AN INTRODUCTION TO ANALYSIS

William R. Wade
University of Tennessee

PRENTICE HALL
Englewood Cliffs, New Jersey 07632

Library of Congress Cataloging-in-Publication Data

Wade, W. R.
 An introduction to analysis / William R. Wade.
   p.   cm.
 Includes bibliographical references and index.
 ISBN 0-13-093089-X
 1. Mathematical analysis. I. Title.
QA300.W25 1995
515—dc20   94–20587
                                           CIP

Acquisitions editor: *George Lobell*
Project manager: *Robert C. Walters, PMI*
Cover design: *Lee Goldstein*
Production coordinator: *Trudy Pisciotti*

 © 1995 by Prentice-Hall, Inc.
A Simon & Schuster Company
Englewood Cliffs, New Jersey 07632

Printed in the United States of America

10 9 8 7 6 5 4 3 2 1

ISBN 0-13-093089-X

PRENTICE-HALL INTERNATIONAL (UK) LIMITED, LONDON
PRENTICE-HALL OF AUSTRALIA PTY. LIMITED, SYDNEY
PRENTICE-HALL CANADA INC. TORONTO
PRENTICE-HALL HISPANOAMERICANA, S.A., MEXICO
PRENTICE-HALL OF INDIA PRIVATE LIMITED, NEW DELHI
PRENTICE-HALL OF JAPAN, INC., TOKYO
SIMON & SCHUSTER ASIA PTE. LTD., SINGAPORE
EDITORA PRENTICE-HALL DO BRASIL, LTDA., RIO DE JANEIRO

CONTENTS

Part I. One-dimensional theory

Chapter 1

THE REAL NUMBER SYSTEM

Chapter 2

CONTINUITY AND DIFFERENTIABILITY ON **R**

iii

Chapter 3

INTEGRABILITY ON **R**

Chapter 4

INFINITE SERIES

Part II. Multidimensional theory

Chapter 5

EUCLIDEAN SPACES

Chapter 6

DIFFERENTIABILITY ON $\mathbf{R}^n$

Chapter 7

INTEGRATION ON $\mathbf{R}^n$

Chapter 8

FUNDAMENTAL THEOREMS OF MULTIVARIABLE CALCULUS

^{e}Chapter 9

FOURIER SERIES

^eChapter 10

STOKES'S THEOREM ON MANIFOLDS

APPENDICES

PREFACE

This text provides a bridge from "sophomore" calculus to graduate courses which use analytic ideas, e.g., real and complex analysis, partial and ordinary differential equations, numerical analysis, fluid mechanics, and differential geometry. It is divided into two parts. The first half, Chapters 1 through 4 together with Appendices A and B, covers one-dimensional theory. The second half, Chapters 5 through 10 together with Appendices C through F, covers multidimensional theory. Separating the one-dimensional material from the n- dimensional material means that certain topics (e.g., convergence of sequences, limits of functions, and uniform continuity) are presented twice, once in $\mathbf{R}$ and once in $\mathbf{R}^n$. Although this is not the most efficient way to present the material, it does give students a second chance to master these difficult concepts.

We have made this text very flexible by dividing it into core material and enrichment material. Enrichment material appears in enrichment sections, marked with an ^{e}, or in core sections, where it is marked with an asterisk. Exercises which use enrichment material are also marked with an asterisk, and the material needed to solve them is cited in the Answers and Hints section. The core material, occupying fewer than 370 pages, can be covered easily in a one year course.

A typical course from this text will cover all core material and some of the enrichment material, chosen at the discretion of the instructor. To make course planning easier, each enrichment section begins with a statement which indicates whether or not that section uses material from any other enrichment section. Since no core material depends on enrichment material, any of the latter can be skipped without loss in the integrity of the course. Thus, instructors can use none, some, or all of this material to supplement their course or to add variety from year to year. For a two-semester course, the first semester should end somewhere in the middle of Chapter 5. For a three-quarter course, the second quarter should begin in the middle of Chapter 4 and end somewhere in the middle of Chapter 6.

A course based on this text is suitable not only for mathematics majors (late undergraduate or first-year graduate), but also for science and engineering majors who need a careful introduction to the concepts and methods of analytic proofs (e.g., those planning to pursue Ph.D.'s in theoretical physics, electrical engineering, or operations research). Such students (especially those specializing in applied mathematics) must be prepared to work in higher dimensions, able to change variables in multiple integrals, optimize functions of several variables, apply the fundamental theorems of vector calculus, and understand the curl, divergence, and gradient operators from both a mechanical and a geometric standpoint.

Since for many students this is the last (for some the only) place to see a rigorous development of vector calculus, we focus our attention on classical, nitty-gritty analysis. By avoiding abstract concepts such as vector spaces and the Lebesgue integral, we have room for a thorough, comprehensive introduction. We include sections on improper integration, the gamma function, Lagrange multipliers, the Inverse and Implicit Function Theorem, Green's Theorem, Gauss's Theorem, Stokes's Theorem, and a full account of the change of variables formula for multiple Riemann integrals.

We assume the reader has completed a three-semester or four-quarter sequence in elementary calculus. Because many of our students now take their elementary calculus in high school (where theory may be almost nonexistent), we assume that the reader is familiar only with the mechanics of calculus, i.e., can differentiate, integrate, and graph simple functions of the form $y = f(x)$ or $z = f(x, y)$. We also assume the reader has had an introductory course in linear algebra, i.e., can add, multiply, and take determinants of matrices with real entries. (Appendix C, which contains an exposition of all definitions and theorems from linear algebra used in the text, can be used as review if the instructor deems it necessary.)

In the first four chapters, we introduce the key ideas of analysis gradually in the one-dimensional setting. After these foundations have been laid in a familiar setting, the material becomes more abstract in Chapter 5, where we discuss both the algebraic and topological structure of the n-dimensional Euclidean spaces $\mathbf{R}^n$, and limits and continuity of multivariable functions. In the next three chapters, we develop the machinery and theory of multidimensional calculus. Chapter 9 gives a short introduction to Fourier series, including summability and convergence of Fourier series, growth of Fourier coefficients, and uniqueness of trigonometric series. And Chapter 10 gives a short introduction to differentiable manifolds which culminates in a proof of Stokes's Theorem on differentiable manifolds.

Always we emphasize the fact that the concepts and results of analysis are based on simple geometric considerations and on analogies with material already known to the student. The aim is to keep the course from looking like a collection of tricks and to share enough of the motivation behind the mathematics so that students are prepared to construct their own proofs when asked. We begin complicated proofs with a short paragraph (marked STRATEGY:) which shows why the proof works. We precede abstruse definitions with a short paragraph which describes in simple terms the behavior we are trying to isolate. And we include examples to show why each hypothesis of a major theorem is necessary.

Each section contains a collection of exercises designed to allow students the luxury and thrill of discovering mathematics on their own. These exercises range from very elementary (to be sure the student understands the concepts introduced in that section) to more challenging (to give the student practice in using these concepts to expand the theory). To minimize frustration, some of the more difficult exercises have several parts which serve as an outline to a solution of the problem. To keep from producing students who know theory but cannot apply it, each set of exercises contains a mix of computational and theoretical assignments. (Exercises which play a prominent role later in the text are marked with a box. These exercises are an integral part of the course and all of them should be assigned.)

Since many students have difficulty reading and understanding mathematics, we have paid close attention to style and organization. We have consciously limited the vocabulary, kept notation consistent from chapter to chapter, and presented the proofs in a unified style. Individual sections are determined by subject matter, not by length of lecture, so that students can comprehend related results in a larger context. Examples and important remarks are numbered and labeled so that students can read the text in small chunks. (Many of these, included for the student's benefit, need not be covered in class.) Paragraphs are short and focused so that students are not overwhelmed by long-winded explanations. To help students discern between central results and peripheral ones, the word "Theorem" is used relatively sparingly; preliminary results and results which are only used in one section are called Remarks, Lemmas, and Examples.

I wish to thank Mr. P.W. Wade and Professors S. Fridli, G.S. Jordan, J. Long, M.E. Mays, M.S. Osborne, and F.E. Schroeck, who carefully read parts of the manuscript and made many valuable suggestions and corrections. Also, I wish to express my gratitude to Ms. C.K. Wade for several lively discussions of a pedagogical nature which helped shape the organization and presentation of this material.

The Author

Chapter 1

THE REAL NUMBER SYSTEM

You have already had several calculus courses in which you evaluated limits, differentiated functions, and computed integrals. You may even remember some of the major results of calculus, such as the Chain Rule, the Mean Value Theorem, and the Fundamental Theorem of Calculus. Although you are probably less familiar with multivariable calculus, you have taken partial derivatives, computed gradients, and evaluated certain line and surface integrals.

In view of all this, you must be asking: Why another course in calculus? The answer to this question is twofold. Although some proofs may have been presented in earlier courses, it is unlikely that the subtler points (e.g., completeness of the real numbers, uniform continuity, and uniform convergence) were covered. Moreover, the skills you acquired were mostly computational; you were rarely asked to prove anything yourself. This course develops the theory of calculus carefully and rigorously from basic principles and gives you a chance to learn how to construct your own proofs. It also serves as an introduction to analysis, an important branch of mathematics which provides a foundation for numerical analysis, functional analysis, harmonic analysis, differential equations, differential geometry, real analysis, complex analysis, and many other areas of specialization within mathematics.

Every rigorous study of mathematics begins with certain undefined concepts, primitive notions on which the theory is based, and certain postulates, properties which are assumed to be true and need no proof. Our study will be based on the primitive notions of set and real numbers and on four postulates (containing a total of eighteen different properties), which will be introduced in the first three sections of this chapter.

1.1 ORDERED FIELD AXIOMS

In this section, we explore the algebraic structure of the real number system. We shall use standard set theoretic notation. For example, $\emptyset$ represents the *empty set* (the set with no elements), $a \in A$ means that a is an *element of A*, and $a \notin A$ means that a is *not* an element of A. We can represent a given finite set in two ways. We

1

can list its elements directly, or we can describe it using sentences or equations. For example, the set of solutions to the equation $x^2 = 1$ can be written as

$$\{1, -1\} \quad \text{or} \quad \{x : x^2 = 1\}.$$

A set A is said to be a *subset* of a set B (notation: $A \subseteq B$) if every element of A is also an element of B. If A is a subset of B but there is at least one element $b \in B$ which does not belong to A, we shall call A a *proper subset* of B (notation: $A \subset B$). Two sets A and B are said to be *equal* (notation: $A = B$) if $A \subseteq B$ and $B \subseteq A$. If A and B are not equal, we write $A \neq B$. A set A is said to be *nonempty* if $A \neq \emptyset$.

The *union* of two sets A and B (notation: $A \cup B$) is the set of elements x such that x belongs to A or B or both. The *intersection* of two sets A and B (notation: $A \cap B$) is the set of elements x such that x belongs to both A and B. The *complement* of B *relative to* A (notation: $A \setminus B$, sometimes B^c if A is understood) is the set of elements x such that x belongs to A but does not belong to B. For example,

$$\{-1, 0, 1\} \cup \{1, 2\} = \{-1, 0, 1, 2\}, \qquad \{-1, 0, 1\} \cap \{1, 2\} = \{1\},$$

$$\{1, 2\} \setminus \{-1, 0, 1\} = \{2\} \quad \text{and} \quad \{-1, 0, 1\} \setminus \{1, 2\} = \{-1, 0\}.$$

Let X and Y be sets. The *Cartesian product* of X and Y is the set of *ordered pairs* defined by

$$X \times Y := \{(x, y) : x \in X, y \in Y\}.$$

(The symbol := means "equal by definition" or "is defined to be.") Two points $(x, y), (z, w) \in X \times Y$ are said to be *equal* if $x = z$ and $y = w$.

Let X and Y be sets. A *relation* on $X \times Y$ is any subset of $X \times Y$. Let $\mathcal{R}$ be a relation on $X \times Y$. The *domain* of $\mathcal{R}$ is the collection of $x \in X$ such that (x, y) belongs to $\mathcal{R}$. When $(x, y) \in \mathcal{R}$, we shall frequently write $x\mathcal{R}y$.

A *function* f from X into Y (notation: $f : X \to Y$) is a relation on $X \times Y$ whose domain is X (notation: $X = \text{Dom}(f)$) such that for each $x \in X$ there is one and only one $y \in Y$ which satisfies $(x, y) \in f$. In this case we say that f is *defined* on X, and call y the *value* of f at x (notation: $y = f(x)$ or $f : x \longmapsto y$). Notice that by the definition of equality of ordered pairs, two functions f and g are equal if and only if they have the same domain and values, i.e., $\text{Dom}(f) = \text{Dom}(g)$ and $f(x) = g(x)$ for all $x \in \text{Dom}(f)$.

We shall denote the set of real numbers by $\mathbf{R}$. We shall assume that $\mathbf{R}$ is a field, i.e., that $\mathbf{R}$ satisfies the following postulate.

POSTULATE 1. [THE FIELD AXIOMS]. There are functions $+$ and $\cdot$, defined on $\mathbf{R}^2 := \mathbf{R} \times \mathbf{R}$, which satisfy the following properties for every $a, b, c \in \mathbf{R}$:

The Closure Properties. $a + b$ and $a \cdot b$ belong to $\mathbf{R}$.

The Associative Properties. $a + (b + c) = (a + b) + c$ and $a \cdot (b \cdot c) = (a \cdot b) \cdot c$.

The Commutative Properties. $a + b = b + a$ and $a \cdot b = b \cdot a$.

The Distributive Law. $a \cdot (b + c) = a \cdot b + a \cdot c$.

Existence of the Additive Identity. There is a unique element $0 \in \mathbf{R}$ such that $0 + a = a$ for all $a \in \mathbf{R}$.

Existence of the Multiplicative Identity. There is a unique element $1 \in \mathbf{R}$ such that $1 \neq 0$ and $1 \cdot a = a$ for all $a \in \mathbf{R}$.

Existence of Additive Inverses. Given $x \in \mathbf{R}$, there is a unique element $-x \in \mathbf{R}$ such that

$$x + (-x) = 0.$$

Existence of Multiplicative Inverses. Given $x \in \mathbf{R} \setminus \{0\}$, there is a unique element $x^{-1} \in \mathbf{R}$ such that

$$x \cdot (x^{-1}) = 1.$$

We note in passing that the word "unique" can be dropped from the statement of Postulate 1 (see Appendix A).

We shall frequently denote $a + (-b)$ by $a - b$, $a \cdot b$ by ab, a^{-1} by $\dfrac{1}{a}$ or $1/a$, and $a \cdot b^{-1}$ by $\dfrac{a}{b}$ or a/b. Notice that by the existence of additive and multiplicative inverses, the equation $x + a = 0$ can be solved for each $a \in \mathbf{R}$ and the equation $ax = 1$ can be solved for each $a \in \mathbf{R}$ provided $a \neq 0$.

From these few properties (i.e., from Postulate 1), one can derive all the usual algebraic laws of real numbers, including the following:

(1)
$$(-1)^2 = 1,$$

(2)
$$0 \cdot a = 0, \quad -a = (-1) \cdot a, \quad -(-a) = a, \quad a \in \mathbf{R},$$

(3)
$$-(a - b) = b - a, \quad a, b \in \mathbf{R},$$

and

(4)
$$a, b \in \mathbf{R} \quad \text{and} \quad ab = 0 \quad \text{imply} \quad a = 0 \quad \text{or} \quad b = 0.$$

We want to keep our attention sharply focused on analysis. Since the proofs of algebraic laws like these lie more in algebra than analysis (see Appendix A), we will not prove them here. In fact, with the exception of the absolute value and the Binomial Formula, we accept all the material usually presented in a high school algebra course as given (including the quadratic formula and graphs of the conic sections), and will use this material without further explanation as the need arises.

The real number system $\mathbf{R}$ contains certain special subsets: the set of *natural numbers*

$$\mathbf{N} := \{1, 2, \dots\},$$

obtained by beginning with 1 and successively adding 1's to form $2 := 1 + 1$, $3 := 2 + 1$, etc.; the set of *integers*

$$\mathbf{Z} := \{\cdots - 2, -1, 0, 1, 2, \ldots\}$$

(*Zahlen* is German for integer); the set of *rationals* (or fractions or quotients)

$$\mathbf{Q} := \left\{\frac{m}{n} : m \in \mathbf{Z} \text{ and } n \in \mathbf{N}\right\};$$

and the set of *irrationals*

$$\mathbf{Q}^c := \mathbf{R} \setminus \mathbf{Q}.$$

Equality in $\mathbf{Q}$ is defined by

$$\frac{m}{n} = \frac{p}{q} \quad \text{if and only if} \quad mq = np.$$

Notice that each of the sets $\mathbf{N}, \mathbf{Z}, \mathbf{Q}$, and $\mathbf{R}$ is a subset of the next. We shall see that each of these subsets is proper, i.e.,

$$\mathbf{N} \subset \mathbf{Z} \subset \mathbf{Q} \subset \mathbf{R}.$$

We make two additional assumptions about $\mathbf{N}$ and $\mathbf{Z}$: $\mathbf{N}$ and $\mathbf{Z}$ satisfy the Closure Properties, e.g., if $n, m \in \mathbf{Z}$ then $n + m$ and nm belong to $\mathbf{Z}$; and, given $n \in \mathbf{Z}$, one and only one of the following statements holds: $n \in \mathbf{N}$, $-n \in \mathbf{N}$, or $n = 0$.

These additional assumptions and Postulate 1 can be used to prove that $\mathbf{Q}$ also satisfies Postulate 1 (see Exercise 6). We notice in passing that none of the other special subsets of $\mathbf{R}$ satisfies Postulate 1. $\mathbf{N}$ satisfies all but three of the properties in Postulate 1: $\mathbf{N}$ has no additive identity (since $0 \notin \mathbf{N}$), $\mathbf{N}$ has no additive inverses (e.g., $-1 \notin \mathbf{N}$), and only one of the nonzero elements of $\mathbf{N}$ (namely, 1) has a multiplicative inverse. $\mathbf{Z}$ satisfies all but one of the properties in Postulate 1: only two nonzero elements of $\mathbf{Z}$ have multiplicative inverses (namely, 1 and -1). $\mathbf{Q}^c$ satisfies all but four of the properties in Postulate 1: $\mathbf{Q}^c$ does not have an additive identity (since $0 \notin \mathbf{R} \setminus \mathbf{Q}$), does not have a multiplicative identity (since $1 \notin \mathbf{R} \setminus \mathbf{Q}$), and it does not satisfy either closure property. Indeed, since $\sqrt{2}$ is irrational (see Remark 5 in Section 1.3), the sum of irrationals may be rational ($\sqrt{2} + (-\sqrt{2}) = 0$) and the product of irrationals may be rational ($\sqrt{2} \cdot \sqrt{2} = 2$).

Postulate 1 is sufficient to derive all algebraic laws of $\mathbf{R}$ but it does not completely describe the real number system. The set of real numbers also has an order relation, i.e., a concept of "less than."

POSTULATE 2. [THE ORDER AXIOMS]. There is a relation $<$ on $\mathbf{R} \times \mathbf{R}$ which has the following properties:

The Trichotomy Property. Given $a, b \in \mathbf{R}$, one and only one of the following statements holds:

$$a < b, \quad b < a, \quad \text{or} \quad a = b.$$

The Transitive Property.

$$a < b \quad \text{and} \quad b < c \quad \text{imply} \quad a < c.$$

The Additive Property.

$$a < b \quad \text{and} \quad c \in \mathbf{R} \quad \text{imply} \quad a + c < b + c.$$

The Multiplicative Properties.

$$a < b \quad \text{and} \quad c > 0 \quad \text{imply} \quad ac < bc,$$

and

$$a < b \quad \text{and} \quad c < 0 \quad \text{imply} \quad bc < ac.$$

By $b > a$ we shall mean $a < b$. By $a \le b$ and $b \ge a$ we shall mean $a < b$ or $a = b$. If $a < b$ and $b < c$, we shall write $a < b < c$.

We shall call a number $a \in \mathbf{R}$ *nonnegative* if $a \ge 0$ and *positive* if $a > 0$. Postulate 2 has a slightly simpler formulation using the set of positive elements as a primitive concept (see Exercise 11). We have introduced Postulate 2 as above because these are the properties we use most often.

Notice that any subset of $\mathbf{R}$ satisfies Postulate 2. Thus $\mathbf{Q}$ also satisfies Postulates 1 and 2. The remaining two postulates, introduced in Sections 1.2 and 1.3 below, identify properties which $\mathbf{Q}$ does not possess. In particular, these four postulates distinguish $\mathbf{R}$ from each of its special subsets $\mathbf{N}$, $\mathbf{Z}$, $\mathbf{Q}$, and $\mathbf{Q}^c$. These postulates actually characterize $\mathbf{R}$; i.e., $\mathbf{R}$ is the only set which satisfies Postulates 1 through 4.

Using all four postulates, one can define a function $f : x \longmapsto x^\alpha$ for any $x > 0$ and $\alpha \in \mathbf{R}$ (see Exercise 5 in Section 3.3) so that the following properties hold: $x^0 = 1$, $x^\alpha > 0$, $x^\alpha \cdot x^\beta = x^{\alpha+\beta}$, and $(x^\alpha)^\beta = x^{\alpha \cdot \beta}$ for all $\alpha, \beta \in \mathbf{R}$ and all $x > 0$, and if $\alpha = n \in \mathbf{N}$ then $x^n = x \cdot \ldots \cdot x$ (there are n factors here). Because it would be impractical to wait until Chapter 3 to use x^α for examples, we shall accept these properties as given and use them as the need arises. We shall also accept, as given, the *trigonometric functions* (whose formulas are) represented by $\sin x$, $\cos x$, $\tan x$, $\cot x$, $\sec x$, $\csc x$, the *exponential function* e^x and its inverse, the *natural logarithm*

$$\log x := \int_1^x \frac{dt}{t},$$

defined and real-valued for each $x \in (0, \infty)$. Although this last function is denoted by $\ln x$ in elementary calculus texts, most analysts denote it, as we did just now, by $\log x$. We will follow this practice throughout this text.

Notice that $x^\alpha \cdot x^{-\alpha} = x^{\alpha-\alpha} = x^0 = 1$. By the uniqueness of multiplicative inverses, it follows that $x^{-\alpha} = (x^\alpha)^{-1} = 1/x^\alpha$ for all $\alpha \in \mathbf{R}$ and $x > 0$.

If $\alpha = 1/m$ for some $m \in \mathbf{N}$, we shall denote x^α by $\sqrt[m]{x}$. (We shall also write $\sqrt{x}$ for $\sqrt[2]{x}$.) Hence

$$(\sqrt[m]{x})^n = (x^{1/m})^n = x^{n/m} = (x^n)^{1/m} = \sqrt[m]{x^n}$$

for all $m \in \mathbf{N}$, $n \in \mathbf{Z}$, and $x > 0$. In particular, $(\sqrt[m]{x})^m = \sqrt[m]{x^m} = x$ for all $x > 0$ and $m \in \mathbf{N}$. These properties hold for $x \geq 0$ and $\alpha \neq 0$ if we define $0^\alpha = 0$ for $\alpha \neq 0$. The symbol 0^0 is left undefined because it is indeterminate (see Example 4 in Section 2.5). Notice that since x^α is positive when $x > 0$, $\sqrt[m]{x} \geq 0$ for all $x \geq 0$ and $m \in \mathbf{N}$.

Postulates 1 and 2 can be used to derive all the usual algebraic laws regarding real numbers and inequalities (e.g., see implications (5) through (9) below). Since arguments based on inequalities are of fundamental importance to analysis, we begin to supply details of proofs at this stage.

What is a proof? Every mathematical result (for us this includes examples, remarks, lemmas, and theorems) has hypotheses and a conclusion. There are three main methods of proof: mathematical induction, direct deduction, and contradiction.

Mathematical induction, a special method for proving statements that depend on positive integers, will be covered in Section 1.3.

To construct a *deductive proof* we assume the hypotheses to be true and proceed step by step to the conclusion. Each step is justified by a hypothesis, a definition, a postulate, or a mathematical result which has already been proved. (Actually, this is usually the way we write a proof. When constructing your own proofs, you may find it helpful to work forward from the hypotheses as far as you can and then work backward from the conclusion, trying to meet in the middle.)

To construct a *proof by contradiction* we assume the hypotheses to be true, the conclusion to be false, and work step by step deductively until a *contradiction* occurs, i.e., a statement which is obviously false or which is contrary to the assumptions made. At this point the proof by contradiction is complete. The phrase "suppose to the contrary" always indicates a proof by contradiction (e.g., see the proof of Theorem 1.2 below).

Here are some examples of deductive proofs. (Note: The symbol ■ indicates that the proof or solution is complete.)

Example 1. If $a, b \in \mathbf{R}$, prove

$$(5) \qquad\qquad a \neq 0 \quad \text{implies} \quad a^2 > 0.$$

PROOF. Suppose $a \neq 0$. By the Trichotomy Property, either $a > 0$ or $a < 0$.

Case 1. $a > 0$. Multiply both sides of this inequality by a. By the first Multiplicative Property, we obtain $a^2 = a \cdot a > 0 \cdot a = 0$.

Case 2. $a < 0$. Multiply both sides of this inequality by a. Since $a < 0$, it follows from the second Multiplicative Property that $a^2 = a \cdot a > 0 \cdot a = 0$. ■

Example 2. Prove $-1 < 0 < 1$.

PROOF. Since $1 \neq 0$, it follows from Example 1 that $1 = 1^2 > 0$. Adding -1 to both sides of this inequality, we conclude that $0 = 1 - 1 > 0 - 1 = -1$. ■

Example 3. If $a, b \in \mathbf{R}$, prove

$$(6) \qquad 0 < a < 1 \quad \text{implies} \quad 0 < a^2 < a, \quad \text{and} \quad a > 1 \quad \text{implies} \quad a^2 > a.$$

PROOF. Suppose $0 < a < 1$. Multiply both sides of this inequality by a. By the first Multiplicative Property,

$$0 = 0 \cdot a < a \cdot a = a^2 < 1 \cdot a = a.$$

On the other hand, if $a > 1$, then $a > 0$ by Example 1 and the Transitive Property. Multiplying $a > 1$ by a, we conclude that $a^2 = a \cdot a > 1 \cdot a = a$. ∎

Similarly (see Exercise 4), we can prove

$$(7) \qquad\qquad 0 \le a < b \quad \text{and} \quad 0 \le c < d \quad \text{imply} \quad ac < bd,$$

$$(8) \qquad\qquad 0 \le a < b \quad \text{implies} \quad 0 \le a^2 < b^2 \quad \text{and} \quad 0 \le \sqrt{a} < \sqrt{b},$$

and

$$(9) \qquad\qquad 0 < a < b \quad \text{implies} \quad \frac{1}{a} > \frac{1}{b} > 0.$$

Although it may seem both pedantic and unnecessary to include proofs of such well known (yes, perhaps even obvious) laws, we include them here for several reasons. We want this book to be reasonably self-contained, because this will make it easier for you to begin to construct your own proofs. We want the first proofs you see in this book to be easily understood, because they deal with familiar properties which are unobscured by new concepts. And we want to form a habit of proving all statements, even "obvious" statements like these. The reason for this hard headed approach is that some "obvious" statements are false. For example, some students think it obvious that any continuous function must be differentiable at some point. Others think it obvious that if every rational in $[0, 1]$ is covered by a small interval, then the sum of the lengths of those intervals must exceed 1. We shall see that both these statements, and many others equally "obvious," are false. In particular, we harbor a skepticism that demands proofs of all statements, even the "obvious" ones.

Much of analysis deals with estimation (of error, of growth, of volume, etc.), in which inequalities and the following concept play a central role.

DEFINITION 1.1. The *absolute value* of a number $a \in \mathbf{R}$ is the number

$$|a| = \begin{cases} a & a \ge 0 \\ -a & a < 0. \end{cases}$$

Here are three important properties that the absolute value satisfies.

Remark 1. *The absolute value is positive definite, i.e., for all $a \in \mathbf{R}$, $|a| \ge 0$ with $|a| = 0$ if and only if $a = 0$.*

PROOF. If $a \ge 0$ then $|a| = a \ge 0$. If $a < 0$ then by Definition 1.1 and the second Multiplicative Property, $|a| = -a = (-1)a > 0$. Thus $|a| \ge 0$ for all $a \in \mathbf{R}$.

If $|a| = 0$ then by definition $\pm a = 0$. Hence $a = 1 \cdot a = (\pm 1)^2 a = (\pm 1)(\pm a) = (\pm 1) \cdot 0 = 0$. Thus $|a| = 0$ implies $a = 0$. Conversely, $|0| = 0$ by definition. ∎

Remark 2. *The absolute value is multiplicative, i.e.,* $|ab| = |a|\,|b|$ *for all* $a, b \in \mathbf{R}$.

PROOF. We consider four cases.

Case 1. $a = 0$ or $b = 0$. Then $ab = 0$, so by definition, $|ab| = 0 = |a|\,|b|$.

Case 2. $a > 0$ and $b > 0$. By the first Multiplicative Property, $ab > 0^2 = 0$. Hence by definition $|ab| = ab = |a|\,|b|$.

Case 3. $a > 0$ and $b < 0$, or, $b > 0$ and $a < 0$. By symmetry, we may suppose $a > 0$ and $b < 0$. (That is, if we can prove it for $a > 0$ and $b < 0$, then by reversing the roles of a and b, we can prove it for $a < 0$ and $b > 0$.) By the second Multiplicative Property, $ab < 0$. Hence by Definition 1.1,

$$|ab| = -(ab) = (-1)(ab) = a((-1)b) = a(-b) = |a|\,|b|.$$

Case 4. $a < 0$ and $b < 0$. By the second Multiplicative Property, $ab > 0$. Hence by Definition 1.1,

$$|ab| = ab = (-1)^2(ab) = (-a)(-b) = |a|\,|b|. \quad ∎$$

The following result is useful when solving inequalities involving absolute value signs.

THEOREM 1.1. *Let* $a \in \mathbf{R}$ *and* $M \geq 0$. *Then* $|a| \leq M$ *if and only if* $-M \leq a \leq M$.

PROOF. Suppose first that $|a| \leq M$. Multiplying by -1, we have $-|a| \geq -M$. Thus

$$-M \leq -|a| = a < 0 \leq M$$

if $a < 0$, and

$$-M \leq 0 \leq a = |a| \leq M$$

if $a \geq 0$. Therefore, $-M \leq a \leq M$.

Conversely, if $-M \leq a \leq M$ then $a \leq M$ and $-M \leq a$. Multiplying the second inequality by -1, we have $-a \leq M$. Consequently, $|a| = a \leq M$ if $a \geq 0$, and $|a| = -a \leq M$ if $a < 0$. ∎

Note: In a similar way we can prove that $|a| < M$ if and only if $-M < a < M$.

The following result (which is equivalent to the Trichotomy Property) will be used many times in this and subsequent chapters.

THEOREM 1.2. *Let* $x, y, a \in \mathbf{R}$.

    i) $x < y + \varepsilon$ *for all* $\varepsilon > 0$ *if and only if* $x \leq y$.

    ii) $x > y - \varepsilon$ *for all* $\varepsilon > 0$ *if and only if* $x \geq y$.

    iii) $|a| < \varepsilon$ *for all* $\varepsilon > 0$ *if and only if* $a = 0$.

PROOF. i) Suppose to the contrary that $x < y + \varepsilon$ for all $\varepsilon > 0$ but $x > y$. Set $\varepsilon_0 = x - y > 0$ and observe that $y + \varepsilon_0 = x$. Hence by the Trichotomy Property, $y + \varepsilon_0$ cannot be greater than x. This contradicts the hypothesis for $\varepsilon = \varepsilon_0$. Thus $x \leq y$.

Conversely, suppose $x \leq y$ and $\varepsilon > 0$ is given. Either $x < y$ or $x = y$. If $x < y$ then $x + 0 < y + 0 < y + \varepsilon$ by the Additive and Transitive Properties. If $x = y$ then $x < y + \varepsilon$ by the Additive Property. Thus $x < y + \varepsilon$ for all $\varepsilon > 0$ in either case. This completes the proof of part i).

ii) Suppose $x > y - \varepsilon$ for all $\varepsilon > 0$. By the second Multiplicative Property, this is equivalent to $-x < -y + \varepsilon$, hence by part i), equivalent to $-x \leq -y$. Multiplying this inequality by -1, we conclude that $x \geq y$.

iii) Suppose $|a| < \varepsilon$ for all $\varepsilon > 0$. By Theorem 1.1, this is equivalent to $-\varepsilon < a < \varepsilon$. It follows from parts i) and ii), that $0 \leq a \leq 0$. We conclude by the Trichotomy Property that $a = 0$. ∎

Some students mistakenly mix absolute values and the Additive Property to conclude that $b < c$ implies $|a + b| < |a + c|$. It is important from the beginning to recognize that this implication is false unless both $a + b$ and $a + c$ are nonnegative. For example, if $a = 1$, $b = -5$, and $c = -1$, then $b < c$ but $|a + b| = 4$ is not less than $|a + c| = 0$.

A correct way to estimate using absolute value signs usually involves one of the following fundamental inequalities.

THEOREM 1.3 [TRIANGLE INEQUALITIES]. *Let $a; b \in \mathbf{R}$. Then*

$$|a + b| \leq |a| + |b|, \quad |a - b| \geq |a| - |b|, \quad and \quad \big| \, |a| - |b| \, \big| \leq |a - b|.$$

PROOF. To prove the first inequality, notice that $|x| \leq |x|$ holds for any $x \in \mathbf{R}$. Thus Theorem 1.1 implies $-|a| \leq a \leq |a|$ and $-|b| \leq b \leq |b|$. Adding these inequalities we obtain

$$-(|a| + |b|) \leq a + b \leq |a| + |b|.$$

Hence by Theorem 1.1 again, $|a + b| \leq |a| + |b|$.

The second inequality follows immediately from the first, since

$$|a| - |b| = |a - b + b| - |b| \leq |a - b| + |b| - |b| = |a - b|.$$

To prove the third inequality, notice by Theorem 1.1 that we need to show

$$-|a - b| \leq |a| - |b| \leq |a - b|.$$

The right-hand inequality has already been proved. Hence by Remark 2,

$$|b| - |a| \leq |b - a| = |-1| \, |a - b| = |a - b|.$$

Multiplying this last inequality by -1 we conclude that

$$-|a - b| \leq |a| - |b|. \quad ∎$$

A *closed interval* is a set of the form

$$[a, b] := \{x \in \mathbf{R} : a \leq x \leq b\}, \qquad [a, \infty) := \{x \in \mathbf{R} : a \leq x\},$$

$$(-\infty, b] := \{x \in \mathbf{R} : x \leq b\}, \quad \text{or} \quad (-\infty, \infty) := \mathbf{R}.$$

A closed interval $[a, b]$ is called *degenerate* if $a = b$, i.e., $[a, b] = \{a\}$, and *nondegenerate* if $a < b$. An *open interval* is a set of the form

$$(a, b) := \{x \in \mathbf{R} : a < x < b\}, \quad (a, \infty) := \{x \in \mathbf{R} : a < x\},$$

$$(-\infty, b) := \{x \in \mathbf{R} : x < b\}, \quad \text{or} \quad (-\infty, \infty) := \mathbf{R}.$$

By an *interval* we mean a closed interval, an open interval, or a set of the form

$$[a, b) := \{x \in \mathbf{R} : a \leq x < b\} \quad \text{or} \quad (a, b] := \{x \in \mathbf{R} : a < x \leq b\}.$$

An interval I is said to be *bounded* if it has the form $[a, b]$, (a, b), $[a, b)$, or $(a, b]$ for some $-\infty < a \leq b < \infty$, in which case the numbers a, b will be called the *endpoints* of I. All other intervals will be called *unbounded*.

Analysis has a strong geometric flavor. Geometry enters the picture because the real number system can be identified with the real line in such a way that $a < b$ if and only if a lies to the left of b (see Figures 1.1, 1.2, and 1.3 below). This gives us a way of translating analytic results on $\mathbf{R}$ into geometric results on the number line, and vice versa. We close with several examples.

The absolute value is closely linked to the idea of length. The *length* of a bounded interval I with endpoints a, b is defined to be $|I| := |b - a|$. And the *distance* between any two points $a, b \in \mathbf{R}$ is defined by $|a - b|$.

Inequalities can be interpreted as statements about intervals. By Theorem 1.1, $|a| \leq M$ if and only if a belongs to the closed interval $[-M, M]$. And, by Theorem 1.2, a belongs to the open intervals $(-\varepsilon, \varepsilon)$ for all $\varepsilon > 0$ if and only if $a = 0$.

We will use this point of view in Chapters 2 through 4 to give geometric interpretations to the calculus of functions defined on $\mathbf{R}$, and in Chapters 5 through 8 to extend this calculus to functions defined on the Euclidean spaces $\mathbf{R}^n$.

EXERCISES

In each of the following exercises, verify the given statement carefully, proceeding step by step. Validate each step which involves an inequality by using some statement found in this section.

$\boxed{1}$. **This exercise is used in Section 4.3.** The *positive part* of an $a \in \mathbf{R}$ is defined by

$$a^+ := \frac{|a| + a}{2}$$

and the *negative part* by

$$a^- := \frac{|a| - a}{2}.$$

a) Prove that $a = a^+ - a^-$ and $|a| = a^+ + a^-$.

b) Prove that

$$a^+ = \begin{cases} a & a \geq 0 \\ 0 & a \leq 0 \end{cases} \quad \text{and} \quad a^- = \begin{cases} 0 & a \geq 0 \\ -a & a \leq 0. \end{cases}$$

2. Solve each of the following inequalities for $x \in \mathbf{R}$.

   a) $|x - 2| < 5$. b) $|1 - x| < 4$. c) $|x^2 - x - 1| < x^2$. d) $|x^2 + x| < 2$.

3. Suppose $a, b, c \in \mathbf{R}$ and $a \leq b$.

   a) Prove that $a + c \leq b + c$.

   b) If $c \geq 0$, prove $a \cdot c \leq b \cdot c$.

4. Prove (7), (8), and (9). Show that each of these statements is false if the hypothesis $a \geq 0$ is removed.

5. a) Prove if $0 < a < 1$ and $b = 1 - \sqrt{1 - a}$, then $0 < b < a$.

   b) Prove if $a > 2$ and $b = 1 + \sqrt{a - 1}$, then $2 < b < a$.

   c) The *arithmetic mean* of $a, b \in \mathbf{R}$ is $A(a, b) = (a + b)/2$ and the *geometric mean* of $a, b \in [0, \infty)$ is $G(a, b) = \sqrt{ab}$. If $0 \leq a \leq b$, prove $a \leq G(a, b) \leq A(a, b) \leq b$. Prove that $G(a, b) = A(a, b)$ if and only if $a = b$.

6. a) Interpreting a rational m/n as $m \cdot n^{-1} \in \mathbf{R}$, use Postulate 1 to prove

$$\frac{m}{n} + \frac{p}{q} = \frac{mq + np}{nq}, \quad \frac{m}{n} \cdot \frac{p}{q} = \frac{mp}{nq}, \quad -\frac{m}{n} = \frac{-m}{n}, \quad \text{and} \quad \left(\frac{\ell}{n}\right)^{-1} = \frac{n}{\ell}$$

   for $m, n, p, q, \ell \in \mathbf{Z}$ and $n, q, \ell \neq 0$.

   b) Prove that Postulate 1 holds with $\mathbf{Q}$ in place of $\mathbf{R}$.

   c) Prove that the sum of a rational and an irrational is always irrational. What can you say about the product of a rational and an irrational?

   d) Let $m/n, p/q \in \mathbf{R}$ with $n, q > 0$. Prove

$$\frac{m}{n} < \frac{p}{q} \quad \text{if and only if} \quad mq < np.$$

   (Restricting this observation to $\mathbf{Q}$ gives a definition of "$<$" on $\mathbf{Q}$.)

7. a) Prove that $|x| \leq 1$ implies $|x^2 - 1| \leq 2|x - 1|$.

   b) Prove that $-1 \leq x \leq 2$ implies $|x^2 + x - 2| \leq 4|x - 1|$.

   c) Prove that $|x| \leq 1$ implies $|x^2 - x - 2| \leq 3|x + 1|$.

   d) Prove that $|x - 1| < 1$ implies $|x^3 + x - 2| < 8|x - 1|$.

8. For each of the following, find all values of $n \in \mathbf{N}$ which satisfy the given inequality.

   a)
$$\frac{1 - n}{1 - n^2} < 0.01.$$

   b)
$$\frac{n^2 + 2n + 3}{2n^3 + 5n^2 + 8n + 3} < 0.025.$$

c)
$$\frac{n-1}{n^3 - n^2 + n - 1} < 0.002.$$

9. Prove that
$$(a_1 b_1 + a_2 b_2)^2 \le (a_1^2 + a_2^2)(b_1^2 + b_2^2)$$
for all $a_1, a_2, b_1, b_2 \in \mathbf{R}$.

10. Suppose $x, a, y, b \in \mathbf{R}$, $|x - a| < \varepsilon$, and $|y - b| < \varepsilon$ for some $\varepsilon > 0$.
   a) Prove $|xy - ab| < (|a| + |b|)\varepsilon + \varepsilon^2$.
   b) Prove $|x^2 y - a^2 b| < \varepsilon(|a|^2 + 2|ab|) + \varepsilon^2(|b| + 2|a|) + \varepsilon^3$.

11. a) Let $\mathbf{P}$ represent the collection of positive real numbers. Prove that $\mathbf{P}$ satisfies the following two properties.

   i) For each $x \in \mathbf{R}$, one and only one of the following hold:
   $$x \in \mathbf{P}, \quad -x \in \mathbf{P}, \quad \text{or} \quad x = 0.$$

   ii) Given $x, y \in \mathbf{P}$, both $x + y$ and $x \cdot y$ belong to $\mathbf{P}$.

   b) Suppose $\mathbf{R}$ contains a subset $\mathbf{P}$ (not necessarily the set of positive numbers) which satisfies properties i) and ii). Define $x \prec y$ by $y - x \in \mathbf{P}$. Prove that Postulate 2 holds with $\prec$ in place of $<$.

1.2   THE WELL-ORDERING PRINCIPLE

In this section we introduce the Well-Ordering Principle, a postulate which distinguishes the set $\mathbf{N}$ from the sets $\mathbf{Z}$, $\mathbf{Q}$, and $\mathbf{R}$. We use it to establish the Principle of Induction and prove the Binomial Formula, a result which shows how to expand powers of a binomial expression, i.e., an expression of the form $a + b$.

The Well-Ordering Principle is different from the preceding two postulates in a fundamental way. Postulates 1 and 2 were statements about the algebraic structure of $\mathbf{R}$, namely, about finite sums and products of elements of $\mathbf{R}$. Postulate 3 is a statement about the "direction" of $\mathbf{N}$ under the order relation $<$, namely, about existence of least elements of subsets of $\mathbf{N}$. Before we state the Well-Ordering Principle, we make precise what we mean by a least element.

DEFINITION 1.2. A number x is a *least element* of a set $E \subset \mathbf{R}$ if $x \in E$ and $x \le a$ for all $a \in E$.

Note: Because French mathematicians (e.g., Borel, Jordan, and Lebesgue) did fundamental work on the connection between analysis and set theory, and *ensemble* is French for set, analysts frequently use E to represent a general set.

POSTULATE 3. [THE WELL-ORDERING PRINCIPLE]. Every nonempty subset of $\mathbf{N}$ has a least element.

Notice that the Well-Ordering Principle is not satisfied by the number systems $\mathbf{Z}$, $\mathbf{Q}$, and $\mathbf{R}$ since none of these systems contains a least element.

Our first application of Postulate 3 is a characterization of $\mathbf{N}$ in $\mathbf{Z}$.

Remark 1. *Suppose $n \in \mathbf{Z}$. Then $n \in \mathbf{N}$ if and only if $n > 0$.*

PROOF. By Example 2 in Section 1.1, $-1 < 0 < 1$. Suppose $E := \{n \in \mathbf{N} : n \leq 0\}$ is nonempty. By the Well-Ordering Principle, E has a least element, say x. Since $1 > 0$ and $x \in \mathbf{N}$, x is a sum of 1's which contains at least two terms. Thus $x - 1$ is a sum of 1's which contains at least one term, i.e., $x - 1 \in \mathbf{N}$. Since x is a least element of E, $x - 1 > 0$. On the other hand, adding -1 to the inequality $x \leq 0$, we have by the Additive Property and the Transitive Property that $x - 1 \leq -1 < 0$. This contradiction proves that $n > 0$ for all $n \in \mathbf{N}$.

Conversely, suppose $n \in \mathbf{Z}$, $n > 0$, but $n \notin \mathbf{N}$. Then $-n \in \mathbf{N}$ and by what we just proved, $-n > 0$. Adding n to both sides of this inequality, we obtain $0 > n$, a contradiction. ∎

Our second application of the Well-Ordering Principle is the *Principle of Mathematical Induction.*

THEOREM 1.4. *Suppose for each $n \in \mathbf{N}$ that $A(n)$ is a proposition (i.e., a verbal statement or formula) which satisfies the following properties: $A(1)$ is true, and if $A(n_0)$ is true for some $n_0 \in \mathbf{N}$, then $A(n_0 + 1)$ is also true. Then $A(n)$ is true for all $n \in \mathbf{N}$.*

PROOF. Suppose the theorem is false. Then the set $E = \{n \in \mathbf{N} : A(n) \text{ is false}\}$ is nonempty. Hence by Postulate 3, E has a least element, say x.

Since $x \in E \subseteq \mathbf{N}$ and $A(1)$ is true, we know that $x \geq 2$. Thus $x - 1 \geq 1$, so $x - 1$ belongs to $\mathbf{N}$. But $x - 1 < x$ and x is a least element of E. Consequently, $A(x - 1)$ is true. Applying the hypothesis to $n_0 = x - 1$, we see that $A(x) = A(n_0 + 1)$ must also be true, i.e, $x \notin E$, a contradiction. ∎

Recall that if $x_0, x_1, \ldots, x_n$ are real numbers and $0 \leq j \leq n$, then

$$\sum_{k=j}^{n} x_k := x_j + x_{j+1} + \cdots + x_n$$

denotes the sum of the x_k's as k ranges from j to n. The following examples illustrate the fact that the Principle of Mathematical Induction can be used to prove a variety of statements involving integers.

Example 1. Prove

$$\sum_{k=1}^{n} (3k - 1)(3k + 2) = 3n^3 + 6n^2 + n$$

for $n \in \mathbf{N}$.

PROOF. Let $A(n)$ represent the statement

$$\sum_{k=1}^{n} (3k - 1)(3k + 2) = 3n^3 + 6n^2 + n.$$

For $n = 1$ the left side of this equation is $2 \cdot 5$ and the right side is $3 + 6 + 1$. Therefore, $A(1)$ is true. Suppose $A(n)$ is true for some $n \geq 1$. Then

$$\sum_{k=1}^{n+1}(3k-1)(3k+2) = (3n+2)(3n+5) + \sum_{k=1}^{n}(3k-1)(3k+2)$$

$$= (3n+2)(3n+5) + 3n^3 + 6n^2 + n = 3n^3 + 15n^2 + 22n + 10.$$

On the other hand, a direct calculation reveals

$$3(n+1)^3 + 6(n+1)^2 + (n+1) = 3n^3 + 15n^2 + 22n + 10.$$

Therefore, $A(n+1)$ is true when $A(n)$ is. We conclude by induction that $A(n)$ holds for all $n \in \mathbf{N}$. ∎

Example 2. Prove that $n \in \mathbf{Z}$ and $n > 0$ implies $n \geq 1$.

PROOF. By Remark 1, it suffices to prove that $n \geq 1$ for all $n \in \mathbf{N}$. The statement is obvious for $n = 1$. If $n \geq 1$ for some $n \in \mathbf{N}$, then by the Additive Property, $n + 1 \geq 1 + 1 > 1$. It follows from the Transitive Property that $n + 1 \geq 1$. We conclude by induction that $n \geq 1$ holds for all $n \in \mathbf{N}$. ∎

It is interesting to note that the Principle of Mathematical Induction is equivalent to the Well-Ordering Principle. Indeed, by the proof of Theorem 1.4, the Well-Ordering Principle implies the Principle of Mathematical Induction. Conversely, suppose the Principle of Mathematical Induction holds but there is a nonempty set $E \subset \mathbf{N}$ which has no least element. Since E is nonempty, let $x_1 \in E$. Suppose $x_{n-1} \in E$ has been chosen. Since E has no least element, there is an $x_n \in E$ such that $x_n < x_{n-1}$. By induction, we can choose $x_n \in E$ for all $n \in \mathbf{N}$ such that $x_1 > x_2 > \ldots$ Since $E \subset \mathbf{N}$, choose $N \in \mathbf{N}$ such that $x_1 = N$. By construction, $x_n - x_{n-1} < 0$, hence by Example 2, $x_n - x_{n-1} \leq -1$ for all $n \in \mathbf{N}$. It follows that

$$x_{N+1} - N = x_{N+1} - x_1 = (x_{N+1} - x_N) + (x_N - x_{N-1}) + \cdots + (x_2 - x_1)$$
$$\leq (-1) + (-1) + \cdots + (-1) = -N.$$

In particular, $x_{N+1} \leq N - N = 0$ so $x_{N+1} \notin \mathbf{N}$, a contradiction.

Two formulas encountered early in an algebra course are the perfect square and cube formulas:

$$(a+b)^2 = a^2 + 2ab + b^2 \quad \text{and} \quad (a+b)^3 = a^3 + 3a^2 b + 3ab^2 + b^3.$$

Our next application of the Principle of Mathematical Induction is a generalization of these formulas from $n = 2$ or 3 to arbitrary $n \in \mathbf{N}$.

Recall that Pascal's Triangle is the triangular array of integers whose rows begin and end with 1's with the property that an interior entry on any row is obtained by adding the two numbers in the preceding row immediately above that entry. Thus the first few rows of Pascal's Triangle are:

$$1$$

$$1 \quad 1$$

$$1 \quad 2 \quad 1$$

$$1 \quad 3 \quad 3 \quad 1$$

$$1 \quad 4 \quad 6 \quad 4 \quad 1$$

$$1 \quad 5 \quad 10 \quad 10 \quad 5 \quad 1$$

$$1 \quad 6 \quad 15 \quad 20 \quad 15 \quad 6 \quad 1$$

Notice that the third and fourth rows are precisely the coefficients which appeared in the perfect square and cube formulas above.

One can write down a formula for each entry in each row of the Pascal Triangle. The first (and only) entry in the first row is

$$\binom{0}{0} := 1.$$

Using the notation $0! := 1$ and $n! := 1 \cdot 2 \cdots (n-1) \cdot n$ for $n \in \mathbf{N}$, define the *binomial coefficient n over k* by

$$\binom{n}{k} = \frac{n!}{(n-k)!k!}$$

for $0 \le k \le n$ and $n = 0, 1, \ldots$.

The following result shows that the binomial coefficient n over k is a formula for the $(k+1)$–st entry in the $(n+1)$–st row of Pascal's Triangle.

Lemma . *If $n, k \in \mathbf{N}$ and $1 \le k \le n$, then*

$$\binom{n+1}{k} = \binom{n}{k-1} + \binom{n}{k}.$$

PROOF. By definition,

$$\binom{n}{k-1} + \binom{n}{k} = \frac{n!\,k}{(n-k+1)!k!} + \frac{n!(n-k+1)}{(n-k+1)!k!}$$

$$= \frac{n!(n+1)}{(n-k+1)!k!} = \binom{n+1}{k}. \quad \blacksquare$$

Binomial coefficients can be used to expand the nth power of a sum of two terms.

THEOREM 1.5 [BINOMIAL FORMULA]. *If $a, b \in \mathbf{R}$ and $n \in \mathbf{N}$, then*

$$(a + b)^n = \sum_{k=0}^{n} \binom{n}{k} a^k b^{n-k}.$$

PROOF. The proof is by induction on n. The formula is obvious for $n = 1$. Suppose the formula is true for some $n \in \mathbf{N}$. Then by the inductive hypothesis and Postulate 1,

$$(a + b)^{n+1} = (a + b)(a + b)^n$$

$$= (a + b) \left(\sum_{k=0}^{n} \binom{n}{k} a^k b^{n-k} \right)$$

$$= \left(\sum_{k=0}^{n} \binom{n}{k} a^{k+1} b^{n-k} \right) + \left(\sum_{k=0}^{n} \binom{n}{k} a^k b^{n-k+1} \right)$$

$$= \left(a^{n+1} + \sum_{k=1}^{n} \binom{n}{k-1} a^k b^{n-k+1} \right) + \left(b^{n+1} + \sum_{k=1}^{n} \binom{n}{k} a^k b^{n-k+1} \right)$$

$$= a^{n+1} + \sum_{k=1}^{n} \left(\binom{n}{k-1} + \binom{n}{k} \right) a^k b^{n-k+1} + b^{n+1}.$$

Hence it follows from the lemma that

$$(a + b)^{n+1} = a^{n+1} + \sum_{k=1}^{n} \binom{n+1}{k} a^k b^{n-k+1} + b^{n+1} = \sum_{k=0}^{n+1} \binom{n+1}{k} a^k b^{n-k+1},$$

i.e., the formula is true for $n + 1$. We conclude by induction that the formula holds for all $n \in \mathbf{N}$. ∎

EXERCISES

$\boxed{1}$. **This exercise is used in Sections 1.4, 1.7, and 3.1.** Prove the following formulas hold for all $n \in \mathbf{N}$.

a) $\displaystyle\sum_{k=1}^{n} k = \frac{n(n+1)}{2}$,   b) $\displaystyle\sum_{k=1}^{n} k^2 = \frac{n(n+1)(2n+1)}{6}$,

c) $\displaystyle\sum_{k=1}^{n} \frac{a-1}{a^k} = 1 - \frac{1}{a^n}$, $a \neq 0$,   d) $\displaystyle\sum_{k=1}^{n} (2k-1)^2 = \frac{n(4n^2 - 1)}{3}$.

2. Use the Binomial Formula to prove each of the following.

a) $2^n = \sum_{k=0}^{n} \binom{n}{k}$ for all $n \in \mathbf{N}$.

b) $(a + b)^n \geq a^n + na^{n-1} b$ for all $n \in \mathbf{N}$ and $a, b \geq 0$.

c) $(1 + 1/n)^n \geq 2$ for all $n \in \mathbf{N}$.

3. Let $n \in \mathbf{N}$. Write

$$\frac{(x+h)^n - x^n}{h}$$

as a sum, none of whose terms has an h in the denominator.

4. a) Suppose $0 < x_1 < 1$ and $x_{n+1} = 1 - \sqrt{1 - x_n}$ for $n \in \mathbf{N}$. Prove $0 < x_{n+1} < x_n < 1$ holds for all $n \in \mathbf{N}$.

   b) Suppose $x_1 \geq 2$ and $x_{n+1} = 1 + \sqrt{x_n - 1}$ for $n \in \mathbf{N}$. Prove $2 \leq x_{n+1} \leq x_n \leq x_1$ holds for all $n \in \mathbf{N}$.

5. Suppose $0 < x_1 < 2$ and $x_{n+1} = \sqrt{2 + x_n}$ for $n \in \mathbf{N}$. Prove $0 < x_n < x_{n+1} < 2$ holds for all $n \in \mathbf{N}$.

6. Prove that each of the following inequalities holds for all $n \in \mathbf{N}$.

$$n < 2^n, \qquad n^2 \leq 2^n + 1, \quad \text{and} \quad n^3 \leq 3^n.$$

$\boxed{7}$. **This exercise is used in Section 1.5.** Prove that $0 \leq a < b$ implies $0 \leq a^n < b^n$ and $0 \leq \sqrt[n]{a} < \sqrt[n]{b}$ for all $n \in \mathbf{N}$.

8. In the next section we shall prove that the square root of an integer m is rational if and only if $m = k^2$ for some $k \in \mathbf{N}$. Assume this result is true.

   a) Prove that $\sqrt{n+3} + \sqrt{n}$ is rational for some $n \in \mathbf{N}$ if and only if $n = 1$.

   b) Find all $n \in \mathbf{N}$ such that $\sqrt{n+7} + \sqrt{n}$ is rational.

1.3 THE COMPLETENESS AXIOM

In this section we introduce the last of four postulates which describe $\mathbf{R}$. To formulate this postulate, which distinguishes $\mathbf{Q}$ from $\mathbf{R}$, we need the following concepts.

DEFINITION 1.3. Let $E \subset \mathbf{R}$ be nonempty.

   i) The set E is said to be *bounded above* if there is an $M \in \mathbf{R}$ such that $a \leq M$ for all $a \in E$.

   ii) A number M is called an *upper bound* of the set E if $a \leq M$ for all $a \in E$.

   iii) A number s is called a *supremum* of the set E if s is an upper bound of E and if $s \leq M$ for all upper bounds M of E. (In this case we shall say that E has a supremum s and shall write $s = \sup E$.)

Notice by iii) that a supremum of a set E (when it exists) is the smallest (or least) upper bound of E.

Remark 1. *A given set can have many upper bounds.*

PROOF. If M_0 is an upper bound for a set E, then so is M for any $M > M_0$. ∎

Remark 2. *If a set E has a supremum, then it has only one supremum.*

PROOF. Let s_1 and s_2 be suprema of the same set E. Then both s_1 and s_2 are upper bounds of E, whence by definition, $s_1 \leq s_2$ and $s_2 \leq s_1$. We conclude by the Trichotomy Property that $s_1 = s_2$. ∎

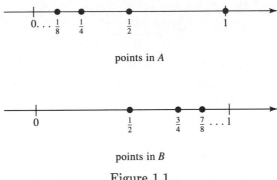

points in A

points in B

Figure 1.1

Note: This proof illustrates a general principle. When asked to prove $a = b$, it is often easier to verify $a \leq b$ and $b \leq a$.

By definition, a least element of a set E always belongs to E. The following result shows that this is not the case for the supremum of a set.

Remark 3. *The supremum of a set E may or may not belong to E.*

Indeed, the sets $A = \{1, \frac{1}{2}, \frac{1}{4}, \frac{1}{8}, \dots\}$ and $B = \{0, \frac{1}{2}, \frac{3}{4}, \frac{7}{8}, \dots\}$ both have the same supremum (see Figure 1.1), namely 1, but B does not contain 1. (This is a heuristic argument, not a proof. A formal proof that $\sup A = \sup B$ can be given using Theorem 1.7 below.)

In this last example, the supremum of B was "near" B even though it did not belong to B. We formalize this relationship in the following fundamental result about suprema.

THEOREM 1.6 [APPROXIMATION PROPERTY FOR SUPREMA]. *If E has a supremum and $\varepsilon > 0$ is any positive number, then there is a point $a \in E$ such that*

$$\sup E - \varepsilon < a \leq \sup E.$$

PROOF. Suppose the theorem is false. Then there is an $\varepsilon_0 > 0$ such that no element of E lies between $s_0 := \sup E - \varepsilon_0$ and $\sup E$. It follows that $a \leq s_0$ for all $a \in E$, i.e., s_0 is an upper bound of E. Thus, by Definition 1.3, $\sup E \leq s_0 = \sup E - \varepsilon_0$. Subtracting $\sup E$ from both sides of this inequality, we conclude that $\varepsilon_0 \leq 0$, a contradiction. ∎

The existence of suprema is the last major assumption about **R** we make.

POSTULATE 4. [THE COMPLETENESS AXIOM]. If E is a nonempty subset of **R** which is bounded above, then E has a supremum.

We shall use this property many times. Our first two applications deal with the distribution of integers and rationals among real numbers.

THEOREM 1.7 [ARCHIMEDEAN PRINCIPLE]. *Given positive real numbers a and b, there is an integer $n \in \mathbf{N}$ such that $b < na$.*

STRATEGY: The idea behind the proof is simple. The Completeness Axiom coupled with the Approximation Property gives us a method for finding "largest" natural numbers in certain sets. If k_0 were the largest natural number which satisfies $k_0 a \leq b$, then $n = (k_0 + 1)$ would satisfy $na > b$. Since no such k_0 exists when $b < a$, we need to consider this case separately. Here are the details.

PROOF. If $b < a$ set $n = 1$. If $a \leq b$, consider the set $E = \{k \in \mathbf{N} : ka \leq b\}$. E is nonempty since $1 \in E$. Since $ka \leq b$ for all $k \in E$ and $a > 0$, it follows from the Multiplicative Property that $k \leq b/a$ for all $k \in E$, i.e., E is bounded above by b/a. Thus, by the Completeness Axiom, E has a supremum, say $s = \sup E$.

By the Approximation Property (applied to $\varepsilon = 1/2$), there is a $k_0 \in E$ such that $s - 1/2 < k_0 \leq s$. Adding 1 to this inequality we obtain $s + 1/2 < k_0 + 1 \leq s + 1$. In particular, $\sup E = s < s + 1/2 < k_0 + 1$. Set $n = k_0 + 1$ and notice that $n \in \mathbf{N}$. Since $n > \sup E$, n cannot belong to E. Thus $na > b$. ∎

THEOREM 1.8 [DENSITY OF RATIONALS] . *If $a, b \in \mathbf{R}$ satisfy $a < b$, then there is a $q \in \mathbf{Q}$ such that $a < q < b$.*

STRATEGY: To motivate the proof, consider the special case $a = 1/4$ and $b = 1/3$. We want to find a fraction $q = m/n$ such that $1/4 < m/n < 1/3$. No such m exists if $1 \leq n \leq 6$ because the fractions p/n are spaced too far apart. If n is large enough, however, so that some of the fractions p/n belong to the interval $(1/4, 1/3)$ (e.g., $n = 7$), then an acceptable value for m is $m = k_0 - 1$, where k_0 is the smallest integer satisfying $b \leq k_0/n$. For the general case, some p/n should belong to (a, b) if n satisfies

$$\frac{1}{n} < b - a.$$

We begin our formal proof at this point.

PROOF. Since $b - a > 0$, it follows from (9) that $(b - a)^{-1} > 0$. Hence by the Archimedean Principle, there is an $n \in \mathbf{N}$ which satisfies $n > (b - a)^{-1}$.

Case 1. $b > 0$. Consider the set $E = \{k \in \mathbf{N} : b \leq k/n\}$. By the Archimedean Principle, E is nonempty. Hence by the Well-Ordering Principle, E has a least element, say k_0. Set $m = k_0 - 1$ and $q = m/n$. Since $m < k_0$ and k_0 is a least element of E, $m \notin E$. This can happen two ways. Either $m \leq 0$ or $b > m/n = q$. In either case we obtain $q < b$. On the other hand, since $k_0 \in E$ implies $b \leq k_0/n$, it follows from the choice of n that

$$a = b - (b - a) < \frac{k_0}{n} - \frac{1}{n} = \frac{k_0 - 1}{n} = q.$$

Case 2. $b \leq 0$. Choose (by the Archimedean Principle) a $k \in \mathbf{N}$ such that $k + b > 0$. By Case 1, there is an $r \in \mathbf{Q}$ such that $k + a < r < k + b$. Therefore, $q := r - k$ belongs to $\mathbf{Q}$ and satisfies the inequality $a < q < b$. ∎

Here is another application of the Archimedean Principle to the distribution of numbers.

Remark 4. *If $x > 1$ and $x \notin \mathbf{N}$, then there is an $n \in \mathbf{N}$ such that $n < x < n + 1$.*

PROOF.   By the Archimedean Principle, the set $E = \{m \in \mathbf{N} : x < m\}$ is nonempty. Hence by the Well-Ordering Principle, E has a least element, say m_0.

Set $n = m_0 - 1$. Since $m_0 \in E$, $n + 1 = m_0 > x$. Since m_0 is least, $n = m_0 - 1 \le x$. Since $x \notin \mathbf{N}$, we also have $n \ne x$. Therefore, $n < x < n + 1$. ∎

Using this last result, we can prove that the set of irrationals is nonempty.

Remark 5. *If $n \in \mathbf{N}$ is not a perfect square (i.e., if there is no $m \in \mathbf{N}$ such that $n = m^2$), then $\sqrt{n}$ is irrational.*

PROOF. Suppose to the contrary that $n \in \mathbf{N}$ is not a perfect square but $\sqrt{n} \in \mathbf{Q}$, i.e., $\sqrt{n} = p/q$ for some $p, q \in \mathbf{N}$. Choose by Remark 4 an integer $m_0 \in \mathbf{N}$ such that

$$(10) \qquad\qquad m_0 < \sqrt{n} < m_0 + 1.$$

Consider the set $E := \{k \in \mathbf{N} : k\sqrt{n} \in \mathbf{Z}\}$. Since $q\sqrt{n} = p$ we know that E is nonempty. Thus by the Well-Ordering Principle, E has a least element, say n_0.

Set $x = n_0(\sqrt{n} - m_0)$. By (10), $0 < \sqrt{n} - m_0 < 1$. Multiplying this inequality by n_0, we find that

$$(11) \qquad\qquad 0 < x < n_0.$$

Since n_0 is a least element of E, it follows from (11) that $x \notin E$. On the other hand,

$$x\sqrt{n} = n_0(\sqrt{n} - m_0)\sqrt{n} = n_0 n - m_0 n_0 \sqrt{n} \in \mathbf{Z}$$

since $n_0 \in E$. Moreover, since $x > 0$ and $x = n_0\sqrt{n} - n_0 m_0$ is the difference of two integers, $x \in \mathbf{N}$. Thus $x \in E$, a contradiction. ∎

For some applications, we also need the following concepts.

DEFINITION 1.4. Let $E \subset \mathbf{R}$ be nonempty.

    i) The set E is said to be *bounded below* if there is an $m \in \mathbf{R}$ such that $a \ge m$ for all $a \in E$.

    ii) A number m is called a *lower bound* of the set E if $a \ge m$ for all $a \in E$.

    iii) A number t is called an *infimum* of the set E if t is a lower bound of E and if $t \ge m$ for all lower bounds m of E. In this case we shall say that *E has an infimum t* and write $t = \inf E$.

    iv) E is said to be *bounded* if it is bounded above and below.

When the set E is finite, i.e., has finitely many points, we shall frequently write $\max E$ for $\sup E$ and $\min E$ for $\inf E$.

(Some authors call the supremum the *least upper bound* and the infimum the *greatest lower bound*. We will not use this terminology because it is somewhat old-fashioned and because it confuses some students, since the **least** upper bound of a given set is always greater than or equal to the **greatest** lower bound.)

To relate suprema to infima, we define the *reflection* of a set $E \subseteq \mathbf{R}$ by

$$-E := \{x : x = -a \text{ for some } a \in E \}.$$

For example, $-(1,2] = [-2,-1)$.

The following result shows that the supremum of a set is the same as the negative of its reflection's infimum. This can be used to prove a completeness axiom for infima (see Exercise 6).

THEOREM 1.9. *Let $E \subseteq \mathbf{R}$ be nonempty.*

i) *E has a supremum if and only if $-E$ has an infimum, in which case*

$$\sup E = -\inf(-E).$$

ii) *E has an infimum if and only if $-E$ has a supremum, in which case*

$$\inf E = -\sup(-E).$$

PROOF. The proofs of these statements are similar. We prove only the first statement.

Suppose E has a supremum s and set $t = -s$. Since s is an upper bound for E, $s \geq a$ for all $a \in E$, so $-s \leq -a$ for all $a \in E$. Therefore, t is a lower bound of $-E$. Suppose that m is any lower bound of $-E$. Then $m \leq -a$ for all $a \in E$, so $-m$ is an upper bound of E. Since s is the supremum of E, it follows that $s \leq -m$, i.e., $t = -s \geq m$. Thus t is the infimum of $-E$ and $\sup E = s = -t = -\inf(-E)$.

Conversely, suppose $-E$ has an infimum t. By definition, $t \leq -a$ for all $a \in E$. Thus $-t$ is an upper bound for E. Since E is nonempty, E has a supremum by the Completeness Axiom. ∎

This remark allows us to obtain information about infima from results about suprema and vice versa (see the proof of Theorem 1.10ii and Exercises 5 and 6).

Here is an important property satisfied by suprema and infima.

THEOREM 1.10 [MONOTONE PROPERTY]. *Suppose that $A \subseteq B$ are nonempty subsets of $\mathbf{R}$.*

i) *If B has a supremum, then $\sup A \leq \sup B$.*
ii) *If B has an infimum, then $\inf A \geq \inf B$.*

PROOF. i) Since $A \subseteq B$, any upper bound of B is an upper bound of A. Therefore, $\sup B$ is an upper bound of A. It follows from the Completeness Axiom that $\sup A$ exists, and from Definition 1.3 that $\sup A \leq \sup B$.

ii) Clearly, $-A \subseteq -B$. Thus by part i) and Theorem 1.9,

$$\inf A = -\sup(-A) \geq -\sup(-B) = \inf B. \quad ∎$$

EXERCISES

1. Find the infimum and supremum of each of the following sets.

   a) $E = \{4, 3, 2, 1, 8, 7, 6, 5\}$.

   b) $E = \{x \in \mathbf{R} : x^2 - 3x - 5 = 0\}$.

   c) $E = [a, b)$, where $a < b$ are real numbers.

   d) $E = \{p/q \in \mathbf{Q} : p^2 < 2q^2 \text{ and } p, q > 0\}$.

   e) $E = \{x \in \mathbf{R} : x = 1 + (-1)^n \text{ for some } n \in \mathbf{N}\}$.

   f) $E = \{x \in \mathbf{R} : x = 1/n - (-1)^n \text{ for some } n \in \mathbf{N}\}$.

   g) $E = \{1 + (-1)^n/n : n \in \mathbf{N}\}$.

2. Let $x_n \in \mathbf{R}$ and suppose that there is an $M \in \mathbf{R}$ such that $|x_n| \leq M$ for $n \in \mathbf{N}$. Prove that $s_n = \sup\{x_n, x_{n+1}, \dots\}$ defines a real number for each $n \in \mathbf{N}$ and $s_1 \geq s_2 \geq \dots$. Prove an analogous result about $t_n = \inf\{x_n, x_{n+1}, \dots\}$.

$\boxed{3}$. [DENSITY OF IRRATIONALS] **This exercise is used in Section 2.2.**
Prove that if $a < b$ are real numbers, then there is an irrational $\xi \in \mathbf{R}$ such that $a < \xi < b$.

4. Prove that for each $a \in \mathbf{R}$ and each $n \in \mathbf{N}$ there exists a rational r_n such that $|a - r_n| < 1/n$.

$\boxed{5}$. [APPROXIMATION PROPERTY FOR INFIMA] **This exercise is used in many sections including 1.4 and 3.1.**

   a) By modifying the proof of Theorem 1.6, prove that if a set $E \subset \mathbf{R}$ has an infimum and $\varepsilon > 0$ is any positive number, then there is a point $a \in E$ such that $\inf E + \varepsilon > a \geq \inf E$.

   b) Give a second proof of the Approximation Property for Infima by using Theorem 1.9.

6. a) Prove that a lower bound of a set need not be unique but the infimum of a given set E is unique.

   b) Prove that if E is a nonempty subset of $\mathbf{R}$ which is bounded below, then E has a infimum.

7. a) Prove that if x is an upper bound of a set $E \subseteq \mathbf{R}$ and $x \in E$, then x is the supremum of E.

   b) Make and prove an analogous statement for the infimum of E.

   c) Show by example that the converse of each of these statements is false.

8. Show that if E is a nonempty bounded subset of $\mathbf{Z}$, then both $\sup E$ and $\inf E$ belong to E.

9. Prove that if a and b are real numbers and $0 \leq a < b$, then there exist $n, m \in \mathbf{N}$ such that $a < m/10^n < b$.

10. a) Prove that $\mathbf{N}$ is bounded below but not bounded above.

   b) Prove that $\mathbf{Z}$ is neither bounded above nor below.

11. Suppose $E, A, B \subset \mathbf{R}$ and $E = A \cup B$. Prove that if E has a supremum and both A and B are nonempty, then $\sup A$ and $\sup B$ both exist, and $\sup E$ is one of the numbers $\sup A$ or $\sup B$.

1.4 SEQUENCES

An *infinite sequence* (more briefly, a *sequence*) is a function whose domain is **N**. A sequence f whose *terms* are $x_n := f(n)$ will be denoted by $x_1, x_2, \ldots$, $\{x_n\}_{n \in \mathbf{N}}$, $\{x_n\}_{n=1}^{\infty}$, or $\{x_n\}$. Thus $1, 1/2, 1/4, 1/8, \ldots$ represents the sequence $\{1/2^{n-1}\}_{n \in \mathbf{N}}$, $-1, 1, -1, 1, \ldots$ represents the sequence $\{(-1)^n\}_{n \in \mathbf{N}}$, and $1, 2, 3, 4, \ldots$ represents the sequence $\{n\}_{n \in \mathbf{N}}$.

For us the terms of an infinite sequence will be numbers, sets, functions, or vectors. We shall call a sequence *real* if each of its terms is a real number. It is important not to confuse the sequence $\{x_n\}_{n \in \mathbf{N}}$ with the set $\{x_n : n \in \mathbf{N}\}$; these are two entirely different concepts. For example, as sequences, $1, 2, 3, 4, \ldots$ is different from $2, 1, 3, 4, \ldots$ but as sets, $\{1, 2, 3, 4, \ldots\}$ is identical with $\{2, 1, 3, 4, \ldots\}$. Again, the sequence $1, -1, 1, -1, \ldots$ is infinite but the set $\{(-1)^n : n \in \mathbf{N}\}$ has only two points.

The limit concept is one of the fundamental building blocks of analysis. We begin our discussion of limit with the simplest case, the limit of real sequences.

Recall from elementary calculus that a real sequence $\{x_n\}$ converges to a number x if x_n gets near x as n gets large; i.e., the distance $|x - x_n|$ gets small as n gets large. Thus, by Theorem 1.2, given any $\varepsilon > 0$ (no matter how small), $|x_n - x|$ gets smaller than ε as n gets large. This leads us to a formal definition of the limit of a sequence.

DEFINITION 1.5. A sequence of real numbers $\{x_n\}$ is said to *converge* to a real number x if given $\varepsilon > 0$ there is an $N \in \mathbf{N}$ such that

$$n \geq N \quad \text{implies} \quad |x_n - x| < \varepsilon.$$

(This definition is illustrated in Figure 1.2 for the case $N = 5$.)

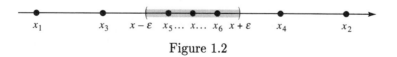

Figure 1.2

We shall use the following phrases and notation interchangeably: a) $\{x_n\}$ converges to x, b) x_n converges to x, c) $x = \lim_{n \to \infty} x_n$, d) $x_n \to x$ as $n \to \infty$, e) the limit of $\{x_n\}$ exists and equals x.

Notice that by definition, a sequence x_n converges to x if and only if $|x_n - x| \to 0$ as $n \to \infty$. In particular, a sequence $x_n \to 0$ if and only if $|x_n| \to 0$ as $n \to \infty$.

We shall occasionally refer to the implication "$n \geq N \quad \text{implies} \quad |x_n - x| \leq \varepsilon$" as "$x_n - x$ is small when n is large." This means that by choosing N large enough we can be sure that $|x_n - x|$ is less than any prescribed positive quantity for all $n \geq N$.

Here are two concrete examples of how to prove that a sequence has a limit.

Example 1. Prove that $1/n \to 0$ as $n \to \infty$.

PROOF. Let $\varepsilon > 0$. Choose by the Archimedean Principle an $N \in \mathbf{N}$ such that $N > 1/\varepsilon$. If $n \geq N$, then $1/n \leq 1/N < \varepsilon$. Consequently,

$$n \geq N \quad \text{implies} \quad \left| \frac{1}{n} - 0 \right| = \left| \frac{1}{n} \right| < \varepsilon. \quad \blacksquare$$

Example 2. Given $a \in \mathbf{R}$, with $|a| < 1$, prove $a^n \to 0$ as $n \to \infty$.

PROOF. Since the result is trivial for $a = 0$ and since $a^n \to 0$ if and only if $|a|^n = |a^n| \to 0$, as $n \to \infty$, we may suppose that $0 < a < 1$. Then $1/a > 1$ so $\delta := 1/a - 1$ is positive. Adding 1 to both sides of this identity and raising it to the nth power, we see by the Binomial Formula that

$$a^{-n} = (1 + \delta)^n = 1 + n\delta + \cdots + \delta^n \geq n\delta.$$

Hence by (9),

(12) $$a^n \leq \frac{1}{n\delta}.$$

Let $\varepsilon > 0$ be given. By the Archimedean Principle, choose an $N \in \mathbf{N}$ such that $N\delta > 1/\varepsilon$. Suppose $n \geq N$. Then $n\delta \geq N\delta > 1/\varepsilon$, i.e.,

$$\frac{1}{n\delta} < \varepsilon.$$

Hence it follows from (12) and our simplifying assumption that $0 < a^n < \varepsilon$ for $n \geq N$. We conclude that $n \geq N$ implies $|a^n| < \varepsilon$, i.e., $a^n \to 0$ as $n \to \infty$. $\blacksquare$

Although the preceding proof introduced estimating techniques which will prove useful in several other contexts, it is more complicated than necessary because we relied only on Definition 1.5. (At the end of this section, return to this example and devise at least two proofs simpler than the one given here. Thus show that the limit theorems proved below can be used to obtain simple, elegant solutions to some limit problems.)

To develop more efficient methods for evaluating limits, we investigate some consequences of Definition 1.5.

Remark 1. *A sequence can have at most one limit.*

PROOF. Suppose x_n converges to both x and y. By definition, given $\varepsilon > 0$ there are integers N_1 and N_2 such that $n \geq N_1$ implies $|x_n - x| < \varepsilon/2$ and $n \geq N_2$ implies $|x_n - y| < \varepsilon/2$. Let $N = \max\{N_1, N_2\}$. By the choice of N_1 and N_2, $n \geq N$ implies both $|x_n - x| < \varepsilon/2$ and $|x_n - y| < \varepsilon/2$. Thus it follows from the Triangle Inequality that

$$|x - y| \leq |x - x_n| + |x_n - y| < \varepsilon,$$

i.e., $|x - y| < \varepsilon$ for all $\varepsilon > 0$. We conclude by Theorem 1.2 that $x = y$.

DEFINITION 1.6. Let $\{x_n\}$ be a sequence of real numbers.

i) $\{x_n\}$ is said to be *bounded above* if there is an $M \in \mathbf{R}$ such that

$$x_n \leq M$$

for all $n \in \mathbf{N}$.

ii) $\{x_n\}$ is said to be *bounded below* if there is an $m \in \mathbf{R}$ such that

$$x_n \geq m$$

for all $n \in \mathbf{N}$.

iii) $\{x_n\}$ is said to be *bounded* if it is bounded above and below.

Notice that $\{x_n\}$ is bounded if and only if there is a $C \in \mathbf{R}$ such that $|x_n| \leq C$ for all $n \in \mathbf{N}$, in which case we shall say that $\{x_n\}$ is bounded by C. Also notice that $\{x_n\}$ is bounded above if and only if $\{-x_n\}$ is bounded below.

Is there a relationship between convergent sequences and bounded sequences?

THEOREM 1.11. *Every convergent sequence in* $\mathbf{R}$ *is bounded.*

STRATEGY: The idea behind the proof is simple (see Figure 1.2). Suppose $x_n \to x$ as $n \to \infty$. By definition, for large N the sequence $x_N, x_{N+1}, \ldots$ is near x, hence must be bounded. Since no finite sequence $x_1, \ldots, x_N$ is unbounded, it would follow that the whole sequence is bounded. We now make this precise.

PROOF. Given $\varepsilon = 1$ there is an $N \in \mathbf{N}$ such that $n \geq N$ implies $|x_n - x| \leq 1$. Hence by the Triangle Inequality, $|x_n| \leq 1 + |x|$ for $n \geq N$. On the other hand, if $1 \leq n \leq N$, then
$$|x_n| \leq M := \max\{|x_1|, |x_2|, \ldots, |x_N|\}.$$
Therefore, $|x_n| \leq 1 + |x| + M$ for all $n \in \mathbf{N}$. ∎

The following example shows that the converse of Theorem 1.11 is false.

Example 3. Prove that $\{(-1)^n\}_{n \in \mathbf{N}}$ is bounded but does not converge.

PROOF. It is obvious that $(-1)^n$ is bounded by 1. Suppose $(-1)^n \to x$ as $n \to \infty$ for some $x \in \mathbf{R}$. Given $\varepsilon = 1/2$ there is an $N \in \mathbf{N}$ such that $n \geq N$ implies $|(-1)^n - x| < \varepsilon$. For n odd this implies $|1 + x| = |-1-x| < 1/2$ and for n even this implies $|1 - x| < 1/2$. Hence

$$2 = |1 + 1| \leq |1 - x| + |1 + x| < \frac{1}{2} + \frac{1}{2} = 1,$$

a contradiction. ∎

It is sometimes possible to decide that a given sequence converges by comparing it with other sequences whose convergence properties are already known (see Example 4 below). The following is the first of many theorems that address this issue.

THEOREM 1.12 [SQUEEZE THEOREM]. *Suppose $\{x_n\}$, $\{y_n\}$, and $\{w_n\}$ are sequences of real numbers.*

i) *If $x_n \to x$ and $y_n \to x$ as $n \to \infty$, and*

$$x_n \leq w_n \leq y_n, \qquad n \geq N_0$$

for some $N_0 \in \mathbf{N}$, then $w_n \to x$ as $n \to \infty$.

ii) *If $x_n \to 0$ as $n \to \infty$ and $\{y_n\}$ is bounded, then $x_n y_n \to 0$ as $n \to \infty$.*

PROOF. i) Let $\varepsilon > 0$. By hypothesis and Theorem 1.1, choose $N_1, N_2 \in \mathbf{N}$ such that $n \geq N_1$ implies $-\varepsilon \leq x_n - x \leq \varepsilon$ and $n \geq N_2$ implies $-\varepsilon \leq y_n - x \leq \varepsilon$. Set $N = \max\{N_0, N_1, N_2\}$. If $n \geq N$ we have by hypothesis and the choice of N_1 and N_2 that

$$x - \varepsilon \leq x_n \leq w_n \leq y_n \leq x + \varepsilon,$$

i.e., $|w_n - x| \leq \varepsilon$ for $n \geq N$. We conclude that $w_n \to x$ as $n \to \infty$.

ii) Suppose $x_n \to 0$ and there is an $M > 0$ such that $|y_n| \leq M$ for $n \in \mathbf{N}$. Let $\varepsilon > 0$ and choose an $N \in \mathbf{N}$ such that $n \geq N$ implies $|x_n| \leq \varepsilon/M$. Then $n \geq N$ implies

$$|x_n y_n| \leq M \frac{\varepsilon}{M} = \varepsilon.$$

We conclude that $x_n y_n \to 0$ as $n \to \infty$. ∎

Notice that in the proofs of Remark 1 and Theorem 1.12, several choices of N_j were made before N was chosen to be their maximum. It is clear that by this same process, if $N_1, \ldots, N_q$ have been chosen so that for each j a property $\mathcal{P}_j$ holds when $n > N_j$ and if $N = \max\{N_1, \ldots, N_q\}$, then all q properties $\mathcal{P}_1, \ldots, \mathcal{P}_q$ hold simultaneously when $n > N$. We shall use this device frequently below, but rarely write N explicitly as a maximum of integers N_j again.

The following example shows how the Squeeze Theorem can be used to find the limit of a complicated sequence by ignoring some of its "less important" factors.

Example 4. Find $\lim_{n \to \infty} \pi^{-n} \cos(n^3 - n^2 + n - 13)$.

SOLUTION. The factor $\cos(n^3 - n^2 + n - 13)$ looks intimidating, but it is superfluous for finding the limit of this sequence. Indeed, since $|\cos x| \leq 1$ for all $x \in \mathbf{R}$ the sequence $\{\cos(n^3 - n^2 + n - 13)\}$ is bounded by 1. Since $\pi > 3$ implies $1/\pi < 1/3 < 1$, $\pi^{-n} \to 0$ as $n \to \infty$ by Example 2. We conclude from Theorem 1.12ii that $\pi^{-n} \cos(n^3 - n^2 + n - 13) \to 0$ as $n \to \infty$. ∎

The Squeeze Theorem can also be used to construct convergent sequences with certain properties. To illustrate how this works, we now prove a result which connects suprema and infima with convergent sequences.

THEOREM 1.13. *Let $E \subset \mathbf{R}$. If E has a supremum (respectively, an infimum), then there is a sequence $x_n \in E$ such that $x_n \to \sup E$ (respectively, $x_n \to \inf E$) as $n \to \infty$.*

PROOF. Suppose E has a supremum. For each $n \in \mathbf{N}$, choose by the Approximation Property an $x_n \in E$ such that $\sup E - 1/n < x_n \leq \sup E$. Then by the

Squeeze Theorem and Example 1, $x_n \to \sup E$ as $n \to \infty$. Similarly, there is a sequence $y_n \in E$ such that $y_n \to \inf E$. ∎

Here is another result which helps to evaluate limits of specific sequences. This one works by viewing complicated sequences in terms of simpler components.

THEOREM 1.14. *Suppose $\{x_n\}$ and $\{y_n\}$ are real sequences and $\alpha \in \mathbf{R}$. If $\{x_n\}$ and $\{y_n\}$ are convergent, then*

i)
$$\lim_{n\to\infty} (x_n + y_n) = \lim_{n\to\infty} x_n + \lim_{n\to\infty} y_n,$$

ii)
$$\lim_{n\to\infty} (\alpha x_n) = \alpha \lim_{n\to\infty} x_n$$

and

iii)
$$\lim_{n\to\infty} (x_n y_n) = \left(\lim_{n\to\infty} x_n\right)\left(\lim_{n\to\infty} y_n\right).$$

If, in addition, $\lim_{n\to\infty} y_n \neq 0$, then

iv)
$$\lim_{n\to\infty} \frac{x_n}{y_n} = \frac{\lim_{n\to\infty} x_n}{\lim_{n\to\infty} y_n}.$$

(In particular, all these limits exist.)

PROOF. Suppose $x_n \to x$ and $y_n \to y$ as $n \to \infty$.

i) Let $\varepsilon > 0$ and choose $N \in \mathbf{N}$ such that $n \geq N$ implies $|x_n - x| < \varepsilon/2$ and $|y_n - y| < \varepsilon/2$. Thus $n \geq N$ implies

$$|(x_n + y_n) - (x + y)| \leq |x_n - x| + |y_n - y| < \frac{\varepsilon}{2} + \frac{\varepsilon}{2} = \varepsilon.$$

ii) It suffices to show $\alpha x_n - \alpha x \to 0$ as $n \to \infty$. But $x_n - x \to 0$ as $n \to \infty$, hence, by the Squeeze Theorem, $\alpha(x_n - x) \to 0$ as $n \to \infty$.

iii) By Theorem 1.11, the sequence $\{x_n\}$ is bounded. Hence, by the Squeeze Theorem the sequences $\{x_n(y_n - y)\}$ and $\{(x_n - x)y\}$ both converge to 0. Since

$$x_n y_n - xy = x_n(y_n - y) + (x_n - x)y,$$

it follows from part i) that $x_n y_n \to xy$ as $n \to \infty$. A similar argument establishes part iv). (See Exercise 3.) ∎

Theorem 1.14 can be used to evaluate limits of sums, products, and quotients of convergent sequences. Here is a typical example.

Example 5. Find $\lim_{n\to\infty} (n^3 + n^2 - 1)/(1 - 3n^3)$.

SOLUTION. Multiplying the numerator and denominator by $1/n^3$, we find

$$\frac{n^3 + n^2 - 1}{1 - 3n^3} = \frac{1 + (1/n) - (1/n^3)}{(1/n^3) - 3}.$$

By Example 1 and Theorem 1.14iii, $1/n^k = (1/n)^k \to 0$, as $n \to \infty$, for any $k \in \mathbf{N}$. Thus

$$\lim_{n \to \infty} \frac{n^3 + n^2 - 1}{1 - 3n^3} = \frac{1 + 0 - 0}{0 - 3} = -\frac{1}{3}. \quad \blacksquare$$

Theorem 1.14 shows how the limit sign interacts with the algebraic structure of $\mathbf{R}$. (Namely, it says that the limit of a sum (product, quotient) is the sum (product, quotient) of the limits.) The following theorem shows how the limit sign interacts with the order structure of $\mathbf{R}$.

THEOREM 1.15 [COMPARISON THEOREM]. *If $\{x_n\}$ and $\{y_n\}$ are convergent sequences which satisfy*

(13) $$x_n \leq y_n, \qquad n \geq N$$

for some $N \in \mathbf{N}$, then

$$\lim_{n \to \infty} x_n \leq \lim_{n \to \infty} y_n.$$

PROOF. Suppose to the contrary that (13) holds but $x = \lim_{n \to \infty} x_n$ is greater than $y = \lim_{n \to \infty} y_n$. Set $\varepsilon = (x - y)/2$. Choose $N_1 > N$ such that $|x_n - x| < \varepsilon$ and $|y_n - y| < \varepsilon$ for $n \geq N_1$. Then for such an n,

$$x_n > x - \varepsilon = x - \left(\frac{x - y}{2}\right) = y + \left(\frac{x - y}{2}\right) = y + \varepsilon > y_n,$$

which contradicts (13). $\blacksquare$

One way to remember this result is that it says the limit of an inequality is the inequality of the limits, provided these limits exist. We shall call this process "taking the limit of an inequality." Since $x_n < y_n$ implies $x_n \leq y_n$, the Comparison Theorem contains the following corollary: if $\{x_n\}$ and $\{y_n\}$ are convergent real sequences, then

$$x_n < y_n, \quad n \geq N, \quad \text{implies} \quad \lim_{n \to \infty} x_n \leq \lim_{n \to \infty} y_n.$$

In particular, if $x_n < M$ for n large and $\{x_n\}$ converges, then $\lim_{n \to \infty} x_n \leq M$. It is important to notice that these results are false if, in the conclusion, $\leq$ is replaced by $<$. For example,

$$\frac{1}{n^2} < \frac{1}{n} \text{ but } \lim_{n \to \infty} \frac{1}{n^2} = \lim_{n \to \infty} \frac{1}{n} = 0.$$

EXERCISES

1. Suppose that $x_n \neq 0$ converges to 0. Show that $1/x_n$ is not bounded.

$\boxed{2}$. **This exercise is used in Section 1.7.**
   Interpret a decimal expansion $0.a_1a_2\ldots$ as

$$0.a_1a_2\cdots = \lim_{n\to\infty} \sum_{k=1}^{n} \frac{a_k}{10^k}.$$

   Prove $0.5 = .4999\ldots$ and $1 = .999\ldots$ (See Exercise 1c in Section 1.2.)

3. Prove Theorem 1.14iv.

4. Suppose that $x_n \geq 0$ and $x_n \to x$ as $n \to \infty$. Prove $\sqrt{x_n} \to \sqrt{x}$ as $n \to \infty$. (For the case $x = 0$ you may wish to use (8) in Section 1.1.)

5. Prove that each of the following sequences converges to zero.
   a) $x_n = \sin((n^4 + n + 1)/(n^2 + 1))/n$.
   b) $x_n = n/(n^2 + 1)$.
   c) $x_n = (\sqrt{2n} + 1)/(n + 1)$.
   d) $x_n = n/2^n$.

6. Find the limit (if it exists) of each of the following sequences.
   a) $x_n = (1 + n - 3n^2)/(3 - 2n + n^2)$.
   b) $x_n = (n^3 + n - 5)/(5n^3 + n - 1)$.
   c) $x_n = \sqrt{2n^2 - 1}/(n + 1)$.
   d) $x_n = \sqrt{n+1} - \sqrt{n}$.

$\boxed{7}$. **This exercise will be used many times from Section 2.3 onward.**
   a) Let $a \in \mathbf{R}$. Show that if $x_n = a$ for all $n \in \mathbf{N}$, then $x_n \to a$ as $n \to \infty$.
   b) If I is a closed interval, $x_n \in I$ for all $n \in \mathbf{N}$, and $x_n \to x$, prove $x \in I$.
   c) Show by example that b) is false if the hypothesis "closed" is omitted.

8. Using the result in Exercise 4, show the following.
   a) Suppose $x_1 \geq 0$ and $x_{n+1} = \sqrt{2 + x_n}$ for $n \in \mathbf{N}$. If $x_n \to x$ as $n \to \infty$, prove $x = 2$.
   b) Suppose $0 \leq x_1 \leq 1$ and $x_{n+1} = 1 - \sqrt{1 - x_n}$ for $n \in \mathbf{N}$. If $x_n \to x$ as $n \to \infty$, prove $x = 0$ or 1.

9. Prove that given $x \in \mathbf{R}$ there is a sequence $r_n \in \mathbf{Q}$ such that $r_n \to x$ as $n \to \infty$.

$\boxed{10}$. **This exercise is used in Section 1.7.**
   a) Suppose $0 \leq y < 1/10^n$ for some integer $n \geq 0$. Prove there is an integer $0 \leq w \leq 9$ such that

$$\frac{w}{10^{n+1}} \leq y \leq \frac{w}{10^{n+1}} + \frac{1}{10^{n+1}}.$$

   b) Prove that given $x \in [0, 1)$ and $n \in \mathbf{N}$, there exist integers $0 \leq x_k \leq 9$, $k = 1, \ldots, n$, such that

$$\sum_{k=1}^{n} \frac{x_k}{10^k} \leq x < \sum_{k=1}^{n} \frac{x_k}{10^k} + \frac{1}{10^n}.$$

c) Prove that given $x \in [0, 1)$ there exist integers $0 \leq x_k \leq 9$, $k \in \mathbf{N}$, such that

$$x = \lim_{n \to \infty} \sum_{k=1}^{n} \frac{x_k}{10^k}.$$

(Note: The numbers x_k are called *digits* of x and $0.x_1 x_2 \ldots$ is called a *decimal expansion* of x. Unless x is a rational number whose denominator is of the form $2^i 5^j$ for some integers $i \geq 0$, $j \geq 0$, this expansion is unique; i.e., there is only one sequence of integers $\{x_k\}$ which satisfies part c). On the other hand, if x is a rational number whose denominator is of the form $2^i 5^j$ then there are two sequences $\{x_k\}$ which satisfy part c), one which satisfies $x_k = 0$ for large k and one which satisfies $x_k = 9$ for large k (see Exercise 2 above). We shall identify the second sequence by saying it *terminates* in 9's.)

11. Suppose $E \subset \mathbf{R}$ is a nonempty bounded set and $\sup E \notin E$. Prove there exists a sequence $\{x_n\}$ which converges to $\sup E$ such that $x_n \in E$ and $x_n < x_{n+1}$ for all $n \in \mathbf{N}$.

1.5 THE BOLZANO–WEIERSTRASS THEOREM

The title of this section refers to a fundamental result about real sequences, namely, that it is always possible to extract a convergent sequence from any bounded real sequence. The first step we take toward this result is to examine the class of monotone sequences, a class for which the Bolzano–Weierstrass Theorem is especially transparent. Later, we shall use this special case to prove the general result.

DEFINITION 1.7. Let $\{x_n\}_{n \in \mathbf{N}}$ be a sequence of real numbers.

i) $\{x_n\}$ is said to be *increasing* (respectively, *strictly increasing*) if $x_1 \leq x_2 \leq \ldots$ (respectively, $x_1 < x_2 < \ldots$).

ii) $\{x_n\}$ is said to be *decreasing* (respectively, *strictly decreasing*) if $x_1 \geq x_2 \geq \ldots$ (respectively, $x_1 > x_2 > \ldots$).

iii) $\{x_n\}$ is said to be *monotone* if it is either increasing or decreasing.

(Some authors call decreasing sequences *nonincreasing* and increasing sequences *nondecreasing*.)

If $\{x_n\}$ is increasing (respectively, decreasing) and converges to x, we shall write $x_n \uparrow x$ (respectively, $x_n \downarrow x$), as $n \to \infty$. Clearly, every strictly increasing sequence is increasing and every strictly decreasing sequence is decreasing. Also, $\{x_n\}$ is increasing if and only if the sequence $\{-x_n\}_{n \in \mathbf{N}}$ is decreasing.

The following result shows that for monotone sequences, convergence and boundedness are equivalent.

THEOREM 1.16 [MONOTONE CONVERGENCE THEOREM].

i) *If* $\{x_n\}_{n \in \mathbf{N}}$ *is increasing and bounded above and* $x = \sup\{x_n : n \in \mathbf{N}\}$, *then* $x_n \uparrow x$ *as* $n \to \infty$.

ii) *If* $\{x_n\}_{n \in \mathbf{N}}$ *is decreasing and bounded below and* $y = \inf\{x_n : n \in \mathbf{N}\}$, *then* $x_n \downarrow y$ *as* $n \to \infty$.

PROOF. i) Suppose $\{x_n\}$ is increasing and bounded above. By the Completeness Axiom, the supremum x exists. Let $\varepsilon > 0$. By the Approximation Property, choose $N \in \mathbf{N}$ such that

$$x - \varepsilon < x_N \leq x.$$

Since $x_N \leq x_n$ for $n \geq N$ and $x_n \leq x$ for all $n \in \mathbf{N}$, it follows that $x - \varepsilon < x_n \leq x$ for all $n \geq N$. In particular, $x_n \uparrow x$ as $n \to \infty$.

ii) If $\{x_n\}$ is decreasing with infimum y, then $\{-x_n\}$ is increasing with supremum $-y$ (see Theorem 1.9). Hence, by part i) and Theorem 1.14ii,

$$y = -(-y) = -\lim_{n \to \infty}(-x_n) = \lim_{n \to \infty} x_n. \quad \blacksquare$$

The Monotone Convergence Theorem is used most often to show that a limit exists. Once existence has been established, it is often easy to find the value of that limit by using Theorems 1.14 and 1.15. The following example illustrates this fact.

Example 1. For each $x > 0$, prove $x^{1/n} \to 1$ as $n \to \infty$.

PROOF. We consider three cases.

Case 1. $x = 1$. Then $x^{1/n} = 1$ for all $n \in \mathbf{N}$ and it follows that $x^{1/n} \to 1$ as $n \to \infty$.

Case 2. $0 < x < 1$. We first show that $\{x^{1/n}\}$ is increasing and bounded above. Indeed, $\{x^{1/n}\}$ is bounded above by 1 since by Exercise 7 in Section 1.2, $0 < x^{1/n} < 1$ for all $n \in \mathbf{N}$. If $\{x^{1/n}\}$ is not increasing, then there is an $n \in \mathbf{N}$ such that $x^{1/n} > x^{1/(n+1)}$. Taking the $n(n+1)$-st power of this inequality, we obtain $x^{n+1} > x^n$, i.e., $x > 1$, a contradiction. Thus $\{x^{1/n}\}$ is increasing and bounded above. Hence, by Theorems 1.16 and 1.15, $a := \lim_{n \to \infty} x^{1/n}$ exists and satisfies $a \leq 1$, and $0 < x^{1/n} \leq a$, i.e., $0 < x \leq a^n$, holds for all $n \in \mathbf{N}$. To show $a = 1$, suppose to the contrary that $a < 1$. Since $a^n \to 0$ as $n \to \infty$ (see Example 2 in Section 1.4), it follows from $0 < x \leq a^n$ and the Squeeze Theorem that $x = 0$, a contradiction.

Case 3. $x > 1$. Then $0 < 1/x < 1$. Hence, by Theorem 1.14 and the cases already considered, we have

$$\lim_{n \to \infty} x^n = \lim_{n \to \infty} \frac{1}{1/x^n} = \frac{1}{\lim_{n \to \infty}(1/x)^n} = 1. \quad \blacksquare$$

Next, we introduce the following monotone property for sequences of sets.

DEFINITION 1.8. A sequence of sets $\{I_n\}_{n \in \mathbf{N}}$ is said to be *nested* if

$$I_1 \supseteq I_2 \supseteq \cdots$$

In Chapters 2 and 5, we shall use this concept to study continuous functions. Here, we use it to prove the Bolzano–Weierstrass Theorem. All of these applications depend in a fundamental way on the following result.

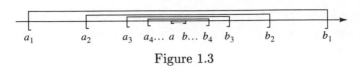

Figure 1.3

THEOREM 1.17 [NESTED INTERVAL PROPERTY]. *If $\{I_n\}_{n\in\mathbf{N}}$ is a nested sequence of nonempty closed bounded intervals, then*

$$K = \bigcap_{n\in\mathbf{N}} I_n := \{x : x \in I_n \text{ for all } n \in \mathbf{N}\}$$

contains at least one number. Moreover, if the lengths of these intervals satisfy $|I_n| \to 0$ as $n \to \infty$, then K contains exactly one number.

PROOF. Let $I_n = [a_n, b_n]$. Since $\{I_n\}$ is nested, the real sequence $\{a_n\}$ is increasing and bounded above by b_1, and $\{b_n\}$ is decreasing and bounded below by a_1 (see Figure 1.3). Thus by Theorem 1.16, there exist $a, b \in \mathbf{R}$ such that $a_n \uparrow a$ and $b_n \downarrow b$ as $n \to \infty$. Since $a_n \le b_n$ for all $n \in \mathbf{N}$, it also follows from the Comparison Theorem that $a_n \le a \le b \le b_n$. Hence, a number x belongs to I_n for all $n \in \mathbf{N}$ if and only if $a \le x \le b$. This proves that $K = [a, b]$.

Suppose now that $|I_n| \to 0$ as $n \to \infty$. Then $b_n - a_n \to 0$ as $n \to \infty$ and we have by Theorem 1.14 that $b - a = 0$. In particular, $K = [a, a] = \{a\}$ contains exactly one number. ∎

Remark 1. *The Nested Interval Property is not true if "closed" is omitted.*

PROOF. The intervals $I_n = (0, 1/n)$, $n \in \mathbf{N}$, are bounded and nested but not closed. If there were an $x \in I_n$ for all $n \in \mathbf{N}$, then $0 < x < 1/n$, i.e., $n < 1/x$ for all $n \in \mathbf{N}$. Since this contradicts the Archimedean Principle, it follows that the intervals I_n have no point in common. ∎

Remark 2. *The Nested Interval Property is not true if "bounded" is omitted.*

PROOF. The intervals $I_n = [n, \infty)$, $n \in \mathbf{N}$ are closed and nested but not bounded. Again, they have no point in common. ∎

We spoke earlier of extracting a convergent sequence from a bounded sequence. The following definition explains exactly what we mean by "extracting" one sequence from another.

DEFINITION 1.9. *By a subsequence of a sequence $\{x_n\}_{n\in\mathbf{N}}$, we shall mean a sequence of the form $\{x_{n_k}\}_{k\in\mathbf{N}}$, where each $n_k \in \mathbf{N}$ and $n_1 < n_2 < \cdots$*

For example, $\{(-1)^{n+1}\}_{n\in\mathbf{N}}$ is a subsequence of $\{\sin(n\pi/2)\}_{n\in\mathbf{N}}$ because if $n_k = (2k - 1)$, then $n_1 < n_2 < \cdots$ and $\sin(n_k\pi/2) = (-1)^{k+1}$ for $k \in \mathbf{N}$.

If $\{x_n\}$ is convergent, then the x_n's get near some x as n gets large. Since n_k gets large as k does, it comes as no surprise that any subsequence of a convergent sequence also converges.

THEOREM 1.18. *If $\{x_n\}_{n\in\mathbf{N}}$ converges to x and $\{x_{n_k}\}_{k\in\mathbf{N}}$ is any subsequence of $\{x_n\}_{n\in\mathbf{N}}$, then x_{n_k} converges to x as $k \to \infty$.*

PROOF. Let $\varepsilon > 0$ and choose $N \in \mathbf{N}$ such that $n \geq N$ implies $|x_n - x| < \varepsilon$. Since $n_k \in \mathbf{N}$ and $n_1 < n_2 < \ldots$, it is clear that $n_k \geq k$ for all $k \in \mathbf{N}$. Hence $k \geq N$ implies $|x_{n_k} - x| < \varepsilon$, i.e., $x_{n_k} \to x$ as $k \to \infty$. ∎

Notice that $\{(-1)^{2k}\}_{k\in\mathbf{N}}$ is a subsequence of $\{(-1)^n\}_{n\in\mathbf{N}}$. Since $(-1)^{2k} = 1$, it follows that the bounded, nonconvergent sequence $\{(-1)^n\}_{n\in\mathbf{N}}$ has a convergent subsequence. This is a general principle.

THEOREM 1.19 [THE BOLZANO–WEIERSTRASS THEOREM]. *Every bounded sequence of real numbers has a convergent subsequence.*

PROOF. We begin with a general observation. Let $\{x_n\}$ be any sequence. If $E = A \cup B$ are sets and E contains x_n for infinitely many values of n, then at least one of the sets A or B also contains x_n for infinitely many values of n. (If not, then E contains x_n for only finitely many n, a contradiction.)

Let $\{x_n\}$ be a bounded sequence. Choose $a, b \in \mathbf{R}$ such that $x_n \in [a, b]$ for all $n \in \mathbf{N}$, and set $I_0 = [a, b]$. Divide I_0 into two halves, say $I' = [a, (a + b)/2]$ and $I'' = [(a + b)/2, b]$. Since $I_0 = I' \cup I''$, at least one of these half intervals contains x_n for infinitely many n. Call it I_1 and choose $n_1 > 1$ such that $x_{n_1} \in I_1$. Notice that $|I_1| = |I_0|/2 = (b - a)/2$.

Suppose closed intervals $I_0 \supset I_1 \supset \cdots \supset I_m$ and natural numbers $n_1 < n_2 < \cdots < n_m$ have been chosen such that for each $0 \leq k \leq m$,

$$(14) \qquad |I_k| = \frac{b - a}{2^k}, \quad x_{n_k} \in I_k, \quad \text{and} \quad x_n \in I_k \quad \text{for infinitely many } n.$$

To choose I_{m+1}, divide $I_m = [a_m, b_m]$ into two halves, say $I' = [a_m, (a_m + b_m)/2]$ and $I'' = [(a_m + b_m)/2, b_m]$. Since $I_m = I' \cup I''$, at least one of these half intervals contains x_n for infinitely many n. Call it I_{m+1} and choose $n_{m+1} > n_m$ such that $x_{n_{m+1}} \in I_{m+1}$. Since

$$|I_{m+1}| = \frac{|I_m|}{2} = \frac{b - a}{2^{m+1}},$$

it follows by induction that there is a nested sequence $\{I_k\}_{k\in\mathbf{N}}$ of nonempty closed bounded intervals which satisfy (14) for all $k \in \mathbf{N}$.

By the Nested Interval Property, there is an $x \in \mathbf{R}$ which belongs to I_k for all $k \in \mathbf{N}$. Since $x \in I_k$, we have by (14) that

$$|x_{n_k} - x| \leq |I_k| \leq \frac{b - a}{2^k}$$

for all $k \in \mathbf{N}$. Hence, by the Squeeze Theorem, $x_{n_k} \to x$ as $k \to \infty$. ∎

Our next application of the Nested Interval Property is in connection with the following concept.

DEFINITION 1.10. A sequence $\{x_n\}$ is said to be a *Cauchy sequence* if given $\varepsilon > 0$ there is an $N \in \mathbf{N}$ such that

$$n, m \geq N \quad \text{implies} \quad |x_n - x_m| < \varepsilon.$$

If a sequence converges to some number x, then the terms of that sequence must eventually get near x, in particular, near each other. This observation leads us to the following result.

Remark 3. *Every convergent sequence is a Cauchy sequence.*

PROOF. Suppose $x_n \to x$ as $n \to \infty$. Let $\varepsilon > 0$ and choose $N \in \mathbf{N}$ such that

$$n \geq N \quad \text{implies} \quad |x_n - x| < \frac{\varepsilon}{2}.$$

Then $n, m \geq N$ implies $|x_n - x_m| \leq |x_n - x| + |x - x_m| < \varepsilon$. ∎

We shall prove (see Theorem 1.20 below) that the converse of Remark 3 also holds. Here is an outline of the proof. It is easy to see that every Cauchy sequence $\{x_n\}$ is bounded, hence by the Bolzano–Weierstrass Theorem, has a convergent subsequence. Since the x_n's of a Cauchy sequence get near each other as $n \to \infty$, we expect that if a subsequence of $\{x_n\}$ converges, then the whole sequence also converges. We begin our formal proof at this point.

Lemma. *If $\{x_n\}$ is Cauchy and some subsequence x_{n_k} converges to x as $k \to \infty$, then $x_n \to x$ as $n \to \infty$.*

PROOF. Let $\varepsilon > 0$. Choose $N_1 \in \mathbf{N}$ such that

$$n, m \geq N_1 \quad \text{implies} \quad |x_n - x_m| < \frac{\varepsilon}{2}$$

and $N_2 \in \mathbf{N}$ such that

$$k \geq N_2 \quad \text{implies} \quad |x_{n_k} - x| < \frac{\varepsilon}{2}.$$

Fix $k \geq N_2$ such that $n_k \geq N_1$. Then

$$|x_n - x| \leq |x_n - x_{n_k}| + |x_{n_k} - x| < \varepsilon$$

for all $n \geq N_1$. Thus $x_n \to x$ as $n \to \infty$. ∎

THEOREM 1.20 [CAUCHY]. *A real sequence converges if and only if it is Cauchy.*

PROOF. By Remark 3, every convergent sequence is Cauchy.

Conversely, suppose $\{x_n\}$ is Cauchy. Choose $N \in \mathbf{N}$ such that $|x_N - x_m| < 1$ for all $m \geq N$. By the Triangle Inequality

$$|x_m| < 1 + |x_N| \quad \text{for} \quad m \geq N.$$

Therefore, $\{x_n\}$ is bounded by $M = \max\{|x_1|, |x_2|, \ldots, |x_{N-1}|, 1 + |x_N|\}$. It follows from the Bolzano–Weierstrass Theorem that $\{x_n\}$ has a convergent subsequence. In particular, we conclude from the lemma that $\{x_n\}$ converges. ∎

It is usually easier to show a sequence is Cauchy than to show it converges. Hence by Theorem 1.20, Cauchy sequences give us a powerful method for proving that a given sequence converges. This method is widely used because, as the following example shows, we can often prove that a sequence is Cauchy even when we have no idea what its limit is.

Example 2. Prove that any real sequence $\{x_n\}$ which satisfies

$$|x_n - x_{n+1}| \le \frac{1}{2^n}, \qquad n \in \mathbf{N}$$

is convergent.

PROOF. If $m > n$, then

$$
\begin{aligned}
|x_n - x_m| &= |x_n - x_{n+1} + x_{n+1} - x_{n+2} + \cdots + x_{m-1} - x_m| \\
&\le |x_n - x_{n+1}| + |x_{n+1} - x_{n+2}| + \cdots + |x_{m-1} - x_m| \\
&\le \frac{1}{2^n} + \cdots + \frac{1}{2^{m-1}} \\
&= \frac{1}{2^{n-1}} \sum_{k=1}^{m-n} \frac{1}{2^k} = \frac{1}{2^{n-1}} \left(1 - \frac{1}{2^{m-n}} \right).
\end{aligned}
$$

(This last step uses Exercise 1c in Section 1.2 for $a = 2$.) It follows that $|x_n - x_m| < 1/2^{n-1}$ for all integers $m > n \ge 1$. But given $\varepsilon > 0$ we can choose $N \in \mathbf{N}$ so large that $n \ge N$ implies $1/2^{n-1} < \varepsilon$. Thus $\{x_n\}$ is Cauchy, hence converges to some real number. ∎

The following result shows that a sequence is not necessarily Cauchy just because x_n is near x_{n+1} for large n.

Remark 4. *A sequence which satisfies $x_{n+1} - x_n \to 0$ is not necessarily Cauchy.*

PROOF. Consider the sequence $x_n := \log n$. By basic properties of logarithms (see Exercise 4 in Section 3.3),

$$x_{n+1} - x_n = \log(n + 1) - \log n = \log((n + 1)/n) \to \log 1 = 0$$

as $n \to \infty$. $\{x_n\}$ cannot be Cauchy, however, because it does not converge; in fact, it is not even bounded as $n \to \infty$. ∎

EXERCISES

1. Prove that

$$x_n = \frac{(n^2 + 20n + 35)\sin(n^3)}{n^2 + n + 1}$$

   has a convergent subsequence.

2. Let $\{x_n\}$ be a real sequence. Suppose for each $\varepsilon > 0$ there is an $N \in \mathbf{N}$ such that $m > n \geq N$ implies $\left| \sum_{k=n}^{m} x_k \right| < \varepsilon$. Prove that

$$\lim_{n \to \infty} \sum_{k=1}^{n} x_k$$

   exists.

3. Let $\{x_n\}$ be a real sequence. Suppose there is an $a > 1$ such that

$$|x_{k+1} - x_k| \leq a^{-k}$$

   for all $k \in \mathbf{N}$. Prove that $x_n \to x$ for some $x \in \mathbf{R}$.

4. Suppose $0 < x_1 < 1$ and $x_{n+1} = 1 - \sqrt{1 - x_n}$ for $n \in \mathbf{N}$. Prove $x_n \downarrow 0$ as $n \to \infty$ and $x_{n+1}/x_n \to 1/2$, as $n \to \infty$. (This is Exercise 4.3 in Apostol [1].)

5. Let $0 < x_1 \leq 3$ and $x_{n+1} = \sqrt{2x_n + 3}$ for $n \in \mathbf{N}$. Prove that $x_n \uparrow 3$ as $n \to \infty$.

6. Suppose $x_1 \geq 2$ and $x_{n+1} = 1 + \sqrt{x_n - 1}$ for $n \in \mathbf{N}$. Prove that $x_n \downarrow 2$ as $n \to \infty$. What happens when $1 \leq x_1 < 2$?

7. Prove

$$\lim_{n \to \infty} x^{1/(2n-1)} = \begin{cases} 1 & x > 0 \\ 0 & x = 0 \\ -1 & x < 0. \end{cases}$$

8. Suppose $x_0 \in \mathbf{R}$ and

$$x_n = \frac{1 + x_{n-1}}{2}$$

   for $n \in \mathbf{N}$. Prove $x_n \to 1$ as $n \to \infty$.

9. Let $0 < y_1 < x_1$ and set

$$x_{n+1} = \frac{x_n + y_n}{2} \quad \text{and} \quad y_{n+1} = \sqrt{x_n y_n} \qquad n \in \mathbf{N}.$$

   a) Prove $0 < y_n < x_n$ for all $n \in \mathbf{N}$.

   b) Prove y_n is increasing and bounded above, and x_n is decreasing and bounded below.

   c) Prove that $0 < x_{n+1} - y_{n+1} < (x_1 - y_1)/2^n$ for $n \in \mathbf{N}$.

   d) Prove that $\lim_{n \to \infty} x_n = \lim_{n \to \infty} y_n$. (This common value is called the *arithmetic-geometric* mean of x_1 and y_1.)

10. Suppose $x_0 = 1$, $y_0 = 0$,

$$x_n = x_{n-1} + 2y_{n-1}$$

and

$$y_n = x_{n-1} + y_{n-1}$$

for $n \in \mathbf{N}$. Prove that $x_n^2 - 2y_n^2 = \pm 1$ for $n \in \mathbf{N}$ and

$$\frac{x_n}{y_n} \to \sqrt{2} \quad \text{as } n \to \infty.$$

11. [Archimedes] Suppose $x_0 = 2\sqrt{3}$, $y_0 = 3$,

$$x_n = \frac{2x_{n-1}y_{n-1}}{x_{n-1} + y_{n-1}}$$

and

$$y_n = \sqrt{x_n y_{n-1}}$$

for $n \in \mathbf{N}$.

a) Prove that $x_n \downarrow x$ and $y_n \uparrow y$, as $n \to \infty$, for some $x, y \in \mathbf{R}$.

b) Prove $x = y$ and

$$3.14155 < x < 3.14161.$$

(The actual value of x is π.)

1.6   THE EXTENDED REAL NUMBER SYSTEM

The sequence $\{\log n\}_{n \in \mathbf{N}}$ fails to converge in a different way than $\{n(-1)^n\}_{n \in \mathbf{N}}$ does. (Indeed, the terms $\log n$ get steadily larger as $n \to \infty$ but the terms $n(-1)^n$ bounce back and forth between large positive values and large negative values.) It is sometimes convenient to emphasize this difference. We do this by adding two symbols, $+\infty$ and $-\infty$, to the real number system $\mathbf{R}$. We shall frequently denote $+\infty$ by ∞.

DEFINITION 1.11.   Let $\{x_n\}$ be a sequence of real numbers.

i) $\{x_n\}$ is said to *diverge* to $+\infty$ (notation: $x_n \to +\infty$ as $n \to \infty$ or $\lim_{n \to \infty} x_n = +\infty$) if given $M > 0$ there is an $N \in \mathbf{N}$ such that

$$n \geq N \quad \text{implies} \quad x_n > M.$$

ii) $\{x_n\}$ is said to *diverge* to $-\infty$ (notation: $x_n \to -\infty$ as $n \to \infty$ or $\lim_{n \to \infty} x_n = -\infty$) if given $M > 0$ there is an $N \in \mathbf{N}$ such that

$$n \geq N \quad \text{implies} \quad x_n < -M.$$

By an *extended real number* x we mean either $x \in \mathbf{R}$, $x = \infty$, or $x = -\infty$. With appropriate modifications, many of the results proved in Sections 1.4 and 1.5 can be generalized to include extended real numbers. Two such results we shall use frequently are the Monotone Convergence Theorem and the Squeeze Theorem (see Exercise 4).

Theorem 1.14 now takes on the following form.

THEOREM 1.21. *Suppose $\{x_n\}$ and $\{y_n\}$ are real sequences such that $x_n \to +\infty$ (respectively, $x_n \to -\infty$) as $n \to \infty$.*

  i) *If $y_n \geq -M_0$ (respectively, $y_n \leq M_0$) for some $M_0 > 0$ and all $n \in \mathbf{N}$, then*

$$\lim_{n\to\infty} (x_n + y_n) = +\infty \text{ (respectively, } \lim_{n\to\infty} (x_n + y_n) = -\infty).$$

  ii) *If $\alpha > 0$, then*

$$\lim_{n\to\infty} (\alpha x_n) = +\infty \text{ (respectively, } \lim_{n\to\infty} (\alpha x_n) = -\infty).$$

  iii) *If $y_n > M_0$ for some $M_0 > 0$ and all $n \in \mathbf{N}$, then*

$$\lim_{n\to\infty} (x_n y_n) = +\infty \text{ (respectively, } \lim_{n\to\infty} (x_n y_n) = -\infty).$$

  iv) *If $\{y_n\}$ is bounded and $x_n \neq 0$, then*

$$\lim_{n\to\infty} \frac{y_n}{x_n} = 0.$$

PROOF. We suppose for simplicity that $x_n \to \infty$ as $n \to \infty$.

i) By hypothesis, $y_n \geq -M_0$. Let $M > 0$ and set $M_1 = M_0 + M$. Since $x_n \to \infty$, choose $N \in \mathbf{N}$ such that $n \geq N$ implies $x_n > M_1$. Then $n \geq N$ implies $x_n + y_n > M_1 - M_0 = M$.

ii) Let $M > 0$ and set $M_1 = M/\alpha$. Choose $N \in \mathbf{N}$ such that $n \geq N$ implies $x_n > M_1$. Then $n \geq N$ implies $\alpha x_n > \alpha M_1 = M$.

iii) Let $M > 0$ and set $M_1 = M/M_0$. Choose $N \in \mathbf{N}$ such that $n \geq N$ implies $x_n > M_1$. Then $n \geq N$ implies $x_n y_n > M_1 M_0 = M$.

iv) Let $\varepsilon > 0$. Choose $M_0 \in \mathbf{R}$ such that $|y_n| \leq M_0$ and $M_1 \in \mathbf{R}$ so large that $M_0/M_1 < \varepsilon$. Choose $N \in \mathbf{N}$ such that $n \geq N$ implies $x_n > M_1$. Then $n \geq N$ implies

$$\left|\frac{y_n}{x_n}\right| = \frac{|y_n|}{x_n} < \frac{M_0}{M_1} < \varepsilon. \quad \blacksquare$$

If we adopt the following conventions

$$x + \infty = \infty, \quad x - \infty = -\infty, \qquad x \in \mathbf{R},$$

$$\infty + \infty = \infty, \quad -\infty - \infty = -\infty,$$

$$x \cdot \infty = \infty, \quad x \cdot (-\infty) = -\infty, \qquad x > 0,$$

$$x \cdot \infty = -\infty, \quad x \cdot (-\infty) = \infty, \qquad x < 0,$$

$$\infty \cdot \infty = (-\infty) \cdot (-\infty) = \infty, \quad \text{and} \quad \infty \cdot (-\infty) = (-\infty) \cdot \infty = -\infty,$$

then Theorem 1.21 contains the following corollary.

COROLLARY 1.22. *Let $\{x_n\}$, $\{y_n\}$ be real sequences, and α, x, y be extended real numbers. If $x_n \to x$ and $y_n \to y$, as $n \to \infty$, then*

$$\lim_{n\to\infty} (x_n + y_n) = x + y$$

(provided the right side is not of the form $\infty - \infty$), and

$$\lim_{n\to\infty} (\alpha x_n) = \alpha x, \qquad \lim_{n\to\infty} (x_n y_n) = xy$$

(provided none of these products is of the form $0 \cdot \pm\infty$).

We have avoided the cases $\infty - \infty$ and $0 \cdot \pm\infty$. These and other "indeterminate forms" will be covered by L'Hôpital's Rule in Section 2.5.

By using the symbols $+\infty$ and $-\infty$, we can extend operations of suprema and infima to arbitrary sets $E \subseteq \mathbf{R}$. Indeed, set $\sup E = +\infty$ if E is unbounded above and $\inf E = -\infty$ if E is unbounded below. Notice that $\sup E$ (respectively, $\inf E$) is finite only when E is bounded above (respectively, bounded below). Also notice that under the convention $-\infty \le a$ and $a \le \infty$ for all $a \in \mathbf{R}$, the Monotone Property still holds; i.e., if $A \subseteq B$ are sets of real numbers, then $\sup A \le \sup B$ and $\inf A \ge \inf B$.

In some situations (for example, the Root Test in Section 4.3), we shall use the following generalization of limits.

DEFINITION 1.12. Let $\{x_n\}$ be a real sequence. Then the *limit supremum* of $\{x_n\}$ is the extended real number

(15)
$$\limsup_{n\to\infty} x_n := \lim_{n\to\infty} (\sup_{k\ge n} x_k),$$

and the *limit infimum* of $\{x_n\}$ is the extended real number

$$\liminf_{n\to\infty} x_n := \lim_{n\to\infty} (\inf_{k\ge n} x_k).$$

Before we proceed, we must show that the limits in Definition 1.12 exist as extended real numbers. To this end, let $\{x_n\}$ be a sequence of real numbers and consider the sequences

$$s_n = \sup_{k\ge n} x_k := \sup\{x_k : k \ge n\} \quad \text{and} \quad t_n = \inf_{k\ge n} x_k := \inf\{x_k : k \ge n\}.$$

Each s_n and t_n is an extended real number and, by the Monotone Property, s_n is a decreasing sequence and t_n an increasing sequence of extended real numbers. In particular, there exist extended real numbers s and t such that $s_n \downarrow s$ and $t_n \uparrow t$ as $n \to \infty$. These extended real numbers are, by Definition 1.12, the limit infimum and limit supremum of the sequence $\{x_n\}$.

Here are two examples of how to compute limits supremum and limits infimum.

Example 1. Find $\limsup_{n\to\infty} x_n$ and $\liminf_{n\to\infty} x_n$ if $x_n = (-1)^n$.

SOLUTION.   Since $\sup_{k \geq n}(-1)^k = 1$ for all $n \in \mathbf{N}$, it is clear that $\limsup_{n \to \infty} x_n = 1$. Similarly, $\liminf_{n \to \infty} x_n = -1$. ∎

Example 2. Find $\limsup_{n \to \infty} x_n$ and $\liminf_{n \to \infty} x_n$ if $x_n = 1 + 1/n$.

SOLUTION. Since $\sup_{k \geq n}(1 + 1/k) = 1 + 1/n$ for all $n \in \mathbf{N}$, $\limsup_{n \to \infty} x_n = 1$. Since $\inf_{k \geq n}(1 + 1/k) = 1$ for all $n \in \mathbf{N}$, $\liminf_{n \to \infty} x_n = 1$. ∎

These examples suggest that there is a connection between limits supremum, limits infimum, and convergent subsequences. The next several results make this connection clear.

THEOREM 1.23. *Let $\{x_n\}$ be a sequence of real numbers and*

$$s = \limsup_{n \to \infty} x_n, \qquad t = \liminf_{n \to \infty} x_n.$$

Then there are subsequences $\{x_{n_k}\}_{k \in \mathbf{N}}$ and $\{x_{\ell_j}\}_{j \in \mathbf{N}}$ such that $x_{n_k} \to s$ as $k \to \infty$ and $x_{\ell_j} \to t$ as $j \to \infty$.

PROOF. We will prove the result for the limit supremum. A similar argument establishes the result for the limit infimum.

Let $s_n = \sup_{k \geq n} x_k$ and observe that $s_n \downarrow s$ as $n \to \infty$.

Case 1. $s = \infty$. Then by definition $s_n = \infty$ for all $n \in \mathbf{N}$. Since $s_1 = \infty$, there is an $n_1 \in \mathbf{N}$ such that $x_{n_1} > 1$. Since $s_{n_1 + 1} = \infty$, there is an $n_2 \geq n_1 + 1 > n_1$ such that $x_{n_2} > 2$. Continuing in this manner, we can choose a subsequence $\{x_{n_k}\}$ such that $x_{n_k} > k$ for all $k \in \mathbf{N}$. Hence, it follows from the Squeeze Theorem that $x_{n_k} \to \infty = s$ as $k \to \infty$.

Case 2. $s = -\infty$. Since $s_n \geq x_n$ for all $n \in \mathbf{N}$, it follows from the Squeeze Theorem that $x_n \to -\infty = s$ as $n \to \infty$.

Case 3. $-\infty < s < \infty$. Set $n_0 = 0$. By Theorem 1.6 (the Approximation Property), there is an integer $n_1 \in \mathbf{N}$ such that $s_{n_0 + 1} - 1 < x_{n_1} \leq s_{n_0 + 1}$. Similarly, there is an integer $n_2 \geq n_1 + 1 > n_1$ such that $s_{n_1 + 1} - 1/2 < x_{n_2} \leq s_{n_1 + 1}$. Continuing in this manner, we can choose integers $n_1 < n_2 < \ldots$ such that

$$(16) \qquad\qquad s_{n_{k-1} + 1} - \frac{1}{k} < x_{n_k} \leq s_{n_{k-1} + 1}$$

for $k \in \mathbf{N}$. Since $s_{n_{k-1} + 1} \to s$ as $k \to \infty$, we conclude by the Squeeze Theorem that $x_{n_k} \to s$ as $k \to \infty$. ∎

This observation leads directly to a characterization of limits in terms of limits infimum and limits supremum.

THEOREM 1.24. *Let $\{x_n\}$ be a real sequence and x be an extended real number. Then $x_n \to x$ as $n \to \infty$ if and only if*

$$(17) \qquad\qquad \limsup_{n \to \infty} x_n = \liminf_{n \to \infty} x_n = x.$$

PROOF. Suppose $x_n \to x$ as $n \to \infty$. Then $x_{n_k} \to x$ as $k \to \infty$ for all subsequences $\{x_{n_k}\}$. Hence, by Theorem 1.23, $\limsup_{n \to \infty} x_n = x$ and $\liminf_{n \to \infty} x_n = x$, i.e., (17) holds.

Conversely, suppose (17) holds.

Case 1. $x = \pm\infty$. By considering $\pm x_n$ we may suppose that $x = \infty$. Thus given $M > 0$ there is an $N \in \mathbf{N}$ such that $\inf_{k \geq N} x_k > M$. It follows that $x_n > M$ for all $n \geq N$, i.e., $x_n \to \infty$ as $n \to \infty$.

Case 2. $-\infty < x < \infty$. Let $\varepsilon > 0$. Choose $N \in \mathbf{N}$ such that

$$\sup_{k \geq N} x_k - x < \frac{\varepsilon}{2} \quad \text{and} \quad x - \inf_{k \geq N} x_k < \frac{\varepsilon}{2}.$$

Let $n, m \geq N$ and suppose for simplicity that $x_n > x_m$. Then

$$|x_n - x_m| = x_n - x_m \leq \sup_{k \geq N} x_k - x + x - \inf_{k \geq N} x_k < \frac{\varepsilon}{2} + \frac{\varepsilon}{2} = \varepsilon.$$

Thus $\{x_n\}$ is Cauchy and converges to some finite real number. But by Theorem 1.23, some subsequence of $\{x_n\}$ converges to x. We conclude that $x_n \to x$ as $n \to \infty$. ∎

Theorem 1.23 also leads to the following geometric interpretation of limits supremum and limits infimum.

Remark 1. *Let $\{x_n\}$ be a sequence of real numbers. Then $\limsup_{n \to \infty} x_n$ (respectively, $\liminf_{n \to \infty} x_n$) is the largest value (respectively, the smallest value) to which some subsequence of $\{x_n\}$ can converge. Namely, if $x_{n_k} \to x$ as $k \to \infty$, then*

$$(18) \qquad \liminf_{n \to \infty} x_n \leq x \leq \limsup_{n \to \infty} x_n.$$

PROOF. Suppose $x_{n_k} \to x$ as $k \to \infty$. Fix $N \in \mathbf{N}$ and choose K so large that $k \geq K$ implies $n_k \geq N$. Clearly,

$$\inf_{j \geq N} x_j \leq x_{n_k} \leq \sup_{j \geq N} x_j$$

for all $k \geq K$. Taking the limit of this inequality as $k \to \infty$, we obtain

$$\inf_{j \geq N} x_j \leq x \leq \sup_{j \geq N} x_j.$$

Taking the limit of this last inequality as $N \to \infty$ and applying Definition 1.12, we obtain (18). ∎

We end this section with several other properties of limits supremum and limits infimum.

Remark 2. *If $\{x_n\}$ is any sequence of real numbers, then*

$$\liminf_{n \to \infty} x_n \leq \limsup_{n \to \infty} x_n.$$

PROOF. Since $\inf_{k \geq n} x_k \leq \sup_{k \geq n} x_k$ for all $n \in \mathbf{N}$, this inequality follows from Theorem 1.15 (the Comparison Theorem). ∎

The following result is an immediate consequence of Definition 1.12, the Comparison Theorem, and the Monotone Convergence Theorem.

Remark 3. *A real sequence $\{x_n\}$ is bounded above if and only if $\limsup_{n\to\infty} x_n < \infty$, and is bounded below if and only if $\liminf_{n\to\infty} x_n > -\infty$.*

The following result shows we can take limits supremum and limits infimum of inequalities.

THEOREM 1.25. *If $x_n \leq y_n$ for n large, then*

$$(19) \qquad \limsup_{n\to\infty} x_n \leq \limsup_{n\to\infty} y_n \quad \text{and} \quad \liminf_{n\to\infty} x_n \leq \liminf_{n\to\infty} y_n.$$

PROOF. If $x_k \leq y_k$ for $k \geq N$, then $\sup_{k\geq n} x_k \leq \sup_{k\geq n} y_k$ and $\inf_{k\geq n} x_k \leq \inf_{k\geq n} y_k$ for any $n \geq N$. Taking the limit of these inequalities as $n \to \infty$, we obtain (19). ∎

EXERCISES

1. Find the limit infimum and the limit supremum of each of the following sequences.

 a) $x_n = 3 - (-1)^n$.
 b) $x_n = \cos(n\pi/2)$.
 c) $x_n = (-1)^{n+1} + (-1)^n/n$.
 d) $x_n = \sqrt{1+n^2}/(2n-5)$.
 e) $x_n = y_n/n$, where $\{y_n\}$ is any bounded sequence.
 f) $x_n = n(1+(-1)^n) + n^{-1}((-1)^n - 1)$.
 g) $x_n = (n^3 + n^2 - n + 1)/(n^2 + 2n + 5)$.

2. Suppose $\{x_n\}$ and $\{y_n\}$ are real sequences. Prove that

$$-\limsup_{n\to\infty} x_n = \liminf_{n\to\infty}(-x_n)$$

 and

$$-\liminf_{n\to\infty} x_n = \limsup_{n\to\infty}(-x_n).$$

3. Let $\{x_n\}$ be a real sequence and $r \in \mathbf{R}$.

 a) Prove that

$$\limsup_{n\to\infty} x_n < r \quad \text{implies} \quad x_n < r$$

 for n large.
 b) Prove that

$$\limsup_{n\to\infty} x_n > r \quad \text{implies} \quad x_n > r$$

 for infinitely many $n \in \mathbf{N}$.

4. a) Prove that every monotone sequence of real numbers has a limit in the extended real number sense.
 b) Suppose x is an extended real number and $\{x_n\}$, $\{y_n\}$, and $\{w_n\}$ are real sequences. If $x_n \to x$ and $y_n \to x$, as $n \to \infty$, and $x_n \leq w_n \leq y_n$ for $n \in \mathbf{N}$, prove that $w_n \to x$ as $n \to \infty$.

5. Suppose $\{x_n\}$ and $\{y_n\}$ are real sequences.

   a) Prove that

$$\liminf_{n\to\infty} x_n + \liminf_{n\to\infty} y_n \le \liminf_{n\to\infty}(x_n + y_n)$$
$$\le \limsup_{n\to\infty} x_n + \liminf_{n\to\infty} y_n$$
$$\le \limsup_{n\to\infty}(x_n + y_n) \le \limsup_{n\to\infty} x_n + \limsup_{n\to\infty} y_n,$$

   provided none of these sums is of the form $\infty - \infty$.

   b) Show that if $\lim_{n\to\infty} x_n$ exists, then

$$\liminf_{n\to\infty}(x_n + y_n) = \lim_{n\to\infty} x_n + \liminf_{n\to\infty} y_n$$

   and

$$\limsup_{n\to\infty}(x_n + y_n) = \lim_{n\to\infty} x_n + \limsup_{n\to\infty} y_n.$$

   c) Show by examples that each of the inequalities in part a) can be strict.

6. Let $\{x_n\}$ and $\{y_n\}$ be real sequences.

   a) Suppose that $x_n \ge 0$ and $y_n \ge 0$ for each $n \in \mathbf{N}$. Prove

$$\limsup_{n\to\infty}(x_n y_n) \le \left(\limsup_{n\to\infty} x_n\right)\left(\limsup_{n\to\infty} y_n\right),$$

   provided the product on the right is not of the form $0 \cdot \infty$. Show by example that this inequality can be strict.

   b) Suppose $x_n \le 0 \le y_n$ for $n \in \mathbf{N}$. Prove

$$\left(\liminf_{n\to\infty} x_n\right)\left(\limsup_{n\to\infty} y_n\right) \le \liminf_{n\to\infty}(x_n y_n),$$

   provided none of these products is of the form $0 \cdot \infty$.

7. Suppose that $x_n \ge 0$ and $y_n \ge 0$ for all $n \in \mathbf{N}$. Prove that if $x_n \to x$ as $n \to \infty$ (x may be an extended real number), then

$$\limsup_{n\to\infty}(x_n y_n) = x \limsup_{n\to\infty} y_n$$

   provided none of these products is of the form $0 \cdot \infty$.

8. Prove that

$$\limsup_{n\to\infty} x_n = \inf_{n\in\mathbf{N}}\left(\sup_{k\ge n} x_k\right) \quad \text{and} \quad \liminf_{n\to\infty} x_n = \sup_{n\in\mathbf{N}}\left(\inf_{k\ge n} x_k\right)$$

   for any real sequence $\{x_n\}$.

9. Suppose $x_n \ge 0$ for $n \in \mathbf{N}$. Under the interpretation $1/0 = \infty$ and $1/\infty = 0$,

prove

$$\limsup_{n\to\infty}\left(\frac{1}{x_n}\right) = \frac{1}{\liminf_{n\to\infty} x_n} \quad \text{and} \quad \liminf_{n\to\infty}\left(\frac{1}{x_n}\right) = \frac{1}{\limsup_{n\to\infty} x_n}.$$

10. Let $x_n \in \mathbf{R}$. Prove that $x_n \to 0$ as $n \to \infty$ if and only if

$$\limsup_{n\to\infty} |x_n| \leq 0.$$

1.7 FUNCTIONS, COUNTABILITY AND THE ALGEBRA OF SETS

In this section we examine the role functions play in distinguishing one kind of infinite set from another and use this point of view to obtain more information about the special subsets of $\mathbf{R}$ introduced in Section 1.1. We also introduce the algebra of sets and examine what happens to it under images and inverse images by functions. (Discussion of the algebra of functions is postponed until Section 2.1.)

We begin with some preliminary remarks. For the first half of this course, most of the functions we consider will be *real-valued functions of a real variable*, i.e., functions $f : E \to \mathbf{R}$ where $E \subseteq \mathbf{R}$. When such a function f is given by a formula, the *domain* of f is defined to be the largest subset of $\mathbf{R}$ on which f is defined and real-valued. For example, if $f(x) = 1/x$, then $\mathrm{Dom}\,(f) = \{x \in \mathbf{R} : x \neq 0\}$. We will also use the usual abuse of notation by representing such functions by their formulas, e.g., calling $1/x$ a function.

We assume you are familiar with the trigonometric functions (whose formulas are) represented by $\sin x$, $\cos x$, $\tan x$, $\cot x$, $\sec x$, $\csc x$, the natural logarithm $\log x$ and its inverse e^x, and the power functions x^α which are defined by

$$x^\alpha := e^{\alpha \log x}, \qquad x > 0, \quad \alpha \in \mathbf{R}.$$

We also assume you can differentiate algebraic combinations of these functions using the basic formulas

$$(\sin x)' = \cos x, \quad (\cos x)' = -\sin x, \quad (\tan x)' = \sec^2 x, \quad (e^x)' = e^x, \quad x \in \mathbf{R},$$

$$(\log x)' = \frac{1}{x}, \quad \text{and} \quad (x^\alpha)' = \alpha x^{\alpha-1}, \quad x > 0, \quad \alpha \in \mathbf{R}.$$

(For a derivation of these identities based on fundamental properties, see Exercise 7 in Section 2.4 and Exercises 4 and 5 in Section 3.3.) Even with these assumptions, we shall repeat some material from elementary calculus.

Frequently, one or more real-valued functions of a real variable will be defined *implicitly* by a relation on $\mathbf{R}^2 := \mathbf{R} \times \mathbf{R}$. If the relation is simple enough, formulas for such functions can be produced by solving the relation for one of its variables. The next two examples illustrate this principle.

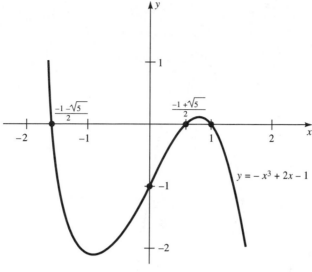

Figure 1.4

Example 1. Find the real-valued function $y = f(x)$ implicitly defined by the relation

$$x^3 + 3e^y = 2x - 1.$$

SOLUTION. Solving for e^y, we have $e^y = (-x^3 + 2x - 1)/3$. Taking the logarithm of this identity, we obtain

$$y = f(x) = \log(-x^3 + 2x - 1) - \log 3.$$

To find the domain of f, recall that $\log x$ is defined (and real-valued) only when $x > 0$. Since the roots of $-x^3 + 2x - 1$ are $x = 1, (-1 \pm \sqrt{5})/2$, the domain of f is $(-\infty, (-1 - \sqrt{5})/2) \cup ((-1 + \sqrt{5})/2, 1)$ (see Figure 1.4). ∎

Example 2. Find a formula for all continuous real-valued functions $y = f(x)$ implicitly defined by

$$x^2 + y^2 = 2x + y - 1.$$

(Note: A formal discussion of continuity appears in Section 2.2. For now, it is enough to recall that a function is continuous on an interval when its graph there is unbroken. The reason for including continuity here is that this relation actually determines infinitely many functions implicitly, but only two continuous ones.)

SOLUTION. Clearly, $y^2 - y + (x^2 - 2x + 1) = 0$ is a quadratic in y. By the quadratic formula,

$$y = \frac{1 \pm \sqrt{1 - 4(x^2 - 2x + 1)}}{2} = \frac{1 \pm \sqrt{-4x^2 + 8x - 3}}{2}.$$

Thus, the given relation determines two continuous functions,

$$f(x) = \frac{1 + \sqrt{-4x^2 + 8x - 3}}{2} \quad \text{and} \quad g(x) = \frac{1 - \sqrt{-4x^2 + 8x - 3}}{2}.$$

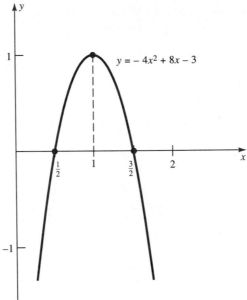

Figure 1.5

Since these functions will be real-valued only when $-4x^2 + 8x - 3 \geq 0$ and since this last quadratic has roots $x = 1/2, 3/2$, the domains of f and g are $[1/2, 3/2]$ (see Figure 1.5). ∎

There is a simple geometric explanation of the fact that the relation in Example 2 determines two continuous functions f and g. By completing the square, we see that this relation describes a circle centered at $(1, 1/2)$ of radius $1/2$:

$$(x - 1)^2 + (y - \frac{1}{2})^2 = \frac{1}{4}.$$

Thus, f represents the upper semicircle and g represents the lower semicircle.

Let $f : X \to Y$. Although each $x \in X$ is assigned a unique $y \in Y$, there is nothing in the definition of functions which keeps two x's from being assigned to the same y, and nothing which says every $y \in Y$ corresponds to some $x \in X$. Functions which satisfy these additional properties are important enough to warrant separate terminology.

DEFINITION 1.13. Let f be a function from a set X into a set Y.

i) f is said to be *one–to–one* (1–1) on X if

$$x_1, x_2 \in X \quad \text{and} \quad f(x_1) = f(x_2) \quad \text{imply} \quad x_1 = x_2.$$

ii) f is said to take X *onto* Y if for each $y \in Y$ there is an $x \in X$ such that $y = f(x)$.

For example, the function $f(x) = x^2$ is 1–1 from $[0, \infty)$ onto $[0, \infty)$ but not 1–1 on any open interval containing 0. Here is a simple, useful characterization of functions which are 1–1 from a set X onto a set Y.

THEOREM 1.26. *Let X and Y be sets and $f : X \to Y$. Then f is 1–1 from X onto Y if and only if there is a unique function g from Y onto X which satisfies*

$$(18) \qquad\qquad f(g(y)) = y, \qquad y \in Y,$$

and

$$(19) \qquad\qquad g(f(x)) = x, \qquad x \in X.$$

PROOF. Suppose f is 1–1 and onto. Let $y \in Y$ and choose a unique $x \in X$ such that $f(x) = y$. Define $g(y) = x$. By definition, (18) and (19) are satisfied. Moreover, it is clear that g takes Y onto X.

Conversely, suppose there is a function g from Y onto X which satisfies (18) and (19). If $x_1, x_2 \in X$ and $f(x_1) = f(x_2)$, then it follows from (19) that $x_1 = g(f(x_1)) = g(f(x_2)) = x_2$. Thus f is 1–1 on X. If $y \in Y$ and $x = g(y)$, then (18) implies that $f(x) = f(g(y)) = y$. Thus f takes X onto Y.

Finally, suppose h is another function which satisfies (18) and (19), and $y \in Y$. Choose $x \in X$ such that $f(x) = y$. Then by (19),

$$h(y) = h(f(x)) = x = g(f(x)) = g(y),$$

i.e., $h = g$ on Y. It follows that the function g is unique. ∎

If f is 1–1 from a set X onto a set Y, we shall say that *f has an inverse function*. We shall call the function g given in Theorem 1.26 the *inverse* of f, and denote it by f^{-1}. (Note: This is different from the function $(f(x))^{-1} := 1/f(x)$.) Notice by (18) and (19) that

$$f(f^{-1}(y)) = y \quad \text{and} \quad f^{-1}((f(x)) = x$$

for all $y \in Y$ and $x \in X$.

Let f be a real-valued function of a real variable. If f has an inverse function f^{-1} and $y = f(x)$, we have by definition that $(x, f(x)) = (f^{-1}(y), y)$. Hence, the graph of $y = f^{-1}(x)$ is a reflection of the graph of $y = f(x)$ about the line $y = x$ (see Figure 1.6).

The following example shows that the inverse function $y = f^{-1}(x)$ can sometimes be found by treating $y = f(x)$ as a relation which implicitly defines $x = f^{-1}(y)$.

Example 3. Prove $f(x) = e^x - e^{-x}$ is 1–1 on **R** and find a formula for f^{-1}.

SOLUTION. Since $f'(x) = e^x + e^{-x}$ is always positive, the function f is strictly increasing on **R** (see Theorem 2.19); i.e., $a < b$ implies $f(a) < f(b)$. In particular, f is 1–1 on **R**.

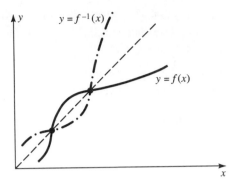

Figure 1.6

To find f^{-1}, let $y = e^x - e^{-x}$. Multiplying this equation by e^x and collecting all nonzero terms on one side of the equation, we have

$$e^{2x} - ye^x - 1 = 0,$$

a quadratic in e^x. By the quadratic formula,

$$e^x = \frac{y \pm \sqrt{y^2 + 4}}{2}.$$

Since e^x is always positive, the minus sign must be discarded. Taking the logarithm of this last identity, we obtain $x = \log(y + \sqrt{y^2 + 4}) - \log 2$. Therefore,

$$f^{-1}(x) = \log(x + \sqrt{x^2 + 4}) - \log 2. \quad \blacksquare$$

Functions which have inverses can be used to "count" infinite sets. Before we make a formal definition, let us examine what it means to count a finite set of objects E. When we count E we assign a number $n \in \mathbf{N}$ to each object in E; i.e., we construct a function f from a subset of $\mathbf{N}$ to E. For example, if E has three objects, then the "counting" function takes $\{1, 2, 3\}$ to E. Now in order to count E properly, we must be careful to avoid two pitfalls. We must not count any element of E more than once (i.e., f must be 1–1), and we cannot miss any element of E (i.e., f must take $\{1, 2, 3\}$ onto E). Accordingly, we make the following definition.

DEFINITION 1.14. Let E be a set.

  i) E is said to be *finite* if $E = \emptyset$ or if there is an $n \in \mathbf{N}$ and a 1–1 function from $\{1, 2, \ldots, n\}$ onto E.
  ii) E is said to be *countably infinite* if there is a 1–1 function from $\mathbf{N}$ onto E.
  iii) E is said to be *countable* if E is either finite or countably infinite.
  iv) E is said to be *uncountable* if E is not countable.

Loosely speaking, a set is countably infinite if it has the same number of elements as $\mathbf{N}$, finite if it has less, uncountable if it has more. The next result shows that two infinite sets can have the same number of elements even though one is a proper subset of the other. (In fact, this property can be used as a definition of "infinite set.")

Remark 1. *The set of even integers $E = \{2, 4, \dots\}$ is countably infinite, i.e., has the same number of elements as* **N**.

PROOF. Let $f(k) = 2k$. It is clear that f takes **N** onto E. Since $2k = 2j$ implies $k = j$, it is also clear that f is 1–1. Thus E is countably infinite. ∎

The following result shows that not every infinite set is countably infinite.

Remark 2 [CANTOR'S DIAGONALIZATION ARGUMENT]. *The open interval $(0, 1)$ is uncountable.*

PROOF. Suppose to the contrary that there is a 1–1 function f from **N** onto the interval $(0, 1)$. Write the numbers $f(j)$, $j \in$ **N**, in decimal notation using the finite expansion when possible, i.e.,

$$f(1) = 0.\alpha_{11}\alpha_{12}\dots,$$

$$f(2) = 0.\alpha_{21}\alpha_{22}\dots,$$

$$f(3) = 0.\alpha_{31}\alpha_{32}\dots,$$

$$\dots,$$

where α_{ij} represents the jth digit in the decimal expansion of $f(i)$ and none of these expansions terminates in 9's (see the note following Exercise 10 in Section 1.4). Since f is onto, given any real number $x \in (0, 1)$, there is a $j \in$ **N** such that $f(j) = x$. Consider the number x whose decimal expansion is given by $0.\beta_1\beta_2\dots,$ where

$$\beta_k := \begin{cases} 8 & \text{if } \alpha_{kk} \neq 8 \text{ or } 9 \\ 0 & \text{if } \alpha_{kk} = 8 \text{ or } 9. \end{cases}$$

Since we have used the finite expansion when possible, each number from the interval $(0, 1)$ has a unique decimal expansion. By construction, the decimal expansions of x and $f(j)$ differ at the jth place. Therefore, $x \neq f(j)$, a contradiction. ∎

It is natural to ask about the countability of the sets **Z**, **Q**, and **R**. To answer these questions, we prove the following result.

THEOREM 1.27. *If $A \subseteq B$ and B is countable, then A is countable.*

PROOF. If A is finite, then A is countable by definition. If A is infinite, then B is infinite. Thus by hypothesis, there is a 1–1 function f from **N** onto B. Since $A \subseteq B$ and f is onto, the set $\{k \in$ **N** $: f(k) \in A\}$ is nonempty. Let k_1 be the least element of this set. Since A is infinite, the set $\{k \in$ **N** $: f(k) \in A \setminus \{f(k_1)\}\}$ is nonempty, so it has a least element, say k_2. By induction, there exist integers $k_1 < k_2 < \dots$ such that k_j is the least element of the set

$$\{k \in \mathbf{N} : f(k) \in A \setminus \{f(k_1), f(k_2), \dots f(k_{j-1})\}\}.$$

Since each k_j is least, it is clear that

(20) $$A = \{f(k_1), f(k_2), \dots\}.$$

Consider the function $g : \mathbf{N} \to B$ defined by $g(j) = f(k_j)$, $j \in \mathbf{N}$. Since f is 1–1, g is also 1–1. By (20), g takes $\mathbf{N}$ onto A. Hence, A is countably infinite by Definition 1.14. $\blacksquare$

Remark 3. *If A is uncountable and $A \subseteq B$, then B is uncountable. In particular, $\mathbf{R}$ is uncountable.*

PROOF. If B were countable, then A would also be countable by Theorem 1.27. This proves the first statement. Since the interval $(0, 1)$ is uncountable (by Remark 2) and a subset of $\mathbf{R}$, it follows that $\mathbf{R}$ is uncountable. $\blacksquare$

We now prove that a countable union of countable sets is countable.

THEOREM 1.28. *Let $A_1, A_2, \ldots$ be sets and*

$$E = \bigcup_{j \in \mathbf{N}} A_j := \bigcup_{j=1}^{\infty} A_j := \{x : x \in A_j \quad \text{for some } j \in \mathbf{N}\}.$$

If each A_j is countable, then so is E.

PROOF. By Theorem 1.27, we may suppose that each A_j is countably infinite and $A_j \bigcap A_k = \emptyset$ for $k \neq j$. Thus, for each $j \in \mathbf{N}$ we can write

$$A_j = \{a_{jk} : k \in \mathbf{N}\}$$

with $a_{jk} = a_{pq}$ if and only if $j = p$ and $k = q$.

List the elements of E as an infinite matrix with elements of A_j forming the jth row. This suggests that E can be "counted" in rising diagonals (follow the arrows in Figure 1.7).

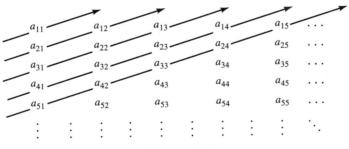

Figure 1.7

To explicitly construct a 1–1 function from E onto $\mathbf{N}$ which embodies this suggestion, set $f(a_{11}) = 1$, $f(a_{21}) = 2$, and $f(a_{12}) = 3$. To describe what happens to the Nth rising diagonal, notice that its elements are given by

$$a_{N,1}, a_{N-1,2}, a_{N-2,3}, \ldots, a_{2,N-1}, a_{1,N},$$

i.e., it contains N elements a_{jk} whose indices satisfy $j + k = N + 1$. Since the sum of integers $1 + 2 + \cdots + (N - 1)$ is given by $(N - 1)N/2$ (see Exercise 1a in Section

1.2), there are $(N-1)N/2$ elements in the first $N-1$ rising diagonals of E. Thus, define f on the Nth rising diagonal by

$$f(a_{jk}) = \frac{(N-1)N}{2} + k = \frac{(j+k-1)(j+k)}{2} + k.$$

By construction, f is a 1–1 function from E onto $\mathbf{N}$; i.e., f^{-1} is a 1–1 function from $\mathbf{N}$ onto E. In particular, E is countably infinite. ∎

Remark 4. *The sets $\mathbf{Z}$ and $\mathbf{Q}$ are countably infinite.*

PROOF. $\mathbf{Z} = \mathbf{N} \cup (-\mathbf{N}) \cup \{0\}$ and $\mathbf{Q} = \bigcup_{n=1}^{\infty} \{p/n : p \in \mathbf{Z}\}$. ∎

Remark 5. *The set of irrationals is uncountable.*

PROOF. If $\mathbf{R} \setminus \mathbf{Q}$ were countable, then $\mathbf{R} = (\mathbf{R} \setminus \mathbf{Q}) \cup \mathbf{Q}$ would also be countable, a contradiction of Remark 3. ∎

Theorem 1.28 had something to say about a countable union of sets. In Chapter 5, we need to deal with uncountable unions and intersections. Here is some notation which will prove useful in this regard. A collection of sets $\mathcal{E}$ is said to be *indexed by* a set A if there is a function F from A onto $\mathcal{E}$. In this case A is called the *index set* of $\mathcal{E}$, and we shall represent $F(\alpha)$ by E_α. In particular, we shall represent a collection of sets indexed by A by

$$\mathcal{E} = \{E_\alpha\}_{\alpha \in A}.$$

Notice, then, that a sequence of sets is a collection of sets indexed by $\mathbf{N}$.

DEFINITION 1.15. Let $\mathcal{E} = \{E_\alpha\}_{\alpha \in A}$ be a collection of sets.
i) The *union* of the collection $\mathcal{E}$ is the set

$$\bigcup_{\alpha \in A} E_\alpha := \{x : x \in E_\alpha \quad \text{for some } \alpha \in A\}.$$

ii) The *intersection* of the collection $\mathcal{E}$ is the set

$$\bigcap_{\alpha \in A} E_\alpha := \{x : x \in E_\alpha \quad \text{for all } \alpha \in A\}.$$

There is an easy way to get from unions to intersections and vice versa.

THEOREM 1.29 [DEMORGAN'S LAWS]. *Let X be a set and $\{E_\alpha\}_{\alpha \in A}$ be a collection of subsets of X. If for each $E \subseteq X$ the symbol E^c represents the set $X \setminus E$, then*

(21) $$\left(\bigcup_{\alpha \in A} E_\alpha \right)^c = \bigcap_{\alpha \in A} E_\alpha^c$$

and

$$(22) \qquad \left(\bigcap_{\alpha \in A} E_\alpha \right)^c = \bigcup_{\alpha \in A} E_\alpha^c$$

PROOF. Suppose x belongs to the left side of (21), i.e., $x \in X$ and $x \notin \bigcup_{\alpha \in A} E_\alpha$. By definition, $x \in X$ and $x \notin E_\alpha$ for all $\alpha \in A$. Hence $x \in E_\alpha^c$ for all $\alpha \in A$, i.e., x belongs to the right side of (21). These steps are reversible. This verifies (21). A similar argument verifies (22). ∎

The following concepts will be used frequently in subsequent chapters.

DEFINITION 1.16. Let X and Y be sets and $f : X \to Y$. The *image* of a set $E \subseteq X$ under f is the set

$$f(E) := \{ y \in Y : y = f(x) \text{ for some } x \in E \}.$$

The *inverse image* of a set $E \subseteq Y$ under f is the set

$$(23) \qquad f^{-1}(E) := \{ x \in X : f(x) = y \text{ for some } y \in E \}.$$

Notice that equation (23) makes sense whether f is 1–1 or not; i.e., f need not be 1–1 for $f^{-1}(E)$ to be defined. In particular, $f^{-1}(E)$ is the inverse image of E under f, not the image of E under the inverse function f^{-1}. In fact, the inverse function f^{-1} exists on $f(X)$ if and only if the inverse image $f^{-1}(\{y\})$ contains at most one point for all $y \in Y$.

By definition, $f(f^{-1}(E)) = E$ for any $E \subseteq f(X)$ and $f^{-1}(f(E)) \supseteq E$ for any $E \subseteq \text{Dom } f$. This last inequality is strict unless f is 1–1 (see Exercise 6 below). For example, if $f : \mathbf{R} \to \mathbf{R}$ is defined by $f(x) = x^2$ and $E = [0, 1)$, then $f(E) = [0, 1)$, but $f^{-1}(f(E)) = (-1, 1)$.

The following result, which plays a prominent role in Chapters 5 and 7, describes images and inverse images of unions and intersections of sets.

THEOREM 1.30. *Let X and Y be sets and $f : X \to Y$.*

i) If $\{E_\alpha\}_{\alpha \in A}$ is a collection of subsets of X, then

$$f \left(\bigcup_{\alpha \in A} E_\alpha \right) = \bigcup_{\alpha \in A} f(E_\alpha) \quad \text{and} \quad f \left(\bigcap_{\alpha \in A} E_\alpha \right) \subseteq \bigcap_{\alpha \in A} f(E_\alpha).$$

Moreover, if $B \subseteq C \subseteq X$ then

$$f(C \setminus B) \supseteq f(C) \setminus f(B).$$

ii) If $\{E_\alpha\}_{\alpha \in A}$ is a collection of subsets of Y, then

$$f^{-1} \left(\bigcup_{\alpha \in A} E_\alpha \right) = \bigcup_{\alpha \in A} f^{-1}(E_\alpha) \quad \text{and} \quad f^{-1} \left(\bigcap_{\alpha \in A} E_\alpha \right) = \bigcap_{\alpha \in A} f^{-1}(E_\alpha).$$

Moreover, if $B \subseteq C \subseteq Y$ then

$$f^{-1}(C \setminus B) = f^{-1}(C) \setminus f^{-1}(B).$$

PROOF. By definition, $y \in f(\cup_{\alpha \in A} E_\alpha)$ if and only if $y = f(x)$ for some $x \in E_\alpha$ and $\alpha \in A$. This is equivalent to $y \in \cup_{\alpha \in A} f(E_\alpha)$. Similarly, $y \in f(\cap_{\alpha \in A} E_\alpha)$ if and only if $y = f(x)$ for some $x \in \cap_{\alpha \in A} E_\alpha$. This implies that for all $\alpha \in A$ there is an $x_\alpha \in E_\alpha$ such that $y = f(x_\alpha)$. Therefore, $y \in \cap_{\alpha \in A} f(E_\alpha)$. Finally, if $y \in f(C) \setminus f(B)$, then $y = f(c)$ for some $c \in C$ but $y \neq f(b)$ for any $b \in B$. It follows that $y \in f(C \setminus B)$. This proves part i). A similar argument proves part ii). ∎

The set inequalities in part i) are strict unless f is 1–1 (see Exercise 6 below). For example, if $f : \mathbf{R} \to \mathbf{R}$ is defined by $f(x) = x^2$, $E_1 = \{1\}$, and $E_2 = \{-1\}$, then $f(E_1 \cap E_2) = \emptyset$ and $f(E_1) \cap f(E_2) = \{1\}$.

EXERCISES

1. Prove that each of the following functions f is 1–1 on E and find a formula for f^{-1}.

   a) $f(x) = 3x - 7$, $E = \mathbf{R}$.
   b) $f(x) = e^{1/x}$, $E = (0, \infty)$.
   c) $f(x) = \tan x$, $E = (-\pi/2, \pi/2)$.
   d) $f(x) = x^2 + 3x - 6$, $E = [-3/2, \infty)$.
   e) $f(x) = 3x - |x| + |x - 2|$, $E = \mathbf{R}$.

2. Each of the following relations implicitly defines one or more continuous, real-valued functions $y = f(x)$. In each case, find formulas for these functions and their domains.

   a) $x^3 + y^3 = 1$.
   b) $e^{xy} + 1 = x$.
   c) $x^2 + xy + y^2 = 2$.
   d) $x^3 + 2x^2 + 2y^2 - x = 2$.

3. Find $f(E)$ and $f^{-1}(E)$ for each of the following.

   a) $f(x) = 1 - 5x$, $E = (-3, 1)$,
   b) $f(x) = x^2$, $E = [-1, 4]$,
   c) $f(x) = x^2 + x$, $E = [-2, 1)$,
   d) $f(x) = \log(x^2 + x + 1)$, $E = (1/2, 5]$,
   e) $f(x) = \sin x$, $E = [0, \infty)$.

4. Prove that the set of odd integers $\{1, 3, \dots\}$ is countably infinite.

5. Give a simple description of each of the following sets.

   a) $\displaystyle\bigcup_{x \in [0,1]} [x - 1, x + 1]$,   b) $\displaystyle\bigcap_{x \in [0,1]} [x - 1, x + 1]$

c) $\bigcup_{k \in \mathbf{N}} [0, 1/k],$ d) $\bigcap_{k \in \mathbf{N}} [0, 1/k].$

6. Let X, Y be sets and $f : X \to Y$. Prove that the following equivalent.

   a) f is 1–1 on X,
   b) $f(A \setminus B) = f(A) \setminus f(B)$ for all $B \subseteq A \subseteq X$,
   c) $f^{-1}(f(E)) = E$ for all $E \subseteq X$,
   d) $f(A \cap B) = f(A) \cap f(B)$ for all $A, B \subseteq X$.

7. Prove (22).

8. Prove Theorem 1.30ii.

9. A number $x_0 \in \mathbf{R}$ is called *algebraic of degree* n if it is the root of a polynomial of the form $P(x) = a_n x^n + a_{n-1} x^{n-1} + \cdots + a_1 x + a_0$ where $a_j \in \mathbf{Z}$ and $a_n \neq 0$. A number x_0 which is not algebraic is called *transcendental*.

   a) Prove that if $n \in \mathbf{N}$ and $q \in \mathbf{Q}$, then n^q is algebraic.
   b) Prove that for each $n \in \mathbf{N}$ the collection of algebraic numbers of degree n is countable.
   c) Prove that the collection of transcendental numbers is uncountable. (Two famous transcendental numbers are π and e. For more information on transcendental numbers and their history, see Kline [5].)

Chapter 2

CONTINUITY AND DIFFERENTIABILITY ON **R**

2.1 LIMITS

In the next three chapters we will study the calculus of real-valued functions of a real variable. Since this calculus is based on the limit concept, we begin by examining the theory of limits. To unify the presentation of one-sided and two-sided limits, we introduce the following idea.

DEFINITION 2.1. Let $E \subseteq \mathbf{R}$ be a nonempty set. A number $x_0 \in \mathbf{R}$ is called a *cluster point* of E if $(x_0 - \delta, x_0 + \delta) \cap E$ contains infinitely many points for each $\delta > 0$.

The etymology of the phrase *cluster point* is obvious. A cluster point of a set E is a point near which E "clusters."

Notice that by definition, x_0 is a cluster point of E if and only if there is a sequence $x_n \in E \setminus \{x_0\}$ such that $x_n \to x_0$ as $n \to \infty$. Thus, no finite set has cluster points. On the other hand, a set may have infinitely many cluster points. Indeed, by the Density of Rationals (Theorem 1.8), every point of **R** is a cluster point of **Q**.

Here are two other examples of sets and their cluster points.

Example 1. Show that 0 is the only cluster point of the set

$$E = \left\{ \frac{1}{n} : n \in \mathbf{N} \right\}.$$

SOLUTION. By Theorem 1.7 (the Archimedean Principle), given $\delta > 0$ there is an $N \in \mathbf{N}$ such that $1/N < \delta$. Since $n \geq N$ implies $1/n \leq 1/N$, it follows that $(-\delta, \delta) \cap E$ contains infinitely many points. Thus 0 is a cluster point of E.

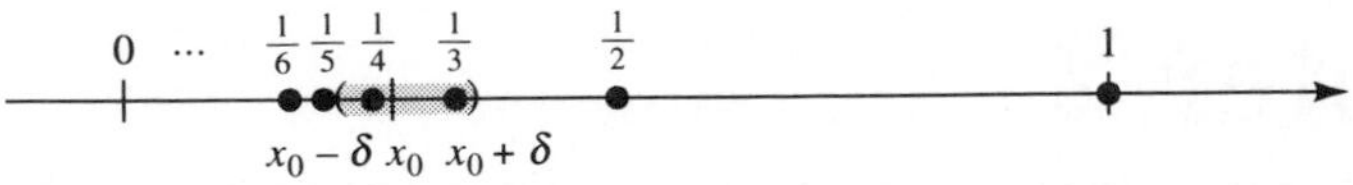

Figure 2.1

On the other hand, if $x_0 \neq 0$ and $\delta < |x_0|$, then $(x_0 - \delta, x_0 + \delta) \cap E$ contains at most finitely many points (see Figure 2.1). Thus, x_0 is not a cluster point of E. ∎

Example 2. Show every point in the interval $[0, 1]$ is a cluster point of the open interval $(0, 1)$.

SOLUTION. Let $x_0 \in [0, 1]$ and $\delta > 0$. Then $x_0 + \delta > 0$ and $x_0 - \delta < 1$. In particular, $(x_0 - \delta, x_0 + \delta) \cap (0, 1)$ is a nondegenerate interval (a, b). But (a, b) contains infinitely many points, e.g., $(a+b)/2$, $(2a+b)/3$, $(3a+b)/4$, $\ldots$ Therefore, x_0 is a cluster point of $(0, 1)$. ∎

To define limits of functions, recall from elementary calculus that a function $f(x)$ converges to L as x approaches x_0 if $f(x)$ is near L when x is near x_0. We wish to introduce a slightly more general idea here. We want to define convergence of $f(x)$ as x approaches x_0 *through* a set E to mean $f(x)$ is near L when x is near x_0 *and x belongs to E*. Here is the formal definition. (Note: We define limits only at cluster points x_0 of E so that implication (1) is not vacuous; i.e., there are always points $x \in E$ which satisfy $0 < |x - x_0| < \delta$ no matter how small δ gets.)

DEFINITION 2.2. Let $E \subseteq \mathbf{R}$, x_0 be a cluster point of E, $f : E \to \mathbf{R}$, and $L \in \mathbf{R}$. Then $f(x)$ is said to *converge to L, as x approaches x_0 through E*, if given $\varepsilon > 0$ there is a $\delta > 0$ such that

(1) $$0 < |x - x_0| < \delta \quad \text{and} \quad x \in E \quad \text{imply} \quad |f(x) - L| < \varepsilon.$$

In this case we write

$$L = \lim_{\substack{x \to x_0 \\ x \in E}} f(x)$$

and call L the *limit* of $f(x)$ as x approaches x_0 through E.

One advantage of this general definition is that both two-sided and one-sided limits can be discussed at the same time. There are other theoretical advantages as well (see the proof of Theorem 4.29, for example).

As defined, it appears that the limit of $f(x)$ as x approaches x_0 through a set E depends on E. The following two remarks show this is the case for some sets but not all.

Remark 1. *In general, the limit of $f(x)$ as $x \to x_0$ through E depends on E.*

PROOF. Consider the function

$$f(x) = \begin{cases} \sin \dfrac{1}{x} & x \neq 0 \\ 0 & x = 0. \end{cases}$$

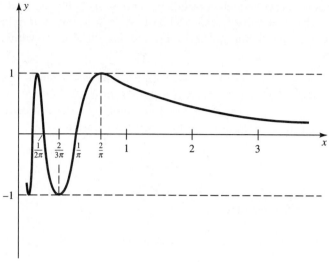

Figure 2.2

Notice that 0 is a cluster point of each of the sets

$$E_1 = \left\{ \frac{1}{n\pi} : n \in \mathbf{N} \right\}, \quad E_2 = \left\{ \frac{2}{(4n+1)\pi} : n \in \mathbf{N} \right\},$$

and

$$E_3 = \left\{ \frac{2}{(4n+3)\pi} : n \in \mathbf{N} \right\}.$$

Since $f(x) = 0$ for $x \in E_1$, $f(x) = 1$ for $x \in E_2$, and $f(x) = -1$ for $x \in E_3$ (see Figure 2.2), we have

$$\lim_{\substack{x \to 0 \\ x \in E_j}} f(x) = \begin{cases} 0 & j = 1 \\ 1 & j = 2 \\ -1 & j = 3. \end{cases} \blacksquare$$

Remark 1 is false when E is an open interval.

Remark 2. *If I is an open interval which contains x_0, and $f(x) \to L$ as $x \to x_0$ through I, then*

$$\lim_{\substack{x \to x_0 \\ x \in E}} f(x) = \lim_{\substack{x \to x_0 \\ x \in I}} f(x)$$

for all sets E which have x_0 as a cluster point. In particular, the limit of a f at x_0 through E does not depend on E.

PROOF. Suppose E is a set with cluster point x_0, and let $\varepsilon > 0$. Since $f(x) \to L$ as $x \to x_0$ through I, choose $\delta_1 > 0$ such that

$$0 < |x - x_0| < \delta_1 \quad \text{and} \quad x \in I \quad \text{imply} \quad |f(x) - L| < \varepsilon.$$

Since I is an open interval which contains x_0, choose $\delta_2 > 0$ such that $|x - x_0| < \delta_2$ implies $x \in I$. Set $\delta = \min\{\delta_1, \delta_2\}$. Then $0 < |x - x_0| < \delta$ implies $x \in I$ and $|f(x) - L| < \varepsilon$. In particular, $|f(x) - L| < \varepsilon$ for all $x \in E$ which satisfy $0 < |x - x_0| < \delta$. ∎

Thus if $f : I \to \mathbf{R}$ converges to L as x approaches x_0 through I, where I is an open interval which contains x_0, then the phrase *through I* is redundant. In this case, we shall simplify the terminology and notation by saying f *converges to L as x approaches x_0*, and writing either $f(x) \to L$ as $x \to x_0$, or

$$L = \lim_{x \to x_0} f(x).$$

Notice that by the proof of Remark 2, $f(x) \to L$ as $x \to x_0$ if and only if given $\varepsilon > 0$ there is a $\delta > 0$ such that

$$(2) \qquad\qquad 0 < |x - x_0| < \delta \quad \text{implies} \quad |f(x) - L| < \varepsilon.$$

The following result shows that the value of a limit as $x \to x_0$ does not in general depend on the value of the function at x_0.

Remark 3. *Let $E \subseteq \mathbf{R}$ and x_0 be a cluster point of E. If $f(x) = g(x)$ for all $x \in E \setminus \{x_0\}$ and $f(x) \to L$ as $x \to x_0$ through E, then $g(x)$ also has a limit as $x \to x_0$ through E, and*

$$\lim_{\substack{x \to x_0 \\ x \in E}} g(x) = \lim_{\substack{x \to x_0 \\ x \in E}} f(x).$$

This follows directly from Definition 2.2. (In fact, this is why the condition $0 < |x - x_0|$ appears in both (1) and (2) above.)

There is a close connection between limits of functions defined on **R** and limits of sequences in **R**.

THEOREM 2.1 [Sequential Characterization of Limits]. *Suppose $E \subseteq$ **R**, x_0 is a cluster point of E, and $f : E \to \mathbf{R}$. Then*

$$L = \lim_{\substack{x \to x_0 \\ x \in E}} f(x)$$

exists if and only if $f(x_n) \to L$ as $n \to \infty$ for every sequence $x_n \in E \setminus \{x_0\}$ which converges to x_0 as $n \to \infty$.

PROOF. Suppose f converges to L as x approaches x_0 through E. Then given $\varepsilon > 0$ there is a $\delta > 0$ such that (1) holds. If $x_n \in E \setminus \{x_0\}$ converges to x_0 as $n \to \infty$, then choose an $N \in \mathbf{N}$ such that $n \geq N$ implies $|x_n - x_0| < \delta$. Since $x_n \in E$ and $x_n \neq x_0$, it follows from (1) that $|f(x_n) - L| < \varepsilon$ for all $n \geq N$. Therefore, $f(x_n) \to L$ as $n \to \infty$.

Conversely, suppose $f(x_n) \to L$ as $n \to \infty$ for every sequence $x_n \in E \setminus \{x_0\}$ which converges to x_0. If f does not converge to L as x approaches x_0 through E, then there is an $\varepsilon > 0$ (call it ε_0) such that the implication "$0 < |x - x_0| < \delta$ and

$x \in E$ implies $|f(x) - f(x_0)| < \varepsilon_0$" does not hold for any $\delta > 0$. Thus, for each $\delta = 1/n$, $n \in \mathbf{N}$, there is a point $x_n \in E$ which satisfies $0 < |x_n - x_0| < 1/n$ and $|f(x_n) - L| \geq \varepsilon_0$. Now the first condition and the Squeeze Theorem imply that $x_n \to x_0$ so by hypothesis, $f(x_n) \to L$, as $n \to \infty$. In particular, $|f(x_n) - L| < \varepsilon_0$ for n large, a contradiction. ∎

(Note: Theorem 2.1 holds because $\mathbf{R}$ is a metric space (see Theorem 5.44). Although this result is false for more general topological spaces, one can prove an analogous result which applies to all topological spaces by using nets or filters.)

Theorem 2.1 allows us to translate results about real sequences to results about limits of functions of a real variable. The next three theorems illustrate this principle. (We supply details of the proof only for the first theorem. The other two are proved in an analogous manner.)

Before stating these results, we need additional notation. Suppose f and g are real-valued functions of a real variable. The *sum* of $f + g$ of f and g is defined by

$$(f + g)(x) := f(x) + g(x), \qquad \mathrm{Dom}\,(f + g) = \mathrm{Dom}\,(f) \cap \mathrm{Dom}\,(g);$$

the *scalar product* αf of a scalar $\alpha \in \mathbf{R}$ with f by

$$(\alpha f)(x) := \alpha f(x), \qquad \mathrm{Dom}\,(\alpha f) = \mathrm{Dom}\,(f);$$

the *product* fg of f and g by

$$(fg)(x) := f(x)g(x), \qquad \mathrm{Dom}\,(fg) = \mathrm{Dom}\,(f) \cap \mathrm{Dom}\,(g);$$

and the *quotient* f/g of f and g by

$$\left(\frac{f}{g}\right)(x) := \frac{f(x)}{g(x)}, \qquad \mathrm{Dom}\,(f/g) = \{x : x \in \mathrm{Dom}\,(f) \cap \mathrm{Dom}\,(g) \text{ and } g(x) \neq 0\}.$$

THEOREM 2.2. *Suppose $E \subseteq \mathbf{R}$, x_0 is a cluster point of E, $\alpha \in \mathbf{R}$, and $f, g : E \to \mathbf{R}$. If $f(x)$ and $g(x)$ converge as x approaches x_0 through E, then so do $(f + g)(x)$, $(fg)(x)$, $(\alpha f)(x)$, and $(f/g)(x)$ (when the limit of $g(x)$ is nonzero). In fact,*

$$\lim_{\substack{x \to x_0 \\ x \in E}} (f + g)(x) = \lim_{\substack{x \to x_0 \\ x \in E}} f(x) + \lim_{\substack{x \to x_0 \\ x \in E}} g(x),$$

$$\lim_{\substack{x \to x_0 \\ x \in E}} (\alpha f)(x) = \alpha \lim_{\substack{x \to x_0 \\ x \in E}} f(x),$$

$$\lim_{\substack{x \to x_0 \\ x \in E}} (fg)(x) = \lim_{\substack{x \to x_0 \\ x \in E}} f(x) \lim_{\substack{x \to x_0 \\ x \in E}} g(x),$$

and (when the limit of $g(x)$ is nonzero)

$$\lim_{\substack{x \to x_0 \\ x \in E}} \left(\frac{f}{g}\right)(x) = \lim_{\substack{x \to x_0 \\ x \in E}} f(x) / \lim_{\substack{x \to x_0 \\ x \in E}} g(x).$$

PROOF. Let
$$L := \lim_{\substack{x \to x_0 \\ x \in E}} f(x) \quad \text{and} \quad M := \lim_{\substack{x \to x_0 \\ x \in E}} g(x).$$

If $x_n \in E \setminus \{x_0\}$ converges to x_0, then by Theorem 2.1, $f(x_n) \to L$ and $g(x_n) \to M$ as $n \to \infty$. By Theorem 1.14i, $f(x_n) + g(x_n) \to L + M$ as $n \to \infty$. Since this holds for any sequence $x_n \in E \setminus \{x_0\}$ which converges to x_0, we conclude by Theorem 2.1 that

$$\lim_{\substack{x \to x_0 \\ x \in E}} (f + g)(x) = L + M = \lim_{\substack{x \to x_0 \\ x \in E}} f(x) + \lim_{\substack{x \to x_0 \\ x \in E}} g(x).$$

Similarly, the other rules follow directly from Theorem 1.14ii through iv. ∎

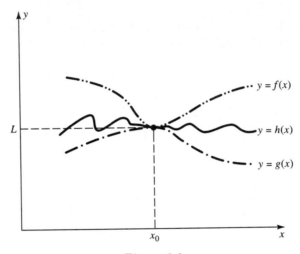

Figure 2.3

The following result is illustrated in Figure 2.3.

THEOREM 2.3 [SQUEEZE THEOREM FOR FUNCTIONS]. *Suppose $E \subseteq$ **R**, x_0 is a cluster point of E, and $f, g, h : E \to$ **R**.*

i) *If $f(x) \leq h(x) \leq g(x)$ for all $x \in E$, and*

$$\lim_{\substack{x \to x_0 \\ x \in E}} f(x) = \lim_{\substack{x \to x_0 \\ x \in E}} g(x) = L,$$

then the limit of h exists, as $x \to x_0$ through E, and

$$\lim_{\substack{x \to x_0 \\ x \in E}} h(x) = L.$$

ii) *If $|g(x)| \leq M$ for all $x \in E$ and $f(x) \to 0$ as $x \to x_0$ through E, then*

$$\lim_{\substack{x \to x_0 \\ x \in E}} f(x)g(x) = 0.$$

THEOREM 2.4 [COMPARISON THEOREM FOR FUNCTIONS]. *Suppose $E \subseteq \mathbf{R}$, x_0 is a cluster point of E, and $f, g : E \to \mathbf{R}$ satisfy*

$$f(x) \le g(x), \qquad x \in E.$$

If f and g have a limit as x approaches x_0 through E, then

$$\lim_{\substack{x \to x_0 \\ x \in E}} f(x) \le \lim_{\substack{x \to x_0 \\ x \in E}} g(x).$$

We shall refer to this last result as taking the limit of an inequality.

Sometimes, to show a function has a limit we apply Definition 2.2 directly.

Example 3. Suppose $f(x) = mx + b$ for some $m, b \in \mathbf{R}$. Prove

$$f(x_0) = \lim_{x \to x_0} f(x)$$

for all $x_0 \in \mathbf{R}$.

PROOF. Given $\varepsilon > 0$, set $\delta = \varepsilon / (|m| + 1)$. If $|x - x_0| < \delta$, then

$$|f(x) - mx_0 - b| = |m|\,|x - x_0| < |m|\delta < \varepsilon.$$

Thus by definition, $f(x) \to f(x_0)$ as $x \to x_0$. ∎

In other situations, we use one of the limit theorems, e.g., Theorems 2.1, 2.2, or 2.3.

Example 4. Prove

$$\lim_{x \to 1} \frac{x - 1}{3x + 1} = 0.$$

PROOF. By Example 3, $x - 1 \to 0$ and $3x + 1 \to 4$ as $x \to 1$. Hence, by Theorem 2.2,

$$\lim_{x \to 1} \frac{x - 1}{3x + 1} = \frac{0}{4} = 0. \quad ∎$$

Let I be a nondegenerate interval and x_0 be the left endpoint of I (respectively, the right endpoint of I). If $f : I \to \mathbf{R}$ converges to L as x approaches x_0 through I, then we shall say that f *converges to L as x approaches x_0 from the right* (respectively, *from the left*) and call L the *right-hand limit* (respectively, the *left-hand limit*) of f at x_0. We shall denote the right-hand limit (when it exists) by

(3) $$f(x_0+) := \lim_{x \to x_0+} f(x)$$

and the left-hand limit (when it exists) by

$$f(x_0-) := \lim_{x \to x_0-} f(x).$$

Thus, L is the right-hand limit (respectively, the left-hand limit) of f at x_0 if and only if given $\varepsilon > 0$ there is a $\delta > 0$ such that $0 < x - x_0 < \delta$ (respectively, $0 < x_0 - x < \delta$) implies $|f(x) - L| < \varepsilon$. The method used in Remark 2 above can be modified to show that if x_0 is a left (respectively, right) endpoint of an interval I, then the limit of a function as $x \to x_0$ through I does not depend on I. Thus, these "one-sided limits" are well defined.

Example 5. If $f(x) = \sqrt{x}$ prove

$$\lim_{x \to 0+} f(x) = 0.$$

PROOF. Let $\varepsilon > 0$ and set $\delta = \varepsilon^2$. If $0 < x < \delta$, then $|f(x)| = \sqrt{|x|} < \sqrt{\delta} = \varepsilon$. ∎

The following result shows that for functions of a real variable, one-sided limits can be used to characterize two-sided limits.

THEOREM 2.5. *The limit*

$$\lim_{x \to x_0} f(x)$$

exists and equals L if and only if

$$(4) \qquad\qquad L = \lim_{x \to x_0+} f(x) = \lim_{x \to x_0-} f(x).$$

PROOF. If the limit of $f(x)$ exists as $x \to x_0$, then by Remark 2 the left and right limits of $f(x)$ exist as $x \to x_0$ and (4) holds.

Conversely, suppose (4) holds. Then given $\varepsilon > 0$ there exist $\delta_j > 0$, $j = 1$ and 2, such that $0 < x - x_0 < \delta_1$ (respectively, $0 < x_0 - x < \delta_2$) implies

$$|f(x) - L| < \varepsilon.$$

Set $\delta = \min\{\delta_1, \delta_2\}$. Then it is clear that (2) holds. Hence, $f(x) \to L$ as $x \to x_0$. ∎

The definition of limits can be expanded to include extended real numbers. We say that $f(x) \to L$ as $x \to \infty$ (respectively, as $x \to -\infty$) if there exists a $c > 0$ such that $f : (c, \infty) \to \mathbf{R}$ (respectively, $f : (-\infty, -c) \to \mathbf{R}$) and given $\varepsilon > 0$ there is an $M > 0$ such that $x > M$ (respectively, $x < -M$) implies $|f(x) - L| < \varepsilon$. In this case we shall write

$$\lim_{x \to \infty} f(x) = L \qquad (\text{respectively,} \lim_{x \to -\infty} f(x) = L).$$

We say that $f(x) \to \infty$ (respectively, $f(x) \to -\infty$) as $x \to x_0$ if there is an open interval I containing x_0 such that $f : I \to \mathbf{R}$ and given $M > 0$ there is a $\delta > 0$ such that $0 < |x - x_0| < \delta$ implies $f(x) > M$ (respectively, $f(x) < -M$). In this case we shall write

$$\lim_{x \to x_0} f(x) = \infty \qquad (\text{respectively,} \lim_{x \to x_0} f(x) = -\infty).$$

Obvious modifications define $f(x) \to \pm\infty$ as $x \to x_0+$ and $x \to x_0-$, and $f(x) \to \pm\infty$ as $x \to \pm\infty$.

Example 6. Prove $1/x \to 0$ as $x \to \infty$.

PROOF. Given $\varepsilon > 0$, set $M = 1/\varepsilon$. If $x > M$, then $1/x < 1/M = \varepsilon$. Thus $1/x \to 0$ as $x \to \infty$. ∎

Example 7. Prove

$$\lim_{x \to 1-} \frac{x+2}{2x^2 - 3x + 1} = -\infty.$$

PROOF. As x converges to 1 from the left, $2x^2 - 3x + 1$ is negative and converges to 0. (Observe $2x^2 - 3x + 1$ is a parabola opening upward with roots $1/2$ and 1.) Therefore, given $M > 0$ choose $\delta \in (0,1)$ such that $1 - \delta < x < 1$ implies $-2/M < 2x^2 - 3x + 1 < 0$, i.e., $1/(2x^2 - 3x + 1) < -M/2$. Notice $0 < x < 1$ also implies $2 < x + 2 < 3$. Therefore,

$$\frac{x+2}{2x^2 - 3x + 1} < -M$$

for all $1 - \delta < x < 1$. ∎

Remark 4. *The Sequential Characterization of Limits holds for "infinite" limits.*

For example, we shall prove that $f(x) \to \infty$ as $x \to x_0$ if and only if $f(x_n) \to \infty$ for any sequence x_n which converges to x_0 and satisfies $x_n \neq x_0$ for $n \in \mathbf{N}$. Indeed, if $f(x) \to \infty$ as $x \to x_0$ and $x_n \to x_0$ as $n \to \infty$, $x_n \neq x_0$, then given $M > 0$ there is a $\delta > 0$ such that $0 < |x - x_0| < \delta$ implies $f(x) > M$, and there is an $N \in \mathbf{N}$ such that $n \geq N$ implies $|x_n - x_0| < \delta$. Consequently, $n \geq N$ implies $f(x_n) > M$, i.e., $f(x_n) \to \infty$ as $n \to \infty$. Conversely, if $f(x)$ does not converge to ∞, then there are numbers $M_0 > 0$ and $x_n \in \mathbf{R}$ such that $|x_n - x_0| < 1/n$ but $f(x_n) \leq M_0$ for all $n \in \mathbf{N}$. The first condition implies $x_n \to x_0$ but the second condition implies that $f(x_n)$ does not converge to ∞ as $n \to \infty$.

By using the Sequential Characterization of Limits, we can prove limit theorems which are function analogues of the results at the end of Section 1.6. We leave this to the reader and will use these results as the need arises.

These limit theorems can be used to evaluate infinite limits.

Example 8. Prove

$$\lim_{x \to \infty} \frac{2x^2 - 1}{1 - x^2} = -2.$$

PROOF. Since the limit of a product is the product of the limits, we have by Example 6 that $1/x^m \to 0$ as $x \to \infty$ for any $m \in \mathbf{N}$. Multiplying numerator and denominator of the expression above by $1/x^2$ we have

$$\lim_{x \to \infty} \frac{2x^2 - 1}{1 - x^2} = \lim_{x \to \infty} \frac{2 - 1/x^2}{-1 + 1/x^2}$$

$$= \frac{\lim_{x \to \infty}(2 - 1/x^2)}{\lim_{x \to \infty}(-1 + 1/x^2)} = \frac{2}{-1} = -2.$$

EXERCISES

1. Find all cluster points of each of the following sets.

   a) $E = \mathbf{R} \setminus \mathbf{Q}$.
   b) $E = [a, b)$, $a, b \in \mathbf{R}$, $a < b$.
   c) $E = \{(-1)^n n : n \in \mathbf{N}\}$.
   d) $E = \{x_n : n \in \mathbf{N}\}$, where $x_n \to x$ as $n \to \infty$.
   e) $E = \{1, 1, 2, 1, 2, 3, 1, 2, 3, 4, \dots\}$.

2. Evaluate the limit

$$\lim_{\substack{x \to 0 \\ x \in E}} f(x)$$

   in each of the following cases.

   a) $f(x) = \cos(1/x)$, $E = \{x \in \mathbf{R} : x = 2/((2n+1)\pi) \text{ for some } n \in \mathbf{N}\}$.
   b) $f(x) = \sin(\sqrt{x} + \pi/2)$, $E = (0, \infty)$.
   c) $f(x) = |x|/x$, $E = \{-1/n : n \in \mathbf{N}\}$.
   d) $f(x) = 1/(x^2 - 1)$, $E = \mathbf{Q}$.
   e) $f(x) = (x^2 + 2x)/\sqrt[3]{x}$, $E = \mathbf{R} \setminus \mathbf{Q}$.

3. Evaluate the following limits when they exist.

   a)
   $$\lim_{x \to 1} \frac{x^2 + x - 2}{x^3 - x}.$$

   b)
   $$\lim_{x \to \infty} \frac{3x^2 - 13x + 4}{1 - x - x^2}.$$

   c)
   $$\lim_{x \to 0+} \frac{x + 1}{x^2 - 2x}.$$

   d)
   $$\lim_{x \to \sqrt{\pi}} \frac{\sqrt{\pi - x^2}}{x + \pi}.$$

   e)
   $$\lim_{x \to 1-} \frac{x^3 - 3x + 2}{x^3 - 1}.$$

   f)
   $$\lim_{x \to \infty} x \sin x.$$

   g)
   $$\lim_{x \to 1} \frac{x^n - 1}{x - 1}, \qquad n \in \mathbf{N}.$$

4. **This exercise is used in Section 2.2**. Recall that a *polynomial of degree n* is a function of the form

$$P(x) = a_n x^n + a_{n-1} x^{n-1} + \cdots + a_1 x + a_0$$

where $a_j \in \mathbf{R}$ for $j = 0, 1, \ldots, n$ and $a_n \neq 0$.

a) Prove $\lim_{x \to x_0} x^n = x_0^n$ for $n = 0, 1, \ldots$

b) Prove that if P is a polynomial, then

$$\lim_{x \to x_0} P(x) = P(x_0)$$

for every $x_0 \in \mathbf{R}$.

c) Suppose that P is a polynomial and $P(x_0) > 0$. Prove $P(x)/(x - x_0) \to \infty$ as $x \to x_0+$, $P(x)/(x - x_0) \to -\infty$ as $x \to x_0-$, but

$$\lim_{x \to x_0} \frac{P(x)}{x - x_0}$$

does not exist.

5. a) Prove Theorem 2.3ii.

 b) Prove that $\lim_{x \to 0} x \sin(1/x) = 0$.

6. Suppose $E \subseteq \mathbf{R}$ has cluster point x_0 and $f : E \to \mathbf{R}$.

 a) Prove that if

$$L = \lim_{\substack{x \to x_0 \\ x \in E}} f(x)$$

exists, then $|f(x)| \to |L|$ as $x \to x_0$ through E.

 b) Show that the converse of a) is false.

7. **This exercise is used in Sections 2.2 and 3.2**. Let $f : E \to \mathbf{R}$. For each $x \in E$, define the *positive part* of f by

$$f^+(x) = \frac{|f(x)| + f(x)}{2}$$

and the *negative part* of f by

$$f^-(x) = \frac{|f(x)| - f(x)}{2}$$

a) Prove $f^+(x) \geq 0$, $f^-(x) \geq 0$, $f(x) = f^+(x) - f^-(x)$ and $|f(x)| = f^+(x) + f^-(x)$ hold for all $x \in E$. (Compare with Exercise 1 in Section 1.1.)

b) Prove that if

$$L = \lim_{\substack{x \to x_0 \\ x \in E}} f(x)$$

exists, then $f^+(x) \to L^+$ and $f^-(x) \to L^-$ as $x \to x_0$ through E.

$\boxed{8}$. **This exercise is used in Sections 2.2 and 3.2.** Let E be a nonempty subset of **R** and suppose $f, g : E \to \mathbf{R}$. For each $x \in E$ define

$$(f \vee g)(x) := \max\{f(x), g(x)\} \quad \text{and} \quad (f \wedge g)(x) := \min\{f(x), g(x)\}.$$

a) Prove

$$(f \vee g)(x) = \frac{(f + g)(x) + |(f - g)(x)|}{2}$$

and

$$f \wedge g = \frac{(f + g)(x) - |(f - g)(x)|}{2}$$

for all $x \in E$.

b) Prove that if

$$L = \lim_{\substack{x \to x_0 \\ x \in E}} f(x) \quad \text{and} \quad M = \lim_{\substack{x \to x_0 \\ x \in E}} g(x)$$

exist, then $(f \vee g)(x) \to L \vee M$ and $(f \wedge g)(x) \to L \wedge M$ as $x \to x_0$ through E.

9. Suppose $f : [a, \infty) \to \mathbf{R}$ for some $a \in \mathbf{R}$. Prove that $f(x) \to L$ as $x \to \infty$ if and only if $f(x_n) \to L$ for any sequence $x_n \in (a, \infty)$ which converges to ∞ as $n \to \infty$.

10. Suppose that $f : [0, 1] \to \mathbf{R}$ and $f(x_0) = \lim_{x \to x_0} f(x)$ for all $x_0 \in [0, 1]$. Prove that $f(q) = 0$ for all $q \in \mathbf{Q} \cap [0, 1]$ if and only if $f(x) = 0$ for all $x \in [0, 1]$.

2.2 CONTINUITY

In elementary calculus, a function was called continuous at x_0 if $x_0 \in \operatorname{Dom} f$ and $f(x) \to f(x_0)$ as $x \to x_0$. Here, we modify the concept of continuity to incorporate the more general limit introduced in Definition 2.2 (see Remark 1 below). In order to include all points of E, whether they are cluster points or not, we define continuity using the ε–δ formulation.

DEFINITION 2.3. Let E be a nonempty subset of **R** and $f : E \to \mathbf{R}$.

i) f is said to be *continuous at a point* $x_0 \in E$ if given $\varepsilon > 0$ there is a $\delta > 0$ (which in general depends on ε and x_0) such that

(5) $$|x - x_0| < \delta \quad \text{and} \quad x \in E \quad \text{imply} \quad |f(x) - f(x_0)| < \varepsilon.$$

ii) f is said to be *continuous on* E (notation: $f : E \to \mathbf{R}$ is continuous) if f is continuous at every $x \in E$.

iii) f is said to be *continuous* if it is continuous on its domain $\operatorname{Dom}(f)$.

Comparing Definitions 2.2 and 2.3, we observe that the following result is true.

Remark 1. *Let $E \subseteq \mathbf{R}$ and x_0 be a cluster point of E. A function $f : E \to \mathbf{R}$ is continuous at $x_0 \in E$ if and only if*

$$f(x_0) = \lim_{\substack{x \to x_0 \\ x \in E}} f(x).$$

By Remark 2 in Section 2.1 and Definition 2.3 above, if f is continuous at a point $x_0 \in I$ for some open interval I, then f is continuous at $x_0 \in E$ for any set E which contains x_0. In this case, we shall simply say that f is continuous at x_0. Hence, Remark 1 contains the following result.

Remark 2. *Let f be a real-valued function of a real variable. Then f is continuous at x_0 if and only if*

$$f(x_0) = \lim_{x \to x_0} f(x).$$

The following result is a corollary of Theorem 2.2 and Remark 1.

THEOREM 2.6. *Let $f, g : E \to \mathbf{R}$ be functions of a real variable. If f, g are continuous at a point $x_0 \in E$ (respectively, continuous on a set E), then so are $f + g$, fg, and αf (for any $\alpha \in \mathbf{R}$). Moreover, f/g is continuous at $x_0 \in E$ when $g(x_0) \neq 0$ (respectively, on E when $g(x) \neq 0$ for all $x \in E$).*

Similarly, it follows from Exercises 7 and 8 in Section 2.1 that if f, g are continuous at a point x_0 or on a set E, then so are $|f|$, f^+, f^-, $f \vee g$ and $f \wedge g$. We also notice by Exercise 4 in Section 2.1, that every polynomial is continuous on $\mathbf{R}$.

Many complicated functions can be broken into simpler pieces using sums, products, quotients, and the following operation.

DEFINITION 2.4. *If f is defined at a point x_0 and g is defined at $f(x_0)$, then the* composition *of g with f is defined by*

$$(g \circ f)(x_0) := g(f(x_0)).$$

The following result contains information about when a limit sign and something else (in this case, the computation of a function) can be interchanged. We shall return to this theme many times and identify conditions under which one can interchange any two of the following objects: limits, integrals, derivatives, infinite summations, and computation of a function (see especially Sections 4.4, 4.5, and 6.1).

THEOREM 2.7. *Let f and g be real-valued functions of a real variable and E be a subset of $\mathbf{R}$.*

i) *If $f(x) \to L$ as $x \to x_0$ through E and g is continuous at L, then*

$$\lim_{\substack{x \to x_0 \\ x \in E}} (g \circ f)(x) = g \left(\lim_{\substack{x \to x_0 \\ x \in E}} f(x) \right).$$

ii) *If f is continuous at $x_0 \in E$ and g is continuous at $f(x_0)$, then $g \circ f$ is continuous at $x_0 \in E$.*

PROOF. It suffices to prove part i). Suppose $x_n \in E$, $x_n \to x_0$ as $n \to \infty$, and $x_n \neq x_0$ for $n \in \mathbf{N}$. By Theorem 2.1 (the Sequential Characterization of Limits), $f(x_n) \to L$ and $g(f(x_n)) \to g(L)$ as $n \to \infty$. Therefore,

$$\lim_{n \to \infty} (g \circ f)(x_n) = g(L) = g \left(\lim_{\substack{x \to x_0 \\ x \in E}} f(x) \right).$$

Hence, by Theorem 2.1 again, $(g \circ f)(x) \to g(L)$ as $x \to x_0$ through E. ∎

DEFINITION 2.5. Let E be a nonempty subset of **R**. A function $f : E \to \mathbf{R}$ is said to be *bounded* on E if there is an $M \in \mathbf{R}$ such that $|f(x)| \leq M$ for all $x \in E$.

The following result shows that continuous functions on a closed, bounded interval are always bounded.

THEOREM 2.8 [EXTREME VALUE THEOREM]. *If f is continuous on a closed bounded interval $[a, b]$, then f is bounded on $[a, b]$. Moreover, if*

$$M = \sup_{x \in [a,b]} f(x) \quad and \quad m = \inf_{x \in [a,b]} f(x),$$

then there exist points $x_m, x_M \in [a, b]$ such that

$$f(x_M) = M \quad and \quad f(x_m) = m.$$

PROOF. Suppose first that f is not bounded on $[a, b]$. Then there exist $x_n \in [a, b]$ such that

$$(6) \qquad\qquad |f(x_n)| > n, \qquad n \in \mathbf{N}.$$

Since $[a, b]$ is bounded, the sequence $\{x_n\}_{n \in \mathbf{N}}$ is bounded. Hence, by Theorem 1.19 (the Bolzano–Weierstrass Theorem), $\{x_n\}$ has a convergent subsequence, say $x_{n_k} \to x_0$ as $k \to \infty$. By Theorem 1.15 (the Comparison Theorem), $x_0 \in [a, b]$, hence $f(x_0) \in \mathbf{R}$. Nevertheless, substituting n_k for n in (6) and letting $k \to \infty$, we have $|f(x_0)| = \infty$, a contradiction. Hence, the function f is bounded on $[a, b]$.

We have proved that both M and m are finite real numbers. To show there is an $x_M \in [a, b]$ such that $f(x_M) = M$, suppose to the contrary that $f(x) < M$ for all $x \in [a, b]$. Then the function

$$g(x) = \frac{1}{M - f(x)}$$

is continuous, hence, bounded on $[a, b]$. In particular, there is a $C > 0$ such that $|g(x)| = g(x) \leq C$. It follows that

$$(7) \qquad\qquad f(x) \leq M - \frac{1}{C}$$

for all $x \in [a, b]$. Taking the supremum of (7) over all $x \in [a, b]$, we obtain $M \leq M - 1/C < M$, a contradiction. Hence, there is an $x_M \in [a, b]$ such that $f(x_M) = M$. A similar argument proves that there is an $x_m \in [a, b]$ such that $f(x_m) = m$. ∎

We shall call the value M (respectively, m) *the maximum* (respectively, *the minimum*) of f on $[a, b]$.

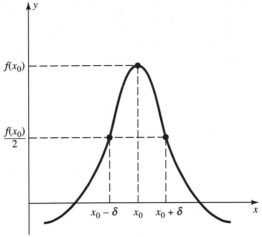

Figure 2.4

Remark 3. *The interval $[a, b]$ in the Extreme Value Theorem is both closed and bounded. The Extreme Value Theorem is false if either "closed" or "bounded" is dropped.*

PROOF. The interval $(0, 1)$ is bounded but not closed, and the function $f(x) = 1/x$ is continuous and unbounded on $(0, 1)$. The interval $[0, \infty)$ is closed but not bounded, and the function $f(x) = x$ is continuous and unbounded on $[0, \infty)$. ∎

A useful conceptualization of functions which are continuous on an interval is that their graphs have no holes or jumps (see Theorem 2.9 below). Our proof of this fact is based on the following elementary observation.

Lemma [SIGN PRESERVING PROPERTY] . *Let f be a real-valued function of a real variable. If f is continuous at a point x_0 and $f(x_0) > 0$, then there are positive numbers ε and δ such that*

$$|x - x_0| < \delta \quad implies \quad f(x) > \varepsilon.$$

STRATEGY: The idea behind this proof is simple. If $f(x_0) > 0$, then $f(x) > f(x_0)/2$ for x near x_0 (see Figure 2.4). Here are the details.

PROOF. By Remark 2, given $\varepsilon = f(x_0)/2$ choose $\delta > 0$ such that $|x - x_0| < \delta$ implies $|f(x) - f(x_0)| < \varepsilon$. It follows that

$$-\frac{f(x_0)}{2} < f(x) - f(x_0) < \frac{f(x_0)}{2}.$$

Solving the left-hand inequality, we see that $f(x) > f(x_0)/2 = \varepsilon$ holds for all $|x - x_0| < \delta$. ∎

A real number y_0 is said to *lie between* two numbers c and d if $c \le y_0 \le d$ or $d \le y_0 \le c$.

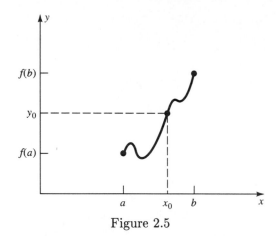

Figure 2.5

THEOREM 2.9 [INTERMEDIATE VALUE THEOREM]. *Let $[a, b]$ be a closed bounded interval and $f : [a, b] \to$* **R** *be continuous. If $f(a) \neq f(b)$ and y_0 is a real number which lies between $f(a)$ and $f(b)$, then there is an $x_0 \in [a, b]$ such that $f(x_0) = y_0$.*

PROOF. We may suppose that $f(a) \leq y_0 \leq f(b)$. Consider the set

$$E = \{x \in [a, b] : f(x) \leq y_0\}.$$

Since $a \in E$ and $E \subseteq [a, b]$, E is a nonempty bounded subset of **R**. Hence, by the Completeness Axiom, $x_0 := \sup E$ is a finite real number. We guess that $f(x_0) = y_0$ (see Figure 2.5).

To prove this guess is correct, use Theorem 1.13 to choose $x_n \in E$ such that $x_n \to x_0$ as $n \to \infty$. Since $E \subseteq [a, b]$, $x_0 \in [a, b]$ (see Exercise 7b in Section 1.4). Hence, by continuity of f and the definition of E, we have $f(x_0) = \lim_{n\to\infty} f(x_n) \leq y_0$.

To show $f(x_0) = y_0$, suppose to the contrary that $f(x_0) < y_0$. Then $y_0 - f(x)$ is a continuous function whose value at $x = x_0$ is positive. Hence, by the lemma, we can choose positive numbers ε and δ such that $y_0 - f(x) > \varepsilon > 0$ for $|x - x_0| < \delta$. In particular, any x which satisfies $x_0 < x < x_0 + \delta$ also satisfies $f(x) < y_0$, a contradiction of the fact that $x_0 = \sup E$. $\blacksquare$

If f fails to be continuous at a point x_0, we say that f is *discontinuous* at x_0 and call x_0 a *point of discontinuity* of f. How badly can a function behave near a point of discontinuity? The following examples can be interpreted as answers to this question. (See also Exercise 9 in Section 5.7.)

Example 1. Prove that the function

$$f(x) = \begin{cases} \dfrac{|x|}{x} & x \neq 0 \\ 1 & x = 0 \end{cases}$$

is continuous on $(-\infty, 0)$ and $[0, \infty)$, discontinuous at 0, and that both $f(0+)$ and $f(0-)$ exist.

PROOF. Since $f(x) = 1$ for $x > 0$, it is clear that $f(0+) = 1$ exists and $f(x) \to f(x_0)$ as $x \to x_0$ for any $x_0 > 0$. In particular, f is continuous on $[0, \infty)$. Similarly, $f(0-) = -1$ and f is continuous on $(-\infty, 0)$. Finally, since $f(0+) \neq f(0-)$, the limit of $f(x)$ as $x \to 0$ does not exist by Theorem 2.5. Therefore, f is not continuous at 0. ∎

Example 2. Assuming $\sin x$ is continuous on $\mathbf{R}$, prove that the function

$$f(x) = \begin{cases} \sin \dfrac{1}{x} & x \neq 0 \\ 1 & x = 0 \end{cases}$$

is continuous on $(-\infty, 0)$ and $(0, \infty)$, discontinuous at 0, and neither $f(0+)$ nor $f(0-)$ exists. (See Figure 2.2.)

PROOF. The function $1/x$ is continuous for $x \neq 0$ by Theorem 2.2. Hence, by Theorem 2.7, $f(x) = \sin(1/x)$ is continuous on $(-\infty, 0)$ and $(0, \infty)$. To prove $f(0+)$ does not exist, let $x_n = 2/((2n+1)\pi)$, and observe (see Appendix B) that $\sin(x_n) = (-1)^n$, $n \in \mathbf{N}$. Since $x_n \downarrow 0$ but $(-1)^n$ does not converge, it follows from Theorem 2.1 (the Sequential Characterization of Limits) that $f(0+)$ does not exist. A similar argument proves $f(0-)$ does not exist. ∎

Example 3. The *Dirichlet function* is defined on $\mathbf{R}$ by

$$f(x) := \begin{cases} 1 & x \in \mathbf{Q} \\ 0 & x \notin \mathbf{Q}. \end{cases}$$

Prove that every point $x \in \mathbf{R}$ is a point of discontinuity of f. (Such functions are called *nowhere continuous*.)

PROOF. By Theorem 1.8 and Exercise 3 in Section 1.3 (Density of Rationals and Irrationals), given any $x_0 \in \mathbf{R}$ and $\delta > 0$ we can choose $x_1 \in \mathbf{Q}$ and $x_2 \in \mathbf{R} \setminus \mathbf{Q}$ such that $|x_i - x_0| < \delta$ for $i = 1, 2$. Since $f(x_1) = 1$ and $f(x_2) = 0$, f cannot be continuous at x_0. ∎

Example 4. Prove that the function

$$f(x) = \begin{cases} \dfrac{1}{q} & x = \dfrac{p}{q} \in \mathbf{Q} \quad \text{(in reduced form)} \\ 0 & x \notin \mathbf{Q}. \end{cases}$$

is continuous at every irrational in the interval $(0, 1)$ but discontinuous at every rational in $(0, 1)$.

PROOF. Let x_0 be a rational in $(0, 1)$ and suppose that f is continuous at x_0. If x_n is a sequence of irrationals which converges to x_0, then $f(x_n) \to f(x_0)$; i.e., $f(x_0) = 0$. But $f(x_0) \neq 0$ by definition. Hence, f is discontinuous at every rational in $(0, 1)$.

Let x_0 be an irrational in $(0, 1)$. We must show $f(x_n) \to f(x_0)$ for every sequence $x_n \in (0, 1)$ which satisfies $x_n \to x_0$ as $n \to \infty$. We may suppose that $x_n \in \mathbf{Q}$ and write $x_n = p_n/q_n$ for $n \in \mathbf{N}$ in reduced form. Since $f(x_0) = 0$, it suffices to show $q_n \to \infty$ as $n \to \infty$. Suppose to the contrary that there exist integers $n_1 < n_2 < \cdots$ such that $|q_{n_k}| \leq M < \infty$ for $k \in \mathbf{N}$. Since $x_{n_k} \in (0,1)$, it follows that the set

$$E := \left\{ x_{n_k} = \frac{p_{n_k}}{q_{n_k}} : k \in \mathbf{N} \right\}$$

contains only a finite number of points. Hence, the limit of any sequence in E must belong to E, a contradiction since x_0 is such a limit and is irrational. ∎

To see how counterintuitive Example 4 is, try to draw a graph of $y = f(x)$. Stranger things can happen.

Remark 4. *The composition of two functions* $g \circ f$ *can be nowhere continuous, even though* f *is discontinuous only on* **Q** *and* g *is discontinuous at only one point.*

PROOF. Let f be the function given in Example 4 and set

$$g(x) = \begin{cases} 1 & x \neq 0 \\ 0 & x = 0. \end{cases}$$

Clearly,

$$(g \circ f)(x) = \begin{cases} 1 & x \in \mathbf{Q} \\ 0 & x \notin \mathbf{Q}. \end{cases}$$

Hence, $g \circ f$ is the Dirichlet function, nowhere continuous by Example 3. ∎

In view of Example 4 and Remark 4, we must be skeptical of proofs which rely exclusively on geometric intuition. And although we shall use geometric intuition to suggest a method of proof for many results in subsequent chapters, these suggestions will always be followed by a careful rigorous proof which contains no fuzzy reasoning based on pictures or sketches no matter how plausible they seem.

EXERCISES

For these exercises, assume that $\sin x$, $\cos x$, *and* e^x *are continuous on* **R**.

1. Use limit theorems to show that each of the following functions is continuous on $[0,1]$.

   a) $$f(x) = xe^{x^2} + 5.$$

   b) $$f(x) = \frac{1-x}{1+x}.$$

c)
$$f(x) = \begin{cases} \sqrt{x}\sin\dfrac{1}{x} & x \neq 0 \\ 0 & x = 0. \end{cases}$$

d)
$$f(x) = \sqrt{1 - x}.$$

e)
$$f(x) = \frac{\sin(e^x)}{x^2 + x - 6}.$$

2. Let
$$f(x) = \begin{cases} \cos\dfrac{1}{x} & x \neq 0 \\ 0 & x = 0. \end{cases}$$

a) Prove f is continuous on $(0, \infty)$ and $(-\infty, 0)$ but discontinuous at 0.

b) Suppose $g : [0, 2/\pi] \to \mathbf{R}$ is continuous on $(0, 2/\pi)$ and that there is a positive constant $C > 0$ such that $|g(x)| < C\sqrt{x}$ for all $x \in (0, 2/\pi)$. Prove $f(x)g(x)$ is continuous on $[0, 2/\pi]$.

3. If $f : [a, b] \to \mathbf{R}$ is continuous, prove that $\sup_{x\in[a,b]} |f(x)|$ is finite.

4. Suppose f is a real-valued function of a real variable. If f is continuous at x_0 with $f(x_0) < M$ for some $M \in \mathbf{R}$, prove there is an open interval I containing x_0 such that $f(x) < M$ for all $x \in I$.

5. Show that there exist nowhere continuous functions f and g whose sum $f + g$ is continuous on $\mathbf{R}$. Show that the same is true for the product of functions.

6. Let (a, b) be an open interval containing a point x_0, $f, g : (a, b) \to \mathbf{R}$, and f be continuous at x_0.

a) Prove that g is continuous at x_0 if and only if $f + g$ is continuous at x_0.

b) Make and prove an analogous statement for the product fg. Show by example that the hypothesis about f you added cannot be dropped.

7. Suppose that $f : [a, b] \to \mathbf{R}$ is continuous. Prove there is a closed interval $[c, d]$ such that $1/(\alpha - f(x))$ is continuous on $[a, b]$ for all $\alpha \notin [c, d]$.

8 . **This exercise is used in Section 5.1.**
Suppose $f : \mathbf{R} \to \mathbf{R}$ satisfies $f(x + y) = f(x) + f(y)$ for each $x, y \in \mathbf{R}$.

a) Show that $f(nx) = nf(x)$ for all $x \in \mathbf{R}$ and $n \in \mathbf{Z}$.

b) Prove $f(qx) = qf(x)$ for all $x \in \mathbf{R}$ and $q \in \mathbf{Q}$.

c) Prove that f is continuous at 0 if and only if f is continuous on $\mathbf{R}$.

d) Prove that if f is continuous at 0, then there is an $m \in \mathbf{R}$ such that $f(x) = mx$ for all $x \in \mathbf{R}$.

9. Suppose that $f : \mathbf{R} \to (0, \infty)$ satisfies $f(x + y) = f(x)f(y)$. Modifying the outline in Exercise 8, show that if f is continuous at 0, then there is an $a \in (0, \infty)$ such that $f(x) = a^x$ for all $x \in \mathbf{R}$. [Note: You may assume that the function a^x is continuous on **R**.]

10. If $f : \mathbf{R} \to \mathbf{R}$ is continuous and

$$\lim_{x \to \infty} f(x) = \lim_{x \to -\infty} f(x) = \infty,$$

prove that f has a minimum on **R**; i.e., there is an $x_m \in \mathbf{R}$ such that

$$f(x_m) = \inf_{x \in \mathbf{R}} f(x) < \infty.$$

2.3 UNIFORM CONTINUITY

The following concept is very important and will be used in the next chapter to relate continuity to integrability.

DEFINITION 2.6. Let E be a nonempty subset of **R** and $f : E \to \mathbf{R}$. Then f is said to be *uniformly continuous* on E (notation: $f : E \to \mathbf{R}$ is uniformly continuous) if given $\varepsilon > 0$ there is a $\delta > 0$ such that

(8) $|x - x_0| < \delta$ and $x, x_0 \in E$ imply $|f(x) - f(x_0)| < \varepsilon.$

Example 1. Prove $f(x) = x^2$ is uniformly continuous on the interval $(0, 1)$.

Proof. Given $\varepsilon > 0$, set $\delta = \varepsilon/3$. If $x, x_0 \in (0, 1)$, then $|x + x_0| \leq |x| + |x_0| \leq 2$. Therefore, if $x, x_0 \in (0, 1)$ and $|x - x_0| < \delta$, then

$$|f(x) - f(x_0)| = |x^2 - x_0^2| = |x - x_0|\,|x + x_0| \leq 2|x - x_0| \leq \frac{2\varepsilon}{3} < \varepsilon. \quad \blacksquare$$

Example 2. Show that $f(x) = x^2$ is not uniformly continuous on **R**.

Proof. Suppose to the contrary that f is uniformly continuous on **R**. Then there is a $\delta > 0$ such that $|x - x_0| < \delta$ implies $|f(x) - f(x_0)| < 1$ for all $x, x_0 \in \mathbf{R}$. By the Archimedean Principle, choose $n \in \mathbf{N}$ so large that $n\delta > 1$. Set $x_0 = n$ and $x = n + \delta/2$. Then $|x - x_0| < \delta$ and

$$1 > |f(x) - f(x_0)| = |x^2 - x_0^2| = n\delta + \frac{\delta^2}{4} > n\delta > 1.$$

This contradiction proves that f is not uniformly continuous on **R**. $\blacksquare$

The definitions of continuity and uniform continuity are very similar. In fact, the only difference is that for a continuous function, the parameter δ may depend on both ε and x_0, whereas for a uniformly continuous function, δ must be chosen independently of x_0. In particular, every function uniformly continuous on E is also continuous on E. We shall see that the converse of this statement is false unless some restriction is made on E.

Here is a key which unlocks the difference between continuity and uniform continuity.

Lemma. *Suppose* $f : E \to \mathbf{R}$ *is uniformly continuous and* $x_n \in E$ *for* $n \in \mathbf{N}$. *If* $\{x_n\}_{n \in \mathbf{N}}$ *is Cauchy, then* $\{f(x_n)\}_{n \in \mathbf{N}}$ *is Cauchy.*

PROOF. Let $\varepsilon > 0$ and choose $\delta > 0$ such that (8) holds. Since $\{x_n\}$ is Cauchy, choose $N \in \mathbf{N}$ such that $n, m \geq N$ implies $|x_n - x_m| < \delta$. Then $n, m \geq N$ implies $|f(x_n) - f(x_m)| < \varepsilon$. ∎

Notice that $f(x) = 1/x$ is continuous on $(0, 1)$ and $x_n = 1/n$ is Cauchy but $f(x_n)$ is not. In particular, $1/x$ is continuous but not uniformly continuous on the open interval $(0, 1)$. Notice how the graph of $y = 1/x$ corroborates this fact. Indeed, as x_0 gets closer to 0, the value of δ gets smaller (compare δ_1 to δ_0 in Figure 2.6), hence, cannot be chosen independently of x_0.

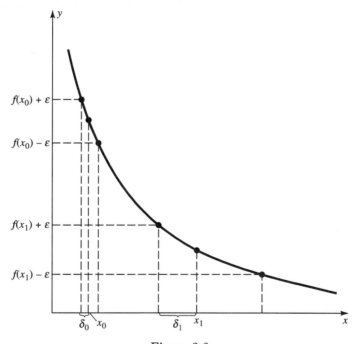

Figure 2.6

Thus, for functions on the open interval $(0, 1)$, continuity and uniform continuity are different. We shall see that this is not the case for functions on the closed bounded interval $[0, 1]$. Indeed, we will prove that for a large class of sets which includes every closed bounded interval, continuity and uniform continuity are equivalent (see Theorem 2.10 and Corollary 5.29).

First, we introduce the "large class of sets."

DEFINITION 2.7. A set $K \subset \mathbf{R}$ is called *sequentially compact* if every sequence $x_n \in K$ has a convergent subsequence whose limit belongs to K.

Observe that by Theorem 1.19 (the Bolzano–Weierstrass Theorem) and Exercise 7 in Section 1.4, every closed, bounded interval is sequentially compact. On the other hand, no open interval is sequentially compact. Indeed, if (a, b) is a bounded open interval, then the sequence $x_n := a + 1/n$ belongs to (a, b) for n large and has limit a which does not belong to (a, b). On the other hand, if (a, b) is unbounded, then (a, b) contains either $\{n\}$ or $\{-n\}$ for n large. Since neither of these sequences has a convergent subsequence it follows that (a, b) is not sequentially compact.

Next, we prove that for functions on sequentially compact sets, continuity and uniform continuity are equivalent.

THEOREM 2.10. *Let $K \subset$ **R** be sequentially compact. If $f : K \to$ **R** is continuous on K, then f is uniformly continuous on K. In particular, if f is continuous on a closed bounded interval $[a, b]$, then f is uniformly continuous on $[a, b]$.*

PROOF. Suppose to the contrary that f is continuous but not uniformly continuous on K. Then there is an $\varepsilon_0 > 0$ and points $x_n, y_n \in K$ such that $|x_n - y_n| < 1/n$ and

$$(9) \qquad\qquad |f(x_n) - f(y_n)| > \varepsilon_0, \qquad n \in \mathbf{N}.$$

Since K is sequentially compact, the sequence $\{x_n\}$ has a convergent subsequence $x_{n_k} \to x \in K$ as $k \to \infty$. Similarly, the sequence $\{y_{n_k}\}_{k \in \mathbf{N}}$ has a convergent subsequence $y_{n_{k_j}} \to y \in K$ as $j \to \infty$. Since $x_{n_{k_j}} \to x$ as $j \to \infty$ and f is continuous, it follows from (9) that $|f(x) - f(y)| \geq \varepsilon_0$, i.e., $f(x) \neq f(y)$. But $|x_n - y_n| < 1/n$ for all $n \in \mathbf{N}$ so Theorem 1.12 (the Squeeze Theorem) implies $x = y$. Therefore, $f(x) = f(y)$, a contradiction. ∎

We now characterize uniform continuity on bounded open intervals.

THEOREM 2.11. *Let (a, b) be a bounded open interval and $f : (a, b) \to$ **R**. Then f is uniformly continuous on (a, b) if and only if f can be continuously extended to $[a, b]$, i.e., if and only if there is a continuous function $g : [a, b] \to$ **R** which satisfies*

$$(10) \qquad\qquad f(x) = g(x), \qquad x \in (a, b).$$

PROOF. Suppose f is uniformly continuous on (a, b). Let $x_n \in (a, b)$ converge to b as $n \to \infty$. Then $\{x_n\}$ is Cauchy, hence by the lemma, so is $\{f(x_n)\}$. In particular,

$$g(b) := \lim_{n \to \infty} f(x_n)$$

exists. This value does not change if we use a different sequence to approximate b. Indeed, let $y_n \in (a, b)$ be another sequence which converges to b as $n \to \infty$. Given $\varepsilon > 0$, choose $\delta > 0$ such that (8) holds for $E = (a, b)$. Since $x_n - y_n \to 0$, choose $N \in \mathbf{N}$ so that $n \geq N$ implies $|x_n - y_n| < \delta$. It follows from (8) that $|f(x_n) - f(y_n)| < \varepsilon$ for $n \geq N$. Taking the limit of this inequality as $n \to \infty$, we obtain

$$|\lim_{n \to \infty} f(x_n) - \lim_{n \to \infty} f(y_n)| \leq \varepsilon$$

for all $\varepsilon > 0$. It follows from Theorem 1.2 that

$$\lim_{n \to \infty} f(x_n) = \lim_{n \to \infty} f(y_n).$$

Thus, $g(b)$ is well defined. A similar argument defines $g(a)$.

Set $g(x) = f(x)$ for $x \in (a, b)$. Then g is defined on $[a, b]$, satisfies (10), and is continuous on $[a, b]$ by the Sequential Characterization of Limits. Thus, f can be "continuously extended" to g as required.

Conversely, suppose that there is a function g continuous on $[a, b]$ which satisfies (10). By Theorem 2.10, g is uniformly continuous on $[a, b]$; hence, g is uniformly continuous on (a, b). We conclude that f is uniformly continuous on (a, b). ∎

EXERCISES

1. Suppose that K_1, K_2 are sequentially compact subsets of $\mathbf{R}$. Prove that $K_1 \cup K_2$ and $K_1 \cap K_2$ are sequentially compact.
2. Prove that each of the following functions is uniformly continuous on $(0, 1)$.

a)
$$f(x) = \frac{x^3 - 1}{x - 1}.$$

b)
$$f(x) = x \sin \frac{1}{x}.$$

c) $f(x)$ is any polynomial.

d)
$$f(x) = \frac{\sin x}{x}.$$

e) $f(x) = x^2 \log x.$

You may use L'Hôpital's Rule (see Theorem 2.18 below) on parts d) and e).
3. Find all real α such that $x^\alpha \sin(1/x)$ is uniformly continuous on the open interval $(0, 1)$.
4. a) Suppose $f : [0, \infty) \to \mathbf{R}$ is continuous and $f(x) \to 0$ as $x \to \infty$. Prove f is uniformly continuous on $[0, \infty)$.
   b) Prove that $f(x) = 1/(x^2 + 1)$ is uniformly continuous on $\mathbf{R}$.

5. a) Let I be a bounded interval. Prove that if $f : I \to \mathbf{R}$ is uniformly continuous on I, then f is bounded on I.
   b) Prove a) may be false if I is unbounded.

6. Suppose $\alpha \in \mathbf{R}$, E is a nonempty subset of $\mathbf{R}$, and $f, g : E \to \mathbf{R}$ are uniformly continuous on E.

    a) Prove that $f + g$ and αf are uniformly continuous on E.

    b) Suppose f, g are bounded on E. Prove fg is uniformly continuous on E.

    c) Show there exist functions f, g uniformly continuous on $\mathbf{R}$ such that fg is not uniformly continuous on $\mathbf{R}$.

    d) Suppose f is bounded on E and that there is a positive constant ε_0 such that $g(x) \geq \varepsilon_0$ for all $x \in E$. Prove that f/g is uniformly continuous on E.

    e) Show there exist functions f, g, uniformly continuous on the interval $(0, 1)$, with $g(x) > 0$ for all $x \in (0, 1)$, such that f/g is not uniformly continuous on $(0, 1)$.

    f) Prove that if f, g are uniformly continuous on a closed interval $[a, b]$ and $g(x) \neq 0$ for $x \in [a, b]$, then f/g is uniformly continuous on $[a, b]$.

7. Let $E \subseteq \mathbf{R}$. A function $f : E \to \mathbf{R}$ is said to be increasing on E if $x_1, x_2 \in E$ and $x_1 < x_2$ imply $f(x_1) \leq f(x_2)$. Suppose f is increasing and bounded on an open bounded interval (a, b).

    a) Prove that $f(a+)$ and $f(b-)$ both exist and are finite.

    b) Prove that f is continuous on (a, b) if and only if f is uniformly continuous on (a, b).

    c) Show b) is false if f is unbounded. Indeed, find an increasing function $g : (0, 1) \to \mathbf{R}$ which is continuous on $(0, 1)$ but not uniformly continuous on $(0, 1)$.

8. Suppose that f is continuous on $[0, 1]$ and set

$$I_k = \left[\frac{k-1}{2^n}, \frac{k}{2^n} \right]$$

for $k = 1, 2, \ldots, 2^n$. Prove that given $\varepsilon > 0$ there is an $N \in \mathbf{N}$ such that $n \geq N$ implies

$$\sup_{x \in I_k} f(x) - \inf_{x \in I_k} f(x) < \varepsilon, \qquad k = 1, 2, \ldots, 2^n.$$

9. Prove that a polynomial of degree n is uniformly continuous on $\mathbf{R}$ if and only if $n = 0$ or 1.

2.4 DIFFERENTIABILITY

Many problems of applied mathematics can be solved by finding the slope of a tangent line. Although *tangent line* is easy to define in classical geometry where lines and circles are the only curves, to define tangent line for more complicated curves we must use the limit process.

Let f be a real-valued function of a real variable. Lines passing through two points on the graph of $y = f(x)$ are called *chords*. The slope of the chord passing through the points $(x, f(x))$ and $(x_0, f(x_0))$ is given by $(f(x) - f(x_0))/(x - x_0)$. We shall say that $y = f(x)$ *has a tangent line* at the point $(x_0, f(x_0))$ if f is defined

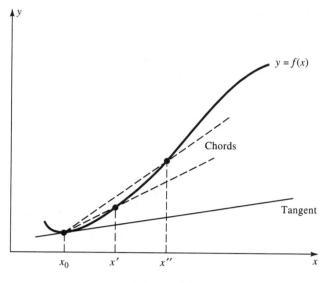

Figure 2.7

in an open interval containing x_0 and the limit of the slopes of these chords exist as $x \to x_0$ (see Figure 2.7). This limit is called the *slope of the tangent line* to $y = f(x)$ at $(x_0, f(x_0))$ and denoted by

$$f'(x_0) := \lim_{x \to x_0} \frac{f(x) - f(x_0)}{x - x_0}.$$

The following definition introduces the standard terminology for functions whose graphs have tangent lines.

DEFINITION 2.8. Let f be a real-valued function of a real variable.

   i) f is said to be *differentiable* at a point $x_0 \in \mathbf{R}$ if f is defined on some open interval containing x_0 and $f'(x_0)$ exists.

   ii) f is said to be *differentiable* if it is differentiable at each point in its domain.

The limit $f'(x_0)$ (when it exists) is called the *derivative* of f at x_0. We note by a change of variables that the derivative can also be defined by

$$f'(x_0) = \lim_{h \to 0} \frac{f(x_0 + h) - f(x_0)}{h}.$$

If f is differentiable at each point in a set E, then the derivative is a function on E. This function is denoted several ways:

$$D_x f = \frac{df}{dx} = f^{(1)} = f'.$$

When $y = f(x)$, we also use the notation dy/dx or y' for f'. Higher-order derivatives are defined recursively; i.e., if $n \in \mathbf{N}$, then $f^{(n+1)}(x) := (f^{(n)}(x))'$, provided these

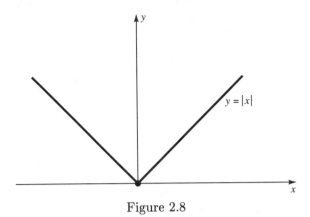

Figure 2.8

derivatives exist. Higher-order derivatives are also denoted several ways including $D_x^n f$, $d^n f/dx^n$, $f^{(n)}$, and by $d^n y/dx^n$ and $y^{(n)}$, when $y = f(x)$. The *second derivatives* $f^{(2)}$ (respectively, $y^{(2)}$) are usually written as f'' (respectively, y''). When they exist at some point x_0, we shall say that f is *twice differentiable* at x_0.

There is a simple connection between differentiability and continuity.

THEOREM 2.12. *If f is differentiable at x_0, then f is continuous at x_0.*

PROOF. If f is defined at x and x_0, and $x \neq x_0$, then

$$(11) \qquad f(x) = \frac{f(x) - f(x_0)}{x - x_0}(x - x_0) + f(x_0).$$

By hypothesis, the right side of this identity converges to $f'(x_0) \cdot 0 + f(x_0) = f(x_0)$ as $x \to x_0$. Taking the limit of (11) as $x \to x_0$, we conclude that

$$\lim_{x \to x_0} f(x) = f(x_0),$$

i.e., f is continuous at x_0. ∎

Remark 1. *The converse of Theorem 2.12 is false.*

PROOF. The function $f(x) = |x|$ is continuous at $x = 0$. But

$$\lim_{h \to 0+} \frac{f(h) - f(0)}{h} = 1 \quad \text{and} \quad \lim_{h \to 0-} \frac{f(h) - f(0)}{h} = -1.$$

Therefore, $f(x)$ is not differentiable at $x = 0$. ∎

This example reflects the conventional wisdom about differentiable and continuous functions. The graph of a differentiable function on an interval is "smooth" with no corners, cusps, or kinks. The graph of a continuous function on an interval is unbroken (has no holes or jumps), but may have corners, cusps, or kinks. In particular, $f(x) = |x|$ is continuous but not differentiable at $x = 0$ because its graph is unbroken and has a corner at the point $(0,0)$ (see Figure 2.8).

It is natural to ask how differentiation (i.e., the process of taking a derivative) interacts with the algebra of functions. The next two results answer this question.

THEOREM 2.13. *If f, g are differentiable at x_0 and $\alpha \in \mathbf{R}$, then $f + g$, fg, αf, and f/g (when $g(x_0) \neq 0$) are all differentiable at x_0. In fact,*

SUM RULE
$$(f + g)'(x_0) = f'(x_0) + g'(x_0),$$

$$(\alpha f)'(x_0) = \alpha f'(x_0),$$

PRODUCT RULE
$$(fg)'(x_0) = f(x_0)g'(x_0) + f'(x_0)g(x_0),$$

and, when $g(x_0) \neq 0$,

QUOTIENT RULE
$$\left(\frac{f}{g}\right)'(x_0) = \frac{g(x_0)f'(x_0) - f(x_0)g'(x_0)}{g^2(x_0)}.$$

PROOF. The first two rules follow directly from Definition 2.8 and Theorem 2.2. The product and quotient rules are easy to prove using Definition 2.8 and Theorems 2.2 and 2.12. For example, it is clear that

$$\frac{f(x)g(x) - f(x_0)g(x_0)}{x - x_0} = \frac{f(x)g(x) - f(x_0)g(x) + f(x_0)g(x) - f(x_0)g(x_0)}{x - x_0}$$

$$= g(x)\frac{f(x) - f(x_0)}{x - x_0} + f(x_0)\frac{g(x) - g(x_0)}{x - x_0}.$$

Taking the limit as $x \to x_0$, we obtain

$$(fg)'(x_0) = g(x_0)f'(x_0) + f(x_0)g'(x_0). \quad \blacksquare$$

The algebraic identity (11) leads to a simple but effective characterization of differentiability using slopes of chords.

Lemma. *Let (a, b) be an open interval, $x_0 \in (a, b)$ and $f : (a, b) \to \mathbf{R}$. Then f is differentiable at x_0 if and only if there is a function $f^* : (a, b) \to \mathbf{R}$ continuous at x_0, such that*

(12)
$$f(x) = f^*(x)(x - x_0) + f(x_0)$$

for all $x \in (a, b)$, in which case $f^*(x_0) = f'(x_0)$.

PROOF. Suppose f is differentiable at x_0 and set

$$f^*(x) := \begin{cases} \dfrac{f(x) - f(x_0)}{x - x_0} & x \in (a, b) \setminus \{x_0\} \\ f'(x_0) & x = x_0. \end{cases}$$

Then (12) holds by definition, and since $f'(x_0)$ exists, f^* is continuous at x_0.

Conversely, suppose that (12) holds for some function f^* continuous at x_0. Then

$$\frac{f(x) - f(x_0)}{x - x_0} = f^*(x)$$

for $x \in (a, b)$, $x \neq x_0$. Take the limit of this identity as $x \to x_0$. Since f^* is continuous at x_0 we conclude that $f'(x_0) = f^*(x_0)$. $\quad \blacksquare$

Our first application of this characterization is a proof of the Chain Rule (see also Exercise 9 below).

THEOREM 2.14 [CHAIN RULE]. *Let f, g be real-valued functions of a real variable. If f is differentiable at x_0 and g is differentiable at $f(x_0)$, then $g \circ f$ is differentiable at x_0 and*

$$(g \circ f)'(x_0) = g'(f(x_0))f'(x_0).$$

PROOF. By hypothesis, f is defined on some open interval (a, b) which contains x_0 and g is defined on some open interval (c, d) which contains $f(x_0)$. Choose by the lemma functions f^*, defined on (a, b) and continuous at x_0, and g^*, defined on (c, d) and continuous at $f(x_0)$, such that

$$f^*(x_0) = f'(x_0), \quad f(x) = f^*(x)(x - x_0) + f(x_0), \qquad x \in (a, b)$$

and

$$(13) \quad g^*(f(x_0)) = g'(f(x_0)), \quad g(y) = g^*(y)(y - f(x_0)) + g(f(x_0)), \qquad y \in (c, d).$$

Since f is continuous at x_0 we may assume (by making (a, b) smaller if necessary) that $f(x) \in (c, d)$ for all $x \in (a, b)$.

Fix $x \in (a, b)$. Apply (13) to $y = f(x)$ and (12) to x to write

$$(g \circ f)(x) = g(f(x)) = g^*(f(x))(f(x) - f(x_0)) + g(f(x_0))$$
$$= g^*(f(x))f^*(x)(x - x_0) + (g \circ f)(x_0).$$

Set $h^*(x) = g^*(f(x))f^*(x)$ for $x \in (a, b)$. Since f is continuous at x_0 and g^* is continuous at $f(x_0)$, it is clear that h^* is continuous at x_0. Moreover,

$$h^*(x_0) = g^*(f(x_0))f^*(x_0) = g'(f(x_0))f'(x_0).$$

We conclude by the lemma that $(g \circ f)'(x_0) = g'(f(x_0))f'(x_0).$ ∎

We close this section with a brief discussion of what it means for a function to be differentiable *on* a set (as opposed to being differentiable at every point *in* a set). This concept will be used in Sections 3.3, 3.6, and 6.1.

Notice that by Definition 2.8, a function which is differentiable at a point x_0 must be defined in an open interval containing x_0, i.e., defined on both sides of x_0. This restriction unnaturally narrows the collection of differentiable functions. For example, the function $f(x) = x^{3/2}$ is smooth and has unique tangent line at $x_0 = 0$, but $f'(0)$ does not exist (because $f(x)$ is not defined for $x < 0$). To enlarge the concept of differentiability to include functions like this, we introduce the following definition.

DEFINITION 2.9. Let E be a set such that each $x_0 \in E$ is a cluster point of E.

    i) A function $f : E \to \mathbf{R}$ is said to be *differentiable on* E if

$$f'_E(x_0) := \lim_{\substack{x \to x_0 \\ x \in E}} \frac{f(x) - f(x_0)}{x - x_0}$$

    exists and is finite for every $x_0 \in E$.

    ii) f is said to be *continuously differentiable* on E if f'_E exists and is continuous on E.

Let E be a set such that each $x_0 \in E$ is a cluster point of E. (In particular, E could be any nondegenerate interval.) For each $n \in \mathbf{N}$, we shall denote the collection of functions f whose nth derivatives exist and are continuous on E by $\mathcal{C}^n(E)$. (Thus $\mathcal{C}^1(E)$ is precisely the collection of functions which are continuously differentiable on E.) We shall denote the collection of functions f which belong to $\mathcal{C}^n(E)$ for all $n \in \mathbf{N}$ by $\mathcal{C}^\infty(E)$. When E is an interval, we shall drop the outer set of parentheses; e.g., we shall write $\mathcal{C}^n[a,b]$ for $\mathcal{C}^n([a,b])$.

By modifying the proof of Theorem 2.12, we can show that if f is differentiable on E, then f is continuous on E. Thus, $\mathcal{C}^m(E) \subset \mathcal{C}^n(E)$ when $m > n$.

It is important to notice that a function which is differentiable on two sets is not necessarily differentiable on their union.

Remark 2. $f(x) = |x|$ *is differentiable on* $[0,1]$ *and on* $[-1,0]$ *but not on* $[-1,1]$.

PROOF. By Remark 1, f is differentiable on $[-1,0) \cup (0,1]$ but not differentiable at $x = 0$. However,

$$f'_{[0,1]}(0) = \lim_{h \to 0+} \frac{|h|}{h} = 1 \quad \text{and} \quad f'_{[-1,0]}(0) = \lim_{h \to 0-} \frac{|h|}{h} = -1.$$

Therefore, f is differentiable on $[0,1]$ and on $[-1,0]$. ∎

We shall use Definition 2.9 only when E is an interval. Notice that by the argument used to prove Remark 2 in Section 2.1, if E is an open interval, then f'_E exists on E if and only if $f'(x)$ exists for each $x \in E$, in which case, $f'_E = f'$. Thus, for open intervals we can drop the subscript E. We will usually drop the subscript for closed intervals too, using the notation

$$f'(a) := \lim_{h \to 0+} \frac{f(a+h) - f(a)}{h} \quad \text{and} \quad f'(b) := \lim_{h \to 0-} \frac{f(b+h) - f(b)}{h}$$

when f is differentiable on $[a,b]$.

The following example shows that Definition 2.9 enlarges the collection of differentiable functions.

Example 1. The function $f(x) = x^{3/2}$ is differentiable on $[0,\infty)$ and $f'(x) = 3\sqrt{x}/2$ for all $x \in [0,\infty)$.

PROOF. By the Power Rule (see Exercise 6 below), $f'(x) = 3\sqrt{x}/2$ for all $x \in (0,\infty)$. And by definition,

$$f'(0) = \lim_{h \to 0+} \frac{h^{3/2} - 0}{h} = \lim_{h \to 0+} \sqrt{h} = 0. \quad ∎$$

The following example shows that $\mathcal{C}^1(\mathbf{R})$ is a proper subset of the collection of differentiable functions.

Example 2. The function

$$f(x) = \begin{cases} x^2 \sin(1/x) & x \neq 0 \\ 0 & x = 0 \end{cases}$$

is differentiable on **R** but not continuously differentiable on any interval which contains the origin.

PROOF. By definition,

$$f'(0) = \lim_{h \to 0} h \sin\left(\frac{1}{h}\right) = 0 \quad \text{and} \quad f'(x) = 2x \sin\left(\frac{1}{x}\right) - \cos\left(\frac{1}{x}\right)$$

for $x \neq 0$. Thus, f is differentiable on **R** but $\lim_{x \to 0} f'(x)$ does not exist. In particular, f' is not continuous on any interval which contains the origin. ∎

EXERCISES

1. For each of the following functions, find all x for which $f'(x)$ exists and find a formula for f'.
   a) $f(x) = (x^3 - 2x^2 + 3x)/\sqrt{x}$.
   b) $f(x) = 1/(x^2 + x - 1)$.
   c) $f(x) = x^x$.
   d) $f(x) = |x^3 + 2x^2 - x - 2|$.

2. Suppose f and g are differentiable at 2 and 3 with $f'(2) = a$, $f'(3) = b$, $g'(2) = c$, and $g'(3) = d$. If $f(2) = 1$, $f(3) = 2$, $g(2) = 3$, and $g(3) = 4$, evaluate each of the following derivatives.
   a) $(fg)'(2)$.
   b) $(f/g)'(3)$.
   c) $(g \circ f)'(3)$.
   d) $(f \circ g)'(2)$.

3. Suppose that

$$f_\alpha(x) = \begin{cases} |x|^\alpha \sin\dfrac{1}{x} & x \neq 0 \\ 0 & x = 0. \end{cases}$$

Show that $f(x)$ is continuous at $x = 0$ when $\alpha > 0$ and differentiable at $x = 0$ when $\alpha > 1$. Graph these functions for $\alpha = 1$ and $\alpha = 2$ and give a geometric interpretation of your results.

4. Suppose that $f : (0, \infty) \to$ **R** satisfies $f(x) - f(y) = f(x/y)$ for all $x, y \in (0, \infty)$ and $f(1) = 0$.
   a) Prove that f is continuous on $(0, \infty)$ if and only if f is continuous at 1.
   b) Prove that f is differentiable on $(0, \infty)$ if and only if f is differentiable at 1.
   c) Prove that if f is differentiable at 1, then $f'(x) = f'(1)/x$ for all $x \in (0, \infty)$.
      (Note: If $f'(1) = 1$, then $f(x) = \log x$.)

5. Let (a, b) be an open interval, $f : (a, b) \to$ **R**, and $x_0 \in (a, b)$. The function f is said to have a *local maximum* at x_0 if there is a $\delta > 0$ such that $f(x_0) \geq f(x)$ for all $|x - x_0| < \delta$.
   a) If f has a local maximum at x_0, prove

$$\limsup_{h \to 0+} \frac{f(x_0 + h) - f(x_0)}{h} \leq 0 \quad \text{and} \quad \liminf_{h \to 0-} \frac{f(x_0 + h) - f(x_0)}{h} \geq 0.$$

b) If f is differentiable at x_0 and has a local maximum at x_0, prove $f'(x_0) = 0$.

c) Make and prove analogous statements for local minima.

d) Show by example that the converses of the statements in parts b) and c) are false. Namely, find an f such that $f'(0) = 0$ but f has neither a local maximum nor a local minimum at 0.

$\boxed{6}$. **This exercise is used in Section 3.3.**

a) Prove $(x^n)' = nx^{n-1}$ for $n \in \mathbf{N}$ and $x \in \mathbf{R}$.

b) Prove that if $f(x) = x^{m/n}$ for some $m, n \in \mathbf{N}$, then $y = f(x)$ is differentiable and satisfies $ny^{n-1}y' = mx^{m-1}$ for every $x \in (0, \infty)$.

c) [POWER RULE] Prove x^q is differentiable on $(0, \infty)$ for every $q \in \mathbf{Q}$ and $(x^q)' = qx^{q-1}$.

7. Using elementary geometry and the definition of $\sin x$, $\cos x$, one can show for every $x, y \in \mathbf{R}$ (see Appendix B) that

(i) $$|\sin x| \le 1, \qquad |\cos x| \le 1, \qquad \sin(0) = 0, \qquad \cos(0) = 1,$$

(ii) $$\sin(-x) = -\sin x, \qquad \cos(-x) = \cos x,$$

(iii) $$\sin^2 x + \cos^2 x = 1, \qquad \cos x = 1 - 2\sin^2\left(\frac{x}{2}\right),$$

(iv) $$\sin(x \pm y) = \sin x \cos y \pm \cos x \sin y.$$

Moreover, if x is measured in radians then

(v) $$\cos x = \sin\left(\frac{\pi}{2} - x\right), \qquad \sin x = \cos\left(\frac{\pi}{2} - x\right),$$

and

(vi) $$0 < x \cos x < \sin x < x, \qquad 0 < x \le \frac{\pi}{2}.$$

Using these properties, prove each of the following statements.

a) The functions $\sin x$ and $\cos x$ are continuous at 0.

b) The functions $\sin x$ and $\cos x$ are continuous on $\mathbf{R}$.

c) The limits

$$\lim_{x \to 0} \frac{\sin x}{x} = 1 \quad \text{and} \quad \lim_{x \to 0} \frac{1 - \cos x}{x} = 0$$

exist.

d) The function $\sin x$ is differentiable on $\mathbf{R}$ with $(\sin x)' = \cos x$.

e) The functions $\cos x$ and $\tan x := \sin x / \cos x$ are differentiable on $\mathbf{R}$ with $(\cos x)' = -\sin x$ and $(\tan x)' = \sec^2 x$.

8. Suppose $n \in \mathbf{N}$ and f, g are real-valued functions of a real variable whose nth derivatives $f^{(n)}$, $g^{(n)}$ exist at a point x_0. Prove Leibniz's generalization of the Product Rule:

$$(fg)^{(n)}(x_0) = \sum_{k=0}^{n} \binom{n}{k} f^{(k)}(x_0) g^{(n-k)}(x_0).$$

9. Consider the following outline to a proof of the Chain Rule. Let $y = f(x)$, $y_0 = f(x_0)$, and observe that $y \to y_0$ as $x \to x_0$. Thus

$$\lim_{x \to x_0} \frac{g \circ f(x) - g \circ f(x_0)}{x - x_0} = \lim_{x \to x_0} \frac{g(f(x)) - g(f(x_0))}{f(x) - f(x_0)} \frac{f(x) - f(x_0)}{x - x_0}$$

$$= \left(\lim_{y \to y_0} \frac{g(y) - g(y_0)}{y - y_0} \right) \left(\lim_{x \to x_0} \frac{f(x) - f(x_0)}{x - x_0} \right)$$

$$= g'(y_0) f'(x_0) = g'(f(x_0)) f'(x_0).$$

   a) Find the flaw in this argument.

   b) Write down a statement which this argument does prove.

2.5   THE MEAN VALUE THEOREM

Suppose f is differentiable on (a, b). Since the graph of f on (a, b) has a tangent at each of its points, it seems likely that the slope of the chord through the points $(a, f(a))$ and $(b, f(b))$ equals the slope $f'(x_0)$ for some value of $x_0 \in (a, b)$ (see Figure 2.9 below). This result, called the Mean Value Theorem, and several of its consequences are proved in this section. We begin with a special case.

Lemma [ROLLE'S THEOREM]. *Suppose f is continuous on a closed bounded non-degenerate interval $[a, b]$ and differentiable on (a, b). If $f(a) = f(b)$, then $f'(x_0) = 0$ for some $x_0 \in (a, b)$.*

   PROOF. By the Extreme Value Theorem, f has a finite maximum M and a finite minimum m on $[a, b]$. If $M = m$, then f is constant on (a, b) and $f'(x) = 0$ for all $x \in (a, b)$.

   Suppose $M \neq m$. Since $f(a) = f(b)$, f must assume one of the values M or m at some point $x_0 \in (a, b)$. By symmetry, we may suppose that $f(x_0) = M$. (That is, if we can prove the theorem when $f(x_0) = M$, then a similar proof establishes the theorem when $f(x_0) = m$.) Since M is the maximum of f on $[a, b]$, we have

$$f(x_0 + h) - f(x_0) \leq 0$$

for all h which satisfy $x_0 + h \in (a, b)$. In the case $h > 0$ this implies

$$f'(x_0) = \lim_{h \to 0+} \frac{f(x_0 + h) - f(x_0)}{h} \leq 0,$$

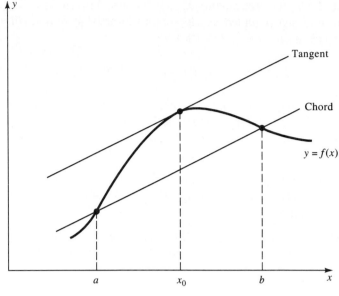

Figure 2.9

and in the case $h < 0$ this implies

$$f'(x_0) = \lim_{h \to 0-} \frac{f(x_0 + h) - f(x_0)}{h} \geq 0.$$

It follows that $f'(x_0) = 0$. ∎

Remark 1. *The continuity hypothesis in Rolle's Theorem cannot be relaxed at even one point in $[a, b]$.*

PROOF. The function

$$f(x) = \begin{cases} x & x \in [0, 1) \\ 0 & x = 1 \end{cases}$$

is continuous on $[0, 1)$ and differentiable on $(0, 1)$, $f(0) = f(1) = 0$, but $f'(x)$ is never zero. ∎

Remark 2. *The differentiability hypothesis in Rolle's Theorem cannot be relaxed at even one point in (a, b).*

PROOF. The function $f(x) = |x|$ is continuous on $[-1, 1]$ and differentiable on $(-1, 1) \setminus \{0\}$, $f(-1) = f(1)$, but $f'(x)$ is never zero. ∎

We shall use Rolle's Theorem to obtain several useful results. The first is a generalization of the Mean Value Theorem also known as *Cauchy's Mean Value Theorem*. (For another generalization of the Mean Value Theorem, see Taylor's Formula (Theorem 4.41).)

THEOREM 2.15 [GENERALIZED MEAN VALUE THEOREM]. *Suppose f, g are continuous on a closed bounded nondegenerate interval $[a, b]$ and differentiable on (a, b). Then there is an $x_0 \in (a, b)$ such that*

$$f'(x_0)(g(b) - g(a)) = g'(x_0)(f(b) - f(a)).$$

PROOF.   Set $h(x) = f(x)(g(b) - g(a)) - g(x)(f(b) - f(a))$. Since $h'(x) = f'(x)(g(b) - g(a)) - g'(x)(f(b) - f(a))$, it is clear that h is continuous on $[a, b]$, differentiable on (a, b), and $h(a) = h(b)$. Thus, by Rolle's Theorem, $h'(x_0) = 0$ for some $x_0 \in (a, b)$. ∎

If we apply Theorem 2.15 to the special case $g(x) = x$, we obtain the following result. (For a geometric interpretation of this result, see the opening paragraph of this section and Figure 2.9.)

THEOREM 2.16 [MEAN VALUE THEOREM]. *If f is continuous on a closed bounded nondegenerate interval $[a, b]$ and differentiable on (a, b), then*

$$f(b) - f(a) = f'(x_0)(b - a)$$

for some $x_0 \in (a, b)$.

The following inequality is an application of the Mean Value Theorem.

THEOREM 2.17 [BERNOULLI'S INEQUALITY]. *Let α be a positive real number and $\delta \geq -1$. If $0 < \alpha \leq 1$, then*

$$(1 + \delta)^\alpha \leq 1 + \alpha\delta,$$

and if $\alpha \geq 1$, then

$$(1 + \delta)^\alpha \geq 1 + \alpha\delta.$$

PROOF.   The proofs of these inequalities are similar. We present the details only for the case $0 < \alpha \leq 1$.

Let $f(x) = x^\alpha$. By the Mean Value Theorem

$$f(1 + \delta) = f(1) + \alpha\delta x_0^{\alpha - 1}$$

for some x_0 between 1 and $1 + \delta$. If $\delta > 0$, then $x_0 > 1$. Since $0 < \alpha \leq 1$, it follows that $x_0^{\alpha - 1} < 1$ (see Exercise 5 in Section 3.3), hence $\delta x_0^{\alpha - 1} < \delta$. On the other hand, if $-1 \leq \delta \leq 0$, then $x_0^{\alpha - 1} \geq 1$ and again $\delta x_0^{\alpha - 1} \leq \delta$. Therefore,

$$(1 + \delta)^\alpha = f(1 + \delta) \leq f(1) + \alpha\delta x_0^{\alpha - 1} \leq 1 + \alpha\delta. \quad ∎$$

Another application of the Mean Value Theorem is the following technique for evaluating limits of certain quotients.

THEOREM 2.18 [L'HÔPITAL'S RULE]. *Let x_0 be an extended real number and I be an open interval which either contains x_0 or has x_0 as an endpoint. Suppose that f and g are differentiable on I, and $g(x) \neq 0 \neq g'(x)$ for all $x \in I$. Suppose further that*

$$A := \lim_{\substack{x \to x_0 \\ x \in I}} f(x) = \lim_{\substack{x \to x_0 \\ x \in I}} g(x)$$

is either 0 or ∞. If

$$B := \lim_{\substack{x \to x_0 \\ x \in I}} \frac{f'(x)}{g'(x)}$$

exists as an extended real number, then

$$\lim_{\substack{x \to x_0 \\ x \in I}} \frac{f(x)}{g(x)} = \lim_{\substack{x \to x_0 \\ x \in I}} \frac{f'(x)}{g'(x)}.$$

PROOF. Let $x, y \in I$ with y between x and x_0 and apply the Generalized Mean Value Theorem to choose a point x^* between x and y such that

(14)
$$\frac{f(y) - f(x)}{g(y) - g(x)} = \frac{f'(x^*)}{g'(x^*)}.$$

Let $\varepsilon > 0$ and suppose for a moment that $|B| < \infty$. If $|x_0| < \infty$, choose $\delta > 0$ such that

$$0 < |t - x_0| < \delta \quad \text{and} \quad t \in I \quad \text{imply} \quad \left| \frac{f'(t)}{g'(t)} - B \right| < \varepsilon.$$

If $|x_0| = \infty$, suppose for simplicity that $x_0 = \infty$ and choose $M > 0$ so large that

$$t > M \quad \text{implies} \quad t \in I \quad \text{and} \quad \left| \frac{f'(t)}{g'(t)} - B \right| < \varepsilon.$$

Hence, for each fixed $x \in I$ which satisfies $0 < |x - x_0| < \delta$ (when $|x_0| < \infty$) or $x > M$ (when $x_0 = \infty$), we have by (14) that

(15)
$$B - \varepsilon < \frac{f(y) - f(x)}{g(y) - g(x)} < B + \varepsilon$$

for all y which lie between x and x_0.

Case 1. $A = 0$. Take the limit of (15) as $y \to x_0$ with $y \in I$. Since $A = 0$, we obtain

$$B - \varepsilon \leq \frac{f(x)}{g(x)} \leq B + \varepsilon$$

for all $x \in I$ which satisfy $0 < |x - x_0| < \delta$ (when $|x_0| < \infty$) or $x > M$ (when $x_0 = \infty$). Thus by definition, $f(x)/g(x) \to B$ as $x \to x_0$ through I.

Case 2. $A = \infty$. Then we can make δ smaller or M larger so that both $g(y)$ and $g(y) - g(x)$ are positive. Multiplying (15) by $g(y) - g(x)$ and dividing by $g(y)$, we obtain

(16) $\qquad \dfrac{f(x)}{g(y)} + (B - \varepsilon)\left(1 - \dfrac{g(x)}{g(y)}\right) < \dfrac{f(y)}{g(y)} < \dfrac{f(x)}{g(y)} + (B + \varepsilon)\left(1 - \dfrac{g(x)}{g(y)}\right)$

for all $y \in I$ which satisfy $0 < |y - x_0| < \delta$ or $y > M$. Since $A = \infty$, we have $g(x)/g(y) \to 0$ and $f(x)/g(y) \to 0$ as $y \to x_0$ through I. Thus by (16) we can choose $\delta > 0$ smaller or $M > 0$ larger (if necessary) so that

$$B - 2\varepsilon \le \frac{f(y)}{g(y)} \le B + 2\varepsilon$$

for all $y \in I$ which satisfy $0 < |y - x_0| < \delta$ or $y > M$. We conclude by definition that $f(y)/g(y) \to B$ as $y \to x_0$ through I. This proves L'Hôpital's Rule when $|B| < \infty$. A similar proof handles the case when $|B| = \infty$ (see Exercise 10). ∎

L'Hôpital's Rule can be used to compare the relative rates of growth of two functions. For example, the next example shows that as $x \to \infty$, e^x converges to ∞ much faster that x^2 does. In particular, $x^2 \le e^x$ for large positive x.

Example 1. Prove $\lim_{x \to \infty} x^2/e^x = 0$.

PROOF. Since the limits of x^2/e^x and x/e^x are of the form ∞/∞, we apply L'Hôpital's Rule twice to verify

$$\lim_{x \to \infty} \frac{x^2}{e^x} = \lim_{x \to \infty} \frac{2x}{e^x} = \lim_{x \to \infty} \frac{2}{e^x} = 0. \quad ∎$$

For each subsequent application of L'Hôpital's Rule, it is important to check that the hypotheses still hold. For example,

$$\lim_{x \to 0} \frac{x^2}{x^2 + \sin x} = \lim_{x \to 0} \frac{2x}{2x + \cos x} = 0 \ne 1 = \lim_{x \to 0} \frac{2}{2 - \sin x}.$$

Notice that the middle limit is not of the form $0/0$.

It is also important that the denominators $g(x)$ and $g'(x)$ have no sequence of roots in the interval I which converges to x_0. For example, the limit of the function

$$\frac{x + \sin x \cos x}{x \sin x + \cos x}$$

is of the form ∞/∞ and does not exist (since $\sin x$ and $\cos x$ oscillate between -1 and 1). If we blindly apply L'Hôpital's Rule, however, we mistakenly conclude that this function *does* have a limit, since the ratio of its derivatives satisfies

$$\lim_{x \to \infty} \frac{2\cos^2 x}{x \cos x} = 0.$$

L'Hôpital's Rule can be used to evaluate limits of the form $0 \cdot \infty$.

Example 2. Find $\lim_{x \to 0} x \log x$.

SOLUTION. By writing x as $1/(1/x)$, we see that the limit in question is of the form ∞/∞. Hence, by L'Hôpital's Rule,

$$\lim_{x \to 0} x \log x = \lim_{x \to 0} \frac{\log x}{1/x} = \lim_{x \to 0} \frac{1/x}{-1/x^2} = 0. \quad \blacksquare$$

The next two examples show that L'Hôpital's Rule can also be used to evaluate limits of the form 1^∞ and 0^0.

Example 3. Prove that the sequence $(1 + 1/n)^n$ is increasing, as $n \to \infty$, and its limit e satisfies $e \leq 3$ and $\log e = 1$.

PROOF. The sequence $(1 + 1/n)^n$ is increasing, since by Bernoulli's Inequality,

$$\left(1 + \frac{1}{n}\right)^{n/(n+1)} \leq \left(1 + \frac{1}{n+1}\right).$$

To prove that this sequence is bounded above, observe by the Binomial Formula that

$$\left(1 + \frac{1}{n}\right)^n = \sum_{k=0}^{n} \binom{n}{k} \left(\frac{1}{n}\right)^k.$$

Now,

$$\binom{n}{k} \left(\frac{1}{n}\right)^k = \frac{n(n-1)\dots(n-k+1)}{n^k} \cdot \frac{1}{k!} \leq \frac{1}{k!} \leq \frac{1}{2^{k-1}}$$

for all $k \in \mathbf{N}$. It follows from Exercise 1c) in Section 1.2 that

$$\left(1 + \frac{1}{n}\right)^n \leq 1 + 1 + \sum_{k=1}^{n} \frac{1}{2^k} = 3 - \frac{1}{2^n} < 3$$

for all $n \in \mathbf{N}$. Hence, by the Monotone Convergence Theorem, the limit defining e exists.

To verify $\log e = 1$, use L'Hôpital's Rule:

$$\log e = \lim_{n \to \infty} \frac{\log(1 + 1/n)}{1/n} = \lim_{n \to \infty} \frac{(n/(n+1))(-1/n^2)}{-1/n^2} = 1. \quad \blacksquare$$

Example 4. Find $L = \lim_{x \to 1} (\log x)^{1-x}$.

SOLUTION. Since $\log L = \lim_{x \to 1} (1 - x) \log \log x$ (if this limit exists) is of the form $0 \cdot \infty$, we have by L'Hôpital's Rule that

$$\log L = \lim_{x \to 1} \frac{\log \log x}{1/(1-x)} = \lim_{x \to 1} \frac{1/(x \log x)}{1/(1-x)^2} = -\lim_{x \to 1} \frac{1-x}{\log x} = \lim_{x \to 1} \frac{1}{1/x} = 1.$$

Hence, $L = e^1 = e$. ∎

EXERCISES

1. Evaluate the following limits.

a)
$$\lim_{x \to 0} \frac{\sin(3x)}{x}.$$

b)
$$\lim_{x \to 0+} \frac{\cos x - e^x}{\log(1 + x^2)}.$$

c)
$$\lim_{x \to 0} \left(\frac{x}{\sin x} \right)^{1/x^2}.$$

d)
$$\lim_{x \to 0+} x^x.$$

e)
$$\lim_{x \to 1} \frac{\log x}{\sin(\pi x)}.$$

f)
$$\lim_{x \to \infty} x \left(\arctan x - \frac{\pi}{2} \right).$$

(For the derivative of $\arctan x$ see Exercise 8 in Section 2.6.)

2. Assuming e^x is differentiable on **R**, prove that

$$f(x) = \begin{cases} \dfrac{x}{1 + e^{1/x}} & x \neq 0 \\ 0 & x = 0 \end{cases}$$

is differentiable on $[0, \infty)$. Is f differentiable at 0?

3. **This exercise is used in Sections 4.7 and 7.5.** Assume that e^x is differentiable on **R** with $(e^x)' = e^x$.

   a) Show that the derivative of

$$f(x) = \begin{cases} e^{-1/x^2} & x \neq 0 \\ 0 & x = 0 \end{cases}$$

   exists and is continuous on **R** with $f'(0) = 0$.

   b) Do analogous statements hold for $f^{(n)}(x)$ when $n = 2, 3, \ldots$?

4. **This exercise is used in Sections 3.4, 4.3, and elsewhere.**

   a) Using $(e^x)' = e^x$, $(\log x)' = 1/x$, and $x^\alpha = e^{\alpha \log x}$, show that $(x^\alpha)' = \alpha x^{\alpha-1}$ for all $x > 0$.

   b) Let $\alpha > 0$. Prove that $\log x \le x^\alpha$ for x large. Prove that there exists a constant C_α such that $\log x \le C_\alpha x^\alpha$ for all $x \in [1, \infty)$, $C_\alpha \to \infty$ as $\alpha \to 0+$, and $C_\alpha \to 0$ as $\alpha \to \infty$.

   c) Obtain an analogue of b) valid for e^x and x^α in place of $\log x$ and x^α.

5. Suppose f is differentiable on an open interval (a, b) with f' bounded on (a, b). Prove f is uniformly continuous on (a, b).

6. a) Suppose f is differentiable on (a, b), continuous on $[a, b]$, and $f(a) = f(b) = 0$. Prove that if $f(c) > 0$ for some $c \in (a, b)$, then there exist $x_1, x_2 \in (a, b)$ such that
$$f'(x_1) > f'(x_2).$$

   b) Suppose f is twice differentiable on (a, b) and that there are points $x_1 < x_2 < x_3$ in (a, b) such that $f(x_1) > f(x_2)$ and $f(x_3) > f(x_2)$. Prove that there is a $c \in (a, b)$ such that $f''(c) > 0$.

7. Let (a, b) be an open interval, $f : (a, b) \to \mathbf{R}$, and $x_0 \in (a, b)$. The function f is said to have a *proper local maximum* at x_0 if there is a $\delta > 0$ such that $f(x_0) > f(x)$ for all $0 < |x - x_0| < \delta$.

   a) If f is differentiable on (a, b) and has a proper local maximum at x_0, prove that $f'(x_0) = 0$ and that given $\delta > 0$, there exist $x_1 < x_0 < x_2$ such that $f'(x_1) > 0$, $f'(x_2) < 0$ and $|x_j - x_0| < \delta$ for $j = 1, 2$.

   b) Make and prove an analogous statement for a proper local minimum.

8. Let f be differentiable on $(0, \infty)$ and suppose $L = \lim_{x \to \infty} f'(x)$ exists and is finite. Prove that if $\lim_{n \to \infty} f(n)$ exists and is finite, then $L = 0$.

9. Suppose $f : [a, b] \to \mathbf{R}$ is continuous and increasing (see Exercise 7 in Section 2.3). Prove that $\sup f(E) = f(\sup E)$ for every nonempty set $E \subseteq [a, b]$.

10. Prove L'Hôpital's Rule for the case $|B| = \infty$ by using (14) and a modified version of (15).

2.6 MONOTONE FUNCTIONS AND THE INVERSE FUNCTION THEOREM

Monotone functions (i.e., those which either increase or decrease on their domain) are important from both a theoretical and a practical point of view (see Theorem 3.15). In this section we study monotone functions and the role they play in the Inverse Function Theorem.

DEFINITION 2.10. Let E be a nonempty subset of $\mathbf{R}$ and $f : E \to \mathbf{R}$.

   i) f is said to be *increasing* (respectively, *strictly increasing*) on E if $x_1, x_2 \in E$ and $x_1 < x_2$ imply $f(x_1) \le f(x_2)$ (respectively, $f(x_1) < f(x_2)$).

   ii) f is said to be *decreasing* (respectively, *strictly decreasing*) on E if $x_1, x_2 \in E$ and $x_1 < x_2$ imply $f(x_1) \ge f(x_2)$ (respectively, $f(x_1) > f(x_2)$).

iii) f is said to be *monotone* (respectively, *strictly monotone*) on E if f is either decreasing or increasing (respectively, either strictly decreasing or strictly increasing) on E.

Thus, although $f(x) = x^2$ is strictly monotone on $[0, 1]$, and on $[-1, 0]$, it is not monotone on $[-1, 1]$.

The derivative gives a simple method for finding where a differentiable function is monotone.

THEOREM 2.19. *Suppose f is continuous on a closed bounded nondegenerate interval $[a, b]$ and differentiable on (a, b).*

  i) *If $f'(x) > 0$ (respectively, $f'(x) < 0$) for all $x \in (a, b)$, then f is strictly increasing (respectively, strictly decreasing) on $[a, b]$.*

  ii) *If $f'(x) = 0$ for all $x \in (a, b)$, then f is constant on $[a, b]$.*

PROOF. Let $a \leq x_1 < x_2 \leq b$. By the Mean Value Theorem, there is an $x_0 \in (a, b)$ such that

$$f(x_2) - f(x_1) = f'(x_0)(x_2 - x_1).$$

Thus, $f(x_2) > f(x_1)$ when $f'(x_0) > 0$ and $f(x_2) < f(x_1)$ when $f'(x_0) < 0$. This proves part i).

To prove part ii), let $a \leq x \leq b$. By the Mean Value Theorem and hypothesis there is an $x_0 \in (a, b)$ such that

$$f(x) - f(a) = f'(x_0)(x - a) = 0.$$

Thus, $f(x) = f(a)$ for all $x \in [a, b]$. ∎

Remark 1. *If f and g are continuous on a closed bounded nondegenerate interval $[a, b]$, differentiable on (a, b), and $f'(x) = g'(x)$ for all $x \in (a, b)$, then $f - g$ is constant on $[a, b]$.*

PROOF. Apply Theorem 2.19ii to the function $f - g$. ∎

Recall (see Examples 2 and 3 in Section 2.2) that there exist functions which have neither right nor left limits at a given point. The following result shows that monotone functions never behave this badly.

Lemma. *Suppose f is increasing on $[a, b]$.*

  i) *If $x_0 \in [a, b)$ then $f(x_0+)$ exists and $f(x_0) \leq f(x_0+)$.*

  ii) *If $x_0 \in (a, b]$ then $f(x_0-)$ exists and $f(x_0-) \leq f(x_0)$.*

PROOF. Fix $x_0 \in (a, b]$. By symmetry it suffices to show that $f(x_0-)$ exists and satisfies $f(x_0-) \leq f(x_0)$. Set $E = \{f(x) : a < x < x_0\}$ and $s = \sup E$. Since f is increasing, $f(x_0)$ is an upper bound of E. Hence, s is a finite real number which satisfies $s \leq f(x_0)$. Given $\varepsilon > 0$, choose by the Approximation Property an $x_1 \in (a, x_0)$ such that $s - \varepsilon < f(x_1) \leq s$. Since f is increasing,

$$s - \varepsilon < f(x_1) \leq f(x) \leq s$$

for all $x_1 < x < x_0$. Therefore, $f(x_0-)$ exists and satisfies $f(x_0-) = s \leq f(x_0)$. ∎

We have seen (Example 3 in Section 2.2) that a function can be nowhere continuous, i.e., can have uncountably many points of discontinuity. How many points of discontinuity can a monotone function have?

*THEOREM 2.20. *If f is monotone on an interval I, then f has at most countably many points of discontinuity on I.*

PROOF. Without loss of generality, we may suppose that f is increasing. Since the countable union of countable sets is countable (Theorem 1.28), it suffices to show that the set of points of discontinuity of f can be written as a countable union of countable sets. Since **R** is the union of closed intervals $[-n, n]$, $n \in \mathbf{N}$, we may suppose that I is a closed bounded interval $[a, b]$.

Let E represent the set of points of discontinuity of f on (a, b). By the lemma, $f(x-) \leq f(x) \leq f(x+)$ for all $x \in (a, b)$. Thus, f is discontinuous at such an x if and only if $f(x+) - f(x-) > 0$. It follows that

$$E = \bigcup_{j=1}^{\infty} A_j,$$

where for each $j \in \mathbf{N}$, $A_j := \{x \in \mathbf{R} : f(x+) - f(x-) \geq 1/j\}$. We will complete the proof by showing each A_j is finite.

Suppose to the contrary that A_{j_0} is infinite for some j_0. Set $y_0 := j_0(f(b) - f(a))$ and observe that since f is finite-valued on I, y_0 is a finite real number. On the other hand, since A_{j_0} is infinite, there exist $x_1 < x_2 < \dots$ in $[a, b]$ such that $f(x_k+) - f(x_k-) \geq 1/j_0$ for $k \in \mathbf{N}$. Since f is monotone, it follows that

$$f(b) - f(a) \geq \sum_{k=1}^{n}(f(x_k+) - f(x_k-)) \geq \frac{n}{j_0},$$

i.e., $y_0 = j_0(f(b) - f(a)) \geq n$ for all $n \in \mathbf{N}$. Taking the limit of this last inequality as $n \to \infty$, we see that $y_0 = +\infty$. With this contradiction, the proof of the theorem is complete. ∎

Let f be a real-valued function of a real variable. Recall (see Figure 1.6) that if f has an inverse function f^{-1}, then the graph of $y = f^{-1}(x)$ is a reflection of the graph of $y = f(x)$ about the line $y = x$. Thus, it is not difficult to imagine that f^{-1} is as smooth as f. This is the subject of the next two theorems.

THEOREM 2.21. *If f is 1–1 and continuous on a closed bounded interval $[a, b]$, then f is strictly monotone on $[a, b]$ and f^{-1} is continuous and strictly monotone on the interval whose endpoints are $f(a)$ and $f(b)$.*

PROOF. We may suppose that $a < b$. There are two cases: either $f(a) < f(b)$ or $f(a) > f(b)$. We supply the details of the proof only for the first case. The proof for the second case is similar.

We first show that if $f(a) < f(b)$, then f is strictly increasing on $[a, b]$. Indeed, suppose $x_1, x_2 \in (a, b)$ and $x_1 < x_2$. If $f(x_1) \geq f(b)$ then $f(b)$ lies between $f(a)$ and $f(x_1)$. Hence, by the Intermediate Value Theorem, there is an $x_0 \in [a, x_1]$ which satisfies $f(x_0) = f(b)$, a contradiction of the fact that f is 1–1. Similarly, $f(x_1) \leq f(a)$ also leads to a contradiction. Therefore, $f(a) < f(x_1) < f(b)$. Applying this same argument with x_1 in place of a and x_2 in place of x_1, we see that $f(x_1) < f(x_2) < f(b)$. This proves that f is strictly increasing on $[a, b]$.

Next, let $c = f(a)$ and $d = f(b)$. By the Intermediate Value Theorem, f takes $[a, b]$ onto $[c, d]$. Hence f is 1–1 from $[a, b]$ onto $[c, d]$ and it follows from Theorem 1.26 that f has an inverse function f^{-1} on $[c, d]$. Suppose f^{-1} is not strictly increasing on $[c, d]$. Then we can choose $y_1 < y_2$ such that

$$x_1 := f^{-1}(y_1) \geq x_2 := f^{-1}(y_2).$$

But f is increasing, so $y_1 = f(x_1) \geq f(x_2) = y_2$, a contradiction. Thus, f^{-1} is strictly increasing on $[c, d]$.

By symmetry, it remains to show that $f^{-1}(y_0+) = f^{-1}(y_0)$ for each $y_0 \in [c, d)$. Fix such a y_0 and let $\varepsilon > 0$. Choose $x_0 \in [a, b)$ such that $f(x_0) = y_0$ and choose $0 < \varepsilon_0 < \varepsilon$ so small that $x_0 + \varepsilon_0 \in [a, b)$. Set $\delta = f(x_0 + \varepsilon_0) - f(x_0)$ and suppose that $0 \leq y - y_0 < \delta$. If $x = f^{-1}(y)$, then the choices of y and δ imply $y_0 \leq y < y_0 + \delta$, i.e.,

$$f(x_0) \leq f(x) < f(x_0 + \varepsilon_0).$$

Since f^{-1} is strictly increasing, it follows that $x_0 \leq x < x_0 + \varepsilon_0$, i.e., $0 \leq x - x_0 < \varepsilon_0$. Therefore,

$$0 \leq f^{-1}(y) - f^{-1}(y_0) < \varepsilon_0 < \varepsilon$$

for all $0 \leq y - y_0 < \delta$. We conclude that $f^{-1}(y_0+)$ exists and equals $f^{-1}(y_0)$. ∎

THEOREM 2.22 [INVERSE FUNCTION THEOREM]. *Let f be 1–1 and continuous on an open interval I. If $f'(x_0)$ exists and is nonzero for some $x_0 \in I$, then f^{-1} is differentiable at $y_0 = f(x_0)$ and*

$$(f^{-1})'(y_0) = \frac{1}{f'(x_0)}.$$

PROOF. Choose $a, b \in \mathbf{R}$ such that $x_0 \in (a, b)$ and $[a, b] \subset I$. By Theorem 2.21, f is strictly monotone, say strictly increasing on $[a, b]$, and f^{-1} exists, is continuous, and strictly increasing on $[f(a), f(b)]$. Since $x_0 \in (a, b)$, choose $y_0 \in (f(a), f(b))$ such that $y_0 = f(x_0)$ and let $h \neq 0$ be so small that $f^{-1}(y_0 + h)$ is defined. Set $x = f^{-1}(y_0 + h)$ and observe that

$$f(x) - f(x_0) = y_0 + h - y_0 = h.$$

Since f^{-1} is continuous, $x \to x_0$ if and only if $h \to 0$. Therefore,

$$\lim_{h \to 0} \frac{f^{-1}(y_0 + h) - f^{-1}(y_0)}{h} = \lim_{x \to x_0} \frac{x - x_0}{f(x) - f(x_0)} = \frac{1}{f'(x_0)}. \quad ∎$$

This theorem is usually presented in elementary calculus texts in a form more easily remembered: if $y = f(x)$ and $x = f^{-1}(y)$, then

$$\frac{dx}{dy} = \frac{1}{dy/dx}.$$

Notice that by using this formula, we do not need to solve explicitly for f^{-1} to be able to compute $(f^{-1})'$ (see Exercises 2, 3, and 8).

Although a differentiable function might not be continuously differentiable, the following result shows that its derivative does satisfy an intermediate value theorem. (The statement of this theorem and its proof come from Rudin [11][1].)

*THEOREM 2.23 [DUHAMEL]. *Suppose f is differentiable on $[a, b]$ with $f'(a) \neq f'(b)$. If y_0 is a real number which lies between $f'(a)$ and $f'(b)$, then there is an $x_0 \in [a, b]$ such that $f'(x_0) = y_0$.*

PROOF. Let $c = (a + b)/2$,

$$\alpha(t) = \begin{cases} a & a \leq t \leq c \\ 2t - b & c \leq t \leq b \end{cases} \quad \text{and} \quad \beta(t) = \begin{cases} 2t - a & a \leq t \leq c \\ b & c \leq t \leq b \end{cases}$$

Then α, β are continuous on $[a, b]$ with $a \leq \alpha(t) < \beta(t) \leq b$ and

$$\beta(t) - \alpha(t) = \begin{cases} 2(t - a) & a \leq t \leq c \\ 2(b - t) & c \leq t \leq b. \end{cases}$$

Define

$$g(t) = \frac{f(\beta(t)) - f(\alpha(t))}{\beta(t) - \alpha(t)}.$$

Clearly, g is continuous on (a, b) and can be continuously extended to $[a, b]$ because

$$\lim_{t \to a+} g(t) = \lim_{t \to a+} \frac{f(2t - a) - f(a)}{2(t - a)} = f'(a)$$

and similarly, $\lim_{t \to b-} g(t) = f'(b)$. Hence, by Theorem 2.9 (the Intermediate Value Theorem), there is a $t_0 \in [a, b]$ such that $g(t_0) = y_0$. On the other hand, by Theorem 2.16 (the Mean Value Theorem), there is an x_0 between $\alpha(t_0)$ and $\beta(t_0)$ such that

$$g(t_0) = \frac{f(\beta(t_0)) - f(\alpha(t_0))}{\beta(t_0) - \alpha(t_0)} = f'(x_0).$$

We conclude that $f'(x_0) = y_0$. ∎

---

[1]Walter Rudin, <u>Principles of Mathematical Analysis</u>, 3rd ed. (New York: McGraw-Hill Book Co., 1976). Reprinted with permission of McGraw- Hill Book Co.

EXERCISES

1. a) Find all $a \in \mathbf{R}$ such that $x^3 + ax^2 + 3x + 15$ is strictly increasing near $x = 1$.
   b) Find all $a \in \mathbf{R}$ such that $ax^2 + 3x + 5$ is strictly increasing on the interval $(1, 2)$.
   c) Find where $f(x) = 2|x - 1| + 5\sqrt{x^2 + 9}$ is strictly increasing and where $f(x)$ is strictly decreasing.

2. Let $f(x) = x^2 e^{x^2}$, $x \in \mathbf{R}$.
   a) Show that f^{-1} exists and is differentiable on $(0, \infty)$.
   b) Compute $(f^{-1})'(e)$.

3. Suppose that f' exists and is continuous on an open interval (a, b) with $f'(x) \neq 0$ for all $x \in (a, b)$.
   a) Prove f is 1–1 on (a, b) and takes (a, b) onto some open interval (c, d).
   b) Show that $(f^{-1})'$ exists and is continuous on (c, d).
   c) Using the function $f(x) = x^3$, show that b) is false if the assumption $f'(x) \neq 0$ fails to hold for some $x \in (a, b)$.

4. Suppose f is 1–1 and continuous on a closed bounded nondegenerate interval $[a, b]$ and differentiable on (a, b). Show that if $f'(x) \geq \varepsilon_0 > 0$ for $x \in (a, b)$, then $(f^{-1})'$ exists and is bounded on $(f(a), f(b))$.

5. Let $[a, b]$ be a closed bounded nondegenerate interval. Find all functions f which satisfy the following conditions for some fixed $\alpha > 0$: f is continuous and 1–1 on $[a, b]$, $f'(x) \neq 0$ and $f'(x) = \alpha(f^{-1})'(f(x))$ for all $x \in (a, b)$.

6. Let I be an interval and $n \in \mathbf{N}$. Show that if $f_j : I \to \mathbf{R}$ are monotone functions and $f = \sum_{j=1}^{n} \alpha_j f_j$ for some $\alpha_j \in \mathbf{R}$, then f has at most countably many points of discontinuity on I.

*7. Suppose f is differentiable at every point in a closed bounded interval $[a, b]$. Prove that if f' is 1–1 on $[a, b]$, then f' is strictly monotone on $[a, b]$.

8. Using the Inverse Function Theorem, prove $(\arcsin x)' = 1/\sqrt{1 - x^2}$ for $x \in (-1, 1)$ and $(\arctan x)' = 1/(1 + x^2)$ for $x \in (-\infty, \infty)$.

Chapter 3

INTEGRABILITY ON **R**

3.1 THE RIEMANN INTEGRAL

In this chapter we discuss integration of real-valued functions of a real variable. We begin our discussion by introducing the following terminology.

DEFINITION 3.1. Let $[a, b]$ be a closed bounded interval.

    i) A *partition* of $[a, b]$ is a set of points $P = \{x_0, x_1, \ldots x_n\}$ such that

$$a = x_0 < x_1 < \cdots < x_n = b.$$

    ii) The *norm* of a partition $P = \{x_0, x_1, \ldots x_n\}$ is the number

$$\|P\| = \max_{1 \le j \le n} |x_j - x_{j-1}|.$$

    iii) A *refinement* of a partition $P = \{x_0, x_1, \ldots x_n\}$ is a partition Q of $[a, b]$ which satisfies $Q \supseteq P$. In this case we say that Q is *finer* than P.

Example 1. [THE DYADIC PARTITION]. Prove that for each $n \in \mathbf{N}$, $P_n = \{j/2^n : j = 0, 1, \ldots, 2^n\}$ is a partition of the interval $[0, 1]$ and P_m is finer than P_n when $m > n$.

    PROOF. Fix $n \in \mathbf{N}$. If $x_j = j/2^n$, then $0 = x_0 < x_1 < \cdots < x_{2^n} = 1$. Thus, P_n is a partition of $[0, 1]$. Let $m > n$ and set $p = m - n$. If $0 \le j \le 2^n$, then $j/2^n = j2^p/2^m$ and $0 \le j2^p \le 2^m$. Thus, P_m is finer than P_n. ∎

    It is clear that by definition, if P and Q are partitions of $[a, b]$, then $P \cup Q$ is finer than both P and Q. (Note that "finer" does not rule out the possibility that $P \cup Q = Q$, which would be the case if Q were a refinement of P.) And if Q is a refinement of P, then $\|Q\| \le \|P\|$. We shall use these observations often.

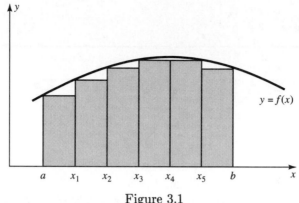

Figure 3.1

Let f be nonnegative on an interval $[a, b]$. You may recall that the integral of f over an interval $[a, b]$ (when this integral exists) is the area of the region bounded by the curves $y = f(x)$, $y = 0$, $x = a$, and $x = b$. This area, A, can be approximated by rectangles whose bases lie in $[a, b]$ and whose heights approximate f (see Figure 3.1). If the tops of these rectangles lie above the curve $y = f(x)$, the resulting approximation is larger than A. If the tops of these rectangles lie below the curve $y = f(x)$, the resulting approximation is smaller than A (see Figure 3.1). Hence, we make the following definition.

DEFINITION 3.2. Let $[a, b]$ be a closed bounded interval, $P = \{x_0, x_1, \ldots x_n\}$ be a partition of $[a, b]$, and $f : [a, b] \to \mathbf{R}$ be bounded.

i) The *upper Riemann sum* of f over P is the number

$$U(f, P) := \sum_{j=1}^{n} M_j(f)(x_j - x_{j-1}),$$

where

$$M_j(f) := \sup_{x \in [x_{j-1}, x_j]} f(x).$$

ii) The *lower Riemann sum* of f over P is the number

$$L(f, P) := \sum_{j=1}^{n} m_j(f)(x_j - x_{j-1}),$$

where

$$m_j(f) := \inf_{x \in [x_{j-1}, x_j]} f(x).$$

(Notice that since f is bounded, the numbers $M_j(f)$ and $m_j(f)$ exist and are finite.)

Some upper and lower Riemann sums can be evaluated with the help of the following elementary observation.

Lemma 1. *If $g : \mathbf{N} \to \mathbf{R}$, then*

$$\sum_{k=m}^{n} (g(k+1) - g(k)) = g(n+1) - g(m)$$

for all $n \geq m$ in $\mathbf{N}$.

PROOF. By the Associative Law,

$$\sum_{k=m}^{n} (g(k+1) - g(k)) = (g(m+1) - g(m))$$

$$+ (g(m+2) - g(m+1)) + \cdots + (g(n+1) - g(n))$$
$$= g(n+1) - g(m). \quad \blacksquare$$

We shall refer to this algebraic identity by saying the sum *telescopes* to $g(n+1) - g(m)$.

Before we define what it means for a function to be integrable, we make several elementary observations concerning upper and lower sums.

Remark 1. *If $f(x) = \alpha$ is constant on $[a, b]$, then*

$$U(f, P) = L(f, P) = \alpha(b - a)$$

for all partitions P of $[a, b]$.

PROOF. Since $M_j(f) = m_j(f) = \alpha$ for all j, the sums $U(f, P)$ and $L(f, P)$ telescope to $\alpha(b - a)$. $\quad \blacksquare$

Remark 2. *$L(f, P) \leq U(f, P)$ for all partitions P and all bounded functions f.*

PROOF. By definition, $m_j(f) \leq M_j(f)$ for all j. $\quad \blacksquare$

The next two results show that as the partitions get finer, the upper and lower Riemann sums get nearer each other.

Remark 3. *If $P = \{x_0, x_1, \ldots x_n\}$ is a partition of $[a, b]$ and $Q = \{c\} \bigcup P$ for some $c \in (a, b)$, then*
$$L(f, P) \leq L(f, Q) \leq U(f, Q) \leq U(f, P).$$

PROOF. By symmetry and Remark 2, we need only show $U(f, Q) \leq U(f, P)$. We may suppose that $c \notin P$. Hence, choose an index j_0 such that $x_{j_0-1} < c < x_{j_0}$. By definition, it is clear that

$$U(f, Q) - U(f, P) = M^{(\ell)}(c - x_{j_0-1}) + M^{(r)}(x_{j_0} - c) - M(x_{j_0} - x_{j_0-1}),$$

where

$$M^{(\ell)} = \sup_{x \in [x_{j_0-1}, c]} f(x), \quad M^{(r)} = \sup_{x \in [c, x_{j_0}]} f(x), \quad \text{and} \quad M = \sup_{x \in [x_{j_0-1}, x_{j_0}]} f(x).$$

By the Monotone Property of Suprema, $M^{(\ell)}$ and $M^{(r)}$ are both less than or equal to M. Therefore,

$$U(f,Q) - U(f,P) \leq M(c - x_{j_0-1}) + M(x_{j_0} - c) - M(x_{j_0} - x_{j_0-1}) = 0. \quad \blacksquare$$

Lemma 2. i) *If P is any partition of $[a,b]$ and Q is a refinement of P, then*

$$L(f,P) \leq L(f,Q) \leq U(f,Q) \leq U(f,P).$$

ii) *If P and Q are any partitions of $[a,b]$ then*

$$L(f,P) \leq U(f,Q).$$

PROOF. i) This part follows immediately from Remark 3 since if Q is finer than P, then Q can be obtained from P in a finite number of steps by adding one point at a time.

ii) Since $P \cup Q$ is a refinement of P and Q, it follows from part i) that

$$L(f,P) \leq L(f,P \cup Q) \quad \text{and} \quad U(f,P \cup Q) \leq U(f,Q)$$

for any pair of partitions P,Q, whether Q is a refinement of P or not. We conclude by Remark 2 that

$$L(f,P) \leq L(f,P \cup Q) \leq U(f,P \cup Q) \leq U(f,Q). \quad \blacksquare$$

We now use the connection between area and integration to motivate the definition of "integrable." Suppose $f(x)$ is nonnegative on $[a,b]$ and the region bounded by the curves $y = f(x)$, $y = 0$, $x = a$, and $x = b$ has a well-defined area A. By Definition 3.2, every upper Riemann sum is an overestimate of A, and every lower Riemann sum is an underestimate of A (see Figure 3.1). Since the estimates $U(f,P)$ and $L(f,P)$ should get nearer to A as P gets finer, the differences $U(f,P) - L(f,P)$ should get smaller. (The shaded area in Figure 3.2 represents the difference $U(f,P) - L(f,P)$ for a particular P.) This leads us to the following definition (see also Exercise 6 below).

DEFINITION 3.3. Let $[a,b]$ be a closed bounded nondegenerate interval. A function $f : [a,b] \to \mathbf{R}$ is said to be *(Riemann) integrable* on $[a,b]$ if f is bounded on $[a,b]$, and given $\varepsilon > 0$ there is a partition P of $[a,b]$ such that $U(f,P) - L(f,P) < \varepsilon$.

Notice that this definition makes sense whether or not f is nonnegative. The connection between nonnegative functions and area was only a convenient vehicle to motivate Definition 3.3.

Also notice that by Lemma 2, $U(f,P) - L(f,P) = |U(f,P) - L(f,P)|$ for all partitions P. Hence, $U(f,P) - L(f,P) < \varepsilon$ is equivalent to $|U(f,P) - L(f,P)| < \varepsilon$.

This section provides a good illustration of how mathematics works. The connection between area and integration leads directly to Definition 3.3. This definition, however, is not easy to apply in concrete situations. Thus, we search for conditions which imply integrability *and* are easy to apply. In view of Figure 3.2, it seems reasonable that a function is integrable if its graph does not jump around too much (so that it can be covered by thinner and thinner rectangles). Since the graph of a continuous function does not jump at all, we are lead to the following simple criterion sufficient (but not necessary) for integrability.

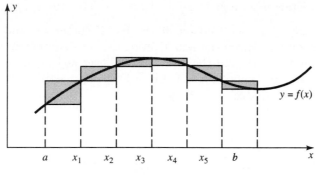

Figure 3.2

THEOREM 3.1. *If f is continuous on a closed bounded interval $[a, b]$, then f is integrable on $[a, b]$.*

PROOF. Let $\varepsilon > 0$. Since f is uniformly continuous on $[a, b]$, choose $\delta > 0$ such that

(1) $$|x - y| < \delta \quad \text{implies} \quad |f(x) - f(y)| < \frac{\varepsilon}{b - a}.$$

Let $P = \{x_0, x_1, \ldots x_n\}$ be any partition of $[a, b]$ which satisfies $\|P\| < \delta$. Fix an index j and notice, by the Extreme Value Theorem, that there are points x_m and x_M in $[x_{j-1}, x_j]$ such that

$$f(x_m) = m_j(f) \quad \text{and} \quad f(x_M) = M_j(f).$$

Since $\|P\| < \delta$, we also have $|x_M - x_m| < \delta$. Hence by (1), $M_j(f) - m_j(f) < \varepsilon/(b-a)$. In particular,

$$U(f, P) - L(f, P) = \sum_{j=1}^{n} (M_j(f) - m_j(f))(x_j - x_{j-1}) < \frac{\varepsilon}{b - a} \sum_{j=1}^{n} (x_j - x_{j-1}) = \varepsilon.$$

(The last step comes from telescoping.) ∎

Although the converse of Theorem 3.1 is false (see Exercise 7 in Section 3.2 and Exercises 3 and 8 below), there is a close connection between integrability and continuity. Indeed, we shall see (Theorem 5.33) that a function is integrable if and only if it has relatively few discontinuities. This principle is illustrated by the following two examples because the nonintegrable function in Example 2 is nowhere continuous (hence has many discontinuities) but the integrable function in Example 3 has only one discontinuity (hence has few discontinuities).

Example 2. The Dirichlet function

$$f(x) = \begin{cases} 1 & x \in \mathbf{Q} \\ 0 & x \notin \mathbf{Q} \end{cases}$$

is not Riemann integrable on $[0, 1]$.

PROOF. Clearly, f is bounded on $[0, 1]$. By Theorem 1.8 and Exercise 3 in Section 1.3 (Density of Rationals and Irrationals), the supremum of f over any nondegenerate interval is 1 and the infimum of f over any nondegenerate interval is 0. Therefore, $U(f, P) - L(f, P) = 1 - 0 = 1$ for any partition P of the interval $[0, 1]$, i.e., f is not integrable on $[0, 1]$. ∎

Example 3. The function

$$f(x) = \begin{cases} 0 & 0 \le x < 1/2 \\ 1 & 1/2 \le x \le 1 \end{cases}$$

is integrable on $[0, 1]$.

PROOF. Since $U(f, P) - L(f, P) \le \|P\|$ for all partitions P of $[0, 1]$, it is clear that f is integrable on $[0, 1]$. ∎

We have defined integrability, but not the value of the integral. We remedy this situation by using the Riemann sums $U(f, P)$ and $L(f, P)$ to define upper and lower integrals.

DEFINITION 3.4. Let $[a, b]$ be a closed bounded nondegenerate interval and $f : [a, b] \to \mathbf{R}$ be bounded.

   i) The *upper integral* of f on $[a, b]$ is the number

$$(U) \int_a^b f(x)\, dx := \inf\{U(f, P) : P \text{ is a partition of } [a, b]\}.$$

   ii) The *lower integral* of f on $[a, b]$ is the number

$$(L) \int_a^b f(x)\, dx := \sup\{L(f, P) : P \text{ is a partition of } [a, b]\}.$$

Although a bounded function might not be integrable (see Example 2 above), the following result shows that the upper and lower integrals of a bounded function always exist.

Remark 4. *If $f : [a, b] \to \mathbf{R}$ is bounded, then its upper and lower integrals exist and are finite, and satisfy*

$$(L) \int_a^b f(x)\, dx \le (U) \int_a^b f(x)\, dx.$$

PROOF. By Lemma 2, $L(f, P) \le U(f, Q)$ for all partitions P and Q of $[a, b]$. Taking the supremum of this inequality over all partitions P of $[a, b]$, we have

$$(L) \int_a^b f(x)\, dx \le U(f, Q),$$

i.e., the lower integral exists and is finite. Taking the infimum of this last inequality over all partitions Q of $[a, b]$, we conclude that the upper integral is also finite, and greater than or equal to the lower integral. ∎

Suppose f is bounded and nonnegative on $[a, b]$. Since the upper and lower sums of f approximate the "area" of the region bounded by the curves $y = f(x)$, $y = 0$, $x = a$, and $x = b$, we guess that f is integrable if and only if the upper and lower integrals of f are equal. The following result shows this guess is true whether or not f is nonnegative.

THEOREM 3.2. *Let* $[a, b]$ *be a closed bounded nondegenerate interval and* $f :$ $[a, b] \to \mathbf{R}$ *be bounded. Then* f *is integrable on* $[a, b]$ *if and only if*

$$(2) \qquad (L) \int_a^b f(x)\, dx = (U) \int_a^b f(x)\, dx,$$

in which case we define the integral *of* f *on* $[a, b]$ *to be the number*

$$\int_a^b f(x)\, dx := (U) \int_a^b f(x)\, dx = (L) \int_a^b f(x)\, dx.$$

PROOF. Suppose f is integrable. Let $\varepsilon > 0$ and choose a partition P of $[a, b]$ such that

$$(3) \qquad U(f, P) - L(f, P) < \varepsilon.$$

By definition, $(U) \int_a^b f(x)\, dx \leq U(f, P)$ and the opposite inequality holds for the lower integral and the lower sum $L(f, P)$. Therefore, it follows from Remark 4 and (3) that

$$\left| (U) \int_a^b f(x)\, dx - (L) \int_a^b f(x)\, dx \right| = (U) \int_a^b f(x)\, dx - (L) \int_a^b f(x)\, dx$$

$$\leq U(f, P) - L(f, P) < \varepsilon.$$

Since this is valid for all $\varepsilon > 0$, (2) holds as promised.

Conversely, suppose (2) holds. Let $\varepsilon > 0$ and choose, by the Approximation Property, partitions P_1 and P_2 of $[a, b]$ such that

$$U(f, P_1) < (U) \int_a^b f(x)\, dx + \frac{\varepsilon}{2}$$

and

$$L(f, P_2) > (L) \int_a^b f(x)\, dx - \frac{\varepsilon}{2}.$$

Set $P = P_1 \cup P_2$. Since P is a refinement of both P_1 and P_2, it follows from Lemma 2, the choices of P_1 and P_2, and (2) that

$$U(f, P) - L(f, P) \leq U(f, P_1) - L(f, P_2)$$

$$\leq (U) \int_a^b f(x)\, dx + \frac{\varepsilon}{2} - (L) \int_a^b f(x)\, dx + \frac{\varepsilon}{2} = \varepsilon. \quad \blacksquare$$

Since the integral has been defined only on nondegenerate intervals $[a, b]$, we have tacitly assumed that $a < b$. We shall use the conventions

$$\int_b^a f(x)\, dx = -\int_a^b f(x)\, dx \quad \text{and} \quad \int_a^a f(x)\, dx = 0$$

to extend the integral to the cases $a > b$ and $a = b$. In particular, if $f(x)$ is integrable and nonpositive on $[a, b]$, then the area of the region bounded by the curves $y = f(x)$, $y = 0$, $x = a$, and $x = b$ is given by $\int_b^a f(x)\, dx$.

In the next section we shall use the machinery of upper and lower sums to prove several familiar theorems about the Riemann integral. We close this section with one more result which reinforces the connection between integration and area.

THEOREM 3.3. *If $f(x) = \alpha$ is constant on $[a, b]$, then*

$$\int_a^b f(x)\, dx = \alpha(b - a).$$

PROOF. By Theorem 3.1, f is integrable on $[a, b]$. Hence, it follows from Theorem 3.2 and Remark 1 that

$$\int_a^b f(x)\, dx = (U) \int_a^b f(x)\, dx = \inf_P U(f, P) = \alpha(b - a). \quad \blacksquare$$

EXERCISES

1. Let $P_n = \{j/n : j = 0, 1, \ldots, n\}$. For each of the following functions, compute the upper and lower sums $U(f, P_n)$, $L(f, P_n)$ and show directly that

   (*) $$\lim_{n \to \infty} L(f, P_n) = \lim_{n \to \infty} U(f, P_n).$$

   (For parts a) and b), you may wish to use Exercise 1 in Section 1.2.)

   a) $$f(x) = \begin{cases} x & 0 \leq x < 1 \\ 0 & \text{otherwise.} \end{cases}$$

   b) $$f(x) = \begin{cases} x^2 & 0 \leq x < 1 \\ 0 & \text{otherwise.} \end{cases}$$

c)
$$f(x) = \begin{cases} 1 & 0 \le x < 1/2 \\ 2 & 1/2 \le x < 1 \\ 0 & \text{otherwise.} \end{cases}$$

d) Prove that any bounded f which satisfies (*) is integrable on $[0, 1]$.

2. Let f be integrable on $[a, b]$ and E be a finite subset of $[a, b]$. Show that if g is a bounded function which satisfies $g(x) = f(x)$ for all $x \in [a, b] \setminus E$, then g is integrable on $[a, b]$ and

$$\int_a^b g(x)\, dx = \int_a^b f(x)\, dx.$$

3. Let $E = \{1/n : n \in \mathbf{N}\}$. Prove that the function

$$f(x) = \begin{cases} 1 & x \in E \\ 0 & \text{otherwise} \end{cases}$$

is integrable on $[0, 1]$. What is the value of $\int_0^1 f(x)\, dx$?

$\boxed{4}$. **This exercise is used in Section 9.2.** Suppose $[a, b]$ is a closed bounded nondegenerate interval and $f : [a, b] \to \mathbf{R}$ is bounded.

a) Prove that if f is continuous at $x_0 \in [a, b]$ and $f(x_0) \ne 0$, then

$$(L) \int_a^b |f(x)|\, dx > 0.$$

b) Show that if f is continuous on $[a, b]$, then $\int_a^b |f(x)|\, dx = 0$ if and only if $f(x) = 0$ for all $x \in [a, b]$.

c) Does this result hold if the absolute values are removed? If it does, prove it. If it does not, provide a counterexample.

5. Suppose f is continuous on a nondegenerate interval $[a, b]$. Show that

$$\int_a^c f(x)\, dx = 0$$

for all $c \in [a, b]$ if and only if $f(x) = 0$ for all $x \in [a, b]$. (Compare with Exercise 4 and notice that f need not be nonnegative here.)

6. Let f be bounded on a nondegenerate interval $[a, b]$. Prove that f is integrable on $[a, b]$ if and only if given $\varepsilon > 0$ there is a partition P_ε of $[a, b]$ such that

$$P \supseteq P_\varepsilon \quad \text{implies} \quad |U(f, P) - L(f, P)| < \varepsilon.$$

$\boxed{7}$. **This exercise is used in Section 7.3.** Let f, g be bounded on $[a, b]$.

a) Prove

$$(U) \int_a^b (f(x) + g(x))\, dx \le (U) \int_a^b f(x)\, dx + (U) \int_a^b g(x)\, dx$$

and

$$(L) \int_a^b (f(x) + g(x)) \, dx \geq (L) \int_a^b f(x) \, dx + (L) \int_a^b g(x) \, dx.$$

b) Prove

$$(U) \int_a^b f(x) \, dx = (U) \int_a^c f(x) \, dx + (U) \int_c^b f(x) \, dx$$

and

$$(L) \int_a^b f(x) \, dx = (L) \int_a^c f(x) \, dx + (L) \int_c^b f(x) \, dx$$

for $a < c < b$.

$\boxed{8}$. **This exercise is used in Sections 3.5 and 4.8.** Suppose $f : [a, b] \to$ **R**.

a) If f is increasing on $[a, b]$ and $P = \{x_0, \ldots, x_n\}$ is any partition of $[a, b]$, prove that

$$\sum_{j=1}^n (M_j(f) - m_j(f))(x_j - x_{j-1}) \leq (f(b) - f(a)) \|P\|.$$

b) Prove that if f is monotone on $[a, b]$, then f is integrable on $[a, b]$.
Note: By Theorem 2.20, f has only countably many (i.e., relatively few) discontinuities on $[a, b]$. This has nothing to do with the proof of part b), but points out a general principle which will be discussed in Section 5.7.

3.2   RIEMANN SUMS

There is another definition of the Riemann integral frequently found in elementary calculus texts.

DEFINITION 3.5. Let $P = \{x_0, x_1, \ldots, x_n\}$ be a partition of $[a, b]$ and $f : [a, b] \to$ **R**.

i) A *Riemann sum* of f with respect to P is a sum of the form

$$\sum_{j=1}^n f(t_j)(x_j - x_{j-1}),$$

where the choice of $t_j \in [x_{j-1}, x_j]$ is arbitrary.

ii) The Riemann sums of f are said to *converge to $I(f)$ as $\|P\| \to 0$* if given $\varepsilon > 0$ there is a partition P_ε of $[a, b]$ such that

$$P \supseteq P_\varepsilon \quad \text{implies} \quad \left| \sum_{j=1}^n f(t_j)(x_j - x_{j-1}) - I(f) \right| < \varepsilon$$

for all choices of $t_j \in [x_{j-1}, x_j]$, $j = 1, 2, \ldots, n$. In this case we shall use the notation

$$I(f) = \lim_{\|P\| \to 0} \sum_{j=1}^{n} f(t_j)(x_j - x_{j-1}).$$

The following result shows that for bounded functions this definition of the Riemann integral is the same as the one using upper and lower integrals.

THEOREM 3.4. *Let $[a, b]$ be a closed bounded nondegenerate interval and $f : [a, b] \to \mathbf{R}$ be bounded. Then f is Riemann integrable on $[a, b]$ if and only if*

$$I(f) = \lim_{\|P\| \to 0} \sum_{j=1}^{n} f(t_j)(x_j - x_{j-1})$$

exists, in which case

$$I(f) = \int_a^b f(x)\,dx.$$

PROOF. Suppose f is integrable on $[a, b]$ and $\varepsilon > 0$. By the Approximation Property, there is a partition P_ε of $[a, b]$ such that

(4) $$L(f, P_\varepsilon) > \int_a^b f(x)\,dx - \varepsilon \quad \text{and} \quad U(f, P_\varepsilon) < \int_a^b f(x)\,dx + \varepsilon.$$

Let $P = \{x_0, x_1, \ldots x_n\} \supseteq P_\varepsilon$. Then (4) holds with P in place of P_ε. But $m_j(f) \leq f(t_j) \leq M_j(f)$ for any choice of $t_j \in [x_{j-1}, x_j]$. Hence,

$$\int_a^b f(x)\,dx - \varepsilon < L(f, P) \leq \sum_{j=1}^{n} f(t_j)(x_j - x_{j-1}) \leq U(f, P) < \int_a^b f(x)\,dx + \varepsilon.$$

In particular,

$$\left| \sum_{j=1}^{n} f(t_j)(x_j - x_{j-1}) - \int_a^b f(x)\,dx \right| < \varepsilon$$

for all partitions $P \supseteq P_\varepsilon$ and all choices of $t_j \in [x_{j-1}, x_j]$, $j = 1, 2, \ldots, n$.

Conversely, suppose the Riemann sums of f converge to $I(f)$. Let $\varepsilon > 0$ and choose a partition $P = \{x_0, x_1, \ldots x_n\}$ of $[a, b]$ such that

(5) $$\left| \sum_{j=1}^{n} f(t_j)(x_j - x_{j-1}) - I(f) \right| < \frac{\varepsilon}{3}$$

for all choices of $t_j \in [x_{j-1}, x_j]$. By the Approximation Property, choose $t_j, u_j \in [x_{j-1}, x_j]$ such that

$$f(t_j) - f(u_j) > M_j(f) - m_j(f) - \frac{\varepsilon}{3(b-a)}.$$

By (5) and telescoping, we have

$$U(f,P) - L(f,P) = \sum_{j=1}^{n}(M_j(f) - m_j(f))(x_j - x_{j-1})$$

$$< \sum_{j=1}^{n}(f(t_j) - f(u_j))(x_j - x_{j-1}) + \frac{\varepsilon}{3(b-a)}\sum_{j=1}^{n}(x_j - x_{j-1})$$

$$\leq \left|\sum_{j=1}^{n}f(t_j)(x_j - x_{j-1}) - I(f)\right|$$

$$+ \left|I(f) - \sum_{j=1}^{n}f(u_j)(x_j - x_{j-1})\right| + \frac{\varepsilon}{3(b-a)}\sum_{j=1}^{n}(x_j - x_{j-1})$$

$$< \frac{2\varepsilon}{3} + \frac{\varepsilon}{3} = \varepsilon.$$

Therefore, f is integrable on $[a,b]$. ∎

The next two results show that Riemann integrals of complicated functions can be broken into simpler pieces.

THEOREM 3.5 [LINEAR PROPERTY]. *If f, g are integrable on $[a,b]$ and $\alpha \in$* **R**, *then $f + g$ and αf are integrable on $[a,b]$. In fact,*

$$(6) \qquad \int_a^b (f(x) + g(x))\,dx = \int_a^b f(x)\,dx + \int_a^b g(x)\,dx$$

and

$$(7) \qquad \int_a^b (\alpha f(x))\,dx = \alpha \int_a^b f(x)\,dx.$$

PROOF. Let $\varepsilon > 0$ and choose P_ε such that for any partition $P = \{x_0, x_1, \ldots x_n\} \supseteq P_\varepsilon$ of $[a,b]$ and any choice of $t_j \in [x_{j-1}, x_j]$, we have

$$\left|\sum_{j=1}^{n}f(t_j)(x_j - x_{j-1}) - \int_a^b f(x)\,dx\right| < \frac{\varepsilon}{2}$$

and

$$\left|\sum_{j=1}^{n}g(t_j)(x_j - x_{j-1}) - \int_a^b g(x)\,dx\right| < \frac{\varepsilon}{2}.$$

By the Triangle Inequality,

$$\left|\sum_{j=1}^{n}f(t_j)(x_j - x_{j-1}) + \sum_{j=1}^{n}g(t_j)(x_j - x_{j-1}) - \int_a^b f(x)\,dx - \int_a^b g(x)\,dx\right| < \varepsilon$$

for any choice of $t_j \in [x_{j-1}, x_j]$. Hence, (6) follows directly from Theorem 3.4.
Similarly, if P_ε is chosen so that

$$P \supseteq P_\varepsilon \quad \text{implies} \quad \left| \sum_{j=1}^{n} f(t_j)(x_j - x_{j-1}) - \int_a^b f(x)\, dx \right| < \frac{\varepsilon}{|\alpha| + 1},$$

then

$$\left| \sum_{j=1}^{n} \alpha f(t_j)(x_j - x_{j-1}) - \alpha \int_a^b f(x)\, dx \right| < |\alpha| \frac{\varepsilon}{|\alpha| + 1} < \varepsilon$$

for any choice of $t_j \in [x_{j-1}, x_j]$. We conclude by Theorem 3.4 that (7) holds.   ∎

THEOREM 3.6. *If f is integrable on $[a, b]$, then f is integrable on each subinterval $[c, d]$ of $[a, b]$. Moreover,*

$$(8) \qquad\qquad \int_a^b f(x)\, dx = \int_a^c f(x)\, dx + \int_c^b f(x)\, dx$$

for all $c \in (a, b)$.

PROOF. Let $\varepsilon > 0$ and choose a partition P of $[a, b]$ such that

$$(9) \qquad\qquad U(f, P) - L(f, P) < \varepsilon.$$

Let $P_1 = (P \cap [a, c]) \cup \{c\}$ and notice that P_1 is a partition of $[a, c]$. Since $P \cup \{c\}$ is a refinement of P, we have by (9) that

$$U(f, P_1) - L(f, P_1) \leq U(f, P) - L(f, P) < \varepsilon.$$

Therefore, f is integrable on $[a, c]$. A similar argument proves that f is integrable on any subinterval $[c, d]$ of $[a, b]$.

To verify (8), suppose P is any partition of $[a, b]$. Let $P_0 = P \cup \{c\}$, $P_1 = P_0 \cap [a, c]$, and $P_2 = P_0 \cap [c, b]$. Then $P_0 = P_1 \cup P_2$ and by definition

$$U(f, P) \geq U(f, P_0) = U(f, P_1) + U(f, P_2)$$

$$\geq (U) \int_a^c f(x)\, dx + (U) \int_c^b f(x)\, dx = \int_a^c f(x)\, dx + \int_c^b f(x)\, dx.$$

(This last equality follows from the fact that f is integrable on both $[a, c]$ and $[c, b]$.)
Taking the infimum of

$$U(f, P) \geq \int_a^c f(x)\, dx + \int_c^b f(x)\, dx$$

over all partitions P of $[a, b]$, we obtain

$$\int_a^b f(x)\, dx = (U) \int_a^b f(x)\, dx \geq \int_a^c f(x)\, dx + \int_c^b f(x)\, dx.$$

A similar argument using lower integrals shows

$$\int_a^b f(x)\,dx \le \int_a^c f(x)\,dx + \int_c^b f(x)\,dx. \quad \blacksquare$$

Using the conventions

$$\int_a^b f(x)\,dx = -\int_b^a f(x)\,dx \quad \text{and} \quad \int_a^a f(x)\,dx = 0,$$

it is easy to see that (8) holds whether or not c lies between a and b (see Exercise 4).

THEOREM 3.7 [COMPARISON THEOREM FOR INTEGRALS]. *If f,g are integrable on $[a,b]$ and $f(x) \le g(x)$ for all $x \in [a,b]$, then*

$$\int_a^b f(x)\,dx \le \int_a^b g(x)\,dx.$$

In particular, if $m \le f(x) \le M$ for $x \in [a,b]$, then

$$m(b-a) \le \int_a^b f(x)\,dx \le M(b-a).$$

PROOF. Let P be a partition of $[a,b]$. By hypothesis, $M_j(f) \le M_j(g)$ whence $U(f,P) \le U(g,P)$. It follows that

$$\int_a^b f(x)\,dx = (U)\int_a^b f(x)\,dx \le U(g,P)$$

for all partitions P of $[a,b]$. Taking the infimum of this inequality over all partitions P of $[a,b]$, we obtain

$$\int_a^b f(x)\,dx \le \int_a^b g(x)\,dx.$$

If $m \le f(x) \le M$, then (by what we just proved and by Theorem 3.3)

$$m(b-a) = \int_a^b m\,dx \le \int_a^b f(x)\,dx \le \int_a^b M\,dx = M(b-a). \quad \blacksquare$$

We shall use the following result nearly every time we need to estimate an integral.

THEOREM 3.8. *If f is integrable on $[a,b]$, then $|f|$ is integrable on $[a,b]$ and*

$$\left| \int_a^b f(x)\,dx \right| \le \int_a^b |f(x)|\,dx.$$

PROOF. Let $P = \{x_0, x_1, \ldots x_n\}$ be a partition of $[a, b]$. We claim that

(10) $$M_j(|f|) - m_j(|f|) \le M_j(f) - m_j(f)$$

holds for $j = 1, 2, \ldots, n$. Indeed, let $x, y \in [x_{j-1}, x_j]$. If $f(x)$, $f(y)$ have the same sign, say both are nonnegative, then

$$|f(x)| - |f(y)| = f(x) - f(y) \le M_j(f) - m_j(f).$$

If $f(x)$, $f(y)$ have opposite signs, say $f(x) \ge 0 \ge f(y)$, then $m_j(f) \le 0$, hence

$$|f(x)| - |f(y)| = f(x) + f(y) \le M_j(f) + 0 \le M_j(f) - m_j(f).$$

Thus (10) holds in any event.

Let $\varepsilon > 0$ and choose a partition P of $[a, b]$ such that $U(f, P) - L(f, P) < \varepsilon$. Since (10) implies $U(|f|, P) - L(|f|, P) \le U(f, P) - L(f, P)$, it follows that

$$U(|f|, P) - L(|f|, P) < \varepsilon.$$

Thus $|f|$ is integrable on $[a, b]$. Since $-|f(x)| \le f(x) \le |f(x)|$ holds for any $x \in [a, b]$, we conclude by Theorem 3.7 that

$$-\int_a^b |f(x)| \, dx \le \int_a^b f(x) \, dx \le \int_a^b |f(x)| \, dx. \quad \blacksquare$$

By Theorem 3.5, the sum of integrable functions is integrable. What about the product?

COROLLARY 3.9. *If f and g are Riemann integrable on $[a, b]$, then so is fg.*

PROOF. Suppose for a moment that the square of any integrable function is integrable. Then by hypothesis, f^2, g^2, and $(f + g)^2$ are integrable on $[a, b]$. Since

$$fg = \frac{(f + g)^2 - f^2 - g^2}{2}$$

it follows from Theorem 3.5 that fg is integrable on $[a, b]$.

It remains to prove that f^2 is integrable on $[a, b]$. Since $M_j(f^2) = (M_j(|f|))^2$ and $m_j(f^2) = (m_j(|f|))^2$, it is clear that

$$\begin{aligned}
M_j(f^2) - m_j(f^2) &= (M_j(|f|))^2 - (m_j(|f|))^2 \\
&= (M_j(|f|) + m_j(|f|))(M_j(|f|) - m_j(|f|)) \\
&\le 2M(M_j(|f|) - (m_j(|f|)),
\end{aligned}$$

where $M = \sup_{x \in [a, b]} |f(x)|$. Multiplying this inequality by $(x_j - x_{j-1})$ and summing over $j = 1, 2, \ldots, n$, we have

$$U(f^2, P) - L(f^2, P) \le 2M(U(|f|, P) - L(|f|, P)).$$

Hence, it follows from Theorem 3.8 that f^2 is integrable on $[a, b]$. $\quad \blacksquare$

We close this section with two integral analogues of the Mean Value Theorem.

THEOREM 3.10 [FIRST MEAN VALUE THEOREM FOR INTEGRALS]. *Suppose that f and g are integrable on $[a, b]$ with $g(x) \geq 0$ for all $x \in [a, b]$. If*

$$m = \inf_{x \in [a,b]} f(x) \quad \text{and} \quad M = \sup_{x \in [a,b]} f(x),$$

then there is a number $c \in [m, M]$ such that

$$\int_a^b f(x)g(x)\,dx = c \int_a^b g(x)\,dx.$$

In particular, if f is continuous on $[a, b]$, then there is an $x_0 \in [a, b]$ which satisfies

$$\int_a^b f(x)g(x)\,dx = f(x_0) \int_a^b g(x)\,dx.$$

PROOF. Since $g \geq 0$ on $[a, b]$, Theorem 3.7 implies

$$m \int_a^b g(x)\,dx \leq \int_a^b f(x)g(x)\,dx \leq M \int_a^b g(x)\,dx.$$

If $\int_a^b g(x)\,dx = 0$, then $\int_a^b f(x)g(x)\,dx = 0$ and there is nothing to prove. Otherwise, set

$$c = \frac{\int_a^b f(x)g(x)\,dx}{\int_a^b g(x)\,dx}$$

and note that $c \in [m, M]$. If f is continuous, then (by the Intermediate Value Theorem) we can choose $x_0 \in [a, b]$ such that $f(x_0) = c$. ∎

Before we state the Second Mean Value Theorem we introduce an idea which will be used in the next section to prove the Fundamental Theorem of Calculus. If f is integrable on $[a, b]$, then f can be used to define a new function

$$F(x) := \int_a^x f(t)\,dt, \qquad x \in [a, b].$$

Example 1. Find $F(x) = \int_0^x f(t)\,dt$ if

$$f(x) = \begin{cases} 1 & x \geq 0 \\ -1 & x < 0. \end{cases}$$

SOLUTION. By Theorem 3.3,

$$F(x) = \int_0^x f(t)\,dt = \begin{cases} x & x \geq 0 \\ -x & x < 0. \end{cases}$$

Hence, $F(x) = |x|$. ∎

Notice in Example 1 that the integral F of f is continuous even though f itself is not. The following result shows this is a general principle.

THEOREM 3.11. *If f is Riemann integrable on $[a, b]$, then $F(x) = \int_a^x f(t)\,dt$ exists and is continuous on $[a, b]$.*

PROOF. By Theorem 3.6, $F(x)$ exists for all $x \in [a, b]$. To prove that F is continuous on $[a, b]$, it suffices to show that $F(x+) = F(x)$ for all $x \in [a, b)$ and $F(x-) = F(x)$ for all $x \in (a, b]$. Fix $x_0 \in [a, b)$. By definition, f is bounded on $[a, b]$. Thus, choose $M \in \mathbf{R}$ such that $|f(t)| \le M$ for all $t \in [a, b]$. Let $\varepsilon > 0$ and set $\delta = \varepsilon/M$. If $0 \le x - x_0 < \delta$, then by Theorem 3.8,

$$|F(x) - F(x_0)| = \left| \int_{x_0}^x f(t)\,dt \right| \le \int_{x_0}^x |f(t)|\,dt \le M|x - x_0| < \varepsilon.$$

Hence, $F(x_0+) = F(x_0)$. A similar argument shows that $F(x_0-) = F(x_0)$ for all $x_0 \in (a, b]$ ∎

THEOREM 3.12 [SECOND MEAN VALUE THEOREM FOR INTEGRALS]. *Suppose f, g are integrable on $[a, b]$ with f increasing and g nonnegative on $[a, b]$. If $m \le f(a+)$ and $M \ge f(b-)$, then there is an $x_0 \in [a, b]$ such that*

$$\int_a^b f(x)g(x)\,dx = m \int_a^{x_0} g(x)\,dx + M \int_{x_0}^b g(x)\,dx.$$

In particular, if f is also nonnegative on $[a, b]$, then there is an $x_0 \in [a, b]$ which satisfies

$$\int_a^b f(x)g(x)\,dx = M \int_{x_0}^b g(x)\,dx.$$

PROOF. Let

$$F(x) = m \int_a^x g(t)\,dt + M \int_x^b g(t)\,dt$$

for $x \in [a, b]$ and observe by Theorem 3.11 that F is continuous on $[a, b]$. Since f is increasing and g is nonnegative, we have $mg(t) \le f(t)g(t) \le Mg(t)$ for all $t \in (a, b)$. Hence, the Comparison Theorem (Theorem 3.7) implies

$$F(b) = m \int_a^b g(t)\,dt \le \int_a^b f(t)g(t)\,dt \le M \int_a^b g(t)\,dt = F(a).$$

Since F is continuous, it follows from the Intermediate Value Theorem that there is an $x_0 \in (a, b)$ such that

$$F(x_0) = \int_a^b f(t)g(t)\,dt.$$

This proves the first statement. The second statement follows from the first since we may use $m = 0$ when $f \ge 0$. ∎

When $g(x) = 1$ and $f(x) \ge 0$, these mean value theorems have simple geometric interpretations. Indeed, let A represent the area bounded by the curves $y = f(x)$,

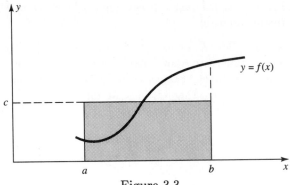

Figure 3.3

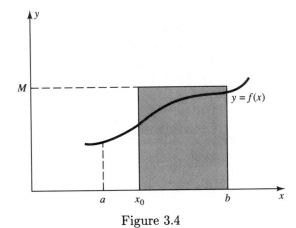

Figure 3.4

$y = 0$, $x = a$, and $x = b$. By the First Mean Value Theorem, there is a $c \in (a, b)$ such that the area of the rectangle of height c and base $b - a$ equals A (see Figure 3.3). And by the Second Mean Value Theorem, given $M \geq f(b-)$ there is an $x_0 \in [a, b]$ such that the area of the rectangle of height M and base $b - x_0$ equals A (see Figure 3.4).

EXERCISES

1. Using the connection between integrals and area, evaluate each of the following integrals.

    a) $$\int_0^1 |x - 0.5| \, dx.$$

    b) $$\int_0^a \sqrt{a^2 - x^2} \, dx, \qquad a > 0.$$

c)
$$\int_{-2}^{2} (|x + 1| + |x|) \, dx.$$

d)
$$\int_{a}^{b} (3x + 1) \, dx, \qquad a < b.$$

2. Prove that if f and g are integrable on $[a, b]$, then so are $f \vee g$ and $f \wedge g$ (see Exercise 8 in Section 2.1).

3. Prove that if f is integrable on $[0, 1]$ and $\beta > 0$, then

$$\lim_{n \to \infty} n^\alpha \int_{0}^{1/n^\beta} f(x) \, dx = 0$$

for all $\alpha < \beta$.

4. Suppose that $a < b < c$ and f is integrable on $[a, c]$. Prove

$$\int_{a}^{b} f(x) \, dx = \int_{a}^{c} f(x) \, dx + \int_{c}^{b} f(x) \, dx.$$

5. a) Suppose that $g_n \geq 0$ is a sequence of integrable functions which satisfies

$$\lim_{n \to \infty} \int_{a}^{b} g_n(x) \, dx = 0.$$

Show that if $f : [a, b] \to \mathbf{R}$ is integrable on $[a, b]$, then

$$\lim_{n \to \infty} \int_{a}^{b} f(x) g_n(x) \, dx = 0.$$

b) Prove that if f is integrable on $[0, 1]$, then

$$\lim_{n \to \infty} \int_{0}^{1} x^n f(x) \, dx = 0.$$

$\boxed{6}$. **This exercise is used in Section 7.3.**

a) Prove that if f is integrable on $[0, 1]$, then

$$\int_{0}^{1} f(x) \, dx = \lim_{n \to \infty} \sum_{k=0}^{n} \int_{1/2^{k+1}}^{1/2^k} f(x) \, dx.$$

b) Suppose f is integrable on $[a, b]$, $x_0 = a$, and x_n is a sequence of numbers in $[a, b]$ such that $x_n \uparrow b$ as $n \to \infty$. Prove

$$\int_{a}^{b} f(x) \, dx = \lim_{n \to \infty} \sum_{k=0}^{n} \int_{x_k}^{x_{k+1}} f(x) \, dx.$$

7. Let $f : [a, b] \to \mathbf{R}$, $a = x_0 < x_1 < \cdots < x_n = b$, and suppose that $f(x_k+)$ exists and is finite for $k = 0, 1, \ldots, n-1$ and $f(x_k-)$ exists and is finite for $k = 1, \ldots, n$. Show that if f is continuous on each subinterval (x_{k-1}, x_k), then f is integrable on $[a, b]$ and

$$\int_a^b f(x)\, dx = \sum_{k=1}^n \int_{x_{k-1}}^{x_k} f(x)\, dx.$$

8. Let f be continuous on a closed bounded nondegenerate interval $[a, b]$ and set

$$M = \sup_{x \in [a,b]} |f(x)|.$$

a) Prove that if $M > 0$, then given $\varepsilon > 0$ there is a nondegenerate interval $I \subset [a, b]$ such that

$$(M - \varepsilon)^n |I| \le \int_a^b |f(x)|^n\, dx \le M^n (b - a).$$

b) Prove

$$\lim_{n \to \infty} \left(\int_a^b |f(x)|^n\, dx \right)^{1/n} = M.$$

3.3 THE FUNDAMENTAL THEOREM OF CALCULUS

Let f be integrable on $[a, b]$ and $F(x) = \int_0^x f(t)\, dt$. By Theorem 3.11, F is continuous on $[a, b]$. The next result shows that if f is continuous, then F is continuously differentiable.

THEOREM 3.13 [FUNDAMENTAL THEOREM OF CALCULUS]. *Let $[a, b]$ be a closed bounded nondegenerate interval and $f : [a, b] \to \mathbf{R}$.*

i) *If f is continuous on $[a, b]$ and $F(x) = \int_a^x f(t)\, dt$, then $F \in \mathcal{C}^1[a, b]$ and*

$$\frac{d}{dx} \int_a^x f(t)\, dt = F'(x) = f(x)$$

for each $x \in [a, b]$.

ii) *If f is differentiable on $[a, b]$ and f' is integrable on $[a, b]$, then*

$$\int_a^x f'(t)\, dt = f(x) - f(a)$$

for each $x \in [a, b]$.

PROOF. Let

$$F(x) = \int_a^x f(t)\, dt, \qquad x \in [a,b].$$

By symmetry, it suffices to show that if $f(x_0+) = f(x_0)$ for some $x_0 \in [a,b)$, then

(11)
$$\lim_{h \to 0+} \frac{F(x_0 + h) - F(x_0)}{h} = f(x_0)$$

(see Definition 2.9). Let $\varepsilon > 0$ and choose a $\delta > 0$ such that $0 \le t - x_0 < \delta$ implies $|f(t) - f(x_0)| < \varepsilon$. Fix $0 < h < \delta$. Notice that by Theorem 3.6,

$$F(x_0 + h) - F(x_0) = \int_{x_0}^{x_0 + h} f(t)\, dt$$

and that by Theorem 3.3,

$$f(x_0) = \frac{1}{h} \int_{x_0}^{x_0 + h} f(x_0)\, dt.$$

Therefore,

$$\frac{F(x_0 + h) - F(x_0)}{h} - f(x_0) = \frac{1}{h} \int_{x_0}^{x_0 + h} (f(t) - f(x_0))\, dt.$$

Since $0 < h < \delta$, it follows from Theorem 3.8 and the choice of δ that

$$\left| \frac{F(x_0 + h) - F(x_0)}{h} - f(x_0) \right| \le \frac{1}{h} \int_{x_0}^{x_0 + h} |f(t) - f(x_0)|\, dt \le \varepsilon.$$

This verifies (11) and the proof of part i) is complete.

To prove part ii) we may suppose that $x = b$. Let $\varepsilon > 0$. Since f' is integrable, choose a partition $P = \{x_0, x_1, \ldots, x_n\}$ of $[a,b]$ such that

$$\left| \sum_{j=1}^{n} f'(t_j)(x_j - x_{j-1}) - \int_a^b f'(x)\, dx \right| < \varepsilon$$

for any choice of points $t_j \in [x_{j-1}, x_j]$. Use the Mean Value Theorem to choose points $t_j \in [x_{j-1}, x_j]$ such that $f(x_j) - f(x_{j-1}) = f'(t_j)(x_j - x_{j-1})$. It follows by telescoping that

$$\left| f(b) - f(a) - \int_a^b f'(t)\, dt \right| = \left| \sum_{j=1}^{n} (f(x_j) - f(x_{j-1})) - \int_a^b f'(t)\, dt \right| < \varepsilon. \quad \blacksquare$$

By the Fundamental Theorem of Calculus, integration is the inverse of differentiation in the following sense. If f' is integrable, then

$$\int_a^b f'(x)\, dx = f(x) \Big|_a^b := f(b) - f(a).$$

In particular,

$$\int_a^b x^\alpha \, dx = \frac{x^{\alpha+1}}{\alpha+1} \Big|_a^b$$

for each $\alpha \geq 0$, and for each $\alpha < 0$ provided $\alpha \neq -1$ and $[a, b]$ does not contain 0 (see Exercise 6 in Section 2.4 and Exercise 5e below). (This result is sometimes called the *Power Rule*.)

The Fundamental Theorem of Calculus can be used to evaluate many integrals.

Example 1. Find $\int_0^1 (3x - 2)^2 \, dx$.

SOLUTION. Since $(3x - 2)^2 = 9x^2 - 12x + 4$, we have by the Power Rule that

$$\int_0^1 (3x - 2)^2 \, dx = 3x^3 - 6x^2 + 4x \Big|_0^1 = 1. \quad \blacksquare$$

Example 2. Find $\int_0^{\pi/2} (1 + \sin x) \, dx$.

SOLUTION. Since $(\cos x)' = -\sin x$, we have by the Fundamental Theorem of Calculus that

$$\int_0^{\pi/2} (1 + \sin x) \, dx = x - \cos x \Big|_0^{\pi/2} = \frac{\pi}{2} + 1. \quad \blacksquare$$

Combining the Product Rule and the Fundamental Theorem of Calculus, we have another tool for evaluating integrals.

THEOREM 3.14 [INTEGRATION BY PARTS]. *Suppose that f, g are differentiable on $[a, b]$ with f', g' integrable on $[a, b]$. Then*

$$\int_a^b f'(x)g(x) \, dx = f(b)g(b) - f(a)g(a) - \int_a^b f(x)g'(x) \, dx.$$

PROOF. By the Product Rule, $(f(x)g(x))' = f'(x)g(x) + f(x)g'(x)$ for $x \in [a, b]$. Since f, g are continuous on $[a, b]$ and f', g' are integrable on $[a, b]$, it follows that $(fg)'$ is a sum of products of integrable functions, hence integrable on $[a, b]$. Thus, by the Fundamental Theorem of Calculus,

$$f(b)g(b) - f(a)g(a) = \int_a^b f'(x)g(x) \, dx + \int_a^b f(x)g'(x) \, dx. \quad \blacksquare$$

Integration by parts can be used to reduce the exponent n on an expression of the form $(ax + b)^n f(x)$ when f is integrable.

Example 3. Find $\int_0^{\pi/2} x \sin x$.

SOLUTION. Let $g(x) = x$ and $f'(x) = \sin x$. Then $g'(x) = 1$ and we may set $f(x) = -\cos x$. Hence, by Theorem 3.14,

$$\int_0^{\pi/2} x \sin x = -x \cos x \Big|_0^{\pi/2} - \int_0^{\pi/2} (-\cos x)\,dx = \sin x \Big|_0^{\pi/2} = 1. \quad \blacksquare$$

Integration by parts is also very effective on integrals involving products of polynomials and logarithms.

Example 4. Find $\int_1^3 \log x\,dx$.

SOLUTION. Let $g(x) = \log x$ and $f'(x) = 1$. Then $g'(x) = 1/x$ and we may set $f(x) = x$. Hence, by Theorem 3.14,

$$\int_1^3 \log x\,dx = x \log x \Big|_1^3 - \int_1^3 dx = 3\log 3 - 2. \quad \blacksquare$$

Complicated problems can frequently be reduced to simpler ones by changing variables. The following result shows how to change variables in a Riemann integral.

THEOREM 3.15 [CHANGE OF VARIABLES]. *Let ϕ be continuously differentiable on a closed bounded interval $[a, b]$ with $\phi(a) \leq \phi(b)$. If*

$$(12) \qquad\qquad f \text{ is continuous on } \phi([a,b]),$$

or if

$$(13) \qquad \phi \text{ is strictly increasing on } [a, b] \text{ and } f \text{ is integrable on } [\phi(a), \phi(b)],$$

then

$$(14) \qquad\qquad \int_{\phi(a)}^{\phi(b)} f(t)\,dt = \int_a^b f(\phi(x))\phi'(x)\,dx.$$

PROOF. Suppose first that (12) holds. By the Fundamental Theorem of Calculus, the functions

$$G(x) := \int_a^x f(\phi(t))\phi'(t)\,dt, \quad x \in [a,b], \quad \text{and} \quad F(u) := \int_{\phi(a)}^u f(t)\,dt, \quad u \in \phi([a,b]),$$

satisfy $G'(x) = f(\phi(x))\phi'(x)$ and $F'(u) = f(u)$. Hence, by the Chain Rule,

$$\frac{d}{dx}(G(x) - F(\phi(x))) = 0$$

for all $x \in [a, b]$. It follows from Theorem 2.19ii that $G(x) - F(\phi(x))$ is constant on $[a, b]$. Evaluation at $x = a$ shows that this constant is zero. Thus, $G(x) = F(\phi(x))$ for all $x \in [a, b]$, in particular, when $x = b$. This proves (14) under hypothesis (12).

STRATEGY: The theorem is more difficult to prove under hypotheses (13), but the idea behind the proof is simple. Since ϕ is increasing, if $P = \{x_0, x_1, \ldots x_n\}$ is a partition of $[a, b]$, then $\widetilde{P} = \{\phi(x_0), \ldots, \phi(x_n)\}$ is a partition of $I := [\phi(a), \phi(b)]$. By the Intermediate Value Theorem, given $u_j \in [\phi(x_{j-1}), \phi(x_j)]$, there is an $s_j \in [x_{j-1}, x_j]$ such that $u_j = \phi(s_j)$. It follows from the Mean Value Theorem that

$$(15) \qquad \sum_{j=1}^{n} f(u_j)(\phi(x_j) - \phi(x_{j-1})) = \sum_{j=1}^{n} f(\phi(s_j))\phi'(c_j)(x_j - x_{j-1})$$

for some $c_j \in [x_{j-1}, x_j]$. Therefore, a Riemann sum of f on $\widetilde{P}$ is almost a Riemann sum of $(f \circ \phi) \cdot \phi'$ on P. Since ϕ' is continuous, not much is changed if we replace c_j by s_j, i.e., make the right side of (15) exactly a Riemann sum on P. Hence, the integral of f on I should equal the integral of $(f \circ \phi) \cdot \phi'$ on $[a, b]$. Here are the details.

Let $\varepsilon > 0$. Since f is bounded there is an $M \in (0, \infty)$ such that $|f(x)| \le M$ for all $x \in I$. Since ϕ' is uniformly continuous on $[a, b]$, choose $\delta > 0$ such that

$$|\phi'(s_j) - \phi'(c_j)| < \frac{\varepsilon}{2M(b-a)},$$

i.e.,

$$(16) \qquad |f(\phi(s_j))\phi'(s_j) - f(\phi(s_j))\phi'(c_j)| < \frac{\varepsilon}{2(b-a)}$$

for all $s_j, c_j \in [a, b]$ with $|s_j - c_j| < \delta$.

Next, since f is integrable on I, choose a partition $\widetilde{P}_\varepsilon = \{w_0, w_1, \ldots, w_m\}$ of I such that if $\widetilde{P} = \{t_0, t_1, \ldots t_n\}$ is finer than $\widetilde{P}_\varepsilon$, then

$$(17) \qquad \left| \sum_{j=1}^{n} f(u_j)(t_j - t_{j-1}) - \int_{\phi(a)}^{\phi(b)} f(t)\, dt \right| < \frac{\varepsilon}{2}$$

for any $u_j \in [t_{j-1}, t_j]$. By Theorem 2.21, ϕ^{-1} exists, is continuous, and is strictly increasing on I. Hence,

$$P_\varepsilon = \{\phi^{-1}(w_0), \phi^{-1}(w_1), \ldots, \phi^{-1}(w_m)\}$$

is a partition of $[a, b]$ and, by making $\widetilde{P}_\varepsilon$ finer if necessary, we may suppose $\|P_\varepsilon\| < \delta$.

Let $P = \{x_0, x_1, \ldots, x_n\}$ be any partition of $[a, b]$ finer than P_ε and let $s_j \in [x_{j-1}, x_j]$. For each $j \in \{1, \ldots, n\}$, set $u_j = \phi(s_j)$, $t_j = \phi(x_j)$, and choose $c_j \in [x_{j-1}, x_j]$ by the Mean Value Theorem such that

$$\phi(x_j) - \phi(x_{j-1}) = \phi'(c_j)(x_j - x_{j-1}).$$

Then $\widetilde{P} = \{t_0, t_1, \ldots, t_n\}$ is a partition of I finer than $\widetilde{P}_\varepsilon$ and (17) holds. It follows from (15), (16), (17), and the choice of the t_j's that

$$\left| \sum_{j=1}^{n} f(\phi(s_j))\phi'(s_j)(x_j - x_{j-1}) - \int_{\phi(a)}^{\phi(b)} f(t)\, dt \right|$$

$$\leq \left| \sum_{j=1}^{n} (f(\phi(s_j))\phi'(s_j) - f(\phi(s_j))\phi'(c_j))(x_j - x_{j-1}) \right|$$

$$+ \left| \sum_{j=1}^{n} f(u_j)(\phi(x_j) - \phi(x_{j-1})) - \int_{\phi(a)}^{\phi(b)} f(t)\, dt \right|$$

$$< \frac{\varepsilon}{2} + \left| \sum_{j=1}^{n} f(u_j)(t_j - t_{j-1}) - \int_{\phi(a)}^{\phi(b)} f(t)\, dt \right| < \varepsilon$$

for any choice of $s_j \in [x_{j-1}, x_j]$. Hence, by Theorem 3.4, $(f \circ \phi) \cdot \phi'$ is integrable on $[a, b]$ and (14) holds. ∎

The following example illustrates a typical application of the Change of Variables Formula.

Example 5. Find $\int_0^1 e^{\sqrt{x+1}}/\sqrt{x+1}\, dx$.

SOLUTION. Let $u = \sqrt{x+1}$ and observe that

$$u'(x) = \frac{1}{2\sqrt{x+1}}.$$

Therefore,

$$\int_0^1 \frac{e^{\sqrt{x+1}}}{\sqrt{x+1}}\, dx = 2 \int_0^1 u'(x)e^{u(x)}\, dx = 2\int_1^{\sqrt{2}} e^u\, du = 2(e^{\sqrt{2}} - e). \quad ∎$$

It is interesting to note that hypothesis (12) does not require that ϕ be 1–1. This observation is used in the following example.

Example 6. Evaluate

$$\int_{-1}^{1} xf(x^2)\, dx$$

for any f continuous on $[0, 1]$.

SOLUTION. Let $\phi(x) = x^2$ and observe that f is continuous on $\phi([-1, 1]) = [0, 1]$. Hence, by Theorem 3.15,

$$\int_{-1}^{1} xf(x^2)\, dx = \frac{1}{2}\int_{-1}^{1} f(\phi(x))\phi'(x)\, dx = \frac{1}{2}\int_1^1 f(t)\, dt = 0. \quad ∎$$

EXERCISES

1. Compute each of the following integrals.

   a)
   $$\int_{-3}^{3} |x^2 + x - 2| \, dx.$$

   b)
   $$\int_{1}^{4} \frac{\sqrt{x} - 1}{\sqrt{x}} \, dx.$$

   c)
   $$\int_{0}^{1} (3x + 1)^{99} \, dx.$$

   d)
   $$\int_{1}^{e} x \log x \, dx.$$

   e)
   $$\int_{0}^{\pi/2} e^x \sin x \, dx.$$

   f)
   $$\int_{0}^{1} \sqrt{\frac{4x^2 - 4x + 1}{x^2 - x + 3}} \, dx.$$

2. Use the First Mean Value Theorem for Integrals to prove the following version of the Mean Value Theorem for Derivatives. If $f \in C^1[a, b]$, then there is an $x_0 \in (a, b)$ such that
   $$f(b) - f(a) = (b - a)f'(x_0).$$

3. a) If $g : \mathbf{R} \to \mathbf{R}$ is continuous, find
   $$\frac{d}{dt} \int_{0}^{t} g(x - t) \, dx.$$

   b) Prove that if $F(x) = \int_{0}^{x} e^{-x^2 t^2} \, dt$, then
   $$xF'(x) + F(x) = 2xe^{-x^4}$$

   for all $x \in \mathbf{R}$.

   c) If $f(x) = \int_{0}^{x^3} e^{t^2} \, dt$, show
   $$6 \int_{0}^{1} x^2 f(x) \, dx - 2 \int_{0}^{1} e^{x^2} \, dx = 1 - e.$$

$\boxed{4}$. **This exercise is used in Sections 3.4 and 4.1.** Define $L : (0, \infty) \to \mathbf{R}$ by

$$L(x) = \int_1^x \frac{dt}{t}.$$

a) Prove L is differentiable and strictly increasing on $(0, \infty)$, with $L'(x) = 1/x$ and $L(1) = 0$.

b) Prove that $L(x) \to \infty$ as $x \to \infty$ and $L(x) \to -\infty$ as $x \to 0+$. (You may wish to prove

$$L(2^n) > \sum_{k=1}^{2^n - 1} \frac{1}{k+1} > \sum_{k=0}^{n-1} \left(\sum_{j=2^k}^{2^{k+1}-1} \frac{1}{2^{k+1}} \right) = \frac{n}{2}$$

for all $n \in \mathbf{N}$.)

c) Using the fact that $(x^q)' = qx^{q-1}$ for $x > 0$ and $q \in \mathbf{Q}$ (see Exercise 6 in Section 2.4), prove $L(x^q) = qL(x)$ for all $q \in \mathbf{Q}$ and $x > 0$.

d) Prove $L(xy) = L(x) + L(y)$ for all $x, y \in (0, \infty)$.

e) Let $e = \lim_{n \to \infty}(1 + 1/n)^n$. Use L'Hôpital's Rule to show that $L(e) = 1$. ($L(x)$ is the *natural logarithm* function $\log x$.)

$\boxed{5}$. **This exercise was used in Section 2.5.** Let $E = L^{-1}$, where L is defined in Exercise 4.

a) Use the Inverse Function Theorem to show that E is differentiable and strictly increasing on $\mathbf{R}$ with $E'(x) = E(x)$, $E(0) = 1$, and $E(1) = e$.

b) Prove $E(x) \to \infty$ as $x \to \infty$ and $E(x) \to 0$ as $x \to -\infty$.

c) Prove $E(xq) = (E(x))^q$ and $E(q) = e^q$ for all $q \in \mathbf{Q}$ and $x \in \mathbf{R}$.

d) Prove $E(x + y) = E(x)E(y)$ for all $x, y \in \mathbf{R}$.

e) For each $\alpha \in \mathbf{R}$ define $e^\alpha = E(\alpha)$. Let $x > 0$ and define $x^\alpha = e^{\alpha \log x} := E(\alpha L(x))$. Prove that $0 < x < y$ implies $x^\alpha < y^\alpha$ for $\alpha > 0$ and $x^\alpha > y^\alpha$ for $\alpha < 0$. Also prove

$$x^{\alpha + \beta} = x^\alpha x^\beta, \qquad x^{-\alpha} = \frac{1}{x^\alpha}, \qquad \text{and} \quad (x^\alpha)' = \alpha x^{\alpha - 1}$$

for all $\alpha, \beta \in \mathbf{R}$ and $x > 0$.

6. a) Suppose g is integrable and nonnegative on $[1, 3]$ with $\int_1^3 g(x)\, dx = 1$. Prove

$$\frac{1}{\pi} \int_1^9 g(\sqrt{x})\, dx < 2.$$

b) Suppose h is integrable and nonnegative on $[1, 11]$ with $\int_1^{11} h(x)\, dx = 3$. Prove that

$$\int_0^2 h(1 + 3x + 3x^2 - x^3)\, dx \leq 1.$$

7. Suppose g is differentiable on $[a, b]$ and g' is integrable on $[a, b]$.

   a) Prove that if f is continuously differentiable and increasing on [a,b] and g is positive on (a, b) with $g(b) = g(a) = 0$, then

   $$\int_a^b f(x)g'(x)\,dx = 0$$

   if and only if f is constant on $[a, b]$.

   b) Show that a) is false if "g is positive on $[a, b]$" is replaced by "g is nonnegative on $[a, b]$ and positive on some subinterval of (a, b)."

8. By Exercise 4 in Section 3.1, if f is nonnegative and continuous on a nondegenerate interval $[a, b]$ and

   $$\int_a^c f(x)\,dx + \int_c^b f(x)\,dx = 0$$

   for some $c \in (a, b)$, then $f(x) = 0$ for all $x \in [a, b]$. Now prove the following result. If f is continuous on $[a, b]$ and

   $$\alpha \int_a^c f(x)\,dx + \beta \int_c^b f(x)\,dx = 0$$

   for all $c \in (a, b)$ and some $\alpha \neq \beta$, then $f(x) = 0$ for all $x \in [a, b]$.

9. Let $-\pi/2 \leq x \leq \pi/2$.

   a) Use $0 \leq \cos x \leq 1$ and the Comparison Theorem for integrals to prove $0 \leq \sin x \leq x$.

   b) For each nonnegative integer m, set

   $$s_m(x) := \sum_{k=0}^m \frac{(-1)^k x^{2k+1}}{(2k+1)!} \quad \text{and} \quad c_m(x) := \sum_{k=0}^m \frac{(-1)^k x^{2k}}{(2k)!}.$$

   Prove[1]

   $$s_{2n+1}(x) \leq \sin x \leq s_{2n}, \quad s_{2n+1} \leq \sin x \leq s_{2n+2},$$

   $$c_{2n+1}(x) \leq \cos x \leq c_{2n}(x), \quad \text{and} \quad c_{2n+1}(x) \leq \cos x \leq c_{2n+2}(x)$$

   hold for $n = 0, 1, 2, \ldots$ and $x \in \mathbf{R}$.

10. Suppose $f : [a, b] \to \mathbf{R}$ is continuously differentiable and 1–1 on $[a, b]$. Prove that

    $$\int_a^b f(x)\,dx + \int_{f(a)}^{f(b)} f^{-1}(x)\,dx = bf(b) - af(a).$$

---

[1]This exercise is due to Deng Bo ("A Simple Derivation of the Maclaurin Series for Sine and Cosine," *American Mathematical Monthly*, vol. 97 (1990), 836.

$\boxed{11}$. **This exercise is used in Section 7.4.** Suppose ϕ is continuously differentiable on a closed bounded interval $[a, b]$ with $\phi'(x) \neq 0$ for all $x \in [a, b]$. Prove that if $[c, d] = \phi([a, b])$ and f is integrable on $[c, d]$, then

$$\int_c^d f(t)\, dt = \int_a^b f(\phi(x)) \, |\phi'(x)|\, dx.$$

3.4 IMPROPER RIEMANN INTEGRATION

To extend the Riemann integral to unbounded intervals or unbounded functions, we begin with an elementary observation.

Remark 1. *If f is integrable on $[a, b]$, then*

$$\int_a^b f(x)\, dx = \lim_{c \to a+} \left(\lim_{d \to b-} \int_c^d f(x)\, dx \right).$$

PROOF. By Theorem 3.11,

$$F(x) = \int_a^x f(t)\, dt$$

is continuous on $[a, b]$. Thus

$$\int_a^b f(x)\, dx = F(b) - F(a) = \lim_{c \to a+} (\lim_{d \to b-} (F(d) - F(c))) = \lim_{c \to a+} (\lim_{d \to b-} \int_c^d f(x)\, dx). \quad \blacksquare$$

This leads to the following generalization of the Riemann integral.

DEFINITION 3.6. Let (a, b) be an open (possibly unbounded) interval and $f : (a, b) \to \mathbf{R}$.

i) f is said to be *locally integrable* on (a, b) if f is integrable on each closed subinterval $[c, d]$ of (a, b).

ii) f is said to be *improperly integrable* on (a, b) if f is locally integrable on (a, b) and if

(18)
$$\int_a^b f(x)\, dx := \lim_{c \to a+} \left(\lim_{d \to b-} \int_c^d f(x)\, dx \right)$$

exists and is finite. This limit is called the *improper (Riemann) integral* of f over (a, b).

Remark 2. *The order of the limits in (18) does not matter. In particular, if the limit in (18) exists, then*

$$\int_a^b f(x)\, dx = \lim_{d \to b-} \left(\lim_{c \to a+} \int_c^d f(x)\, dx \right).$$

PROOF. Let $x_0 \in (a, b)$ be fixed. By Theorems 3.6 and 2.2,

$$\lim_{c \to a+} \left(\lim_{d \to b-} \int_c^d f(x)\,dx \right) = \lim_{c \to a+} \left(\int_c^{x_0} f(x)\,dx + \lim_{d \to b-} \int_{x_0}^d f(x)\,dx \right)$$

$$= \lim_{c \to a+} \int_c^{x_0} f(x)\,dx + \lim_{d \to b-} \int_{x_0}^d f(x)\,dx$$

$$= \lim_{d \to b-} \left(\lim_{c \to a+} \int_c^d f(x)\,dx \right). \blacksquare$$

Thus, we shall use the notation

$$\lim_{c \to a+, d \to b-} \int_c^d f(x)\,dx$$

to represent the limit in (18). If the integral is not improper at one of the endpoints; e.g., if f is Riemann integrable on closed subintervals of $(a, b]$, we shall say that f is improperly integrable on $(a, b]$ and simplify the notation even further by writing

$$\int_a^b f(x)\,dx = \lim_{c \to a+} \int_c^b f(x)\,dx.$$

The following example shows that an improperly integrable function need not be bounded.

Example 1. Show that $f(x) = 1/\sqrt{x}$ is improperly integrable on $(0, 1]$.

SOLUTION. By definition,

$$\int_0^1 \frac{1}{\sqrt{x}}\,dx = \lim_{a \to 0+} \int_a^1 \frac{1}{\sqrt{x}}\,dx = \lim_{a \to 0+} (2 - 2\sqrt{a}) = 2. \blacksquare$$

The following example shows that a function can be improperly integrable on an unbounded interval.

Example 2. Show that $f(x) = 1/x^2$ is improperly integrable on $[1, \infty)$.

SOLUTION. By definition,

$$\int_1^\infty \frac{1}{x^2}\,dx = \lim_{d \to \infty} \int_1^d \frac{1}{x^2}\,dx = \lim_{d \to \infty} \left(1 - \frac{1}{d} \right) = 1. \blacksquare$$

Because an improper integral is a limit of Riemann integrals, many of the results we proved earlier in this chapter have analogues for the improper integral. The next two results illustrate this principle.

THEOREM 3.16. *If f, g are improperly integrable on (a, b) and $\alpha, \beta \in \mathbf{R}$, then $\alpha f + \beta g$ is improperly integrable on (a, b) and*

$$\int_a^b (\alpha f(x) + \beta g(x))\, dx = \alpha \int_a^b f(x)\, dx + \beta \int_a^b g(x)\, dx.$$

PROOF. By Theorem 3.5 (the Linear Property for Riemann Integrals),

$$\int_c^d (\alpha f(x) + \beta g(x))\, dx = \alpha \int_c^d f(x)\, dx + \beta \int_c^d g(x)\, dx$$

for all $a < c < d < b$. Taking the limit as $c \to a+$ and $d \to b-$ finishes the proof. ∎

THEOREM 3.17 [COMPARISON THEOREM FOR IMPROPER INTEGRALS]. *Suppose f, g are locally integrable on (a, b). If $0 \le f(x) \le g(x)$ for $x \in (a, b)$, and g is improperly integrable on (a, b), then f is improperly integrable on (a, b) and*

$$\int_a^b f(x)\, dx \le \int_a^b g(x)\, dx.$$

PROOF. Fix $c \in (a, b)$. Let $F(d) = \int_c^d f(x)\, dx$ and $G(d) = \int_c^d g(x)\, dx$ for $d \in [c, b)$. By the Comparison Theorem for Integrals, $F(d) \le G(d)$. Since $f \ge 0$, the function F is increasing on $[c, b]$, hence $F(b-)$ exists. Thus, by definition, f is improperly integrable on (c, b) and

$$\int_c^b f(x)\, dx = F(b-) \le G(b-) = \int_c^b g(x)\, dx.$$

A similar argument works for the case $c \to a+$. ∎

This test is frequently used in conjunction with the following inequalities: $|\sin x| \le |x|$ for all $x \in \mathbf{R}$ (see Appendix B); for any $\alpha > 0$ there exists a constant $B_\alpha > 1$ such that $|x^\alpha| \le e^x$ and $|\log x| \le x^\alpha$ for all $x \ge B_\alpha$ (see Exercise 4 in Section 2.5). Here are two typical examples.

Example 3. Prove that $f(x) = |\sin x / \sqrt{x^3}|$ is improperly integrable on $(0, 1]$.

PROOF. Since $0 \le f(x) = |\sin x / \sqrt{x^3}| \le |x|/x^{3/2} = 1/\sqrt{x}$ on $(0, 1]$ and this last function is improperly integrable on $(0, 1]$ by Example 1, it follows from the Comparison Test that $f(x)$ is improperly integrable on $(0, 1]$. ∎

Example 4. Prove that $f(x) = |\log x / \sqrt[4]{x^9}|$ is improperly integrable on $[1, \infty)$.

PROOF. Since f is continuous on $(0, \infty)$, f is integrable on $[1, C]$ for any $C \in \mathbf{R}$. Since $0 \le f(x) = |\log x / \sqrt[4]{x^9}| \le x^{1/4}/x^{9/4} = 1/x^2$ for $x \ge C := B_{1/4}$, and this last function is improperly integrable on $[1, \infty)$ by Example 2, it follows from the Comparison Test that $f(x)$ is improperly integrable on $[1, \infty)$. ∎

The Comparison Theorem for improper integrals can also be used to prove certain product functions are integrable.

Remark 3. *If f is bounded and locally integrable on (a, b) and $|g|$ is improperly integrable on (a, b), then $|fg|$ is improperly integrable on (a, b).*

PROOF. Let $M = \sup_{x \in (a,b)} |f(x)|$. Then $0 \le |f(x)g(x)| \le M|g(x)|$ for all $x \in (a, b)$. Hence, by Theorem 3.17, $|fg|$ is improperly integrable on (a, b). ∎

For the Riemann integral, we proved that $|f|$ is integrable when f is (see Theorem 3.8). This is not the case for the improper integral (see Example 5 below). For this reason we introduce the following concepts.

DEFINITION 3.7. Let $f : (a, b) \to$ **R**.

 i) f is said to be *absolutely integrable* on (a, b) if $|f|$ is improperly integrable on (a, b).
 ii) f is said to be *conditionally integrable* on (a, b) if f is improperly integrable but not absolutely integrable on (a, b).

The following result, an analogue of Theorem 3.8 for absolutely integrable functions, shows that absolutely integrability is stronger than improper integrability.

THEOREM 3.18. *If f is absolutely integrable on (a, b), then f is improperly integrable on (a, b) and*

$$\left| \int_a^b f(x)\,dx \right| \le \int_a^b |f(x)|\,dx.$$

PROOF. Since $0 \le |f(x)| + f(x) \le 2|f(x)|$, we have by Theorem 3.17 that $|f| + f$ is improperly integrable on $[a, b]$. Hence by Theorem 3.16, so is $f = (|f| + f) - |f|$. Moreover,

$$\left| \int_c^d f(x)\,dx \right| \le \int_c^d |f(x)|\,dx$$

for every $a < c < d < b$. We finish the proof by taking the limit of this last inequality as $c \to a+$ and $d \to b-$. ∎

The converse of Theorem 3.18, however, is false.

Example 5. Prove that the function $\sin x / x$ is conditionally integrable on $[1, \infty)$.

PROOF. Integrating by parts, we have

$$\int_1^d \frac{\sin x}{x}\,dx = -\frac{\cos x}{x}\Big|_1^d - \int_1^d \frac{\cos x}{x^2}\,dx$$

$$= \cos(1) - \frac{\cos d}{d} - \int_1^d \frac{\cos x}{x^2}\,dx.$$

Since $1/x^2$ is absolutely integrable on $[1, \infty)$, it follows from Remark 3 that $\cos x / x^2$ is absolutely integrable on $[1, \infty)$. Therefore, $\sin x / x$ is improperly integrable on $[1, \infty)$ and

$$\int_1^\infty \frac{\sin x}{x}\,dx = \cos(1) - \int_1^\infty \frac{\cos x}{x^2}\,dx.$$

To show $\sin x / x$ is not absolutely integrable on $[1, \infty)$, notice that

$$\int_1^{n\pi} \frac{|\sin x|}{x}\, dx \geq \sum_{k=2}^n \int_{(k-1)\pi}^{k\pi} \frac{|\sin x|}{x}\, dx$$

$$\geq \sum_{k=2}^n \frac{1}{k\pi} \int_{(k-1)\pi}^{k\pi} |\sin x|\, dx$$

$$= \sum_{k=2}^n \frac{2}{k\pi} = \frac{2}{\pi} \sum_{k=2}^n \frac{1}{k}$$

for each $n \in \mathbf{N}$. Since

$$\sum_{k=2}^n \frac{1}{k} \geq \sum_{k=2}^n \int_k^{k+1} \frac{1}{x}\, dx = \int_2^{n+1} \frac{1}{x}\, dx = \log(n+1) - \log 2 \to \infty$$

as $n \to \infty$, it follows from the Squeeze Theorem that

$$\lim_{n \to \infty} \int_1^{n\pi} \frac{|\sin x|}{x}\, dx = \infty.$$

Thus, $\sin x / x$ is not absolutely integrable on $[1, \infty)$. ∎

EXERCISES

1. Evaluate the following improper integrals.

   a) $$\int_1^\infty \frac{1+x}{x^3}\, dx.$$

   b) $$\int_{-\infty}^\infty \frac{1}{1+x^2}\, dx.$$

   c) $$\int_0^{\pi/2} \frac{\cos x}{\sqrt[3]{\sin x}}\, dx.$$

   d) $$\int_0^\infty \sqrt{x}\, e^{-\sqrt{x}}\, dx.$$

2. For each of the following, find all values of $p \in \mathbf{R}$ for which f is improperly integrable on I.
   a) $f(x) = 1/x^p$, $I = (1, \infty)$.
   b) $f(x) = 1/x^p$, $I = (0, 1)$.
   c) $f(x) = 1/(x \log^p x)$, $I = (e, \infty)$.
   d) $f(x) = 1/(1 + x^p)$, $I = (0, \infty)$.

3. Show that for each $p > 0$, $\sin x/x^p$ is improperly integrable on $[1, \infty)$ and $\cos x/\log^p x$ is improperly integrable on (e, ∞).

4. Decide which of the following functions are improperly integrable on I and which are not.

   a) $f(x) = \sin x$, $I = (0, \infty)$.
   b) $f(x) = 1/x^2$, $I = [-1, 1]$.
   c) $f(x) = x^{-1}\sin(x^{-1})$, $I = (1, \infty)$.
   d) $f(x) = \log x$, $I = (0, 1)$.
   e) $f(x) = (1 - \cos x)/x^2$, $I = (0, \infty)$.

5. Use the examples provided by Exercise 2b to show that the product of two improperly integrable functions might not be improperly integrable.

6. Suppose that f, g are nonnegative and locally integrable on $[a, b)$ and

$$L = \lim_{x \to b-} \frac{f(x)}{g(x)}$$

   exists.

   a) Show that if $0 \le L < \infty$ and g is improperly integrable on $[a, b)$, then so is f.
   b) Show that if $0 < L \le \infty$ and f is not improperly integrable on $[a, b)$, then neither is g.

7. a) Suppose f is improperly integrable on $[0, \infty)$. Prove that if $L = \lim_{x \to \infty} f(x)$ exists, then $L = 0$.

   b) Let

$$f(x) = \begin{cases} 1 & n \le x < n + 2^{-n}, \ n \in \mathbf{N}, \\ 0 & \text{otherwise.} \end{cases}$$

   Prove that f is improperly integrable on $[0, \infty)$ but $\lim_{x \to \infty} f(x)$ does not exist.

8. Prove that if f is absolutely integrable on $[1, \infty)$, then

$$\lim_{n \to \infty} \int_1^\infty f(x^n)\, dx = 0.$$

9. Assuming $e = \lim_{n \to \infty} \sum_{k=0}^n 1/k!$ (see Example 2 in Section 4.7), prove

$$\lim_{n \to \infty} \left(\frac{1}{n!} \int_1^\infty x^n e^{-x}\, dx \right) = 1.$$

10. a) Prove

$$\int_0^{\pi/2} e^{-a \sin x}\, dx \le \frac{2}{a}$$

   for all $a > 0$.

   b) Sharpen your estimates in part a) to prove that the inequality holds with $29/(19a)$ in place of $2/a$.

   c) What happens if $\cos x$ replaces $\sin x$?

^{e}3.5 FUNCTIONS OF BOUNDED VARIATION   *This section uses no material from any other enrichment section.*

In this section we study functions which do not wiggle too much. These functions, which play a prominent role in the theory of Fourier series (see Sections 9.3 and 9.4) and probability theory, are important tools for theoretical as well as applied mathematics.

Let $\phi : [a, b] \to \mathbf{R}$. To measure how much ϕ wiggles on an interval $[a, b]$, set

$$V(\phi, P) = \sum_{j=1}^{n} |\phi(x_j) - \phi(x_{j-1})|$$

for each partition $P = \{x_0, x_1, \ldots, x_n\}$ of $[a, b]$. The *variation* of ϕ is defined by

(19) $\text{Var}\,(\phi) := \sup\{V(\phi, P) : P \text{ is a partition of } [a, b]\}.$

DEFINITION 3.8. Let $[a, b]$ be a closed bounded nondegenerate interval and $\phi : [a, b] \to \mathbf{R}$. Then ϕ is said to be of *bounded variation* on $[a, b]$ if $\text{Var}\,(\phi) < \infty$.

The following three remarks show how the collection of functions of bounded variation is related to other collections of functions we have studied.

Remark 1. If $\phi \in C^1[a, b]$, then ϕ is of bounded variation on $[a, b]$.

PROOF. Let $P = \{x_0, x_1, \ldots, x_n\}$ be a partition of $[a, b]$. By the Extreme Value Theorem, there is an $M > 0$ such that $|\phi'(x)| \leq M$ for all $x \in [a, b]$. Therefore, it follows from the Mean Value Theorem that for each k between 1 and n there is a point c_k between x_{k-1} and x_k such that

$$|\phi(x_k) - \phi(x_{k-1})| = |\phi'(c_k)|(x_k - x_{k-1}) \leq M(x_k - x_{k-1}).$$

By telescoping, we obtain $V(\phi, P) \leq M(b-a)$ for any partition P of $[a, b]$. Therefore,

$$\text{Var}\,(\phi) \leq M(b - a). \quad \blacksquare$$

Remark 2. *If ϕ is monotone on $[a, b]$, then ϕ is of bounded variation on $[a, b]$. However, there exist functions of bounded variation which are not monotone.*

PROOF. Let ϕ be increasing on $[a, b]$ and $P = \{x_0, x_1, \ldots, x_n\}$ be a partition of $[a, b]$. Then by telescoping,

$$\sum_{j=1}^{n} |\phi(x_j) - \phi(x_{j-1})| = \sum_{j=1}^{n} (\phi(x_j) - \phi(x_{j-1}))$$
$$= \phi(x_n) - \phi(x_0) = \phi(b) - \phi(a) =: M < \infty.$$

Thus, $\text{Var}\,(f) \leq M$. On the other hand, by Remark 1, $\phi(x) = x^2$ is of bounded variation on $[-1, 1]$. $\blacksquare$

Remark 3. *If ϕ is of bounded variation on $[a,b]$, then ϕ is bounded on $[a,b]$. However, there exist bounded functions which are not of bounded variation.*

PROOF. Let $x \in [a,b]$ and note by definition that

$$|\phi(x) - \phi(a)| \le |\phi(x) - \phi(a)| + |\phi(b) - \phi(x)| \le \mathrm{Var}\,(\phi).$$

Hence, by the Triangle Inequality,

$$|\phi(x)| \le |\phi(a)| + \mathrm{Var}\,(\phi).$$

To find a bounded function which is not of bounded variation, consider

$$\phi(x) := \begin{cases} \sin(1/x) & x \ne 0 \\ 0 & x = 0. \end{cases}$$

Clearly, ϕ is bounded by 1. On the other hand, if

$$x_j = \begin{cases} 0 & j = 0 \\ \dfrac{2}{(n-j)\pi} & 0 < j < n, \end{cases}$$

then

$$\sum_{j=1}^{n} |\phi(x_j) - \phi(x_{j-1})| = 2n \to \infty$$

as $n \to \infty$. Thus ϕ is not of bounded variation on $[0, 2/\pi]$. ∎

The following result and Exercise 3 below are partial answers to the question: Is the class of functions of bounded variation preserved by algebraic operations?

THEOREM 3.19. *If ϕ and ψ are of bounded variation on a closed bounded interval $[a,b]$, then so are $\phi + \psi$ and $\phi - \psi$.*

PROOF. Let $a = x_0 < x_1 < \cdots < x_n = b$. Then

$$\sum_{j=1}^{n} |\phi(x_j) \pm \psi(x_j) - (\phi(x_{j-1}) \pm \psi(x_{j-1}))|$$

$$\le \sum_{j=1}^{n} |\phi(x_j) - \phi(x_{j-1})| + \sum_{j=1}^{n} |\psi(x_j) - \psi(x_{j-1})|$$

$$\le \mathrm{Var}\,(\phi) + \mathrm{Var}\,(\psi).$$

Therefore, $\mathrm{Var}\,(\phi \pm \psi) \le \mathrm{Var}\,(\phi) + \mathrm{Var}\,(\psi)$. ∎

It turns out that there is a close connection between functions of bounded variation and monotone functions (see Corollary 3.21 below). To make this connection clear, we introduce the following concept.

DEFINITION 3.9. Let ϕ be of bounded variation on a closed bounded interval $[a, b]$. The *total variation* of ϕ is the function Φ defined on $[a, b]$ by

$$\Phi(x) := \sup\{\sum_{j=1}^{k} |\phi(x_j) - \phi(x_{j-1})| : \{x_0, x_1, \ldots, x_k\} \text{ is a partition of } [a, x]\}.$$

THEOREM 3.20. Let ϕ be of bounded variation on $[a, b]$ and Φ be its total variation. Then

  i) $|\phi(y) - \phi(x)| \leq \Phi(y) - \Phi(x)$ for all $a \leq x < y \leq b$,
 ii) Φ and $\Phi - \phi$ are increasing on $[a, b]$, and
iii) $\text{Var}\,(\phi) \leq \text{Var}\,(\Phi)$.

PROOF. i) Let $x < y$ belong to $[a, b]$ and $\{x_0, x_1, \ldots, x_k\}$ be a partition of $[a, x]$. Then $\{x_0, x_1, \ldots, x_k, y\}$ is a partition of $[a, y]$, and we have by Definition 3.9 that

$$\sum_{j=1}^{k} |\phi(x_j) - \phi(x_{j-1})| \leq \sum_{j=1}^{k} |\phi(x_j) - \phi(x_{j-1})| + |\phi(y) - \phi(x)| \leq \Phi(y).$$

Taking the supremum of this inequality over all partitions $\{x_0, x_1, \ldots, x_k\}$ of $[a, x]$, we obtain

$$\Phi(x) \leq \Phi(x) + |\phi(y) - \phi(x)| \leq \Phi(y).$$

ii) By the Monotone Property of Suprema, Φ is increasing on $[a, b]$. To show $\Phi - \phi$ is increasing, suppose $a \leq x < y \leq b$. By part i),

$$\phi(y) - \phi(x) \leq |\phi(y) - \phi(x)| \leq \Phi(y) - \Phi(x).$$

Therefore, $\Phi(x) - \phi(x) \leq \Phi(y) - \phi(y)$.

iii) Let $P = \{x_0, x_1, \ldots, x_n\}$ be a partition of $[a, b]$. By part i) and Definition 3.8,

$$\sum_{j=1}^{n} |\phi(x_j) - \phi(x_{j-1})| \leq \sum_{j=1}^{n} |\Phi(x_j) - \Phi(x_{j-1})| \leq \text{Var}\,(\Phi).$$

Taking the supremum of this inequality over all partitions P of $[a, b]$, we obtain $\text{Var}\,(\phi) \leq \text{Var}\,(\Phi)$. ∎

COROLLARY 3.21. Let $[a, b]$ be a closed bounded interval. Then ϕ is of bounded variation on $[a, b]$ if and only if there exist increasing functions f, g on $[a, b]$ such that

$$\phi(x) = f(x) - g(x), \qquad x \in [a, b].$$

PROOF. Suppose ϕ is of bounded variation, let Φ represent the total variation of ϕ, $f = \Phi$, and $g = \Phi - \phi$. By Theorem 3.20 f and g are increasing and by construction, $\phi = f - g$.

Conversely, suppose $\phi = f - g$ for some increasing f, g on $[a, b]$. Then by Remark 2 and Theorem 3.19, ϕ is of bounded variation on $[a, b]$. ∎

In particular, if f is of bounded variation on $[a, b]$ then

i) $f(x+)$ exists for each $x \in [a, b)$ and $f(x-)$ exists for each $x \in (a, b]$ (see the Lemma in Section 2.6),

ii) f has at most countably many points of discontinuity in $[a, b]$ (see Theorem 2.20), and

iii) f is integrable on $[a, b]$ (see Exercise 8 in Section 3.1).

EXERCISES

1. a) Show $4k/(4k^2 - 1) > 1/k$ for $k \in \mathbf{N}$.

b) Prove

$$\sum_{k=1}^{2^n-1} \frac{1}{k} > \sum_{k=0}^{n-1} \left(\sum_{j=2^k}^{2^{k+1}-1} \frac{1}{2^{k+1}} \right) = \frac{n}{2}$$

for $n \in \mathbf{N}$.

c) Prove that

$$\phi(x) = \begin{cases} x^2 \sin \dfrac{1}{x^2} & x \neq 0 \\ 0 & x = 0 \end{cases}$$

is not of bounded variation on $[0, 1]$.

2. a) Show $(8k^2 + 2)/(4k^2 - 1)^2 < 1/k^2$ for $k = 2, 3, \ldots$

b) Prove

$$\sum_{k=1}^{n} \frac{1}{k^2} \leq 1 + \sum_{k=1}^{n-1} \left(\frac{1}{k} - \frac{1}{k+1} \right) = 2 - \frac{1}{n}$$

for $n \in \mathbf{N}$.

c) Prove that

$$\phi(x) = \begin{cases} x^2 \sin \dfrac{1}{x} & x \neq 0 \\ 0 & x = 0 \end{cases}$$

is of bounded variation on any bounded interval $[a, b]$.

$\boxed{3}$. **This exercise is used in Section 9.3.** Suppose ϕ and ψ are of bounded variation on a closed bounded interval $[a, b]$.

a) Prove that $\alpha\phi$ is of bounded variation on $[a, b]$ for every $\alpha \in \mathbf{R}$.

b) Prove that $\phi\psi$ is of bounded variation on $[a, b]$.

c) If there is an $\varepsilon_0 > 0$ such that

$$\phi(x) \geq \varepsilon_0, \qquad x \in [a, b],$$

prove that $1/\phi$ is of bounded variation on $[a, b]$.

4. Suppose ϕ is of bounded variation on a closed bounded interval $[a, b]$. Prove that ϕ is continuous on (a, b) if and only if ϕ is uniformly continuous on (a, b).

5. a) If ϕ is continuous on a closed bounded nondegenerate interval $[a, b]$, differentiable on (a, b), and ϕ' is bounded on (a, b), prove that ϕ is of bounded variation on $[a, b]$.

    b) Show that $\phi(x) = \sqrt[3]{x}$ is of bounded variation on $[-1, 1]$ but ϕ' is unbounded at some point in $(-1, 1)$.

6. Let P be a polynomial of degree N.

    a) Show that P is of bounded variation on any closed bounded interval $[a, b]$.

    b) Obtain an estimate for $\text{Var}(P)$ on $[a, b]$ using values of the derivative $P'(x)$ at no more than N points.

7. Let ϕ be a function of bounded variation on $[a, b]$ and Φ be its total variation function. Prove that if Φ is continuous at some point $x_0 \in (a, b)$, then ϕ is continuous at x_0.

$\boxed{8}$. **This exercise is used in Section 9.4.** If f is integrable on $[a, b]$, prove

$$F(x) = \int_a^x f(t) \, dt$$

is of bounded variation on $[a, b]$.

9. Suppose f' exists and is integrable on $[a, b]$. Prove that f is of bounded variation and

$$\text{Var}(f) = \int_a^b |f'(x)| \, dx.$$

What happens to this result if f' is bounded but not necessarily integrable?

e **3.6   CONVEX FUNCTIONS**   *This section uses no material from any other enrichment section.*

In this section we examine another collection of functions which is important for certain applications, especially for Fourier analysis, functional analysis, numerical analysis, and probability theory.

DEFINITION 3.10. Let I be an interval and $f : I \to \mathbf{R}$.

    i) f is said to be *convex* on I if

$$f(\alpha x + (1 - \alpha)y) \leq \alpha f(x) + (1 - \alpha)f(y)$$

    for all $0 \leq \alpha \leq 1$ and all $x, y \in I$.

    ii) f is said to be *concave* on I if $-f$ is convex on I.

Notice that by definition, a function f is convex on an interval I if and only if f is convex on every closed subinterval of I.

It is easy to check that $f(x) = mx + b$ is both convex and concave on any interval (see also Exercise 3) but in general it is difficult to apply Definition 3.10 directly. For this reason, we include the following simple geometric characterizations of convexity.

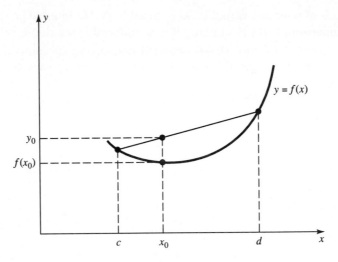

Figure 3.5

Remark 1. *Let I be an interval and $f : I \to$* **R**. *Then f is convex on I if and only if given any $[c, d] \subseteq I$, the chord through the points $(c, f(c))$, $(d, f(d))$ lies on or above the graph of $y = f(x)$ for all $x \in [c, d]$.* (See Figure 3.5.)

PROOF. Suppose f is convex on I and $x_0 \in [c, d]$. Choose $0 \le \alpha \le 1$ such that $x_0 = \alpha c + (1 - \alpha)d$. The chord from $(c, f(c))$ to $(d, f(d))$ has slope $(f(d) - f(c))/(d - c)$. Hence, the point on this chord which has the form (x_0, y_0) must satisfy $y_0 = \alpha f(c) + (1 - \alpha)f(d)$. Since f is convex, it follows that $f(x_0) \le y_0$, i.e., the point (x_0, y_0) lies on or above the point $(x_0, f(x_0))$. A similar argument establishes the reverse implication. ∎

Thus, both $f(x) = |x|$ and $f(x) = x^2$ are convex on any interval.

Remark 2. *A function f is convex on an open interval (a, b) if and only if the slope of the chord always increases, i.e.,*

$$a < c < x < d < b \quad \text{implies} \quad \frac{f(x) - f(c)}{x - c} \le \frac{f(d) - f(x)}{d - x}.$$

PROOF. Fix $a < c < x < d < b$ and let $\lambda(x)$ be the equation of the chord to f through the points $(c, f(c))$ and $(d, f(d))$. If f is convex, then $f(x) \le \lambda(x)$ (see Figure 3.6). Therefore,

$$\frac{f(x) - f(c)}{x - c} \le \frac{\lambda(x) - \lambda(c)}{x - c} = \frac{\lambda(d) - \lambda(x)}{d - x} \le \frac{f(d) - f(x)}{d - x}.$$

Conversely, if f is not convex, then $\lambda(x) < f(x)$ for some $x \in (c, d)$. It follows that

$$\frac{f(x) - f(c)}{x - c} > \frac{\lambda(x) - \lambda(c)}{x - c} = \frac{\lambda(d) - \lambda(x)}{d - x} > \frac{f(d) - f(x)}{d - x}.$$

Therefore, the slope of the chord decreases. ∎

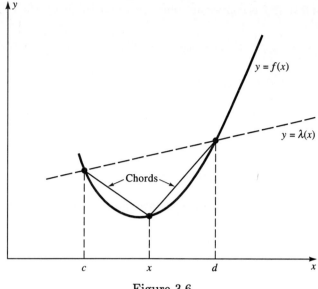

Figure 3.6

This leads us to a characterization of differentiable convex functions.

THEOREM 3.22. *Suppose f is differentiable on (a,b). Then f is convex on (a,b) if and only if f' is increasing on (a,b).*

PROOF. Suppose f is convex and that $c, d \in (a,b)$ satisfy $c < d$. Choose h so small that $a < c - h$, $c + h < d$, and $d + h < b$. Then by Remark 2,

$$\frac{f(c+h) - f(c)}{h} \leq \frac{f(d+h) - f(d)}{h}.$$

In particular, $f'(c) \leq f'(d)$.

Conversely, suppose f' is increasing. Let $a < c < x < d < b$ and use the Mean Value Theorem to choose x_0 (between c and x) and x_1 (between x and d) such that

$$\frac{f(x) - f(c)}{x - c} = f'(x_0) \quad \text{and} \quad \frac{f(d) - f(x)}{d - x} = f'(x_1).$$

Since $x_0 < x_1$ it follows that $f'(x_0) \leq f'(x_1)$. In particular, we conclude by Remark 2 that f is convex on (a,b). ∎

Combining Theorems 2.19 and 3.22, we obtain the usual convexity criterion in terms of the second derivative: If f is twice differentiable on (a,b), then f is convex on (a,b) if and only if $f''(x) \geq 0$ for all $x \in (a,b)$. In particular, convexity is what elementary calculus texts call *concave upward* and concavity is what elementary calculus texts call *concave downward*.

On open intervals, convex functions are always continuous. (The statements and proofs of the next two results come from Zygmund [15].)

THEOREM 3.23. *If f is convex on some open interval (a, b), then f is continuous on (a, b).*

PROOF. Let $x_0 \in (a, b)$. By symmetry, it suffices to show that $f(x) \to f(x_0)$ as $x \to x_0+$.

Let $a < c < x_0 < x < d < b$, $y = g(x)$ represent the equation of the chord through $(c, f(c))$, $(x_0, f(x_0))$, and $y = h(x)$ represent the equation of the chord through $(x_0, f(x_0))$, $(d, f(d))$. Since f is convex, we have by Remark 1 that $f(x) \le h(x)$. Since $f(x_0)$ lies on or below the chord from $(c, f(c))$ to $(x, f(x))$, we also have that $g(x) \le f(x)$. Consequently,

$$g(x) \le f(x) \le h(x), \qquad x \in (x_0, b).$$

Both chords $y = g(x)$ and $y = h(x)$ pass through the point $(x_0, f(x_0))$, so $g(x) \to f(x_0)$ and $h(x) \to f(x_0)$ as $x \to x_0+$. Hence, it follows from the Squeeze Theorem that $f(x) \to f(x_0)$ as $x \to x_0+$. ∎

Theorem 3.23 does not hold for closed intervals $[a, b]$. Indeed, the function

$$f(x) := \begin{cases} 0 & 0 \le x < 1 \\ 1 & x = 1 \end{cases}$$

is convex on $[0, 1]$ but not continuous there.

Recall that a function f is said to have a *proper maximum* (respectively, *proper minimum*) at x_0 if there exists a $\delta > 0$ such that $f(x) < f(x_0)$ (respectively, $f(x) > f(x_0)$) for all $0 < |x - x_0| < \delta$. As far as proper extrema are concerned, convex functions behave like strictly increasing functions.

THEOREM 3.24. i) *If f is convex on an open interval (a, b), then f has no proper maximum on (a, b).*

ii) *If f is convex on $[0, \infty)$ and has a proper minimum, then $f(x) \to \infty$ as $x \to \infty$.*

PROOF. i) Suppose $x_0 \in (a, b)$ and $f(x_0)$ is a proper maximum of f. Then there exist $c < x_0 < d$ such that $f(x) < f(x_0)$ for $c < x < d$. In particular, the chord through $(c, f(c))$, $(d, f(d))$ must lie below $f(x_0)$ for c, d near x_0, a contradiction.

ii) Suppose $x_0 \in (a, b)$ and $f(x_0)$ is a proper minimum of f. Fix $x_1 > x_0$. Let $y = g(x)$ represent the equation of the chord through $(x_0, f(x_0))$ and $(x_1, f(x_1))$. Since $f(x_0)$ is a proper minimum, $f(x_1) > f(x_0)$, hence g has positive slope. Moreover, by the proof of Theorem 3.21, $g(x) \le f(x)$ for all $x \in (x_1, \infty)$. Since $g(x) \to \infty$ as $x \to \infty$, we conclude that $f(x) \to \infty$ as $x \to \infty$. ∎

What about differentiability of convex functions? To answer this question we introduce the following concepts (compare with Definition 2.9).

DEFINITION 3.11. Let $f : (a, b) \to \mathbf{R}$ and $x \in (a, b)$.

i) f is said to have a *right-hand derivative* at x if

$$D_R f(x) := \lim_{h \to 0+} \frac{f(x + h) - f(x)}{h},$$

exists as an extended real number.

ii) f is said to have a *left-hand derivative* at x if

$$D_L f(x) := \lim_{h \to 0-} \frac{f(x+h) - f(x)}{h},$$

exists as an extended real number.

The following remark is a simple consequence of the definition of differentiability and the characterization of two-sided limits by one-sided limits (see Theorem 2.5).

Remark 3. *A function f is differentiable at x if and only if both the left-hand and right-hand derivatives of f exist and are finite at x, in which case*

$$f'(x) = D_R f(x) = D_L f(x).$$

The next result shows that the left-hand and right-hand derivatives of a convex function are remarkably well-behaved.

THEOREM 3.25. *Let f be convex on an open interval (a, b). Then the left-hand and right-hand derivatives of f exist, are increasing on (a, b), and satisfy*

$$-\infty < D_L f(x) \le D_R f(x) < \infty$$

for all $x \in (a, b)$.

PROOF. Let $h < 0$ and notice that the slope of the chord through the points $(x, f(x))$ and $(x+h, f(x+h))$ is $(f(x+h) - f(x))/h$. By Remark 2, these slopes increase as $h \to 0-$. Since increasing functions have a limit (which may be $+\infty$), it follows that $D_L f(x)$ exists and satisfies $-\infty < D_L f(x) \le \infty$. Similarly, $D_R f(x)$ exists and satisfies $-\infty \le D_R f(x) < \infty$. Remark 2 also implies

$$(20) \qquad\qquad D_L f(x) \le D_R f(x).$$

Hence, both numbers are finite and by symmetry it remains to show that $D_R f(x)$ is increasing on (a, b).

Let $x_1 < u < t < x_2$ be points which belong to (a, b). Then

$$\frac{f(u) - f(x_1)}{u - x_1} \le \frac{f(x_2) - f(t)}{x_2 - t}.$$

Taking the limit of this inequality as $u \to x_1+$ and $t \to x_2-$, we conclude by (20) that

$$(21) \qquad\qquad D_R f(x_1) \le D_L f(x_2) \le D_R f(x_2). \quad \blacksquare$$

The next proof uses Theorem 2.20, an optional result from Section 2.6.

*COROLLARY. *If f is convex on an open interval (a, b), then f is differentiable at all but countably many points of (a, b); i.e., there is a countable set $E \subset (a, b)$ such that $f'(x)$ exists for all $x \in (a, b) \setminus E$.*

PROOF. Let E be the set where either $D_L f(x)$ or $D_R f(x)$ is discontinuous. By Theorems 3.25 and 2.20, the set E is countable. Suppose $x_0 \in (a, b) \setminus E$ and $x < x_0$. By (21),
$$D_R f(x) \le D_L f(x_0) \le D_R f(x_0).$$

Let $x \to x_0$. Since both $D_L f(x)$ and $D_R f(x)$ are continuous at x_0, we obtain $D_R f(x_0) \le D_L f(x_0) \le D_R f(x_0)$. In particular, $f'(x_0)$ exists for all $x_0 \in (a, b) \setminus E$. ∎

How useful is a statement about $f'(x)$ which holds for all but countably many points x? We address this question by proving a generalization of Theorem 2.19. (The proof here uses Theorem 2.23, an optional result from Section 2.6.)

*THEOREM 3.26. *Suppose f is continuous on a closed bounded nondegenerate interval $[a, b]$ and differentiable on (a, b). If $f'(x) \ge 0$ for all but countably many $x \in (a, b)$, then f is increasing on $[a, b]$.*

PROOF. Suppose $f'(x_1) < 0$ for some $x_1 \in (a, b)$ and let $y \in (f'(x_1), 0)$. By Theorem 2.23 (the Intermediate Value Theorem for derivatives), there is an $x = x(y) \in (a, b)$ such that $f'(x) = y < 0$. It follows that if $f'(x) < 0$ for one $x \in (a, b)$, then $f'(x) < 0$ for uncountably many $x \in (a, b)$, a contradiction. Therefore, $f'(x) \ge 0$ for all $x \in (a, b)$, hence by Theorem 2.19, f is increasing on (a, b). ∎

COROLLARY. *If f is continuous on a closed bounded nondegenerate interval $[a, b]$ and differentiable on (a, b) with $f'(x) = 0$ for all but countably many $x \in (a, b)$, then f is constant on (a, b).*

We close this section by proving a fundamental inequality concerning convex functions and integration.

THEOREM 3.27 [JENSEN'S INEQUALITY]. *Let ϕ be convex on a closed bounded interval $[a, b]$ and $f : [0, 1] \to [a, b]$. If f and $\phi \circ f$ are integrable on $[0, 1]$, then*

$$(22) \qquad \phi \left(\int_0^1 f(x)\, dx \right) \le \int_0^1 (\phi \circ f)(x)\, dx.$$

PROOF. Set
$$c = \int_0^1 f(x)\, dx$$

and observe that

$$(23) \qquad \phi \left(\int_0^1 f(x)\, dx \right) = \phi(c) + s \left(\int_0^1 f(x)\, dx - c \right)$$

for all $s \in$ **R**. (Note: Since $a \le f(x) \le b$ for each $x \in [0, 1]$, c must belong to the interval $[a, b]$ by the Comparison Theorem for Integrals. Thus $\phi(c)$ is defined.)

Let

$$s = \sup_{x \in [a,c)} \frac{\phi(c) - \phi(x)}{c - x}.$$

By Remark 2, $s \le (\phi(u) - \phi(c))/(u - c)$ for all $u \in (c, b]$, i.e.,

(24) $$\phi(c) + s(u - c) \le \phi(u)$$

for all $u \in [c, b]$. On the other hand, if $u \in [a, c)$, we have by the definition of s that

$$s \ge \frac{\phi(c) - \phi(u)}{c - u}.$$

Thus (24) holds for all $u \in [a, b]$. Applying (24) to $u = f(x)$, we obtain

$$\phi(c) + s(f(x) - c) \le (\phi \circ f)(x).$$

Integrating this inequality as x runs from 0 to 1, we obtain

$$\phi(c) + s \left(\int_0^1 f(x)\, dx - c \right) \le \int_0^1 (\phi \circ f)(x)\, dx.$$

Combining this inequality with (23), we conclude that (22) holds. ∎

EXERCISES

1. Suppose f, g are convex on an interval I. Prove that $f + g$ and cf are convex on I for any $c \ge 0$.
2. Suppose that f_n is a sequence of functions convex on an interval I and

$$f(x) = \lim_{n \to \infty} f_n(x)$$

   exists for each $x \in I$. Prove that f is convex on I.
3. Prove that a function f is both convex and concave on I if and only if there exist $m, b \in \mathbf{R}$ such that $f(x) = mx + b$ for $x \in I$.
4. Prove that $f(x) = x^p$ is convex on $[0, \infty)$ for $p \ge 1$, and concave on $[0, \infty)$ for $0 < p \le 1$.
5. Show that if f is increasing on $[a, b]$, then

$$F(x) = \int_a^x f(t)\, dt$$

   is convex on $[a, b]$.
   (Recall that by Exercise 8 in Section 3.1, f is integrable on $[a, b]$.)
6. Let f be continuous on a closed bounded interval $[a, b]$ and suppose $D_R f(x)$

exists for all $x \in (a, b)$.

a) Show that if $f(b) < y_0 < f(a)$, then

$$x_0 := \sup\{x \in [a, b] : f(x) > y_0\}$$

satisfies $f(x_0) = y_0$ and $D_R f(x_0) \leq 0$.

b) Prove that if $f(b) < f(a)$, then there are uncountably many points x which satisfy $D_R f(x) \leq 0$.

c) Prove that if $D_R f(x) > 0$ for all but countably many points $x \in (a, b)$, then f is increasing on $[a, b]$.

d) Prove that if $D_R f(x) \geq 0$ and $g(x) = f(x) + x/n$ for some $n \in \mathbf{N}$, then $D_R g(x) > 0$.

e) Prove that if $D_R f(x) \geq 0$ for all but countably many points $x \in (a, b)$, then f is increasing on $[a, b]$.

7. Suppose $f : [0, 1] \to [a, b]$ is integrable on $[0, 1]$.

a) Prove

$$e^{\int_0^1 f(x)\, dx} \leq \int_0^1 e^{f(x)}\, dx.$$

b) Prove

$$\left(\int_0^1 |f(x)|^r \, dx \right)^{1/r} \leq \int_0^1 |f(x)|\, dx$$

for all $0 < r \leq 1$.

c) Prove

$$\int_0^1 |f(x)|\, dx \leq \left(\int_0^1 |f(x)|^p \, dx \right)^{1/p}$$

for all $p \geq 1$.

8. If $f : [a, b] \to \mathbf{R}$ is integrable on $[a, b]$, prove

$$\int_a^b |f(x)|\, dx \leq (b - a)^{1/2} \left(\int_a^b f^2(x)\, dx \right)^{1/2}.$$

Chapter 4

INFINITE SERIES

Infinite series are one of the most widely used tools of analysis. They are used to approximate numbers and functions. (Series of Ramanujan type have been used to compute more than a billion digits of the decimal expansion of π.) They are used to approximate solutions of differential equations. (You may have used power series to solve ordinary differential equations with nonconstant coefficients.) They even form the basis for some very practical applications including pattern recognition (e.g., reading zip codes), image enhancement (e.g., removing raindrop clutter from a radar scan), and data compression (e.g., reducing the space needed to store data from a comprehensive census of a large country). Other applications of infinite series can be found in Section 4.8 below. In view of the variety of these applications, it should come as no surprise that the subject matter of this chapter is of fundamental importance.

4.1 INTRODUCTION

Let $\{a_k\}_{k \in \mathbf{N}}$ be a sequence of numbers or functions. We shall call an expression of the form

$$(1) \qquad \sum_{k=1}^{\infty} a_k$$

an *infinite series* with *terms* a_k. (No convergence is assumed at this point. This is merely a formal expression.)

DEFINITION 4.1. Let $S = \sum_{k=1}^{\infty} a_k$ be an infinite series whose terms a_k belong to $\mathbf{R}$.

   i) The *partial sums of S of order* n are the numbers, defined for each $n \in \mathbf{N}$, by

$$s_n := \sum_{k=1}^{n} a_k.$$

ii) S is said to *converge* if its sequence of partial sums $\{s_n\}$ converges to some $s \in \mathbf{R}$ as $n \to \infty$. In this case we shall write

(2)
$$\sum_{k=1}^{\infty} a_k = s$$

and call s the *sum* or *value* of the series $\sum_{k=1}^{\infty} a_k$.

iii) S is said to *diverge* if its sequence of partial sums $\{s_n\}$ does not converge as $n \to \infty$. When s_n diverges to ∞ as $n \to \infty$, we shall also write

$$\sum_{k=1}^{\infty} a_k = \infty.$$

You are already familiar with one type of infinite series, decimal expansions. Every decimal expansion of a number $x \in (0, 1)$ is a series of the form $\sum_{k=1}^{\infty} x_k/10^k$, where the x_k's are integers between 0 and 9. For example, when we write $1/3 = 0.333\ldots$ we mean

$$\frac{1}{3} = \sum_{k=1}^{\infty} \frac{3}{10^k}.$$

In particular, $.3, .33, .333, \ldots$ are approximations to $1/3$ which get closer and closer to $1/3$ as more terms of the decimal expansion are taken.

One way to determine if a given series converges is to find a formula for its partial sums simple enough so that we can decide whether or not they converge. Here are two simple examples.

Example 1. Prove $\sum_{k=1}^{\infty} 2^{-k} = 1$.

PROOF. By induction, we can show that the partial sums $s_n = \sum_{k=1}^{n} 1/2^k$ satisfy $s_n = 1 - 2^{-n}$ for $n \in \mathbf{N}$. Thus, $s_n \to 1$ as $n \to \infty$. ∎

Example 2. Prove $\sum_{k=1}^{\infty} (-1)^k$ diverges.

PROOF. The partial sums $s_n = \sum_{k=1}^{n} (-1)^k$ satisfy

$$s_n = \begin{cases} -1 & \text{if } n \text{ is odd} \\ 0 & \text{if } n \text{ is even.} \end{cases}$$

Thus, s_n does not converge as $n \to \infty$. ∎

Another way to show a series diverges is to estimate its partial sums.

Example 3. [THE HARMONIC SERIES]. Prove that $1/k$ converges but $\sum_{k=1}^{\infty} 1/k = \infty$.

PROOF. By the Archimedean Principle, the sequence $1/k$ converges to zero (see Example 1 in Section 1.4). On the other hand, by the Comparison Theorem for Integrals,

$$\sum_{k=1}^{n} \frac{1}{k} \geq \sum_{k=1}^{n} \int_{k}^{k+1} \frac{1}{x}\, dx = \int_{1}^{n+1} \frac{1}{x}\, dx = \log(n+1).$$

We conclude that $s_n \to \infty$ as $n \to \infty$. ∎

Note: This example shows that the terms of a divergent series may converge. In particular, a series does not converge just because its terms converge.

There are many methods for deciding whether or not a given series converges. The following method is simple but useful in a variety of situations.

THEOREM 4.1 [DIVERGENCE TEST]. *Let* $\{a_k\}_{k \in \mathbf{N}}$ *be a sequence of real numbers.*

i) *If* $\sum_{k=1}^{\infty} a_k$ *converges, then* $a_k \to 0$ *as* $k \to \infty$.
ii) *The series* $\sum_{k=1}^{\infty} a_k$ *diverges when* a_k *does not converge to zero.*

PROOF. It suffices to prove part i). Suppose $\sum_{k=1}^{\infty} a_k$ converges to some $s \in \mathbf{R}$. By definition, the sequence of partial sums s_n converges to s as $n \to \infty$. Therefore, $a_k = s_k - s_{k-1} \to s - s = 0$ as $k \to \infty$. ∎

It is important to realize from the beginning that Theorem 4.1 is a test for divergence, not a test for convergence. Indeed, the harmonic series shows that a series can diverge even when its terms converge to zero.

Finding the sum of a convergent series is usually difficult. The following two results show this is not the case for certain kinds of series.

THEOREM 4.2 [TELESCOPIC SERIES]. *If* $\{a_k\}$ *is a convergent real sequence, then*

$$\sum_{k=1}^{\infty}(a_k - a_{k+1}) = a_1 - \lim_{k \to \infty} a_k.$$

PROOF. By telescoping, we have

$$s_n := \sum_{k=1}^{n}(a_k - a_{k+1}) = a_1 - a_{n+1}.$$

Hence, $s_n \to a_1 - \lim_{k \to \infty} a_k$ as $n \to \infty$. ∎

THEOREM 4.3 [GEOMETRIC SERIES]. *The series* $\sum_{k=1}^{\infty} x^k$ *converges if and only if* $|x| < 1$, *in which case*

$$\sum_{k=1}^{\infty} x^k = \frac{x}{1-x}.$$

(See also Exercise 1 below.)

PROOF. If $|x| \geq 1$, then $\sum_{k=1}^{\infty} x^k$ diverges by the Divergence Test. If $|x| < 1$, then set $s_n = \sum_{k=1}^{n} x^k$ and observe by telescoping that

$$(1-x)s_n = (1-x)(x + x^2 + \cdots + x^n)$$
$$= x + x^2 + \cdots + x^n \ - x^2 - x^3 - \cdots - x^{n+1} = x - x^{n+1}.$$

Hence,

$$s_n = \frac{x}{1-x} - \frac{x^{n+1}}{1-x}$$

for all $n \in \mathbf{N}$. Since $x^{n+1} \to 0$ as $n \to \infty$ for all $|x| < 1$ (see Example 2 in Section 1.4), we conclude that $s_n \to x/(1 - x)$ as $n \to \infty$. $\blacksquare$

Note: In everyday speech, the words *sequence* and *series* are considered synonyms. Example 3 shows that in mathematics, this is not the case. In particular, you must not apply a result valid for sequences to series and vice versa. Nevertheless, because convergence of an infinite series is defined in terms of convergence of its sequence of partial sums, any result about sequences contains a result about infinite series. The following three theorems illustrate this principle.

THEOREM 4.4 [THE CAUCHY CRITERION]. *Let $\{a_k\}$ be a real sequence. Then the infinite series $\sum_{k=1}^{\infty} a_k$ converges if and only if given $\varepsilon > 0$ there is an $N \in \mathbf{N}$ such that*

$$m > n \geq N \quad implies \quad \left| \sum_{k=n}^{m} a_k \right| < \varepsilon.$$

PROOF. Let s_n represent the sequence of partial sums of $\sum_{k=1}^{\infty} a_k$. By Cauchy's Theorem (Theorem 1.20), s_n converges if and only if given $\varepsilon > 0$ there is an $N \in \mathbf{N}$ such that $m, n \geq N$ imply $|s_m - s_{n-1}| < \varepsilon$. Since

$$s_m - s_{n-1} = \sum_{k=n}^{m} a_k$$

for all integers $m > n > 1$, the proof is complete. $\blacksquare$

COROLLARY 4.5. *Let $\{a_k\}$ be a real sequence. Then the infinite series $\sum_{k=1}^{\infty} a_k$ converges if and only if given $\varepsilon > 0$ there is an $N \in \mathbf{N}$ such that*

$$n \geq N \quad implies \quad \left| \sum_{k=n}^{\infty} a_k \right| < \varepsilon.$$

THEOREM 4.6. *Let $\{a_k\}$ and $\{b_k\}$ be real sequences. If $\sum_{k=1}^{\infty} a_k$ and $\sum_{k=1}^{\infty} b_k$ are convergent series, then*

$$\sum_{k=1}^{\infty} (a_k + b_k) = \sum_{k=1}^{\infty} a_k + \sum_{k=1}^{\infty} b_k$$

and

$$\sum_{k=1}^{\infty} (\alpha a_k) = \alpha \sum_{k=1}^{\infty} a_k$$

for any $\alpha \in \mathbf{R}$.

PROOF. Both identities are corollaries of Theorem 1.14; we provide the details only for the first identity.

Let s_n represent the partial sums of $\sum_{k=1}^{\infty} a_k$ and t_n represent the partial sums of $\sum_{k=1}^{\infty} b_k$. By the Commutative Property of real addition,

$$\sum_{k=1}^{n}(a_k + b_k) = s_n + t_n, \qquad n \in \mathbf{N}.$$

Taking the limit of this identity as $n \to \infty$, we conclude by Theorem 1.14 that

$$\sum_{k=1}^{\infty}(a_k + b_k) = \lim_{n\to\infty} s_n + \lim_{n\to\infty} t_n = \sum_{k=1}^{\infty} a_k + \sum_{k=1}^{\infty} b_k. \quad \blacksquare$$

EXERCISES

1. Show that

$$\sum_{k=n}^{\infty} x^k = \frac{x^n}{1 - x}$$

for $|x| < 1$ and $n = 0, 1, \ldots$.

2. Prove that each of the following series converges and find its value.

a) $\displaystyle\sum_{k=1}^{\infty} \frac{(-1)^{k+1}}{\pi^k}.$ b) $\displaystyle\sum_{k=1}^{\infty} \frac{(-1)^k + 4}{5^k}.$

c) $\displaystyle\sum_{k=1}^{\infty} \frac{3^k}{7^{k-1}}.$ d) $\displaystyle\sum_{k=0}^{\infty} 2^k e^{-k}.$

3. Represent each of the following series as a telescopic series and find its value.

a) $\displaystyle\sum_{k=1}^{\infty} \frac{1}{k(k+1)}.$ b) $\displaystyle\sum_{k=2}^{\infty} \log\left(\frac{k(k+2)}{(k+1)^2}\right).$

c) $\displaystyle\sum_{k=1}^{\infty} \sqrt[k]{\frac{\pi}{4}\left(1 - \left(\frac{\pi}{4}\right)^{j_k}\right)}$, where $j_k = -1/(k(k+1))$ for $k \in \mathbf{N}$.

4. Find all $x \in \mathbf{R}$ for which

$$\sum_{k=1}^{\infty} 3(x^k - x^{k-1})(x^k + x^{k-1})$$

converges. For each such x, find the value of this series.

5. Prove that each of the following series diverges.

a) $\displaystyle\sum_{k=1}^{\infty} \cos\left(\frac{1}{k^2}\right).$ b) $\displaystyle\sum_{k=1}^{\infty} \left(1 - \frac{1}{k}\right)^k.$ c) $\displaystyle\sum_{k=1}^{\infty} \frac{k+1}{k^2}.$

6. a) Prove that if $\sum_{k=1}^{\infty} a_k$ converges, then its partial sums s_n are bounded.

   b) Show that the converse of part a) is false. Namely, show that a series $\sum_{k=1}^{\infty} a_k$ may have bounded partial sums and still diverge.

7. Let $\{b_k\}$ be a real sequence and $b \in \mathbf{R}$.

   a) Suppose there is an $N \in \mathbf{N}$ such that $|b - b_k| \le M$ for all $k \ge N$. Prove that

   $$\left| nb - \sum_{k=1}^{n} b_k \right| \le \sum_{k=1}^{N} |b_k - b| + M(n - N)$$

   for all $n > N$.

   b) Prove that if $b_k \to b$ as $k \to \infty$, then

   $$\frac{b_1 + b_2 + \cdots + b_n}{n} \to b$$

   as $n \to \infty$.

   c) Show that the converse of b) is false.

8. A series $\sum_{k=0}^{\infty} a_k$ is said to be *Cesàro summable* to L if

   $$\sigma_n := \sum_{k=0}^{n-1} \left(1 - \frac{k}{n} \right) a_k$$

   converges to L as $n \to \infty$.

   a) Let $s_n = \sum_{k=0}^{n-1} a_k$. Prove that

   $$\sigma_n = \frac{s_1 + \cdots + s_n}{n}$$

   for each $n \in \mathbf{N}$.

   b) Prove that if $a_k \in \mathbf{R}$ and $\sum_{k=0}^{\infty} a_k = L$ converges, then $\sum_{k=0}^{\infty} a_k$ is Cesàro summable to L.

   c) Prove that $\sum_{k=0}^{\infty} (-1)^k$ is Cesàro summable to $1/2$, hence the converse of b) is false.

   d) [TAUBER]. Prove that if $a_k \ge 0$ for $k \in \mathbf{N}$ and $\sum_{k=0}^{\infty} a_k$ is Cesàro summable to L, then $\sum_{k=0}^{\infty} a_k = L$.

9. a) Suppose $\{a_k\}$ is a decreasing sequence of real numbers. Prove that if $\sum_{k=1}^{\infty} a_k$ converges, then $k a_k \to 0$ as $k \to \infty$.

   b) Let $s_n = \sum_{k=1}^{n} (-1)^{k+1}/k$ for $n \in \mathbf{N}$. Prove that s_{2n} is strictly increasing, s_{2n+1} is strictly decreasing, and $s_{2n+1} - s_{2n} \to 0$ as $n \to \infty$.

   c) Prove that part a) is false if "decreasing" is removed.

10. Suppose $a_k \ge 0$ for k large and $\sum_{k=1}^{\infty} a_k$ converges. Prove

$$\lim_{j \to \infty} \sum_{k=1}^{\infty} \frac{a_k}{j + k} = 0.$$

4.2 TESTS FOR CONVERGENCE

Although we obtained exact values in the previous section for telescopic series and geometric series, finding exact values of a given series is frequently difficult, if not impossible. Fortunately, for many applications it is not as important to be able to find the value of a series as it is to know that the series converges. When it does converge, we can use its partial sums to approximate its value as accurately as we wish (up to the limitations of whatever computing device we are using). Therefore, in this section we include several tests which can be used to decide whether a given series converges or whether it diverges, and to determine how accurately its value is approximated by its nth partial sums.

Let $\mathcal{P}_k$ be a statement which depends on $k \in \mathbf{N}$. We shall say that $\mathcal{P}_k$ holds for *large* k if there is an $N \in \mathbf{N}$ such that $\mathcal{P}_k$ is true for $k \geq N$.

The partial sums of a divergent series may be bounded (like $\sum_{k=1}^{\infty}(-1)^k$) or unbounded (like $\sum_{k=1}^{\infty} 1/k$). When the terms of a divergent series are nonnegative, the former cannot happen.

THEOREM 4.7. *Suppose $a_k \geq 0$ for $k \geq N$. Then $\sum_{k=1}^{\infty} a_k$ converges if and only if its sequence of partial sums $\{s_n\}$ is bounded, in which case,*

$$s := \sup\{s_n : n \geq N\} = \sum_{k=1}^{\infty} a_k.$$

PROOF. Set $s_n = \sum_{k=1}^{n} a_k$ for $n \in \mathbf{N}$. If $\sum_{k=1}^{\infty} a_k$ converges, then s_n converges as $n \to \infty$. Since every convergent sequence is bounded (Theorem 1.11), $\sum_{k=1}^{\infty} a_k$ has bounded partial sums.

Conversely, suppose $|s_n| \leq M$ for $n \in \mathbf{N}$. Since $a_k \geq 0$ for $k \geq N$, s_n is an increasing sequence when $n \geq N$. Since any increasing bounded sequence converges to its supremum (see Theorem 1.16), it follows that $\sum_{k=1}^{\infty} a_k$ converges to s. ∎

If $a_k \geq 0$ for large k, we shall write $\sum_{k=1}^{\infty} a_k < \infty$ when the series is convergent and $\sum_{k=1}^{\infty} a_k = \infty$ when the series is divergent.

In some cases, integration can be used to test convergence of a series. The idea behind this test is that

$$\int_1^{\infty} f(x)\, dx = \sum_{k=1}^{\infty} \int_k^{k+1} f(x)\, dx \approx \sum_{k=1}^{\infty} f(k)$$

when f is almost constant on each interval $[k, k+1]$. This will surely be the case for large k if $f(k) \downarrow 0$ as $k \to \infty$ (see Figure 4.1). This observation leads us to the following test.

THEOREM 4.8. *Suppose $f : [1, \infty) \to \mathbf{R}$ is positive and decreasing on $[1, \infty)$.*

i) *For all $n \in \mathbf{N}$,*

$$f(n) \leq \sum_{k=1}^{n} f(k) - \int_1^n f(x)\, dx \leq f(1).$$

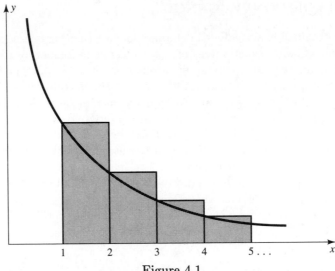

Figure 4.1

ii) [INTEGRAL TEST]. $\sum_{k=1}^{\infty} f(k)$ *converges if and only if* f *is improperly integrable on* $[1, \infty)$, *i.e., if and only if*

$$\int_1^{\infty} f(x)\, dx < \infty.$$

iii) *If* $\sum_{k=1}^{\infty} f(k)$ *converges, then*

$$0 \le \sum_{k=1}^{n} f(k) + \int_n^{\infty} f(x)\, dx - \sum_{k=1}^{\infty} f(k) \le f(n)$$

for all $n \in \mathbf{N}$.

PROOF. i) Let $s_n = \sum_{k=1}^{n} f(k)$ and $t_n = \int_1^n f(x)\, dx$ for $n \in \mathbf{N}$. Since f is decreasing, f is locally integrable on $[1, \infty)$ and $f(k+1) \le f(x) \le f(k)$ for all $x \in [k, k+1]$. Hence, by the Comparison Theorem for Integrals,

$$f(k+1) \le \int_k^{k+1} f(x)\, dx \le f(k)$$

for $k \in \mathbf{N}$. Summing over $k = 1, \ldots, n-1$, we obtain

$$s_n - f(1) = \sum_{k=2}^{n} f(k) \le \int_1^n f(x)\, dx = t_n \le \sum_{k=1}^{n-1} f(k) = s_n - f(n)$$

for all $n \ge N$. This proves part i).

ii) By the inequality just derived, $\{s_n\}$ is bounded if and only if $\{t_n\}$ is. Since $f(x) \ge 0$ implies both s_n and t_n are increasing sequences, it follows from the

Monotone Convergence Theorem that s_n converges if and only if t_n converges, as $n \to \infty$. This proves part ii).

iii) Let $u_k = s_k - t_k$ for $k \in \mathbf{N}$, and observe, since f is decreasing, that

$$0 \le u_k - u_{k+1} = \int_k^{k+1} f(x)\, dx - f(k+1) \le f(k) - f(k+1).$$

Summing these inequalities over $k \ge n$ and telescoping, we have

$$0 \le u_n - \lim_{j \to \infty} u_j = \sum_{k=n}^{\infty} (u_k - u_{k+1}) \le \sum_{k=n}^{\infty} (f(k) - f(k+1)) = f(n).$$

Since $u_j \to \sum_{k=1}^{\infty} f(k) - \int_1^{\infty} f(x)\, dx$ as $j \to \infty$, we conclude that

$$0 \le \sum_{k=1}^{n} f(k) + \int_n^{\infty} f(x)\, dx - \sum_{k=1}^{\infty} f(k) \le f(n). \quad \blacksquare$$

The next two examples show how this result can be used to estimate both divergent and convergent series.

Example 1. Prove that there exist numbers $C_n \in (0,1)$ such that

$$\sum_{k=1}^{n} \frac{1}{k} = \log n + C_n$$

for all $n \in \mathbf{N}$.

PROOF. Clearly, $f(x) = 1/x$ is positive, decreasing, and locally integrable on $[1, \infty)$. Hence, by Theorem 4.8i,

$$\frac{1}{n} \le \sum_{k=1}^{n} \frac{1}{k} - \int_1^n \frac{1}{x}\, dx = \sum_{k=1}^{n} \frac{1}{k} - \log n \le 1. \quad \blacksquare$$

Example 2. Prove $\sum_{k=1}^{\infty} k e^{-k^2}$ converges and estimate its value to an accuracy of 10^{-3}.

PROOF. Let $f(x) = x e^{-x^2}$. Since $f'(x) = e^{-x^2}(1 - 2x^2) \le 0$ for $x \ge 1$, f is decreasing on $[1, \infty)$. Since

$$\int_1^{\infty} x e^{-x^2}\, dx = \frac{1}{2} \int_1^{\infty} e^{-u}\, du = \frac{1}{2e} < \infty,$$

it follows from the Integral Test that $\sum_{k=1}^{\infty} k e^{-k^2}$ converges. To estimate the value s of this series, notice that $f(2) = 0.036631$ and $f(3) = 0.000370$. Therefore, by Theorem 4.8iii, s is approximately equal to

$$\sum_{k=1}^{3} k e^{-k^2} + \int_3^{\infty} x e^{-x^2}\, dx = \frac{1}{e} + \frac{2}{e^4} + \frac{3}{e^9} + \frac{1}{2e^9} \approx 0.4049427$$

with an error no more than 0.000370. $\blacksquare$

COROLLARY 4.9 [p–SERIES TEST]. *The series*

$$(3) \qquad \sum_{k=1}^{\infty} \frac{1}{k^p}$$

converges if and only if $p > 1$.

PROOF. If $p = 1$ or $p \leq 0$, the series diverges. If $p > 0$ and $p \neq 1$, set $f(x) = x^{-p}$ and observe that $f'(x) = -px^{-p-1} < 0$ for all $x \in [1, \infty)$. Hence, f is nonnegative and decreasing on $[1, \infty)$. Since

$$\int_1^{\infty} \frac{dx}{x^p} = \lim_{n \to \infty} x^{1-p} \Big|_1^n = \lim_{n \to \infty} \frac{n^{1-p} - 1}{1 - p}$$

has a finite limit if and only if $1 - p < 0$, it follows from the Integral Test that (3) converges if and only if $p > 1$. ∎

The Integral Test, which requires that f satisfies some very restrictive hypotheses, has limited applications. The following test can be used in a much broader context.

THEOREM 4.10 [COMPARISON TEST]. *Suppose $0 \leq a_k \leq b_k$ for large k.*
  i) *If $\sum_{k=1}^{\infty} b_k < \infty$, then $\sum_{k=1}^{\infty} a_k < \infty$.*
  ii) *If $\sum_{k=1}^{\infty} a_k = \infty$, then $\sum_{k=1}^{\infty} b_k = \infty$.*

PROOF. By hypothesis, choose $N \in \mathbf{N}$ so large that $0 \leq a_k \leq b_k$ for $k > N$. Set $s_n = \sum_{k=1}^{n} a_k$ and $t_n = \sum_{k=1}^{n} b_k$, $n \in \mathbf{N}$. Then $0 \leq s_n - s_N \leq t_n - t_N$ for all $n \geq N$. Since N is fixed, it follows that s_n is bounded when t_n is, and t_n is unbounded when s_n is. Apply Theorem 4.7 and the proof of the theorem is complete. ∎

The Comparison Test is used to compare one series with another whose convergence properties are already known. Although there is no simple algorithm for this process, the idea is to examine the terms of the given series, ignoring the superfluous factors, and to replace the more complicated factors by simpler ones. Here is a typical example.

Example 3. Determine whether the series

$$(4) \qquad \sum_{k=1}^{\infty} \frac{3k}{k^2 + k} \sqrt{\frac{\log k}{k}}$$

converges or diverges.

SOLUTION. The kth term of this series can be written by using three factors:

$$\frac{1}{k} \frac{3k}{k+1} \sqrt{\frac{\log k}{k}}.$$

The factor $3k/(k+1)$ is bounded by 3 and can be ignored. Since $\log k \leq \sqrt{k}$ for large k (see Exercise 4 in Section 2.5), the factor $\sqrt{\log k/k}$ satisfies

$$\sqrt{\frac{\log k}{k}} \leq \sqrt{\frac{\sqrt{k}}{k}} = \frac{1}{\sqrt[4]{k}}$$

for large k. Therefore, the terms of (4) are bounded by $3/k^{5/4}$. Since $\sum_{k=1}^{\infty} 3/k^{5/4}$ converges by the p–Series Test, it follows from the Comparison Test that (4) converges. ∎

The Comparison Test may not be easy to apply to a given series, even when we know which series it should be compared with. For situations like this, the following test is usually more efficient.

THEOREM 4.11 [LIMIT COMPARISON TEST]. *Suppose $a_k, b_k \geq 0$ for large k. If*

$$0 < \lim_{n \to \infty} \frac{a_n}{b_n} < \infty,$$

then $\sum_{k=1}^{\infty} a_k$ converges if and only if $\sum_{k=1}^{\infty} b_k$ converges.
(See also Exercise 8 below.)

PROOF. Let $c = \lim_{n \to \infty} a_n/b_n$. Then there is an $N \in \mathbf{N}$ such that

$$\frac{c}{2} b_k < a_k < \frac{3c}{2} b_k$$

for $k \geq N$. Hence, the result follows immediately from the Comparison Test and Theorem 4.6. ∎

In general, the Limit Comparison Test is used to replace a series $\sum_{k=1}^{\infty} a_k$ by $\sum_{k=1}^{\infty} b_k$ when $a_k \approx b_k$ for k large. For example, to determine whether or not the series

$$S := \sum_{k=1}^{\infty} \frac{k}{\sqrt{4k^4 + k^2 + 5k}}$$

converges, notice that its terms are approximately $1/(2k)$ for k large. This leads us to compare S with the harmonic series $\sum_{k=1}^{\infty} 1/k$. Since the harmonic series diverges and

$$\frac{k/(\sqrt{4k^4 + k^2 + 5k})}{1/k} = \frac{k^2}{\sqrt{4k^4 + k^2 + 5k}} \to \frac{1}{2} \neq 0$$

as $k \to \infty$, it follows from the Limit Comparison Test that S diverges.

Here is another application of the Limit Comparison Test.

Example 4. Let $a_k \to 0$ as $k \to \infty$. Prove that $\sum_{k=1}^{\infty} \sin |a_k|$ converges if and only if $\sum_{k=1}^{\infty} |a_k|$ converges.

PROOF. By L'Hôpital's Rule,

$$\lim_{k \to \infty} \frac{\sin |a_k|}{|a_k|} = \lim_{x \to 0+} \frac{\sin x}{x} = 1.$$

Hence, by the Limit Comparison Test, $\sum_{k=1}^{\infty} \sin |a_k|$ converges if and only if $\sum_{k=1}^{\infty} |a_k|$ converges. ∎

We have been discussing series whose terms are nonnegative. We now turn our attention to series whose terms are of mixed sign.

If $f : [1, N] \to \mathbf{R}$ for some $N \in \mathbf{N}$, then the *summation* $\sum_{k=1}^{N-1} f(k)$ is an approximation to $\int_1^N f(x)\, dx$ and the *finite difference* $f(k+1) - f(k)$ is an approximation to $f'(k)$ for $k = 1, 2, \ldots, N-1$. In particular, summation is an analogue of integration and finite difference is an analogue of differentiation. In this context, the following formula can be interpreted as a discrete analogue of integration by parts.

THEOREM 4.12 [ABEL'S FORMULA]. *Let $\{a_k\}_{k \in \mathbf{N}}$ and $\{b_k\}_{k \in \mathbf{N}}$ be real sequences, and for each pair of integers $n \geq m \geq 1$ set*

$$A_{n,m} = \sum_{k=m}^{n} a_k, \qquad n \geq m.$$

Then

$$\sum_{k=m}^{n} a_k b_k = A_{n,m} b_n - \sum_{k=m}^{n-1} A_{k,m}(b_{k+1} - b_k)$$

for all integers $n > m \geq 1$.

PROOF. Since $A_{k,m} - A_{(k-1),m} = a_k$ for $k > m$ and $A_{m,m} = a_m$, we have

$$\sum_{k=m}^{n} a_k b_k = a_m b_m + \sum_{k=m+1}^{n} (A_{k,m} - A_{(k-1),m}) b_k$$

$$= a_m b_m + \sum_{k=m+1}^{n} A_{k,m} b_k - \sum_{k=m}^{n-1} A_{k,m} b_{k+1}$$

$$= a_m b_m + A_{n,m} b_n - a_m b_{m+1} - \sum_{k=m+1}^{n-1} A_{k,m}(b_{k+1} - b_k)$$

$$= A_{n,m} b_n - \sum_{k=m}^{n-1} A_{k,m}(b_{k+1} - b_k). \blacksquare$$

Our first application of Abel's Formula is a test for convergence of series whose terms have the form $a_k b_k$, where $\{b_k\}$ is monotone and convergent.

THEOREM 4.13 [ABEL'S TEST]. *Let $a_k, b_k \in \mathbf{R}$ for $k \in \mathbf{N}$. If $\sum_{k=1}^{\infty} a_k$ converges and b_k is a convergent, monotone sequence, then $\sum_{k=1}^{\infty} a_k b_k$ converges.*

PROOF. It suffices to show that the partial sums of $\sum_{k=1}^{\infty} a_k b_k$ are Cauchy. Let $\varepsilon > 0$. We suppose without loss of generality that $b_k \downarrow b$, as $k \to \infty$, for some $b \in \mathbf{R}$. Since b_k converges, choose $M \in \mathbf{R}$ such that $|b_k| \le M$ for $k \in \mathbf{N}$. Since $\sum_{k=1}^{\infty} a_k$ converges, choose $N \in \mathbf{N}$ so large that

$$(6) \qquad k > m \ge N \quad \text{implies} \quad |A_{k,m}| := \left| \sum_{j=m}^{k} a_j \right| < \frac{\varepsilon}{4M}.$$

Let $n > m \ge N$. By Abel's Formula and (6),

$$(7) \qquad \left| \sum_{k=m}^{n} a_k b_k \right| \le |A_{n,m}| |b_n| + \sum_{k=m}^{n-1} |A_{k,m}| |b_{k+1} - b_k|$$

$$\le \frac{\varepsilon}{4} + \frac{\varepsilon}{4M} \sum_{k=m}^{n-1} |b_{k+1} - b_k|.$$

Since $\{b_k\}$ is decreasing, $|b_{k+1} - b_k| = (b_k - b_{k+1})$ and

$$\sum_{k=m}^{n-1} |b_{k+1} - b_k| = b_m - b_n$$

telescopes. Hence, it follows from (7) that

$$\left| \sum_{k=m}^{n} a_k b_k \right| < \frac{\varepsilon}{4} + \frac{\varepsilon}{2} < \varepsilon$$

for $n > m \ge N$. ∎

Example 5. Show

$$\sum_{k=1}^{\infty} \left(1 + \frac{1}{k} \right)^k \left(\frac{\log(k+1)}{\sqrt{k+1}} - \frac{\log k}{\sqrt{k}} \right)$$

converges.

PROOF. By L'Hôpital's Rule, $\log k / \sqrt{k} \to 0$ as $k \to \infty$, so the telescopic series

$$\sum_{k=1}^{\infty} \left(\frac{\log(k+1)}{\sqrt{k+1}} - \frac{\log k}{\sqrt{k}} \right)$$

converges. By Example 3 in Section 2.5, $(1+1/k)^k \uparrow e$ as $k \to \infty$. Hence, it follows from Abel's Test that the original series converges. ∎

Closely related is the following result.

THEOREM 4.14 [DIRICHLET'S TEST]. *Let $a_k, b_k \in \mathbf{R}$ for $k \in \mathbf{N}$. If the sequence of partial sums $s_n = \sum_{k=1}^{n} a_k$ is bounded and $b_k \downarrow 0$ as $k \to \infty$, then $\sum_{k=1}^{\infty} a_k b_k$ converges.*

PROOF. Choose $M \in \mathbf{R}$ such that

$$|s_n| = \left| \sum_{k=1}^{n} a_k \right| \leq M, \qquad n \in \mathbf{N}.$$

By the Triangle Inequality,

$$(8) \qquad |A_{n,m}| = \left| \sum_{k=m}^{n} a_k \right| = |s_n - s_{m-1}| \leq 2M$$

for $n > m > 1$.

Let $\varepsilon > 0$ and choose $N \in \mathbf{N}$ so that $|b_k| < \varepsilon/(4M)$ for $k \geq N$. Since $\{b_k\}$ is decreasing, we find by Abel's Formula and (8) that

$$\left| \sum_{k=m}^{n} a_k b_k \right| \leq |A_{n,m}| |b_n| + \sum_{k=m}^{n-1} |A_{k,m}| (b_k - b_{k+1})$$

$$< \frac{\varepsilon}{2} + 2M(b_m - b_n) < \varepsilon$$

for all $n > m \geq N$. ∎

Example 6. Prove $S(x) = \sum_{k=1}^{\infty} \sin(kx)/k$ converges for each $x \in \mathbf{R}$.

PROOF. Since $\phi(x) = \sin(kx)$ is periodic of period 2π (i.e., $\phi(x + 2\pi) = \phi(x)$ for all $x \in \mathbf{R}$) and has value identically zero when $x = 0$ or 2π, we need only show that $S(x)$ converges for each $x \in (0, 2\pi)$. By Dirichlet's Test, it suffices to show

$$(9) \qquad \widetilde{D}_n(x) := \sum_{k=1}^{n} \sin(kx) \qquad n \in \mathbf{N}$$

is a bounded sequence for each fixed $x \in (0, 2\pi)$.

This proof, originally discovered by Dirichlet, involves a clever trick which leads to a formula for $\widetilde{D}_n$. Indeed, applying a sum angle formula (see Appendix B) and telescoping, we have

$$2\sin(\frac{x}{2})\widetilde{D}_n(x) = \sum_{k=1}^{n} 2\sin(\frac{x}{2})\sin(kx)$$

$$= \sum_{k=1}^{n} \left(\cos((k - \frac{1}{2})x) - \cos((k + \frac{1}{2})x) \right)$$

$$= \cos(\frac{x}{2}) - \cos((n + \frac{1}{2})x).$$

Therefore,

$$| \tilde{D}_n(x) | = \left| \frac{\cos(\frac{x}{2}) - \cos\left((n + \frac{1}{2})x\right)}{2\sin(\frac{x}{2})} \right| \leq \frac{1}{|\sin(\frac{x}{2})|}$$

for all $n \in \mathbf{N}$. ∎

The following special case of Dirichlet's Test is widely used.

COROLLARY 4.15 [ALTERNATING SERIES TEST]. *If $a_k \downarrow 0$ as $k \to \infty$, then*

$$\sum_{k=1}^{\infty} (-1)^k a_k$$

converges. In fact, if $s = \sum_{k=1}^{\infty}(-1)^k a_k$ and $s_n = \sum_{k=1}^{n} a_k$, then

$$0 < |s - s_n| < a_{n+1}$$

for all $n \in \mathbf{N}$.

PROOF. Since the partial sums of $\sum_{k=1}^{\infty}(-1)^k$ are bounded, $\sum_{k=1}^{\infty}(-1)^k a_k$ converges by Dirichlet's Test.

To estimate $s - s_n$, suppose first that n is even, say $n = 2m$. Then

$$0 \geq (-a_{2m+1} + a_{2m+2}) + (-a_{2m+3} + a_{2m+4}) + \cdots$$
$$= \sum_{k=2m+1}^{\infty} (-1)^k a_k = s - s_n$$
$$= -a_{2m+1} + (a_{2m+2} - a_{2m+3}) + (a_{2m+4} - a_{2m+5}) + \cdots$$
$$\geq -a_{2m+1},$$

i.e., $0 \geq s - s_n \geq -a_{n+1}$. A similar argument proves that $0 \leq s - s_n \leq a_{n+1}$ when n is odd. ∎

This result gives an estimate of the rate of approximation to an alternating series by its partial sums.

Example 7. For each $\alpha \in \mathbf{R}$, prove that the series $\sum_{k=1}^{\infty}(-1)^k k/(k^2 + \alpha)$ converges. If s_n represents its nth partial sum and s its value, find an n so large that s_n approximates s to an accuracy of 10^{-2}.

PROOF. Let $f(x) = x/(x^2 + \alpha)$ and note that $f(x) \to 0$ as $x \to \infty$. Since $f'(x) = (\alpha - x^2)/(x^2 + \alpha)^2$ is negative for $x > \sqrt{|\alpha|}$, it follows that $k/(k^2 + \alpha) \downarrow 0$ as $k \to \infty$. Hence, the given series converges by the Alternating Series Test.

By Corollary 4.15, s_n will estimate s to an accuracy of 10^{-2} if $f(n) < 10^{-2}$, i.e., if $n^2 - 100n + \alpha > 0$. When $\alpha > 50^2$, this last quadratic has no real roots; hence, the inequality is always satisfied and we may choose $n = 1$. When $\alpha \leq 50^2$, the quadratic has roots $50 \pm \sqrt{50^2 - \alpha}$. Hence, choose any n which satisfies $n > 50 + \sqrt{50^2 - \alpha}$. ∎

EXERCISES

1. Prove that each of the following series diverges.

   a) $\displaystyle\sum_{k=2}^{\infty} \frac{1}{k \log k (\log \log k)^p}$, $p \le 1$.   b) $\displaystyle\sum_{k=1}^{\infty} \frac{1}{\log^p(k+1)}$, $p > 0$.

   c) $\displaystyle\sum_{k=1}^{\infty} \left(\frac{\sqrt{k}-1}{\sqrt{k}} \right)^{\sqrt{k}}$.   d) $\displaystyle\sum_{k=1}^{\infty} \frac{k^2 + 2k + 3}{k^3 - 2k^2 + \sqrt{2}}$.   e) $\displaystyle\sum_{k=1}^{\infty} \frac{\sqrt[k]{k}}{k}$.

2. Prove that the following series converge. For each alternating series, let s_n represent its partial sums, s its value, and find an n so large that s_n approximates s to an accuracy of 10^{-2}.

   a) $\displaystyle\sum_{k=1}^{\infty} (-1)^k \left(\frac{\pi}{2} - \arctan k \right)$.   b) $\displaystyle\sum_{k=1}^{\infty} \frac{(-1)^k k^2}{2^k}$.   c) $\displaystyle\sum_{k=1}^{\infty} \frac{\log k}{k^p}$, $p > 1$.

   d) $\displaystyle\sum_{k=1}^{\infty} \frac{k^2 - \sqrt{k}}{k^4 + k^2 + 1}$.   e) $\displaystyle\sum_{k=1}^{\infty} \frac{(-1)^k}{k^2} \frac{2 \cdot 4 \cdots (2k)}{1 \cdot 3 \cdots (2k-1)}$.

3. a) Find all $p \ge 0$ such that the following series converges.

$$\sum_{k=1}^{\infty} \frac{1}{k \log^p(k+1)}.$$

   b) For each such p, prove that the partial sums of this series s_n and its value s satisfy
$$|s - s_n| \le \frac{n+p-1}{n(p-1)} \left(\frac{1}{\log^{p-1}(n)} \right)$$
   for all $n \ge 2$.

4. Suppose $a_k \to 0$. Prove that $\sum_{k=1}^{\infty} a_k$ converges if and only if the series $\sum_{k=1}^{\infty} (a_{2k} + a_{2k+1})$ converges.

5. Suppose $a_k \in [0,1)$ and $a_k \to 0$ as $k \to \infty$. Prove that $\sum_{k=1}^{\infty} \arcsin a_k$ converges if and only if $\sum_{k=1}^{\infty} a_k$ converges.

6. Prove that
$$\sum_{k=1}^{\infty} a_k \cos(kx)$$

converges for every $x \in (0, 2\pi)$ and every $a_k \downarrow 0$. What happens when $x = 0$?

7. Suppose $a_k \downarrow 0$ as $k \to \infty$. Prove that

$$\sum_{k=1}^{\infty} a_k \sin((2k+1)x)$$

converges for all $x \in \mathbf{R}$.

8. Suppose $a_k, b_k \geq 0$ for $k \in \mathbf{N}$.

a) If

$$\lim_{n \to \infty} \frac{a_n}{b_n} = 0$$

and $\sum_{k=1}^{\infty} b_k$ converges, prove $\sum_{k=1}^{\infty} a_k$ converges.

b) If

$$\lim_{n \to \infty} \frac{a_n}{b_n} = \infty$$

and $\sum_{k=1}^{\infty} b_k$ diverges, prove $\sum_{k=1}^{\infty} a_k$ diverges.

9. Show that under the hypotheses of Dirichlet's Test,

$$\sum_{k=1}^{\infty} a_k b_k = \sum_{k=1}^{\infty} s_k(b_k - b_{k+1}).$$

10. Suppose $\{a_k\}$ and $\{b_k\}$ are real sequences such that $a_k \to 0$ as $k \to \infty$,

$$\sum_{k=1}^{\infty} |a_{k+1} - a_k| < \infty, \quad \text{and} \quad \left| \sum_{k=1}^{n} b_k \right| \leq M \qquad n \in \mathbf{N}.$$

Prove that $\sum_{k=1}^{\infty} a_k b_k$ converges.

11. Suppose $a, b \in \mathbf{R}$ satisfy $b/a \in \mathbf{R} \setminus \mathbf{Z}$. Find all $q > 0$ such that

$$\sum_{k=1}^{\infty} \frac{1}{(ak+b)q^k}$$

converges.

12. Suppose $\sum_{k=1}^{\infty} a_k$ converges. Prove that if $b_k \uparrow \infty$ and $\sum_{k=1}^{\infty} a_k b_k$ converges, then

$$b_m \sum_{k=m}^{\infty} a_k \to 0$$

as $m \to \infty$.

(Note: The exercise is trivial if $a_k \geq 0$ for large k, but this is not an assumption and has nothing to do with the proof.)

4.3 ABSOLUTE CONVERGENCE

In this section we investigate what happens to a convergent series when its terms are replaced by their absolute values. We begin by defining the following concepts.

DEFINITION 4.2. Let $S = \sum_{k=1}^{\infty} a_k$ be an infinite series.

   i) S is said to *converge absolutely* if $\sum_{k=1}^{\infty} |a_k| < \infty$.

   ii) S is said to *converge conditionally* if S converges but not absolutely.

The Cauchy Criterion gives us the following test for absolute convergence.

Remark 1. *A series $\sum_{k=1}^{\infty} a_k$ converges absolutely if and only if given $\varepsilon > 0$ there is an $N \in \mathbf{N}$ such that*

$$(10) \qquad m > n \geq N \quad implies \quad \sum_{k=n}^{m} |a_k| < \varepsilon.$$

As was the case for improper integrals, absolute convergence is stronger than convergence.

Remark 2. *If $\sum_{k=1}^{\infty} a_k$ converges absolutely, then $\sum_{k=1}^{\infty} a_k$ converges, but not conversely. In particular, there exist conditionally convergent series.*

PROOF. Given $\varepsilon > 0$, choose $N \in \mathbf{N}$ so that (10) holds. Then

$$\left| \sum_{k=n}^{m} a_k \right| \leq \sum_{k=n}^{m} |a_k| < \varepsilon$$

for $m > n \geq N$. Hence, by the Cauchy Criterion, $\sum_{k=1}^{\infty} a_k$ converges. On the other hand, $\sum_{k=1}^{\infty} (-1)^k / k$ converges by the Alternating Series Test, but does not converge absolutely since the harmonic series diverges. ∎

Absolutely convergent series behave much better than conditionally convergent series (see Theorems 4.16 through 4.18 below). To facilitate our discussion, recall (see Exercise 1 in Section 1.1) that the *positive and negative parts* of an $a \in \mathbf{R}$ are defined by

$$a^+ := \frac{|a| + a}{2} = \begin{cases} a & a \geq 0 \\ 0 & a < 0 \end{cases}$$

and

$$a^- := \frac{|a| - a}{2} = \begin{cases} 0 & a \geq 0 \\ -a & a < 0. \end{cases}$$

Notice that

$$(11) \qquad a^+ \geq 0, \quad a^- \geq 0,$$

and

$$(12) \qquad a = a^+ - a^-, \qquad |a| = a^+ + a^-$$

for all $a \in \mathbf{R}$.

THEOREM 4.16. i) *If $\sum_{k=1}^{\infty} a_k$ converges absolutely, then so do $\sum_{k=1}^{\infty} a_k^+$ and $\sum_{k=1}^{\infty} a_k^-$. In fact,*

$$\sum_{k=1}^{\infty} |a_k| = \sum_{k=1}^{\infty} a_k^+ + \sum_{k=1}^{\infty} a_k^- \quad \text{and} \quad \sum_{k=1}^{\infty} a_k = \sum_{k=1}^{\infty} a_k^+ - \sum_{k=1}^{\infty} a_k^-.$$

ii) *If $\sum_{k=1}^{\infty} a_k$ converges conditionally, then*

$$\sum_{k=1}^{\infty} a_k^+ = \sum_{k=1}^{\infty} a_k^- = \infty.$$

PROOF. By definition, $a_k^+ = (|a_k| + a_k)/2$. Since both $\sum_{k=1}^{\infty} |a_k|$ and $\sum_{k=1}^{\infty} a_k$ converge, it follows from Theorem 4.6 that

$$\sum_{k=1}^{\infty} a_k^+ = \frac{1}{2} \sum_{k=1}^{\infty} |a_k| + \frac{1}{2} \sum_{k=1}^{\infty} a_k$$

converges. Similarly,

$$\sum_{k=1}^{\infty} a_k^- = \frac{1}{2} \sum_{k=1}^{\infty} |a_k| - \frac{1}{2} \sum_{k=1}^{\infty} a_k$$

converges. This proves part i).

Suppose part ii) is false. By symmetry we may suppose that $\sum_{k=1}^{\infty} a_k^+$ converges. Since $\sum_{k=1}^{\infty} a_k$ converges, it follows from (12) that

$$\sum_{k=1}^{\infty} a_k^- = \sum_{k=1}^{\infty} a_k^+ - \sum_{k=1}^{\infty} a_k$$

converges. Thus,

$$\sum_{k=1}^{\infty} |a_k| = \sum_{k=1}^{\infty} a_k^+ + \sum_{k=1}^{\infty} a_k^-$$

converges, a contradiction. ∎

We have "inserted parentheses" (i.e., grouped terms together) to aid evaluation of some series (e.g., to evaluate some telescopic series and to estimate the rate of approximation by the partial sums of an alternating series). This is valid for convergent series (absolutely or conditionally) because if the sequence of partial sums s_n converges to s, then any subsequence s_{n_k} also converges to s. The situation is more complicated when we start changing the order of the terms (compare Theorem 4.17 with Theorem 4.18).

DEFINITION 4.3. A series $\sum_{j=1}^{\infty} b_j$ is called a *rearrangement* of a series $\sum_{k=1}^{\infty} a_k$ if there is a 1–1 function f from $\mathbf{N}$ onto $\mathbf{N}$ such that

$$b_{f(k)} = a_k, \qquad k \in \mathbf{N}.$$

THEOREM 4.17. *If $\sum_{k=1}^{\infty} a_k$ converges absolutely and $\sum_{j=1}^{\infty} b_j$ is any rearrangement of $\sum_{k=1}^{\infty} a_k$, then $\sum_{j=1}^{\infty} b_j$ converges and*

$$\sum_{k=1}^{\infty} a_k = \sum_{j=1}^{\infty} b_j.$$

PROOF. Let $\varepsilon > 0$. Set $s_n = \sum_{k=1}^{n} a_k$, $s = \sum_{k=1}^{\infty} a_k$, and $t_m = \sum_{j=1}^{m} b_j$, $n, m \in \mathbf{N}$. Since $\sum_{k=1}^{\infty} a_k$ converges absolutely, we can choose $N \in \mathbf{N}$ (see Corollary 4.5) such that

$$(13) \qquad \sum_{k=N+1}^{\infty} |a_k| < \frac{\varepsilon}{2}.$$

Thus

$$(14) \qquad |s_N - s| = \left| \sum_{k=N+1}^{\infty} a_k \right| \le \sum_{k=N+1}^{\infty} |a_k| < \frac{\varepsilon}{2}.$$

Let f be a 1–1 function from $\mathbf{N}$ onto $\mathbf{N}$ which satisfies

$$b_{f(k)} = a_k, \qquad k \in \mathbf{N}$$

and set $M = \max\{f(1), \dots, f(N)\}$. Notice that

$$\{a_1, \dots, a_N\} \subseteq \{b_1, \dots, b_M\}.$$

Let $m \ge M$. Then $t_m - s_N$ contains only a_k's whose indices satisfy $k > N$. Thus, it follows from (13) that

$$|t_m - s_N| \le \sum_{k=N+1}^{\infty} |a_k| < \frac{\varepsilon}{2}.$$

Hence, by (14),

$$|t_m - s| \le |t_m - s_N| + |s_N - s| < \frac{\varepsilon}{2} + \frac{\varepsilon}{2} = \varepsilon$$

for $m \ge M$. Therefore,

$$s = \sum_{j=1}^{\infty} b_j. \quad \blacksquare$$

The following result shows that Theorem 4.17 is false for conditionally convergent series.

*THEOREM 4.18. *Let $\sum_{k=1}^{\infty} a_k$ be conditionally convergent and let $x \leq y$ be any pair of extended real numbers. Then there is a rearrangement $\sum_{j=1}^{\infty} b_j$ of $\sum_{k=1}^{\infty} a_k$ whose partial sums s_n satisfy*

(15) $$\liminf_{n\to\infty} s_n = x \quad \text{and} \quad \limsup_{n\to\infty} s_n = y.$$

In particular, given any real number $c \in \mathbf{R}$, there is a rearrangement of $\sum_{k=1}^{\infty} a_k$ which converges to c.

PROOF. Since each a_k^+ and $-a_k^-$ is either a_k or 0, it suffices to show there are integers $0 < k_1 < r_1 < k_2 < r_2 < \dots$ such that if $b_1 = a_1^+$, $b_2 = a_2^+$, ..., $b_{k_1} = a_{k_1}^+$, $b_{k_1+1} = -a_1^-$, ..., $b_{r_1} = -a_{r_1-k_1}^-$, $b_{r_1+1} = a_{k_1+1}^+$, ..., and $s_n = \sum_{j=1}^{n} b_j$, then (15) holds. We suppose for simplicity that x and y are both finite.

STRATEGY: The idea behind the proof is simple. Since $\sum_{k=1}^{\infty} a_k^+ = \sum_{k=1}^{\infty} a_k^- = \infty$ by Theorem 4.16, add enough a_k^+'s until $s_n > y$, then subtract enough a_k^-'s until $s_n < x$ and continue adding and subtracting in such a way that some s_n's get successively closer to y and others to x. We now make this precise.

Since $\sum_{k=1}^{\infty} a_k^+ = \infty$, choose an integer $k_1 \in \mathbf{N}$ least such that

$$s_{k_1} := b_1 + b_2 + \cdots + b_{k_1} := a_1^+ + a_2^+ + \cdots + a_{k_1}^+ > y.$$

Since k_1 is least, $s_{k_1-1} \leq y$, hence $s_{k_1} \leq y + b_{k_1}$. Similarly, since $\sum_{k=1}^{\infty} a_k^- = \infty$ we can choose an integer $r_1 > k_1$ least such that

$$s_{r_1} := b_1 + b_2 + \cdots + b_{r_1} := s_{k_1} - a_1^- - \cdots - a_{r_1-k_1}^- < x,$$

and $s_{r_1} \geq x + b_{r_1}$. Since the $-a_\ell^-$'s are nonpositive, it is clear that $s_\ell \leq s_{k_1} \leq y + b_{k_1}$ for $k_1 < \ell \leq r_1$,. Therefore,

$$s_{k_1} > y \quad \text{and} \quad x + b_{r_1} \leq s_\ell \leq y + b_{k_1}$$

for all $k_1 \leq \ell \leq r_1$. By a similar argument, if $k_2 > r_1$ is least such that $s_{k_2} > y$, then $s_{r_1} < x$ and $x + b_{r_1} \leq s_\ell \leq y + b_{k_2}$ for all $r_1 \leq \ell \leq k_2$. In particular,

$$y < \sup_{k_1 \leq \ell \leq k_2} s_\ell \leq y + \max\{b_{k_1}, b_{k_2}\} \leq y + \sup_{\ell \geq k_1} b_\ell.$$

In the same way, if $r_2 > k_2$ is least such that $s_{r_2} < x$, then

$$x + \inf_{\ell \geq r_1} b_\ell \leq \sup_{r_1 \leq \ell \leq r_2} s_\ell < x.$$

Continuing this process, we generate integers $k_1 < r_1 < k_2 < r_2 < \dots$ such that for each $j \in \mathbf{N}$,

$$y < \sup_{k_j \leq \ell \leq k_{j+1}} s_\ell \leq y + \sup_{\ell \geq k_j} b_\ell \quad \text{and} \quad x + \inf_{\ell \geq r_j} b_\ell \leq \inf_{r_j \leq \ell \leq r_{j+1}} s_\ell < x.$$

The first of these inequalities implies

$$y < \sup_{\ell \geq k_j} s_\ell \leq y + \sup_{\ell \geq k_j} b_\ell.$$

Taking the limit of this inequality as $j \to \infty$, bearing in mind that by the Divergence Test $b_n \to 0$ as $n \to \infty$, we conclude that

$$y \leq \limsup_{n \to \infty} s_n \leq y + \limsup_{n \to \infty} b_n = y.$$

This proves the second half of (15) holds. A similar argument establishes the first half of (15). ∎

Thus, it is important to have tests which identify absolutely convergent series. Since every result about series with nonnegative terms can be applied to the series $\sum_{k=1}^{\infty} |a_k|$, we have already identified five such tests in the previous section. The next four results provide additional tests for absolute convergence.

THEOREM 4.19 [ROOT TEST]. *Let* $a_k \in \mathbf{R}$ *and set* $r = \limsup_{k \to \infty} \sqrt[k]{|a_k|}$.

   i) *If* $r < 1$, *then* $\sum_{k=1}^{\infty} a_k$ *converges absolutely.*
   ii) *If* $1 < r \leq \infty$, *then* $\sum_{k=1}^{\infty} a_k$ *diverges.*

PROOF. Suppose $r < 1$. Let $r < x < 1$ and notice that the geometric series $\sum_{k=1}^{\infty} x^k$ converges. By hypothesis (see Exercise 3a in Section 1.6),

$$\sqrt[k]{|a_k|} < x$$

for large k. Hence, $|a_k| < x^k$ for large k and it follows from the Comparison Test that $\sum_{k=1}^{\infty} |a_k|$ converges. This proves part i).
   Suppose $r > 1$. Then by Exercise 3b in Section 1.6,

$$\sqrt[k]{|a_k|} > 1$$

for infinitely many $k \in \mathbf{N}$. Hence, $|a_k| > 1$ for infinitely many k and it follows from the Divergence Test that $\sum_{k=1}^{\infty} a_k$ diverges. ∎

The following test is weaker than the Root Test (see Exercise 11) but is easier to use when the terms of $\sum_{k=1}^{\infty} a_k$ are made up of products (e.g., factorials).

THEOREM 4.20 [RATIO TEST]. *Let* $a_k \in \mathbf{R}$ *with* $a_k \neq 0$ *for* $k \in \mathbf{N}$. *If*

$$r = \lim_{k \to \infty} \frac{|a_{k+1}|}{|a_k|}$$

exists, then $\sum_{k=1}^{\infty} a_k$ *converges absolutely when* $r < 1$ *and diverges when* $r > 1$.

PROOF. If $r > 1$, then $|a_{k+1}| \geq |a_k|$ for k large and thus a_k cannot converge to zero. Hence, by the Divergence Test, $\sum_{k=1}^{\infty} a_k$ diverges.
   If $r < 1$, then observe for any $x \in (r, 1)$ that

$$\frac{|a_{k+1}|}{|a_k|} < x = \frac{x^{k+1}}{x^k}$$

for k large. Hence, the sequence $|a_k|/x^k$ is decreasing for large k and thus bounded. In particular, there is an $M > 0$ such that $|a_k| \leq Mx^k$ for all $k \in \mathbf{N}$. Since $x < 1$, it follows from the Comparison Test that $\sum_{k=1}^{\infty} |a_k|$ converges. ∎

Remark 3. *The Root and Ratio Tests are inconclusive when $r = 1$.*

For example, under the Ratio Test $\sum_{k=1}^{\infty} 1/k$ and $\sum_{k=1}^{\infty} 1/k^2$ both yield $r = 1$. Nevertheless, the first series diverges whereas the second converges absolutely.

The next result shows that the proofs of the Root and Ratio Tests actually contain information about how accurately the partial sums of certain series approximate their values.

Remark 4. *Suppose $\sum_{k=1}^{\infty} a_k$ converges absolutely and s is the value of $\sum_{k=1}^{\infty} |a_k|$.*

  i) *If there exist numbers $x \in (0, 1)$ and $N \in \mathbf{N}$ such that*

$$\sqrt[k]{|a_k|} \leq x$$

for all $k > N$, then

$$0 \leq s - \sum_{k=1}^{n} |a_k| \leq \frac{x^{n+1}}{1 - x}$$

for all $n \geq N$.

  ii) *If there exist numbers $x \in (0, 1)$ and $N \in \mathbf{N}$ such that*

$$\frac{|a_{k+1}|}{|a_k|} \leq x$$

for $k > N$, then

$$0 \leq s - \sum_{k=1}^{n} |a_k| \leq \frac{|a_N| x^{n-N+1}}{1 - x}$$

for all $n \geq N$.

PROOF. Let $n \geq N$. Since $|a_k| \leq x^k$ for $k > N$, we have, by summing a geometric series, that

$$0 \leq s - \sum_{k=1}^{n} |a_k| = \sum_{k=n+1}^{\infty} |a_k| \leq \sum_{k=n+1}^{\infty} x^k = \frac{x^{n+1}}{1 - x}$$

for all $n \geq N$. This proves part i). The proof of part ii) is left as an exercise. ∎

Example 1. Prove $\sum_{k=1}^{\infty} k^{2k}/(3k^2 + k)^k$ converges absolutely. If s_n represent its nth partial sum and s represents its value, find an n so large that s_n approximates s to an accuracy of 10^{-2}.

SOLUTION. Since

$$\left(\frac{k^{2k}}{(3k^2 + k)^k} \right)^{1/k} = \frac{k^2}{3k^2 + k} \leq \frac{1}{3}$$

for all $k \geq N := 1$, the series converges absolutely by the Root Test. Since $(1/3)^{n+1}/(1 - 1/3) \leq 10^{-2}$ for $n \geq 4$, we conclude by Remark 4i that it takes at most four terms to approximate the value of this series to an accuracy of 10^{-2}. ∎

The following tests can be used in some cases where the Root and Ratio Tests yield $r = 1$ (see Exercise 7).

THEOREM 4.21 [THE LOGARITHM TEST]. *Suppose* $a_k \neq 0$ *for large* k *and*

$$p = \lim_{k \to \infty} \frac{\log(1/|a_k|)}{\log k}$$

exists as an extended real number. If $p > 1$, *then* $\sum_{k=1}^{\infty} a_k$ *converges absolutely. If* $p < 1$, *then* $\sum_{k=1}^{\infty} a_k$ *does not converge absolutely.*

PROOF. If $p > 1$, then $\log(1/|a_k|)$ is eventually positive, i.e., $|a_k| < 1$ for large k. Let $p > q > 1$ and choose $N \in \mathbf{N}$ so that $k \geq N$ implies $|a_k| < 1$ and $\log(1/|a_k|) > q \log k = \log(k^q)$. It follows that $|a_k| < k^{-q}$ for $k \geq N$. Hence, by the Comparison Test, $\sum_{k=1}^{\infty} |a_k|$ converges.

If $0 \leq p < 1$, then a similar argument shows that $|a_k| > 1/k$ for large k. Hence, by the Comparison Test, $\sum_{k=1}^{\infty} |a_k|$ diverges.

Finally, if $p < 0$, then $\log(1/|a_k|) < 0$, i.e., $|a_k| > 1$ for large k. We conclude by the Divergence Test that $\sum_{k=1}^{\infty} |a_k|$ diverges. ∎

*THEOREM 4.22** [RAABE'S TEST]. *Suppose there is a constant* C *and a parameter* p *such that*

(16)
$$\left| \frac{a_{k+1}}{a_k} \right| \leq 1 - \frac{p}{k+C}$$

for large k. *If* $p > 1$, *then* $\sum_{k=1}^{\infty} a_k$ *converges absolutely.*

PROOF. Set $x_k = k + C - 1$ for $k \in \mathbf{N}$ and choose $N \in \mathbf{N}$ such that $x_k > 1$ and (16) hold for $k \geq N$. By the p–Series Test and the Limit Comparison Test,

(17)
$$\sum_{k=N}^{\infty} x_k^{-p} < \infty.$$

By (16) and Bernoulli's Inequality,

$$\left| \frac{a_{k+1}}{a_k} \right| \leq 1 - \frac{p}{x_{k+1}} \leq \left(1 - \frac{1}{x_{k+1}} \right)^p = \frac{x_k^p}{x_{k+1}^p}.$$

Hence, the sequence $\{|a_k| x_k^p\}_{k=N}^{\infty}$, is decreasing and bounded above. In particular, there is an $M > 0$ such that $|a_k| \leq M x_k^{-p}$ for $k \geq N$. We conclude by (17) that $\sum_{k=1}^{\infty} a_k$ converges. ∎

EXERCISES

1. For each of the following series, let s_n represent its partial sums, s represent its value. Prove that s is finite and find an n so large that s_n approximates s to an accuracy of 10^{-2}.

  a) $\sum_{k=1}^{\infty} \frac{1}{k!}$.   b) $\sum_{k=1}^{\infty} \frac{1}{k^k}$.   c) $\sum_{k=1}^{\infty} \frac{2^k}{k!}$.   d) $\sum_{k=1}^{\infty} \left(\frac{k}{k+1} \right)^{k^2}$.

2. Using any test covered in this chapter, find out which of the following series converge absolutely, which converge conditionally, and which diverge.

a) $\displaystyle\sum_{k=1}^{\infty} \frac{(-1)^k k^3}{(k+1)!}$. b) $\displaystyle\sum_{k=1}^{\infty} \frac{(-1)(-3)\dots(1-2k)}{1\cdot 4\dots(3k-2)}$. c) $\displaystyle\sum_{k=1}^{\infty} \frac{(k+1)^k}{p^k k!}$, $p > e$.

d) $\displaystyle\sum_{k=1}^{\infty} \frac{(-1)^{k+1}\sqrt{k}}{k+1}$. e) $\displaystyle\sum_{k=1}^{\infty} \frac{(-1)^k \sqrt{k+1}}{\sqrt{k}\,k^k}$. f) $\displaystyle\sum_{k=1}^{\infty} \left(\frac{\sqrt{k}-1}{\sqrt{k}}\right)^k$.

3. For each of the following, find all values $x \in \mathbf{R}$ for which the given series converges.

$$\text{a) } \sum_{k=1}^{\infty} \frac{x^k}{k}. \qquad \text{b) } \sum_{k=1}^{\infty} \frac{x^{3k}}{2^k}.$$

$$* \text{ c) } \sum_{k=1}^{\infty} \frac{3\cdot 5\dots(2k+1)x^k}{2^k(k+2)!}. \qquad \text{d) } \sum_{k=1}^{\infty} \frac{(x+2)^k}{k\sqrt{k+1}}.$$

4. For each of the following, find all values of $p \in \mathbf{R}$ for which the given series converges absolutely.

a) $\displaystyle\sum_{k=2}^{\infty} \frac{1}{k \log^p k}$. b) $\displaystyle\sum_{k=2}^{\infty} \frac{1}{(\log k)^{p\log k}}$. c) $\displaystyle\sum_{k=2}^{\infty} \frac{1}{\log^p k}$.

d) $\displaystyle\sum_{k=1}^{\infty}(\sqrt{k^{2p}+1}-k^p)$. e) $\displaystyle\sum_{k=1}^{\infty} \frac{k^p}{p^k}$.

5. Using Exercise 9 in Section 3.3, prove that

$$\sin x = \sum_{k=0}^{\infty} \frac{(-1)^k x^{2k+1}}{(2k+1)!} \quad \text{and} \quad \cos x = \sum_{k=0}^{\infty} \frac{(-1)^k x^{2k}}{(2k)!}$$

for all $x \in \mathbf{R}$.

6. Prove Remark 4ii.

7. a) Prove that the Root Test applied to the series

$$\sum_{k=2}^{\infty} \frac{1}{(\log k)^{\log k}}$$

yields $r = 1$. Use the Logarithm Test to prove this series converges.

*b) Prove that the Ratio Test applied to the series

$$\sum_{k=1}^{\infty} \frac{1\cdot 3\cdots(2k-1)}{4\cdot 6\cdots(2k+2)}$$

yields $r = 1$. Use Raabe's Test to prove this series converges.

*8. Suppose that $\{a_k\}$ is a sequence of nonzero real numbers and

$$p = \lim_{k \to \infty} k \left(1 - \left| \frac{a_{k+1}}{a_k} \right| \right)$$

exists as an extended real number. Prove that $\sum_{k=1}^{\infty} a_k$ converges absolutely when $p > 1$.

9. Suppose $a_{kj} \geq 0$ for $k, j \in \mathbf{N}$. Set

$$A_k = \sum_{j=1}^{\infty} a_{kj}$$

for each $k \in \mathbf{N}$, and suppose $\sum_{k=1}^{\infty} A_k$ converges.

a) Prove that

$$\sum_{j=1}^{\infty} \left(\sum_{k=1}^{\infty} a_{kj} \right) \leq \sum_{k=1}^{\infty} \left(\sum_{j=1}^{\infty} a_{kj} \right).$$

b) Show

$$\sum_{j=1}^{\infty} \left(\sum_{k=1}^{\infty} a_{kj} \right) = \sum_{k=1}^{\infty} \left(\sum_{j=1}^{\infty} a_{kj} \right).$$

c) Prove that b) may not hold if a_{kj} has both positive and negative values.
   Hint: Consider

$$a_{kj} = \begin{cases} 1 & j = k \\ -1 & j = k+1 \\ 0 & \text{otherwise.} \end{cases}$$

10. a) Suppose $\sum_{k=1}^{\infty} a_k$ converges absolutely. Prove that $\sum_{k=1}^{\infty} |a_k|^p$ converges for all $p \geq 1$.

   b) Suppose $\sum_{k=1}^{\infty} a_k$ converges conditionally. Prove that $\sum_{k=1}^{\infty} k^p a_k$ diverges for all $p > 1$.

11. a) Let $a_n > 0$ for $n \in \mathbf{N}$. Set $b_0 = b_1 = 0$, $b_2 = \log(a_2/a_1)$, and

$$b_k = \log\left(\frac{a_k}{a_{k-1}}\right) - \log\left(\frac{a_{k-1}}{a_{k-2}}\right) \qquad k = 3, 4, \ldots$$

Prove that if

$$r = \lim_{n \to \infty} \frac{a_{n+1}}{a_n}$$

exists and is positive, then

$$\lim_{n \to \infty} \log(a_n^{1/n}) = \lim_{n \to \infty} \sum_{k=0}^{n} \left(1 - \frac{k}{n} \right) b_k = \sum_{k=0}^{\infty} b_k = \log r.$$

   b) Prove that if $a_n \in \mathbf{R} \setminus \{0\}$ and $|a_{n+1}/a_n| \to r$ as $n \to \infty$, for some $r > 0$, then $|a_n|^{1/n} \to r$ as $n \to \infty$.

4.4 UNIFORM CONVERGENCE OF SEQUENCES

You are familiar with what it means for a sequence of numbers to converge. In this section we examine what it means for a sequence of functions to converge. It turns out there are several different ways to define *convergence* of a sequence of functions. We begin with the simplest way.

DEFINITION 4.4. Let E be a nonempty subset of $\mathbf{R}$. A sequence of functions $f_n : E \to \mathbf{R}$ is said to *converge pointwise* on E (notation: $f_n \to f$ pointwise on E as $n \to \infty$) if $f(x) = \lim_{n \to \infty} f_n(x)$ exists for each $x \in E$.

Because $\{f_n\}$ converges pointwise on a set E if and only if the sequence of real numbers $\{f_n(x)\}$ converges for each $x \in E$, every result about convergence of real numbers contains a result about pointwise convergence of functions. Here is a typical example.

Remark 1. *Let E be a nonempty subset of $\mathbf{R}$ and $f_n : E \to \mathbf{R}$ be a sequence of functions. Then $f_n \to f$ pointwise on E if and only if given $\varepsilon > 0$ and $x \in E$ there is an $N \in \mathbf{N}$* (which may depend on x as well as ε) *such that*

$$n \geq N \quad implies \quad |f_n(x) - f(x)| < \varepsilon.$$

PROOF. By Definition 4.4, $f_n \to f$ pointwise on E if and only if $f_n(x) \to f(x)$ for all $x \in E$. This occurs, by Definition 1.5, if and only if given $\varepsilon > 0$ and $x \in E$ there is an $N \in \mathbf{N}$ such that $n \geq N$ implies $|f_n(x) - f(x)| < \varepsilon$. ∎

Given $f_n \to f$ pointwise on $[a, b]$, it is natural to ask: What does f inherit from f_n? The next four remarks show that, in general, the answer to this question is not much.

Remark 2. *The pointwise limit of continuous (respectively, differentiable) functions is not necessarily continuous (respectively, differentiable).*

PROOF. Let $f_n(x) = x^n$ and

$$f(x) = \begin{cases} 0 & 0 \leq x < 1 \\ 1 & x = 1. \end{cases}$$

Then $f_n \to f$ pointwise on $[0, 1]$ (see Figure 4.4 below), each f_n is continuous and differentiable on $[0, 1]$, but $f(x)$ is neither differentiable nor continuous at $x = 1$. ∎

Remark 3. *The pointwise limit of integrable functions is not necessarily integrable.*

PROOF. Set

$$f_n(x) = \begin{cases} 1 & x = p/n \in \mathbf{Q} \text{ written in reduced form} \\ 0 & \text{otherwise,} \end{cases}$$

for $n \in \mathbf{N}$ and

$$f(x) = \begin{cases} 1 & x \in \mathbf{Q} \\ 0 & \text{otherwise.} \end{cases}$$

Then $f_n \to f$ pointwise on $[0, 1]$, each f_n is integrable on $[0, 1]$ (with integral zero), but f is not integrable on $[0, 1]$ (see Example 2 in Section 3.1). ∎

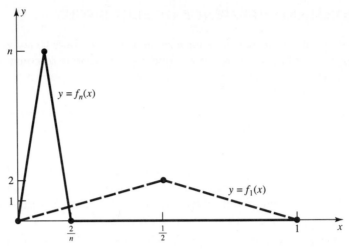

Figure 4.2

Remark 4. *There exist differentiable functions f_n and f such that $f_n \to f$ pointwise on $[0, 1]$ but*

$$(18) \qquad \lim_{n\to\infty} f_n'(x) \neq \left(\lim_{n\to\infty} f_n(x) \right)'$$

for $x = 1$.

PROOF. Let $f_n(x) = x^n/n$ and $f(x) = 0$. Then $f_n \to f$ pointwise on $[0, 1]$, each f_n is differentiable with $f_n'(x) = x^{n-1}$. Thus, the left side of (18) is 1 at $x = 1$ but the right side of (18) is zero. ∎

Remark 5. *There exist continuous functions f_n and f such that $f_n \to f$ pointwise on $[0, 1]$ but*

$$(19) \qquad \lim_{n\to\infty} \int_0^1 f_n(x)\, dx \neq \int_0^1 \left(\lim_{n\to\infty} f_n(x) \right) dx.$$

PROOF. Let $f_1(x) = 1$ and

$$f_n(x) = \begin{cases} n^2 x & 0 < x < 1/n \\ 2n - n^2 x & 1/n \leq x < 2/n \\ 0 & 2/n \leq x \leq 1 \end{cases}$$

for $n = 2, 3, \ldots$ (see Figure 4.2). Then $f_n \to 0$ pointwise on $[0, 1]$ and, since the area of a triangle is one-half base times altitude, $\int_0^1 f_n(x)\, dx = 1$ for each $n \in \mathbf{N}$. Thus, the left side of (19) is 1 but the right side is zero. ∎

In view of the preceding examples, it is clear that pointwise convergence is of limited value for the calculus of limits of sequences. It turns out that the following concept, discovered independently by Stokes, Cauchy, and Weierstrass around 1850, is much more useful in this context.

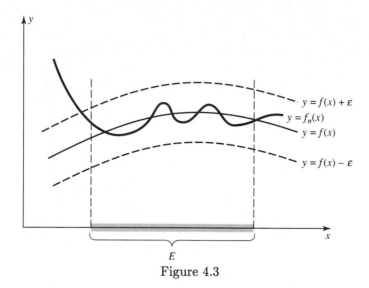

$$y = f(x) + \varepsilon$$
$$y = f_n(x)$$
$$y = f(x)$$
$$y = f(x) - \varepsilon$$

E

Figure 4.3

DEFINITION 4.5. Let E be a nonempty subset of $\mathbf{R}$. A sequence of functions $f_n : E \to \mathbf{R}$ is said to *converge uniformly* on E to a function f (notation: $f_n \to f$ uniformly on E as $n \to \infty$) if given $\varepsilon > 0$ there is an $N \in \mathbf{N}$ such that

$$n \geq N \quad \text{implies} \quad |f_n(x) - f(x)| < \varepsilon$$

for all $x \in E$.

Comparing Definition 4.5 with Remark 1 above, we see that the only difference between uniform convergence and pointwise convergence is that for uniform convergence, the integer N must be chosen independently of x (see Figure 4.3). Notice how this is similar to the difference between uniform continuity and continuity (see the discussion following Example 2 in Section 2.3).

By definition, if f_n converges uniformly on E, then f_n converges pointwise on E. The following example shows that the converse of this statement is false.

Example 1. Prove that $x^n \to 0$ uniformly on $[0, b]$ for any $b < 1$, and pointwise, but not uniformly, on $[0, 1)$.

PROOF. We know that $x^n \to 0$ pointwise on $[0, 1)$. Let $b < 1$. Given $\varepsilon > 0$ choose $N \in \mathbf{N}$ such that $n \geq N$ implies $b^n < \varepsilon$. Then $x \in [0, b]$ and $n \geq N$ imply

$$|x^n| \leq b^n < \varepsilon,$$

i.e., $x^n \to 0$ uniformly for $x \in [0, b]$.

Suppose to the contrary that x^n converges to 0 uniformly on $[0, 1)$. Then given $0 < \varepsilon < 1/2$, there is an $N \in \mathbf{N}$ such that $|x^N| < \varepsilon$ for all $x \in [0, 1)$. On the other hand, since $x^N \to 1$ as $x \to 1-$, we can choose an $x_0 \in (0, 1)$ such that $x_0^N > \varepsilon$ (see Figure 4.4). Thus, $\varepsilon < x_0^N < \varepsilon$, a contradiction. ∎

The next several results show that if $f_n \to f$ or $f_n' \to f'$ uniformly, then f inherits much from f_n.

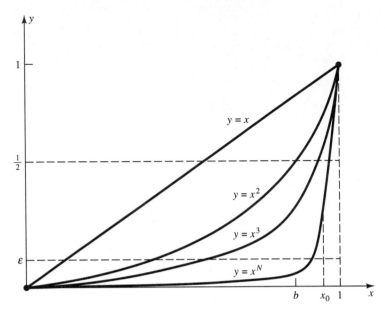

Figure 4.4

THEOREM 4.23. *Suppose $f_n \to f$ uniformly on an open interval (a, b) and $c \in (a, b)$. If each f_n is continuous at c, then f is continuous at c.*

PROOF. Let $\varepsilon > 0$ and choose $N \in \mathbf{N}$ such that

$$n \geq N \quad \text{and} \quad x \in (a, b) \quad \text{imply} \quad |f_n(x) - f(x)| < \frac{\varepsilon}{3}.$$

Since f_N is continuous at c, choose $\delta > 0$ such that

$$|x - c| < \delta \quad \text{and} \quad x \in (a, b) \quad \text{imply} \quad |f_N(x) - f_N(c)| < \frac{\varepsilon}{3}.$$

Suppose $|x - c| < \delta$ and $x \in (a, b)$. Then

$$|f(x) - f(c)| \leq |f(x) - f_N(x)| + |f_N(x) - f_N(c)| + |f_N(c) - f(c)| < \varepsilon.$$

Thus f is continuous at c. ∎

(For a generalization of this result see Exercise 6. For a converse of this result when the sequence f_n is monotone, see Theorem 5.32.)

Here is an important theorem about interchanging a limit sign and an integral sign.

THEOREM 4.24. *Suppose $f_n \to f$ uniformly on a closed interval $[a, b]$. If each f_n is integrable on $[a, b]$, then so is f and*

$$\lim_{n \to \infty} \int_a^b f_n(x) \, dx = \int_a^b \left(\lim_{n \to \infty} f_n(x) \right) \, dx.$$

In fact,

$$\lim_{n\to\infty} \int_a^x f_n(t)\, dt = \int_a^x \left(\lim_{n\to\infty} f_n(t) \right) dt$$

uniformly for $x \in [a, b]$.

PROOF. Let $\varepsilon > 0$ and choose $N \in \mathbf{N}$ such that

(20) $n \geq N$ implies $|f(x) - f_n(x)| < \dfrac{\varepsilon}{3(b - a)}$

for all $x \in [a, b]$. Using this inequality for $n = N$, we see that by the definition of upper and lower sums,

$$U(f - f_N, P) \leq \frac{\varepsilon}{3} \quad \text{and} \quad L(f - f_N, P) \geq -\frac{\varepsilon}{3}$$

for any partition P of $[a, b]$. Since f_N is integrable, choose a partition P such that

$$U(f_N, P) - L(f_N, P) < \frac{\varepsilon}{3}.$$

It follows that

$$U(f, P) - L(f, P) \leq U(f - f_N, P) + U(f_N, P) - L(f_N, P) - L(f_N - f, P)$$
$$< \frac{\varepsilon}{3} + \frac{\varepsilon}{3} + \frac{\varepsilon}{3} = \varepsilon,$$

i.e., f is integrable on $[a, b]$. We conclude by Theorem 3.8 and (20) that

$$\left| \int_a^x f_n(t)\, dt - \int_a^x f(t)\, dt \right| \leq \int_a^x |f_n(t) - f(t)|\, dt \leq \frac{\varepsilon(x - a)}{3(b - a)} < \varepsilon$$

for all $x \in [a, b]$ and $n \geq N$. ∎

Here is a Cauchy Criterion for uniform convergence.

Lemma [UNIFORM CAUCHY CRITERION]. *Let E be a nonempty subset of $\mathbf{R}$ and $f_n : E \to \mathbf{R}$ be a sequence of functions. Then f_n converges uniformly on E if and only if given $\varepsilon > 0$, there is an $N \in \mathbf{N}$ such that*

(21) $n, m \geq N$ implies $|f_n(x) - f_m(x)| < \varepsilon$

for all $x \in E$.

PROOF. Suppose first that $f_n \to f$ uniformly on E as $n \to \infty$. Let $\varepsilon > 0$ and choose $N \in \mathbf{N}$ such that

$$n \geq N \quad \text{implies} \quad |f_n(x) - f(x)| < \frac{\varepsilon}{2}$$

for $x \in E$. Since $|f_n(x) - f_m(x)| \leq |f_n(x) - f(x)| + |f(x) - f_m(x)|$, it is clear that (21) holds.

Conversely, if (21) holds, then $\{f_n(x)\}_{n \in \mathbf{N}}$ is Cauchy for each $x \in E$. Hence, by Cauchy's Theorem for real sequences (Theorem 1.20),

$$f(x) := \lim_{n \to \infty} f_n(x)$$

exists for each $x \in E$. Take the limit of the inequality in (21) as $m \to \infty$. We obtain $|f_n(x) - f(x)| \leq \varepsilon$ for all $n \geq N$ and $x \in E$. Hence, by definition, $f_n \to f$ uniformly on E. ∎

Here is a result about interchanging a limit sign and the derivative sign. (The proof presented here comes from Apostol [1].)

THEOREM 4.25. *Let (a, b) be an open interval and f_n be a sequence of functions which converges at some $x_0 \in (a, b)$. If each f_n is differentiable on (a, b), and f_n' converges uniformly on (a, b) as $n \to \infty$, then f_n converges uniformly on (a, b) and*

$$\lim_{n \to \infty} f_n'(x) = \left(\lim_{n \to \infty} f_n(x) \right)'$$

for each $x \in (a, b)$.

PROOF. Fix $c \in (a, b)$ and define

$$g_n(x) = \begin{cases} \dfrac{f_n(x) - f_n(c)}{x - c} & x \neq c \\ f_n'(c) & x = c \end{cases}$$

for $n \in \mathbf{N}$. Clearly,

(22) $$f_n(x) = f_n(c) + (x - c)g_n(x)$$

for $n \in \mathbf{N}$ and $x \in (a, b)$.

We claim that for any $c \in (a, b)$, the sequence g_n converges uniformly on (a, b). Let $\varepsilon > 0$, $n, m \in \mathbf{N}$, and $x \in (a, b)$ with $x \neq c$. By the Mean Value Theorem, there is a ξ between x and c such that

$$g_n(x) - g_m(x) = \frac{f_n(x) - f_m(x) - (f_n(c) - f_m(c))}{x - c} = f_n'(\xi) - f_m'(\xi).$$

Since f_n' converges uniformly on (a, b), it follows that there is an $N \in \mathbf{N}$ such that

$$n, m \geq N \quad \text{implies} \quad |g_n(x) - g_m(x)| < \varepsilon$$

for $x \in (a, b)$ with $x \neq c$. This implication also holds for $x = c$ because $g_n(c) = f_n'(c)$ for all $n \in \mathbf{N}$. This proves the claim.

To show f_n converges uniformly on (a, b), notice that by the claim, g_n converges uniformly as $n \to \infty$ and (22) holds for $c = x_0$. Since $f_n(x_0)$ converges as $n \to \infty$ by hypothesis, it follows from (22) that f_n converges uniformly on (a, b) as $n \to \infty$.

Fix $c \in (a, b)$. Define f, g on (a, b) by $f(x) := \lim_{n \to \infty} f_n(x)$ and $g(x) := \lim_{n \to \infty} g_n(x)$. We need to show

$$(23) \qquad\qquad f'(c) = \lim_{n \to \infty} f'_n(c).$$

Since each g_n is continuous at c, the claim implies g is continuous at c. Since $g_n(c) = f'_n(c)$, it follows that the right side of (23) can be written as

$$\lim_{n \to \infty} f'_n(c) = \lim_{n \to \infty} g_n(c) = g(c) = \lim_{x \to c} g(x).$$

On the other hand, if $x \neq c$ we have by definition that

$$\frac{f(x) - f(c)}{x - c} = \lim_{n \to \infty} \frac{f_n(x) - f_n(c)}{x - c} = \lim_{n \to \infty} g_n(x) = g(x).$$

Therefore, the left side of (23) reduces to

$$f'(c) = \lim_{x \to c} \frac{f(x) - f(c)}{x - c} = \lim_{x \to c} g(x).$$

This verifies (23), and the proof of the theorem is complete. ∎

EXERCISES

1. a) Prove $x/n \to 0$ uniformly, as $n \to \infty$, on any closed bounded interval $[a, b]$.

   b) Prove $1/(nx) \to 0$ pointwise but not uniformly on $(0, 1)$ as $n \to \infty$.

2. Find

$$\lim_{n \to \infty} \int_0^\pi \frac{\sin x}{\sqrt{nx^2 + a}} \, dx$$

   for any $a \geq 0$.

3. Suppose that $f_n \to f$ and $g_n \to g$, as $n \to \infty$, uniformly on some set $E \subseteq \mathbf{R}$.

   a) Prove that $f_n + g_n \to f + g$ and $\alpha f_n \to \alpha f$, as $n \to \infty$, uniformly on E for all $\alpha \in \mathbf{R}$.

   b) Prove that $f_n g_n \to fg$ pointwise on E.

   c) Prove that if f and g are bounded on E, then $f_n g_n \to fg$ uniformly on E.

   d) Show c) may be false when g is unbounded.

4. Let g be continuous on a closed bounded interval $[a, b]$ with $|g(x)| > 0$ for $x \in [a, b]$. Suppose $f_n \to f$ and $g_n \to g$ as $n \to \infty$, uniformly on $[a, b]$.

   a) Prove $1/g_n$ is defined for large n and $f_n/g_n \to f/g$ uniformly on $[a, b]$ as $n \to \infty$.

   b) Show that a) is false if $[a, b]$ is replaced by (a, b).

5. A sequence of functions f_n is said to be *uniformly bounded* on a set E if there is an $M > 0$ such that

$$|f_n(x)| \leq M$$

for all $x \in E$ and all $n \in \mathbf{N}$. Suppose each f_n is a bounded function on a set E and $f_n \to f$ uniformly on E. Prove $\{f_n\}$ is uniformly bounded on E and f is a bounded function on E.

$\boxed{6}$. **This exercise is used in Section 4.5.** Suppose $f_n \to f$ uniformly on E as $n \to \infty$.

a) Prove that if each f_n is continuous on E, then f is continuous on E.

b) Prove that if each f_n is uniformly continuous on E, then f is uniformly continuous on E.

7. Suppose $b > a > 0$. Prove that

$$\lim_{n \to \infty} \int_a^b \left(1 + \frac{x}{n}\right)^n e^{-x}\, dx = b - a.$$

8. Let $[a, b]$ be a closed bounded interval, $f : [a, b] \to \mathbf{R}$ be bounded, and $g : [a, b] \to \mathbf{R}$ be continuous with $g(a) = g(b) = 0$. Let f_n be a uniformly bounded sequence of functions on $[a, b]$ (see Exercise 5 above). Prove that if $f_n \to f$ uniformly on all closed interval $[c, d] \subset (a, b)$, then $f_n g \to fg$ uniformly on $[a, b]$.

9. Let f_n be integrable on $[0, 1]$ and $f_n \to f$ uniformly on $[0, 1]$. Show that if $b_n \uparrow 1$ as $n \to \infty$, then

$$\lim_{n \to \infty} \int_0^{b_n} f_n(x)\, dx = \int_0^1 f(x)\, dx.$$

10. Let E be a nonempty subset of $\mathbf{R}$ and f be a real-valued function defined on E. Suppose f_n is a sequence of bounded functions on E which converges to f uniformly on E. Prove that

$$\frac{f_1(x) + \cdots + f_n(x)}{n} \to f(x)$$

uniformly on E as $n \to \infty$ (compare with Exercise 7 in Section 4.1).

4.5 UNIFORM CONVERGENCE OF SERIES

In this section we extend the concepts introduced in Section 4.4 from sequences to series.

DEFINITION 4.6. Let f_k be a sequence of functions defined on some set E and set

$$s_n(x) := \sum_{k=1}^n f_k(x), \qquad x \in E, \ n \in \mathbf{N}.$$

i) The series $\sum_{k=1}^\infty f_k$ is said to *converge pointwise* on E if $s_n(x)$ converges pointwise on E as $n \to \infty$.

ii) The series $\sum_{k=1}^{\infty} f_k$ is said to *converge uniformly* on E if $s_n(x)$ converges uniformly on E as $n \to \infty$.

iii) The series $\sum_{k=1}^{\infty} f_k$ is said to *converge absolutely* on E if $\sum_{k=1}^{\infty} |f_k(x)|$ converges for each $x \in E$.

Since convergence of series is defined in terms of convergence of sequences of partial sums, every result about convergence of sequences of functions contains a result about convergence of series of functions. For example, the following result is an immediate consequence of Theorems 4.23 through 4.25.

THEOREM 4.26. *Let $\{f_k\}$ be a sequence of real-valued functions defined on an interval I.*

  i) *Suppose $I = (a, b)$ is open, $c \in I$, and each f_k is continuous at c. If $f = \sum_{k=1}^{\infty} f_k$ converges uniformly on (a, b), then f is continuous at c.*

  ii) *[TERM-BY-TERM INTEGRATION]. Suppose $I = [a, b]$ is closed and bounded, and each f_k is integrable on $[a, b]$. If $f = \sum_{k=1}^{\infty} f_k$ converges uniformly on $[a, b]$, then f is integrable on $[a, b]$ and*

$$\int_a^b \sum_{k=1}^{\infty} f_k(x)\,dx = \sum_{k=1}^{\infty} \int_a^b f_k(x)\,dx.$$

  iii) *[TERM-BY-TERM DIFFERENTIATION]. Suppose $I = (a, b)$ is open, and each f_k is differentiable on I. If $f = \sum_{k=1}^{\infty} f_k$ converges at some $x_0 \in (a, b)$, and $\sum_{k=1}^{\infty} f_k'$ converges uniformly on (a, b), then $\sum_{k=1}^{\infty} f_k$ converges uniformly to f, f is differentiable on (a, b), and*

$$\left(\sum_{k=1}^{\infty} f_k(x) \right)' = \sum_{k=1}^{\infty} f_k'(x)$$

for $x \in (a, b)$.

Here are two much used tests for uniform convergence of series.

THEOREM 4.27 [WEIERSTRASS M–TEST]. *Let f_k be defined on a set E and suppose $M_k \geq 0$ satisfies $\sum_{k=1}^{\infty} M_k < \infty$. If $|f_k(x)| \leq M_k$ for $k \in \mathbf{N}$ and $x \in E$, then $\sum_{k=1}^{\infty} f_k$ converges absolutely and uniformly on E.*

PROOF. Let $\varepsilon > 0$ and use the Cauchy Criterion to choose $N \in \mathbf{N}$ such that

$$m > n \geq N \quad \text{implies} \quad \sum_{k=n}^{m} M_k < \varepsilon.$$

Thus, by hypothesis,

$$\left| \sum_{k=n}^{m} f_k(x) \right| \leq \sum_{k=n}^{m} |f_k(x)| \leq \sum_{k=n}^{m} M_k < \varepsilon$$

for $m > n \geq N$ and $x \in E$. Hence, the partial sums of $\sum_{k=1}^{\infty} f_k$ are uniformly Cauchy and the partial sums of $\sum_{k=1}^{\infty} |f_k(x)|$ are Cauchy for each $x \in E$. ∎

THEOREM 4.28 [DIRICHLET'S TEST FOR UNIFORM CONVERGENCE]. *Let E be a nonempty subset of* $\mathbf{R}$ *and* $f_k, g_k : E \to \mathbf{R}$. *If*

$$\left| \sum_{k=1}^{n} f_k(x) \right| \leq M < \infty$$

for $n \in \mathbf{N}$ *and* $x \in E$, *and if* $g_k \downarrow 0$ *uniformly on* E *as* $k \to \infty$, *then* $\sum_{k=1}^{\infty} f_k g_k$ *converges uniformly on* E.

PROOF. Let

$$F_{n,m}(x) = \sum_{k=m}^{n} f_k(x), \qquad m, n \in \mathbf{N}, \ n \geq m, \ x \in E$$

and fix integers $n > m > 0$. By Abel's Formula and the hypothesis,

$$\left| \sum_{k=m}^{n} f_k(x) g_k(x) \right| = \left| F_{n,m}(x) g_n(x) + \sum_{k=m}^{n-1} F_{k,m}(x)(g_k(x) - g_{k+1}(x)) \right|$$

$$\leq 2M g_n(x) + 2M \sum_{k=m}^{n-1} (g_k(x) - g_{k+1}(x))$$

$$= 2M g_m(x)$$

for all $x \in E$. Since $g_m(x) \to 0$ uniformly on E, as $m \to \infty$, it follows from the uniform Cauchy Criterion that $\sum_{k=1}^{\infty} f_k(x) g_k(x)$ converges uniformly on E. ∎

Here is a typical application of Dirichlet's Test.

Example 1. Prove that if $a_k \downarrow 0$ as $k \to \infty$, then $\sum_{k=0}^{\infty} a_k \cos kx$ converges uniformly on any closed subinterval $[a, b]$ of $(0, 2\pi)$.

PROOF. Let $f_k(x) = \cos kx$ and $g_k(x) = a_k$ for $k \in \mathbf{N}$. By the technique used in Example 6 in Section 4.2, we can show that

$$D_n(x) := \sum_{k=0}^{n} \cos kx = \frac{\sin(\frac{x}{2}) + \sin\left(\left(n + \frac{1}{2}\right)x\right)}{2\sin(\frac{x}{2})}$$

for $n \in \mathbf{N}$ and $x \in (0, 2\pi)$. Hence the partial sums of $\sum_{k=0}^{\infty} f_k(x)$ satisfy

$$|D_n(x)| = \left| \frac{\sin(\frac{x}{2}) + \sin\left(\left(n + \frac{1}{2}\right)x\right)}{2\sin(\frac{x}{2})} \right| \leq \frac{1}{|\sin(\frac{x}{2})|}$$

for $x \in (0, 2\pi)$. If $\delta = \min\{2\pi - b, a\}$ and $x \in [a, b]$, then $\sin(x/2) \geq \sin(\delta/2)$ (see Figure 4.5). Therefore, $\sum_{k=1}^{\infty} a_k \cos kx$ converges uniformly on $[a, b]$ by Dirichlet's Test. ∎

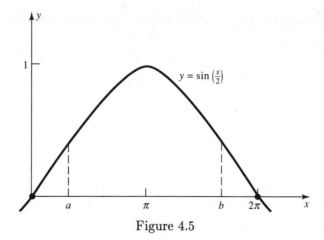

Figure 4.5

This example can be used to show that uniform convergence of a series is not sufficient for term-by-term differentiation. Indeed, although $\sum_{k=1}^{\infty} \cos kx/k$ converges uniformly on $[\pi/2, 3\pi/2]$, its term-by-term derivative $\sum_{k=1}^{\infty}(-\sin kx)$ converges at no point in $[\pi/2, 3\pi/2]$.

A *double series* is a series of numbers or functions of the form

$$\sum_{k=1}^{\infty} \left(\sum_{j=1}^{\infty} a_{kj} \right).$$

Such a double series is said to converge if $\sum_{j=1}^{\infty} a_{kj}$ converges for each $k \in \mathbf{N}$ and

$$\sum_{k=1}^{\infty} \sum_{j=1}^{\infty} a_{kj} := \lim_{N \to \infty} \sum_{k=1}^{N} \left(\sum_{j=1}^{\infty} a_{kj} \right)$$

exists and is finite.

When working with double series, one frequently wants to be able to change the order of summation. We already know that the order of summation can be changed when $a_{kj} \geq 0$ (see Exercise 9 in Section 4.3). We now prove a more general result. (The elegant proof given here, which comes from Rudin [11][1], uses uniform convergence.)

THEOREM 4.29. *Let $a_{kj} \in \mathbf{R}$ for $k, j \in \mathbf{N}$ and suppose*

$$A_j = \sum_{k=1}^{\infty} |a_{kj}| < \infty$$

for each $j \in \mathbf{N}$. If $\sum_{j=1}^{\infty} A_j$ converges, then

$$\sum_{k=1}^{\infty} \sum_{j=1}^{\infty} a_{kj} = \sum_{j=1}^{\infty} \sum_{k=1}^{\infty} a_{kj}.$$

---

[1] Walter Rudin, <u>Principles of Mathematical Analysis</u>, 3rd ed. (New York: McGraw-Hill Book Co., 1976). Reprinted with permission of McGraw-Hill Book Co.

PROOF. Let $E = \{0, 1, \frac{1}{2}, \frac{1}{3}, \dots\}$. For each $j \in \mathbf{N}$, define a function f_j on E by

$$f_j(0) = \sum_{k=1}^{\infty} a_{kj}, \quad f_j(\frac{1}{n}) = \sum_{k=1}^{n} a_{kj}, \qquad n \in \mathbf{N}.$$

By hypothesis, $f_j(0)$ exists and

$$\lim_{\substack{x \to 0 \\ x \in E}} f_j(x) = f_j(0)$$

for each $j \in \mathbf{N}$. Moreover, since $|f_j(x)| \le A_j$ for all $x \in E$ and $j \in \mathbf{N}$, it follows from the Weierstrass M–Test that

$$f(x) := \sum_{j=1}^{\infty} f_j(x)$$

converges uniformly on E. Hence, by Exercise 6 in Section 4.4,

$$\lim_{\substack{x \to 0 \\ x \in E}} f(x) = f(0).$$

Therefore,

$$\sum_{k=1}^{\infty}\sum_{j=1}^{\infty} a_{kj} = \lim_{n \to \infty} \sum_{k=1}^{n}\sum_{j=1}^{\infty} a_{kj} = \lim_{n \to \infty} \sum_{j=1}^{\infty}\sum_{k=1}^{n} a_{kj}$$

$$= \lim_{n \to \infty} \sum_{j=1}^{\infty} f_j(\frac{1}{n}) = \lim_{\substack{x \to 0 \\ x \in E}} f(x) = f(0) = \sum_{j=1}^{\infty}\sum_{k=1}^{\infty} a_{kj}. \quad \blacksquare$$

EXERCISES

1. a) Prove $\sum_{k=1}^{\infty} \cos(kx)/k^2$ converges uniformly on $\mathbf{R}$.
   b) Prove $\sum_{k=1}^{\infty} \sin(x/k^2)$ converges uniformly on any closed bounded interval $[a, b]$.

2. Prove that the geometric series

$$\sum_{k=0}^{\infty} x^k = \frac{1}{1-x}$$

converges uniformly on any closed interval $[a, b] \subset (-1, 1)$.

3. Suppose $a_k \downarrow 0$ as $k \to \infty$. Prove $\sum_{k=1}^{\infty} a_k \sin kx$ converges uniformly on any closed interval $[a, b] \subset (0, 2\pi)$.

4. Let $E(x) = \sum_{k=0}^{\infty} x^k/k!$.

   a) Prove that the series defining $E(x)$ converges uniformly on any closed bounded interval $[a, b]$.

   b) Prove that
   $$\int_a^b E(x)\, dx = E(b) - E(a)$$
   for all $a, b \in \mathbf{R}$.

   c) Prove that the function $y = E(x)$ satisfies the initial value problem
   $$y' - y = 0, \qquad y(0) = 1.$$

   (We shall see in Section 4.7 that $E(x) = e^x$.)

5. Suppose
   $$f(x) = \sum_{k=1}^{\infty} \frac{\cos(kx)}{k^2}.$$

   Prove
   $$\int_0^{\pi/2} f(x)\, dx = \sum_{k=0}^{\infty} \frac{(-1)^k}{(2k+1)^3}.$$

6. Show that
   $$f(x) = \sum_{k=1}^{\infty} \frac{1}{k} \sin\left(\frac{x}{k+1}\right)$$

   converges uniformly on $\mathbf{R}$ to a differentiable function f which satisfies
   $$|f(x)| \le |x| \quad \text{and} \quad |f'(x)| \le 1$$
   for all $x \in \mathbf{R}$.

7. Prove that
   $$\left| \sum_{k=1}^{\infty} (1 - \cos(1/k)) \right| \le 2.$$

8. Suppose $f = \sum_{k=1}^{\infty} f_k$ converges uniformly on a set $E \subseteq \mathbf{R}$. If g_1 is bounded on E and $g_k(x) \ge g_{k+1}(x) \ge 0$ for all $x \in E$ and $k \in \mathbf{N}$, prove that $\sum_{k=1}^{\infty} f_k g_k$ converges uniformly on E.

9. Let $n \ge 0$ be a fixed nonnegative integer and recall that $0! := 1$. The Bessel function of order n is the function defined by
   $$B_n(x) := \sum_{k=0}^{\infty} \frac{(-1)^k}{(k!)(n+k)!} \left(\frac{x}{2}\right)^{n+2k}.$$

   a) Show $B_n(x)$ converges pointwise on $\mathbf{R}$ and uniformly on any closed bounded interval $[a, b]$.

b) Prove that $y = B_n(x)$ satisfies the differential equation

$$x^2 y'' + xy' + (x^2 - n^2)y = 0$$

for $x \in \mathbf{R}$.

c) Prove that

$$(x^n B_n(x))' = x^n B_{n-1}(x)$$

for $n \in \mathbf{N}$ and $x \in \mathbf{R}$.

4.6 POWER SERIES

Polynomials are functions of the form $P(x) = \sum_{k=0}^{n} a_k x^k$ where $a_k \in \mathbf{R}$ and $n \geq 0$. In this section we investigate a natural generalization of polynomials, namely, series of the form $\sum_{k=0}^{\infty} a_k x^k$.

Actually, we shall consider a slightly more general class of series. A *power series* (centered at x_0) is a series of the form

$$\sum_{k=0}^{\infty} a_k (x - x_0)^k.$$

Since such a series is identically a_0 when $x = x_0$, it is clear that a power series always converges at at least one point. The following result shows that this may be the only point.

Remark 1. *There exist power series which converge only at one point.*

PROOF. For each $x \neq 0$, $(k^k |x|^k)^{1/k} = k|x| \to \infty$ as $k \to \infty$. Therefore, by the Root Test, the series $\sum_{k=1}^{\infty} k^k x^k$ converges only at $x = 0$. ∎

Applying the Root Test to a general power series (see the proof of Theorem 4.30 below), we are led naturally to the following idea.

DEFINITION 4.7. The *radius of convergence* of a power series $\sum_{k=0}^{\infty} a_k (x - x_0)^k$ is the extended real number

$$R := \frac{1}{\limsup_{k \to \infty} \sqrt[k]{|a_k|}}.$$

(Here, we interpret $1/\infty = 0$ and $1/0 = \infty$.)

In general, a series of functions can converge at several isolated points. (For example, the series $\sum_{k=1}^{\infty} \sin(kx)$ converges only when $x = n\pi$ for some $n \in \mathbf{Z}$.) The next result shows this cannot happen for power series.

THEOREM 4.30. *Let $S(x)$ be a power series centered at x_0 with radius of convergence R. Then*

i) *$S(x)$ converges absolutely for each $x \in (x_0 - R, x_0 + R)$,*

ii) *$S(x)$ diverges for each $x \notin [x_0 - R, x_0 + R]$, and*

iii) *$S(x)$ converges uniformly on any closed interval $[a, b] \subset (x_0 - R, x_0 + R)$.*

PROOF. Suppose $S(x) = \sum_{k=0}^{\infty} a_k(x - x_0)^k$ has radius of convergence R. Then

$$r := \limsup_{k \to \infty} \sqrt[k]{|a_k(x - x_0)^k|} = \frac{|x - x_0|}{R}.$$

By the Root Test, $S(x)$ converges absolutely if $r < 1$ and diverges if $r > 1$. Thus, $S(x)$ converges absolutely if $|x - x_0| < R$ and diverges if $|x - x_0| > R$. This proves parts i) and ii).

To prove part iii), let $[a, b] \subset (x_0 - R, x_0 + R)$. Choose an $x_1 \in (x_0 - R, x_0 + R)$ such that $x \in [a, b]$ implies $|x - x_0| \le |x_1 - x_0|$ (see Figure 4.6). Set $M_k = |a_k| \, |x_1 - x_0|^k$ and observe by part i) that $\sum_{k=0}^{\infty} M_k$ converges. Since $|a_k(x - x_0)^k| \le M_k$ for $x \in [a, b]$ and $k \in \mathbf{N}$, it follows from the Weierstrass M–Test that $S(x)$ converges uniformly on $[a, b]$. ∎

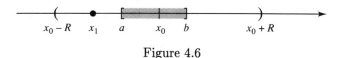

Figure 4.6

The following result provides another way to compute the radius of convergence of a power series (see also Exercise 5 below).

THEOREM 4.31. *If the limit*

$$R = \lim_{k \to \infty} \frac{|a_k|}{|a_{k+1}|}$$

exists, then R is the radius of convergence of the power series $S(x) = \sum_{k=0}^{\infty} a_k(x - x_0)^k$.

PROOF. Repeat the proof of Theorem 4.30, using the Ratio Test instead of the Root Test, to find that $S(x)$ converges absolutely on $(x_0 - R, x_0 + R)$ and diverges for each $x \notin [x_0 - R, x_0 + R]$. By Theorem 4.30, R must be the radius of convergence of $S(x)$. ∎

DEFINITION 4.8. The *interval of convergence* of a power series $S(x)$ is the largest interval on which $S(x)$ converges.

By Theorem 4.30, for a given power series $S(x) = \sum_{k=0}^{\infty} a_k(x - x_0)^k$, there are only three possibilities:

i) $R = \infty$, in which case $S(x)$ converges absolutely on $\mathbf{R}$ and the interval of convergence is $(-\infty, \infty)$,

ii) $R = 0$, in which case $S(x)$ converges only at $x = x_0$ and the interval of convergence is degenerate, and

iii) $0 < R < \infty$, in which case $S(x)$ converges absolutely for $|x - x_0| < R$, diverges for $|x - x_0| > R$, and the interval of convergence has endpoints $x_0 - R$ and $x_0 + R$.

To find the interval of convergence of a power series, one needs to compute the radius of convergence and then check both endpoints $x_0 - R$ and $x_0 + R$.

Example 1. Find the interval of convergence of $S(x) = \sum_{k=1}^{\infty} x^k / \sqrt{k}$.

SOLUTION. By Theorem 4.31,

$$R = \lim_{k \to \infty} \frac{\sqrt{k}}{\sqrt{k+1}} = \sqrt{\lim_{k \to \infty} \frac{k}{k+1}} = 1.$$

Thus, the interval of convergence has endpoints 1 and -1. $S(x)$ diverges at $x = 1$ by the p–Series Test and converges at $x = -1$ by the Alternating Series Test. Thus, the interval of convergence of $S(x)$ is $[-1, 1)$. ∎

Remark 2. *The interval of convergence may contain none, one, or both its endpoints.*

PROOF. By Theorem 4.31, the radius of convergence of each of the series

$$\sum_{k=1}^{\infty} x^k, \qquad \sum_{k=1}^{\infty} \frac{x^k}{k}, \qquad \sum_{k=1}^{\infty} \frac{x^k}{k^2}$$

is 1, but by the Divergence Test, the Alternating Series Test, and the p–Series Test, the intervals of convergence of these series are $(-1, 1)$, $[-1, 1)$, and $[-1, 1]$, respectively. ∎

We now pass from convergence properties of power series to the calculus of power series. The next several results answer the question: What properties (e.g., continuity, differentiability, integrability) does the limit of a power series satisfy?

THEOREM 4.32. If $f(x) = \sum_{k=0}^{\infty} a_k (x - x_0)^k$ is a power series with positive radius of convergence R, then f is continuous on $(x_0 - R, x_0 + R)$.

PROOF. Let $x \in (x_0 - R, x_0 + R)$ and choose $a, b \in \mathbf{R}$ such that $x \in [a, b] \subset (x_0 - R, x_0 + R)$. By Theorems 4.30 and 4.23, f is continuous on $[a, b]$ hence at x. ∎

The following result shows that continuity of the limit extends to the endpoints when they belong to the interval of convergence.

THEOREM 4.33 [ABEL'S THEOREM]. Let $f(x) = \sum_{k=0}^{\infty} a_k (x - x_0)^k$ have a positive, finite radius of convergence R. If $f(x)$ converges at $x = x_0 + R$ (respectively, $x = x_0 - R$), then $f(x)$ is continuous and converges uniformly on $[x_0, x_0 + R]$ (respectively, on $[x_0 - R, x_0]$).

PROOF. By symmetry, it suffices to consider the right-hand endpoint. Thus, suppose $f(x)$ converges at $x = x_0 + R$ and fix $x_1 \in (x_0, x_0 + R]$. Set $b_k = a_k R^k$ and $c_k = (x_1 - x_0)^k / R^k$ for $k \in \mathbf{N}$. By hypothesis, $\sum_{k=1}^{\infty} b_k$ is convergent. Hence, given $\varepsilon > 0$, there is an integer $N > 1$ such that

$$k > m \geq N \quad \text{implies} \quad \left| \sum_{j=m}^{k} b_j \right| < \frac{\varepsilon}{c_1}.$$

Since $0 < x_1 - x_0 \le R$, the sequence $\{c_k\}$ is decreasing. Applying Abel's Formula and telescoping, we have

$$\left| \sum_{k=m}^{n} a_k(x_1 - x_0)^k \right| = \left| \sum_{k=m}^{n} b_k c_k \right|$$

$$= \left| c_n \sum_{k=m}^{n} b_k + \sum_{k=m}^{n-1} (c_k - c_{k+1}) \sum_{j=m}^{k} b_j \right|$$

$$< c_n \frac{\varepsilon}{c_1} + (c_m - c_n) \frac{\varepsilon}{c_1} = c_m \frac{\varepsilon}{c_1}.$$

Since $c_m \le c_1$ it follows that

$$\left| \sum_{k=m}^{n} a_k(x_1 - x_0)^k \right| < \varepsilon$$

for all $x_1 \in (x_0, x_0 + R]$. Since this inequality also holds for $x_1 = x_0$, we conclude that $\sum_{k=0}^{\infty} a_k(x - x_0)^k$ converges uniformly on $[x_0, x_0 + R]$. ∎

Remark 3. *If a power series $S(x) = \sum_{k=0}^{\infty} a_k(x - x_0)^k$ converges at some $x_1 > x_0$, then $S(x)$ converges uniformly on $[x_0, x_1]$ and absolutely on $[x_0, x_1)$. It need not converge absolutely at $x = x_1$.*

PROOF. By Theorems 4.30 and 4.33, $S(x)$ converges uniformly on $[x_0, x_1]$ and absolutely on $[x_0, x_1)$. The power series $\sum_{k=1}^{\infty} (-x)^k / k$ converges at $x = 0, 1$ but not absolutely at $x = 1$. ∎

To discuss differentiability of the limit of a power series, we first show that the radius of convergence of a power series is not changed by term-by-term differentiation (see also Exercise 7 in Section 1.6).

Lemma. *If $a_n \in \mathbf{R}$ for $n \in \mathbf{N}$, then*

$$\limsup_{n \to \infty} \sqrt[n]{n|a_n|} = \limsup_{n \to \infty} \sqrt[n]{|a_n|}.$$

PROOF. Let $\varepsilon > 0$. Since $\sqrt[n]{n} \to 1$ as $n \to \infty$, choose $N \in \mathbf{N}$ so that

$$n \ge N \quad \text{implies} \quad 1 - \varepsilon < \sqrt[n]{n} < 1 + \varepsilon.$$

Multiply this inequality by $\sqrt[n]{|a_n|}$ and take the limit supremum as $n \to \infty$. We obtain

$$(1 - \varepsilon) \limsup_{n \to \infty} \sqrt[n]{|a_n|} \le \limsup_{n \to \infty} \sqrt[n]{n|a_n|} \le (1 + \varepsilon) \limsup_{n \to \infty} \sqrt[n]{|a_n|}.$$

Take the limit of this inequality as $\varepsilon \to 0$ and the proof of the lemma is complete. ∎

We use this result to prove that each power series with a positive radius of convergence is term-by-term differentiable.

THEOREM 4.34. *If* $f(x) = \sum_{k=0}^{\infty} a_k(x - x_0)^k$ *is a power series with positive radius of convergence R, then*

$$f'(x) = \sum_{k=1}^{\infty} k a_k(x - x_0)^{k-1}$$

for $x \in (x_0 - R, x_0 + R)$.

PROOF. Let $x \in (x_0 - R, x_0 + R)$. Choose a closed subinterval $[a, b]$ of $(x_0 - R, x_0 + R)$ such that $x \in (a, b)$. By the lemma, the radius of convergence of the derived series $\sum_{k=1}^{\infty} k a_k(x - x_0)^{k-1}$ is also R. Moreover, by Theorem 4.30 the derived series converges uniformly on $[a, b]$. We conclude by Theorem 4.25 that the series $f(x)$ is term-by-term differentiable on (a, b), hence at x. ∎

Recall that $\mathcal{C}^{\infty}(a, b)$ represents the set of functions f such that $f^{(k)}$ exists and is continuous on (a, b) for all $k \in \mathbf{N}$. The following result generalizes Theorem 4.34.

COROLLARY 4.35. *If* $f(x) = \sum_{k=0}^{\infty} a_k(x - x_0)^k$ *has a positive radius of convergence R, then $f \in \mathcal{C}^{\infty}(x_0 - R, x_0 + R)$ and*

(24)
$$f^{(k)}(x) = \sum_{n=k}^{\infty} \frac{n!}{(n-k)!} a_n(x - x_0)^{n-k}$$

for $x \in (x_0 - R, x_0 + R)$ *and* $k \in \mathbf{N}$.

PROOF. The proof is by induction on k. By Theorem 4.34 and the fact that $0! := 1$, (24) holds for $k = 1$ and $x \in (x_0 - R, x_0 + R)$. If (24) holds for some $k \in \mathbf{N}$ and all $x \in (x_0 - R, x_0 + R)$, then $f^{(k)}$ is a power series with radius of convergence R. It follows from Theorem 4.34 that

$$f^{(k+1)}(x) = (f^{(k)}(x))' = \left(\sum_{n=k}^{\infty} \frac{n!}{(n-k)!} a_n(x - x_0)^{n-k} \right)'$$

$$= \sum_{n=k+1}^{\infty} \frac{n!}{(n-k-1)!} a_n(x - x_0)^{n-k-1}$$

for all $x \in (x_0 - R, x_0 + R)$. Hence, (24) holds for $k + 1$ in place of k. ∎

The following result shows that each power series with a positive radius of convergence can also be integrated term-by-term.

THEOREM 4.36. *Let* $f(x) = \sum_{k=0}^{\infty} a_k(x - x_0)^k$ *be a power series and $a < b$ be real numbers.*

i) *If $f(x)$ converges at $x = a$ and $x = b$, then*

$$\int_a^b f(x)\, dx = \sum_{k=0}^{\infty} a_k \int_a^b (x - x_0)^k\, dx.$$

ii) If $f(x)$ converges on $[a, b)$ and if $\sum_{k=0}^{\infty} a_k(b - x_0)^{k+1}/(k + 1)$ converges, then f is improperly integrable on $[a, b)$ and

$$\int_a^b f(x)\,dx = \sum_{k=0}^{\infty} a_k \int_a^b (x - x_0)^k\,dx.$$

PROOF. i) By Abel's Theorem, $f(x)$ converges uniformly on $[a, b]$. Hence, by Theorem 4.24, $f(x)$ is term-by-term integrable on $[a, b]$.

ii) Let $a \le t < b$ and set $A = \sum_{k=0}^{\infty} a_k(a - x_0)^{k+1}/(k + 1)$. By part i),

$$\int_a^t f(x)\,dx = \sum_{k=0}^{\infty} a_k \int_a^t (x - x_0)^k\,dx = \sum_{k=0}^{\infty} \frac{a_k}{k + 1}(t - x_0)^{k+1} - A.$$

The left-most term of this last difference is a power series which by hypothesis converges at $t = b$. Thus, by the definition of improper integration and Abel's Theorem,

$$\int_a^b f(x)\,dx = \lim_{t \to b-} \int_a^t f(x)\,dx$$

$$= \lim_{t \to b-} \sum_{k=0}^{\infty} \frac{a_k}{k + 1}(t - x_0)^{k+1} - A$$

$$= \sum_{k=0}^{\infty} \frac{a_k}{k + 1}(b - x_0)^{k+1} - A = \sum_{k=0}^{\infty} a_k \int_a^b (x - x_0)^k\,dx. \ \blacksquare$$

The following result shows that the product of two power series is a power series. (For a result on the division of power series, see Taylor [13], p. 619.)

THEOREM 4.37. If $f(x) = \sum_{k=0}^{\infty} a_k x^k$ and $g(x) = \sum_{k=0}^{\infty} b_k x^k$ converge on $(-r, r)$ and

$$c_k = \sum_{j=0}^{k} a_j b_{k-j}, \qquad k = 0, 1, \ldots,$$

then $\sum_{k=0}^{\infty} c_k x^k$ converges on $(-r, r)$ and converges to $f(x)g(x)$.

PROOF. Fix $x \in (-r, r)$ and for each $n \in \mathbf{N}$, set

$$f_n(x) = \sum_{k=0}^{n} a_k x^k, \qquad g_n(x) = \sum_{k=0}^{n} b_k x^k, \quad \text{and} \quad h_n(x) = \sum_{k=0}^{n} c_k x^k.$$

By changing the order of summation, we see that

$$h_n(x) = \sum_{k=0}^{n} \sum_{j=0}^{k} a_j b_{k-j} x^j x^{k-j} = \sum_{j=0}^{n} a_j x^j \sum_{k=j}^{n} b_{k-j} x^{k-j}$$

$$= \sum_{j=0}^{n} a_j x^j g_{n-j}(x) = g(x) f_n(x) + \sum_{j=0}^{n} a_j x^j (g_{n-j}(x) - g(x)).$$

Thus, it suffices to show

$$\lim_{n\to\infty} \sum_{j=0}^{n} a_j x^j (g_{n-j}(x) - g(x)) = 0.$$

Let $\varepsilon > 0$. Since $g_n(x)$ converges as $n \to \infty$, choose $M > 0$ such that

$$|g_{n-j}(x) - g(x)| \le M$$

for all integers $n > j > 0$. Since $C = \sum_{k=0}^{\infty} |a_k x^k|$ is finite, choose $N \in \mathbf{N}$ such that

$$\ell \ge N \quad \text{implies} \quad |g_\ell(x) - g(x)| < \frac{\varepsilon}{2C} \quad \text{and} \quad \sum_{j=N+1}^{\infty} |a_j x^j| < \frac{\varepsilon}{2M}.$$

Let $n > 2N$. Then

$$\left| \sum_{j=0}^{n} a_j x^j (g_{n-j}(x) - g(x)) \right|$$

$$= \left| \sum_{j=0}^{N} a_j x^j (g_{n-j}(x) - g(x)) + \sum_{j=N+1}^{n} a_j x^j (g_{n-j}(x) - g(x)) \right|$$

$$< \frac{\varepsilon}{2C} \sum_{j=0}^{N} |a_j x^j| + M \sum_{j=N+1}^{n} |a_j x^j|$$

$$< \frac{\varepsilon}{2} + \frac{\varepsilon}{2} = \varepsilon. \ \blacksquare$$

COROLLARY 4.38. *If $\sum_{k=0}^{\infty} a_k$ and $\sum_{k=0}^{\infty} b_k$ converge and*

$$c_k = \sum_{j=0}^{k} a_j b_{k-j}, \qquad k = 0, 1, \ldots,$$

then $\sum_{k=0}^{\infty} c_k$ converges to $(\sum_{k=0}^{\infty} a_k)(\sum_{k=0}^{\infty} b_k)$.

PROOF. Repeat the proof of Theorem 4.37 with $x = 1$. $\blacksquare$

We close this section with some optional material on finding exact values of convergent power series. Namely, we show how term-by-term differentiation and integration can be used in conjunction with geometric series to obtain simple formulas for certain kinds of power series. Such formulas are called *closed forms*.

Example 2. Find a closed form of the power series

$$f(x) = \sum_{k=1}^{\infty} k x^k.$$

SOLUTION. Since the interval of convergence of this power series is $(-1, 1)$, we have by Theorem 4.36 and Theorem 4.3 that

$$\int_0^x \frac{f(t)}{t}\, dt = \sum_{k=1}^\infty k \int_0^x t^{k-1}\, dt = \sum_{k=1}^\infty x^k = \frac{x}{1-x}$$

for each $x \in (-1, 1)$. (Note that $f(x)/x$ is defined at $x = 0$ and has value 1.) Hence, by the Fundamental Theorem of Calculus,

$$\frac{f(x)}{x} = \left(\frac{x}{1-x}\right)' = \frac{1}{(1-x)^2}$$

and it follows that

$$f(x) = \frac{x}{(1-x)^2}, \qquad x \in (-1, 1). \quad \blacksquare$$

*Example 3. Find a closed form of the power series

$$g(x) = \sum_{k=0}^\infty \frac{x^k}{k+1}.$$

SOLUTION. Since the interval of convergence of this power series is $[-1, 1)$, we have by Theorem 4.34 that

$$(xg(x))' = \sum_{k=0}^\infty \left(\frac{x^{k+1}}{k+1}\right)' = \sum_{k=0}^\infty x^k = \frac{1}{1-x}$$

for $x \in (-1, 1)$. Hence, by the Fundamental Theorem of Calculus,

$$xg(x) = \int_0^x \frac{dt}{1-t} = -\log(1-x)$$

for $x \in (-1, 1)$. Since $g(-1)$ exists and $\log(1-x)$ is continuous at $x = -1$, we conclude by Abel's Theorem that

$$g(x) = -\frac{\log(1-x)}{x}, \qquad x \in [-1, 1), \quad x \neq 0. \quad \blacksquare$$

EXERCISES

1. Find the interval of convergence of each of the following power series.

a) $\displaystyle\sum_{k=0}^\infty \frac{x^k}{2^k}.$ b) $\displaystyle\sum_{k=0}^\infty ((-1)^k + 3)^k (x-1)^k.$

c) $\displaystyle\sum_{k=1}^{\infty} \log\left(\frac{k+1}{k}\right) x^k.$ d) $\displaystyle\sum_{k=1}^{\infty} \frac{1 \cdot 3 \dots (2k-1)}{(k+1)!} x^{2k}.$

*2. Find a closed form for each of the following series and the largest set on which this formula is valid.

a) $\displaystyle\sum_{k=1}^{\infty} 3x^{3k-1}.$ b) $\displaystyle\sum_{k=2}^{\infty} kx^{k-2}.$ c) $\displaystyle\sum_{k=1}^{\infty} \frac{2k}{k+1}(1-x)^k.$ d) $\displaystyle\sum_{k=0}^{\infty} \frac{x^{3k}}{k+1}.$

3. Use Theorems 4.34 and 4.37 to give two different proofs of the following identity:

$$\frac{1}{(1-x)^2} = \sum_{k=0}^{\infty} (k+1)x^k \qquad x \in (-1,1).$$

4. Show that if $\sum_{k=1}^{\infty} a_k x^k$ has radius of convergence R and $a_k \neq 0$ for large k, then

$$\liminf_{k\to\infty} \left|\frac{a_k}{a_{k+1}}\right| \leq R \leq \limsup_{k\to\infty} \left|\frac{a_k}{a_{k+1}}\right|.$$

5. Suppose $|a_k| \leq |b_k|$ for large k. Prove that if $\sum_{k=1}^{\infty} b_k x^k$ converges on an open interval I, then $\sum_{k=1}^{\infty} a_k x^k$ also converges on I. Is this result true if "open" is omitted?

6. Suppose $\{a_k\}_{k=0}^{\infty}$ is a bounded sequence of real numbers.

   a) Prove

$$f_{x_0}(x) = \sum_{k=0}^{\infty} a_k(x-x_0)^k$$

   has a positive radius of convergence.

   b) Prove that $f_{0.5}$ is integrable on $[0,1]$ and

$$\int_0^1 f_{0.5}(x)\,dx = \sum_{k=0}^{\infty} \frac{a_{2k}}{(2k+1)2^{2k}}.$$

7. A series $\sum_{k=0}^{\infty} a_k$ is said to be *Abel summable* to L if

$$\lim_{r\to 1-} \sum_{k=0}^{\infty} a_k r^k = L.$$

   a) Prove that if $\sum_{k=0}^{\infty} a_k$ converges to L, then $\sum_{k=0}^{\infty} a_k$ is Abel summable to L.
   b) Find the Abel sum of $\sum_{k=0}^{\infty}(-1)^k$.

*8. Prove

$$f(x) = \sum_{k=0}^{\infty} \left(\frac{x}{(-1)^k + 4} \right)^k$$

is differentiable on $(-3, 3)$ and

$$|f'(x)| \leq \frac{3}{(3-x)^2}$$

for $0 \leq x < 3$.

9. Suppose $a_k \downarrow 0$ as $k \to \infty$. Prove that given $\varepsilon > 0$ there is a $\delta > 0$ such that

$$\left| \sum_{k=0}^{\infty} (-1)^k a_k (x^k - y^k) \right| < \varepsilon$$

for all $x, y \in [0, 1]$ which satisfy $|x - y| < \delta$.

10. a) Prove the following weak form of Stirling's Formula (compare with Theorem 7.23).

$$\frac{n^n}{e^{n-1}} < n! < \frac{n^{n+1}}{e^{n-1}}.$$

b) Find all $x \in \mathbf{R}$ for which the power series

$$\sum_{k=0}^{\infty} \frac{n^n}{n!} x^n$$

converges absolutely.

4.7 ANALYTIC FUNCTIONS

In this section we study functions which can be represented by power series. (For a discussion of how to represent functions by trigonometric series instead of power series, see Chapter 9.) We begin with the following definition.

DEFINITION 4.9. A real-valued function f is said to be *analytic* on an open interval (a, b) if given $x_0 \in (a, b)$ there is a power series centered at x_0 which converges to f near x_0, i.e., there exist coefficients $\{a_k\}_{k=0}^{\infty}$ and points $c, d \in (a, b)$ such that $c < x_0 < d$ and

$$f(x) = \sum_{k=0}^{\infty} a_k (x - x_0)^k$$

for all $x \in (c, d)$.

We shall develop several techniques for showing that a given function is analytic. To simplify statements of results, we shall use the conventions $f^{(0)} := f$ and $0! := 1$.

First, it is important to realize that if f can be represented by a power series, then the coefficients of that power series can be computed using derivatives of f.

THEOREM 4.39 [UNIQUENESS]. *Let $c < d$ be extended real numbers, $x_0 \in (c, d)$, and $f : (c, d) \to \mathbf{R}$. If $f(x) = \sum_{k=0}^{\infty} a_k (x - x_0)^k$ for each $x \in (c, d)$, then $f \in \mathcal{C}^{\infty}(c, d)$ and*

$$a_k = \frac{f^{(k)}(x_0)}{k!}, \qquad k = 0, 1, \dots$$

PROOF. Clearly, $f(x_0) = a_0$. Fix $k \in \mathbf{N}$. By hypothesis, the radius of convergence R of the power series $\sum_{k=0}^{\infty} a_k (x - x_0)^k$ is positive and $(c, d) \subseteq (x_0 - R, x_0 + R)$. Hence, by Corollary 4.35, $f \in \mathcal{C}^{\infty}(c, d)$ and

$$(25) \qquad f^{(k)}(x) = \sum_{n=k}^{\infty} \frac{n!}{(n-k)!} a_n (x - x_0)^{n-k}$$

for $x \in (c, d)$. Apply this to $x = x_0$. The terms on the right side of (25) are zero when $n > k$ and $k! a_k$ when $n = k$. Hence, $f^{(k)}(x_0) = k! a_k$ for each $k \in \mathbf{N}$. ∎

In particular, if f is analytic on (a, b), then for each $x_0 \in (a, b)$ there is only one power series centered at x_0 which represents f near x_0. This power series has a special name.

DEFINITION 4.10. Let $f \in \mathcal{C}^{\infty}(a, b)$ and $x_0 \in (a, b)$.

i) The *Taylor expansion* (or *Taylor series*) of f centered at x_0 is the series

$$\sum_{k=0}^{\infty} \frac{f^{(k)}(x_0)}{k!} (x - x_0)^k.$$

(No convergence is implied or assumed.)

ii) The *remainder term of order n* of the Taylor expansion of f centered at x_0 is the function

$$R_n(x) = R_n^{f, x_0}(x) := f(x) - \sum_{k=0}^{n-1} \frac{f^{(k)}(x_0)}{k!} (x - x_0)^k.$$

Note: The Taylor expansion of f centered at $x_0 = 0$ is usually called the *Maclaurin expansion* (or *Maclaurin series*) of f.

The following result summarizes Corollary 4.35, Definition 4.9, and Theorem 4.39.

Remark 1. *If f is analytic on (a, b), then $f \in \mathcal{C}^{\infty}(a, b)$. Conversely, a given $f \in \mathcal{C}^{\infty}(a, b)$ is analytic on (a, b) if and only if for each $x_0 \in (a, b)$ the Taylor expansion of f centered at x_0 converges to $f(x)$ for all x in some open interval containing x_0.*

The next remark shows that not every $\mathcal{C}^{\infty}$ function is analytic.

Remark 2. [CAUCHY]. *The function*

$$f(x) = \begin{cases} e^{-1/x^2} & x \neq 0 \\ 0 & x = 0 \end{cases}$$

belongs to $C^\infty(-\infty, \infty)$ *but* f *is not analytic on any interval which contains* $x = 0$.

PROOF. It is easy to see (Exercise 3 in Section 2.5) that $f \in C^\infty(-\infty, \infty)$ and $f^{(k)}(0) = 0$ for all $k \in \mathbf{N}$. Thus, the Taylor expansion of f about the point $x_0 = 0$ is identically zero but $f(x) = 0$ only when $x = 0$. ∎

Fortunately, functions like this are rare. One of the aims of this section is to prove that many of the classical C^∞ functions used in elementary calculus are analytic on their domain.

First, we use Theorem 4.39 to reformulate the problem of deciding whether or not a given function is analytic.

THEOREM 4.40. *A function* $f \in C^\infty(a, b)$ *is analytic on* (a, b) *if and only if given* $x_0 \in (a, b)$ *there is an interval* (c, d) *containing* x_0 *such that the remainder term* $R_n^{f, x_0}(x)$ *converges to zero for all* $x \in (c, d)$.

PROOF. By Theorem 4.39, f is analytic on (a, b) if and only if given $x_0 \in (a, b)$ there is an interval (c, d) containing x_0 such that the Taylor expansion of f centered at x_0 converges to f pointwise on (c, d). By Definition 4.10, this happens if and only if $R_n^{f, x_0} \to 0$, as $n \to \infty$, for every $x \in (c, d)$. ∎

Thus, to decide whether a given $f \in C^\infty(a, b)$ is analytic on (a, b) we need only estimate the corresponding remainder term. We shall prove two results (see Theorems 4.41 and 4.43 below) which can be used to estimate the remainder term in concrete situations.

Since $R_1^{f, x_0} = f(x) - f(x_0)$, the remainder term of order 1 can always be estimated using the Mean Value Theorem. The following result shows how to use the Generalized Mean Value Theorem to estimate the remainder term of order n.

THEOREM 4.41 [TAYLOR'S FORMULA]. *Let* $n \in \mathbf{N}$, $f : (a, b) \to \mathbf{R}$, *and suppose* $f^{(n)}$ *exists on* (a, b). *Then given* $x, x_0 \in (a, b)$, *there is a number* c *between* x *and* x_0 *such that*

$$R_n^{f, x_0}(x) = \frac{f^{(n)}(c)}{n!}(x - x_0)^n.$$

In particular,

$$f(x) = \sum_{k=0}^{n-1} \frac{f^{(k)}(x_0)}{k!}(x - x_0)^k + \frac{f^{(n)}(c)}{n!}(x - x_0)^n$$

for some number c *between* x *and* x_0.

PROOF. Without loss of generality, suppose $x_0 < x$. We shall apply the Generalized Mean Value Theorem to the functions

$$F(t) = \frac{(x - t)^n}{n!} \quad \text{and} \quad G(t) = R_n^{f, t}(x) = f(t) - \sum_{k=0}^{n-1} \frac{f^{(k)}(t)}{k!}(x - t)^k,$$

defined for $t \in (a, b)$.

Notice by the Chain Rule that

$$(26) \qquad\qquad F'(t) = -\frac{(x-t)^{n-1}}{(n-1)!}$$

for $t \in \mathbf{R}$. Also notice that

$$\frac{d}{dt}\left(\frac{f^{(k)}(t)}{k!}(x-t)^k\right) = \frac{f^{(k+1)}(t)}{k!}(x-t)^k - \frac{f^{(k)}(t)}{(k-1)!}(x-t)^{k-1}$$

for $t \in (a, b)$ and $k \in \mathbf{N}$. By telescoping, we obtain

$$(27) \qquad\qquad G'(t) = -\frac{f^{(n)}(t)}{(n-1)!}(x-t)^{n-1}$$

for $t \in (a, b)$. Thus, F and G are differentiable on (x_0, x) and continuous on $[x_0, x]$.

By the Generalized Mean Value Theorem and the fact that $F(x) = G(x) = 0$, there is a number $c \in (x_0, x)$ such that

$$-F(x_0)G'(c) = (F(x) - F(x_0))G'(c) = (G(x) - G(x_0))F'(c) = -G(x_0)F'(c).$$

Hence, it follows from (26) and (27) that

$$\frac{(x-x_0)^n}{n!}\left(\frac{f^{(n)}(c)(x-c)^{n-1}}{(n-1)!}\right) = R_n^{f,x_0}(x)\frac{(x-c)^{n-1}}{(n-1)!}.$$

Solving this equation for R_n^{f,x_0} completes the proof. ∎

The following theorem, a corollary of Taylor's Formula, is the first of several results which identify conditions on the derivatives $f^{(n)}$ of an $f \in C^\infty$ sufficient for f to be analytic on an interval (a, b).

THEOREM 4.42. *Let $f \in C^\infty(a, b)$. If there is an $M > 0$ such that*

$$\left| f^{(n)}(x) \right| \le M^n$$

for all $x \in (a, b)$ and $n \in \mathbf{N}$, then f is analytic on (a, b). In fact, for each $x_0 \in (a, b)$,

$$f(x) = \sum_{k=0}^{\infty} \frac{f^{(k)}(x_0)}{k!}(x-x_0)^k$$

holds for all $x \in (a, b)$.

PROOF. Fix $x \in (a, b)$ and set $C = \max\{M|a-x_0|, M|b-x_0|\}$. By Theorem 4.41,

$$|R_n(x)| := |R_n^{f,x_0}(x)| \le \frac{M^n|x-x_0|^n}{n!} \le \frac{C^n}{n!}$$

for all $n \in \mathbf{N}$. But $C^n/n! \to 0$ as $n \to \infty$ for any $C \in \mathbf{R}$. Thus, by the Squeeze Theorem, the remainder term $R_n(x)$ converges to zero for every $x \in (a, b)$. ∎

Example 1. Prove that $\sin x$ and $\cos x$ are analytic on $\mathbf{R}$ and have Maclaurin expansions

$$\sin x = \sum_{k=0}^{\infty} \frac{(-1)^k x^{2k+1}}{(2k+1)!}, \qquad \cos x = \sum_{k=0}^{\infty} \frac{(-1)^k x^{2k}}{(2k)!}.$$

PROOF. Set $f(x) = \sin x$. It is easy to see that

$$f^{(n)}(x) = \begin{cases} \sin x & n = 4j, \\ \cos x & n = 4j+1, \\ -\sin x & n = 4j+2, \\ -\cos x & n = 4j+3 \end{cases}$$

for $j \in \mathbf{N}$, i.e.,

$$f^{(n)}(0) = \begin{cases} (-1)^k & n = 2k-1 \\ 0 & n = 2k \end{cases}$$

for $k \in \mathbf{N}$. Hence, by Theorem 4.42, $f(x) = \sin x$ is analytic on $\mathbf{R}$ and its Maclaurin expansion has the promised form. A similar argument verifies the result for $\cos x$. ∎

Example 2. Prove that e^x is analytic on $\mathbf{R}$ and has Maclaurin expansion

$$e^x = \sum_{k=0}^{\infty} \frac{x^k}{k!}.$$

PROOF. Fix $C > 0$ and set $M = e^C$. If $f(x) = e^x$, then $f^{(n)}(x) = e^x$ for all $x \in \mathbf{R}$. Hence, $f^{(n)}(x) \leq M \leq M^n$ for $n = 0, 1, \dots$ and $x \in [-C, C]$. It follows from Theorem 4.42 that the Maclaurin series of f converges to f on $[-C, C]$. Since $f^{(n)}(0) = 1$ for all $n \in \mathbf{N}$, and $C > 0$ was arbitrary, we conclude that $\sum_{k=0}^{\infty} x^k/k!$ converges to e^x for all $x \in \mathbf{R}$. ∎

In Examples 1 and 2, we found the Taylor expansion of a given f by computing the derivatives of f and estimating the remainder term. The next three examples show that by using Theorems 4.34, 4.36, 4.37, and 4.39, some Taylor expansions can be found without computing the derivatives of f.

Example 3. Prove $\arctan x$ is analytic on $(-1, 1)$ and has Maclaurin expansion

$$\arctan x = \sum_{k=0}^{\infty} \frac{(-1)^k x^{2k+1}}{2k+1}.$$

PROOF. For each $0 < x < 1$, the geometric series $\sum_{k=0}^{\infty} (-1)^k t^{2k}$ converges uniformly on $[-x, x]$ to $1/(1+t^2)$ (see Exercise 1 in Section 4.1). Thus, by Theorem 4.36

$$\arctan x = \int_0^x \frac{dt}{1+t^2} = \int_0^x \sum_{k=0}^{\infty} (-1)^k t^{2k} \, dt = \sum_{k=0}^{\infty} \frac{(-1)^k x^{2k+1}}{2k+1}$$

By uniqueness, this is the Maclaurin expansion of $\arctan x$. ∎

Example 4. Prove $\arctan x/(1-x)$ is analytic on $(-1,1)$ and find its Maclaurin expansion.

PROOF. By Theorem 4.37, for each $|x| < 1$,

$$\left(\frac{\arctan x}{1-x}\right) = \left(\sum_{k=0}^{\infty} x^k\right)\left(\sum_{k=0}^{\infty} \frac{(-1)^k x^{2k+1}}{2k+1}\right)$$

$$= \sum_{k=1}^{\infty}\left(\sum_{j\in A_k} \frac{(-1)^j}{2j+1}\right) x^k,$$

where $A_k := \{j \in \mathbf{N} : 0 \le j \le (k-1)/2\}$. ∎

Example 5. Prove $\log x$ is analytic on $(0,2)$ with Taylor expansion centered at $x_0 = 1$ given by

$$\log x = \sum_{k=1}^{\infty} \frac{(-1)^{k+1}}{k}(x-1)^k.$$

PROOF. By Theorem 4.36, for each $x \in (0,2)$,

$$\log x = \int_1^x \frac{dt}{t} = \int_1^x \frac{dt}{1-(1-t)}$$

$$= \int_1^x \sum_{k=0}^{\infty} (1-t)^k \, dt = \sum_{k=1}^{\infty} \frac{(-1)^{k+1}}{k}(x-1)^k. \ \blacksquare$$

In some situations it is useful to have an integral form of the remainder term. This requires a slightly stronger hypothesis than Theorem 4.41 but can yield a sharper estimate.

THEOREM 4.43 [LAGRANGE]. *Let $n \in \mathbf{N}$. If $f \in C^n(a,b)$, then*

$$R_n(x) := R_n^{f,x_0}(x) = \frac{1}{(n-1)!} \int_{x_0}^x (x-t)^{n-1} f^{(n)}(t) \, dt$$

for all $x, x_0 \in (a,b)$.

PROOF. The proof is by induction on n. If $n = 1$, the formula holds by the Fundamental Theorem of Calculus.

Suppose the formula holds for some $n \in \mathbf{N}$. Since

$$R_{n+1}(x) = R_n(x) - \frac{f^{(n)}(x_0)}{n!}(x-x_0)^n$$

and

$$\frac{(x-x_0)^n}{n!} = \frac{1}{(n-1)!} \int_{x_0}^x (x-t)^{n-1} \, dt$$

it follows that

$$(28) \qquad R_{n+1}(x) = \frac{1}{(n-1)!} \int_{x_0}^{x} (x-t)^{n-1} \left(f^{(n)}(t) - f^{(n)}(x_0) \right) dt.$$

Let $u(t) = f^{(n)}(t) - f^{(n)}(x_0)$, $v(t) = (x-t)^n/n$ and integrate the right side of (28) by parts. Since $u(x_0) = 0$ and $v(x) = 0$, we have

$$R_{n+1}(x) = -\frac{1}{(n-1)!} \int_{x_0}^{x} u'(t)v(t)\, dt = \frac{1}{n!} \int_{x_0}^{x} (x-t)^n f^{(n+1)}(t)\, dt.$$

Hence, the formula holds for $n+1$. ∎

The rest of this section contains some additional (but optional) material on analytic functions.

Lagrange's Theorem gives us another condition on the derivatives of f sufficient to conclude that f is analytic.

*THEOREM 4.44 [BERNSTEIN]. *If $f \in C^{\infty}(a,b)$ and $f^{(n)}(x) \geq 0$ for all $x \in (a,b)$ and $n \in \mathbf{N}$, then f is analytic on (a,b). In fact, if $x_0 \in (a,b)$ and $f^{(n)}(x) \geq 0$ for $x \in [x_0, b)$ and $n \in \mathbf{N}$, then*

$$(29) \qquad f(x) = \sum_{k=0}^{\infty} \frac{f^{(k)}(x_0)}{k!} (x - x_0)^k$$

for all $x \in [x_0, b)$.

PROOF. Fix $x_0 < x < b$ and $n \in \mathbf{N}$. Use Lagrange's Theorem and a change of variables $t = xu$ to write

$$(30) \qquad R_n(x) = R_n^{f,x_0}(x) = \frac{x^n}{(n-1)!} \int_{x_0/x}^{1} (1-u)^{n-1} f^{(n)}(xu)\, du.$$

Since $f^{(n)} \geq 0$, (30) implies $R_n(x) \geq 0$. On the other hand, by definition and hypothesis,

$$R_n(x) = f(x) - \sum_{k=0}^{n-1} \frac{f^{(k)}(x_0)}{k!} (x - x_0)^k \leq f(x).$$

Therefore,

$$(31) \qquad 0 \leq R_n(x) \leq f(x)$$

for all $x \in (x_0, b)$.

Let $b_0 \in (x_0, b)$ and notice that it suffices to verify (29) for $x_0 \leq x \leq b_0$. (We introduce the parameter b_0 in order to handle the cases $b \in \mathbf{R}$ and $b = \infty$ simultaneously.) Since $R_n(x_0) = 0$ for all $n \in \mathbf{N}$, we need only show that $R_n(x) \to 0$ as $n \to \infty$ for each $x \in (x_0, b_0]$.

By hypothesis, $f^{(n+1)}(t) \geq 0$ for $t \in [x_0, b)$, so $f^{(n)}$ is increasing on $[x_0, b)$. Since $x < b_0 < b$, we have by (30) and (31) that

$$0 \leq R_n(x) = \frac{1}{(n-1)!} \int_{x_0/x}^{1} x^n (1-u)^{n-1} f^{(n)}(xu)\, du$$

$$\leq \frac{1}{(n-1)!} \int_{x_0/b_0}^{1} x^n (1-u)^{n-1} f^{(n)}(b_0 u)\, du$$

$$= \left(\frac{x}{b_0}\right)^n R_n(b_0).$$

Since $x/b_0 < 1$, $(x/b_0)^n \to 0$ as $n \to \infty$. Moreover by (31), $R_n(b_0) \leq f(b_0)$. We conclude by the Squeeze Theorem that $R_n(x) \to 0$ as $n \to \infty$. ∎

We shall now use Bernstein's Theorem to generalize the Binomial Formula (compare Theorem 1.5 with Theorem 4.45 below). First we introduce some notation. Let $\alpha \in \mathbf{R}$ and $k \in \mathbf{N}$. The *generalized binomial coefficient* α *over* k is defined by

$$\binom{\alpha}{k} := \begin{cases} \dfrac{\alpha(\alpha-1)\dots(\alpha-k+1)}{k!} & k \neq 0 \\ 1 & k = 0. \end{cases}$$

Notice that when $\alpha \in \mathbf{N}$, these generalized binomial coefficients coincide with the usual binomial coefficients, because in this case $\binom{\alpha}{k} = 0$ for $k > \alpha$.

Next, we show that the center of a power series can be changed within its interval of convergence.

*Lemma 1. *Suppose I is an open interval centered at c and*

$$f(x) = \sum_{k=0}^{\infty} a_k (x-c)^k, \qquad x \in I.$$

If $x_0 \in I$ and $r > 0$ satisfy $(x_0 - r, x_0 + r) \subseteq I$, then

$$f(x) = \sum_{k=0}^{\infty} \frac{f^{(k)}(x_0)}{k!}(x - x_0)^k$$

for all $x \in (x_0 - r, x_0 + r)$.

PROOF. By making the change of variables $w = x - c$, we may suppose that $c = 0$ and $I = (-R, R)$, i.e., that

$$f(x) = \sum_{k=0}^{\infty} a_k x^k, \qquad x \in (-R, R).$$

Suppose $(x_0 - r, x_0 + r) \subseteq (-R, R)$ and fix $x_0 \in (x_0 - r, x_0 + r)$. By hypothesis and the Binomial Formula,

$$(32) \quad f(x) = \sum_{k=0}^{\infty} a_k x^k = \sum_{k=0}^{\infty} a_k ((x - x_0) + x_0)^k = \sum_{k=0}^{\infty} a_k \sum_{j=0}^{k} \binom{k}{j} x_0^{k-j} (x - x_0)^j.$$

Since $\sum_{k=0}^{\infty} a_k y^k$ converges absolutely at $y := |x - x_0| + |x_0| < R$, we have

$$\sum_{k=0}^{\infty} \left| a_k \sum_{j=0}^{k} \binom{k}{j} x_0^{k-j} (x - x_0)^j \right| \leq \sum_{k=0}^{\infty} |a_k| \sum_{j=0}^{k} \binom{k}{j} |x_0|^{k-j} |x - x_0|^j$$

$$= \sum_{k=0}^{\infty} |a_k| (|x - x_0| + |x_0|)^k < \infty.$$

Hence, by (32), Theorem 4.29, and Corollary 4.35,

$$f(x) = \sum_{k=0}^{\infty} a_k \sum_{j=0}^{k} \binom{k}{j} x_0^{k-j} (x - x_0)^j$$

$$= \sum_{j=0}^{\infty} \left(\sum_{k=j}^{\infty} \binom{k}{j} a_k x_0^{k-j} \right) (x - x_0)^j$$

$$= \sum_{j=0}^{\infty} \left(\sum_{k=j}^{\infty} \frac{k!}{(k-j)!} a_k (x_0 - 0)^{k-j} \right) \frac{(x - x_0)^j}{j!}$$

$$= \sum_{j=0}^{\infty} \frac{f^{(j)}(x_0)}{j!} (x - x_0)^j. \quad \blacksquare$$

Our final preliminary result is a special case of the Binomial Theorem.

*Lemma 2. If $\beta > 0$, then $f(t) = (1 - t)^{-\beta}$ is analytic on $(-\infty, 1)$ and

$$(33) \qquad\qquad f(t) = \sum_{k=0}^{\infty} \binom{-\beta}{k} (-1)^k t^k$$

for all $|t| < 1$.

PROOF. Since $f^{(n)}(t) = \beta(\beta + 1) \dots (\beta + n - 1)(1 - t)^{-\beta - n}$, $f^{(n)}(t) \geq 0$ for all $n \in \mathbf{N}$ and $t < 1$. Hence, by Bernstein's Theorem, f is analytic on $(-\infty, 1)$ and the Taylor series of f centered at $c = -1$ converges to f on $[-1, 1)$. Hence, by Lemma 1, the Maclaurin series of f converges to f on $(-1, 1)$. Since

$$f^{(n)}(0) = (-1)^n \binom{-\beta}{n} n!$$

it follows that (33) holds for all $|t| < 1$. $\quad \blacksquare$

*THEOREM 4.45 [THE BINOMIAL SERIES]. *If $\alpha \in \mathbf{R}$ and $|x| < 1$, then*

$$(34) \qquad\qquad (1+x)^\alpha = \sum_{k=0}^{\infty} \binom{\alpha}{k} x^k.$$

In particular, $(1+x)^\alpha$ is analytic on $(-1, 1)$ for all $\alpha \in \mathbf{R}$.

PROOF. We first show that (34) holds for $\alpha < 0$. Indeed, fix $\alpha < 0$, $x \in (-1, 1)$, and set $\beta = -\alpha$, $t = -x$. Then by Lemma 2,

$$(1+x)^\alpha = (1-t)^{-\beta} = \sum_{k=0}^{\infty} \binom{-\beta}{k}(-1)^k t^k = \sum_{k=0}^{\infty} \binom{\alpha}{k} x^k.$$

Next, we show that if (34) holds for some $\alpha \in \mathbf{R}$, then (34) holds for $\alpha + 1$. (Since we have already verified (34) for $\alpha < 0$, this proves that (34) holds for $\alpha < 1$, $\alpha < 2$, ..., i.e., for all $\alpha \in \mathbf{R}$.) Suppose (34) holds for some $\alpha \in \mathbf{R}$. Integrate (34) term-by-term to obtain

$$\frac{(1+x)^{\alpha+1}}{\alpha+1} = \sum_{k=0}^{\infty} \binom{\alpha}{k} \frac{x^{k+1}}{k+1} + C.$$

Evaluating the constant of integration by setting $x = 0$, we see that

$$(35) \qquad\qquad (1+x)^{\alpha+1} = (\alpha+1)\sum_{k=0}^{\infty} \binom{\alpha}{k} \frac{x^{k+1}}{k+1} + 1.$$

But by definition,

$$\binom{\alpha}{k}\frac{\alpha+1}{k+1} = \binom{\alpha+1}{k+1}.$$

Therefore, it follows from (35) that

$$(1+x)^{\alpha+1} = \sum_{k=0}^{\infty} \binom{\alpha+1}{k+1} x^{k+1} + 1 = \sum_{k=0}^{\infty} \binom{\alpha}{k} x^k. \quad \blacksquare$$

We close this section by showing that an analytic function cannot be extended in an arbitrary way to produce another analytic function.

First, we prove that if a power series centered at x_0 is zero on one side of x_0, then it is zero on both sides of x_0.

*Lemma 3. *Suppose f, g are analytic on an open interval (c, d) and $x_0 \in (c, d)$. If $f(x) = g(x)$ for $x \in (c, x_0)$, then there is a $\delta > 0$ such that $f(x) = g(x)$ for all $x \in (x_0 - \delta, x_0 + \delta)$.*

PROOF. By Theorem 4.39 and Definition 4.9, there is a $\delta > 0$ such that

$$(36) \qquad f(x) = \sum_{k=0}^{\infty} \frac{f^{(k)}(x_0)}{k!}(x - x_0)^k \quad \text{and} \quad g(x) = \sum_{k=0}^{\infty} \frac{g^{(k)}(x_0)}{k!}(x - x_0)^k$$

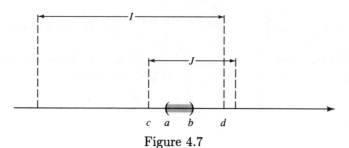

Figure 4.7

for all $x \in (x_0 - \delta, x_0 + \delta)$. By hypothesis, f, g are continuous at x_0 and

$$f(x_0) = \lim_{x \to x_0-} f(x) = \lim_{x \to x_0-} g(x) = g(x_0).$$

Similarly, $f^{(k)}(x_0) = g^{(k)}(x_0)$ for $k \in \mathbf{N}$. We conclude from (36) that $f(x) = g(x)$ for all $x \in (x_0 - \delta, x_0 + \delta)$. ∎

*THEOREM 4.46 [ANALYTIC CONTINUATION]. *Suppose I and J are open intervals, f is analytic on I, g is analytic on J, and $a < b$ are points in $I \cap J$. If $f(x) = g(x)$ for $x \in (a, b)$, then $f(x) = g(x)$ for all $x \in I \cap J$.*

PROOF. We assume for simplicity that I and J are bounded intervals. Since $I \cap J \neq \emptyset$, choose $c, d \in \mathbf{R}$ such that $I \cap J = (c, d)$ (see Figure 4.7).

Consider the set $E = \{t \in (a, d) : f(x) = g(x) \text{ for all } x \in (a, t)\}$. By our assumption, $d < \infty$ and by hypothesis $b \in E$. Thus, E is bounded and nonempty. Let $x_0 = \sup E$. If $x_0 < d$, then by Lemma 3 there is a $\delta > 0$ such that $f(x) = g(x)$ for all $x \in (x_0 - \delta, x_0 + \delta)$. This contradicts the choice of x_0. Therefore, $x_0 = d$, i.e., $f(x) = g(x)$ for all $x \in (a, d)$. A similar argument proves $f(x) = g(x)$ for all $x \in (c, b)$. ∎

EXERCISES

1. Prove that each of the following functions is analytic on $\mathbf{R}$ and find its Maclaurin expansion.

   a) $\cos(3x)$.   b) 2^x.   c) $\cos^2 x$.   d) $\sin^2 x + \cos^2 x$.   e) $x^3 e^{x^2}$.

2. Prove that each of the following functions is analytic on $(-1, 1)$ and find its Maclaurin expansion.

   a) $\log(1 - x)$.   b) $\dfrac{x^2}{1 - x^3}$.   c) $\dfrac{e^x}{1 - x}$.   d) $\dfrac{x^3}{(1 - x)^2}$.   *e) $\arcsin x$.

3. Prove that each of the following functions is analytic on $\mathbf{R}$ and find its Maclaurin expansion.

   a) $(x^2 - 1)e^x$.   b) $e^x \cos x$.   c) $\dfrac{\sin x}{e^x}$.

d) $f(x) := \begin{cases} (a^x - 1)/x & x \neq 0 \\ 1 & x = 0 \end{cases}$ (where $a > 0$ fixed).

4. For each of the following functions, find its Taylor expansion centered at $x_0 = 1$ and determine the largest interval on which it converges.

$\quad$ a) $\log_{10} x$. $\quad$ b) $x^2 + 2x - 1$. $\quad$ c) e^x.

5. Let $f \in C^\infty(a, b)$. Prove that f is analytic on (a, b) if and only if f' is analytic on (a, b).

6. a) Prove

$$|\delta + \sin(\delta + \pi)| \leq \frac{\delta^3}{3!}$$

$\quad$ for all $\delta > 0$.

$\quad$ b) Prove that if $|x - \pi| \leq \delta$, then $|x + \sin x - \pi| \leq \delta^3/3!$.

7. Suppose $f \in C^\infty(-\infty, \infty)$ and

$$\lim_{n\to\infty} \frac{1}{n!} \int_0^a x^n f^{(n+1)}(a - x)\, dx = 0$$

for all $a \in \mathbf{R}$. Prove f is analytic on $(-\infty, \infty)$ and

$$f(x) = \sum_{k=0}^\infty \frac{f^{(k)}(0)}{k!} x^k, \qquad x \in \mathbf{R}.$$

8. a) Prove that

$$\left| \int_0^1 e^{x^2}\, dx - \sum_{k=0}^{n-1} \frac{1}{(2k+1)k!} \right| \leq \frac{3}{n!}$$

$\quad$ for $n \in \mathbf{N}$.

$\quad$ b) Show that

$$2.9253 < \int_{-1}^1 e^{x^2}\, dx < 2.9254.$$

*9. Suppose f is analytic on $(-\infty, \infty)$ and

$$\int_a^b |f(x)|\, dx = 0$$

for some $a \neq b$ in $\mathbf{R}$. Prove that $f(x) = 0$ for all $x \in \mathbf{R}$.

*10. Prove that

$$\left(\sum_{k=1}^\infty |a_k|^\beta \right)^{1/\beta} \leq \sum_{k=1}^\infty |a_k|$$

for all $a_k \in \mathbf{R}$ and all $\beta > 1$.

^{e}**4.8 APPLICATIONS**   *This section uses no material from any other enrichment section.*

The theory of infinite series is a potent tool for both pure and applied mathematics. In this section we give several examples to back up this claim.

We begin with a nontrivial theorem from number theory. Recall that an integer $n \geq 2$ is called *prime* if the only factors of n in $\mathbf{N}$ are 1 and n. Also recall that given $n \in \mathbf{N}$ there are primes $p_1, p_2, \ldots, p_k$ and exponents $\alpha_1, \alpha_2, \ldots, \alpha_k$ such that

$$n = p_1^{\alpha_1} p_2^{\alpha_2} \cdots p_k^{\alpha_k}.$$

THEOREM 4.47 [EUCLID'S THEOREM; EULER'S PROOF]. *There are infinitely many primes in* $\mathbf{N}$.

PROOF. Suppose to the contrary that $p_1, p_2, \ldots, p_k$ represent all the primes of $\mathbf{N}$. Fix $N \in \mathbf{N}$ and set $\alpha = \sup\{\alpha_1, \ldots, \alpha_k\}$, where this supremum is taken over all α_j's which satisfy

$$n = p_1^{\alpha_1} p_2^{\alpha_2} \cdots p_k^{\alpha_k}$$

for some $n \leq N$. Since every integer $j \in [1, N]$ must have the form $j = p_1^{e_1} \cdots p_k^{e_k}$ for some choice of integers $0 \leq e_i \leq \alpha$, we have

$$\left(1 + \frac{1}{p_1} + \cdots + \frac{1}{p_1^{\alpha}}\right) \left(1 + \frac{1}{p_2} + \cdots + \frac{1}{p_2^{\alpha}}\right) \cdots \left(1 + \frac{1}{p_k} + \cdots + \frac{1}{p_k^{\alpha}}\right)$$

$$= \sum_{1 \leq e_i \leq \alpha} 1 \cdot \frac{1}{p_1^{e_1}} \cdots \frac{1}{p_k^{e_k}} \geq \sum_{j=1}^{N} \frac{1}{j}.$$

On the other hand, for each integer $i \in [1, k]$, we have by Theorem 4.3 that

$$1 + \frac{1}{p_i} + \cdots + \frac{1}{p_i^{\alpha}} \leq \sum_{k=1}^{\infty} \left(\frac{1}{p_i}\right)^k = \frac{p_i}{p_i - 1}.$$

Consequently,

$$\sum_{j=1}^{N} \frac{1}{j} \leq \left(\frac{p_1}{p_1 - 1}\right) \cdots \left(\frac{p_k}{p_k - 1}\right) = M < \infty.$$

Taking the limit of this inequality as $N \to \infty$, we conclude that $\sum_{j=1}^{\infty} 1/j \leq M < \infty$, a contradiction. ∎

Our next application, a result used to approximate roots of twice differentiable functions, shows that if an initial guess x_0 is close enough to a root of a suitably well-behaved function f, then the sequence x_n generated by (37) converges to a root of f.

THEOREM 4.48 [NEWTON]. *Suppose $f : [a, b] \to \mathbf{R}$ is continuous on $[a, b]$ and $f(c) = 0$ for some $c \in (a, b)$. If f'' exists and is bounded on (a, b) and there is an $\varepsilon_0 > 0$ such that*

$$|f'(x)| \geq \varepsilon_0$$

for all $x \in (a, b)$, then there is a closed interval $I \subseteq (a, b)$ containing c such that given $x_0 \in I$, the sequence $\{x_n\}_{n \in \mathbf{N}}$ defined by

$$(37) \qquad\qquad x_n = x_{n-1} - \frac{f(x_{n-1})}{f'(x_{n-1})} \qquad n \in \mathbf{N},$$

satisfies $x_n \in I$ and $x_n \to c$ as $n \to \infty$.

PROOF. Choose $M > 0$ such that $|f''(x)| \leq M$ for $x \in (a, b)$. Choose $r_0 \in (0, 1)$ so small that $I = [c - r_0, c + r_0]$ is a subinterval of (a, b) and $r_0 < \varepsilon_0/M$. Suppose $x_0 \in I$ and define the sequence $\{x_n\}$ by (37). Set $r := r_0 M/\varepsilon_0$ and observe by the choice of r_0 that $r < 1$. Thus, it suffices to show that

$$(38) \qquad\qquad |x_n - c| \leq r^n |x_0 - c|$$

and

$$(39) \qquad\qquad |x_n - c| < r_0$$

hold for all $n \in \mathbf{N}$.

The proof is by induction on n. Clearly, (38) and (39) hold for $n = 0$. Fix $n \in \mathbf{N}$ and suppose

$$(40) \qquad\qquad |x_{n-1} - c| \leq r^{n-1} |x_0 - c|$$

and

$$(41) \qquad\qquad |x_{n-1} - c| < r_0.$$

Use Taylor's Formula to choose a point ξ between c and x_{n-1} such that

$$-f(x_{n-1}) = f(c) - f(x_{n-1}) = f'(x_{n-1})(c - x_{n-1}) + \frac{1}{2} f''(\xi)(c - x_{n-1})^2.$$

Since (37) implies $-f(x_{n-1}) = f'(x_{n-1})(x_n - x_{n-1})$, it follows that

$$f'(x_{n-1})(x_n - c) = \frac{1}{2} f''(\xi)(c - x_{n-1})^2.$$

Solving this equation for $x_n - c$, we have by the choice of M and ε_0 that

$$(42) \qquad\qquad |x_n - c| = \left| \frac{f''(\xi)}{2f'(x_{n-1})} \right| |x_{n-1} - c|^2 < \frac{M}{2\varepsilon_0} |x_{n-1} - c|^2.$$

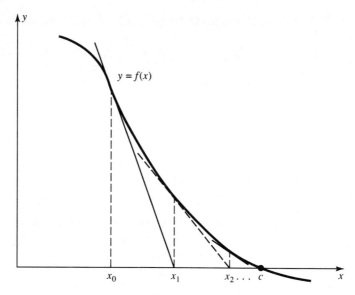

Figure 4.8

Since $M/\varepsilon_0 < 1/r_0$, it follows from (42) and (41) that

$$|x_n - c| < \frac{M}{\varepsilon_0}|x_{n-1} - c|^2 < \frac{1}{r_0}|r_0|^2 = r_0.$$

This proves (39). Again, by (42), (40), and the choice of r, we have

$$|x_n - c| \le \frac{M}{\varepsilon_0}(r^{n-1}|x_0 - c|)^2 = \frac{r}{r_0}(r^{2n-2}|x_0 - c|^2) < r^{2n-1}|x_0 - c|.$$

Since $r < 1$ and $2n - 1 \ge n$ imply $r^{2n-1} \le r^n$, we conclude that

$$|x_n - c| < r^{2n-1}|x_0 - c| \le r^n|x_0 - c|. \quad \blacksquare$$

Notice if x_{n-1} and x_n satisfy (37), then x_n is the x–intercept of the tangent line to $y = f(x)$ at the point $(x_{n-1}, f(x_{n-1}))$ (see Exercise 4). Thus, Newton's method is based on a simple geometric principle (see Figure 4.8). Also notice that by (42), this method converges very rapidly. Indeed, the number of decimal places of accuracy nearly doubles with each successive approximation.

As a general rule, it is extremely difficult to show a particular number is irrational. Infinite series can be used to give an easy proof that certain kinds of numbers are irrational.

THEOREM 4.49 [HERMITE]. *The number e is irrational.*

PROOF. Suppose to the contrary that $e = p/q$ for some $p, q \in \mathbf{N}$. By Example 2 in Section 4.7,

$$\frac{q}{p} = e^{-1} = \sum_{k=0}^{\infty} \frac{(-1)^k}{k!}.$$

Breaking this sum into two pieces and multiplying by $(-1)^{p+1}p!$, we have

$$x := (-1)^{p+1}\left(q(p-1)! - \sum_{k=0}^{p}\frac{(-1)^k p!}{k!}\right) = y := \sum_{k=p+1}^{\infty}(-1)^{k+p+1}\frac{p!}{k!}.$$

Since $p!/k! \in \mathbf{N}$ for all integers $k \le p$, the number x must be an integer. On the other hand,

$$y = \frac{1}{p+1} - \frac{1}{(p+1)(p+2)} + \frac{1}{(p+1)(p+2)(p+3)} - \cdots$$

lies between $1/(p+1)$ and $1/(p+1) - 1/(p+1)(p+2)$. Therefore, y is a number which satisfies $0 < y < 1$. In particular, $x \ne y$, a contradiction. ∎

We know that a continuous function can fail to be differentiable at one point (e.g., $f(x) = |x|$). Hence, it is not difficult to see that given any finite set of points E, there is a continuous function which fails to be differentiable at every point in E. We shall now show that there is a continuous function which fails to be differentiable at any point in $\mathbf{R}$. Once again, here is a clear indication that although we use sketches to motivate proofs and to explain results, we cannot rely on sketches to give a complete picture of the general situation.

THEOREM 4.50 [WEIERSTRASS]. *There is a function f continuous on $\mathbf{R}$ which is not differentiable at any point in $\mathbf{R}$.*
Note: Such functions are called *nowhere differentiable*.

PROOF. Let

$$f_0(x) = \begin{cases} x & 0 \le x < 1/2 \\ 1-x & 1/2 \le x < 1 \end{cases}$$

and extend f to $\mathbf{R}$ by periodicity of period 1, i.e., so that $f_0(x) = f_0(x+1)$ for all $x \in \mathbf{R}$ (see Figure 4.9). Set $f_k(x) = f_0(2^k x)/2^k$ for $x \in \mathbf{R}$ and $k \in \mathbf{N}$ and consider the function

$$f(x) = \sum_{k=0}^{\infty} f_k(x), \qquad x \in \mathbf{R}.$$

Normalizing f_k by 2^k has two consequences. First, since $f_0'(y) = \pm 1$ for each y which satisfies $2y \notin \mathbf{Z}$, it is easy to see that

(43) $f_k'(y) = \pm 1$ for each y which satisfies $2^{k+1}y \notin \mathbf{Z}$.

Second, by the Weierstrass M–Test, f converges uniformly, hence, is continuous on $\mathbf{R}$.

Since f is periodic of period 1, it suffices to show that f is not differentiable at any $x \in [0,1)$. Suppose to the contrary that f is differentiable at some $x \in [0,1)$. For each $n \in \mathbf{N}$, choose $p \in \mathbf{Z}$ such that $x \in [\alpha_n, \beta_n)$ for $\alpha_n = p/2^n$ and $\beta_n = (p+1)/2^n$. Notice that if h is any function differentiable at x, then

$$h'(x) = \lim_{n \to \infty} \frac{h(\beta_n) - h(\alpha_n)}{\beta_n - \alpha_n}.$$

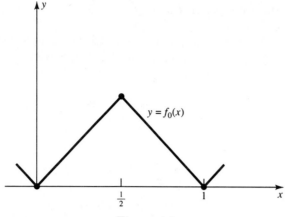

$$y = f_0(x)$$

Figure 4.9

Hence, by (43),

$$(44) \qquad \frac{f_k(\beta_n) - f_k(\alpha_n)}{\beta_n - \alpha_n} = f_k'(x) = \pm 1$$

for each $n > k$. On the other hand, since $f_0(y) = 0$ if and only if $y \in \mathbf{Z}$, it is clear that $f_k(\beta_n) = f_k(\alpha_n) = 0$ for $k \geq n$. Therefore,

$$f(\beta_n) = \sum_{k=0}^{n-1} f_k(\beta_n) \quad \text{and} \quad f(\alpha_n) = \sum_{k=0}^{n-1} f_k(\alpha_n).$$

Since $f'(x)$ exists, it follows from (44) that

$$f'(x) = \lim_{n \to \infty} \frac{f(\beta_n) - f(\alpha_n)}{\beta_n - \alpha_n} = \lim_{n \to \infty} \sum_{k=0}^{n-1} \frac{f_k(\beta_n) - f_k(\alpha_n)}{\beta_n - \alpha_n} = \sum_{k=0}^{\infty} f_k'(x).$$

Hence, by the Divergence Test, $\pm 1 = f_k'(x) \to 0$ as $k \to \infty$, a contradiction. ∎

EXERCISES

1. Using a calculator and Theorem 4.48, approximate all real roots of $f(x) = x^3 + 3x^2 + 4x + 1$ to five decimal places.

2. a) Using the proof of Theorem 4.48, prove that (38) holds if $r/2$ replaces r.
   b) Use part a) to estimate the difference $|x_4 - \pi|$, where $x_0 = 3$, $f(x) = \sin x$, and x_n is defined by (37). Evaluate x_4 directly and verify that x_4 is actually closer than our theory predicts.

3. Prove that given any $n \in \mathbf{N}$, there is a function $f \in C^n(\mathbf{R})$ such that $f^{(n+1)}(x)$ does not exist for any $x \in \mathbf{R}$.

4. Prove that if x_{n-1}, x_n satisfy (37), then x_n is the x–intercept of the tangent line

to $y = f(x)$ at the point $(x_{n-1}, f(x_{n-1}))$.

5. Prove that $\cos(1)$ is irrational.

6. Suppose $f : \mathbf{R} \to \mathbf{R}$. If f'' exists and is bounded on $\mathbf{R}$, and there is an $\varepsilon_0 > 0$ such that $|f'(x)| \geq \varepsilon_0$ for all $x \in \mathbf{R}$, prove that there exists a $\delta > 0$ such that if $|f(x_0)| \leq \delta$ for some $x_0 \in \mathbf{R}$, then f has a root; i.e., there $f(c) = 0$ for some $c \in \mathbf{R}$.

   Hint: Choose r_0 as in the proof of Theorem 4.48, define $\{x_n\}$ by (37), and find a δ so that $|f(x_0)| \leq \delta$ implies $|x_n - x_{n-1}| < r_0^{n+1}$.

Chapter 5

EUCLIDEAN SPACES

The world we live in is at least four-dimensional: three spatial dimensions together with the time dimension. Moreover, certain problems from engineering, physics, chemistry, and economics force us to consider even higher dimensions. For example, guidance systems for missiles frequently require as many as 100 variables (longitude, latitude, altitude, velocity, time after launch, pitch, yaw, fuel on board, etc.). Another example, the state of a gas in a closed container, can best be described by a function of $6m$ variables, where m is the number of molecules in the system. (Six enters the picture because each molecule of gas is described by three space variables and three momentum variables.) Thus, there are practical reasons for studying functions of more than one variable.

For each $n \in \mathbf{N}$, set

$$\mathbf{R}^n := \{(x_1, x_2, \ldots, x_n) : x_j \in \mathbf{R} \text{ for } j = 1, 2, \ldots, n\}.$$

Elements $\boldsymbol{x} = (x_1, x_2, \ldots, x_n)$ of $\mathbf{R}^n$ are called *points* or *vectors* or *ordered n-tuples* and each number x_j is called the *jth coordinate* or *component* of $\boldsymbol{x}$. When $n = 2$ (respectively, $n = 3$), we usually denote the components of $\boldsymbol{x}$ by x, y (respectively, by x, y, z).

The reader has already encountered the sets $\mathbf{R}^n$ for small n. $\mathbf{R}^1 = \mathbf{R}$ is the real line; we shall call its elements *scalars*. $\mathbf{R}^2$ is the xy plane used to graph functions of the form $y = f(x)$. And $\mathbf{R}^3$ is the xyz space used to graph functions of the form $z = f(x, y)$.

In preceding chapters we developed the calculus of one-variable functions. In this chapter we shall lay foundations for a calculus of functions of several variables, i.e., functions defined on some subset of $\mathbf{R}^n$. Much of what we intend to do has been anticipated in Chapters 1 and 2. When there is no significant difference between the one-dimensional case and the multidimensional case, we shall leave the details as exercises.

5.1 ALGEBRAIC STRUCTURE OF $\mathbf{R}^n$

We began our study of one-dimensional calculus by examining the algebraic structure of $\mathbf{R}$. In this section we examine the algebraic structure of $\mathbf{R}^n$.

DEFINITION 5.1. Let $\boldsymbol{x} = (x_1, \ldots, x_n)$, $\boldsymbol{y} = (y_1, \ldots, y_n) \in \mathbf{R}^n$ be vectors and $\alpha \in \mathbf{R}$ be a scalar.

   i) $\boldsymbol{x}, \boldsymbol{y}$ are said to be *equal* if their components are equal, i.e., if $x_j = y_j$ for $j = 1, 2, \ldots, n$.

   ii) The *zero vector* is the vector whose components are all zero, i.e., $\boldsymbol{0} := (0, 0, \ldots, 0)$.

   iii) The *usual basis* of $\mathbf{R}^n$ is the collection $\{\boldsymbol{e}_1, \ldots, \boldsymbol{e}_n\}$, where $\boldsymbol{e}_j$ is the point in $\mathbf{R}^n$ whose jth coordinate is 1 and all other coordinates are 0. (In $\mathbf{R}^2$ or $\mathbf{R}^3$, $\boldsymbol{e}_1$ is denoted by $\mathbf{i}$, $\boldsymbol{e}_2$ is denoted by $\mathbf{j}$, and, in $\mathbf{R}^3$, $\boldsymbol{e}_3$ is denoted by $\mathbf{k}$. Thus, in $\mathbf{R}^3$, $\mathbf{i} := (1, 0, 0)$, $\mathbf{j} := (0, 1, 0)$, and $\mathbf{k} := (0, 0, 1)$.)

   iv) The *sum* of $\boldsymbol{x}$ and $\boldsymbol{y}$ is the vector
$$\boldsymbol{x} + \boldsymbol{y} := (x_1 + y_1, x_2 + y_2, \ldots, x_n + y_n).$$

   v) The *difference* of $\boldsymbol{x}$ and $\boldsymbol{y}$ is the vector
$$\boldsymbol{x} - \boldsymbol{y} := (x_1 - y_1, x_2 - y_2, \ldots, x_n - y_n).$$

   vi) The *product* of a scalar α and a vector $\boldsymbol{x}$ is the vector
$$\alpha\boldsymbol{x} := (\alpha x_1, \alpha x_2, \ldots, \alpha x_n).$$

   vii) The (*Euclidean*) *dot product* (or *scalar product* or *inner product*) of $\boldsymbol{x}$ and $\boldsymbol{y}$ is the scalar
$$\boldsymbol{x} \cdot \boldsymbol{y} := x_1 y_1 + x_2 y_2 + \cdots + x_n y_n.$$

   viii) Two nonzero vectors $\boldsymbol{x}, \boldsymbol{y}$ are said to be *orthogonal* if $\boldsymbol{x} \cdot \boldsymbol{y} = 0$.

Let $\boldsymbol{x} = (x_1, \ldots, x_n) \in \mathbf{R}^n$. By definition,
$$\boldsymbol{x} = \sum_{j=1}^{n} x_j \boldsymbol{e}_j.$$

Hence, any vector in $\mathbf{R}^2$ can be written as $x\mathbf{i} + y\mathbf{j}$ and any vector in $\mathbf{R}^3$ can be written as $x\mathbf{i} + y\mathbf{j} + z\mathbf{k}$. Also notice that if $i \neq j$, then $\boldsymbol{e}_i \cdot \boldsymbol{e}_j = 0$. Thus, the usual basis is an *orthogonal basis*.

By a *Euclidean space* we mean one of the spaces $\mathbf{R}^n$ together with the Euclidean dot product. We shall refer to $\mathbf{R}^n$ as *n-dimensional Euclidean space*. We shall not discuss other bases of $\mathbf{R}^n$ or the more general concept of "vector spaces," which can be introduced using postulates similar in spirit to Postulate 1 in Chapter 1. Instead, we will introduce just enough algebraic machinery in $\mathbf{R}^n$ to develop the calculus of multivariable functions. For more information about $\mathbf{R}^n$ and abstract vector spaces, see Noble and Daniel [9].

The algebraic operations introduced on $\mathbf{R}^n$ above are generalizations of addition, subtraction, and multiplication on $\mathbf{R}$. It is natural to ask: Do the usual laws of algebra hold in $\mathbf{R}^n$? An answer to this question is contained in the following result.

THEOREM 5.1. *Let* $\boldsymbol{x}, \boldsymbol{y}, \boldsymbol{z} \in \mathbf{R}^n$ *and* $\alpha, \beta \in \mathbf{R}$. *Then* $\alpha \boldsymbol{0} = \boldsymbol{0}$, $0\boldsymbol{x} = \boldsymbol{0}$, $1\boldsymbol{x} = \boldsymbol{x}$, $\alpha(\beta\boldsymbol{x}) = \beta(\alpha\boldsymbol{x}) = (\alpha\beta)\boldsymbol{x}$, $\alpha(\boldsymbol{x} \cdot \boldsymbol{y}) = (\alpha\boldsymbol{x}) \cdot \boldsymbol{y} = \boldsymbol{x} \cdot (\alpha\boldsymbol{y})$, $\alpha(\boldsymbol{x} + \boldsymbol{y}) = \alpha\boldsymbol{x} + \alpha\boldsymbol{y}$, $\boldsymbol{0} + \boldsymbol{x} = \boldsymbol{x}$, $\boldsymbol{x} - \boldsymbol{x} = \boldsymbol{0}$, $0 \cdot \boldsymbol{x} = 0$, $\boldsymbol{x} + (\boldsymbol{y} + \boldsymbol{z}) = (\boldsymbol{x} + \boldsymbol{y}) + \boldsymbol{z}$, $\boldsymbol{x} + \boldsymbol{y} = \boldsymbol{y} + \boldsymbol{x}$, $\boldsymbol{x} \cdot \boldsymbol{y} = \boldsymbol{y} \cdot \boldsymbol{x}$, *and* $\boldsymbol{x} \cdot (\boldsymbol{y} + \boldsymbol{z}) = \boldsymbol{x} \cdot \boldsymbol{y} + \boldsymbol{x} \cdot \boldsymbol{z}$.

PROOF. These properties are direct consequences of Definition 5.1 and corresponding properties of real numbers. We will prove the associative property for vector addition, and leave the proof of the rest of these properties as an exercise.

By definition and the Associative Property of Addition on $\mathbf{R}$ (see Postulate 1 in Section 1.1),

$$
\begin{aligned}
\boldsymbol{x} + (\boldsymbol{y} + \boldsymbol{z}) &= (x_1, \ldots, x_n) + (y_1 + z_1, \ldots, y_n + z_n) \\
&= (x_1 + (y_1 + z_1), \ldots, x_n + (y_n + z_n)) \\
&= ((x_1 + y_1) + z_1, \ldots, (x_n + y_n) + z_n) = (\boldsymbol{x} + \boldsymbol{y}) + \boldsymbol{z}. \quad \blacksquare
\end{aligned}
$$

There are many ways to measure the distance between two points in $\mathbf{R}^n$, but only one "Euclidean" way.

DEFINITION 5.2. *Let* $\boldsymbol{x}, \boldsymbol{y} \in \mathbf{R}^n$.

i) The *sup–norm* of a vector $\boldsymbol{x} \in \mathbf{R}^n$ is the scalar

$$
\|\boldsymbol{x}\|_\infty := \max\{x_1, \ldots, x_n\}.
$$

ii) The *(Euclidean) norm* of a vector $\boldsymbol{x} \in \mathbf{R}^n$ is the scalar

$$
\|\boldsymbol{x}\| := \sqrt{\boldsymbol{x} \cdot \boldsymbol{x}}.
$$

iii) The *(Euclidean) distance* between two points $\boldsymbol{a}, \boldsymbol{b} \in \mathbf{R}^n$ to be $\|\boldsymbol{a} - \boldsymbol{b}\|$.

(Note: The subscript ∞ is frequently used for supremum norms because the supremum of a continuous function on a closed bounded interval can be computed by a certain limit as $p \to \infty$—see Exercise 8 in Section 3.2.)

Since $\boldsymbol{x} \cdot \boldsymbol{x} = x_1^2 + x_2^2 + \cdots + x_n^2$, it is clear that

$$
\|\boldsymbol{x}\| = \sqrt{x_1^2 + x_2^2 + \cdots + x_n^2}.
$$

Hence, by the Pythagorean Theorem, the norm of a vector $\boldsymbol{x} \in \mathbf{R}^2$ is the distance between the points $\boldsymbol{x}$ and $\boldsymbol{0}$. Thus, Definition 5.2iii is consistent with the usual notion of distance in $\mathbf{R}^2$. (For a relationship between the Euclidean norm and the sup–norm, see Theorem 5.3v below.)

Here is a fundamental inequality which relates the absolute value of a dot product of two vectors to the product of their norms. (Some authors call this the Cauchy–Schwarz–Bunyakovsky Inequality.)

THEOREM 5.2 [CAUCHY–SCHWARZ INEQUALITY]. *If* $x, y \in \mathbf{R}^n$, *then*

$$|x \cdot y| \le \|x\| \, \|y\|.$$

PROOF. The inequality is trivial when $y = 0$. Suppose $y \ne 0$. By definition,

$$(1) \qquad 0 \le \|x - ty\|^2 = (x - ty) \cdot (x - ty) = \|x\|^2 - 2t(x \cdot y) + t^2 \|y\|^2$$

holds for any scalar t. Substituting the value $t = (x \cdot y)/\|y\|^2$ into (1), we obtain

$$0 \le \|x\|^2 - t(x \cdot y) = \|x\|^2 - \frac{(x \cdot y)^2}{\|y\|^2},$$

i.e., $0 \le \|x\|^2 - (x \cdot y)^2/\|y\|^2$. Solving this inequality for $(x \cdot y)^2$, we conclude that

$$(x \cdot y)^2 \le \|x\|^2 \|y\|^2. \quad \blacksquare$$

The analogy between $\mathbf{R}$ and $\mathbf{R}^n$ is reinforced further by an analogy between the absolute value and the norm (compare the following result with Theorem 1.3 and the remarks preceding it).

THEOREM 5.3. *Let* $x, y \in \mathbf{R}^n$. *Then*

    i) $\|x\| \ge 0$ *with equality only when* $x = 0$,
    ii) $\|\alpha x\| = |\alpha| \, \|x\|$ *for all scalars* α,
    iii) [TRIANGLE INEQUALITIES]. $\|x + y\| \le \|x\| + \|y\|$ *and* $\|x - y\| \ge \|x\| - \|y\|$,
    iv) $\|x\| \le \sum_{j=1}^{n} |x_j|$, *and*
    v) $|x_j| \le \|x\| \le \sqrt{n} \, \|x\|_\infty$ *for each* $j = 1, 2, \ldots, n$.

PROOF. Statements i) and ii) are obvious.

To prove the first inequality in part iii), observe that by definition, Theorem 5.1, and the Cauchy–Schwarz Inequality,

$$\|x + y\|^2 = (x + y) \cdot (x + y) = x \cdot x + 2x \cdot y + y \cdot y$$
$$= \|x\|^2 + 2x \cdot y + \|y\|^2 \le \|x\|^2 + 2\|x\| \, \|y\| + \|y\|^2 = (\|x\| + \|y\|)^2.$$

The second inequality follows directly from the first (see the proof of Theorem 1.3).

Part v) is trivial because

$$|x_\ell|^2 \le \|x\|^2 = x_1^2 + \cdots + x_n^2 \le n(\max_{1 \le j \le n} |x_j|)^2.$$

To prove part iv), observe that

$$(|x_1| + \cdots + |x_n|)^2 = |x_1|^2 + \cdots + |x_n|^2 + 2 \sum_{(i,j) \in A} |x_i| \, |x_j|,$$

where $A = \{(i,j) : 1 \le i,j \le n \text{ and } i < j\}$. Since

$$\sum_{(i,j) \in A} |x_i| \, |x_j| \ge 0,$$

we conclude that

$$\|\boldsymbol{x}\|^2 = x_1^2 + \cdots + x_n^2 \le (|x_1| + \cdots + |x_n|)^2. \quad \blacksquare$$

We have referred to elements of $\mathbf{R}^n$ as vectors or points. For engineers, a vector is a directed line segment which begins at a point $\boldsymbol{a}$ and ends at a point $\boldsymbol{b}$. How can we reconcile these two points of view? We say that two "engineering" vectors are equivalent if they have the same length and point in the same direction. Thus, the engineering vector from $\boldsymbol{a}$ to $\boldsymbol{b}$ is equivalent to the engineering vector which begins at $\boldsymbol{0}$ and ends at $\boldsymbol{b} - \boldsymbol{a}$ which can in turn be identified with the point $\boldsymbol{b} - \boldsymbol{a}$. In this way, we can identify any engineering vector with a point in $\mathbf{R}^n$.

In general, we make no distinction between vectors and points, but in each situation we adopt the interpretation which proves most useful. For example, thinking of $\mathbf{R}^n$ as a collection of vectors emanating from the origin, we say that $\boldsymbol{a} \ne \boldsymbol{0}$ is *parallel* to $\boldsymbol{b} \ne \boldsymbol{0}$ if there is a scalar $t \in \mathbf{R}$ such that $\boldsymbol{a} = t\boldsymbol{b}$. On the other hand, thinking of $\mathbf{R}^n$ as a collection of points we define the *line segment* from $\boldsymbol{a}$ to $\boldsymbol{b}$ to be the set of points

$$L(\boldsymbol{a};\boldsymbol{b}) = \{\phi(t) = (1-t)\boldsymbol{a} + t\boldsymbol{b} : t \in [0,1]\}.$$

Notice that $\phi(0) = \boldsymbol{a}$ and $\phi(1) = \boldsymbol{b}$.

The identification of vectors with points is a powerful tool for at least three reasons.

First, it provides us with geometric interpretations of most of the concepts and results which appear in this chapter and the next. To illustrate this principle, let us examine the concepts and results introduced above. For simplicity, we consider only the case $n = 2$.

Let $\boldsymbol{a} = (a_1, a_2)$ and $\boldsymbol{b} = (b_1, b_2) \in \mathbf{R}^2$. We shall call the set

$$\mathcal{P} = \{(x,y) = u(a_1, a_2) + v(b_1, b_2) : u, v \in [0,1]\}$$

the *parallelogram associated* with $\boldsymbol{a}$ and $\boldsymbol{b}$. The sum $\boldsymbol{a} + \boldsymbol{b}$ is the vector which begins at the origin and ends at the opposite vertex of $\mathcal{P}$. The difference $\boldsymbol{a} - \boldsymbol{b}$ is equivalent to the engineering vector which begins at $\boldsymbol{b}$ and ends at $\boldsymbol{a}$ (see Figure 5.1). The norm of $\boldsymbol{a}$ is the length or magnitude of the vector which begins at $\boldsymbol{0}$ and ends at $\boldsymbol{a}$. In this context, the Triangle Inequality $\|\boldsymbol{a} + \boldsymbol{b}\| \le \|\boldsymbol{a}\| + \|\boldsymbol{b}\|$ states that the length of one side of a triangle (namely, the triangle whose vertices are $\boldsymbol{0}$, $\boldsymbol{a}$, and $\boldsymbol{a} + \boldsymbol{b}$) is less than or equal to the sum of the lengths of its other two sides.

Second, the identification of vectors with points can be used to construct proofs. For example, the proof of the Cauchy–Schwarz Inequality, namely the mysterious choice of t above, is based on a simple geometric concept. Inequality (1) is trivial

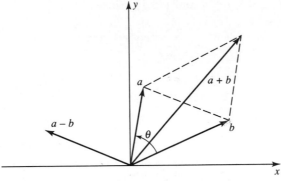

Figure 5.1

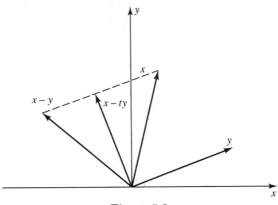

Figure 5.2

for large t but is sharpest when $\|\boldsymbol{x} - t\boldsymbol{y}\|$ is minimal. To identify a value of t which minimizes $\|\boldsymbol{x} - t\boldsymbol{y}\|$, look at the case $n = 2$ and use the vector interpretation. The set of points $\boldsymbol{x} - t\boldsymbol{y}$, $t \in [0, 1]$, is a line segment from $\boldsymbol{x}$ to $\boldsymbol{x} - \boldsymbol{y}$, i.e., a line segment which is parallel to the vector $\boldsymbol{y}$ (see Figure 5.2). The length of the vector $\boldsymbol{x} - t\boldsymbol{y}$ is minimized when $\boldsymbol{x} - t\boldsymbol{y}$ is orthogonal to $\boldsymbol{y}$, i.e.,

$$0 = (\boldsymbol{x} - t\boldsymbol{y}) \cdot \boldsymbol{y} = \boldsymbol{x} \cdot \boldsymbol{y} - t\boldsymbol{y} \cdot \boldsymbol{y} = \boldsymbol{x} \cdot \boldsymbol{y} - t\|\boldsymbol{y}\|^2.$$

It follows that $t = \boldsymbol{x} \cdot \boldsymbol{y} / \|\boldsymbol{y}\|^2$. In particular, the choice of t in the proof of Theorem 5.2 is a natural one under the interpretation of $\mathbf{R}^n$ as a collection of vectors.

And third of all, the identification of vectors and points can be used to extend concepts from $\mathbf{R}^2$ or $\mathbf{R}^3$ to $\mathbf{R}^n$. Here are several examples.

Let $\boldsymbol{a}, \boldsymbol{b}$ be nonzero vectors in $\mathbf{R}^2$ and let $\theta \in [0, \pi]$ be the angle between $\boldsymbol{a}$ and $\boldsymbol{b}$ (see Figure 5.1). By the Law of Cosines (see Appendix B),

$$\|\boldsymbol{a} - \boldsymbol{b}\|^2 = \|\boldsymbol{a}\|^2 + \|\boldsymbol{b}\|^2 - 2\|\boldsymbol{a}\| \, \|\boldsymbol{b}\| \cos \theta.$$

Since Theorem 5.1 implies

$$\|\boldsymbol{a} - \boldsymbol{b}\|^2 = (\boldsymbol{a} - \boldsymbol{b}) \cdot (\boldsymbol{a} - \boldsymbol{b}) = \|\boldsymbol{a}\|^2 - 2\boldsymbol{a} \cdot \boldsymbol{b} + \|\boldsymbol{b}\|^2,$$

it follows that

$$-2\boldsymbol{a} \cdot \boldsymbol{b} = -2\|\boldsymbol{a}\| \|\boldsymbol{b}\| \cos \theta,$$

i.e.,

$$(2) \qquad\qquad \cos \theta = \frac{\boldsymbol{a} \cdot \boldsymbol{b}}{\|\boldsymbol{a}\| \|\boldsymbol{b}\|}.$$

Motivated by this identity, we define the *angle* between two nonzero vectors $\boldsymbol{a}, \boldsymbol{b} \in \mathbf{R}^n$ (for any $n \in \mathbf{N}$) to be the number $\theta \in [0, \pi]$ defined by (2). Notice that by the Cauchy–Schwarz Inequality, the right side of (2) always belongs to the interval $[-1, 1]$. Hence, for each pair of nonzero vectors $\boldsymbol{a}, \boldsymbol{b} \in \mathbf{R}^n$, there is a unique angle $\theta \in [0, \pi]$ which satisfies (2). Also notice that this definition is consistent with Definition 5.1viii, since for $\theta \in [0, \pi]$, $\theta = \pi/2$ if and only if $\cos \theta = 0$.

By an *open ball* in $\mathbf{R}^n$ we mean a set of the form

$$B_r(\boldsymbol{a}) := \{\boldsymbol{x} \in \mathbf{R}^n : \|\boldsymbol{x} - \boldsymbol{a}\| < r\}$$

for some $r > 0$ and $\boldsymbol{a} \in \mathbf{R}^n$. The point $\boldsymbol{a}$ is called the *center* and r is called the *radius* of $B_r(\boldsymbol{a})$. (We shall call an open ball *rational* if its radius r and every component a_j of its center $\boldsymbol{a}$ is a rational number.) Notice that by definition, $B_r(\boldsymbol{a})$ is the set of points $\boldsymbol{x}$ such that the distance from $\boldsymbol{x}$ to the center $\boldsymbol{a}$ is less than r units. If $\boldsymbol{a} = (a_1, a_2, \ldots, a_n)$, then $\boldsymbol{x} = (x_1, \ldots, x_n) \in B_r(\boldsymbol{a})$ if and only if

$$(x_1 - a_1)^2 + (x_2 - a_2)^2 + \cdots + (x_n - a_n)^2 < r^2.$$

By a *hyperplane* in $\mathbf{R}^n$ we mean a set of the form

$$\Pi_{\boldsymbol{b}}(\boldsymbol{a}) := \{\boldsymbol{x} \in \mathbf{R}^n : (\boldsymbol{x} - \boldsymbol{a}) \cdot \boldsymbol{b} = 0\}$$

for some $\boldsymbol{a} \in \mathbf{R}^n$ and some nonzero vector $\boldsymbol{b}$ in $\mathbf{R}^n$. (We shall call a hyperplane in $\mathbf{R}^3$ a *plane*.) The vector $\boldsymbol{b}$ is called a *normal* (*vector*) of the hyperplane $\Pi_{\boldsymbol{b}}(\boldsymbol{a})$. Notice that by definition, $\Pi_{\boldsymbol{b}}(\boldsymbol{a})$ is the set of all points $\boldsymbol{x}$ such that $\boldsymbol{x} - \boldsymbol{a}$ and $\boldsymbol{b}$ are orthogonal. (Several such points $\boldsymbol{x}$ are shown in Figure 5.3.) A hyperplane Π is said to *pass through* a point $\boldsymbol{x}$ if $\boldsymbol{x} \in \Pi$. Thus, $\Pi_{\boldsymbol{b}}(\boldsymbol{a})$ always passes through the point $\boldsymbol{a}$. An *equation* of a hyperplane Π is an expression of the form $F(\boldsymbol{x}) = 0$ such that $F : \mathbf{R}^n \to \mathbf{R}$ and Π passes through $\boldsymbol{x}$ if and only if $F(\boldsymbol{x}) = 0$. Hence, an equation of the hyperplane $\Pi_{\boldsymbol{b}}(\boldsymbol{a})$ is given by

$$b_1 x_1 + b_2 x_2 + \cdots + b_n x_n = d,$$

where $\boldsymbol{b} = (b_1, \ldots, b_n)$ and $d = b_1 a_1 + b_2 a_2 + \cdots + b_n a_n$. In particular, planes in $\mathbf{R}^3$ have equations of the form

$$ax + by + cz = d.$$

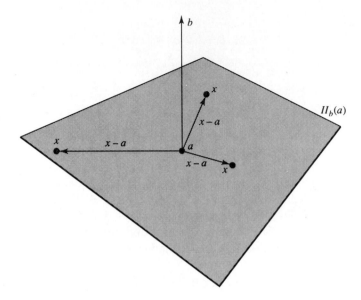

Figure 5.3

Notice we have used "$\mathbf{R}^3$" language instead of "$\mathbf{R}^2$" language (e.g., balls not discs and (hyper)planes not lines) to name objects in $\mathbf{R}^n$. This will be our usual custom unless, as is the case for rectangles, no suitable three-dimensional name exists.

Equations of hyperplanes belong to a special collection of functions on $\mathbf{R}^n$ which plays a prominent role in the chapter on differentiation. This collection is defined as follows.

DEFINITION 5.3. A function $F : \mathbf{R}^n \to \mathbf{R}^m$ is said to be linear if $F(\boldsymbol{x}+\boldsymbol{y}) = F(\boldsymbol{x}) + F(\boldsymbol{y})$ and $F(\alpha\boldsymbol{x}) = \alpha F(\boldsymbol{x})$ for all $\boldsymbol{x},\boldsymbol{y} \in \mathbf{R}^n$ and all scalars α.

A one-variable function $F : \mathbf{R} \to \mathbf{R}$ is linear if and only if $F(x) = mx$ (see Exercise 8 in Section 2.2). Thus, linear functions are easy to describe in the one-dimensional case. To obtain an analogous description of multivariable linear functions, recall that an $m \times n$ matrix B is a rectangular array which has m rows and n columns:

$$B = [b_{ij}]_{m\times n} := \begin{bmatrix} b_{11} & b_{12} & \dots & b_{1n} \\ b_{21} & b_{22} & \dots & b_{2n} \\ \vdots & \vdots & \ddots & \vdots \\ b_{m1} & b_{m2} & \dots & b_{mn} \end{bmatrix}.$$

For us, the *entries* b_{ij} of a matrix B will usually be numbers or real-valued functions. Let $B = [b_{ij}]_{m\times n}$ and $C = [c_{\nu k}]_{p\times q}$ be such matrices. Recall that the *product* of B and a scalar α is defined by

$$\alpha B = [\alpha b_{ij}]_{m\times n},$$

the *sum* of B and C is defined (when $m = p$ and $n = q$) by

$$B + C = [b_{ij} + c_{ij}]_{m\times n},$$

and the *product* of B and C is defined (when $n = p$) by

$$BC = \left[\sum_{\nu=1}^{n} b_{i\nu} c_{\nu j} \right]_{m \times q}.$$

Also recall that most of the usual laws of algebra hold for addition and multiplication of matrices (see Theorem C.1 in Appendix C). One glaring exception is that matrix multiplication is not commutative.

We shall identify points $\pmb{x} = (x_1, x_2, \ldots, x_n) \in \mathbf{R}^n$ with $1 \times n$ row matrices or $n \times 1$ column matrices by setting

$$[\pmb{x}] = [\,x_1 \quad x_2 \quad \ldots \quad x_n\,] \quad \text{or} \quad [\pmb{x}] = [\,x_1 \quad x_2 \quad \ldots \quad x_n\,]^T := \begin{bmatrix} x_1 \\ x_2 \\ \vdots \\ x_n \end{bmatrix}.$$

(Recall (see Appendix C) that B^T represents the transpose of a matrix B.) Abusing the notation slightly, we shall frequently represent the product of an $m \times n$ matrix B and an $n \times 1$ column matrix $[\pmb{x}]$ by $B\pmb{x}$.

The following result shows that the function $\pmb{x} \longmapsto [\pmb{x}]$ takes vector addition to matrix addition, the dot product to matrix multiplication, and scalar multiplication to scalar multiplication.

Remark 1. *If* $\pmb{x}, \pmb{y} \in \mathbf{R}^n$ *and* α *is a scalar, then*

$$[\pmb{x} + \pmb{y}] = [\pmb{x}] + [\pmb{y}], \qquad [\pmb{x} \cdot \pmb{y}] = [\pmb{x}][\pmb{y}]^T, \quad and \quad [\alpha \pmb{x}] = \alpha[\pmb{x}].$$

PROOF. These laws follow immediately from the definitions of addition and multiplication of matrices and vectors. For example,

$$[\pmb{x} + \pmb{y}] = [\,x_1 + y_1 \quad x_2 + y_2 \quad \ldots \quad x_n + y_n\,]$$
$$= [\,x_1 \quad x_2 \quad \ldots \quad x_n\,] + [\,y_1 \quad y_2 \quad \ldots \quad y_n\,] = [\pmb{x}] + [\pmb{y}]. \; \blacksquare$$

We are now prepared to obtain a representation theorem for linear functions on $\mathbf{R}^n$.

THEOREM 5.4. *Let* $B = [b_{ij}]$ *be an* $m \times n$ *matrix whose entries are real numbers and let* $\pmb{e}_1, \ldots, \pmb{e}_n$ *represent the usual basis of* $\mathbf{R}^n$. *If*

(3) $$F(\pmb{x}) = B\pmb{x}, \qquad \pmb{x} \in \mathbf{R}^n,$$

then F *is a linear function from* $\mathbf{R}^n$ *to* $\mathbf{R}^m$ *and*

(4) $$(b_{1j}, b_{2j}, \ldots, b_{mj}) = F(\pmb{e}_j), \qquad j = 1, 2, \ldots, n.$$

Conversely, if $F : \mathbf{R}^n \to \mathbf{R}^m$ is linear and $B = [b_{ij}]_{m \times n}$ has entries defined by (4), then F and B satisfy (3). In particular, for each linear function $F : \mathbf{R}^n \to \mathbf{R}^m$ there is one and only one $m \times n$ matrix B which satisfies (3).

PROOF. By Remark 1 and the distributive law of matrix multiplication (see Theorem C.1),

$$F(\boldsymbol{x} + \boldsymbol{y}) = B[\boldsymbol{x} + \boldsymbol{y}] = B([\boldsymbol{x}] + [\boldsymbol{y}]) = B[\boldsymbol{x}] + B[\boldsymbol{y}] = F(\boldsymbol{x}) + F(\boldsymbol{y})$$

for all $\boldsymbol{x}, \boldsymbol{y} \in \mathbf{R}^n$. Similarly, $F(\alpha \boldsymbol{x}) = B[\alpha \boldsymbol{x}] = B(\alpha[\boldsymbol{x}]) = \alpha B[\boldsymbol{x}] = \alpha F(\boldsymbol{x})$ for all $\boldsymbol{x} \in \mathbf{R}^n$ and $\alpha \in \mathbf{R}$. Thus, F is linear. Moreover, (4) holds by the definition of matrix multiplication.

Conversely, suppose $F : \mathbf{R}^n \to \mathbf{R}^m$ is linear and define B by (4). Then

$$F(\boldsymbol{x}) = F(\sum_{j=1}^{n} x_j \boldsymbol{e}_j)$$

$$= \sum_{j=1}^{n} x_j F(\boldsymbol{e}_j) = \sum_{j=1}^{n} x_j (b_{1j}, b_{2j}, \ldots, b_{mj})$$

$$= (\sum_{j=1}^{n} x_j b_{1j}, \sum_{j=1}^{n} x_j b_{2j}, \ldots, \sum_{j=1}^{n} x_j b_{mj}) = B\boldsymbol{x}. \ \blacksquare$$

The unique matrix B which satisfies (3) is called the *matrix which represents F*.

Notice that by Theorem 5.4, an equation of a hyperplane in $\mathbf{R}^n$ has the form $F(\boldsymbol{x}) = d$ for some linear $F : \mathbf{R}^n \to \mathbf{R}$.

The following result shows that under the identification of linear functions with matrices, function composition is taken to matrix multiplication.

Remark 2. *If $F : \mathbf{R}^n \to \mathbf{R}^m$ and $G : \mathbf{R}^m \to \mathbf{R}^p$ are linear, then so is $G \circ F$. In fact, if B is the $m \times n$ matrix which represents F, and C is the $p \times m$ matrix which represents G, then CB is the matrix which represents $G \circ F$.*

PROOF. Let $\boldsymbol{e}_1, \ldots, \boldsymbol{e}_n$ be the usual basis of $\mathbf{R}^n$, $\boldsymbol{u}_1, \ldots, \boldsymbol{u}_m$ be the usual basis of $\mathbf{R}^m$, and $\boldsymbol{w}_1, \ldots, \boldsymbol{w}_p$ be the usual basis of $\mathbf{R}^p$. If $B = [b_{ij}]_{m \times n}$ represents F and $C = [c_{\nu k}]_{p \times m}$ represents G, then by Definition 5.1iii and Theorem 5.4,

$$\sum_{k=1}^{m} b_{kj} \boldsymbol{u}_k = (b_{1j}, \ldots, b_{mj}) = F(\boldsymbol{e}_j), \qquad j = 1, 2, \ldots, n,$$

and

$$\sum_{\nu=1}^{p} c_{\nu k} \boldsymbol{w}_\nu = (c_{1k}, \ldots, c_{pk}) = G(\boldsymbol{u}_k), \qquad k = 1, 2, \ldots, m.$$

Hence

$$(G \circ F)(\boldsymbol{e}_j) = G(F(\boldsymbol{e}_j)) = G(\sum_{k=1}^{m} b_{kj} \boldsymbol{u}_k) = \sum_{k=1}^{m} b_{kj} G(\boldsymbol{u}_k)$$

$$= \sum_{k=1}^{m} \sum_{\nu=1}^{p} b_{kj} c_{\nu k} \boldsymbol{w}_\nu = (\sum_{k=1}^{m} b_{kj} c_{1k}, \ldots, \sum_{k=1}^{m} b_{kj} c_{pk})$$

for each $1 \leq j \leq n$. Since this last vector is the jth column of the matrix CB, it follows that CB is the matrix which represents $G \circ F$. $\blacksquare$

Since matrix multiplication can be viewed as a generalization of the dot product, the following result is an analogue of the Cauchy–Schwarz Inequality.

THEOREM 5.5. *If* $B = [b_{ij}]_{m \times n}$ *is a matrix with real entries and*

$$\|B\|_\infty := \max\{|b_{ij}| : 1 \leq i \leq m, 1 \leq j \leq n\},$$

then

$$\|B\boldsymbol{x}\| \leq \sqrt{mn}\,\|B\|_\infty\,\|\boldsymbol{x}\|$$

for all $\boldsymbol{x} \in \mathbf{R}^n$.

PROOF. By definition, if $B = [b_{ij}]_{m \times n}$, then

$$B\boldsymbol{x} = \Big(\sum_{j=1}^{n} b_{1j}x_j, \ldots, \sum_{j=1}^{n} b_{mj}x_j\Big).$$

Hence, by Theorem 5.3v and the Cauchy–Schwarz Inequality,

$$\|B\boldsymbol{x}\| = \Big\|\Big(\sum_{j=1}^{n} b_{1j}x_j, \ldots, \sum_{j=1}^{n} b_{mj}x_j\Big)\Big\|$$

$$\leq \sqrt{m}\,\max_{1 \leq i \leq m}\,\Big|\sum_{j=1}^{n} b_{ij}x_j\Big|$$

$$\leq \sqrt{m}\,\max_{1 \leq i \leq m}\,\sqrt{\sum_{j=1}^{n} |b_{ij}|^2}\,\|\boldsymbol{x}\|$$

$$\leq \sqrt{mn}\,\max_{1 \leq i \leq m}\,\max_{1 \leq j \leq n}\,|b_{ij}|\,\|\boldsymbol{x}\| = \sqrt{mn}\,\|B\|_\infty\,\|\boldsymbol{x}\|. \quad\blacksquare$$

The dot product of two vectors is a scalar; hence, the dot product does not satisfy the Closure Property on $\mathbf{R}^n$ for any $n > 1$. For three-dimensional vectors, there is another product which does satisfy the Closure Property.

DEFINITION 5.4. The *cross product* of two vectors $\boldsymbol{x} = (x_1, x_2, x_3)$ and $\boldsymbol{y} = (y_1, y_2, y_3)$ in $\mathbf{R}^3$ is the vector defined by

$$\boldsymbol{x} \times \boldsymbol{y} := (x_2 y_3 - x_3 y_2,\, x_3 y_1 - x_1 y_3,\, x_1 y_2 - x_2 y_1).$$

Using the usual basis $\mathbf{i} = \boldsymbol{e}_1$, $\mathbf{j} = \boldsymbol{e}_2$, $\mathbf{k} = \boldsymbol{e}_3$, and the determinant operator (see Appendix C), we can give the cross product a more easily remembered form:

$$\boldsymbol{x} \times \boldsymbol{y} = \det \begin{bmatrix} \mathbf{i} & \mathbf{j} & \mathbf{k} \\ x_1 & x_2 & x_3 \\ y_1 & y_2 & y_3 \end{bmatrix}.$$

The following result shows that the cross product satisfies some, but not all, of the usual laws of algebra. (Specifically, notice that the cross product satisfies neither the commutative property nor the associative property.)

THEOREM 5.6. *Let $x, y, z \in \mathbf{R}^3$ be vectors and α be a scalar. Then*

i) $$x \times x = 0, \qquad x \times y = -y \times x,$$

ii) $$(\alpha x) \times y = \alpha(x \times y) = x \times (\alpha y),$$

iii) $$x \times (y + z) = (x \times y) + (x \times z),$$

iv) $$(x \times y) \cdot z = x \cdot (y \times z) = \det \begin{bmatrix} x_1 & x_2 & x_3 \\ y_1 & y_2 & y_3 \\ z_1 & z_2 & z_3 \end{bmatrix},$$

v) $$x \times (y \times z) = (x \cdot z)y - (x \cdot y)z,$$

and

vi) $$\|x \times y\|^2 = (x \cdot x)(y \cdot y) - (x \cdot y)^2,$$

PROOF. These properties follow immediately from the definitions. We will prove properties iv) and v) and leave the rest as an exercise.

To prove iv), notice that by definition,

$$(x \times y) \cdot z = (x_2 y_3 - x_3 y_2)z_1 + (x_3 y_1 - x_1 y_3)z_2 + (x_1 y_2 - x_2 y_1)z_3$$
$$= x_1(y_2 z_3 - y_3 z_2) + x_2(y_3 z_1 - y_1 z_3) + x_3(y_1 z_2 - y_2 z_1).$$

Since this last expression is both the scalar $x \cdot (y \times z)$ and the value of the determinant on the right side of iv) (expanded along the first row), this verifies iv).

To prove v), notice that since

$$x \times (y \times z) = (x_1, x_2, x_3) \times (y_2 z_3 - y_3 z_2, y_3 z_1 - y_1 z_3, y_1 z_2 - y_2 z_1),$$

the first component of $x \times (y \times z)$ is

$$x_2 y_1 z_2 - x_2 y_2 z_1 - x_3 y_3 z_1 + x_3 y_1 z_3 = (x_1 z_1 + x_2 z_2 + x_3 z_3)y_1 - (x_1 y_1 + x_2 y_2 + x_3 y_3)z_1.$$

This proves that the first components of $x \times (y \times z)$ and $(x \cdot z)y - (x \cdot y)z$ are equal. A similar argument shows that the second and third components are also equal. ∎

Notice that by Theorem 5.6, $x \cdot (x \times y) = (x \times x) \cdot y = 0$ for all $x, y \in \mathbf{R}^3$. Similarly, $y \cdot (x \times y) = 0$. Thus, $x \times y$ is always orthogonal to both x and y. It follows that the cross product can be used to construct a vector orthogonal to any two nonzero, nonparallel vectors (see Figure 5.4).

By (2), there is a close connection between dot products and cosines. The following result shows there is a similar connection between cross products and sines. (For a connection between cross products and area or volume, see Exercise 7.)

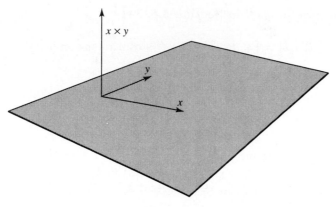

Figure 5.4

Remark 3. *Let $\boldsymbol{x}, \boldsymbol{y}$ be nonzero vectors in $\mathbf{R}^3$ and θ be the angle between $\boldsymbol{x}$ and $\boldsymbol{y}$. Then*

$$\|\boldsymbol{x} \times \boldsymbol{y}\| = \|\boldsymbol{x}\| \, \|\boldsymbol{y}\| \, \sin\theta.$$

PROOF. By Theorem 5.6vi and (2),

$$\|\boldsymbol{x} \times \boldsymbol{y}\|^2 = (\|\boldsymbol{x}\| \, \|\boldsymbol{y}\|)^2 - (\|\boldsymbol{x}\| \, \|\boldsymbol{y}\| \cos\theta)^2$$
$$= (\|\boldsymbol{x}\| \, \|\boldsymbol{y}\|)^2 (1 - \cos^2\theta) = (\|\boldsymbol{x}\| \, \|\boldsymbol{y}\|)^2 \sin^2\theta. \quad\blacksquare$$

EXERCISES

1. a) Find all nonzero vectors orthogonal to $(1, -1, 0)$ which lie in the plane $z = x$.
   b) Find all nonzero vectors orthogonal to the vector $(3, 2, -5)$ whose components sum to 4.
   c) Find an equation of the plane containing the point $(1, 0, 1)$ with normal $(-1, 2, 1)$.
   d) Find an equation of the plane orthogonal to $3x + 2y - 5z = 0$ which passes through the point $(-1, 1, 1)$.

2. Using Postulate 1 in Section 1.1 and Definition 5.1, prove Theorem 5.1.

3. Use the proof of Theorem 5.2 to show that equality in the Cauchy–Schwarz Inequality holds if and only if $\boldsymbol{x} = \boldsymbol{0}$, $\boldsymbol{y} = \boldsymbol{0}$ or $\boldsymbol{x}$ is parallel to $\boldsymbol{y}$.

4. Prove Theorem 5.6, parts i) through iii) and vi).

5. Suppose that $\boldsymbol{a}, \boldsymbol{b}, \boldsymbol{c} \in \mathbf{R}^3$ are three points which do not lie on the same straight line and Π is the plane which contains the points $\boldsymbol{a}, \boldsymbol{b}, \boldsymbol{c}$. Prove that an equation of Π is given by

$$\det \begin{bmatrix} x - a_1 & y - a_2 & z - a_3 \\ b_1 - a_1 & b_2 - a_2 & b_3 - a_3 \\ c_1 - a_1 & c_2 - a_2 & c_3 - a_3 \end{bmatrix} = 0.$$

6. **This exercise is used in Section 5.8.** Let

$$C[a, b] = \{f : [a, b] \to \mathbf{R} : f \text{ is continuous on } [a, b]\}$$

and set

$$\|f\|_\infty := \sup_{x \in [a,b]} |f(x)|.$$

a) Prove that $\|f\|_\infty$ is a finite, nonnegative real number for each $f \in C[a, b]$.
b) Show $\|f\|_\infty = 0$ if and only if $f(x) = 0$ for all $x \in [a, b]$.
c) Prove that $\|\alpha f\|_\infty = |\alpha| \|f\|_\infty$ for all $f \in C[a, b]$ and all scalars α.
d) Prove $\|f + g\|_\infty \le \|f\|_\infty + \|g\|_\infty$ and $\|f - g\|_\infty \ge \|f\|_\infty - \|g\|_\infty$ for all $f, g \in C[a, b]$.

7. **This exercise is used in Appendix E.** Recall that the area of a parallelogram with base b and altitude h is given by bh and the volume of a parallelepiped is given by the area of its base times its altitude.

a) Let $\boldsymbol{a}, \boldsymbol{b} \in \mathbf{R}^3$ be nonzero vectors and $\mathcal{P}$ represent the parallelogram

$$\{(x, y, z) = u\boldsymbol{a} + v\boldsymbol{b} : u, v \in [0, 1]\}.$$

Prove that the area of $\mathcal{P}$ is $\|\boldsymbol{a} \times \boldsymbol{b}\|$.

b) Let $\boldsymbol{a}, \boldsymbol{b}, \boldsymbol{c} \in \mathbf{R}^3$ be nonzero vectors and $\mathcal{P}$ represent the parallelepiped

$$\{(x, y, z) = t\boldsymbol{a} + u\boldsymbol{b} + v\boldsymbol{c} : t, u, v \in [0, 1]\}.$$

Prove that the volume of $\mathcal{P}$ is $|(\boldsymbol{a} \times \boldsymbol{b}) \cdot \boldsymbol{c}|$.

8. Suppose $\{a_k\}$ and $\{b_k\}$ are sequences of real numbers which satisfy

$$\sum_{k=1}^{\infty} a_k^2 < \infty, \qquad \sum_{k=1}^{\infty} b_k^2 < \infty.$$

Prove that the infinite series $\sum_{k=1}^{\infty} a_k b_k$ converges absolutely.

9. [ROTATIONS IN $\mathbf{R}^2$]. **This exercise is used in Section 7.1.** Let

$$B = \begin{bmatrix} \cos\theta & -\sin\theta \\ \sin\theta & \cos\theta \end{bmatrix}$$

for some $\theta \in \mathbf{R}$.

a) Prove that $\|B(x, y)\| = \|(x, y)\|$ for all $(x, y) \in \mathbf{R}^2$.
b) Let $(x, y) \in \mathbf{R}^2$ be a nonzero vector and φ represent the angle between $B(x, y)$ and (x, y). Prove $\cos\varphi = \cos\theta$. Thus, show that B rotates $\mathbf{R}^2$ through an angle θ. (When $\theta > 0$, we shall call B *counterclockwise rotation* about the origin through the angle θ.)

10. The distance from a point $\boldsymbol{x}_0 = (x_0, y_0, z_0)$ to a plane Π in $\mathbf{R}^3$ is defined to be

$$\text{dist}\,(\boldsymbol{x}_0, \Pi) := \begin{cases} 0 & \boldsymbol{x}_0 \in \Pi \\ \|\boldsymbol{v}\| & \boldsymbol{x}_0 \notin \Pi, \end{cases}$$

where $\boldsymbol{v} := (x_0 - x_1, y_0 - y_1, z_0 - z_1)$ for some $(x_1, y_1, z_1) \in \Pi$ and $\boldsymbol{v}$ is orthogonal to Π, i.e., parallel to its normal. Sketch Π and $\boldsymbol{x}_0$ for a typical plane Π and convince yourself that this is the correct definition. Prove that this definition does not depend on the choice of $\boldsymbol{v}$ by showing the distance from $\boldsymbol{x}_0 = (x_0, y_0, z_0)$ to the plane Π described by $ax + by + cz = d$ is

$$\text{dist}\,(\boldsymbol{x}_0, \Pi) = \frac{|ax_0 + by_0 + cz_0 - d|}{\sqrt{a^2 + b^2 + c^2}}.$$

5.2 OPEN SETS AND CLOSED SETS IN $\mathbf{R}^n$

Topology is a very important branch of mathematics which grew out of classical geometry and analysis. It is based on a fundamental concept of *open* set. In the next three sections, we look at the most commonly used definition of this concept in the spaces $\mathbf{R}^n$. (In Section 5.8, we shall see that there are other ways to define *open*, even in $\mathbf{R}$.)

For several results in Chapters 1 and 2 (e.g., the Extreme Value Theorem), it was necessary to distinguish between open intervals and closed intervals. What shall we use to replace these concepts in $\mathbf{R}^n$?

Notice that every point x in an open interval (a, b) is "surrounded" by points in (a, b). This leads us to the following definition.

DEFINITION 5.5. A set $V \subseteq \mathbf{R}^n$ is said to be *open* if given $\boldsymbol{x} \in V$ there is an $\varepsilon > 0$ such that the open ball $B_\varepsilon(\boldsymbol{x})$ is contained in V.

The following result shows that this terminology is consistent as used on balls.

Remark 1. *Every open ball in $\mathbf{R}^n$ is open.*

PROOF. Let $\boldsymbol{x} \in B_r(\boldsymbol{a})$. Using Figure 5.5 for guidance, we set $\varepsilon = r - \|\boldsymbol{x} - \boldsymbol{a}\|$. If $\boldsymbol{y} \in B_\varepsilon(\boldsymbol{x})$, then

$$\|\boldsymbol{y} - \boldsymbol{a}\| \le \|\boldsymbol{y} - \boldsymbol{x}\| + \|\boldsymbol{x} - \boldsymbol{a}\| < \varepsilon + \|\boldsymbol{x} - \boldsymbol{a}\| = r.$$

Therefore, $B_\varepsilon(\boldsymbol{x}) \subseteq B_r(\boldsymbol{a})$ and $B_r(\boldsymbol{a})$ is open by definition. ∎

(Note: Although this proof is valid in any Euclidean space $\mathbf{R}^n$, the choice of ε was suggested by a two-dimensional sketch. We shall find that sketches in $\mathbf{R}^2$ frequently guide us to correct proofs for the general case.)

Since every point x which does not belong to a closed interval $[a, b]$ is "surrounded" by points in $[a, b]^c$, we shall use the following definition to define *closed sets* in $\mathbf{R}^n$.

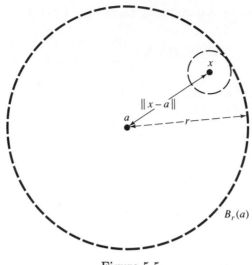

Figure 5.5

DEFINITION 5.6. A set $E \subseteq \mathbf{R}^n$ is said to be *closed* if $E^c := \mathbf{R}^n \setminus E$ is open.

Modifying the proof of Remark 1, we can show that the set $\{\boldsymbol{x} : \|\boldsymbol{x}{-}\boldsymbol{a}\| > r\}$ is open for every $\boldsymbol{a} \in \mathbf{R}^n$ and $r \geq 0$. Therefore, the *closed ball* defined by $\{\boldsymbol{x} : \|\boldsymbol{x} - \boldsymbol{a}\| \leq r\}$ is closed for every $\boldsymbol{a} \in \mathbf{R}^n$ and $r \geq 0$.

Here is another collection of closed sets.

Remark 2. *Every finite set in $\mathbf{R}^n$ is closed.*

PROOF. Suppose $E = \{\boldsymbol{x}_1, \boldsymbol{x}_2, \ldots, \boldsymbol{x}_p\}$. Let $\boldsymbol{x} \in E^c$ and set

$$\varepsilon = \min\{\|\boldsymbol{x} - \boldsymbol{x}_k\| : k = 1, 2, \ldots, p\}.$$

Then $\boldsymbol{x}_k \notin B_\varepsilon(\boldsymbol{x})$ for any $k = 1, 2, \ldots, p$. Therefore, $B_\varepsilon(\boldsymbol{x}) \subseteq E^c$ and E^c is open by definition. ∎

Students sometimes mistakenly believe that every set is either open or closed. Some sets are neither open nor closed (like the interval $[0, 1)$). And, as the following result shows, every Euclidean space contains two special sets which are both open and closed.

Remark 3. *The empty set $\emptyset$ and the whole space $\mathbf{R}^n$ are both open and closed.*

PROOF. Because the empty set contains no points, "every" point $\boldsymbol{x} \in \emptyset$ satisfies $B_\varepsilon(\boldsymbol{x}) \subseteq \emptyset$. (This is called the *vacuous implication*.) Therefore, $\emptyset$ is open and by definition $\mathbf{R}^n = \emptyset^c$ is closed. On the other hand, since $B_\varepsilon(\boldsymbol{x}) \subseteq \mathbf{R}^n$ for all $\boldsymbol{x} \in \mathbf{R}^n$ and all $\varepsilon > 0$, $\mathbf{R}^n$ is open, hence $\emptyset = (\mathbf{R}^n)^c$ is closed. ∎

Collections of open sets and closed sets behave differently under unions and intersections.

THEOREM 5.7. *Let X be a Euclidean space.*

i) *If $\{V_\alpha\}_{\alpha \in A}$ is any collection of open sets in X, then*

$$\bigcup_{\alpha \in A} V_\alpha$$

is open.

ii) *If $\{V_k : k = 1, 2, \ldots, n\}$ is a finite collection of open sets in X, then*

$$\bigcap_{k=1}^{n} V_k := \bigcap_{k \in \{1,2,\ldots,n\}} V_k$$

is open.

iii) *If $\{E_\alpha\}_{\alpha \in A}$ is any collection of closed sets in X, then*

$$\bigcap_{\alpha \in A} E_\alpha$$

is closed.

iv) *If $\{E_k : k = 1, 2, \ldots, n\}$ is a finite collection of closed sets in X, then*

$$\bigcup_{k=1}^{n} E_k := \bigcup_{k \in \{1,2,\ldots,n\}} E_k$$

is closed.

v) *If V is open in X and E is closed in X, then $V \setminus E$ is open and $E \setminus V$ is closed.*

PROOF. i) Let $\boldsymbol{x} \in \bigcup_{\alpha \in A} V_\alpha$. Then $\boldsymbol{x} \in V_\alpha$ for some $\alpha \in A$. Since V_α is open, it follows that there is an $r > 0$ such that $B_r(\boldsymbol{x}) \subseteq V_\alpha$. Thus, $B_r(\boldsymbol{x}) \subseteq \bigcup_{\alpha \in A} V_\alpha$; i.e., this union is open.

ii) Let $\boldsymbol{x} \in \bigcap_{k=1}^{n} V_k$. Then $\boldsymbol{x} \in V_k$ for $k = 1, 2, \ldots, n$. Since each V_k is open, it follows that there are numbers $r_k > 0$ such that $B_{r_k}(\boldsymbol{x}) \subseteq V_k$. Let $r = \min\{r_1, \ldots, r_n\}$. Then $r > 0$ and $B_r(\boldsymbol{x}) \subseteq V_k$ for all $k = 1, 2, \ldots, n$, i.e., $B_r(\boldsymbol{x}) \subseteq \bigcap_{k=1}^{n} V_k$. Hence, this intersection is open.

iii) By DeMorgan's Law (Theorem 1.29) and part i),

$$\left(\bigcap_{\alpha \in A} E_\alpha \right)^c = \bigcup_{\alpha \in A} E_\alpha^c$$

is open, so $\bigcap_{\alpha \in A} E_\alpha$ is closed.

iv) By DeMorgan's Law and part ii),

$$\left(\bigcup_{k=1}^{n} E_k \right)^c = \bigcap_{k=1}^{n} E_k^c$$

is open, so $\bigcup_{k=1}^{n} E_k$ is closed.

v) Since $V \setminus E = V \cap E^c$ and $E \setminus V = E \cap V^c$, the former is open by part ii) and the latter is closed by part iii). ∎

Remark 4. *Statements ii) and iv) of Theorem 5.7 are false if arbitrary collections are used in place of finite collections.*

PROOF. In the Euclidean space $X = \mathbf{R}$,

$$\bigcap_{k \in \mathbf{N}} \left(-\frac{1}{k}, \frac{1}{k} \right) = \{0\}$$

is closed and

$$\bigcup_{k \in \mathbf{N}} \left[\frac{1}{k+1}, \frac{k}{k+1} \right] = (0, 1)$$

is open. ∎

Our next main result (Theorem 5.8) is a technical theorem which shows how many open sets are needed to "cover" a subset of a Euclidean space. First, we prove that every open ball can be "approximated" by a rational ball.

Lemma 1. *Given any open ball $B_r(\boldsymbol{x}) \subset \mathbf{R}^n$, there is a rational open ball $B_q(\boldsymbol{a})$ such that $\boldsymbol{x} \in B_q(\boldsymbol{a})$ and $B_q(\boldsymbol{a}) \subseteq B_r(\boldsymbol{x})$.*

PROOF. Let $B_r(\boldsymbol{x}) \subset \mathbf{R}^n$ be given. To find $\boldsymbol{a}$, use Theorem 1.8 (Density of Rationals) to choose $a_j \in \mathbf{Q}$ such that

$$|x_j - a_j| < \frac{r}{4n}$$

for $j = 1, 2, \ldots, n$, and set $\boldsymbol{a} = (a_1, a_2, \ldots, a_n)$. By Theorem 5.3,

$$\|\boldsymbol{x} - \boldsymbol{a}\| \le \sum_{j=1}^{n} |x_j - a_j| \le n \left(\frac{r}{4n} \right) = \frac{r}{4}.$$

Choose $q \in \mathbf{Q}$ such that $r/4 < q < r/2$. Since $r/4 < q$, we have $\boldsymbol{x} \in B_q(\boldsymbol{a})$. Moreover, if $\boldsymbol{y} \in B_q(\boldsymbol{a})$, then

$$\|\boldsymbol{x} - \boldsymbol{y}\| \le \|\boldsymbol{y} - \boldsymbol{a}\| + \|\boldsymbol{a} - \boldsymbol{x}\| < q + \frac{r}{4} < \frac{r}{2} + \frac{r}{4} < r.$$

Therefore, $B_q(\boldsymbol{a}) \subseteq B_r(\boldsymbol{x})$. ∎

Next, we show that there are relatively few rational balls.

Lemma 2. *The collection $\mathcal{B}$ of rational open balls in $\mathbf{R}^n$ is countable.*

PROOF. By Theorem 1.28, $\mathcal{B}$ is countable since it can be represented as a countable union of countable sets:

$$\mathcal{B} = \bigcup_{a_1 \in \mathbf{Q}} \cdots \bigcup_{a_n \in \mathbf{Q}} \{ B_q(\boldsymbol{a}) : q \in \mathbf{Q}, q > 0 \} . \quad ∎$$

THEOREM 5.8 [LINDELÖF]. *Let E be any subset of $\mathbf{R}^n$. If $\{V_\alpha\}_{\alpha \in A}$ is a collection of open sets and $E \subseteq \cup_{\alpha \in A} V_\alpha$, then there is a countable subset A_0 of A such that*

$$E \subseteq \bigcup_{\alpha \in A_0} V_\alpha.$$

PROOF. Let $\boldsymbol{x} \in E$. By hypothesis, $\boldsymbol{x} \in V_\alpha$ for some $\alpha \in A$. By Lemma 1, there is a ball $B_{\boldsymbol{x}} \in \mathcal{B}$ such that

(5) $$\boldsymbol{x} \in B_{\boldsymbol{x}} \subseteq V_\alpha.$$

The collection $\mathcal{B}$ is countable, hence so is the subcollection

(6) $$\{U_1, U_2, \dots\} := \{B_{\boldsymbol{x}} : \boldsymbol{x} \in E\}.$$

By (5), for each $k \in \mathbf{N}$ there is at least one $\alpha_k \in A$ such that $U_k \subseteq V_{\alpha_k}$. Hence, by (6),

$$E \subseteq \bigcup_{\boldsymbol{x} \in E} B_{\boldsymbol{x}} = \bigcup_{k \in \mathbf{N}} U_k \subseteq \bigcup_{k \in \mathbf{N}} V_{\alpha_k}.$$

Thus, set $A_0 := \{\alpha_k : k \in \mathbf{N}\}$. ∎

By Definition 5.5, a set V is open if each of its points $\boldsymbol{a}$ is "interior" to V, i.e., surrounded by points in V. A given set E may have both "interior" points and "boundary" points, i.e., points which are neither "interior" to V nor "interior" to V^c. For example, a "boundary" point of the closed ball $\{(x, y) : x^2 + y^2 \leq 1\}$ is a point which satisfies $x^2 + y^2 = 1$. Here is a formal definition of the boundary of a set.

DEFINITION 5.7. Let $E \subseteq \mathbf{R}^n$. The *boundary* of E is the set

$$\partial E := \{\boldsymbol{x} \in \mathbf{R}^n : B_r(\boldsymbol{x}) \cap E \neq \emptyset \text{ and } B_r(\boldsymbol{x}) \cap E^c \neq \emptyset \text{ for all } r > 0\}.$$

(We will refer to the last two conditions in the definition of ∂E by saying $B_r(\boldsymbol{x})$ *intersects E and E^c.*)

Boundary points will be used to study integration of multivariable functions. In order to obtain a simple formula for the boundary of a set, we introduce the following concepts.

DEFINITION 5.8. Let $E \subseteq \mathbf{R}^n$.

i) The *interior* of E is the set

$$E^o := \bigcup \{V : V \subseteq E \text{ and } V \text{ is open in } \mathbf{R}^n\}.$$

ii) The *closure* of E is the set

$$\overline{E} := \bigcap \{B : B \supseteq E \text{ and } B \text{ is closed in } \mathbf{R}^n\}.$$

Notice that by Theorem 5.7, the interior of a set is always open and the closure of a set is always closed. The following result shows that E^o is the largest open set contained in E, $\overline{E}$ is the smallest closed set which contains E, and that the boundary of E can be computed using $\overline{E}$ and E^o.

THEOREM 5.9. *Let $E \subseteq \mathbf{R}^n$. Then*

  i) $E^o \subseteq E \subseteq \overline{E}$,

  ii) *if V is open and $V \subseteq E$ then $V \subseteq E^o$,*

  iii) *if C is closed and $C \supseteq E$ then $C \supseteq \overline{E}$, and*

  iv) $\partial E = \overline{E} \setminus E^o$.

PROOF. Since every open set V in the union defining E^o is a subset of E, it is clear that the union of these V's is a subset of E. Thus, $E^o \subseteq E$. A similar argument establishes $E \subseteq \overline{E}$. This proves i).

By Definition 5.8, if V is an open subset of E, then $V \subseteq E^o$ and if C is a closed set containing E, then $\overline{E} \subseteq C$. This proves ii) and iii).

To prove iv), it suffices to show

  v) $\boldsymbol{x} \in \overline{E}$ if and only if $B_r(\boldsymbol{x}) \cap E \neq \emptyset$ for all $r > 0$, and

  vi) $\boldsymbol{x} \notin E^o$ if and only if $B_r(\boldsymbol{x}) \cap E^c \neq \emptyset$ for all $r > 0$.

We will provide the details for v) and leave the proof of vi) as an exercise. Suppose $\boldsymbol{x} \in \overline{E}$ but $B_{r_0}(\boldsymbol{x}) \cap E = \emptyset$ for some $r_0 > 0$. Then $(B_{r_0}(\boldsymbol{x}))^c$ is a closed set which contains E, hence by part iii), $\overline{E} \subseteq (B_{r_0}(\boldsymbol{x}))^c$. It follows that $\overline{E} \cap B_{r_0}(\boldsymbol{x}) = \emptyset$, e.g., $\boldsymbol{x} \notin \overline{E}$, a contradiction. Conversely, suppose $\boldsymbol{x} \notin \overline{E}$. Since $(\overline{E})^c$ is open, there is an $r_0 > 0$ such that $B_{r_0}(\boldsymbol{x}) \subseteq (\overline{E})^c$. In particular, $\emptyset = B_{r_0}(\boldsymbol{x}) \cap \overline{E} \supseteq B_{r_0}(\boldsymbol{x}) \cap E$ for some $r_0 > 0$. ∎

Notice that by parts ii) and iii), $E = E^o$ if and only if E is open and $E = \overline{E}$ if and only if E is closed.

We have introduced topological operations (interior, closure, and boundary) and set operations (union and intersection). How do these operations interact with each other?

THEOREM 5.10. *Let $A, B \subseteq \mathbf{R}^n$. Then*

  i) $$(A \cup B)^o \supseteq A^o \cup B^o, \quad (A \cap B)^o = A^o \cap B^o,$$

  ii) $$\overline{A \cup B} = \overline{A} \cup \overline{B}, \qquad \overline{A \cap B} \subseteq \overline{A} \cap \overline{B},$$

  iii) $\partial(A \cup B) \subseteq \partial A \cup \partial B,$ *and* $\partial(A \cap B) \subseteq (A \cap \partial B) \cup (B \cap \partial A) \cup (\partial A \cap \partial B).$

PROOF. i) Since the union of two open sets is open, $A^o \cup B^o$ is an open subset of $A \cup B$. Hence, by Theorem 5.9ii, $A^o \cup B^o \subseteq (A \cup B)^o$.

Similarly, $(A \cap B)^o \supseteq A^o \cap B^o$. On the other hand, if $V \subset A \cap B$, then $V \subset A$ and $V \subset B$. Thus, $(A \cap B)^o \subseteq A^o \cap B^o$.

ii) Since $\overline{A} \cup \overline{B}$ is closed and contains $A \cup B$, it is clear that by Theorem 5.9iii, $\overline{A \cup B} \subseteq \overline{A} \cup \overline{B}$. Similarly, $\overline{A \cap B} \subseteq \overline{A} \cap \overline{B}$. To prove the reverse inequality for union, suppose $\boldsymbol{x} \notin \overline{A \cup B}$. Then there is a closed set E which contains $A \cup B$ such that $\boldsymbol{x} \notin E$. Since E contains both A and B, it follows that $\boldsymbol{x} \notin \overline{A}$ and $\boldsymbol{x} \notin \overline{B}$. This proves part ii).

iii) Let $\boldsymbol{x} \in \partial(A \cup B)$, i.e., suppose $B_r(\boldsymbol{x})$ intersects $A \cup B$ and $(A \cup B)^c$ for all $r > 0$. Since $(A \cup B)^c = A^c \cap B^c$, it follows that $B_r(\boldsymbol{x})$ intersects both A^c and B^c for all $r > 0$. Thus, $B_r(\boldsymbol{x})$ intersects A and A^c for all $r > 0$ or $B_r(\boldsymbol{x})$ intersects B and B^c for all $r > 0$, i.e., $\boldsymbol{x} \in \partial A \cup \partial B$. This proves the first set inequality in part iii).

To prove the second set inequality, suppose $\boldsymbol{x} \in \partial(A \cap B)$; i.e., suppose $B_r(\boldsymbol{x})$ intersects $A \cap B$ and $(A \cap B)^c$ for all $r > 0$. If $\boldsymbol{x} \in (A \cap \partial B) \cup (B \cap \partial A)$, then there is nothing to prove. If $\boldsymbol{x} \notin (A \cap \partial B) \cup (B \cap \partial A)$, then $\boldsymbol{x} \in (A^c \cup (\partial B)^c) \cap (B^c \cup (\partial A)^c)$. Since $B_r(\boldsymbol{x})$ intersects both A and B for all $r > 0$, it follows that $\boldsymbol{x} \in \partial B$ and $\boldsymbol{x} \in \partial A$, i.e., $\boldsymbol{x} \in \partial A \cap \partial B$. ∎

EXERCISES

1. Show that for all real numbers $a < b$, the sets (a, b), (a, ∞), $(-\infty, b)$ are open, $[a, b]$, $[a, \infty)$, $(-\infty, b]$ are closed, and $[a, b)$, $(a, b]$ are neither open nor closed.
2. Identify which of the following sets are open, which are closed, and which are neither. Find E^o, $\overline{E}$, and ∂E and sketch E in each case.

   a) $\{(x, y) : x^2 + 4y^2 \le 1\}$.
   b) $\{(x, y) : x^2 - 2x + y^2 = 0\} \cup \{(x, 0) : x \in [2, 3]\}$.
   c) $\{(x, y) : y \ge x^2, \ 0 \le y < 1\}$.
   d) $\{(x, y) : x^2 - y^2 < 1, \ -1 < y < 1\}$.

3. Let $s < r$, $V = \{\boldsymbol{x} \in \mathbf{R}^n : s < \|\boldsymbol{x}\| < r\}$, and $E = \{\boldsymbol{x} \in \mathbf{R}^n : s \le \|\boldsymbol{x}\| \le r\}$. Prove that V is open and E is closed.
4. Suppose $A \subseteq B \subseteq \mathbf{R}^n$. Prove that $\overline{A} \subseteq \overline{B}$ and $A^o \subseteq B^o$.
5. **This exercise is used in Section 5.4.** Show that if E is closed in $\mathbf{R}^n$ and $\boldsymbol{a} \notin E$, then
$$\inf_{\boldsymbol{x} \in E} \|\boldsymbol{x} - \boldsymbol{a}\| > 0.$$

6. Suppose $\{V_\alpha\}_{\alpha \in A}$ is a collection of nonempty open sets in $\mathbf{R}^n$ which satisfies $V_\alpha \cap V_\beta = \emptyset$ for all $\alpha \ne \beta$ in A. Prove that A is countable. What happens to this result when "open" is omitted?
7. Prove that if V is open in $\mathbf{R}^n$, then there are open balls $B_1, B_2, \dots$ such that
$$V = \bigcup_{j \in \mathbf{N}} B_j.$$

8. Prove statement vi) which appears in the proof of Theorem 5.9.
9. Show that Theorem 5.10 is best possible in the following sense.

   a) There exist sets A, B in $\mathbf{R}$ such that $\overline{(A \cup B)}^o \ne A^o \cup B^o$.
   b) There exist sets A, B in $\mathbf{R}$ such that $\overline{A \cap B} \ne \overline{A} \cap \overline{B}$.
   c) There exist sets A, B in $\mathbf{R}$ such that $\partial(A \cup B) \ne \partial A \cup \partial B$ and $\partial(A \cap B) \ne (A \cap \partial B) \cup (B \cap \partial A) \cup (\partial A \cap \partial B)$.

10. Let $f : \mathbf{R} \to \mathbf{R}$. Prove that f is continuous on $\mathbf{R}$ if and only if $f^{-1}(I)$ is open in $\mathbf{R}$ for every open interval I. Hint: To show f is continuous at a, consider the

open interval $I = (f(a) - \varepsilon, f(a) + \varepsilon)$.

5.3 SEQUENCES AND COMPACT SETS IN $\mathbf{R}^n$

In this section we examine what it means for a sequence of vectors to be bounded and to converge. We use the one-dimensional case for motivation and for guidance (compare the following with Definitions 1.5 and 1.6).

DEFINITION 5.9. Let $\{\boldsymbol{x}_k\}_{k \in \mathbf{N}}$ be a sequence of points in $\mathbf{R}^n$.

i) $\{\boldsymbol{x}_k\}$ is said to be *bounded* if there is an $M > 0$ such that

$$\|\boldsymbol{x}_k\| \le M$$

for all $k \in \mathbf{N}$.

ii) $\{\boldsymbol{x}_k\}$ is said to *converge* to a point $\boldsymbol{x} \in \mathbf{R}^n$ (notation: $\boldsymbol{x}_k \to \boldsymbol{x}$ as $k \to \infty$) if given $\varepsilon > 0$ there is an $N \in \mathbf{N}$ such that

$$k \ge N \quad \text{implies} \quad \|\boldsymbol{x}_k - \boldsymbol{x}\| < \varepsilon.$$

In this case we call $\boldsymbol{x}$ the *limit* of the sequence $\{\boldsymbol{x}_k\}$.

Notice that by definition, a sequence of points $\boldsymbol{x}_k \in \mathbf{R}^n$ converges to some $\boldsymbol{x} \in \mathbf{R}^n$ if and only if the sequence of numbers $\|\boldsymbol{x}_k - \boldsymbol{x}\|$ converges to 0 in $\mathbf{R}$, as $k \to \infty$.

By modifying proofs which appeared earlier, we can extend many of the results in Chapter 1 to sequences of vectors. Since these modifications amount to little more than replacing absolute values by the norm sign, i.e., $|x - y|$ by $\|\boldsymbol{x} - \boldsymbol{y}\|$, we shall be brief and leave some of the details as exercises.

A given sequence $\{\boldsymbol{x}_k\}$ in $\mathbf{R}^n$ can have at most one limit. For if $\boldsymbol{x}_k$ converges to both $\boldsymbol{x}$ and $\boldsymbol{y}$, then $0 \le \|\boldsymbol{x} - \boldsymbol{y}\| \le \|\boldsymbol{x} - \boldsymbol{x}_k\| + \|\boldsymbol{x}_k - \boldsymbol{y}\| \to 0$ as $k \to \infty$. Hence, by Theorem 1.12 (the Squeeze Theorem), $\|\boldsymbol{x} - \boldsymbol{y}\| = 0$, i.e., $\boldsymbol{x} = \boldsymbol{y}$.

Every convergent sequence is bounded. Indeed, if $\boldsymbol{x}_k \to \boldsymbol{x}$, then choose $N \in \mathbf{N}$ such that $\|\boldsymbol{x}_k - \boldsymbol{x}\| < 1$ for all $k \ge N$. Set

$$M = \max\{1 + \|\boldsymbol{x}\|, \|\boldsymbol{x}_1\|, \ldots, \|\boldsymbol{x}_N\|\}.$$

Since $\|\boldsymbol{x}_k\| \le \|\boldsymbol{x}_k - \boldsymbol{x}\| + \|\boldsymbol{x}\|$, it follows that $\|\boldsymbol{x}_k\| \le M$ for all $k \in \mathbf{N}$, i.e., $\{\boldsymbol{x}_k\}$ is bounded.

We can define subsequences of a sequence of points in $\mathbf{R}^n$ exactly as we did for sequences of real numbers (see Definition 1.9). By modifying the proof of Theorem 1.18, we can show that a sequence $\{\boldsymbol{x}_k\}_{k \in \mathbf{N}}$ in $\mathbf{R}^n$ converges to some $\boldsymbol{x}$ if and only if every subsequence $\{\boldsymbol{x}_{k_j}\}_{j \in \mathbf{N}}$ converges to that same $\boldsymbol{x}$.

The following result shows that evaluation of limits of sequences in $\mathbf{R}^n$ reduces to taking limits of real sequences.

THEOREM 5.11. *Let* $\boldsymbol{x}, \boldsymbol{x}_k \in \mathbf{R}^n$ *for* $k \in \mathbf{N}$. *Denote the* jth *component of* $\boldsymbol{x}_k$ *by* $x_k(j)$ *and the* jth *component of* $\boldsymbol{x}$ *by* $x(j)$. *Then the sequence* $\{\boldsymbol{x}_k\}_{k \in \mathbf{N}}$ *converges to* $\boldsymbol{x}$ *in* $\mathbf{R}^n$, *as* $k \to \infty$, *if and only if for each* $j = 1, 2, \ldots, n$, *the component sequence* $\{x_k(j)\}_{k \in \mathbf{N}}$ *converges to* $x(j)$ *in* $\mathbf{R}$, *as* $k \to \infty$.

PROOF. By Theorem 5.3,

$$|x_k(\ell) - x(\ell)| \le \|\boldsymbol{x}_k - \boldsymbol{x}\| \le \sqrt{n} \max_{1 \le j \le n} |x_k(j) - x(j)|$$

for $\ell = 1, 2, \ldots, n$. Hence, by the Squeeze Theorem, $x_k(j) \to x(j)$ as $k \to \infty$ for all $1 \le j \le n$ if and only if the real sequence $\|\boldsymbol{x}_k - \boldsymbol{x}\| \to 0$ as $k \to \infty$. Since $\|\boldsymbol{x}_k - \boldsymbol{x}\| \to 0$ if and only if $\boldsymbol{x}_k \to \boldsymbol{x}$, as $k \to \infty$, the proof of the theorem is complete. ∎

The following result is an analogue of Theorem 1.14.

THEOREM 5.12. *Let* $\{\boldsymbol{x}_k\}$ *and* $\{\boldsymbol{y}_k\}$ *be sequences in* $\mathbf{R}^n$. *If* $\boldsymbol{x}_k \to \boldsymbol{x}$ *and* $\boldsymbol{y}_k \to \boldsymbol{y}$, *as* $k \to \infty$, *and* $\alpha \in \mathbf{R}$, *then*

$$\alpha \boldsymbol{x}_k \to \alpha \boldsymbol{x}, \quad \boldsymbol{x}_k + \boldsymbol{y}_k \to \boldsymbol{x} + \boldsymbol{y}, \quad \text{and} \quad \boldsymbol{x}_k \cdot \boldsymbol{y}_k \to \boldsymbol{x} \cdot \boldsymbol{y}$$

as $k \to \infty$.

PROOF. By Theorem 5.3,

$$0 \le \|\alpha \boldsymbol{x}_k - \alpha \boldsymbol{x}\| = |\alpha| \, \|\boldsymbol{x}_k - \boldsymbol{x}\|.$$

By hypothesis, this last sequence converges to 0 as $k \to \infty$. Hence, it follows from the Squeeze Theorem that the real sequence $\|\alpha \boldsymbol{x}_k - \alpha \boldsymbol{x}\| \to 0$ as $k \to \infty$. In particular, the vectors $\alpha \boldsymbol{x}_k \to \alpha \boldsymbol{x}$ as $k \to \infty$.

Similarly, the second statement follows from the inequality

$$\|\boldsymbol{x}_k + \boldsymbol{y}_k - (\boldsymbol{x} + \boldsymbol{y})\| \le \|\boldsymbol{x}_k - \boldsymbol{x}\| + \|\boldsymbol{y}_k - \boldsymbol{y}\|.$$

Finally, since every convergent sequence is bounded, choose $M > 0$ such that $\|\boldsymbol{y}_k\| \le M$ for all $k \in \mathbf{N}$. Then it follows from the Cauchy–Schwarz Inequality that

$$\begin{aligned}
|\boldsymbol{x}_k \cdot \boldsymbol{y}_k - \boldsymbol{x} \cdot \boldsymbol{y}| &\le |(\boldsymbol{x}_k - \boldsymbol{x}) \cdot \boldsymbol{y}_k| + |\boldsymbol{x} \cdot (\boldsymbol{y}_k - \boldsymbol{y})| \\
&\le \|\boldsymbol{x}_k - \boldsymbol{x}\| \, \|\boldsymbol{y}_k\| + \|\boldsymbol{x}\| \, \|\boldsymbol{y}_k - \boldsymbol{y}\| \\
&\le M \|\boldsymbol{x}_k - \boldsymbol{x}\| + \|\boldsymbol{x}\| \, \|\boldsymbol{y}_k - \boldsymbol{y}\|.
\end{aligned}$$

Hence, $\boldsymbol{x}_k \cdot \boldsymbol{y}_k \to \boldsymbol{x} \cdot \boldsymbol{y}$ as $k \to \infty$. ∎

By iterating the one-dimensional Bolzano–Weierstrass Theorem, we can extend it to $\mathbf{R}^n$.

THEOREM 5.13 [Bolzano–Weierstrass Theorem for $\mathbf{R}^n$]. *Every bounded sequence in $\mathbf{R}^n$ has a convergent subsequence.*

Proof. Suppose $\{\boldsymbol{x}_k\}$ is bounded in $\mathbf{R}^n$. For each $j \in \{1, \ldots, n\}$, let $x_k(j)$ represent the jth component of the vector $\boldsymbol{x}_k$. By hypothesis, the sequence $\{x_k(j)\}_{k \in \mathbf{N}}$ is bounded in $\mathbf{R}$ for each $j = 1, 2, \ldots, n$.

Let $j = 1$. By the one-dimensional Bolzano–Weierstrass Theorem, there is a sequence of integers $1 \leq k(1,1) < k(1,2) < \cdots$ and a number $x(1)$ such that $x_{k(1,\nu)}(1) \to x(1)$ as $\nu \to \infty$.

Let $j = 2$. Again, since the sequence $\{x_{k(1,\nu)}(2)\}_{\nu \in \mathbf{N}}$ is bounded in $\mathbf{R}$, there is a subsequence $\{k(2,\nu)\}_{\nu \in \mathbf{N}}$ of $\{k(1,\nu)\}_{\nu \in \mathbf{N}}$ and a number $x(2)$ such that $x_{k(2,\nu)}(2) \to x(2)$ as $\nu \to \infty$. Since $\{k(2,\nu)\}_{\nu \in \mathbf{N}}$ is a subsequence of $\{k(1,\nu)\}_{\nu \in \mathbf{N}}$, we also have $x_{k(2,\nu)}(1) \to x(1)$ as $\nu \to \infty$. Thus, $x_{k(2,\nu)}(\ell) \to x(\ell)$ as $\nu \to \infty$ for all $1 \leq \ell \leq j = 2$.

Continuing this process until $j = n$, we choose a subsequence $k_\nu = k(n,\nu)$ and points $x(\ell)$ such that

$$\lim_{\nu \to \infty} x_{k_\nu}(\ell) = x(\ell)$$

for $1 \leq \ell \leq j = n$. Set $\boldsymbol{x} = (x(1), x(2), \ldots, x(n))$. Then by Theorem 5.11, $\boldsymbol{x}_{k_\nu}$ converges to $\boldsymbol{x}$ as $\nu \to \infty$. ∎

DEFINITION 5.10. A sequence of points $\boldsymbol{x}_k \in \mathbf{R}^n$ is said to be *Cauchy* if given $\varepsilon > 0$ there is an $N \in \mathbf{N}$ such that

$$k, m \geq N \quad \text{imply} \quad \|\boldsymbol{x}_k - \boldsymbol{x}_m\| < \varepsilon.$$

The following result shows that convergence and the Cauchy condition are equivalent.

THEOREM 5.14. *A sequence $\{\boldsymbol{x}_k\}$ in $\mathbf{R}^n$ is Cauchy if and only if it converges.*

Proof. Every convergent sequence in $\mathbf{R}^n$ is Cauchy. For, if $\boldsymbol{x}_k \to \boldsymbol{x}$ as $k \to \infty$ then given $\varepsilon > 0$ there is an $N \in \mathbf{N}$ such that $k \geq N$ implies $\|\boldsymbol{x}_k - \boldsymbol{x}\| < \varepsilon/2$. Thus

$$\|\boldsymbol{x}_k - \boldsymbol{x}_m\| \leq \|\boldsymbol{x}_k - \boldsymbol{x}\| + \|\boldsymbol{x} - \boldsymbol{x}_m\| < \frac{\varepsilon}{2} + \frac{\varepsilon}{2} = \varepsilon$$

for all $m, k \geq N$. The converse follows from the Bolzano–Weierstrass Theorem (see Exercise 6). ∎

The following result shows that convergence of sequences could have been defined using open sets instead of ε's. (We shall use this point of view in Section 5.6 to characterize continuous functions by open sets without using ε's and δ's.)

Remark 1. *A sequence $\{\boldsymbol{x}_k\}$ converges to $\boldsymbol{x} \in \mathbf{R}^n$ if and only if for every open set V which contains $\boldsymbol{x}$ there is an $N \in \mathbf{N}$ such that*

$$(7) \qquad\qquad k \geq N \quad \text{implies} \quad \boldsymbol{x}_k \in V.$$

PROOF. Suppose $\boldsymbol{x}_k \to \boldsymbol{x}$ as $k \to \infty$ and V is an open set which contains $\boldsymbol{x}$. By definition, there is an $r > 0$ such that $B_r(\boldsymbol{x}) \subseteq V$. Given $\varepsilon = r$, choose $N \in \mathbf{N}$ such that $k \geq N$ implies $\|\boldsymbol{x}_k - \boldsymbol{x}\| < \varepsilon = r$. Thus, (7) holds.

Conversely, given $\varepsilon > 0$ apply (7) to the open set $V = B_\varepsilon(\boldsymbol{x})$. Thus, choose $N \in \mathbf{N}$ such that $k \geq N$ implies $\boldsymbol{x}_k \in V$, i.e., $\|\boldsymbol{x}_k - \boldsymbol{x}\| < \varepsilon$. By definition, $\boldsymbol{x}_k \to \boldsymbol{x}$ as $k \to \infty$. ∎

The following result shows that sequences can be used to characterize closed sets.

THEOREM 5.15. *Let $E \subseteq \mathbf{R}^n$. Then E is closed if and only if the limit of every convergent sequence $\boldsymbol{x}_k \in E$ satisfies*

$$\lim_{k \to \infty} \boldsymbol{x}_k \in E.$$

PROOF. The theorem is vacuously satisfied if E is the empty set.

Suppose that $E \neq \emptyset$ is closed but some sequence $\boldsymbol{x}_k \in E$ converges to a point $\boldsymbol{x} \notin E$. Since E is closed, E^c is open. Thus, by Remark 1, there is an $N \in \mathbf{N}$ such that $k \geq N$ implies $\boldsymbol{x}_k \notin E$, a contradiction.

Conversely, suppose E is a nonempty set such that every convergent sequence in E has its limit in E. If E is not closed, then by Remark 3 in Section 5.2, $E \neq \mathbf{R}^n$ and by definition, E^c is nonempty and not open. Thus, there is at least one point $\boldsymbol{x} \in E^c$ such that no ball $B_r(\boldsymbol{x})$ is contained in E^c. Let $\boldsymbol{x}_k \in B_{1/k}(\boldsymbol{x}) \cap E$ for $k = 1, 2, \ldots$ Then $\boldsymbol{x}_k \in E$ and, since $\|\boldsymbol{x}_k - \boldsymbol{x}\| < 1/k$ for all $k \in \mathbf{N}$, $\boldsymbol{x}_k \to \boldsymbol{x}$ as $k \to \infty$. Thus, by hypothesis, $\boldsymbol{x} \in E$, a contradiction. ∎

A set $H \subseteq \mathbf{R}^n$ is called *sequentially compact* if every sequence in H has a convergent subsequence which converges to a limit in H. Recall that for real-valued functions on sequentially compact sets, uniform continuity and continuity are equivalent (see Theorem 2.10). Thus far, this result has had limited application because the only sequentially compact sets we identified in $\mathbf{R}$ were closed bounded intervals. Is there a simple characterization of sequentially compact sets which would allow us to decide whether a set is sequentially compact by inspection? The following concept will be used to answer this question.

DEFINITION 5.11. Let $\mathcal{V} = \{V_\alpha\}_{\alpha \in A}$ be a collection of subsets of a Euclidean space $\mathbf{R}^n$ and suppose $E \subseteq \mathbf{R}^n$.

i) $\mathcal{V}$ is said to *cover E* (or be a *covering* of E) if

$$E \subseteq \bigcup_{\alpha \in A} V_\alpha.$$

ii) $\mathcal{V}$ is said to be an *open covering* of E if $\mathcal{V}$ covers E and each V_α is open.

iii) Let $\mathcal{V}$ be a covering of E. $\mathcal{V}$ is said to have a *finite* (respectively, *countable*) *subcovering* if there is a finite (respectively, countable) subset A_0 of A such that $\{V_\alpha\}_{\alpha \in A_0}$ covers E.

Notice that the collections of open intervals

$$\left\{ \left(\frac{1}{k+1}, \frac{k}{k+1} \right) \right\}_{k \in \mathbf{N}} \quad \text{and} \quad \left\{ \left(-\frac{1}{k}, \frac{k+1}{k} \right) \right\}_{k \in \mathbf{N}}$$

are open coverings of the interval $(0, 1)$. The first covering of $(0, 1)$ has no finite subcover but any member of the second covering covers $(0, 1)$.

By Lindelöf's Theorem, any open covering of a set E in $\mathbf{R}^n$ has a countable subcovering. In general, an open covering of a set may or may not have a finite subcovering. Sets which satisfy this special property are important enough to be given a name.

DEFINITION 5.12. A subset H of a Euclidean space X is said to be *compact* if every open covering of H has a finite subcover.

Before we answer the question posed above (concerning recognition of sequentially compact sets), we need to make some elementary observations concerning compact sets in general.

Remark 2. *The empty set and all finite subsets of a Euclidean space are compact.*

PROOF. These statements follow immediately from Definition 5.12. The empty set needs no set to cover it and any finite set H can be covered by finitely many sets, one set for each element in H. ∎

Remark 3. *Let $E \subset H \subseteq \mathbf{R}^n$. If H is compact and E is closed, then E is compact.*

PROOF.   Let $\mathcal{V} = \{V_\alpha\}_{\alpha \in A}$ be an open covering of E. Since $E^c = \mathbf{R}^n \setminus E$ is open, $\mathcal{V} \cup \{E^c\}$ is an open covering of H. Since H is compact, there is a finite set $A_0 \subseteq A$ such that

$$H \subseteq E^c \cup \left(\bigcup_{\alpha \in A_0} V_\alpha \right).$$

Since $E \cap E^c = \emptyset$, it follows that E is covered by $\{V_\alpha\}_{\alpha \in A_0}$. ∎

Remark 4. *Let $H \subset \mathbf{R}^n$. If H is compact, then H is closed.*

PROOF. Suppose H is compact but not closed. By Theorem 5.15, there is a convergent sequence $\boldsymbol{x}_k \in H$ whose limit $\boldsymbol{x}$ does not belong to H. For each $\boldsymbol{y} \in H$, set $r := r(\boldsymbol{y}) := \|\boldsymbol{x} - \boldsymbol{y}\|/2$. Since $\boldsymbol{x}$ does not belong to H, $r > 0$, hence each $B_r(\boldsymbol{y})$ is open and contains $\boldsymbol{y}$; i.e., $\{B_{r(\boldsymbol{y})}(\boldsymbol{y}) : \boldsymbol{y} \in H\}$ is an open covering of H. Since H is compact, we can choose points $\boldsymbol{y}_j$ and radii $r_j = r(\boldsymbol{y}_j)$ such that $\{B_{r_j}(\boldsymbol{y}_j) : j = 1, 2, \ldots, N\}$ covers H.

Let $r := \min\{r_1, \ldots, r_N\}$. Since $\boldsymbol{x}_k \to \boldsymbol{x}$, $\boldsymbol{x}_k \in B_r(\boldsymbol{x})$ for k large. On the other hand, given $\boldsymbol{x}_k \in H \cap B_r(\boldsymbol{x})$, there is a $j \in \mathbf{N}$ such that $\boldsymbol{x}_k \in B_{r_j}(\boldsymbol{y}_j)$. It follows from the choices of r_j and r that

$$r_j \geq \|\boldsymbol{x}_k - \boldsymbol{y}_j\| \geq \|\boldsymbol{x} - \boldsymbol{y}_j\| - \|\boldsymbol{x}_k - \boldsymbol{x}\|$$
$$= 2r_j - \|\boldsymbol{x}_k - \boldsymbol{x}\| > 2r_j - r \geq 2r_j - r_j = r_j,$$

a contradiction. ∎

Compact sets play a prominent role in the study of functions of several variables, partly because of the following result (compare with the Nested Interval Property (Theorem 1.17)).

THEOREM 5.16 [CANTOR'S INTERSECTION THEOREM]. *Let* $H_1, H_2, \ldots$ *be a sequence of nonempty compact sets in* $\mathbf{R}^n$. *If* $H_1 \supseteq H_2 \supseteq \ldots$, *then*

$$\bigcap_{j \in \mathbf{N}} H_j \neq \emptyset.$$

PROOF. Suppose to the contrary that $\bigcap_{j \in \mathbf{N}} H_j = \emptyset$. Then by DeMorgan's Law, $\mathbf{R}^n = \cup_{j \in \mathbf{N}} H_j^c$. Hence, $\{H_j^c\}_{j \in \mathbf{N}}$ is an open covering of H_1. Since H_1 is compact, it follows that there is an $N \in \mathbf{N}$ such that

(8)
$$H_1 \subseteq \bigcup_{j=1}^{N} H_j^c, \quad \text{i.e.,} \quad H_1^c \supseteq \bigcap_{j=1}^{N} H_j.$$

Since the H_j's are nested, it is easy to see by induction that

$$\bigcap_{j=1}^{m} H_j = H_m \subseteq H_1$$

for all $m \in \mathbf{N}$. Combining this observation with (8), we conclude that $H_N \subseteq H_1 \cap H_1^c = \emptyset$, i.e., H_N is empty. This contradicts the fact that each H_j is nonempty. ∎

We are prepared to answer the question posed above.

THEOREM 5.17 [HEINE–BOREL]. *Let* H *be a subset of a Euclidean space* $\mathbf{R}^n$. *Then the following three statements are equivalent.*

i) *H is compact.*
ii) *H is sequentially compact.*
iii) *H is closed and bounded.*

PROOF. *i) implies ii).* Suppose H is compact. Let $\boldsymbol{x}_k \in H$ for $k \in \mathbf{N}$. There are two cases. Either there is an $\boldsymbol{a} \in H$ such that for each $r > 0$, $B_r(\boldsymbol{a})$ contains $\boldsymbol{x}_k$ for infinitely many k's, or for each $\boldsymbol{a} \in H$ there exists an $r_{\boldsymbol{a}} > 0$ such that $B_{r_{\boldsymbol{a}}}(\boldsymbol{a})$ contains $\boldsymbol{x}_k$ for only finitely many k's.

If the second case holds, then

$$H \subseteq \bigcup_{\boldsymbol{a} \in H} B_{r_{\boldsymbol{a}}}(\boldsymbol{a}).$$

Since H is compact, there are points $\boldsymbol{a}_1, \boldsymbol{a}_2, \ldots, \boldsymbol{a}_N$ such that

$$H \subseteq \bigcup_{j=1}^{N} B_{r_{\boldsymbol{a}_j}}(\boldsymbol{a}_j).$$

Since each $B_{r_{\boldsymbol{a}_j}}(\boldsymbol{a}_j)$ contains $\boldsymbol{x}_k$ for only finitely many k's and $\boldsymbol{x}_k \in H$ for all $k \in \mathbf{N}$, it follows that $\mathbf{N}$ is finite, a contradiction. Hence, the second case cannot hold.

Evidently, the first case holds. Let $\boldsymbol{x}_{k_1} \in B_1(\boldsymbol{a})$. Since $B_{1/2}(\boldsymbol{a})$ contains $\boldsymbol{x}_k$ for infinitely many k's, choose $k_2 > k_1$ such that $\boldsymbol{x}_{k_2} \in B_{1/2}(\boldsymbol{a})$. Continuing in this manner, we can choose integers $k_1 < k_2 < \ldots$ such that $\boldsymbol{x}_{k_j} \in B_{1/j}(\boldsymbol{a})$ for $j \in \mathbf{N}$. Since $\|\boldsymbol{x}_{k_j} - \boldsymbol{a}\| < 1/j$, $\boldsymbol{x}_{k_j}$ converges to $\boldsymbol{a}$. Therefore, H is sequentially compact.

ii) implies iii). Suppose H is sequentially compact. To show H is closed, let $\boldsymbol{x}_k \in H$ converge to $\boldsymbol{x}$. Then any subsequence $\boldsymbol{x}_{k_j}$ also converges to $\boldsymbol{x}$. Since H is sequentially compact, it follows that $\boldsymbol{x} \in H$. Hence, by Theorem 5.15, H is closed.

To show H is bounded, suppose to the contrary that there exist points $\boldsymbol{y}_k \in H$ such that $\|\boldsymbol{y}_k\| \geq k$ for $k \in \mathbf{N}$. Since H is sequentially compact, some subsequence $\{\boldsymbol{y}_{k_j}\}_{j \in \mathbf{N}}$ converges, hence, is bounded. It follows that $\|\boldsymbol{y}_{k_j}\| \geq k_j$ is bounded, a contradiction.

iii) implies i). Suppose to the contrary that H is closed and bounded but H is not compact. Let $\mathcal{V}$ be an open covering of H which has no finite subcover of H. By Lindelöf's Theorem, we may suppose that $\mathcal{V} = \{V_k\}_{k \in \mathbf{N}}$, i.e.,

$$(9) \qquad\qquad H \subseteq \bigcup_{k \in \mathbf{N}} V_k.$$

By the choice of $\mathcal{V}$, $\cup_{j=1}^k V_j$ cannot contain H for any $k \in \mathbf{N}$. Thus, we can choose a point

$$(10) \qquad\qquad \boldsymbol{x}_k \in H \setminus \bigcup_{j=1}^k V_j$$

for each $k \in \mathbf{N}$. Since H is bounded, the sequence $\boldsymbol{x}_k$ is bounded. Hence, by the Bolzano–Weierstrass Theorem, there is a subsequence $\boldsymbol{x}_{k_\nu}$ which converges to some $\boldsymbol{x}$ as $\nu \to \infty$. Since H is closed, $\boldsymbol{x} \in H$. Hence by (9), $\boldsymbol{x} \in V_N$ for some $N \in \mathbf{N}$. But V_N is open, hence there is an $M \in \mathbf{N}$ such that $\nu \geq M$ implies $k_\nu > N$ and $\boldsymbol{x}_{k_\nu} \in V_N$. This contradicts (10). We conclude that H is compact. $\blacksquare$

Remark 5. *By Cantor's Intersection Theorem and the Heine–Borel Theorem, if $H_1 \supseteq H_2 \supseteq \ldots$ are closed bounded nonempty sets in $\mathbf{R}^n$, then their intersection is nonempty. Neither closed nor bounded can be omitted from this statement, even in the case $n = 1$.*

PROOF. $[1, \infty) \supset [2, \infty) \supset \ldots$ are closed, nonempty sets which satisfy

$$\bigcap_{k \in \mathbf{N}} [k, \infty) = \emptyset,$$

and $(0, 1) \supset (0, 1/2) \supset \ldots$ are bounded, nonempty sets which satisfy

$$\bigcap_{k \in \mathbf{N}} (0, 1/k) = \emptyset. \quad \blacksquare$$

EXERCISES

1. Find the limit of each of the following vector sequences.

   a)
   $$\boldsymbol{x}_k = \left(\frac{1}{k}, \frac{k - 3k^2}{k + k^2} \right).$$

   b)
   $$\boldsymbol{x}_k = \left(1, \sin \pi k, \cos \frac{1}{k} \right).$$

   c)
   $$\boldsymbol{x}_k = \left(k - \sqrt{k^2 + k}, k^{1/k}, \frac{1}{k} \right).$$

2. Suppose $\boldsymbol{x}_k \to \boldsymbol{0}$ in $\mathbf{R}^n$ as $k \to \infty$ and $\boldsymbol{y}_k$ is bounded in $\mathbf{R}^n$. Prove $\boldsymbol{x}_k \cdot \boldsymbol{y}_k \to 0$ as $k \to \infty$.

3. Given $\boldsymbol{a} \in \mathbf{R}^n$ and $E \subseteq \mathbf{R}^n$, $\boldsymbol{a}$ is called a *cluster point* of E if $E \cap B_r(\boldsymbol{a})$ contains infinitely many points for each $r > 0$. Prove that every bounded infinite subset of $\mathbf{R}^n$ has at least one cluster point.

4. Identify which of the following sets are compact and which are not. If E is not compact, find the smallest compact set H (if there is one) such that $E \subset H$.

   a) $\{1/k : k \in \mathbf{N}\} \cup \{0\}$,
   b) $\{(x, y) \in \mathbf{R}^2 : a \le x^2 + y^2 \le b\}$ for real numbers $0 < a < b$,
   c) $\{(x, y) \in \mathbf{R}^2 : y = \sin(1/x) \text{ for some } x \in (0, 1]\}$,
   d) $\{(x, y, z) \in \mathbf{R}^3 : |xyz| \le 1\}$.

5. Let A, B be compact subsets of $\mathbf{R}^n$. Prove that $A \cup B$ and $A \cap B$ are compact.

6. Prove Theorem 5.14 by modifying the proof of Theorem 1.20.

7. Suppose $E \subseteq \mathbf{R}$ is closed and bounded above. Prove $\sup E \in E$.

8. Let E be a subset of $\mathbf{R}^n$. Prove that E is closed if and only if it contains all of its cluster points. (Cluster points in $\mathbf{R}^n$ are defined in Exercise 3 above.)

9. Let $E_0, E_1, E_2, \ldots$ be nonempty subsets of $\mathbf{R}^n$ with

   $$\overline{E_k} \subset E_{k-1} \qquad k \in \mathbf{N}.$$

   If E_0 is bounded, prove that
   $$\bigcap_{k \in \mathbf{N}} E_k$$
   is nonempty.

10. Let $E \subseteq \mathbf{R}^n$ be closed.

    a) Prove $\partial E \subseteq E$.
    b) Prove that $\partial E = E$ if and only if $E^\circ = \emptyset$.
    c) Show b) is false if E is not closed.

11. Prove that the Bolzano–Weierstrass Theorem does not hold for $\mathcal{C}[a,b]$ and $\|f\|_\infty$ (see Exercise 6 in Section 5.1). Namely, prove that if $f_n(x) = x^n$, then $\|f_n\|_\infty$ is bounded but $\|f_{n_k} - f\|_\infty$ does not converge to zero for any $f \in \mathcal{C}[0,1]$ and any subsequence $\{n_k\}$.

5.4 CONVEX SETS AND CONNECTED SETS IN $\mathbf{R}^n$

We have introduced open sets (analogues of open intervals), closed sets (analogues of closed intervals), and compact sets (analogues of closed bounded intervals). Some theorems from Chapters 2 and 3, however, depend on properties of intervals not yet discussed. For example, the proof of the Mean Value Theorem tacitly used the fact that an interval is convex, i.e., contains the line segment between any two of its points, and the proof of the Intermediate Value Theorem tacitly used the fact that an interval is connected, i.e., is unbroken and all of one piece. In this section we introduce analogues of these concepts and prove several results which will be used to study the calculus of functions of several variables.

DEFINITION 5.13. A set $E \subseteq \mathbf{R}^n$ is said to be *convex* if given $\boldsymbol{a}, \boldsymbol{b} \in E$ the line segment $L(\boldsymbol{a}; \boldsymbol{b})$ is a subset of E.

The following result shows that every Euclidean space contains many convex sets.

Remark 1. *Every open ball in $\mathbf{R}^n$ is convex.*

PROOF. Let $\boldsymbol{a}, \boldsymbol{b} \in B_r(\boldsymbol{c})$ and $\boldsymbol{x} \in L(\boldsymbol{a}; \boldsymbol{b})$. By definition, there is a $t \in [0,1]$ such that $\boldsymbol{x} = (1-t)\boldsymbol{a} + t\boldsymbol{b}$. Thus

$$\|\boldsymbol{x} - \boldsymbol{c}\| = \|(1-t)(\boldsymbol{a} - \boldsymbol{c}) + t(\boldsymbol{b} - \boldsymbol{c})\| < (1-t)r + tr = r,$$

i.e., $\boldsymbol{x} \in B_r(\boldsymbol{c})$. ∎

On the other hand, there are many subsets of a Euclidean space which are not convex. For example, since the line segment from $(-1,1)$ to $(1,1)$ passes through the point $(0,1)$, the set $\{(x,y) : y \le |x|\}$ is not convex.

The next result, which we shall use in Chapter 10, shows that convexity is preserved by the topological operations of interior and closure.

*THEOREM 5.18. *If $E \subseteq \mathbf{R}^n$ is convex, then so are E^o and $\overline{E}$.*

PROOF. We supply the details for E^o. The proof that $\overline{E}$ is convex is easier and left as an exercise. Let

$$V = \{(1-t)\boldsymbol{a} + t\boldsymbol{b} : \boldsymbol{a}, \boldsymbol{b} \in E^o \quad \text{and} \quad 0 \le t \le 1\}.$$

By definition, E^o is convex if and only if $V \subseteq E^o$. Since E^o is the largest open set containing E (see Theorem 5.9), it suffices to show $V \subseteq E$ and V is open.

Since $E^o \subseteq E$ and E is convex, it is clear that $V \subseteq E$. To show V is open, we must show that given $(1-t)\boldsymbol{a} + t\boldsymbol{b} \in V$ there is an $r_0 > 0$ such that $B_{r_0}((1-t)\boldsymbol{a} + t\boldsymbol{b}) \subseteq V$.

Fix $0 \leq t \leq 1$ and $\boldsymbol{a}, \boldsymbol{b} \in E^o$. Since $\boldsymbol{a}, \boldsymbol{b} \in E^o$, choose $r > 0$ so that $B_r(\boldsymbol{a})$ and $B_r(\boldsymbol{b})$ are subsets of E^o.

Case 1. $t = 0$. Then set $r_0 = r$. Since $(1-t)\boldsymbol{a} + t\boldsymbol{b} = \boldsymbol{a}$ and $E^o \subseteq V$ by definition, it follows that $B_{r_0}((1-t)\boldsymbol{a} + t\boldsymbol{b}) = B_r(\boldsymbol{a}) \subset E^o \subseteq V$.

Case 2. $0 < t \leq 1$. Then set $r_0 = tr$ and fix $\boldsymbol{x}_0 \in B_{r_0}((1-t)\boldsymbol{a} + t\boldsymbol{b})$. We claim that $\boldsymbol{x}_0$ belongs to V, i.e., $\boldsymbol{x}_0 = (1-t)\boldsymbol{a}_0 + t\boldsymbol{b}_0$ for some $\boldsymbol{a}_0, \boldsymbol{b}_0 \in E^o$. To prove this, notice that $\boldsymbol{x}_0 = (1-t)\boldsymbol{a} + (\boldsymbol{x}_0 - (1-t)\boldsymbol{a})$. Hence, if

$$\boldsymbol{a}_0 = \boldsymbol{a} \quad \text{and} \quad \boldsymbol{b}_0 = \frac{1}{t}(\boldsymbol{x}_0 - (1-t)\boldsymbol{a})$$

then $\boldsymbol{x}_0 = (1-t)\boldsymbol{a}_0 + t\boldsymbol{b}_0$. Clearly, $\boldsymbol{a}_0 = \boldsymbol{a} \in E^o$. Moreover,

$$\|\boldsymbol{b}_0 - \boldsymbol{b}\| = \left\|\frac{1}{t}(\boldsymbol{x}_0 - (1-t)\boldsymbol{a}) - \boldsymbol{b}\right\| = \frac{1}{t}\|\boldsymbol{x}_0 - (1-t)\boldsymbol{a} - t\boldsymbol{b}\| < \frac{r_0}{t} = r.$$

Therefore, $\boldsymbol{b}_0 \in B_r(\boldsymbol{b}) \subseteq E^o$. We have proved that $B_{r_0}((1-t)\boldsymbol{a} + t\boldsymbol{b}) \subseteq V$. Thus, V is open by definition. ∎

DEFINITION 5.14. Let E be a subset of $\mathbf{R}^n$.

i) A pair of open sets U, V in $\mathbf{R}^n$ is said to *separate* E if $E \subseteq U \cup V$, $U \cap V = \emptyset$, $E \cap U \neq \emptyset$ and $E \cap V \neq \emptyset$.

ii) E is said to be *connected* (in $\mathbf{R}^n$) if E cannot be separated by any pair of open sets U, V in $\mathbf{R}^n$.

Notice that by the vacuous implication, the empty set is connected in any Euclidean space. It is also obvious that if $E = \{\boldsymbol{a}\}$ for some $\boldsymbol{a} \in \mathbf{R}^n$, then E is connected in $\mathbf{R}^n$.

Loosely speaking, a connected set is all in one piece, i.e., cannot be broken into smaller pieces which do not touch each other. Since there are many irrationals between every pair of rationals, it comes as no surprise that the rationals are not connected in $\mathbf{R}$.

Remark 3. $\mathbf{Q}$ *is not connected in* $\mathbf{R}$.

PROOF. The pair $U = \{x \in \mathbf{Q} : x < \sqrt{2}\}$, $V = \{x \in \mathbf{Q} : x > \sqrt{2}\}$ separates $\mathbf{Q}$. ∎

We will find the following concepts useful when dealing with continuous functions of several variables.

DEFINITION 5.15. Let $E \subseteq \mathbf{R}^n$.

i) A set $U \subseteq E$ is said to be *relatively open* in E if there is a set V open in $\mathbf{R}^n$ such that $U = E \cap V$.

ii) A set $A \subseteq E$ is said to be *relatively closed* in E if there is a set C closed in $\mathbf{R}^n$ such that $A = E \cap C$.

The following result shows that relatively open sets can be used to characterize connected sets in $\mathbf{R}^n$.

THEOREM 5.19. *Let $E \subseteq \mathbf{R}^n$. Then E is connected if and only if there are no sets U, V, relatively open in E, such that $U \cap V = \emptyset$, $E = U \cup V$, $U \neq \emptyset$, and $V \neq \emptyset$.*

PROOF. Suppose E is connected but there exist sets U, V, relatively open in E, such that $U \cap V = \emptyset$, $E = U \cup V$, $U \neq \emptyset$, and $V \neq \emptyset$. We will reach a contradiction by constructing a pair of open sets U_0, V_0 which separates the connected set E.

We first show that

$$(11) \qquad\qquad\qquad \overline{U} \cap V = \emptyset.$$

Indeed, since V is relatively open in E, there is a set Ω, open in $\mathbf{R}^n$, such that $V = E \cap \Omega$. Since $U \cap V = \emptyset$, it follows that $U \subset \Omega^c$. This last set is closed in $\mathbf{R}^n$. Therefore,

$$\overline{U} \subseteq \overline{\Omega^c} = \Omega^c,$$

i.e., (11) holds.

Next, we use (11) to construct the set V_0. Set

$$\delta_{\boldsymbol{x}} = \inf\{\|\boldsymbol{x} - \boldsymbol{u}\| : \boldsymbol{u} \in \overline{U}\}, \quad \boldsymbol{x} \in V, \quad \text{and} \quad V_0 = \bigcup_{\boldsymbol{x} \in V} B_{\delta_{\boldsymbol{x}}/2}(\boldsymbol{x}).$$

Clearly, V_0 is open in $\mathbf{R}^n$. Since $\delta_{\boldsymbol{x}} > 0$ for each $\boldsymbol{x} \notin \overline{U}$ (see Exercise 5 in Section 5.2), V_0 contains V, hence $V_0 \cap E \supseteq V$. The reverse inequality also holds since by construction $V_0 \cap U = \emptyset$ and by hypothesis $E = U \cup V$. Therefore, $V_0 \cap E = V$.

Similarly, we can construct an open set U_0 such that $U_0 \cap E = U$ by setting

$$\varepsilon_{\boldsymbol{y}} = \inf\{\|\boldsymbol{v} - \boldsymbol{y}\| : \boldsymbol{v} \in \overline{V}\}, \quad \boldsymbol{y} \in U \quad \text{and} \quad U_0 = \bigcup_{\boldsymbol{y} \in U} B_{\varepsilon_{\boldsymbol{y}}/2}(\boldsymbol{y}).$$

To prove the pair U_0, V_0 separates E, it remains to prove $U_0 \cap V_0 = \emptyset$. Suppose to the contrary that there is a point $\boldsymbol{a} \in U_0 \cap V_0$. Then $\boldsymbol{a} \in B_{\delta_{\boldsymbol{x}}/2}(\boldsymbol{x})$ for some $\boldsymbol{x} \in V$ and $\boldsymbol{a} \in B_{\varepsilon_{\boldsymbol{y}}/2}(\boldsymbol{y})$ for some $\boldsymbol{y} \in U$. We may suppose $\delta_{\boldsymbol{x}} \leq \varepsilon_{\boldsymbol{y}}$. Then

$$\|\boldsymbol{x} - \boldsymbol{y}\| \leq \|\boldsymbol{x} - \boldsymbol{a}\| + \|\boldsymbol{a} - \boldsymbol{y}\| < \frac{\delta_{\boldsymbol{x}}}{2} + \frac{\varepsilon_{\boldsymbol{y}}}{2} \leq \varepsilon_{\boldsymbol{y}}.$$

Therefore, $\|\boldsymbol{x} - \boldsymbol{y}\| < \inf\{\|\boldsymbol{v} - \boldsymbol{y}\| : \boldsymbol{v} \in \overline{V}\}$. Since $\boldsymbol{x} \in V$, this is impossible. We conclude that $U_0 \cap V_0 = \emptyset$.

Conversely, if a pair of open sets U_0, V_0 separates E, then $U := U_0 \cap E$ and $V := V_0 \cap E$ are relatively open in E, $U \cap V = \emptyset$, $E = U \cup V$, $U \neq \emptyset$ and $V \neq \emptyset$. ∎

We close this section by examining connectivity in the Euclidean space $\mathbf{R}$.

THEOREM 5.20. *A nonempty subset E of $\mathbf{R}$ is connected if and only if E is an interval.*

PROOF. Let E be a nonempty connected subset of $\mathbf{R}$. If E contains only one point c, then E is a degenerate interval, i.e., $E = [c]$. If E contains at least two points, then set $a = \inf E$ and $b = \sup E$. Notice that $-\infty \leq a < b \leq \infty$. Suppose for simplicity that $a, b \notin E$, i.e., $E \subseteq (a, b)$. If $E \neq (a, b)$, then there is an $x \in (a, b)$ such that $x \notin E$. By the Approximation Property, $E \cap (a, x) \neq \emptyset$ and $E \cap (x, b) \neq \emptyset$, and by assumption, $E \subseteq (a, x) \cup (x, b)$. Hence, E is separated by the open sets (a, x), (x, b), a contradiction.

Conversely, suppose I is an interval which is not connected. Then there are open sets U, V in $\mathbf{R}$ which separate I, i.e., $I \subseteq U \cup V$ and there are points $x_1 \in I \cap U$ and $x_2 \in I \cap V$. We may suppose that $x_1 < x_2$. Consider the set

$$W = \{t \in I : \text{ the interval } (x_1, t) \text{ satisfies } (x_1, t) \subseteq U\}.$$

Since U is open, $W \neq \emptyset$. Since V is open, $x_2 \notin W$ and W is bounded above by some $c < x_2$. Thus, $x_3 = \sup W$ is a finite number which belongs to $(x_1, c] \subset I$. In particular, either $x_3 \in U$ or $x_3 \in V$.

Suppose $x_3 \in U$. Since U is open and $x_3 > x_1$, we can choose $\delta > 0$ so small that $x_3 - \delta > x_1$ and $(x_3 - \delta, x_3 + \delta) \subset U$. Since $x_3 = \sup W$, we can choose by the Approximation Property a $t \in W$ such that $t > x_3 - \delta$ and $(x_1, t) \subset U$. It follows that $(x_1, x_3 + \delta) = (x_1, t) \cup (x_3 - \delta, x_3 + \delta) \subset U$, i.e., x_3 is not the supremum of W, a contradiction. On the other hand, if $x_3 \in V$, the same reasoning shows us that there is a $\delta > 0$ such that $(x_3 - \delta, x_3 + \delta) \subset V$ and a $t \in W$ such that $t > x_3 - \delta$ and $(x_3 - \delta, t) \subset U$. It follows that $(x_3 - \delta, t) \subset U \cap V$, i.e., $U \cap V \neq \emptyset$, a contradiction. Thus, the pair U, V does not separate I, and I must be connected. ∎

EXERCISES

1. Let $a \leq b$ and $c \leq d$. Prove that the rectangle

$$[a, b] \times [c, d] := \{(x, y) : x \in [a, b], y \in [c, d]\}$$

is closed, convex, and connected.

2. Let A, B be convex sets in $\mathbf{R}^n$.
   a) Prove that $A \cap B$ is convex. What about $A \cup B$?
   b) Prove that the closure of A is convex.

3. Prove that the intersection of connected sets in $\mathbf{R}$ is connected. Show that this is false if "$\mathbf{R}$" is replaced by "$\mathbf{R}^2$."

4. Prove that if $E \subseteq \mathbf{R}$ is connected, then E^o is also connected. Show that this is false if "$\mathbf{R}$" is replaced by "$\mathbf{R}^2$."

5. Suppose that $E \subset \mathbf{R}^n$ is connected and $E \subseteq A \subseteq \overline{E}$. Prove that A is connected.

6. Prove that every convex subset of $\mathbf{R}^n$ is connected. What about the converse of this statement?

7. **This exercise is used in Section 5.6.**   Let $H \subseteq \mathbf{R}^n$.   Prove that H is compact if and only if every cover $\{E_\alpha\}_{\alpha \in A}$ of H, where the E_α's are relatively open in H, has a finite subcover.

8. Suppose $\{E_\alpha\}_{\alpha \in A}$ is a collection of connected sets in a Euclidean space $\mathbf{R}^n$ such

that $\cap_{\alpha \in A} E_\alpha \neq \emptyset$. Prove that

$$E = \bigcup_{\alpha \in A} E_\alpha$$

is connected.

*9. Let I be an interval and $f : I \to \mathbf{R}$. Prove that f is convex on I if and only if

$$E = \{(x, y) : x \in I \quad \text{and} \quad y \geq f(x)\}$$

is convex in $\mathbf{R}^2$.

5.5 LIMITS OF FUNCTIONS ON $\mathbf{R}^n$

In Section 5.3 we obtained multidimensional analogues of most of the results which appeared in Chapter 1 concerning limits of sequences. Now we turn our attention to multidimensional analogues of results which appeared in the first three sections of Chapter 2. Namely, we shall examine limits (this section) and continuity (the next section) of functions of several variables. Again, since the multidimensional case is very similar to the one-dimensional case, we shall be brief and leave some of the details as exercises.

Let $f : E \to \mathbf{R}^m$ for some $E \subseteq \mathbf{R}^n$. The largest subset of $\mathbf{R}^n$ on which f is defined is called the *domain* of f, and will be denoted by $\text{Dom}\,(f)$. Since $f(\boldsymbol{x}) \in \mathbf{R}^m$ for each $\boldsymbol{x} \in \text{Dom}\,(f)$, there are functions $f_j : \text{Dom}\,(f) \to \mathbf{R}$ (called the *component functions* of f) such that $f(\boldsymbol{x}) = (f_1(\boldsymbol{x}), \ldots, f_m(\boldsymbol{x}))$ for each $\boldsymbol{x} \in \text{Dom}\,(f)$. When $m = 1$, f has only one component and we shall call f *real-valued*.

For the most part, domains of functions of one variable were fairly simple, e.g., unions of intervals. For functions of several variables, the domains exhibit much more variety.

Example 1. Find the domain of $f(x, y) = (\log(xy - y + 2x - 2), \sqrt{9 - x^2 - y^2})$.

SOLUTION. This function has two components: $f_1(x, y) = \log(xy - y + 2x - 2)$ and $f_2(x, y) = \sqrt{9 - x^2 - y^2}$. Since the logarithm is real-valued only when its argument is positive, the domain of f_1 is the set of points (x, y) which satisfy

$$0 < xy - y + 2x - 2 = (x - 1)(y + 2).$$

Since the square root function is real-valued if and only if its argument is non-negative, the domain of f_2 is the set of points (x, y) which satisfy $x^2 + y^2 \leq 9$. Thus

$$\text{Dom}\,(f) = \{(x, y) : x^2 + y^2 \leq 9 \text{ and } (x - 1)(y + 2) > 0\}. \quad \blacksquare$$

(We have shaded the region $\text{Dom}\,(f)$ in Figure 5.6. The dashed lines do not belong to $\text{Dom}\,(f)$ but the solid curves do belong to $\text{Dom}\,(f)$.)

Example 2. Find the domain of

$$f(x, y) = (\sqrt{1 - x^2}, \log(x^2 - y^2), \sin x \cos y).$$

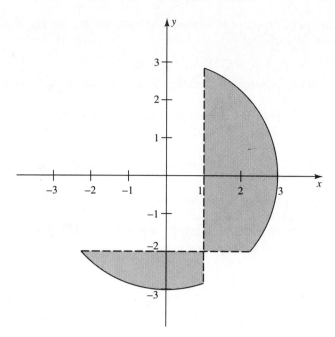

Figure 5.6

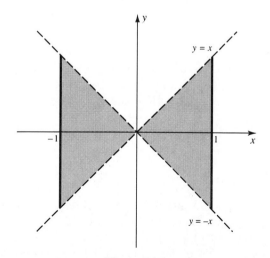

Figure 5.7

SOLUTION. This function has three component functions: $f_1(x, y) = \sqrt{1 - x^2}$, $f_2(x, y) = \log(x^2 - y^2)$, and $f_3(x, y) = \sin x \cos y$. f_1 is real-valued when $1 - x^2 \geq 0$, i.e., $-1 \leq x \leq 1$. f_2 is real-valued when $x^2 - y^2 > 0$, i.e., when $-|x| < y < |x|$. The domain of f_3 is all of $\mathbf{R}^2$. Thus

$$\text{Dom}\,(f) = \{(x, y) : -1 \leq x \leq 1 \text{ and } -|x| < y < |x|\}$$

(see Figure 5.7). ∎

Using the norm $\| \ \|$ in place of absolute values, we extend the limit concept to functions of several variables.

DEFINITION 5.16. Let $E \subseteq \mathbf{R}^n$ and $\boldsymbol{a} \in \mathbf{R}^n$.

  i) $\boldsymbol{a}$ is said to be a *cluster point* of E if $B_r(\boldsymbol{a}) \cap E$ contains infinitely many points for every $r > 0$.

  ii) A function $f : E \to \mathbf{R}^m$ is said to *converge* to $\boldsymbol{L} \in \mathbf{R}^m$, as $\boldsymbol{x} \to \boldsymbol{a}$ *through* E, if $\boldsymbol{a}$ is a cluster point of E and given $\varepsilon > 0$ there is a $\delta > 0$ such that

$$\boldsymbol{x} \in E \quad \text{and} \quad 0 < \|\boldsymbol{x} - \boldsymbol{a}\| < \delta \quad \text{imply} \quad \|f(\boldsymbol{x}) - \boldsymbol{L}\| < \varepsilon.$$

As in the one-dimensional case, the value of f at $\boldsymbol{a}$ does not affect the limit of $f(\boldsymbol{x})$ as $\boldsymbol{x} \to \boldsymbol{a}$ through E, and the limit (if it exists) is unique.

If f converges to $\boldsymbol{L}$ as $\boldsymbol{x} \to \boldsymbol{a}$ through E, we shall say that f *has a limit* as $\boldsymbol{x} \to \boldsymbol{a}$ and write either "$f(\boldsymbol{x}) \to \boldsymbol{L}$ as $\boldsymbol{x} \to \boldsymbol{a}$ through E" or

$$\boldsymbol{L} = \lim_{\substack{\boldsymbol{x} \to \boldsymbol{a} \\ \boldsymbol{x} \in E}} f(\boldsymbol{x}).$$

As in the one-dimensional case, if f converges to $\boldsymbol{L}$ as $\boldsymbol{x} \to \boldsymbol{a}$ through some open ball $B_r(\boldsymbol{a})$, then it converges to $\boldsymbol{L}$ through all sets E for which $\boldsymbol{a}$ is a cluster point. In this case, we shall simply write

$$\boldsymbol{L} = \lim_{\boldsymbol{x} \to \boldsymbol{a}} f(\boldsymbol{x}).$$

Proving that a given multivariable function f has a limit often involves showing that $\|f(\boldsymbol{x}) - \boldsymbol{L}\|$ is *dominated by* (i.e., less than or equal to) some nonnegative function g which satisfies $g(\boldsymbol{x}) \to 0$ as $\boldsymbol{x} \to \boldsymbol{a}$. Here is a typical example.

Example 3. Prove that

$$f(x, y) = \frac{3x^2 y}{x^2 + y^2}$$

converges as $(x, y) \to (0, 0)$.

PROOF. Since the numerator is a polynomial of degree 3 and the denominator is a polynomial of degree 2, we expect the numerator to overpower the denominator, i.e., the limit to be 0 as $(x, y) \to (0, 0)$. To prove this, we must estimate $f(x, y)$ near $(0, 0)$. Since $2|xy| \leq x^2 + y^2$ for all $(x, y) \in \mathbf{R}^2$, it is easy to check that

$$|f(x, y)| \leq \frac{3}{2}|x|$$

for all $(x, y) \neq (0, 0)$. Let $\varepsilon > 0$ and set $\delta = \varepsilon/2$. If $0 < \|(x, y)\| < \delta$, then $|f(x, y)| \leq 2|x| \leq 2\|(x, y)\| < 2\delta/2 = \varepsilon$. Thus, by definition,

$$\lim_{(x, y) \to (0, 0)} f(x, y) = 0. \quad \blacksquare$$

It is important to realize that by Definition 5.16, if f converges to $\boldsymbol{L}$ as $\boldsymbol{x} \to \boldsymbol{a}$, then $\|f(\boldsymbol{x}) - \boldsymbol{L}\|$ is small for all $\boldsymbol{x}$ near $\boldsymbol{a}$. In particular, $f(\boldsymbol{x}) \to \boldsymbol{L}$ as $\boldsymbol{x} \to \boldsymbol{a}$, no matter what path $\boldsymbol{x}$ takes. The next two examples show how to use this observation to prove that a limit does not exist.

Example 4. Prove that the function

$$f(x, y) = \frac{2xy}{x^2 + y^2}$$

has no limit as $(x, y) \to (0, 0)$.

PROOF. Suppose f has a limit L, as $(x, y) \to (0, 0)$. If (x, y) approaches $(0, 0)$ along a vertical path, e.g., if $x = 0$ and $y \downarrow 0$, then $L = 0$ (because $f(0, y) = 0$ for all $y \neq 0$). If (x, y) approaches $(0, 0)$ along a "diagonal" path, e.g., if $y = x$ and $x \downarrow 0$, then $L = 1$ (because $f(x, x) = 1$ for all $x \neq 0$). Since $0 \neq 1$, f has no limit at $(0, 0)$. ∎

The diagonal path was chosen in the solution to Example 4 so that the denominator of $f(x, y)$ would collapse to a single term. This same strategy is used in the next example.

Example 5. Determine whether

$$f(x, y) = \frac{xy^2}{x^2 + y^4}$$

has a limit as $(x, y) \to (0, 0)$.

SOLUTION. The vertical path $x = 0$ gives $f(0, y) = 0$ even before we take the limit as $y \to 0$. On the other hand, the parabolic path $x = y^2$ gives

$$f(y^2, y) = \frac{y^4}{2y^4} = \frac{1}{2} \neq 0.$$

Therefore, f cannot have a limit as $(x, y) \to (0, 0)$. ∎

(Notice that if $y = mx$, then

$$f(x, y) = \frac{m^2 x^3}{x^2 + m^4 x^4} \to 0$$

as $x \to 0$. Thus, Example 5 shows that the two-dimensional limit of a function might not exist even when its limit along every linear path exists and gives the same value.)

Before we look at more examples, we state three theorems about evaluation of limits of functions of several variables.

By modifying the proof of Theorem 2.1, we can prove the following result (see Exercise 4).

THEOREM 5.21 [SEQUENTIAL CHARACTERIZATION OF LIMITS]. *Let $E \subseteq \mathbf{R}^n$, $\boldsymbol{a}$ be a cluster point of E, and $f : E \to \mathbf{R}^m$. Then*

$$\boldsymbol{L} = \lim_{\substack{\boldsymbol{x} \to \boldsymbol{a} \\ \boldsymbol{x} \in E}} f(\boldsymbol{x})$$

exists if and only if $f(\boldsymbol{x}_k) \to \boldsymbol{L}$ for every sequence $\boldsymbol{x}_k \in E \setminus \{\boldsymbol{a}\}$ which converges to $\boldsymbol{a}$ as $k \to \infty$.

Combining Theorems 5.11 and 5.21, we can prove the following result (see Exercise 5).

THEOREM 5.22. *Let $E \subseteq \mathbf{R}^n$, $\boldsymbol{a}$ be a cluster point of E, $f : E \to \mathbf{R}^m$, and $\boldsymbol{L} = (L_1, L_2, \ldots, L_m) \in \mathbf{R}^m$. Then*

$$(12) \qquad\qquad \boldsymbol{L} = \lim_{\substack{\boldsymbol{x} \to \boldsymbol{a} \\ \boldsymbol{x} \in E}} f(\boldsymbol{x})$$

exists if and only if

$$(13) \qquad\qquad L_j = \lim_{\substack{\boldsymbol{x} \to \boldsymbol{a} \\ \boldsymbol{x} \in E}} f_j(\boldsymbol{x})$$

exists for each $j = 1, 2, \ldots, m$.

Thus, evaluation of limits of vector-valued functions reduces to the real-valued case, i.e., the case where the range is one-dimensional.

To state a result about algebraic combinations of functions, we introduce the following terminology. Let $E \subseteq \mathbf{R}^n$ and $f, g : E \to \mathbf{R}^m$. For each $\boldsymbol{x} \in E$, the *product* of a scalar $\alpha \in \mathbf{R}$ with f is defined by

$$(\alpha f)(\boldsymbol{x}) := \alpha f(\boldsymbol{x}),$$

the *sum* of f and g is defined by

$$(f + g)(\boldsymbol{x}) := f(\boldsymbol{x}) + g(\boldsymbol{x}),$$

and the (*Euclidean*) *dot product* of f and g is defined by

$$(f \cdot g)(\boldsymbol{x}) := f(\boldsymbol{x}) \cdot g(\boldsymbol{x}).$$

(Notice that when $m = 1$, the dot product of two functions is the pointwise product defined in Section 2.1.)

By modifying the proof of Theorem 2.2, we can prove the following result (see Exercise 6).

THEOREM 5.23. *Let $E \subseteq \mathbf{R}^n$, $\boldsymbol{a}$ be a cluster point of E, $\alpha \in \mathbf{R}$, and $f, g : E \to \mathbf{R}^m$. If*

$$\lim_{\substack{\boldsymbol{x} \to \boldsymbol{a} \\ \boldsymbol{x} \in E}} f(\boldsymbol{x}) \quad and \quad \lim_{\substack{\boldsymbol{x} \to \boldsymbol{a} \\ \boldsymbol{x} \in E}} g(\boldsymbol{x})$$

exist, then

$$\lim_{\substack{\boldsymbol{x} \to \boldsymbol{a} \\ \boldsymbol{x} \in E}} (\alpha f)(\boldsymbol{x}) = \alpha \lim_{\substack{\boldsymbol{x} \to \boldsymbol{a} \\ \boldsymbol{x} \in E}} f(\boldsymbol{x}),$$

$$\lim_{\substack{\boldsymbol{x} \to \boldsymbol{a} \\ \boldsymbol{x} \in E}} (f + g)(\boldsymbol{x}) = \lim_{\substack{\boldsymbol{x} \to \boldsymbol{a} \\ \boldsymbol{x} \in E}} f(\boldsymbol{x}) + \lim_{\substack{\boldsymbol{x} \to \boldsymbol{a} \\ \boldsymbol{x} \in E}} g(\boldsymbol{x}),$$

$$\lim_{\substack{\boldsymbol{x} \to \boldsymbol{a} \\ \boldsymbol{x} \in E}} (f \cdot g)(\boldsymbol{x}) = \left(\lim_{\substack{\boldsymbol{x} \to \boldsymbol{a} \\ \boldsymbol{x} \in E}} f(\boldsymbol{x}) \right) \cdot \left(\lim_{\substack{\boldsymbol{x} \to \boldsymbol{a} \\ \boldsymbol{x} \in E}} g(\boldsymbol{x}) \right),$$

and

$$\left\| \lim_{\substack{\boldsymbol{x} \to \boldsymbol{a} \\ \boldsymbol{x} \in E}} f(\boldsymbol{x}) \right\| = \lim_{\substack{\boldsymbol{x} \to \boldsymbol{a} \\ \boldsymbol{x} \in E}} \| f(\boldsymbol{x}) \|.$$

Here is a typical application of Theorem 5.23.

Example 6. Prove that the function

$$f(x, y) = \frac{2 + x - y}{1 + 2x^2 + 3y^2}$$

has a limit as $(x, y) \to (0, 0)$.

PROOF. By Exercise 3, $2 + x - y \to 2$ and $1 + 2x^2 + 3y^2 \to 1$ as $(x, y) \to (0, 0)$. Hence, it follows from Theorem 5.23 that

$$\lim_{(x,y) \to (0,0)} \frac{2 + x - y}{1 + 2x^2 + 3y^2} = \frac{2}{1} = 2. \quad \blacksquare$$

When asked whether the limit of a function $f(\boldsymbol{x})$ exists, it is natural to try taking the limit as each variable moves independently. Comparing Examples 4 and 6, we see that this strategy works for some functions but not all. To look at this problem more closely, we introduce the following terminology. Let $E \subseteq \mathbf{R}^2$, (a, b) be a cluster point of E, and $f : E \to \mathbf{R}^m$. The *iterated limits* of f at (a, b) are defined to be

$$\lim_{x \to a} \lim_{y \to b} f(x, y) := \lim_{x \to a} \left(\lim_{y \to b} f(x, y) \right), \quad \lim_{y \to b} \lim_{x \to a} f(x, y) := \lim_{y \to b} \left(\lim_{x \to a} f(x, y) \right),$$

when they exist.

The iterated limits of a given function might not exist. Even when they do, we cannot be sure that the corresponding two-dimensional limit exists. Indeed, although the iterated limits of the function f in Example 4 above exist and are both zero at $(0, 0)$, f has no limit as $(x, y) \to (0, 0)$.

It is also possible for both iterated limits to exist but give different values.

Example 7. Evaluate the iterated limits of

$$f(x, y) = \frac{x^2}{x^2 + y^2}$$

at $(0, 0)$.

SOLUTION. For each $x \neq 0$, $x^2/(x^2 + y^2) \to 1$ as $y \to 0$. Therefore,

$$\lim_{x \to 0} \lim_{y \to 0} \frac{x^2}{x^2 + y^2} = \lim_{x \to 0} \frac{x^2}{x^2} = 1.$$

On the other hand,

$$\lim_{y \to 0} \lim_{x \to 0} \frac{x^2}{x^2 + y^2} = \lim_{y \to 0} \frac{0}{y^2} = 0. \quad \blacksquare$$

This leads us to ask: When are the iterated limits equal? The following result shows that if f has a limit as $(x, y) \to (a, b)$ and the iterated limits exist, then these limits must be equal.

Remark 1. *Suppose I and J are open intervals, $a \in I$, $b \in J$, $\lim_{x \to a} f(x, y_0)$ exists for each $y_0 \in J$, and $\lim_{y \to b} f(x_0, y)$ exists for each $x_0 \in I$. If*

$$L = \lim_{(x, y) \to (a, b)} f(x, y)$$

also exists, then

$$L = \lim_{x \to a} \lim_{y \to b} f(x, y) = \lim_{y \to b} \lim_{x \to a} f(x, y).$$

PROOF. Let

$$g(x) := \lim_{y \to b} f(x, y)$$

for $x \in I$. Given $\varepsilon > 0$, choose $\delta > 0$ such that

$$0 < \|(x, y) - (a, b)\| < \delta \quad \text{implies} \quad |f(x, y) - L| < \varepsilon.$$

Suppose $x \in I$ and $0 < |x - a| < \delta/\sqrt{2}$. Then for any y which satisfies $0 < |y - b| < \delta/\sqrt{2}$, we have $0 < \|(x, y) - (a, b)\| < \delta$ and

$$|g(x) - L| \leq |g(x) - f(x, y)| + |f(x, y) - L| < |g(x) - f(x, y)| + \varepsilon.$$

Taking the limit of this inequality as $y \to b$, we find that $|g(x) - L| \leq \varepsilon$ for all $x \in I$ which satisfy $|x - a| < \delta/\sqrt{2}$. It follows that $g(x) \to L$ as $x \to a$, i.e.,

$$L = \lim_{x \to a} \lim_{y \to b} f(x, y).$$

A similar argument proves that the other iterated limit also exists and equals L. $\blacksquare$

This result gives us another method for showing that the limit of a function f does not exist. Indeed, if the iterated limits of f exist at a point (a, b) and give different values, then f has no limit as $(x, y) \to (a, b)$. Notice by Example 7 that the conclusion of Remark 1 might not hold if the hypothesis "$f(x, y) \to L$ as $(x, y) \to (0, 0)$" is omitted. In particular, if the limit of a function does not exist, we must be careful about the order in which an iterated limit is taken.

EXERCISES

1. Compute the iterated limits at $(0, 0)$ of each of the following functions. Determine which of these functions has a limit as $(x, y) \to (0, 0)$ in $\mathbf{R}^2$ and prove that the limit exists.

a)
$$f(x, y) = \frac{\sin x \sin y}{x^2 + y^2},$$

b)
$$f(x, y) = \frac{x^3 - y^3}{x^2 + y^2},$$

c)
$$f(x, y) = \frac{x^\alpha y^4}{x^2 + y^4}, \quad \alpha > 0,$$

d)
$$f(x, y) = \frac{x^2 + y^4}{x^2 + 2y^4},$$

e)
$$f(x, y) = \frac{x - y}{(x^2 + y^2)^\alpha}, \quad \alpha < \frac{1}{2}.$$

2. For each of the following functions, find the domain of f, prove that the given limit exists, and find the value of that limit. (Note: You can prove that the limit exists without using ε's and δ's—see Example 6.)

a)
$$\lim_{(x,y)\to(1,-1)} \left(\frac{x - 1}{y - 1}, \frac{x^2 + x - 2}{x - 1} \right),$$

b)
$$\lim_{(x,y)\to(0,1)} \left(\frac{y \sin x}{x}, \tan \frac{x}{y}, x^2 + y^2 - xy \right),$$

c)
$$\lim_{(x,y)\to(0,0)} \left(\frac{x^4 + y^4}{x^2 + y^2}, \frac{\sqrt{|xy|}}{\sqrt[3]{x^2 + y^2}} \right),$$

d) $$\lim_{(x,y)\to(1,1)} \left(\frac{x^2-1}{y^2+1}, \frac{x^2y-2xy+y-(x-1)^2}{x^2+y^2-2x-2y+2} \right).$$

3. A *polynomial* on $\mathbf{R}^n$ is a function of the form

$$P(x_1, x_2, \ldots, x_n) = \sum_{j_1=0}^{N_1} \cdots \sum_{j_n=0}^{N_n} a_{j_1,\ldots,j_n} x_1^{j_1} \ldots x_n^{j_n},$$

where $a_{j_1,\ldots,j_n}$ are scalars and $N_1, \ldots, N_n$ are nonnegative integers. Prove that if P is a polynomial on $\mathbf{R}^n$ and $\boldsymbol{a} \in \mathbf{R}^n$, then $\lim_{\boldsymbol{x}\to\boldsymbol{a}} P(\boldsymbol{x}) = P(\boldsymbol{a})$.

4. Prove Theorem 5.21.
5. Prove Theorem 5.22.
6. Prove Theorem 5.23.

5.6 CONTINUOUS FUNCTIONS ON $\mathbf{R}^n$

In this section we discuss continuity and uniform continuity of functions of several variables.

The following definition is the multidimensional analogue of Definition 2.3.

DEFINITION 5.17. Let E be a nonempty subset of $\mathbf{R}^n$ and $f : E \to \mathbf{R}^m$.

 i) f is said to be *continuous* at $\boldsymbol{a} \in E$ if given $\varepsilon > 0$ there is a $\delta > 0$ such that

$$\boldsymbol{x} \in E \quad \text{and} \quad \|\boldsymbol{x} - \boldsymbol{a}\| < \delta \quad \text{imply} \quad \|f(\boldsymbol{x}) - f(\boldsymbol{a})\| < \varepsilon.$$

 ii) f is said to be *continuous* on E (notation: $f : E \to \mathbf{R}^m$ is continuous) if f is continuous at each point $\boldsymbol{a} \in E$.

 iii) f is said to be *continuous* if it is continuous on its domain.

Thus, for any cluster point $\boldsymbol{a}$ of E, f is continuous at $\boldsymbol{a} \in E$ if and only if

$$f(\boldsymbol{a}) = \lim_{\substack{\boldsymbol{x}\to\boldsymbol{a} \\ \boldsymbol{x}\in E}} f(\boldsymbol{x}).$$

Notice that by Theorem 5.22, a vector-valued function f is continuous at a point $\boldsymbol{a} \in E$ (respectively, on a set E) if and only if each component function f_j is continuous at $\boldsymbol{a} \in E$ (respectively, on E). Thus, the theory of continuous vector-valued functions reduces to the real-valued case, i.e., to functions $f : E \to \mathbf{R}$ for some $E \subseteq \mathbf{R}^n$. Also notice by Theorem 5.23 that if f and g are continuous on E, then so are $f + g$, $f \cdot g$, $\|f\|$, and αf for all scalars α.

By modifying the proof of Theorem 2.7, we can prove the following result (see Exercise 4).

THEOREM 5.24. *Let* $E \subseteq \mathbf{R}^n$, $\Omega \subseteq \mathbf{R}^m$, $f : E \to \Omega$, *and* $g : \Omega \to \mathbf{R}^p$. *If* $f(\boldsymbol{x})$ *converges to* $\boldsymbol{L}$ *as* $\boldsymbol{x} \to \boldsymbol{a}$ *through* E *and if* g *is continuous at* $\boldsymbol{L}$, *then*

$$\lim_{\substack{\boldsymbol{x} \to \boldsymbol{a} \\ \boldsymbol{x} \in E}} (g \circ f)(\boldsymbol{x}) = g\left(\lim_{\substack{\boldsymbol{x} \to \boldsymbol{a} \\ \boldsymbol{x} \in E}} f(\boldsymbol{x}) \right).$$

We now look at continuous functions from a different point of view. By Definition 5.17, if f is continuous at $\boldsymbol{a} \in E$ and $\varepsilon > 0$, then there is a $\delta > 0$ such that $f(B_\delta(\boldsymbol{a}) \cap E) \subseteq B_\varepsilon(f(\boldsymbol{a}))$, i.e.,

$$B_\delta(\boldsymbol{a}) \cap E \subseteq f^{-1}(B_\varepsilon(f(\boldsymbol{a})))$$

In particular, the inverse image of an open ball centered at $f(\boldsymbol{a})$ contains a relatively open subset of E "centered at" $\boldsymbol{a}$. This observation can be used to give the following simple but powerful characterization of continuous functions (see also Exercise 2).

THEOREM 5.25. *Let* $E \subseteq \mathbf{R}^n$ *and* $f : E \to \mathbf{R}^m$. *Then* f *is continuous on* E *if and only if* $f^{-1}(V) \cap E$ *is relatively open in* E *for every* V *open in* $\mathbf{R}^m$.

PROOF. Suppose f is continuous on E and V is open in $\mathbf{R}^m$. We may suppose $f^{-1}(V) \cap E$ is nonempty. Let $\boldsymbol{a} \in f^{-1}(V) \cap E$, i.e., $\boldsymbol{a} \in E$ and $f(\boldsymbol{a}) \in V$. Since V is open, choose $\varepsilon > 0$ such that $B_\varepsilon(f(\boldsymbol{a})) \subseteq V$. Since f is continuous at $\boldsymbol{a} \in E$, choose $\delta > 0$ such that

(14) $$B_\delta(\boldsymbol{a}) \cap E \subseteq f^{-1}(B_\varepsilon(f(\boldsymbol{a}))) \subseteq f^{-1}(V).$$

Set

$$U = \bigcup_{\boldsymbol{a} \in f^{-1}(V)} B_\delta(\boldsymbol{a}).$$

Since U is a union of open sets, U is open. Since U is a union of $B_\delta(\boldsymbol{a})$ over all $\boldsymbol{a} \in f^{-1}(V)$, U contains $f^{-1}(V)$, i.e., $U \cap E \supseteq f^{-1}(V) \cap E$. On the other hand, by (14) $U \cap E \subseteq f^{-1}(V) \cap E$. We have proved $f^{-1}(V) \cap E = U \cap E$. Therefore, $f^{-1}(V) \cap E$ is relatively open in E by Definition 5.15.

Conversely, let $\varepsilon > 0$, $\boldsymbol{a} \in E$, and set $V = B_\varepsilon(f(\boldsymbol{a}))$. By hypothesis, $f^{-1}(V) \cap E$ is relatively open in E. Since $\boldsymbol{a} \in f^{-1}(V)$, it follows that there is a $\delta > 0$ such that $E \cap B_\delta(\boldsymbol{a}) \subseteq f^{-1}(V)$, i.e., if $\boldsymbol{x} \in E$ and $\|\boldsymbol{x} - \boldsymbol{a}\| < \delta$ then $\|f(\boldsymbol{x}) - f(\boldsymbol{a})\| < \varepsilon$. ∎

We shall refer to Theorem 5.25 by saying that open sets are invariant under inverse images by continuous functions. It is interesting to notice that closed sets are also invariant under inverse images by continuous functions (see Exercise 2).

This invariance does not extend to compact sets. Indeed, if $f(x) = 1/x$ and $H = [0, 1]$, then f is continuous on $(0, \infty)$ and H is compact, but $f^{-1}(H) = [1, \infty)$ is not compact. The next result shows that compact sets are invariant under **images**, rather than inverse images, by continuous functions.

THEOREM 5.26. *If H is compact in $\mathbf{R}^n$ and $f : H \to \mathbf{R}^m$ is continuous on H, then $f(H)$ is compact.*

PROOF. Suppose $\{V_\alpha\}_{\alpha \in A}$ is an open covering of $f(H)$. By Theorem 1.30,

$$H \subseteq f^{-1}(f(H)) \subseteq f^{-1}\left(\bigcup_{\alpha \in A} V_\alpha\right) = \bigcup_{\alpha \in A} f^{-1}(V_\alpha).$$

Hence, by Theorem 5.25, $\{f^{-1}(V_\alpha)\}_{\alpha \in A}$ is a covering of H whose sets are all relatively open in H. Since H is compact, there are indices $\alpha_1, \alpha_2, \ldots, \alpha_N$ such that

$$H \subseteq \bigcup_{j=1}^{N} f^{-1}(V_{\alpha_j})$$

(see Exercise 7 in Section 5.4). It follows from Theorem 1.30 that

$$f(H) \subseteq f\left(\bigcup_{j=1}^{N} f^{-1}(V_{\alpha_j})\right) = \bigcup_{j=1}^{N}(f \circ f^{-1})(V_{\alpha_j}) = \bigcup_{j=1}^{N} V_{\alpha_j}.$$

Therefore, $f(H)$ is compact. ∎

(Note: Theorem 5.26 does not hold if "compact" is replaced by "open" or "closed." For example, if $f(x) = x^2$ and $V = (-1, 1)$, then f is continuous on $\mathbf{R}$ and V is open, but $f(V) = [0, 1)$ is neither open nor closed.)

The following result shows that connected sets are also invariant under images by continuous functions.

THEOREM 5.27. *If E is connected in $\mathbf{R}^n$ and $f : E \to \mathbf{R}^m$ is continuous on E, then $f(E)$ is connected in $\mathbf{R}^m$.*

PROOF. Suppose $f(E)$ is not connected and let $U, V \subset \mathbf{R}^m$ be a pair of open sets which separates $f(E)$. By Theorem 5.25, $f^{-1}(U) \cap E$ and $f^{-1}(V) \cap E$ are relatively open in E. Since $f(E) \subseteq U \cup V$, we have

$$E = (f^{-1}(U) \cap E) \cup (f^{-1}(V) \cap E).$$

Since $U \cap V = \emptyset$, we also have $f^{-1}(U) \cap f^{-1}(V) = \emptyset$. Thus, $f^{-1}(U) \cap E$, $f^{-1}(V) \cap E$ is a pair of relatively open sets which separates E. Hence, by Theorem 5.19, E cannot be connected, a contradiction. ∎

To illustrate the power of these ideas, compare the proofs of the following two results with those of Theorems 2.8 and 2.9.

THEOREM 5.28 [EXTREME VALUE THEOREM FOR MULTIVARIABLE FUNCTIONS]. *Let H be a nonempty, compact set in a Euclidean space X and suppose $f : H \to \mathbf{R}$ is continuous. Then*

$$M := \sup\{f(\boldsymbol{x}) : \boldsymbol{x} \in H\} \quad \text{and} \quad m := \inf\{f(\boldsymbol{x}) : \boldsymbol{x} \in H\}$$

are finite real numbers and there exist points $\boldsymbol{x}_M, \boldsymbol{x}_m \in H$ such that $M = f(\boldsymbol{x}_M)$ and $m = f(\boldsymbol{x}_m)$.

PROOF. By symmetry, it suffices to prove the result for M. Since H is compact, $f(H)$ is compact. Hence, by the Heine–Borel Theorem, $f(H)$ is closed and bounded. Since $f(H)$ is bounded, M is finite. By the Approximation Property, choose $\boldsymbol{x}_k \in H$ such that $f(\boldsymbol{x}_k) \to M$ as $k \to \infty$. Since $f(H)$ is closed, $M \in f(H)$. Therefore, there is an $\boldsymbol{x}_M \in H$ such that $M = f(\boldsymbol{x}_M)$. ∎

THEOREM 5.29 [INTERMEDIATE VALUE THEOREM FOR MULTIVARIABLE FUNCTIONS]. *Let E be a connected subset of a Euclidean space X. If $f : E \to \mathbf{R}$ is continuous, $f(a) \neq f(b)$ for some $a, b \in E$, and y is a number which lies between $f(a)$ and $f(b)$, then there is an $x \in E$ such that $f(x) = y$.*

PROOF. Since E is connected and f is continuous on E, $f(E)$ is connected in $\mathbf{R}$. Thus, by Theorem 5.20, $f(E)$ is an interval. ∎

We close this section by examining the multidimensional analogue of uniform continuity.

DEFINITION 5.18. Let $E \subseteq \mathbf{R}^n$. A function $f : E \to \mathbf{R}^m$ is said to be *uniformly continuous* on E if given $\varepsilon > 0$ there is a $\delta > 0$ such that

$$\|\boldsymbol{x} - \boldsymbol{a}\| < \delta \quad \text{and} \quad \boldsymbol{x}, \boldsymbol{a} \in E \quad \text{imply} \quad \|f(\boldsymbol{x}) - f(\boldsymbol{a})\| < \varepsilon.$$

As in the one-dimensional case, not every continuous function is uniformly continuous. The following result, an analogue of Theorem 2.10, shows that on compact sets, continuity and uniform continuity are equivalent.

THEOREM 5.30. *Let $H \subseteq \mathbf{R}^n$ be compact and $f : H \to \mathbf{R}^m$. Then f is continuous on H if and only if f is uniformly continuous on H.*

PROOF. By definition, if f is uniformly continuous on H, then f is continuous on H.

Conversely, suppose f is continuous on a compact set H. By the Heine–Borel Theorem, H is sequentially compact. Therefore, f is uniformly continuous on H (see Exercise 3). ∎

EXERCISES

1. Prove that
$$f(x, y) = \begin{cases} e^{-1/|x-y|} & x \neq y \\ 0 & x = y \end{cases}$$
   is continuous on $\mathbf{R}^2$.
2. Let $E \subseteq \mathbf{R}^n$ and $f : E \to \mathbf{R}^m$. Prove that f is continuous on E if and only if $f^{-1}(C) \cap E$ is relatively closed in E for every set C closed in $\mathbf{R}^m$.
3. Let $H \subseteq \mathbf{R}^n$ be sequentially compact and $f : H \to \mathbf{R}^m$. Modify the proof of Theorem 2.10 to show that if f is continuous on H, then f is uniformly

continuous on H.

4. Prove Theorem 5.24.

$\boxed{*5}$. **This exercise is used in Section** ^{e}**5.8.** Let X, Y be Euclidean spaces, H be a compact subset of X, and $f : H \to Y$ be continuous.

a) Show that given $\varepsilon > 0$ there exist points $x_1, \ldots, x_N \in H$ and positive numbers $\delta_1, \ldots, \delta_N$ such that

$$H \subseteq \bigcup_{j=1}^{N} B_{\delta_j}(x_j) \quad \text{and} \quad f(H \cap B_{2\delta_j}(x_j)) \subseteq B_{\varepsilon/2}(f(x_j)).$$

b) Use part a) to give another proof of Theorem 5.30.

$\boxed{6}$. **This exercise is used in Section 8.1.** Let X, Y be Euclidean spaces, $H \subseteq X$, and $f : H \to Y$ be 1–1 on H.

a) Show that $(f^{-1})^{-1}(E) = f(E)$ for all $E \subseteq H$.

b) Use Exercise 2 and part a) to prove that if H is compact and f is 1–1 and continuous on H, then f^{-1} is continuous on $f(H)$.

$\boxed{*7}$. **This exercise is used in Section** ^{e}**5.8.** Let H be a nonempty compact subset of a Euclidean space X.

a) Suppose $f : H \to \mathbf{R}^m$ is continuous. Prove that

$$\|f\|_\infty := \sup_{x \in H} \|f(x)\|$$

is finite and there exists an $x_0 \in H$ such that $\|f(x_0)\| = \|f\|_\infty$.

b) A sequence of functions $f_k : H \to \mathbf{R}^m$ is said to converge uniformly on H to a function $f : H \to \mathbf{R}^m$ if given $\varepsilon > 0$ there is an $N \in \mathbf{N}$ such that

$$k \geq N \quad \text{and} \quad x \in H \quad \text{imply} \quad \|f_k(x) - f(x)\| < \varepsilon.$$

Show that $\|f_k - f\|_\infty \to 0$ as $k \to \infty$ if and only if $f_k \to f$ uniformly on H as $k \to \infty$.

c) Prove that a sequence of functions f_k converges uniformly on H if and only if given $\varepsilon > 0$ there is an $N \in \mathbf{N}$ such that

$$k, j \geq N \quad \text{implies} \quad \|f_k - f_j\|_\infty < \varepsilon.$$

$\boxed{*8}$. **This exercise is used to prove *Corollary 6.17.**

a) A set $E \subseteq \mathbf{R}^n$ is said to be *polygonally connected* if any two points $a, b \in E$ can be connected by a polygonal path in E; i.e., there exist points $x_k \in E$, $k = 1, \ldots, N$, such that $x_0 = a$, $x_N = b$ and $L(x_{k-1}; x_k) \subseteq E$ for $k = 1, \ldots, N$. Prove every polygonally connected set in $\mathbf{R}^n$ is connected.

b) Let $E \subseteq \mathbf{R}^n$ be open and connected and $x_0 \in E$. Let U be the set of points $x \in E$ which can be polygonally connected in E to x_0. Prove that U is open.

c) Prove every open connected set in $\mathbf{R}^n$ is polygonally connected.

*e*5.7   **APPLICATIONS**   *This section requires no material from any other enrichment section.*

We have seen that the topological ideas introduced in this chapter (e.g., closed sets, open sets, compact sets, and connected sets) are powerful theoretical tools. In this section we continue this theme by proving three independent theorems.

First, we use connectivity to characterize the graph of a continuous function.

THEOREM 5.31 [CLOSED GRAPH THEOREM]. *Let $[a, b]$ be a closed interval and $f : [a, b] \to \mathbf{R}$. Then f is continuous on $[a, b]$ if and only if the graph of f is closed and connected in $\mathbf{R}^2$.*

PROOF. For any points $c, d \in [a, b]$, let $\mathcal{G}[c, d]$ represent the graph of $y = f(x)$ for $x \in [c, d]$. Suppose f is continuous on $[a, b]$. The function $x \longmapsto (x, f(x))$ is continuous from $[a, b]$ into $\mathbf{R}^2$ and $[a, b]$ is connected in $\mathbf{R}$. Thus $\mathcal{G}[a, b]$ is connected in $\mathbf{R}^2$ by Theorem 5.27. To prove $\mathcal{G}[a, b]$ is closed, we shall use Theorem 5.15. Let $x_k \in [a, b]$ and $(x_k, f(x_k)) \to (x, y)$ as $k \to \infty$. Then $x_k \to x$ and $f(x_k) \to y$, as $k \to \infty$. Hence, $x \in [a, b]$ and since f is continuous, $f(x_k) \to f(x)$. In particular, the graph of f is closed.

Conversely, suppose the graph of f is closed and connected in $\mathbf{R}^2$. We first show that f satisfies the Intermediate Value Theorem on $[a, b]$. Indeed, suppose to the contrary that there exist $x_1 < x_2$ in $[a, b]$ with $f(x_1) \neq f(x_2)$ and a value y_0 between $f(x_1)$ and $f(x_2)$ such that $f(t) \neq y_0$ for all $t \in [x_1, x_2]$. Suppose for simplicity that $f(x_1) < f(x_2)$. Let $\delta < \min\{|f(x_1) - y_0|, |f(x_2) - y_0|\}$. Since $f(t) \neq y_0$ for any $t \in [x_1, x_2]$, the open sets

$$U = \{(x, y) : x_1 < x < x_2, \ y < y_0\} \cup B_{\delta/2}((x_1, f(x_1)),$$

$$V = \{(x, y) : x_1 < x < x_2, \ y > y_0\} \cup B_{\delta/2}((x_2, f(x_2)))$$

separate $\mathcal{G}[x_1, x_2]$ (see Figure 5.8). Hence, the open sets

$$U \cup \{(x, y) : x < x_1\} \quad \text{and} \quad V \cup \{(x, y) : x > x_2\}$$

separate $\mathcal{G}[a, b]$, a contradiction. Therefore, f satisfies the Intermediate Value Theorem on $[a, b]$.

If f is not continuous on $[a, b]$, then there exist numbers $x_0 \in [a, b]$, $\varepsilon_0 > 0$, and $x_k \in [a, b]$ such that $x_k \to x_0$ and $|f(x_k) - f(x_0)| > \varepsilon_0$. By symmetry, we may suppose that $f(x_k) > f(x_0) + \varepsilon_0$ for infinitely many k's, say

$$f(x_{k_j}) > f(x_0) + \varepsilon_0 > f(x_0), \qquad j \in \mathbf{N}.$$

By the Intermediate Value Theorem, choose c_j between x_{k_j} and x_0 such that $f(c_j) = f(x_0) + \varepsilon_0$. By construction, $(c_j, f(c_j)) \to (x_0, f(x_0) + \varepsilon_0)$ and $c_j \to x_0$ as $j \to \infty$. Hence, the graph of f on $[a, b]$ is not closed. ∎

Next, we obtain a partial converse of Theorem 4.23. A sequence of real-valued functions $\{f_k\}$ is said to be *pointwise increasing* (respectively, *pointwise decreasing*) on a set E if $f_k(\boldsymbol{x}) \leq f_{k+1}(\boldsymbol{x})$ (respectively, $f_k(\boldsymbol{x}) \geq f_{k+1}(\boldsymbol{x})$) for all $\boldsymbol{x} \in E$ and $k \in \mathbf{N}$. A sequence is said to be *pointwise monotone* on E if it is pointwise increasing on E or pointwise decreasing on E.

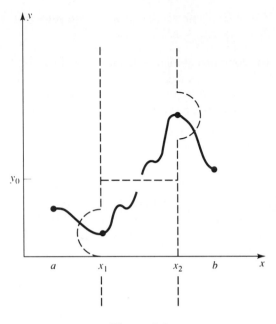

Figure 5.8

THEOREM 5.32 [DINI]. *Suppose $H \subset \mathbf{R}^n$ is compact and $f_k : H \to \mathbf{R}$ is a pointwise monotone sequence of continuous functions. If $f_k \to f$ pointwise on H as $k \to \infty$ and f is continuous on H, then $f_k \to f$ uniformly on H. In particular, if ϕ_k is a pointwise monotone sequence of functions continuous on an interval $[a, b]$ which converges pointwise to a continuous function, then*

$$\lim_{k \to \infty} \int_a^b \phi_k(t)\, dt = \int_a^b \left(\lim_{k \to \infty} \phi_k(t) \right)\, dt.$$

PROOF. By Theorem 4.24, we need only show that $f_k \to f$ uniformly on H. We may suppose that f_k is pointwise increasing.

Let $\varepsilon > 0$. For each $\boldsymbol{x} \in H$, choose $N(\boldsymbol{x}) \in \mathbf{N}$ such that

$$k \geq N(\boldsymbol{x}) \quad \text{implies} \quad |f_k(\boldsymbol{x}) - f(\boldsymbol{x})| < \frac{\varepsilon}{3}.$$

Since f and $f_{N(\boldsymbol{x})}$ are continuous on H, choose an $r = r(\boldsymbol{x}) > 0$ such that

$$\boldsymbol{y} \in H \cap B_r(\boldsymbol{x}) \quad \text{implies} \quad |f(\boldsymbol{x}) - f(\boldsymbol{y})| < \frac{\varepsilon}{3} \quad \text{and} \quad |f_{N(\boldsymbol{x})}(\boldsymbol{x}) - f_{N(\boldsymbol{x})}(\boldsymbol{y})| < \frac{\varepsilon}{3}.$$

Since H is compact, choose $\boldsymbol{x}_j \in H$ and $r_j = r(\boldsymbol{x}_j)$ such that

$$H \subset \bigcup_{j=1}^M B_{r_j}(\boldsymbol{x}_j).$$

Set $N = \max\{N(\boldsymbol{x}_1), \ldots, N(\boldsymbol{x}_M)\}$, let $\boldsymbol{x} \in H$, and suppose $k \geq N$. Since $\boldsymbol{x} \in B_{r_j}(\boldsymbol{x}_j)$ for some $j \in \{1, \ldots, M\}$ and $k \geq N(\boldsymbol{x}_j)$, it follows that

$$
\begin{aligned}
|f(\boldsymbol{x}) - f_k(\boldsymbol{x})| = f(\boldsymbol{x}) - f_k(\boldsymbol{x}) &\leq f(\boldsymbol{x}) - f_{N(\boldsymbol{x}_j)}(\boldsymbol{x}) \\
&\leq |f(\boldsymbol{x}) - f(\boldsymbol{x}_j)| + |f(\boldsymbol{x}_j) - f_{N(\boldsymbol{x}_j)}(\boldsymbol{x}_j)| \\
&\quad + |f_{N(\boldsymbol{x}_j)}(\boldsymbol{x}_j) - f_{N(\boldsymbol{x}_j)}(\boldsymbol{x})| \\
&< \varepsilon.
\end{aligned}
$$

Since this inequality holds for all $\boldsymbol{x} \in H$, we conclude that $f_k \to f$ uniformly on H as $k \to \infty$. ∎

Our final application of topological ideas to analysis is a characterization of Riemann integrability of a function f by the size of the set of points of discontinuity of f. To measure the size of this set, we make the following definition. (Recall that $|I|$ denotes the length of an interval I.)

DEFINITION 5.19. i) A set $E \subset \mathbf{R}$ is said to be of *measure zero* if given $\varepsilon > 0$ there is a countable collection of intervals $\{I_j\}_{j \in \mathbf{N}}$ which covers E such that

$$
\sum_{j=1}^{\infty} |I_j| \leq \varepsilon.
$$

ii) A function $f : [a, b] \to \mathbf{R}$ is said to be *almost everywhere continuous* on $[a, b]$ if the set of points $x \in [a, b]$ where f is discontinuous is a set of measure zero.

Notice that by definition, if E is of measure zero, then every subset of E is also of measure zero. Loosely speaking, a set is of measure zero if it is so small that it can be covered by a sequence of intervals whose total length is as small as we wish.

It is easy to see that a single point $E = \{x\}$ is a set of measure zero. Indeed, $I_1 := (x - \varepsilon/2, x + \varepsilon/2)$, $I_k := \emptyset$ for $k \geq 2$, cover E and have total length ε. Modifying this technique, we can show that any finite set is a set of measure zero (see Remark 1 below). On the other hand, since $[a, b]$ is compact, any open covering of $[a, b]$ has a finite subcover; hence, any covering of $[a, b]$ by open intervals must have total length greater than or equal to $b - a$. Hence, a closed nondegenerate interval cannot be of measure zero.

The following result shows that if a set is small in the set theoretical sense, then it is small in the measure theoretic sense.

Remark 1. *Every countable set of real numbers is a set of measure zero.*

PROOF. We may suppose that E is countably infinite, say $E = \{x_1, x_2, \ldots\}$. Given $\varepsilon > 0$ and $j \in \mathbf{N}$, set

$$
I_j = (x_j - \varepsilon 2^{-j-1}, x_j + \varepsilon 2^{-j-1}).
$$

Then $x_j \in I_j$ and $|I_j| = \varepsilon 2^{-j}$ for $j \in \mathbf{N}$. Therefore, $E \subseteq \cup_{j=1}^{\infty} I_j$ and

$$
\sum_{j=1}^{\infty} |I_j| = \varepsilon \sum_{j=1}^{\infty} \frac{1}{2^j} = \varepsilon. \quad \blacksquare
$$

The converse of Remark 1 is false; i.e., there exist uncountable sets of measure zero (see Exercise 9 below).

The following result shows that the countable union of sets of measure zero is a set of measure zero.

Remark 2. *If $E_1, E_2, \ldots$ is a sequence of sets of measure zero, then*

$$E = \bigcup_{k=1}^{\infty} E_k$$

is also a set of measure zero.

PROOF. Let $\varepsilon > 0$. By hypothesis, given $k \in \mathbf{N}$ we can choose a collection of intervals $\{I_j^{(k)}\}_{j \in \mathbf{N}}$ which covers E_k such that

$$\sum_{j=1}^{\infty} |I_j^{(k)}| < \frac{\varepsilon}{2^k}.$$

Then the collection $\{I_j^{(k)}\}_{k,j \in \mathbf{N}}$ is countable, covers E, and

$$\sum_{k=1}^{\infty} \sum_{j=1}^{\infty} |I_j^{(k)}| \le \sum_{k=1}^{\infty} \frac{\varepsilon}{2^k} = \varepsilon.$$

Consequently, E is of measure zero. ∎

To facilitate our discussion of points of discontinuity, we introduce the following concepts.

DEFINITION 5.20. Let $[a, b]$ be a closed interval and $f : [a, b] \to \mathbf{R}$ be bounded.

i) The *oscillation* of f on an interval J which intersects $[a, b]$ is defined to be

$$\Omega_f(J) := \sup_{x,y \in J \cap [a,b]} (f(x) - f(y)).$$

ii) The *oscillation* of f at a point $t \in [a, b]$ is defined to be

$$\omega_f(t) := \lim_{h \to 0+} \Omega_f((t - h, t + h)),$$

when this limit exists.

Remark 3. *If $f : [a, b] \to \mathbf{R}$ is bounded, then $\omega_f(t)$ exists for all $t \in [a, b]$ and satisfies $0 \le \omega_f(t) < \infty$.*

PROOF. Fix $t \in [a, b]$ and for each interval J, set

$$M_J = \sup_{x \in J \cap [a,b]} f(x), \qquad m_J = \inf_{x \in J \cap [a,b]} f(x).$$

Since $\sup(-f(x)) = -\inf f(x)$, it is obvious that

$$(15) \qquad\qquad \Omega_f(J) = M_J - m_J \geq 0.$$

Suppose for simplicity that $t \in (a, b)$ and choose h_0 so small that $(t - h_0, t + h_0) \subset (a, b)$. For each $0 < h < h_0$, set

$$F(h) = \Omega_f((t - h, t + h)).$$

By the Monotone Property of Suprema, $F(h)$ is increasing on $(0, h_0)$, hence has a finite limit as $h \to 0+$. By (15), $F(h) \geq 0$. Therefore, $\omega_f(t)$ exists, and is both finite and nonnegative. ∎

The next result shows that by using the oscillation function ω_f, one can represent the set of points of discontinuity of any bounded f as a countable union.

Remark 4. *Let $f : [a, b] \to \mathbf{R}$ be bounded. If E represents the set of points of discontinuity of f in $[a, b]$, then*

$$E = \bigcup_{j=1}^{\infty} \left\{ t \in [a, b] : \omega_f(t) \geq \frac{1}{j} \right\}.$$

PROOF. By (15), f is continuous at $t \in [a, b]$ if and only if $\omega_f(t) = 0$. Hence, t belongs to E if and only if $\omega_f(t) > 0$. Since by the Archimedean Principle, $\omega_f(t) > 0$ if and only if $\omega_f(t) \geq 1/j$ for some $j \in \mathbf{N}$, the result follows at once. ∎

We need two technical results about the oscillation of f at a point t.

Lemma 1. *Let $f : [a, b] \to \mathbf{R}$ be bounded. For each $\varepsilon > 0$, the set*

$$H = \{ t \in [a, b] : \omega_f(t) \geq \varepsilon \}$$

is compact.

PROOF. Suppose to the contrary that H is not compact. Then H is nonempty. Since H is bounded, it cannot be closed. Hence, there are points $t_k \in H$ such that $t_k \to t$ as $k \to \infty$ but $t \notin H$. Since $\omega_f(t) < \varepsilon$, it follows that there is an $h_0 > 0$ such that

$$(16) \qquad\qquad \Omega_f((t - h_0, t + h_0)) < \varepsilon.$$

Since $t_k \to t$, choose $N \in \mathbf{N}$ so that

$$\left(t_N - \frac{h_0}{2}, t_N + \frac{h_0}{2} \right) \subset (t - h_0, t + h_0).$$

Then by (16), $\Omega_f((t_N - h_0/2, t_N + h_0/2)) < \varepsilon$. Therefore, $\omega_f(t_N) < \varepsilon$ which contradicts the fact that $t_N \in H$. ∎

Lemma 2. *Let I be a closed bounded interval and $f : I \to \mathbf{R}$ be bounded. If $\varepsilon > 0$ and $\omega_f(t) < \varepsilon$ for all $t \in I$, then there is a $\delta > 0$ such that $\Omega_f(J) < \varepsilon$ for all closed intervals $J \subseteq I$ which satisfy $|J| < \delta$.*

PROOF. For each $t \in I$, choose δ_t such that

(17) $$\Omega_f((t - \delta_t, t + \delta_t)) < \varepsilon.$$

Since $\{(t - \delta_t/2, t + \delta_t/2)\}_{t \in I}$ covers I, choose $t_1, \ldots, t_N$ such that

$$I \subseteq \bigcup_{j=1}^{N} \left(t_j - \frac{\delta_{t_j}}{2}, t_j + \frac{\delta_{t_j}}{2} \right)$$

and set

$$\delta = \min_{1 \leq j \leq N} \frac{\delta_{t_j}}{2}.$$

If $J \subseteq I$, then

$$J \cap \left(t_j - \frac{\delta_{t_j}}{2}, t_j + \frac{\delta_{t_j}}{2} \right) \neq \emptyset$$

for some $j \in \{1, \ldots, N\}$. If J also satisfies $|J| < \delta$, then it follows from $2\delta \leq \delta_{t_j}$ that $J \subseteq (t_j - \delta_{t_j}, t_j + \delta_{t_j})$. In particular, (17) implies

$$\Omega_f(J) \leq \Omega_f((t_j - \delta_{t_j}, t_j + \delta_{t_j})) < \varepsilon. \quad \blacksquare$$

THEOREM 5.33 [LEBESGUE]. *Let $f : [a, b] \to \mathbf{R}$ be bounded. Then f is Riemann integrable on $[a, b]$ if and only if f is almost everywhere continuous on $[a, b]$. In particular, if f is bounded and has countably many points of discontinuity on $[a, b]$, then f is integrable on $[a, b]$.*

PROOF. Let E be the set of points of discontinuity of f in $[a, b]$. Suppose f is integrable but E is not of measure zero. By Remarks 2 and 4, there is a $j_0 \in \mathbf{N}$ such that

$$H := \{t \in [a, b] : \omega_f(t) \geq \frac{1}{j_0}\}$$

is not of measure zero. In particular, there is an $\varepsilon_0 > 0$ such that if $\{I_k\}_{k \in \mathbf{N}}$ is any collection of intervals which covers H, then

(18) $$\sum_{k=1}^{\infty} |I_k| \geq \varepsilon_0.$$

Let $P = \{x_0, \ldots, x_n\}$ be a partition of $[a, b]$. If $(x_{k-1}, x_k) \cap H \neq \emptyset$, then by definition, $M_k(f) - m_k(f) \geq 1/j_0$. Hence,

$$U(f, P) - L(f, P) = \sum_{k=1}^{n} (M_k(f) - m_k(f))(x_k - x_{k-1})$$

$$\geq \sum_{(x_{k-1}, x_k) \cap H \neq \emptyset} (M_k(f) - m_k(f))(x_k - x_{k-1})$$

$$\geq \frac{1}{j_0} \sum_{(x_{k-1}, x_k) \cap H \neq \emptyset} (x_k - x_{k-1}).$$

But $\{[x_{k-1}, x_k] : (x_{k-1}, x_k) \cap H \neq \emptyset\}$ is a collection of intervals which covers H. Hence, it follows from (18) that

$$U(f, P) - L(f, P) \geq \frac{\varepsilon_0}{j_0} > 0.$$

Therefore, f cannot be integrable on $[a, b]$.

Conversely, suppose E is of measure zero. Let $M = \sup_{x \in [a,b]} f(x)$ and $m = \inf_{x \in [a,b]} f(x)$. Given $\varepsilon > 0$, choose $j_0 \in \mathbf{N}$ such that

$$\frac{M - m + b - a}{j_0} < \varepsilon.$$

Since E is of measure zero, so is

$$H = \{t \in [a, b] : \omega_f(t) \geq \frac{1}{j_0}\}.$$

Hence, by Definition 5.19, there exists a collection of intervals which covers H whose lengths sum to a real number less than $1/(2j_0)$. By expanding these intervals slightly, we may suppose there exist open intervals $I_1, I_2, \ldots$ which cover H such that

$$\sum_{\nu=1}^{\infty} |I_\nu| < \frac{1}{j_0}.$$

Hence, by Lemma 1, we can choose $N \in \mathbf{N}$ such that $\{I_1, I_2, \ldots, I_N\}$ covers H and

(19)
$$\sum_{\nu=1}^{N} |I_\nu| < \frac{1}{j_0}.$$

We must find a partition P such that $U(f, P) - L(f, P) < \varepsilon$. The endpoints of the I_ν's form part of this partition. Other points will come from further division of that part of $[a, b]$ not covered by the I_ν's. Indeed, let $I' \subseteq [a, b] \setminus (\cup_{\nu=1}^{N} I_\nu)$. Since the I_ν's cover H, $\omega_f(t) < 1/j_0$ for all $t \in I'$. Hence, by Lemma 2, there is a $\delta > 0$ such that if $J \subseteq I'$ satisfies $|J| < \delta$, then $\Omega_f(J) < 1/j_0$. Subdivide $[a, b] \setminus (\cup_{\nu=1}^{N} I_\nu)$ into intervals J_ℓ, $\ell = 1, \ldots, s$, such that $|J_\ell| < \delta$. Then

(20)
$$\Omega_f(J_\ell) < \frac{1}{j_0}$$

for $\ell = 1, \ldots, s$.

Let $P = \{x_0, x_1, \ldots, x_n\}$ represent the collection of points x such that x is an endpoint of some I_ν or of some J_ℓ. Notice that if $(x_{k-1}, x_k) \cap H \neq \emptyset$, then x_{k-1} and x_k are endpoints of some I_ν, whence by (19),

$$\sum_{(x_{k-1}, x_k) \cap H \neq \emptyset} (M_k(f) - m_k(f))(x_k - x_{k-1}) \leq \frac{M - m}{j_0}.$$

On the other hand, if $(x_{k-1}, x_k) \cap H = \emptyset$, then x_{k-1} and x_k are endpoints of some J_ℓ, whence by (20),

$$\sum_{(x_{k-1}, x_k) \cap H = \emptyset} (M_k(f) - m_k(f))(x_k - x_{k-1}) \leq \frac{1}{j_0} \sum_{k=1}^{n} (x_k - x_{k-1}) = \frac{b-a}{j_0}.$$

Consequently,

$$U(f, P) - L(f, P) = \sum_{k=1}^{n} (M_k(f) - m_k(f))(x_k - x_{k-1}) \leq \frac{M - m + b - a}{j_0} < \varepsilon.$$

We conclude that f is integrable on $[a, b]$. ∎

Recall that if $\alpha > 0$ and $f(x)$ is positive, then

$$f^\alpha(x) := e^{\alpha \log(f(x))}.$$

Suppose f is Riemann integrable. Although Corollary 3.9 implies that f^n is integrable for each $n \in \mathbf{N}$, we have not yet investigated the integrability of noninteger powers of f, e.g., $\sqrt{f}$ and $\sqrt[3]{f}$. The following result shows that Lebesgue's Theorem answers the question of integrability for all positive powers of f, rational or irrational.

COROLLARY 5.34. *If $f : [a, b] \to [0, \infty)$ is Riemann integrable, then so is f^α for every $\alpha > 0$.*

EXERCISES

1. Suppose $f_k : [a, b] \to [0, \infty)$ for $k \in \mathbf{N}$ and

$$f(x) := \sum_{k=1}^{\infty} f_k(x)$$

converges pointwise on $[a, b]$. If f and f_k are continuous on $[a, b]$ for each $k \in \mathbf{N}$, prove that

$$\int_a^b \sum_{k=1}^{\infty} f_k(x) \, dx = \sum_{k=1}^{\infty} \int_a^b f_k(x) \, dx.$$

2. Let E be a compact subset of $\mathbf{R}^n$. Suppose $g, f_k, g_k : E \to \mathbf{R}$ are continuous on E with $g_k \geq 0$ and $f_1 \geq f_2 \cdots \geq f_k \geq 0$ for all $k \in \mathbf{N}$. If $g = \sum_{k=1}^{\infty} g_k$ converges pointwise on E, prove $\sum_{k=1}^{\infty} f_k g_k$ converges uniformly on E.

3. Suppose $f, f_k : \mathbf{R} \to \mathbf{R}$ are continuous and nonnegative. Prove that if $f(x) \to 0$ as $x \to \pm\infty$ and $f_k \uparrow f$ everywhere on $\mathbf{R}$, then $f_k \to f$ uniformly on $\mathbf{R}$.

4. For each of the following functions, find a formula for $\omega_f(t)$.

a)
$$f(x) = \begin{cases} 1 & x \in \mathbf{Q} \\ 0 & x \notin \mathbf{Q}. \end{cases}$$

b)
$$f(x) = \begin{cases} 1 & x \geq 0 \\ 0 & x < 0. \end{cases}$$

c)
$$f(x) = \begin{cases} \sin(1/x) & x \neq 0 \\ 0 & x = 0. \end{cases}$$

5. Prove $(1 - x/k)^k \to e^{-x}$ uniformly on any compact subset of $\mathbf{R}$.

6. Show that if $f : [a, b] \to \mathbf{R}$ is integrable and $g : f([a, b]) \to \mathbf{R}$ is continuous, then $g \circ f$ is integrable on $[a, b]$. (Notice by Remark 4 in Section 2.2 that this result is false if g is allowed even one point of discontinuity.)

7. Using Theorem 4.24 or Theorem 5.32, prove that each of the following limits exists. Find a value for the limit in each case.

a)
$$\lim_{k \to \infty} \int_0^{\pi/2} \sin x \sqrt{\frac{2k}{4k - 3x}} \, dx.$$

b)
$$\lim_{k \to \infty} \int_0^1 x^2 f\left(\frac{k}{k^2 + x}\right) dx,$$

where f is continuously differentiable on $[0, 1]$ and $f'(0) > 0$.

c)
$$\lim_{k \to \infty} \int_0^1 x^3 \cos\left(\frac{\log k + x}{k + x}\right) dx.$$

d)
$$\lim_{k \to \infty} \int_{-1}^1 \left(1 + \frac{x}{k}\right)^k e^x \, dx.$$

8. a) Prove that given $\varepsilon > 0$ there is a sequence of open intervals $\{I_k\}_{k \in \mathbf{N}}$ which covers $[0, 1] \cap \mathbf{Q}$ such that

$$\sum_{k=1}^{\infty} |I_k| < \varepsilon.$$

b) Prove that if $\{I_k\}_{k \in \mathbf{N}}$ is a sequence of open intervals which covers $[0, 1]$, then there is an $N \in \mathbf{N}$ such that

$$\sum_{k=1}^{N} |I_k| \geq 1.$$

9. Let E_1 be the unit interval $[0,1]$ with its middle third $(1/3, 2/3)$ removed, i.e., $E_1 = [0, 1/3] \cup [2/3, 1]$. Let E_2 be E_1 with its middle thirds removed, i.e.,

$$E_2 = [0, 1/9] \cup [2/9, 1/3] \cup [2/3, 7/9] \cup [8/9, 1].$$

Continuing in this manner, generate nested sets E_k such that each E_k is the union of 2^k closed intervals of length $1/3^k$. The *Cantor set* is the set

$$E := \bigcap_{k=1}^{\infty} E_k.$$

Assume that every point $x \in [0,1]$ has a binary expansion and a ternary expansion; i.e., there exist $a_k \in \{0,1\}$ and $b_k \in \{0,1,2\}$ such that

$$x = \sum_{k=1}^{\infty} \frac{a_k}{2^k} = \sum_{k=1}^{\infty} \frac{b_k}{3^k}.$$

(For example, if $x = 1/3$, then $a_{2k-1} = 0$, $a_{2k} = 1$ for all k and either $b_1 = 1$, $b_k = 0$ for $k > 1$ or $b_1 = 0$ and $b_k = 1$ for all $k > 1$.)

a) Prove that E is a nonempty compact set of measure zero.

b) Show that a point $x \in [0,1]$ belongs to E if and only if x has a ternary expansion whose digits satisfy $b_k \neq 1$ for all $k \in \mathbf{N}$.

c) Define $f : E \to [0,1]$ by

$$f\left(\sum_{k=1}^{\infty} \frac{b_k}{3^k}\right) = \sum_{k=1}^{\infty} \frac{b_k/2}{2^k}.$$

Prove f is 1–1 from E onto $[0,1]$; i.e., prove E is uncountable.

d) Extend f from E to $[0,1]$ by making f constant on the middle thirds $E_{k-1} \setminus E_k$. Prove that $f : [0,1] \to [0,1]$ is continuous and increasing. *(Note: The function f is almost everywhere constant on $[0,1]$, i.e., constant off a set of measure zero. Yet, it begins at $f(0) = 0$ and ends at $f(1) = 1$.)*

^{e}5.8 METRIC SPACES   *This section uses no material from any other enrichment section.*

In Sections 5.2 through 5.6 we introduced several topological ideas in the concrete Euclidean space setting. In this section we investigate these ideas from a more general viewpoint.

DEFINITION 5.21. A *metric space* is a set X together with a function $\rho = \rho_X : X \times X \to \mathbf{R}$ (called the *metric* of X), which satisfies the following properties for all $x, y, z \in X$:

POSITIVE DEFINITE $\rho(x,y) \geq 0$ with $\rho(x,y) = 0$ if and only if $x = y$,

SYMMETRIC $\rho(x,y) = \rho(y,x)$,

TRIANGLE INEQUALITY $\rho(x,y) \leq \rho(x,z) + \rho(z,y).$

The concept of a metric space, somewhat more abstract than anything considered thus far, occupies a central place in mathematics and is well worth studying in its own right. We begin with several different examples.

Example 1. R is a metric space with metric $\rho(x, y) = |x - y|$.

PROOF. See Theorem 1.3 and the remarks preceding it. ∎

Example 2. For each $n \in \mathbf{N}$, $\mathbf{R}^n$ is a metric space with metric $\rho(\boldsymbol{x}, \boldsymbol{y}) = \|\boldsymbol{x} - \boldsymbol{y}\|$.

PROOF. See Theorem 5.3. ∎

Example 3. The set $\mathcal{C}[a, b]$ of continuous functions $f : [a, b] \to \mathbf{R}$ is a metric space with metric

$$\rho(f, g) = \sup_{x \in [a,b]} |f(x) - g(x)|.$$

PROOF. See Exercise 6 in Section 5.1. ∎

The following example will be used to give a simple interpretation of relatively open sets (see Corollary 5.36 below.)

Example 4. If X is a metric space with metric ρ and $Y \subset X$, then Y is a metric space with metric ρ. (We shall call such a metric space a *subspace* of X.)

PROOF. If the positive definite property, the symmetric property, and the triangle inequality hold for all $x, y \in X$, then they hold for all $x, y \in Y$. ∎

Example 5. Q is a metric space with metric $\rho(x, y) = |x - y|$.

PROOF. Combine Examples 1 and 4. ∎

By Example 1, **R** is a metric space with metric $\rho(x, y) = |x - y|$. The following example illustrates the fact that a given set can have more than one metric. Hence, to describe a given metric space, we must specify both the set X and the metric ρ.

Example 6. R is a metric space with metric

$$\sigma(x, y) = \begin{cases} 0 & x = y \\ 1 & x \neq y. \end{cases}$$

(This metric is called the *discrete metric*.)

PROOF. The function σ is obviously positive definite and symmetric. To prove σ satisfies the triangle inequality, we consider three cases. If $x = z$, then $\sigma(x, y) = 0 + \sigma(z, y) = \sigma(x, y) + \sigma(z, y)$. A similar equality holds if $y = z$. Finally, if $x \neq z$ and $y \neq z$, then $\sigma(x, y) \leq 1 < 2 = \sigma(x, z) + \sigma(z, y)$. ∎

For the next several pages, let X represent a metric space with metric ρ. Using the analogy between the norm on $\mathbf{R}^n$ and the metric ρ, we can extend most of the concepts and results covered in Sections 5.2 through 5.6 from Euclidean spaces to arbitrary metric spaces. Since this process amounts to little more than replacing norms by the metric, e.g., $\|\boldsymbol{x} - \boldsymbol{y}\|$ by $\rho(x, y)$, we shall leave the details to the reader.

We will be careful, however, to point out where the theory of metric spaces diverges from that of Euclidean spaces.

To make our presentation easier to follow, we will cover the material in roughly the same order that was used in Sections 5.2 through 5.6. Thus, we begin with the following definition.

DEFINITION 5.22. i) An *open ball* in X is a set of the form

$$B_r(a) := \{x \in X : \rho(x, a) < r\},$$

where $a \in X$ and $r > 0$. The point a is called the *center* of $B_r(a)$ and r is called its *radius*.

ii) A set $V \subseteq X$ is said to be *open* if given $a \in V$ there is an $\varepsilon > 0$ such that $B_\varepsilon(a) \subseteq V$.

iii) A set $E \subseteq X$ is said to be *closed* if $E^c := X \setminus E$ is open.

By modifying arguments presented in Section 5.2, we can prove the following results.

THEOREM 5.35. *Let X be a metric space.*

i) *Every open ball in X is open.*

ii) *Every finite subset of X is closed.*

iii) *The empty set $\emptyset$ and the whole space X are both open and closed.*

THEOREM 5.36. *Let X be a metric space.*

i) *If $\{V_\alpha\}_{\alpha \in A}$ is any collection of open sets in X, then*

$$\bigcup_{\alpha \in A} V_\alpha$$

is open.

ii) *If $\{V_k : k = 1, 2, \ldots, n\}$ is a finite collection of open sets in X, then*

$$\bigcap_{k=1}^{n} V_k := \bigcap_{k \in \{1,2,\ldots,n\}} V_k$$

is open.

iii) *If $\{E_\alpha\}_{\alpha \in A}$ is any collection of closed sets in X, then*

$$\bigcap_{\alpha \in A} E_\alpha$$

is closed.

iv) *If $\{E_k : k = 1, 2, \ldots, n\}$ is a finite collection of closed sets in X, then*

$$\bigcup_{k=1}^{n} E_k := \bigcup_{k \in \{1,2,\ldots,n\}} E_k$$

is closed.

v) *If V is open in X and E is closed in X, then $V \setminus E$ is open and $E \setminus V$ is closed.*

The following result shows that given $Y \subset \mathbf{R}^n$, a set $V \subseteq Y$ is relatively open in Y if and only if V is open in the subspace Y (compare with Definition 5.15).

COROLLARY 5.37. *If X is a metric space and Y is a subspace of X, then U is open in Y if and only if there is a set V open in X such that $U = Y \cap V$.*

PROOF. We begin by relating the balls in Y to the balls in X. Let $a \in Y$, $r > 0$, and $B_r(a) = \{x \in X : \rho(x, a) < r\}$. By definition, the open ball in Y centered at a of radius r is the set $\{x \in Y : \rho(x, a) < r\}$, i.e., the set $B_r(a) \cap Y$. Thus, each open ball in Y has the form $B_r(a) \cap Y$ for some $a \in Y$ and $r > 0$.

To prove the theorem, first suppose that U is open in Y. Then given $x \in U$ there is an $r_x > 0$ such that $B_{r_x}(x) \cap Y \subseteq U$. Set

$$V = \bigcup_{x \in U} B_{r_x}(x).$$

By Theorem 5.36, V is open in X. Evidently, $V \cap Y \subseteq U$ and by construction, $U \subseteq V \cap Y$. Thus $U = V \cap Y$.

Conversely, if $U = V \cap Y$ for some open set V in X, then given $x \in U$ there is an $r > 0$ such that $B_r(x) \subseteq V$. It follows that $B_r(x) \cap Y \subseteq U$, i.e., U is open in Y. ∎

The following is the metric space analogue of Definitions 5.7 and 5.8.

DEFINITION 5.23. Let $E \subseteq X$.

i) The *interior* of E is the set

$$E^o := \bigcup \{V : V \subseteq E \text{ and } V \text{ is open in } X\}.$$

ii) The *closure* of E is the set

$$\overline{E} := \bigcap \{B : B \supseteq E \text{ and } B \text{ is closed in } X\}.$$

iii) The *boundary* of E is the set

$$\partial E := \{x \in X : B_r(x) \cap E \neq \emptyset \quad \text{and} \quad B_r(x) \cap E^c \neq \emptyset \text{ for all } r > 0\}.$$

Using arguments presented in Section 5.2, we can prove the following results.

THEOREM 5.38. *Let $E \subseteq X$. Then E^o is the largest open set contained in E, $\overline{E}$ is the smallest closed set containing E, and $\partial E = \overline{E} \setminus E^o$. In particular, $E^o \subseteq E \subseteq \overline{E}$, $E = E^o$ if and only if E is open, and $E = \overline{E}$ if and only if E is closed.*

THEOREM 5.39. *Let $A, B \subseteq X$. Then*

i) $$(A \cup B)^o \supseteq A^o \cup B^o \quad \text{and} \quad (A \cap B)^o = A^o \cap B^o,$$

ii) $$\overline{A \cup B} = \overline{A} \cup \overline{B} \quad \text{and} \quad \overline{A \cap B} \subseteq \overline{A} \cap \overline{B},$$

and

iii) $\partial(A \cup B) \subseteq \partial A \cup \partial B \quad \text{and} \quad \partial(A \cap B) \subseteq (A \cap \partial B) \cup (B \cap \partial A) \cup (\partial A \cap \partial B).$

The following result illustrates the fact that not every topological theorem valid for Euclidean spaces has an analogue for arbitrary metric spaces.

Remark 1. *Lindelöf's Theorem does not hold for arbitrary metric spaces.*

PROOF. Consider the discrete metric space $X = \mathbf{R}$ introduced in Example 6. Since $B_{1/2}(x) = \{x\}$ for each $x \in X$, every point in X is open. Thus, $\{x\}_{x \in X}$ is an uncountable open covering of X which has no proper subcover. ∎

Next, we summarize the theory of sequences in a metric space.

DEFINITION 5.24. Let $\{x_k\}$ be a sequence of points in X.

   i) $\{x_k\}$ is said to be *bounded* if given $y_0 \in X$ there is an $M > 0$ such that $x_k \in B_M(y_0)$, i.e.,

$$\rho(x_k, y_0) < M \qquad k \in \mathbf{N}.$$

   ii) $\{x_k\}$ is said to *converge* (in X) to a point $x \in X$ (notation: $x_k \to x$ as $k \to \infty$) if given $\varepsilon > 0$ there is an $N \in \mathbf{N}$ such that

$$k \geq N \quad \text{implies} \quad \rho(x_k, x) < \varepsilon,$$

in which case we call x the *limit* of the sequence $\{x_k\}$.

By modifying arguments presented in Section 5.3, we can prove the following result.

THEOREM 5.40. *Let X be a metric space.*

   i) *A given sequence $\{x_k\}$ in X can have at most one limit.*
   ii) *Every convergent sequence in a metric space is bounded.*
   iii) *A sequence $\{x_k\}$ in a metric space X converges to some $x \in X$ if and only if every subsequence $\{x_{k_j}\}_{j \in \mathbf{N}}$ converges to that same x.*
   iv) *A sequence $\{x_k\}$ converges to $x \in X$ if and only if for every open set V which contains x there is an $N \in \mathbf{N}$ such that $k \geq N$ implies $x_k \in V$.*
   v) *Let $E \subseteq X$. Then E is closed if and only if the limit of every convergent sequence $x_k \in E$ satisfies*

$$\lim_{k \to \infty} x_k \in E.$$

The following result shows that the Bolzano–Weierstrass Theorem does not hold for arbitrary metric spaces.

Remark 2. *A metric space may contain bounded sequences which have no convergent subsequences.*

PROOF. Let $X = \mathbf{R}$ be the discrete metric space introduced in Example 6. Since $\sigma(0, k) = 1$ for all $k \in \mathbf{N}$, $\{k\}$ is a bounded sequence in X. Suppose there exist integers $k_1 < k_2 < \dots$ and an $x \in X$ such that $k_j \to x$ as $j \to \infty$. Then there is an $N \in \mathbf{N}$ such that $\sigma(k_j, x) < 1/2$ for $j \geq N$, i.e., $k_j = x$ for all $j \geq N$. This contradiction proves that $\{k\}$ has no convergent subsequences. ∎

Cauchy sequences can be defined on arbitrary metric spaces as follows.

DEFINITION 5.25. Let X be a metric space. A sequence of points $x_k \in X$ is said to be *Cauchy* if given $\varepsilon > 0$ there is an $N \in \mathbf{N}$ such that

$$k, m \geq N \quad \text{imply} \quad \rho(x_k, x_m) < \varepsilon.$$

Using the triangle inequality, we can prove that every convergent sequence in a metric space X is Cauchy. The following result shows that for an arbitrary metric space, the converse of this statement can be false.

Remark 3. *The metric space $X = \mathbf{Q}$ introduced in Example 5 contains Cauchy sequences which do not converge.*

PROOF. Choose $q_k \in \mathbf{Q}$ such that $q_k \to \sqrt{2}$. Then $\{q_k\}$ is Cauchy but does not converge in X since $\sqrt{2} \notin X$. ∎

This leads us to the following concept.

DEFINITION 5.26. A metric space X is said to be *complete* if every Cauchy sequence in X converges to some point in X.

By Theorem 5.14, $\mathbf{R}^n$ is complete. By Exercise 7 in Section 5.6, $\mathcal{C}[a, b]$ is also complete. Remark 3 shows that not every metric space is complete.

Sequentially compact and compact sets can be defined on arbitrary metric spaces as follows.

DEFINITION 5.27. Let H be a subset of a metric space X.

  i) H is said to be *sequentially compact* if every sequence in H has a convergent subsequence whose limit belongs to H.
  ii) H is said to be *compact* if every open covering of H has a finite subcover.

By modifying arguments presented in Section 5.3, we can prove the following results.

THEOREM 5.41. *Let X be a metric space.*

  i) *The empty set and all finite subsets of a metric space are compact.*
  ii) *If $E \subset H \subseteq X$, where H is compact and E is closed, then E is compact.*
  iii) *If H is a compact subset of X, then H is closed.*

THEOREM 5.42 [CANTOR'S INTERSECTION THEOREM]. *Let $H_1, H_2, \ldots$ be a sequence of nonempty compact sets in a metric space X. If $H_1 \supseteq H_2 \supseteq \ldots$, then*

$$\bigcap_{j \in \mathbf{N}} H_j \neq \emptyset.$$

THEOREM 5.43. *Let H be a subset of a metric space X.*

  i) *If H is compact, then H is sequentially compact.*
  ii) *If H is sequentially compact, then H is closed and bounded.*

Nevertheless, as the following result shows, the full statement of the Heine–Borel Theorem can be false for an arbitrary metric space.

Remark 4. *There exist metric spaces in which not every closed bounded set is compact.*

PROOF. Let $X = \mathbf{R}$ be the discrete metric space introduced in Example 6. Since $\sigma(0, x) \le 1$ for all $x \in \mathbf{R}$, every subset of X is bounded. Since $x_k \to x$ in X implies $x_k = x$ for large k, every subset of X is closed. Thus, $[0, 1]$ is a closed bounded subset of X. Since $\{x\}_{x \in [0,1]}$ is an uncountable open covering of $[0, 1]$ which has no proper subcover, we conclude that $[0, 1]$ is closed and bounded, but not compact. ∎

For the remainder of this section, let X and Y represent metric spaces with metrics $\rho = \rho_X$ and ρ_Y.

DEFINITION 5.28. Let E be a subset of X.

   i) A point $a \in X$ is called a *cluster point* of E if $E \cap B_r(a)$ contains infinitely many points for each $r > 0$. ·

   ii) Let a be a cluster point of E and $f : E \to Y$. Then f is said to *converge* to an $L \in Y$, as $x \to a$ *through* E, if given $\varepsilon > 0$ there is a $\delta > 0$ such that

$$x \in E \quad \text{and} \quad 0 < \rho_X(x, a) < \delta \quad \text{imply} \quad \rho_Y(f(x), L) < \varepsilon.$$

If f converges to L as $x \to a$ through E, we shall write $f(x) \to L$ as $x \to a$ through E or

$$L = \lim_{\substack{x \to a \\ x \in E}} f(x).$$

If f converges to L as $x \to a$ through some open ball $E = B_r(a)$, we shall simply write

$$L = \lim_{x \to a} f(x).$$

As in the Euclidean case, the value of f at a does not affect the limit L of $f(x)$ as $x \to a$ through E, the limit L (if it exists) is unique, and if f converges to L through some open ball centered at a, then it converges to L through all sets E which have a as a cluster point.

The following result shows that sequences can be used to characterize limits of functions between arbitrary metric spaces.

THEOREM 5.44 [SEQUENTIAL CHARACTERIZATION OF LIMITS]. *Let $E \subseteq X$, a be a cluster point of E, and $f : E \to Y$. Then*

$$L = \lim_{\substack{x \to a \\ x \in E}} f(x)$$

exists if and only if $f(x_k) \to L$ for every sequence $x_k \in E \setminus \{a\}$ which converges to a as $k \to \infty$.

Continuity of functions on metric spaces is defined as follows.

DEFINITION 5.29. Let $E \subseteq X$ and $f : E \to Y$.

   i) f is said to be *continuous* at a point $a \in E$ if given $\varepsilon > 0$ there is a $\delta > 0$ such that

$$x \in E \quad \text{and} \quad \rho_X(x, a) < \delta \quad \text{imply} \quad \rho_Y(f(x), f(a)) < \varepsilon.$$

   ii) f is said to be *continuous* on a nonempty subset E of X if f is continuous at every point $a \in E$.

If f is continuous at $a \in X$, we shall simply say f is continuous at a. Thus, given a cluster point a of X, f is continuous at a if and only if

$$f(a) = \lim_{x \to a} f(x).$$

By modifying arguments presented in Section 5.6, we can prove the following results.

THEOREM 5.45. *Let X, Y, Z be metric spaces, E be a nonempty subset of X and Ω be a nonempty subset of Y. If $f : E \to Y$ converges to L as $x \to a$ through E and if $g : \Omega \to Z$ is continuous at L, then*

$$\lim_{\substack{x \to a \\ x \in E}} (g \circ f)(x) = g \left(\lim_{\substack{x \to a \\ x \in E}} f(x) \right).$$

THEOREM 5.46. *Let X and Y be metric spaces and $f : X \to Y$. Then the following statements are equivalent.*

   i) *f is continuous on X.*
   ii) *$f^{-1}(V)$ is open in X for every V open in Y.*
   iii) *$f^{-1}(E)$ is closed in X for every E closed in Y.*

(Notice how the metric space setting makes this theorem simpler to state because we can avoid referring to the relative topology.)

THEOREM 5.47. *Let X and Y be metric spaces and suppose $f : X \to Y$ is continuous. If X is compact, then $f(X)$ is compact.*

THEOREM 5.48. *Let X, Y be metric spaces and H be compact in X. If $f : H \to Y$ is continuous on H, then $f(H)$ is compact in Y.*

THEOREM 5.49. *Let H be a nonempty, compact set in a metric space X and suppose $f : H \to \mathbf{R}$ is continuous. Then*

$$M := \sup\{f(x) : x \in H\} \quad \text{and} \quad m := \inf\{f(x) : x \in H\}$$

are finite real numbers and there exist points $x_M, x_m \in H$ such that $M = f(x_M)$ and $m = f(x_m)$.

Uniform convergence can be defined on arbitrary metric spaces as follows.

DEFINITION 5.30. A function $f : X \to Y$ is said to be *uniformly continuous* on a set $E \subseteq X$ if given $\varepsilon > 0$ there is a $\delta > 0$ such that

$$\rho_X(x,a) < \delta \quad \text{and} \quad x,a \in E \quad \text{imply} \quad \rho_Y(f(x), f(a)) < \varepsilon.$$

Clearly, if f is uniformly continuous on E, then f is continuous on E. The following result contains a partial converse of this statement.

THEOREM 5.50. *If X is a metric space, H is a compact subset of X, and $f : H \to Y$ is continuous on H, then f is uniformly continuous on H.*

An outline of the proof of this result can be found in Exercise 5 in Section 5.6. We close this section with a discussion of connected sets in arbitrary metric spaces.

DEFINITION 5.31. Let E be a subset of a metric space X.

   i) A pair of open sets U, V in X is said to *separate* E if $E \subseteq U \cup V$, $U \cap V = \emptyset$, $E \cap U \neq \emptyset$ and $E \cap V \neq \emptyset$.

   ii) E is said to be *connected* (in X) if E cannot be separated by any pair of open sets U, V in X.

By modifying arguments presented in Sections 5.4 and 5.6, we can prove the following results.

THEOREM 5.51. *Let X be a metric space.*

   i) *The empty set is connected in X.*

   ii) *Every singleton $\{a\}$ is connected in X.*

THEOREM 5.52. *Let X, Y be metric spaces. If X is connected and $f : X \to Y$ is continuous, then $f(X)$ is connected in Y.*

The following result shows that the union of intersecting connected sets is connected.

THEOREM 5.53. *If $\{E_\alpha\}_{\alpha \in A}$ is a collection of connected sets in a metric space X such that $\cap_{\alpha \in A} E_\alpha \neq \emptyset$, then*

$$E = \bigcup_{\alpha \in A} E_\alpha$$

is connected.

PROOF. Suppose U, V is a pair of open sets which separates E and let $x \in \cap_{\alpha \in A} E_\alpha$. Since $E \subseteq U \cup V$, we may suppose that $x \in U$. Choose $\alpha_0 \in A$ such that $V \cap E_{\alpha_0} \neq \emptyset$. Since $x \in E_{\alpha_0}$, we also have $U \cap E_{\alpha_0} \neq \emptyset$. Therefore, U and V separate E_{α_0}, a contradiction. ∎

DEFINITION 5.32. Let $E \subseteq X$. The *component* of an $x \in E$ is the set

$$E(x) := \bigcup \{B \subseteq E : B \text{ is connected and } x \in B\}.$$

We shall prove (see Theorem 5.54 below) that components can be used to decompose any subset of a metric space into a union of maximal connected subsets. First, we make two elementary observations.

Remark 5. *Given $E \subseteq X$ and $x \in E$, the component $E(x)$ is the largest connected subset of E which contains x.*

PROOF. Since $\{x\}$ is connected, the union defining $E(x)$ is nonempty. Since each set B defining $E(x)$ contains x, this collection has a nonempty intersection. Hence, by Theorem 5.53, the set $E(x)$ is connected. It is the largest connected subset of E containing x since if $B \subseteq E$ is connected and $x \in B$, then by definition, $B \subseteq E(x)$. ∎

Remark 6. *Let $E \subseteq X$ and $a, b \in E$. If $E(a) \cap E(b) \neq \emptyset$, then $E(a) = E(b)$.*

PROOF. If $E(a)$ and $E(b)$ have a point in common, then by Theorem 5.53 $E(a) \cup E(b)$ is connected. By the maximal property (Remark 5), it follows that $E(a) \subseteq E(b)$ and $E(b) \subseteq E(a)$. In particular, $E(a) = E(b)$. ∎

THEOREM 5.54. *Let E be a subset of a metric space X. Then there exist connected sets $\{E_\alpha\}_{\alpha \in A}$ such that $E_\alpha \cap E_\beta = \emptyset$ for $\alpha \neq \beta$ and*

$$E = \bigcup_{\alpha \in A} E_\alpha.$$

Moreover, the E_α's are maximal in the following sense. If B is a nonempty, connected subset of E, then there is a unique $\alpha \in A$ such that $B \subseteq E_\alpha$.

PROOF. By Remarks 5 and 6, $\{E(x)\}_{x \in E}$ is a collection of connected subsets of E which satisfy the maximal condition such that if $x, y \in E$, then either $E(x) \cap E(y) = \emptyset$ or $E(x) = E(y)$. Since each $E(x) \subseteq E$, it is clear that

$$E = \bigcup_{x \in E} E(x).$$

Thus,

$$\{E_\alpha : \alpha \in A\} := \{E(x) : x \in E\}$$

is the collection for which we search. ∎

Connected sets can be used to obtain an intermediate value theorem for real-valued functions on a metric space.

THEOREM 5.55. *Let X be a metric space and E be a connected subset of X. If $f : E \to \mathbf{R}$ is continuous, $f(a) \neq f(b)$ for some $a, b \in E$, and y is a number which lies between $f(a)$ and $f(b)$, then there is an $x \in E$ such that $f(x) = y$.*

PROOF. Since E is connected and f is continuous on E, $f(E)$ is connected in $\mathbf{R}$. Thus by Theorem 5.20, $f(E)$ is an interval. ∎

EXERCISES

1. Prove Theorems 5.34 through 5.38.
2. Prove Theorems 5.39 through 5.42.

3. Prove Theorems 5.43 through 5.49.

4. Prove that the discrete space $X = \mathbf{R}$ is complete.

5. Let X be a complete metric space and Y be a subspace of X. Prove that Y is a complete metric space if and only if Y is closed.

6. Let X be a metric space. Prove that a subset V of X is open if and only if V is a union of open balls in X.

7. Let $X = \mathbf{Q}$ be the metric space introduced in Example 5.

   a) Prove $(\sqrt{2}, \sqrt{3}) \cap \mathbf{Q}$ is closed and bounded but not compact in X.

   b) Show that Theorem 5.42 does not hold if "compact" is replaced by "closed and bounded."

8. Let X be a metric space and $\{x_k\}$, $\{y_k\}$ be sequences in X. If $x_k \to x$ and $y_k \to y$ as $k \to \infty$, prove $\rho(x_k, y_k) \to \rho(x, y)$.

9. Let X and Y be metric spaces and D be a *dense* subspace of X; i.e., $D \subset X$ and $\overline{D} = X$. If Y is complete and $f : D \to Y$ is uniformly continuous on D, prove that f has a uniformly continuous extension to X, i.e., prove that there is a uniformly continuous function $g : X \to Y$ such that $g(x) = f(x)$ for all $x \in D$.

10. A set E in a metric space is called *clopen* if it is both open and closed.

    a) Prove that every metric space has at least two clopen sets.

    b) Prove that a metric space is connected if and only if it contains exactly two clopen sets.

11. Let X be a metric space. Prove that X is connected if and only if every nonempty proper subset of X has a nonempty boundary.

Chapter 6

DIFFERENTIABILITY ON $\mathbf{R}^n$

In this chapter we define what it means for a function of several variables to be differentiable and obtain multidimensional analogues of many results which appeared in Chapter 2.

6.1 PARTIAL DERIVATIVES AND PARTIAL INTEGRALS

The most natural way to define derivatives and integrals of functions of several variables is to allow one variable to move at a time. The corresponding objects, partial derivatives and partial integrals, are the subject of this section. Our main goal is to identify conditions under which partial derivatives, partial integrals, and evaluation of limits commute with each other, e.g., under which the limit of a partial integral is the partial integral of a limit.

We begin with some notation. The *Cartesian product* of a finite collection of sets $E_1, E_2, \ldots, E_n$ is the set of ordered n-tuples defined by

$$E_1 \times E_2 \times \cdots \times E_n := \{(x_1, x_2, \ldots, x_n) : x_j \in E_j \text{ for } j = 1, 2, \ldots, n\}.$$

Thus, the Cartesian product of n subsets of $\mathbf{R}$ is a subset of $\mathbf{R}^n$. By a *rectangle* in $\mathbf{R}^n$ (or an *n-dimensional rectangle*) we mean a Cartesian product of n closed bounded intervals. An n-dimensional rectangle $R = [a_1, b_1] \times \cdots \times [a_n, b_n]$ is called an *n-dimensional cube with side s* if $|b_j - a_j| = s$ for $j = 1, \ldots, n$.

Let $f : \{x_1\} \times \cdots \times \{x_{j-1}\} \times [a, b] \times \{x_{j+1}\} \times \cdots \times \{x_n\} \to \mathbf{R}$. We shall denote the function

$$g(t) := f(x_1, \ldots, x_{j-1}, t, x_{j+1}, \ldots, x_n), \qquad t \in [a, b],$$

by $f(x_1, \ldots, x_{j-1}, \cdot, x_{j+1}, \ldots, x_n)$. If g is integrable on $[a, b]$, then the *partial integral* of f on $[a, b]$ with respect to x_j is defined by

$$\int_a^b f(x_1, \ldots, x_n)\, dx_j := \int_a^b g(t)\, dt.$$

If g is differentiable at some $t_0 \in (a, b)$, then the *partial derivative* (or *first-order partial derivative*) of f at $(x_1, \ldots, x_{j-1}, t_0, x_{j+1}, \ldots x_n)$ with respect to x_j is defined by

$$f_{x_j}(x_1, \ldots, x_{j-1}, t_0, x_{j+1}, \ldots, x_n)$$
$$:= \frac{\partial f}{\partial x_j}(x_1, \ldots, x_{j-1}, t_0, x_{j+1}, \ldots, x_n) := g'(t_0).$$

Thus, the partial derivative f_{x_j} exists at a point $\boldsymbol{a}$ if and only if the limit

$$\frac{\partial f}{\partial x_j}(\boldsymbol{a}) := \lim_{h \to 0} \frac{f(\boldsymbol{a} + h\boldsymbol{e}_j) - f(\boldsymbol{a})}{h}$$

exists. (Some authors use f_j to denote the partial derivative f_{x_j}. To avoid confusing first-order partial derivatives with sequences and components of functions, we will not use this notation.)

We extend partial derivatives to vector-valued functions in the following way. Suppose $f = (f_1, f_2, \ldots, f_m) : V \to \mathbf{R}^m$ and $j \in \{1, 2, \ldots, n\}$, where V is some open subset of $\mathbf{R}^n$. If the first-order partial derivatives with respect to x_j of all components of f exist at $\boldsymbol{a} \in V$, then we define the *first-order partial derivative* of f with respect to x_j to be the vector-valued function

$$f_{x_j}(\boldsymbol{a}) := \frac{\partial f}{\partial x_j}(\boldsymbol{a}) := \left(\frac{\partial f_1}{\partial x_j}(\boldsymbol{a}), \ldots, \frac{\partial f_m}{\partial x_j}(\boldsymbol{a})\right).$$

Higher-order partial derivatives are defined by iteration. For example, the *second-order partial derivative* of f with respect to x_j and x_k is defined by

$$f_{x_j x_k} := \frac{\partial^2 f}{\partial x_k \partial x_j} := \frac{\partial}{\partial x_k}\left(\frac{\partial f}{\partial x_j}\right)$$

when it exists. Second-order partial derivatives are called *mixed* when $j \neq k$.

This brings us to the following important collection of functions.

DEFINITION 6.1. Let V be open in $\mathbf{R}^n$, $f : V \to \mathbf{R}^m$, and $p \in \mathbf{N}$.

    i) f is said to be $\mathcal{C}^p$ on V if each partial derivative of f of order $k \leq p$ exists and is continuous on V.

    ii) f is said to be $\mathcal{C}^\infty$ on V if f is $\mathcal{C}^p$ on V for all $p \in \mathbf{N}$.

Clearly, if f is $\mathcal{C}^p$ on V and $q < p$, then f is $\mathcal{C}^q$ on V.

For simplicity, we shall state all results in this section for the case $n = 2$, using x for x_1 and y for x_2. It is clear that, with appropriate changes in notation, these results also hold for any $n \in \mathbf{N}$.

Since partial derivatives and partial integrals are essentially one-dimensional ideas, each one-dimensional result about derivatives and integrals contains information about partial derivatives and partial integrals. Here are three examples.

By the Product Rule (Theorem 2.13), if f_x and g_x exist, then

$$\frac{\partial}{\partial x}(fg) = f\frac{\partial g}{\partial x} + g\frac{\partial f}{\partial x}.$$

By the Mean Value Theorem (Theorem 2.16), if $f(\cdot, y)$ is continuous on $[a, b]$ and the partial derivative $f_x(\cdot, y)$ exists on (a, b), then there is a point $c \in (a, b)$ (which may depend on y) such that

$$f(b, y) - f(a, y) = (b - a)\frac{\partial f}{\partial x}(c, y).$$

And by the Fundamental Theorem of Calculus (Theorem 3.13), if $f(\cdot, y)$ is continuous on $[a, b]$, then

$$\frac{\partial}{\partial x}\int_a^x f(t, y)\, dt = f(x, y),$$

and if the partial derivative $f_x(\cdot, y)$ exists and is integrable on $[a, b]$, then

$$\int_a^b \frac{\partial f}{\partial x}(x, y)\, dx = f(b, y) - f(a, y).$$

Our first result about the commutation of partial derivatives, partial integrals, and evaluation of limits deals with interchanging two first-order partial derivatives (see also Exercise 9 in Section 6.3).

THEOREM 6.1. *Suppose V is open in $\mathbf{R}^n$, $(a, b) \in V$, and $f : V \to \mathbf{R}$. If f is C^1 on V, and if one of the mixed second partial derivatives of f exists on V and is continuous at the point (a, b), then the other mixed second partial derivative exists at (a, b) and*

$$\frac{\partial^2 f}{\partial y \partial x}(a, b) = \frac{\partial^2 f}{\partial x \partial y}(a, b).$$

PROOF. Suppose f_{yx} exists on V and is continuous at the point (a, b). Choose $r > 0$ such that $B_r(a, b) \subset V$ and set

$$\Delta(h, k) = f(a + h, b + k) - f(a + h, b) - f(a, b + k) + f(a, b)$$

for $|h|, |k| < r/\sqrt{2}$. Apply the Mean Value Theorem twice to choose scalars $s, t \in (0, 1)$ such that

$$\Delta(h, k) = k\frac{\partial f}{\partial y}(a + h, b + tk) - k\frac{\partial f}{\partial y}(a, b + tk) = hk\frac{\partial^2 f}{\partial x \partial y}(a + sh, b + tk).$$

Since this last mixed partial derivative is continuous at the point (a, b), we have

(1) $$\lim_{k \to 0}\lim_{h \to 0}\frac{\Delta(h, k)}{hk} = \frac{\partial^2 f}{\partial x \partial y}(a, b).$$

On the other hand, the Mean Value Theorem also implies that there is a scalar $u \in (0,1)$ such that

$$\Delta(h,k) = f(a+h,b+k) - f(a,b+k) - f(a+h,b) + f(a,b)$$
$$= h\frac{\partial f}{\partial x}(a+uh,b+k) - h\frac{\partial f}{\partial x}(a+uh,b).$$

Hence, it follows from (1) that

$$\lim_{k \to 0} \lim_{h \to 0} \frac{1}{k}\left(\frac{\partial f}{\partial x}(a+uh,b+k) - \frac{\partial f}{\partial x}(a+uh,b)\right)$$
$$= \lim_{k \to 0} \lim_{h \to 0} \frac{\Delta(h,k)}{hk} = \frac{\partial^2 f}{\partial x \partial y}(a,b).$$

Since f_x is continuous on $B_r(a,b)$, we can let $h = 0$ in the first expression. We conclude by definition that

$$\frac{\partial^2 f}{\partial y \partial x}(a,b) = \lim_{k \to 0}\frac{1}{k}\left(\frac{\partial f}{\partial x}(a,b+k) - \frac{\partial f}{\partial x}(a,b)\right) = \frac{\partial^2 f}{\partial x \partial y}(a,b). \quad \blacksquare$$

We shall refer to the conclusion of Theorem 6.1 by saying the first partial derivatives of f *commute*.

The following example shows that Theorem 6.1 is false if the assumption about continuity of the second-order partial derivative is dropped.

Example 1. Prove that

$$f(x,y) = \begin{cases} xy\left(\dfrac{x^2 - y^2}{x^2 + y^2}\right) & (x,y) \neq \mathbf{0} \\ 0 & (x,y) = \mathbf{0} \end{cases}$$

is $\mathcal{C}^1$ on $\mathbf{R}^2$, both mixed second partial derivatives of f exist on $\mathbf{R}^2$, but the first partial derivatives of f do not commute at $(0,0)$, i.e., $f_{xy}(0,0) \neq f_{yx}(0,0)$.

PROOF. By the one-dimensional Product and Quotient Rules,

$$\frac{\partial f}{\partial x}(x,y) = xy\frac{\partial}{\partial x}\left(\frac{x^2 - y^2}{x^2 + y^2}\right) + \frac{\partial}{\partial x}(xy)\left(\frac{x^2 - y^2}{x^2 + y^2}\right)$$
$$= xy\left(\frac{4xy^2}{(x^2 + y^2)^2}\right) + y\left(\frac{x^2 - y^2}{x^2 + y^2}\right)$$

for $(x,y) \neq (0,0)$. Since $2|xy| \leq x^2 + y^2$, we have $|f_x(x,y)| \leq 2|y|$. Therefore, $f_x(x,y) \to 0$ as $(x,y) \to (0,0)$. On the other hand, by definition

$$\frac{\partial f}{\partial x}(0,y) = \lim_{h \to 0} y\left(\frac{h^2 - y^2}{h^2 + y^2}\right) = -y$$

for all $y \in \mathbf{R}$, hence $f_x(0,0) = 0$. This proves that f_x exists and is continuous on $\mathbf{R}^2$ with value $-y$ at $(0, y)$. A similar argument shows that f_y exists and is continuous on $\mathbf{R}^2$ with value x at $(x, 0)$. It follows that the mixed second partial derivatives of f exist on $\mathbf{R}^2$, and

$$\frac{\partial^2 f}{\partial y \partial x}(0,0) = -1 \neq 1 = \frac{\partial^2 f}{\partial x \partial y}(0,0). \quad \blacksquare$$

The following result shows that we can interchange a limit sign and a partial integral sign when the integrand is continuous on a rectangle.

THEOREM 6.2. *Let $H = [a, b] \times [c, d]$ be a rectangle and $f : H \to \mathbf{R}$ be continuous. If*

$$F(y) = \int_a^b f(x, y) \, dx,$$

then F is continuous on $[c, d]$, i.e.,

$$\lim_{\substack{y \to y_0 \\ y \in [c,d]}} \int_a^b f(x, y) \, dx = \int_a^b \lim_{\substack{y \to y_0 \\ y \in [c,d]}} f(x, y) \, dx$$

for all $y_0 \in [c, d]$.

PROOF. For each $y \in [c, d]$, $f(\cdot, y)$ is continuous on $[a, b]$. Hence, by Theorem 3.1, $F(y)$ exists for $y \in [c, d]$.

Fix $y_0 \in [c, d]$ and let $\varepsilon > 0$. Since H is compact, choose $\delta > 0$ such that $\|(x, y) - (z, w)\| < \delta$ and $(x, y), (z, w) \in H$ imply

$$|f(x, y) - f(z, w)| < \frac{\varepsilon}{b - a}.$$

Since $|y - y_0| = \|(x, y) - (x, y_0)\|$, it follows that

$$|F(y) - F(y_0)| \leq \int_a^b |f(x, y) - f(x, y_0)| \, dx < \varepsilon$$

for all $y \in [c, d]$ which satisfy $|y - y_0| < \delta$. We conclude that $F(y) \to F(y_0)$ as $y \to y_0$ through $[c, d]$. $\blacksquare$

The following result shows that we can interchange a derivative and an integral sign when the partial derivative is continuous on a rectangle. We will refer to this process as *differentiating under the integral sign*.

THEOREM 6.3. *Let $H = [a, b] \times [c, d]$ be a rectangle in $\mathbf{R}^2$ and $f : H \to \mathbf{R}$. Suppose that $f(\cdot, y)$ is integrable on $[a, b]$ for each $y \in [c, d]$, and that the partial derivative $f_y(x, \cdot)$ exists on $[c, d]$ for each $x \in [a, b]$. If the two-variable function $f_y(x, y)$ is continuous on H, then*

$$(2) \qquad \frac{d}{dy} \int_a^b f(x, y) \, dx = \int_a^b \frac{\partial f}{\partial y}(x, y) \, dx$$

for all $y \in [c, d]$.

PROOF. Recall that "$f_y(x, \cdot)$ exists on $[c, d]$" means $f_y(x, \cdot)$ exists on (c, d) and

$$f_y(x, c) := \lim_{h \to 0+} \frac{f(x, c + h) - f(x, c)}{h}, \quad f_y(x, d) := \lim_{h \to 0-} \frac{f(x, d + h) - f(x, d)}{h}$$

both exist (see Definition 2.9). Hence, it suffices to show

$$\lim_{h \to 0+} \int_a^b \frac{f(x, y + h) - f(x, y)}{h}\, dx = \int_a^b \frac{\partial f}{\partial y}(x, y)\, dx$$

for $y \in [c, d)$ and

$$\lim_{h \to 0-} \int_a^b \frac{f(x, y + h) - f(x, y)}{h}\, dx = \int_a^b \frac{\partial f}{\partial y}(x, y)\, dx$$

for $y \in (c, d]$. The arguments are similar and we provide the details only for the first identity.

Fix $x \in [a, b]$ and $y \in [c, d)$, and let $h > 0$ be so small that $y + h \in [c, d)$. By the Mean Value Theorem, choose a point $z(x; h)$ between y and $y + h$ such that

$$\frac{f(x, y + h) - f(x, y)}{h} = \frac{\partial f}{\partial y}(x, z(x; h)).$$

Since $z(x; h) \to y$ as $h \to 0+$, it follows from Theorem 6.2 that

$$\frac{d}{dy} \int_a^b f(x, y)\, dx = \lim_{h \to 0+} \int_a^b \frac{\partial f}{\partial y}(x, z(x; h))\, dx = \int_a^b \frac{\partial f}{\partial y}(x, y)\, dx. \quad \blacksquare$$

To see what happens to these results when the improper integral is used, we borrow the following concept from the theory of infinite series.

DEFINITION 6.2. Let $a < b$ be extended real numbers, I be an interval in $\mathbf{R}$, and $f : (a, b) \times I \to \mathbf{R}$. The improper integral

$$\int_a^b f(x, y)\, dx$$

is said to *converge uniformly* on I if $f(\cdot, y)$ is improperly integrable on (a, b) for each $y \in I$, and if given $\varepsilon > 0$ there exist real numbers $A, B \in (a, b)$ such that

$$\left| \int_a^b f(x, y)\, dx - \int_\alpha^\beta f(x, y)\, dx \right| < \varepsilon$$

for all $a < \alpha < A$, $B < \beta < b$, and all $y \in I$.

For most applications, the following simple test for uniform convergence of an improper integral will be used instead of Definition 6.2 (compare with Theorem 4.27).

THEOREM 6.4 [WEIERSTRASS M–TEST]. *Suppose $a < b$ are extended real numbers, I is an interval in $\mathbf{R}$, $f : (a, b) \times I \to \mathbf{R}$, and $f(\cdot, y)$ is locally integrable on the interval (a, b) for each $y \in I$. If there is a function $M : (a, b) \to \mathbf{R}$, absolutely integrable on (a, b), such that*

$$|f(x, y)| \le M(x)$$

for all $x \in (a, b)$ and $y \in I$, then

$$\int_a^b f(x, y)\, dx$$

converges uniformly on I.

PROOF.   By hypothesis and the Comparison Test for improper integrals, $\int_a^b f(x, y)\, dx$ exists and is finite for each $y \in I$. Moreover, since $M(x)$ is improperly integrable on (a, b), there exist real numbers A, B such that $a < A < B < b$ and

$$\int_a^A M(x)\, dx + \int_B^b M(x)\, dx < \varepsilon.$$

Thus, for each $a < \alpha < A < B < \beta < b$ and each $y \in I$, we have

$$\left| \int_a^b f(x, y)\, dx - \int_\alpha^\beta f(x, y)\, dx \right| \le \int_a^\alpha |f(x, y)|\, dx + \int_\beta^b |f(x, y)|\, dx$$

$$\le \int_a^A M(x)\, dx + \int_B^b M(x)\, dx < \varepsilon. \ \blacksquare$$

The following is an improper integral analogue of Theorem 6.2.

THEOREM 6.5. *Suppose $a < b$ are extended real numbers, $c < d$ are finite real numbers, and $f : (a, b) \times [c, d] \to \mathbf{R}$ is continuous. If*

$$F(y) = \int_a^b f(x, y)\, dx$$

converges uniformly on $[c, d]$, then F is continuous on $[c, d]$, i.e.,

$$\lim_{\substack{y \to y_0 \\ y \in [c,d]}} \int_a^b f(x, y)\, dx = \int_a^b \lim_{\substack{y \to y_0 \\ y \in [c,d]}} f(x, y)\, dx$$

for all $y_0 \in [c, d]$.

PROOF.   Let $\varepsilon > 0$ and $y_0 \in [c, d]$. Choose real numbers A, B such that $a < A < B < b$ and

$$\left| F(x) - \int_A^B f(x, y)\, dx \right| < \frac{\varepsilon}{3}$$

for all $y \in [c, d]$. By Theorem 6.2, choose $\delta > 0$ such that

$$\int_A^B |f(x, y) - f(x, y_0)| \, dx < \frac{\varepsilon}{3}$$

for all $y \in [c, d]$ which satisfy $|y - y_0| < \delta$. Then

$$|F(y) - F(y_0)| \leq \left| F(y) - \int_A^B f(x, y) \, dx \right| + \left| \int_A^B (f(x, y) - f(x, y_0)) \, dx \right|$$

$$+ \left| F(y_0) - \int_A^B f(x, y_0) \, dx \right|$$

$$\leq \frac{2\varepsilon}{3} + \int_A^B |f(x, y) - f(x, y_0)| \, dx < \varepsilon$$

for all $y \in [c, d]$ which satisfy $|y - y_0| < \delta$. ∎

The proof of Theorem 6.3 can be modified to prove the following result.

THEOREM 6.6. *Suppose $a < b$ are extended real numbers, $c < d$ are finite real numbers, $f : (a, b) \times [c, d] \to \mathbf{R}$ is continuous, and the improper integral*

$$F(y) = \int_a^b f(x, y) \, dx$$

exists for all $y \in [c, d]$. If $f_y(x, y)$ exists and is continuous on $(a, b) \times [c, d]$ and if

$$\phi(y) = \int_a^b \frac{\partial f}{\partial y}(x, y) \, dx$$

converges uniformly on $[c, d]$, then F is differentiable on $[c, d]$ and $F'(y) = \phi(y)$, i.e.,

$$\frac{d}{dy} \int_a^b f(x, y) \, dx = \int_a^b \frac{\partial f}{\partial y}(x, y) \, dx$$

for all $y \in [c, d]$.

For a result about interchanging two partial integrals, see Theorem 7.10 and Exercise 10 in Section 7.3.

EXERCISES

1. Compute all mixed second-order partial derivatives of each of the following functions and verify that the mixed partial derivatives are equal.

    a) $f(x, y) = xe^y$.   b) $f(x, y) = \cos(xy)$.   c) $f(x, y) = \dfrac{x + y}{x^2 + 1}$.

2. Compute all first-order partial derivatives of each of the following functions and find where they are continuous.

a) $f(x, y) = x^2 + \sin(xy)$.   b) $f(x, y, z) = \dfrac{xy}{1 + z}$.   c) $f(x, y) = \sqrt{x^2 + y^2}$.

3. For each of the following functions, compute f_x and determine where it is continuous.

a)
$$f(x, y) = \begin{cases} \dfrac{x^4 + y^4}{x^2 + y^2} & (x, y) \neq (0, 0) \\ 0 & (x, y) = (0, 0). \end{cases}$$

b)
$$f(x, y) = \begin{cases} \dfrac{x^2 - y^2}{\sqrt[3]{x^2 + y^2}} & (x, y) \neq (0, 0) \\ 0 & (x, y) = (0, 0). \end{cases}$$

4. Prove that every rectangle in $\mathbf{R}^n$ is closed and bounded, hence compact.
5. Suppose $H = [a, b] \times [c, d]$ is a rectangle, $f : H \to \mathbf{R}$ is continuous, and $g : [a, b] \to \mathbf{R}$ is integrable. Prove

$$F(y) = \int_a^b g(x) f(x, y) \, dx$$

is uniformly continuous on $[c, d]$.

6. a) Prove
$$\int_0^1 \frac{\cos(x^2 + y^2)}{\sqrt{x}} \, dx$$

converges uniformly on $(-\infty, \infty)$.
b) Prove $\int_0^\infty e^{-xy} \, dx$ converges uniformly on $[1, \infty)$.
c) Prove $\int_0^\infty ye^{-xy} \, dx$ exists for each $y \in [0, \infty)$, converges uniformly on any $[a, b] \subset (0, \infty)$, but does not converge uniformly on $[0, 1]$.

7. Evaluate each of the following expressions.

a)
$$\lim_{y \to 0} \int_0^1 \cos(x^2 y + xy^2) \, dx$$

b)
$$\frac{d}{dy} \int_{-1}^1 \sqrt{x^2 y^2 + xy + y + 2} \, dx \qquad \text{at } y = 0$$

c)
$$\lim_{y \to 0+} \int_0^1 \frac{x \cos y}{\sqrt[3]{1 - x + y}} \, dx$$

d)
$$\frac{d}{dy} \int_{\pi}^{\infty} \frac{e^{-xy}\sin x}{x}\,dx \qquad \text{at } y = 1$$

*DEFINITION 6.3. The *Laplace transform* of a function $f : (0, \infty) \to \mathbf{R}$ is said to exist at a point $s \in (0, \infty)$ if the integral

$$\mathcal{L}\{f\}(s) := \int_{0}^{\infty} e^{-st} f(t)\,dt$$

converges. (Note: This integral is improper at ∞ and may be improper at 0.)

*8. Prove

a)
$$\mathcal{L}\{1\}(s) = \frac{1}{s}, \qquad s > 0,$$

b)
$$\mathcal{L}\{t^n\}(s) = \frac{n!}{s^{n+1}}, \qquad s > 0,\ n \in \mathbf{N}$$

c)
$$\mathcal{L}\{e^{at}\}(s) = \frac{1}{s - a}, \qquad s > a,\ a \in \mathbf{R}$$

d)
$$\mathcal{L}\{\cos(bt)\}(s) = \frac{s}{s^2 + b^2}, \qquad s > 0,\ b \in \mathbf{R}$$

e)
$$\mathcal{L}\{\sin(bt)\}(s) = \frac{b}{s^2 + b^2}. \qquad s > 0,\ b \in \mathbf{R}$$

*9. Suppose $f : (0, \infty) \to \mathbf{R}$ is continuous and bounded and $\mathcal{L}\{f\}$ exists at some $a \in (0, \infty)$. Let

$$\phi(t) = \int_{0}^{t} e^{-au} f(u)\,du \qquad t \in (0, \infty).$$

a) Prove that

$$\int_{0}^{N} e^{-st} f(t)\,dt = \phi(N)e^{-(s-a)N} + (s - a)\int_{0}^{N} e^{-(s-a)t}\phi(t)\,dt$$

for all $N \in \mathbf{N}$.

b) Prove that the integral $\int_0^\infty e^{-(s-a)t}\phi(t)\,dt$ converges uniformly on $[b,\infty)$ for any $b > a$ and

$$\int_0^\infty e^{-st} f(t)\,dt = (s-a)\int_0^\infty e^{-(s-a)t}\phi(t)\,dt, \qquad s > a.$$

c) Prove that $\mathcal{L}\{f\}$ exists, is continuous on (a,∞), and satisfies

$$\lim_{s\to\infty} \mathcal{L}\{f\}(s) = 0.$$

d) Let $g(t) = tf(t)$ for $t \in (0,\infty)$. Prove that $\mathcal{L}\{f\}$ is differentiable on (a,∞) and

$$\frac{d}{ds}\mathcal{L}\{f\}(s) = -\mathcal{L}\{g\}(s)$$

for all $s \in (a,\infty)$.

e) If in addition, f' is continuous and bounded on $(0,\infty)$, prove that

$$\mathcal{L}(f')(s) = s\mathcal{L}(f)(s) - f(0)$$

for all $s \in (a,\infty)$.

*10. Using Exercises 8 and 9, find the Laplace transforms of the each of the functions te^t, $t\sin\pi t$, and $t^2\cos t$.

6.2   THE DEFINITION OF DIFFERENTIABILITY

For the one-dimensional case, there is a close connection between differentiability and linear functions (see Exercise 2). This leads us to the following definition.

DEFINITION 6.4. Let V be open in $\mathbf{R}^n$, $\boldsymbol{a} \in V$ and $f : V \to \mathbf{R}^m$. Then f is said to be differentiable at $\boldsymbol{a}$ if there is a linear function $T_{\boldsymbol{a}} : \mathbf{R}^n \to \mathbf{R}^m$ (called the *total derivative* of f at $\boldsymbol{a}$) such that

$$(3) \qquad\qquad \lim_{\boldsymbol{h}\to 0} \frac{f(\boldsymbol{a}+\boldsymbol{h}) - f(\boldsymbol{a}) - T_{\boldsymbol{a}}(\boldsymbol{h})}{\|\boldsymbol{h}\|} = \boldsymbol{0}.$$

Notice that since $f(\boldsymbol{a}+\boldsymbol{h})$, $f(\boldsymbol{a})$, and $T_{\boldsymbol{a}}(\boldsymbol{h})$ are vectors in $\mathbf{R}^m$, the numerator of (3) is a vector in $\mathbf{R}^m$ and the limit in (3) is a limit of vectors.

The following result justifies our calling $T_{\boldsymbol{a}}$ the total derivative of f at $\boldsymbol{a}$.

Remark 1. *If f is differentiable at $\boldsymbol{a}$, then there is only one linear function $T_{\boldsymbol{a}}$ which satisfies (3).*

PROOF. Suppose $F : \mathbf{R}^n \to \mathbf{R}^m$ is linear and (3) holds for F in place of $T_{\boldsymbol{a}}$. To prove $F(\boldsymbol{x}) = T_{\boldsymbol{a}}(\boldsymbol{x})$ for all $\boldsymbol{x} \in \mathbf{R}^n$, notice first that since F and $T_{\boldsymbol{a}}$ are linear,

$F(\mathbf{0}) = T_{\boldsymbol{a}}(\mathbf{0}) = \mathbf{0}$. Next, suppose $\boldsymbol{x} \neq \mathbf{0}$ and set $\boldsymbol{h} = u\boldsymbol{x}$ for $u > 0$. Clearly, $\boldsymbol{h} \to \mathbf{0}$ as $u \to 0+$. Since F and $T_{\boldsymbol{a}}$ are linear and satisfy (3), we have

$$
\frac{F(\boldsymbol{x}) - T_{\boldsymbol{a}}(\boldsymbol{x})}{\|\boldsymbol{x}\|} = \frac{uF(\boldsymbol{x}) - uT_{\boldsymbol{a}}(\boldsymbol{x})}{u\,\|\boldsymbol{x}\|} = \frac{F(\boldsymbol{h}) - T_{\boldsymbol{a}}(\boldsymbol{h})}{\|\boldsymbol{h}\|}
$$

$$
= \frac{F(\boldsymbol{h}) - f(\boldsymbol{a}+\boldsymbol{h}) + f(\boldsymbol{a})}{\|\boldsymbol{h}\|} + \frac{f(\boldsymbol{a}+\boldsymbol{h}) - f(\boldsymbol{a}) - T_{\boldsymbol{a}}(\boldsymbol{h})}{\|\boldsymbol{h}\|} \to \mathbf{0}
$$

as $u \to 0+$. Therefore, $F(\boldsymbol{x}) - T_{\boldsymbol{a}}(\boldsymbol{x}) = \|\boldsymbol{x}\| \cdot \mathbf{0} = \mathbf{0}$, i.e., $F = T_{\boldsymbol{a}}$ on $\mathbf{R}^n$. ∎

The total derivative (if it exists) is a linear function on $\mathbf{R}^n$. Since linear functions from $\mathbf{R}^n$ to $\mathbf{R}^m$ can be represented by $m \times n$ matrices (see Theorem 5.4), $T_{\boldsymbol{a}}$ must have a matrix representation. Is there an easy way to compute this matrix? To answer this question, we introduce the following notation. Let V be open in $\mathbf{R}^n$ and $\boldsymbol{a} \in V$. For each function $f : V \to \mathbf{R}^m$ whose first-order partial derivatives exist at $\boldsymbol{a}$, define the *Jacobian matrix* of f by

$$
Df(\boldsymbol{a}) := \left[\frac{\partial f_i}{\partial x_j}(\boldsymbol{a})\right]_{m \times n} = \begin{bmatrix} \dfrac{\partial f_1}{\partial x_1}(\boldsymbol{a}) & \cdots & \dfrac{\partial f_1}{\partial x_n}(\boldsymbol{a}) \\ \vdots & \ddots & \vdots \\ \dfrac{\partial f_m}{\partial x_1}(\boldsymbol{a}) & \cdots & \dfrac{\partial f_m}{\partial x_n}(\boldsymbol{a}) \end{bmatrix}.
$$

If $\{k_1, k_2, \ldots, k_n\} \subseteq \{1, 2, \ldots, m\}$, we shall also use the notation

$$
\frac{\partial(f_{k_1}, \ldots, f_{k_n})}{\partial(x_1, \ldots, x_n)} := \det\left[\frac{\partial f_{k_i}}{\partial x_j}(\boldsymbol{a})\right]_{n \times n} = \det \begin{bmatrix} \dfrac{\partial f_{k_1}}{\partial x_1}(\boldsymbol{a}) & \cdots & \dfrac{\partial f_{k_1}}{\partial x_n}(\boldsymbol{a}) \\ \vdots & \ddots & \vdots \\ \dfrac{\partial f_{k_n}}{\partial x_1}(\boldsymbol{a}) & \cdots & \dfrac{\partial f_{k_n}}{\partial x_n}(\boldsymbol{a}) \end{bmatrix}
$$

when all these partial derivatives exist. (This notation simplifies the statements of some results in Sections 6.5 and 6.6.) Finally, when $n = m$, the *Jacobian* of f at $\boldsymbol{a}$ is defined by

$$
\Delta_f(\boldsymbol{a}) := \frac{\partial(f_1, \ldots, f_n)}{\partial(x_1, \ldots, x_n)} := \det Df(\boldsymbol{a}).
$$

(Note: The Jacobian Δ_f can be interpreted as a change of volumes (see Exercise 7 in Section 7.4) and will play a prominent role in Chapters 7 and 8.)

The following result shows that when f is differentiable, the Jacobian matrix of f is the matrix which represents the total derivative of f.

THEOREM 6.7. *Let V be open in $\mathbf{R}^n$, $\boldsymbol{a} \in V$, and $f : V \to \mathbf{R}^m$. If f is differentiable at $\boldsymbol{a}$, then the first-order partial derivatives of f exist at $\boldsymbol{a}$ and the total derivative of f at $\boldsymbol{a}$ is $Df(\boldsymbol{a})$, i.e., if $T_{\boldsymbol{a}}$ is the total derivative of f at $\boldsymbol{a}$ and B is the matrix which represents $T_{\boldsymbol{a}}$, then $B = Df(\boldsymbol{a})$.*

PROOF. Let $B = [b_{ij}]_{m \times n}$ be the matrix which represents $T_{\boldsymbol{a}}$, i.e., $T_{\boldsymbol{a}}(\boldsymbol{e}_j) = (b_{1j}, \ldots, b_{mj})$ (see Theorem 5.4). Fix $1 \leq j \leq n$, set $\boldsymbol{h} = u\boldsymbol{e}_j$ for some $u > 0$, and observe that

$$\frac{f(a+h) - f(a) - T_a(h)}{\|h\|} = \frac{f(a+ue_j) - f(a)}{u} - T_a(e_j).$$

Taking the limit of this identity as $u \to 0+$, it follows from (3) that

$$\lim_{u \to 0+} \frac{f(a+ue_j) - f(a)}{u} = T_a(e_j) = (b_{1j}, \ldots, b_{mj}).$$

A similar argument shows that the limit of this quotient as $u \to 0-$ exists and also equals $(b_{1j}, \ldots, b_{mj})$. Since a vector-valued function converges if and only if each of its components converges (Theorem 5.22), it follows that the first-order partial derivative of each component f_i with respect to x_j exists at a and satisfies

$$\frac{\partial f_i}{\partial x_j}(a) = b_{ij}$$

for $i = 1, 2, \ldots, m$. In particular, $B = Df(a)$. ∎

Theorem 6.7 shows us that the action of T_a on $\mathbf{R}^n$ is given by

$$T_a(h) = Df(a)(h),$$

where this last expression is the product of the $m \times n$ matrix $Df(a)$ with the $n \times 1$ matrix $[h]$. Combining Theorem 6.7 with Remark 1, we see that if V is open in $\mathbf{R}^n$, $a \in V$, $f : V \to \mathbf{R}^m$, and B is an $m \times n$ matrix which satisfies

$$(4) \qquad \lim_{h \to 0} \frac{f(a+h) - f(a) - Bh}{\|h\|} = \mathbf{0},$$

then f is differentiable at a and $Df(a) = B$. This observation, referred to as the *uniqueness of the total derivative*, will be used many times below.

Abusing the notation slightly, we shall use $Df(a)$ to represent both the Jacobian matrix and the total derivative T_a. This is done to make the multidimensional theory look more like the one-dimensional case. In this regard, Df and Δ_f play the same role that f' played for one-variable functions. Most of this chapter and some of the next is devoted to seeing how many of the one variable results from Chapters 2 and 3 hold for functions of several variables when f' is replaced by Df or Δ_f.

If $n = 1$ or $m = 1$, the Jacobian matrix Df is an $m \times 1$ or $1 \times n$ matrix, hence can be identified with a vector. These vectors have separate notation. If $n = 1$, then the vector identified with

$$Df(a) = \begin{bmatrix} f_1'(a) \\ \vdots \\ f_m'(a) \end{bmatrix}$$

is denoted by $f'(a)$, i.e.,

$$f'(a) := (f_1'(a), \ldots, f_m'(a)).$$

If $m = 1$, then the vector identified with

$$Df(\boldsymbol{a}) = \left[\frac{\partial f}{\partial x_1}(\boldsymbol{a}) \quad \cdots \quad \frac{\partial f}{\partial x_n}(\boldsymbol{a}) \right]$$

is denoted by $\nabla f(\boldsymbol{a})$, i.e.,

$$\nabla f(\boldsymbol{a}) := \left(\frac{\partial f}{\partial x_1}(\boldsymbol{a}), \ldots, \frac{\partial f}{\partial x_n}(\boldsymbol{a}) \right).$$

(∇f is sometimes called the *gradient* of f because it identifies the direction of steep-est ascent. For this connection and a relationship between gradients and directional derivatives, see Exercise 1 in Section 6.3.)

As in the one-dimensional case, differentiability is stronger than continuity.

THEOREM 6.8. *Let V be open in* $\mathbf{R}^n$, $\boldsymbol{a} \in V$ *and* $f : V \to \mathbf{R}^m$. *If f is differentiable at $\boldsymbol{a}$, then f is continuous at $\boldsymbol{a}$.*

PROOF. By (3) and Theorem 6.7, choose $\delta > 0$ such that

$$\|\boldsymbol{h}\| < \delta \quad \text{implies} \quad \|f(\boldsymbol{a} + \boldsymbol{h}) - f(\boldsymbol{a}) - Df(\boldsymbol{a})(\boldsymbol{h})\| < \|\boldsymbol{h}\|.$$

By the Triangle Inequality and Theorem 5.5,

$$\|f(\boldsymbol{a} + \boldsymbol{h}) - f(\boldsymbol{a})\| \le \|Df(\boldsymbol{a})(\boldsymbol{h})\| + \|\boldsymbol{h}\| \le (\sqrt{mn}\,\|Df(\boldsymbol{a})\|_\infty + 1)\|\boldsymbol{h}\|.$$

Therefore, $f(\boldsymbol{a} + \boldsymbol{h}) \to f(\boldsymbol{a})$ as $\boldsymbol{h} \to \boldsymbol{0}$, i.e., f is continuous at $\boldsymbol{a}$. ∎

In particular, any function which fails to be continuous at $\boldsymbol{a}$ cannot be differen-tiable at $\boldsymbol{a}$.

At this point you may be wondering why the definition of differentiability is not simpler. In the one-dimensional case, a function is differentiable if and only if its derivative exists. Working by analogy, we might guess that a function of several variables is differentiable if and only if all its first-order partial derivatives exist. The following example shows that this guess is wrong (see also Exercise 3 below).

Example 1. Prove that the first-order partial derivatives of

$$f(x, y) = \begin{cases} x + y & x = 0 \quad \text{or} \quad y = 0 \\ 1 & \text{otherwise} \end{cases}$$

exist at $(0, 0)$, but f is not differentiable at $(0, 0)$.

PROOF. Since

$$\lim_{x \to 0} f(x, x) = 1 \ne 0 = f(0, 0),$$

it is clear that f is not continuous (hence, not differentiable) at $(0,0)$. On the other hand, the first-order partial derivatives of f exist since

$$f_x(0,0) = \lim_{h \to 0} \frac{f(h,0) - f(0,0)}{h} = 1$$

and similarly, $f_y(0,0) = 1$. ∎

The function in Example 1 failed to be differentiable because it was discontinuous at $(0,0)$. Still looking for a simpler definition of differentiability, we might modify our guess by insisting that if a function is *continuous* and has first-order partial derivatives, then it is differentiable. The following example shows that this guess is also wrong.

Example 2. Prove that

$$f(x,y) = \begin{cases} \dfrac{x^3 - xy^2}{x^2 + y^2} & (x,y) \neq (0,0) \\ 0 & (x,y) = (0,0) \end{cases}$$

is continuous, has first-order partial derivatives everywhere on $\mathbf{R}^2$, but f is not differentiable at $(0,0)$.

PROOF. Clearly, f is continuous and has first-order partial derivatives at every point $(x,y) \neq (0,0)$. What happens at $(0,0)$? Since

$$|f(x,y)| = \frac{|x||x^2 - y^2|}{x^2 + y^2} \le |x|,$$

it follows from the Squeeze Theorem that f is continuous at $(0,0)$ with $f(0,0) = 0$. Moreover, the function f has first-order partial derivatives at $(0,0)$, since

$$\frac{\partial f}{\partial x}(0,0) = \lim_{h \to 0} \frac{f(h,0) - f(0,0)}{h} = \lim_{h \to 0} \frac{h^3}{h^3} = 1,$$

and

$$\frac{\partial f}{\partial y}(0,0) = \lim_{h \to 0} \frac{f(0,h) - f(0,0)}{h} = \lim_{h \to 0} \frac{0}{h^3} = 0.$$

Finally, if f were differentiable at $(0,0)$, then

$$0 = \lim_{(h,k) \to (0,0)} \frac{f(h,k) - f(0,0) - \nabla f(0,0) \cdot (h,k)}{\sqrt{h^2 + k^2}} = \lim_{(h,k) \to (0,0)} \frac{-2hk^2}{(h^2 + k^2)^{3/2}}.$$

But the path $h = k$ gives a limit of $-1/\sqrt{2} \neq 0$ as $h \to 0+$. Thus, f is not differentiable at $(0,0)$. ∎

Although there is no simpler definition of differentiability, there is an uncomplicated condition with wide applicability which *implies* differentiability. Indeed, the following result shows that if f is C^1 on some open set V, then f is differentiable on V.

THEOREM 6.9. *Let V be open in $\mathbf{R}^n$, $\boldsymbol{a} \in V$, and $f : V \to \mathbf{R}^m$. If all first-order partial derivatives of f exist in V and are continuous at $\boldsymbol{a}$, then f is differentiable at $\boldsymbol{a}$.*

PROOF. Since a function converges if and only if each of its components converge (Theorem 5.22), we may suppose $m = 1$. It suffices to show

$$\lim_{\boldsymbol{h} \to \boldsymbol{0}} \frac{f(\boldsymbol{a} + \boldsymbol{h}) - f(\boldsymbol{a}) - \nabla f(\boldsymbol{a}) \cdot \boldsymbol{h}}{\|\boldsymbol{h}\|} = 0.$$

Let $\boldsymbol{a} = (a_1, \ldots, a_n)$. Suppose $\boldsymbol{h} = (h_1, \ldots, h_n) \neq \boldsymbol{0}$ is so small that $\boldsymbol{a} + \boldsymbol{h} \in V$. By telescoping and using the one-dimensional Mean Value Theorem, we can choose numbers c_j between a_j and $a_j + h_j$ such that

$$\begin{aligned} f(\boldsymbol{a} + \boldsymbol{h}) - f(\boldsymbol{a}) &= f(a_1 + h_1, \ldots, a_n + h_n) - f(a_1, a_2 + h_2, \ldots, a_n + h_n) \\ &\quad + \cdots + f(a_1, \ldots, a_{n-1}, a_n + h_n) - f(a_1, \ldots, a_n) \\ &= \sum_{j=1}^{n} h_j \frac{\partial f}{\partial x_j}(a_1, \ldots, a_{j-1}, c_j, a_{j+1} + h_{j+1}, \ldots, a_n + h_n). \end{aligned}$$

Therefore,

$$(5) \qquad\qquad f(\boldsymbol{a} + \boldsymbol{h}) - f(\boldsymbol{a}) - \nabla f(\boldsymbol{a}) \cdot \boldsymbol{h} = \boldsymbol{h} \cdot \boldsymbol{\delta}$$

where $\boldsymbol{\delta} \in \mathbf{R}^n$ is the vector with components

$$\delta_j = \frac{\partial f}{\partial x_j}(a_1, \ldots, a_{j-1}, c_j, a_{j+1} + h_{j+1}, \ldots, a_n + h_n) - \frac{\partial f}{\partial x_j}(a_1, \ldots, a_n).$$

Since the first-order partial derivatives of f are continuous at $\boldsymbol{a}$, $\delta_j \to 0$ for each $1 \leq j \leq n$, i.e., $\|\boldsymbol{\delta}\| \to 0$ as $\boldsymbol{h} \to \boldsymbol{0}$. Moreover, by the Cauchy–Schwarz Inequality and (5),

$$(6) \qquad\qquad 0 \leq \frac{|f(\boldsymbol{a} + \boldsymbol{h}) - f(\boldsymbol{a}) - \nabla f(\boldsymbol{a}) \cdot \boldsymbol{h}|}{\|\boldsymbol{h}\|} = \frac{\|\boldsymbol{h} \cdot \boldsymbol{\delta}\|}{\|\boldsymbol{h}\|} \leq \|\boldsymbol{\delta}\|.$$

It follows from the Squeeze Theorem that the first quotient in (6) converges to 0 as $\boldsymbol{h} \to \boldsymbol{0}$. Thus, f is differentiable at $\boldsymbol{a}$ by definition. ∎

If all first-order partial derivatives of a function f exist and are continuous at a point $\boldsymbol{a}$ (respectively, on an open set V), we shall call f *continuously differentiable* at $\boldsymbol{a}$ (respectively, on V). By Theorem 6.9, every continuously differentiable function is differentiable. In particular, every function which is $\mathcal{C}^p$ on an open set V, for some $1 \leq p \leq \infty$, is continuously differentiable on V.

Although Theorem 6.9 is much simpler to use than Definition 6.4, the following example shows that there are times when Definition 6.4 must be used directly.

Example 3. Prove that

$$f(x, y) = \begin{cases} (x^2 + y^2) \sin \dfrac{1}{\sqrt{x^2 + y^2}} & (x, y) \neq (0, 0) \\ 0 & (x, y) = (0, 0) \end{cases}$$

is differentiable but not continuously differentiable at $(0, 0)$.

PROOF. Clearly, f is C^1, hence differentiable, on $\mathbf{R} \setminus \{(0, 0)\}$. To prove that f is differentiable at $(0, 0)$, we must verify (4) for $\boldsymbol{a} = (0, 0)$ and $B = \nabla f(\boldsymbol{a})$. By definition,

$$f_x(0, 0) = \lim_{t \to 0} \frac{f(t, 0) - f(0, 0)}{t} = \lim_{t \to 0} t \sin \frac{1}{|t|} = 0,$$

and similarly, $f_y(0, 0) = 0$. Thus,

$$\frac{f(h, k) - f(0, 0) - \nabla f(0, 0) \cdot (h, k)}{\|(h, k)\|} = \sqrt{h^2 + k^2} \sin \frac{1}{\sqrt{h^2 + k^2}} \to 0$$

as $(h, k) \to (0, 0)$; i.e., f is differentiable at $(0, 0)$. On the other hand, if $(x, y) \neq (0, 0)$ it follows from the one-dimensional Product Rule that

$$f_x(x, y) = \frac{-x}{\sqrt{x^2 + y^2}} \cos \frac{1}{\sqrt{x^2 + y^2}} + 2x \sin \frac{1}{\sqrt{x^2 + y^2}}.$$

Thus, $f_x(x, 0)$ has no limit as $x \to 0$, and the partial derivative f_x is not continuous at $(0, 0)$. ∎

Combining Theorem 6.9 with Example 3, we see that every continuously differentiable function is differentiable, but not conversely.

These results suggest the following procedure to determine whether a function f is differentiable at a point $\boldsymbol{a}$. Compute all first-order partial derivatives of f at $\boldsymbol{a}$. If one of these does not exist, then f is not differentiable at $\boldsymbol{a}$ (Theorem 6.7). If all first-order partial derivatives exist and are continuous at $\boldsymbol{a}$, then f is differentiable at $\boldsymbol{a}$ (Theorem 6.9). If one of the first-order partial derivatives exists but fails to be continuous at $\boldsymbol{a}$, then use Definition 6.4 directly. This will involve evaluation of a limit of vectors using the methods outlined in Sections 5.5 and 5.6.

We close this section with some additional comments about the definition of differentiability. Definition 6.4 describes differentiability from the analytic point of view. For the one-dimensional case, the derivative has both an analytic description (in terms of linear functions) and a geometric description (in terms of tangent lines). In Section 6.7, we shall prove (for the case $n = 2$ and $m = 1$) that f satisfies (4) for $\boldsymbol{a} = (a, b)$ and $B = \nabla f(a, b)$ if and only if $z = f(x, y)$ has a unique tangent plane at $(a, b, f(a, b))$. Moreover, we shall show that the corresponding differential dz at (a, b) approximates $f(a + \Delta x, b + \Delta y)$ along this tangent plane (see Figure 6.3). Thus, Definition 6.4 captures both the analytic and geometric spirit of the one-dimensional derivative.

For the one-dimensional case, if g is differentiable at a, then $g'(a)$ is the slope of the tangent line to $y = g(x)$ at $x = a$. The following result shows that for the two-dimensional case, if f is differentiable, then $f_x(a, b)$ and $f_y(a, b)$ can be used to compute a normal vector of the tangent plane to $z = f(x, y)$ at $(x, y) = (a, b)$.

Remark 2. *Suppose V is an open subset of $\mathbf{R}^2$ and $f : V \to \mathbf{R}$ is differentiable at $(a, b) \in V$. If Π is a plane passing through the point $(a, b, f(a, b))$ with equation $F(x, y, z) = d$ such that the line $F(x, b, z) = d$ is tangent to the curve $z = f(x, b)$ at $x = a$ and the line $F(a, y, z) = d$ is tangent to the curve $z = f(a, y)$ at $y = b$, then*

$$\boldsymbol{n} = (-f_x(a, b), -f_y(a, b), 1)$$

is normal to Π and

$$z = f_x(a, b)(x - a) + f_y(a, b)(y - b) + f(a, b)$$

is an equation of Π.

PROOF. Since f is differentiable at (a, b), $f(\cdot, b)$ is differentiable at a and $f(a, \cdot)$ is differentiable at b. By one-variable calculus, the slope of the line tangent to the curve $z = f(x, b)$ at $x = a$ is $f_x(a, b)$, so $\boldsymbol{u} = (1, 0, f_x(a, b))$ is a vector parallel to Π. Similarly, $\boldsymbol{v} = (0, 1, f_y(a, b))$ is also a vector parallel to Π. Therefore,

$$\boldsymbol{n} := \boldsymbol{u} \times \boldsymbol{v} = (-f_x(a, b), -f_y(a, b), 1)$$

is normal to Π and

$$\boldsymbol{n} \cdot (x - a, y - b, z - f(a, b)) = 0$$

is an equation of Π. ∎

EXERCISES

1. For each of the following functions, compute the Jacobian matrix Df at all points where it exists.

   a) $f(x, y) = (\sin x, \sqrt{xy}, \cos y)$.   b) $f(s, t, u, v) = (st + u^2, uv - s^2)$.

   c) $f(t) = (\log t, 1/(1 + t))$.   d) $f(r, \theta) = (r \cos \theta, r \sin \theta)$.

2. Prove that $f : \mathbf{R} \to \mathbf{R}$ is differentiable at a (in the sense of Definition 2.8) if and only if there is a linear function $\lambda : \mathbf{R} \to \mathbf{R}$ such that

$$\lim_{h \to 0} \frac{f(a + h) - f(a) - \lambda(h)}{h} = 0.$$

   (Note: By Theorem 5.4, $\lambda : \mathbf{R} \to \mathbf{R}$ is linear if and only if $\lambda(x) = mx$ for some $m \in \mathbf{R}$.)

3. Let V be open in $\mathbf{R}^n$, $\boldsymbol{a} \in V$, and $f, g : V \to \mathbf{R}^m$.

*DEFINITION 6.5. If $\boldsymbol{u}$ is a *unit* vector in $\mathbf{R}^n$, i.e., $\|\boldsymbol{u}\| = 1$, then the *directional derivative* of f at $\boldsymbol{a}$ in the direction $\boldsymbol{u}$ is defined by

$$D_{\boldsymbol{u}}f(\boldsymbol{a}) := \lim_{t \to 0} \frac{f(\boldsymbol{a} + t\boldsymbol{u}) - f(\boldsymbol{a})}{t}$$

when this limit exists.

a) Prove that $D_{\boldsymbol{u}}f(\boldsymbol{a})$ exists for $\boldsymbol{u} = \boldsymbol{e}_k$ if and only if $f_{x_k}(\boldsymbol{a})$ exists, in which case

$$D_{\boldsymbol{e}_k} f(\boldsymbol{a}) = \frac{\partial f}{\partial x_k}(\boldsymbol{a}).$$

b) Show if f has directional derivatives at $\boldsymbol{a}$ in all directions $\boldsymbol{u}$, then the first-order partial derivatives of f exist at $\boldsymbol{a}$. Use Example 1 to show that the converse of this statement is false.

c) Prove that the directional derivatives of

$$f(x, y) = \begin{cases} \dfrac{x^2 y}{x^4 + y^2} & y \neq 0 \\ 0 & y = 0 \end{cases}$$

exist at $(0,0)$ in all directions $\boldsymbol{u}$ but f is neither continuous nor differentiable at $(0,0)$.

4. Prove that $f = (f_1, \ldots, f_m)$ is differentiable at a point $\boldsymbol{a} \in \mathbf{R}^n$ if and only if its components f_j are differentiable at $\boldsymbol{a}$ for $j = 1, 2, \ldots, m$.

5. Let I be an open interval, $a \in I$, and $f, g : I \to \mathbf{R}^m$.

a) Prove that f is differentiable at a if and only if

$$\lim_{h \to 0} \frac{f(a + h) - f(a)}{h}$$

exists, in which case this limit is $f'(a)$. Prove $(\alpha f)'(a) = \alpha f'(a)$ for all scalars α.

b) Prove that if f and g are differentiable at a, then

$$(f + g)'(a) = f'(a) + g'(a) \quad \text{and} \quad (f \cdot g)'(a) = g(a) \cdot f'(a) + f(a) \cdot g'(a).$$

c) Let J be an open interval, $b \in J$, $h : J \to \mathbf{R}$, and $a = h(b)$. Show that if f is differentiable at a and h is differentiable at b, then

$$(f \circ h)'(b) = h'(b) f'(h(b)).$$

6. Let α be a scalar, V be open in $\mathbf{R}^n$, $\boldsymbol{a} \in V$, $f, g : V \to \mathbf{R}^m$, and suppose all first-order partial derivatives of f and g exist at $\boldsymbol{a}$.

a) Prove that the Jacobian matrices $D(f + g)$ and $D(\alpha f)$ satisfy

$$D(f + g)(\boldsymbol{a}) = Df(\boldsymbol{a}) + Dg(\boldsymbol{a}) \quad \text{and} \quad D(\alpha f)(\boldsymbol{a}) = \alpha Df(\boldsymbol{a}).$$

b) Prove that the Jacobian matrix $D(f \cdot g)$ satisfies

$$D(f \cdot g)(\boldsymbol{a}) = f(\boldsymbol{a}) Dg(\boldsymbol{a}) + g(\boldsymbol{a}) Df(\boldsymbol{a}).$$

Write out this last identity for $f(x, y, z) = (x^2, y^2)$ and $g(x, y, z) = (xyz, x + y + z)$, interpreting each product appropriately.

7. Prove that $f(x, y) = \sqrt{|xy|}$ is not differentiable at $(0, 0)$.

8. Prove that

$$f(x, y) = \begin{cases} \dfrac{x^2 + y^2}{\sin \sqrt{x^2 + y^2}} & 0 < \|(x, y)\| < \pi \\ 0 & (x, y) = (0, 0) \end{cases}$$

   is not differentiable at $(0, 0)$.

9. Let $r > 0$, $f : B_r(\mathbf{0}) \to \mathbf{R}$, and suppose there exists an $\alpha > 1$ such that $|f(\boldsymbol{x})| \le \|\boldsymbol{x}\|^\alpha$ for all $\boldsymbol{x} \in B_r(\mathbf{0})$. Prove f is differentiable at $\mathbf{0}$. What happens when $\alpha = 1$?

10. Prove that if $\alpha > 1/2$, then

$$f(x, y) = \begin{cases} (xy)^\alpha \log(x^2 + y^2) & (x, y) \ne (0, 0) \\ 0 & (x, y) = (0, 0) \end{cases}$$

   is differentiable on $\mathbf{R}^2$.

11. Prove that

$$f(x, y) = \begin{cases} \dfrac{x^4 + y^4}{(x^2 + y^2)^\alpha} & (x, y) \ne (0, 0) \\ 0 & (x, y) = (0, 0) \end{cases}$$

   is differentiable on $\mathbf{R}^2$ for all $\alpha < 3/2$.

6.3 DIFFERENTIABILITY THEOREMS

In this section we begin to explore the analogy between Df and f'. Our first result is a simple characterization of differentiability which uses the uniqueness of the total derivative.

THEOREM 6.10. *Let V be open in $\mathbf{R}^n$, $\boldsymbol{a} \in V$ and $f : V \to \mathbf{R}^m$. If all first-order partial derivatives of f exist at $\boldsymbol{a}$, then f is differentiable at $\boldsymbol{a}$ if and only if there is a function $\varepsilon : \mathbf{R}^n \to \mathbf{R}^m$ such that $\varepsilon(\boldsymbol{h}) \to \mathbf{0}$ as $\boldsymbol{h} \to \mathbf{0}$ and*

$$(7) \qquad f(\boldsymbol{a} + \boldsymbol{h}) - f(\boldsymbol{a}) = Df(\boldsymbol{a})(\boldsymbol{h}) + \|\boldsymbol{h}\|\varepsilon(\boldsymbol{h})$$

for $\boldsymbol{h}$ sufficiently small; i.e., there is an $r > 0$ such that (7) holds for $\|\boldsymbol{h}\| < r$.

PROOF. Choose $r > 0$ such that $B_r(\boldsymbol{a}) \subset V$. If f is differentiable, set

$$(8) \qquad \varepsilon(\boldsymbol{h}) = \frac{f(\boldsymbol{a} + \boldsymbol{h}) - f(\boldsymbol{a}) - (Df(\boldsymbol{a}))(\boldsymbol{h})}{\|\boldsymbol{h}\|}$$

for $0 < \|\boldsymbol{h}\| < r$, and $\varepsilon(\boldsymbol{h}) = \mathbf{0}$ otherwise. Clearly, (7) holds for $\boldsymbol{h} = \mathbf{0}$. By (8), (7) also holds for $0 < \|\boldsymbol{h}\| < r$. And by Definition 6.4, $\varepsilon(\boldsymbol{h}) \to \mathbf{0}$ as $\boldsymbol{h} \to \mathbf{0}$.

Conversely, if (7) holds for $\|\boldsymbol{h}\| < r$, then (8) holds for $0 < \|\boldsymbol{h}\| < r$. Since $\varepsilon(\boldsymbol{h}) \to \mathbf{0}$ as $\boldsymbol{h} \to \mathbf{0}$, it follows that f is differentiable at $\boldsymbol{a}$. ∎

The first application of this characterization is an analogue of the rule $(mx)' = m$.

THEOREM 6.11. *If $F : \mathbf{R}^n \to \mathbf{R}^m$ is linear, then F is differentiable everywhere on $\mathbf{R}^n$ and $DF(\boldsymbol{a}) = F$, i.e., if B is the matrix which represents F, then $DF(\boldsymbol{a}) = B$ for all $\boldsymbol{a} \in \mathbf{R}^n$.*

PROOF. Since F is linear, we have

$$F(\boldsymbol{a} + \boldsymbol{h}) - F(\boldsymbol{a}) = F(\boldsymbol{h})$$

for all $\boldsymbol{h} \in \mathbf{R}^n$. Hence, by Theorem 6.10 (with $\varepsilon(\boldsymbol{h}) := \boldsymbol{0}$), $Df(\boldsymbol{a}) = F$ for all $\boldsymbol{a} \in \mathbf{R}^n$. ∎

A second application leads to the following rules for differentiating sums and products.

THEOREM 6.12. *Let V be open in $\mathbf{R}^n$, $\boldsymbol{a} \in V$, α be a scalar, and $f, g : V \to \mathbf{R}^m$. If f and g are differentiable at $\boldsymbol{a}$, then so are $f + g$, αf, and $f \cdot g$. In fact,*

$$(9) \qquad D(f + g)(\boldsymbol{a}) = Df(\boldsymbol{a}) + Dg(\boldsymbol{a}),$$

$$(10) \qquad D(\alpha f)(\boldsymbol{a}) = \alpha Df(\boldsymbol{a}),$$

and

$$(11) \qquad D(f \cdot g)(\boldsymbol{a}) = g(\boldsymbol{a})Df(\boldsymbol{a}) + f(\boldsymbol{a})Dg(\boldsymbol{a}).$$

(The products gDf and fDg which appear in (11) represent matrix multiplication.)

PROOF. The proofs of these rules are similar. We provide the details only for (11). Let

$$(12) \qquad B = g(\boldsymbol{a})Df(\boldsymbol{a}) + f(\boldsymbol{a})Dg(\boldsymbol{a}).$$

Since $g(\boldsymbol{a})$ and $f(\boldsymbol{a})$ are $1 \times m$ matrices and $Df(\boldsymbol{a})$ and $Dg(\boldsymbol{a})$ are $m \times n$ matrices, we see that B is a $1 \times n$ matrix, the right size for $D(f \cdot g)$. By uniqueness of the total derivative, we need only show

$$\lim_{\boldsymbol{h} \to \boldsymbol{0}} \frac{(f \cdot g)(\boldsymbol{a} + \boldsymbol{h}) - (f \cdot g)(\boldsymbol{a}) - B\boldsymbol{h}}{\|\boldsymbol{h}\|} = 0.$$

Choose functions $\varepsilon, \delta : \mathbf{R}^n \to \mathbf{R}^m$ such that $\varepsilon(\boldsymbol{h}), \delta(\boldsymbol{h}) \to 0$ as $\boldsymbol{h} \to \boldsymbol{0}$, and

$$f(\boldsymbol{a} + \boldsymbol{h}) - f(\boldsymbol{a}) = Df(\boldsymbol{a})(\boldsymbol{h}) + \|\boldsymbol{h}\|\varepsilon(\boldsymbol{h}), \qquad g(\boldsymbol{a} + \boldsymbol{h}) - g(\boldsymbol{a}) = Dg(\boldsymbol{a})(\boldsymbol{h}) + \|\boldsymbol{h}\|\delta(\boldsymbol{h})$$

both hold for $\boldsymbol{h}$ sufficiently small. By (12),

$$\begin{aligned}
(f \cdot g)&(\boldsymbol{a} + \boldsymbol{h}) - (f \cdot g)(\boldsymbol{a}) - B\boldsymbol{h} \\
&= (f \cdot g)(\boldsymbol{a} + \boldsymbol{h}) - (f \cdot g)(\boldsymbol{a}) - g(\boldsymbol{a})Df(\boldsymbol{a})(\boldsymbol{h}) + f(\boldsymbol{a})Dg(\boldsymbol{a})(\boldsymbol{h}) \\
&= (f(\boldsymbol{a} + \boldsymbol{h}) - f(\boldsymbol{a}) - Df(\boldsymbol{a})(\boldsymbol{h})) \cdot g(\boldsymbol{a} + \boldsymbol{h}) \\
&\qquad + (Df(\boldsymbol{a})(\boldsymbol{h})) \cdot (g(\boldsymbol{a} + \boldsymbol{h}) - g(\boldsymbol{a})) \\
&\qquad\qquad + f(\boldsymbol{a}) \cdot (g(\boldsymbol{a} + \boldsymbol{h}) - g(\boldsymbol{a}) - Dg(a)(\boldsymbol{h})) \\
&=: T_1(\boldsymbol{h}) + T_2(\boldsymbol{h}) + T_3(\boldsymbol{h}).
\end{aligned}$$

By the uniqueness of the total derivative, we must show $T_j(\boldsymbol{h})/\|\boldsymbol{h}\| \to 0$ as $\boldsymbol{h} \to \boldsymbol{0}$ for $j = 1, 2, 3$. To estimate the first term, observe by the Cauchy–Schwarz Inequality and the choice of ε that

$$|T_1(\boldsymbol{h})| \le \|f(\boldsymbol{a}+\boldsymbol{h}) - f(\boldsymbol{a}) - Df(\boldsymbol{a})(\boldsymbol{h})\| \, \|g(\boldsymbol{a}+\boldsymbol{h})\|$$
$$= \|\boldsymbol{h}\| \, \|\varepsilon(\boldsymbol{h})\| \, \|g(\boldsymbol{a}+\boldsymbol{h})\|.$$

Since g is continuous at $\boldsymbol{a}$ (see Theorem 6.8) and $\varepsilon(\boldsymbol{h}) \to \boldsymbol{0}$ as $\boldsymbol{h} \to \boldsymbol{0}$, it follows that $|T_1(\boldsymbol{h})|/\|\boldsymbol{h}\| \to 0$ as $\boldsymbol{h} \to \boldsymbol{0}$. A similar argument shows that $|T_3(\boldsymbol{h})|/\|\boldsymbol{h}\| \to 0$ as $\boldsymbol{h} \to \boldsymbol{0}$. To estimate the second term, observe by the Cauchy–Schwarz Inequality and Theorem 5.5 that

$$|T_2(\boldsymbol{h})| \le \|(Df(\boldsymbol{a})(\boldsymbol{h})\| \, \|g(\boldsymbol{a}+\boldsymbol{h}) - g(\boldsymbol{a}))\|$$
$$\le \sqrt{mn} \, \|Df(\boldsymbol{a})\|_\infty \, \|\boldsymbol{h}\| \, \|g(\boldsymbol{a}+\boldsymbol{h}) - g(\boldsymbol{a})\|.$$

Thus, $|T_2(\boldsymbol{h})|/\|\boldsymbol{h}\| \le \sqrt{mn} \, \|Df(\boldsymbol{a})\|_\infty \, \|g(\boldsymbol{a}+\boldsymbol{h}) - g(\boldsymbol{a})\| \to 0$ as $\boldsymbol{h} \to \boldsymbol{0}$. We conclude that $f \cdot g$ is differentiable at $\boldsymbol{a}$ and the derivative is B. ∎

Formula (9) is called the *Sum Rule*, (10) is sometimes called the *Homogeneous Rule*, and (11) is called the *Dot Product Rule*.

The following result is a multidimensional analogue of the Chain Rule.

THEOREM 6.13 [CHAIN RULE]. *Let V be open in* $\mathbf{R}^n$, *U be open in* $\mathbf{R}^m$, *$g : V \to \mathbf{R}^m$, $f : U \to \mathbf{R}^p$, $\boldsymbol{a} \in V$, and $g(\boldsymbol{a}) \in U$. If g is differentiable at $\boldsymbol{a}$ and f is differentiable at $g(\boldsymbol{a})$, then $f \circ g$ is differentiable at $\boldsymbol{a}$ and*

$$(13) \qquad\qquad D(f \circ g)(\boldsymbol{a}) = Df(g(\boldsymbol{a}))Dg(\boldsymbol{a}).$$

(The product on the right side of (13) is matrix multiplication.)

PROOF. We must show

$$(14) \qquad \lim_{\boldsymbol{h} \to \boldsymbol{0}} \frac{f(g(\boldsymbol{a}+\boldsymbol{h})) - f(g(\boldsymbol{a})) - Df(g(\boldsymbol{a}))Dg(\boldsymbol{a})(\boldsymbol{h})}{\|\boldsymbol{h}\|} = \boldsymbol{0}.$$

Let $\boldsymbol{b} = g(\boldsymbol{a})$. Choose functions $\varepsilon : \mathbf{R}^n \to \mathbf{R}^m$ and $\delta : \mathbf{R}^m \to \mathbf{R}^p$ such that $\varepsilon(\boldsymbol{h}) \to \boldsymbol{0}$ as $\boldsymbol{h} \to \boldsymbol{0}$, $\delta(\boldsymbol{k}) \to \boldsymbol{0}$ as $\boldsymbol{k} \to \boldsymbol{0}$, and

$$(15) \qquad\qquad g(\boldsymbol{a}+\boldsymbol{h}) - g(\boldsymbol{a}) = Dg(\boldsymbol{a})(\boldsymbol{h}) + \|\boldsymbol{h}\|\varepsilon(\boldsymbol{h}),$$

$$(16) \qquad\qquad f(\boldsymbol{b}+\boldsymbol{k}) - f(\boldsymbol{b}) = Df(\boldsymbol{b})(\boldsymbol{k}) + \|\boldsymbol{k}\|\delta(\boldsymbol{k})$$

both hold for $\boldsymbol{h}$ and $\boldsymbol{k}$ sufficiently small. Fix $\boldsymbol{h} \ne \boldsymbol{0}$ and set $\boldsymbol{k} = g(\boldsymbol{a}+\boldsymbol{h}) - g(\boldsymbol{a})$. Since (16) and (15) imply

$$f(g(\boldsymbol{a}+\boldsymbol{h})) - f(g(\boldsymbol{a})) = f(\boldsymbol{b}+\boldsymbol{k}) - f(\boldsymbol{b}) = Df(\boldsymbol{b})(\boldsymbol{k}) + \|\boldsymbol{k}\|\delta(\boldsymbol{k})$$
$$= Df(\boldsymbol{b})(Dg(\boldsymbol{a})(\boldsymbol{h}) + \|\boldsymbol{h}\|\varepsilon(\boldsymbol{h})) + \|\boldsymbol{k}\|\delta(\boldsymbol{k}),$$

we have

$$f(g(\boldsymbol{a}+\boldsymbol{h})) - f(g(\boldsymbol{a})) - Df(g(\boldsymbol{a}))Dg(\boldsymbol{a})(\boldsymbol{h}) = \|\boldsymbol{h}\|Df(\boldsymbol{b})\varepsilon(\boldsymbol{h}) + \|\boldsymbol{k}\|\delta(\boldsymbol{k})$$
$$=: T_1(\boldsymbol{h}) + T_2(\boldsymbol{h}).$$

By the uniqueness of the total derivative, it suffices to show that $T_j(\boldsymbol{h})/\|\boldsymbol{h}\| \to \boldsymbol{0}$ as $\boldsymbol{h} \to \boldsymbol{0}$ for $j = 1, 2$.

Since $\varepsilon(\boldsymbol{h}) \to \boldsymbol{0}$ as $\boldsymbol{h} \to \boldsymbol{0}$, it is clear that $T_1(\boldsymbol{h})/\|\boldsymbol{h}\| \to \boldsymbol{0}$ as $\boldsymbol{h} \to \boldsymbol{0}$. On the other hand, by (15) and Theorem 5.5, we have

$$\|\boldsymbol{k}\| = \|g(\boldsymbol{a}+\boldsymbol{h}) - g(\boldsymbol{a})\| = \|Dg(\boldsymbol{a})(\boldsymbol{h}) + \|\boldsymbol{h}\|\varepsilon(\boldsymbol{h})\|$$
$$\leq \sqrt{mn}\,\|\boldsymbol{h}\|\,(\|Dg(\boldsymbol{a})\|_\infty + \|\varepsilon(\boldsymbol{h})\|).$$

Thus, $\|\boldsymbol{k}\|/\|\boldsymbol{h}\|$ is bounded for $\boldsymbol{h}$ sufficiently small. Since $\boldsymbol{k} \to \boldsymbol{0}$ (hence $\delta(\boldsymbol{k}) \to \boldsymbol{0}$) as $\boldsymbol{h} \to \boldsymbol{0}$, it follows that $\|T_2(\boldsymbol{h})\|/\|\boldsymbol{h}\| = (\|\boldsymbol{k}\|/\|\boldsymbol{h}\|)\,\|\delta(\boldsymbol{k})\| \to 0$ as $\boldsymbol{h} \to \boldsymbol{0}$. We conclude that $f \circ g$ is differentiable at $\boldsymbol{a}$ and the derivative is $Df(g(\boldsymbol{a}))Dg(\boldsymbol{a})$. ∎

Theorem 6.13 contains many versions of the Chain Rule usually listed separately in elementary calculus texts. Here are two examples.

Example 1. Let $F, G, H : \mathbf{R}^2 \to \mathbf{R}$ be differentiable. If $z = F(x, y)$, $x = G(r, \theta)$, and $y = H(r, \theta)$, show

$$\frac{\partial z}{\partial r} = \frac{\partial z}{\partial x}\frac{\partial x}{\partial r} + \frac{\partial z}{\partial y}\frac{\partial y}{\partial r} \quad \text{and} \quad \frac{\partial z}{\partial \theta} = \frac{\partial z}{\partial x}\frac{\partial x}{\partial \theta} + \frac{\partial z}{\partial y}\frac{\partial y}{\partial \theta}.$$

SOLUTION. Let $z = \psi(r, \theta) = F(G(r, \theta), H(r, \theta))$ and $\phi = (H, G)$. By definition,

$$DF = \begin{bmatrix} \dfrac{\partial z}{\partial x} & \dfrac{\partial z}{\partial y} \end{bmatrix},$$

$$D\psi = \begin{bmatrix} \dfrac{\partial z}{\partial r} & \dfrac{\partial z}{\partial \theta} \end{bmatrix},$$

and

$$D\phi = \begin{bmatrix} \dfrac{\partial x}{\partial r} & \dfrac{\partial x}{\partial \theta} \\ \dfrac{\partial y}{\partial r} & \dfrac{\partial y}{\partial \theta} \end{bmatrix}.$$

Since $\psi = F \circ \phi$, it follows from the Chain Rule that

$$\begin{bmatrix} \dfrac{\partial z}{\partial r} & \dfrac{\partial z}{\partial \theta} \end{bmatrix} = \begin{bmatrix} \dfrac{\partial z}{\partial x} & \dfrac{\partial z}{\partial y} \end{bmatrix} \begin{bmatrix} \dfrac{\partial x}{\partial r} & \dfrac{\partial x}{\partial \theta} \\ \dfrac{\partial y}{\partial r} & \dfrac{\partial y}{\partial \theta} \end{bmatrix}$$
$$= \begin{bmatrix} \dfrac{\partial z}{\partial x}\dfrac{\partial x}{\partial r} + \dfrac{\partial z}{\partial y}\dfrac{\partial y}{\partial r} & \dfrac{\partial z}{\partial x}\dfrac{\partial x}{\partial \theta} + \dfrac{\partial z}{\partial y}\dfrac{\partial y}{\partial \theta} \end{bmatrix}. \;\blacksquare$$

Example 2. Let $f : \mathbf{R}^3 \to \mathbf{R}$ and $\phi, \psi, \sigma : \mathbf{R} \to \mathbf{R}$ be differentiable. If $w = f(x, y, z)$, $x = \phi(t)$, $y = \psi(t)$, and $z = \sigma(t)$, show

$$\frac{dw}{dt} = \frac{\partial w}{\partial x}\frac{dx}{dt} + \frac{\partial w}{\partial y}\frac{dy}{dt} + \frac{\partial w}{\partial z}\frac{dz}{dt}.$$

SOLUTION. Let $w = G(t) := f(\phi(t), \psi(t), \sigma(t))$ and $g := (\phi, \psi, \sigma)$. By definition

$$Df = \left[\begin{array}{ccc} \dfrac{\partial w}{\partial x} & \dfrac{\partial w}{\partial y} & \dfrac{\partial w}{\partial z} \end{array} \right],$$

$$Dg = \left[\begin{array}{c} dx/dt \\ dy/dt \\ dz/dt \end{array} \right] \quad \text{and} \quad DG = \left[\dfrac{dw}{dt} \right].$$

Since $G = f \circ g$, it follows from the Chain Rule that

$$\left[\frac{dw}{dt} \right] = \left[\begin{array}{ccc} \dfrac{\partial w}{\partial x} & \dfrac{\partial w}{\partial y} & \dfrac{\partial w}{\partial z} \end{array} \right] \left[\begin{array}{c} dx/dt \\ dy/dt \\ dz/dt \end{array} \right] = \left[\frac{\partial w}{\partial x}\frac{dx}{dt} + \frac{\partial w}{\partial y}\frac{dy}{dt} + \frac{\partial w}{\partial z}\frac{dz}{dt} \right]. \quad \blacksquare$$

EXERCISES

1. Let V be open in $\mathbf{R}^n$, $\mathbf{a} \in V$, $f : V \to \mathbf{R}$, and suppose f is differentiable at $\mathbf{a}$.

    a) Prove the directional derivative $D_{\mathbf{u}} f(\mathbf{a})$ exists for each $\mathbf{u} \in \mathbf{R}^n$, $\|\mathbf{u}\| = 1$ (see Exercise 3 in Section 6.2), and

$$D_{\mathbf{u}} f(\mathbf{a}) = \nabla f(\mathbf{a}) \cdot \mathbf{u}.$$

    b) If $\nabla f(\mathbf{a}) \neq \mathbf{0}$ and θ represents the angle between $\mathbf{u}$ and $\nabla f(\mathbf{a})$, prove $D_{\mathbf{u}} f(\mathbf{a}) = \|\nabla f(\mathbf{a})\| \cos \theta$.

    c) Show that as $\mathbf{u}$ ranges over all unit vectors in $\mathbf{R}^n$, the maximum of $D_{\mathbf{u}} f(\mathbf{a})$ is $\|\nabla f(\mathbf{a})\|$ and it occurs when $\mathbf{u}$ is parallel to $\nabla f(\mathbf{a})$.

2. Let $F : \mathbf{R}^3 \to \mathbf{R}$ and $f, g, h : \mathbf{R}^2 \to \mathbf{R}$ be $\mathcal{C}^2$ functions. If $w = F(x, y, z)$, where $x = f(p, q)$, $y = g(p, q)$, and $z = h(p, q)$, find formulas for w_p, w_q, and w_{pp}.

3. Let $r > 0$, $\mathbf{a} \in \mathbf{R}^n$, and $g : B_r(\mathbf{a}) \to \mathbf{R}^m$ be differentiable at $\mathbf{a}$.

    a) If $f : B_r(g(\mathbf{a})) \to \mathbf{R}$ is differentiable at $g(\mathbf{a})$, prove that the partial derivatives of $h = f \circ g$ are given by

$$\frac{\partial h}{\partial x_j}(\mathbf{a}) = \nabla f(g(\mathbf{a})) \cdot \frac{\partial g}{\partial x_j}(\mathbf{a})$$

    for $j = 1, 2, \ldots, n$.

    b) If $n = m$ and $f : B_r(g(\mathbf{a})) \to \mathbf{R}^n$ is differentiable at $g(\mathbf{a})$, prove $\Delta_{f \circ g}(\mathbf{a}) = \Delta_f(g(\mathbf{a}))\Delta_g(\mathbf{a})$.

4. Let $f, g : \mathbf{R} \to \mathbf{R}$ be twice differentiable. Prove that $u(x, y) := f(xy)$ satisfies

$$x \frac{\partial u}{\partial x} - y \frac{\partial u}{\partial y} = 0$$

and $v(x, y) := f(x - y) + g(x + y)$ satisfies the *wave equation*; i.e.,

$$\frac{\partial^2 v}{\partial x^2} - \frac{\partial^2 v}{\partial y^2} = 0.$$

5. Let $u : \mathbf{R} \to [0, \infty)$ be differentiable. Prove that

$$F(x, y, z) := u(\sqrt{x^2 + y^2 + z^2})$$

satisfies

$$u'(\sqrt{x^2 + y^2 + z^2}) = \left(\left(\frac{\partial F}{\partial x} \right)^2 + \left(\frac{\partial F}{\partial y} \right)^2 + \left(\frac{\partial F}{\partial z} \right)^2 \right)^{1/2}.$$

6. Let

$$u(x, t) = \frac{e^{-x^2/4t}}{\sqrt{4\pi t}}, \qquad t > 0, \; x \in \mathbf{R}.$$

   a) Prove that u satisfies the *heat equation*; i.e., $u_{xx} - u_t = 0$ for all $t > 0$ and $x \in \mathbf{R}$.

   b) If $a > 0$, prove that $u(x, t) \to 0$, as $t \to 0+$, uniformly for $x \in [a, \infty)$.

7. Suppose I is an open interval and $f : I \to \mathbf{R}^m$ is differentiable on I. If $f(I) \subseteq \partial B_r(\mathbf{0})$ for some fixed $r > 0$, prove that $f(t)$ is orthogonal to $f'(t)$ for all $t \in I$.

8. Suppose $z = F(x, y)$ is differentiable at (a, b), $F_y(a, b) \neq 0$, and I is an open interval containing a. Prove that if $f : I \to \mathbf{R}$ is differentiable at a and $F(x, f(x)) = 0$ for all $x \in I$, then

$$\frac{df}{dx}(a) = \frac{-\dfrac{\partial F}{\partial x}(a, b)}{\dfrac{\partial F}{\partial y}(a, b)}.$$

9. Let $r > 0$, $(a, b) \in \mathbf{R}^2$, $f : B_r(a, b) \to \mathbf{R}$, and suppose that the first-order partial derivatives f_x and f_y exist in $B_r(a, b)$ and are differentiable at (a, b).

   a) Set

$$\Delta(h) = f(a + h, b + h) - f(a + h, b) - f(a, b + h) + f(a, b))$$

   and prove for h sufficiently small that

$$\frac{\Delta(h)}{h} = f_y(a + h, b + th) - f_y(a, b) - \nabla f_y(a, b) \cdot (h, th)$$
$$- (f_y(a, b + th) - f_y(a, b) - \nabla f_y(a, b) \cdot (0, th)) + h f_{yx}(a, b)$$

for some $t \in (0,1)$.

b) Prove that

$$\lim_{h \to 0} \frac{\Delta(h)}{h^2} = f_{yx}(a,b).$$

c) Prove

$$\frac{\partial^2 f}{\partial x \partial y}(a,b) = \frac{\partial^2 f}{\partial y \partial x}(a,b).$$

10. Let $f, g : \mathbf{R}^2 \to \mathbf{R}$ be differentiable and satisfy the *Cauchy–Riemann equations*, i.e.,

$$\frac{\partial f}{\partial x} = \frac{\partial g}{\partial y} \quad \text{and} \quad \frac{\partial f}{\partial y} = -\frac{\partial g}{\partial x}$$

hold on $\mathbf{R}^2$. If $u(r,\theta) = f(r\cos\theta, r\sin\theta)$, and $v(r,\theta) = g(r\cos\theta, r\sin\theta)$, prove

$$\frac{\partial u}{\partial r} = \frac{1}{r}\frac{\partial v}{\partial \theta}, \qquad \frac{\partial v}{\partial r} = -\frac{1}{r}\frac{\partial u}{\partial \theta}.$$

11. Let $f : \mathbf{R}^2 \to \mathbf{R}$ be $\mathcal{C}^2$ on $\mathbf{R}^2$ and set $u(r,\theta) = f(r\cos\theta, r\sin\theta)$. If f satisfies the *Laplace equation*, i.e., if

$$\frac{\partial^2 f}{\partial x^2} + \frac{\partial^2 f}{\partial y^2} = 0,$$

prove

$$\frac{1}{r^2}\frac{\partial^2 u}{\partial \theta^2} + \frac{1}{r}\frac{\partial u}{\partial r} + \frac{\partial^2 u}{\partial r^2} = 0.$$

6.4 THE MEAN VALUE THEOREM AND TAYLOR'S FORMULA

Using Df as a replacement for f', we guess that the multidimensional analogue of the Mean Value Theorem is

$$f(\boldsymbol{x}) - f(\boldsymbol{a}) = Df(\boldsymbol{c})(\boldsymbol{x} - \boldsymbol{a})$$

for some $\boldsymbol{c}$ "between" $\boldsymbol{x}$ and $\boldsymbol{a}$, i.e., $\boldsymbol{c} \in L(\boldsymbol{x}; \boldsymbol{a})$. The following result shows that our guess is wrong for functions $f : \mathbf{R}^n \to \mathbf{R}^m$ when $m > 1$.

Remark 1. *The function $f(t) = (\cos t, \sin t)$ is differentiable on $\mathbf{R}$ and satisfies $f(2\pi) = f(0)$, but there is no $c \in \mathbf{R}$ such that $\nabla f(c) = (0,0)$.*

PROOF. $\nabla f(t) = (-\sin t, \cos t)$ exists and is continuous for $t \in \mathbf{R}$ but $(0,0) \neq (-\sin t, \cos t)$ for $t \in \mathbf{R}$. ∎

The following is a correct version of the Mean Value Theorem for multivariable functions.

THEOREM 6.14 [THE MEAN VALUE THEOREM]. *Let V be open in $\mathbf{R}^n$ and $f : V \to \mathbf{R}^m$ be differentiable on V. If $\boldsymbol{x}, \boldsymbol{a} \in V$ and $L(\boldsymbol{x}; \boldsymbol{a}) \subset V$, then given any $\boldsymbol{u} \in \mathbf{R}^m$ there is a $\boldsymbol{c} \in L(\boldsymbol{x}; \boldsymbol{a})$ such that*

$$\boldsymbol{u} \cdot (f(\boldsymbol{x}) - f(\boldsymbol{a})) = \boldsymbol{u} \cdot (Df(\boldsymbol{c})(\boldsymbol{x} - \boldsymbol{a})).$$

PROOF. Let

$$g(t) = \boldsymbol{a} + t(\boldsymbol{x} - \boldsymbol{a}), \qquad t \in \mathbf{R}$$

and notice by Theorem 6.11 that $g : \mathbf{R} \to \mathbf{R}^m$ is differentiable with $Dg(t) = \boldsymbol{x} - \boldsymbol{a}$ for all $t \in \mathbf{R}$. Since $L(\boldsymbol{x}; \boldsymbol{a}) \subset V$ and V is open, choose $\delta > 0$ such that $g(t) \in V$ for all $t \in I_\delta := (-\delta, 1 + \delta)$. By the Chain Rule,

$$(17) \qquad\qquad D(f \circ g)(t) = Df(g(t))(\boldsymbol{x} - \boldsymbol{a}), \qquad t \in I_\delta.$$

Fix $\boldsymbol{u} \in \mathbf{R}^m$ and consider the function

$$F(t) = \boldsymbol{u} \cdot (f \circ g)(t), \qquad t \in I_\delta.$$

The function F is a real-valued function on I_δ. By (11) (the Dot Product Rule) and (17), F is differentiable on I_δ with

$$F'(t) = \boldsymbol{u} \cdot D(f \circ g)(t) = \boldsymbol{u} \cdot (Df(g(t))(\boldsymbol{x} - \boldsymbol{a})).$$

Hence, by the one-dimensional Mean Value Theorem, there is a $t_0 \in (0, 1)$ such that

$$\boldsymbol{u} \cdot (f(\boldsymbol{x}) - f(\boldsymbol{a})) = F(1) - F(0) = F'(t_0) = \boldsymbol{u} \cdot (Df(g(t_0))(\boldsymbol{x} - \boldsymbol{a})).$$

Thus, set $\boldsymbol{c} = g(t_0)$. ∎

The special case when f is scalar-valued deserves separate mention (see also Exercises 1 and 5).

COROLLARY 6.15. *Let V be open and convex in $\mathbf{R}^n$ and $f : V \to \mathbf{R}$. If f is differentiable on V and $\boldsymbol{a} + \boldsymbol{h}, \boldsymbol{a}$ both belong to V, then there is a $0 < t < 1$ such that*

$$(18) \qquad f(\boldsymbol{a} + \boldsymbol{h}) - f(\boldsymbol{a}) = \sum_{j=1}^{n} \frac{\partial f}{\partial x_j}(\boldsymbol{a} + t\boldsymbol{h})h_j = \nabla f(\boldsymbol{a} + t\boldsymbol{h}) \cdot \boldsymbol{h}.$$

PROOF. Let u be a nonzero scalar, and suppose $\boldsymbol{a} + \boldsymbol{h}, \boldsymbol{a}$ both belong to V. Since V is convex, $L(\boldsymbol{a} + \boldsymbol{h}; \boldsymbol{a}) \subset V$. Hence, by Theorem 6.14,

$$u(f(\boldsymbol{a} + \boldsymbol{h}) - f(\boldsymbol{a})) = u(\nabla f(\boldsymbol{c}) \cdot \boldsymbol{h})$$

for some $\boldsymbol{c} \in L(\boldsymbol{a} + \boldsymbol{h}; \boldsymbol{a})$. Dividing this inequality by u and choosing $t \in (0, 1)$ such that $\boldsymbol{c} = \boldsymbol{a} + t\boldsymbol{h}$, we conclude that (18) holds. ∎

The following two results illustrate typical applications of the Mean Value Theorem.

COROLLARY 6.16. *Let V be an open set in $\mathbf{R}^n$, H be a compact subset of V, and $f : V \to \mathbf{R}^m$ be continuously differentiable on V. If E is a convex subset of H, then there is a constant M (which depends on H and f but not on E) such that*

$$\|f(\boldsymbol{x}) - f(\boldsymbol{a})\| \le M \|\boldsymbol{x} - \boldsymbol{a}\|$$

for all $\boldsymbol{x}, \boldsymbol{a} \in E$.

PROOF. Since H is compact and the entries of Df are continuous on H, it is clear that

$$M := \sqrt{mn} \sup_{\boldsymbol{c} \in H} \|Df(\boldsymbol{c})\|_\infty$$

is finite and depends only on H and f.

Let $\boldsymbol{x}, \boldsymbol{a} \in E$ and $\boldsymbol{u} = f(\boldsymbol{x}) - f(\boldsymbol{a})$. Since E is convex, $L(\boldsymbol{x}; \boldsymbol{a}) \subseteq E$. Hence by Theorem 6.14, there is a $\boldsymbol{c} \in L(\boldsymbol{x}; \boldsymbol{a})$ such that

$$\|f(\boldsymbol{x}) - f(\boldsymbol{a})\|^2 = \boldsymbol{u} \cdot (f(\boldsymbol{x}) - f(\boldsymbol{a})) = \boldsymbol{u} \cdot (Df(\boldsymbol{c})(\boldsymbol{x} - \boldsymbol{a})) = (f(\boldsymbol{x}) - f(\boldsymbol{a})) \cdot (Df(\boldsymbol{c})(\boldsymbol{x} - \boldsymbol{a})).$$

It follows from the Cauchy–Schwarz Inequality and Theorem 5.5 that

$$\|f(\boldsymbol{x}) - f(\boldsymbol{a})\|^2 \le \sqrt{mn} \, \|f(\boldsymbol{x}) - f(\boldsymbol{a})\| \, \|Df(\boldsymbol{c})\|_\infty \, \|\boldsymbol{x} - \boldsymbol{a}\|.$$

If $\|f(\boldsymbol{x}) - f(\boldsymbol{a})\| = 0$, there is nothing to prove. Otherwise, we can divide the inequality above by $\|f(\boldsymbol{x}) - f(\boldsymbol{a})\|$ to obtain

$$\|f(\boldsymbol{x}) - f(\boldsymbol{a})\| \le \sqrt{mn} \|Df(\boldsymbol{c})\|_\infty \, \|\boldsymbol{x} - \boldsymbol{a}\| \le M \|\boldsymbol{x} - \boldsymbol{a}\|. \quad \blacksquare$$

*COROLLARY 6.17. *Suppose V is an open connected set in $\mathbf{R}^n$ and $f : V \to \mathbf{R}^m$ is differentiable on V. If $Df(\boldsymbol{c}) = O$ for all $\boldsymbol{c} \in V$, then f is constant on V.*

PROOF. Fix $\boldsymbol{a} \in V$ and let $\boldsymbol{x} \in V$. Since V is open and connected, V is polygonally connected (see Exercise 8 in Section 5.6). Thus, there exist points $\boldsymbol{x}_0 = \boldsymbol{a}, \boldsymbol{x}_1, \ldots, \boldsymbol{x}_k = \boldsymbol{x}$ such that $L(\boldsymbol{x}_{j-1}; \boldsymbol{x}_j) \subset V$ for $j = 1, 2, \ldots, k$ (see Figure 6.1). Let $\boldsymbol{u} = f(\boldsymbol{x}) - f(\boldsymbol{a})$ and choose by Theorem 6.14 points $\boldsymbol{c}_j \in L(\boldsymbol{x}_{j-1}; \boldsymbol{x}_j)$ such that

$$\boldsymbol{u} \cdot (f(\boldsymbol{x}_j) - f(\boldsymbol{x}_{j-1})) = \boldsymbol{u} \cdot (Df(\boldsymbol{c}_j)(\boldsymbol{x}_j - \boldsymbol{x}_{j-1})) = 0$$

for $j = 1, 2, \ldots, k$. Summing over j and telescoping, we see by the choice of $\boldsymbol{u}$ that

$$0 = \sum_{j=1}^{k} \boldsymbol{u} \cdot (f(\boldsymbol{x}_j) - f(\boldsymbol{x}_{j-1})) = \boldsymbol{u} \cdot (f(\boldsymbol{x}) - f(\boldsymbol{a})) = \|f(\boldsymbol{x}) - f(\boldsymbol{a})\|^2.$$

Therefore, $f(\boldsymbol{x}) = f(\boldsymbol{a})$. $\blacksquare$

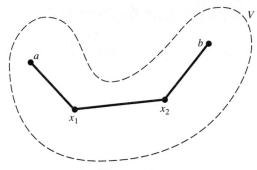

Figure 6.1

To obtain a multidimensional version of Taylor's Formula, we need to define higher-order differentials. Let $p \geq 1$, V be open in $\mathbf{R}^n$, $\boldsymbol{a} \in V$, and $f : V \to \mathbf{R}$. If the $(p-1)$-st order partial derivatives of f exist on V and are differentiable at $\boldsymbol{a}$, then the pth *total differential* of f at $\boldsymbol{a}$ and $\boldsymbol{h} = (h_1, \ldots, h_n)$ is the expression

$$D^{(p)} f(\boldsymbol{a}; \boldsymbol{h}) = \sum_{i_1=1}^{n} \cdots \sum_{i_p=1}^{n} \frac{\partial^p f}{\partial x_{i_1} \ldots \partial x_{i_p}}(\boldsymbol{a}) h_{i_1} \ldots h_{i_p}.$$

Notice that

$$D^{(p)} f(\boldsymbol{a}; \boldsymbol{h}) = D^{(1)}(D^{(p-1)} f)(\boldsymbol{a}; \boldsymbol{h})$$

$$= \sum_{j=1}^{n} \frac{\partial}{\partial x_j} \left(\sum_{i_1=1}^{n} \cdots \sum_{i_{p-1}=1}^{n} \frac{\partial^{p-1} f}{\partial x_{i_1} \ldots \partial x_{i_{p-1}}}(\boldsymbol{a}) h_{i_1} \ldots h_{i_{p-1}} \right) h_j$$

for $p > 1$.

The first total differential of f is closely related to the total derivative of f, since by definition,

$$D^{(1)} f(\boldsymbol{a}; \boldsymbol{h}) := \sum_{j=1}^{n} \frac{\partial f}{\partial x_j}(\boldsymbol{a}) h_j = \nabla f(\boldsymbol{a}) \cdot \boldsymbol{h} = Df(\boldsymbol{a})(\boldsymbol{h}).$$

For the case $n = 2$, this differential has a simple geometric interpretation (see Figure 6.3 below).

Although total differentials look messy to evaluate, when f is a sufficiently smooth function of two variables they are relatively easy to calculate using binomial coefficients (see the next example and Exercise 4 below).

Example 1. Suppose $f : V \to \mathbf{R}$ is $\mathcal{C}^2$ on V. Find a formula for the second total differential of f at (a, b).

SOLUTION. By definition,

$$D^{(2)} f((a, b); (h, k)) = h^2 \frac{\partial^2 f}{\partial x^2}(a, b) + hk \frac{\partial^2 f}{\partial x \partial y}(a, b) + hk \frac{\partial^2 f}{\partial y \partial x}(a, b) + k^2 \frac{\partial^2 f}{\partial y^2}(a, b).$$

But by Theorem 6.1, $f_{xy}(a, b) = f_{yx}(a, b)$. Therefore,

$$D^{(2)} f((a, b); (h, k)) = h^2 \frac{\partial^2 f}{\partial x^2}(a, b) + 2hk \frac{\partial^2 f}{\partial x \partial y}(a, b) + k^2 \frac{\partial^2 f}{\partial y^2}(a, b). \quad \blacksquare$$

Thus, the second total differential of $f(x, y) = (xy)^2$ is $D^{(2)} f((x, y); (h, k)) = 2y^2 h^2 + 8xyhk + 2x^2 k^2$.

THEOREM 6.18 [TAYLOR'S FORMULA]. *Let* $p \in \mathbf{N}$, *V be an open set in* $\mathbf{R}^n$, *$\boldsymbol{x}, \boldsymbol{a} \in V$, $f : V \to \mathbf{R}$, and suppose that the partial derivatives of f order $p - 1$ exist on V. If each $(p - 1)$-st order partial derivative of f is differentiable on V and $L(\boldsymbol{x}; \boldsymbol{a}) \subset V$, then there is a point $\boldsymbol{c} \in L(\boldsymbol{x}; \boldsymbol{a})$ such that*

$$f(\boldsymbol{x}) = f(\boldsymbol{a}) + \sum_{k=1}^{p-1} \frac{1}{k!} D^{(k)} f(\boldsymbol{a}; \boldsymbol{h}) + \frac{1}{p!} D^{(p)} f(\boldsymbol{c}, \boldsymbol{h})$$

for $\boldsymbol{h} := \boldsymbol{x} - \boldsymbol{a}$.

PROOF. Let $\boldsymbol{h} = \boldsymbol{x} - \boldsymbol{a}$. As in the proof of Theorem 6.14, choose $\delta > 0$ so small that $\boldsymbol{a} + t\boldsymbol{h} \subset V$ for $t \in I_\delta := (-\delta, 1 + \delta)$. The function $F(t) = f(\boldsymbol{a} + t\boldsymbol{h})$ is differentiable on I_δ. In fact, by the Chain Rule,

$$F'(t) = Df(\boldsymbol{a} + t\boldsymbol{h})(\boldsymbol{h}) = \sum_{k=1}^{n} \frac{\partial f}{\partial x_k}(\boldsymbol{a} + t\boldsymbol{h}) \, h_k.$$

An easy proof by induction shows that

$$F^{(j)}(t) = \sum_{i_1=1}^{n} \cdots \sum_{i_j=1}^{n} \frac{\partial^j f}{\partial x_{i_1} \cdots \partial x_{i_j}}(\boldsymbol{a} + t\boldsymbol{h}) \, h_{i_1} \cdots h_{i_j}$$

for $j = 1, 2, \ldots, p$. Thus,

(19) $$F^{(j)}(0) = D^{(j)} f(\boldsymbol{a}; \boldsymbol{h}) \quad \text{and} \quad F^{(p)}(t) = D^{(p)} f(\boldsymbol{a} + t\boldsymbol{h}; \boldsymbol{h})$$

for $j = 1, \ldots, p - 1$ and $t \in I_\delta$.

We have proved that $F : I_\delta \to \mathbf{R}$ has a derivative of order p everywhere on $I_\delta \supset [0, 1]$. Therefore, by the one-dimensional Taylor Formula and (19),

$$f(\boldsymbol{x}) - f(\boldsymbol{a}) = F(1) - F(0) = \sum_{j=1}^{p-1} \frac{1}{j!} F^{(j)}(0) + \frac{1}{p!} F^{(p)}(t)$$

$$= \sum_{j=1}^{p-1} \frac{1}{j!} D^{(j)} f(\boldsymbol{a}; \boldsymbol{h}) + \frac{1}{p!} D^{(p)} f(\boldsymbol{a} + t\boldsymbol{h}; \boldsymbol{h})$$

for some $t \in (0, 1)$. Thus, set $\boldsymbol{c} = \boldsymbol{a} + t\boldsymbol{h}$. $\blacksquare$

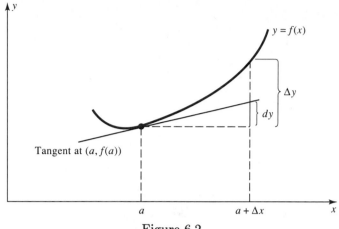

Figure 6.2

Applied mathematicians and physicists use a different notation for the first total differential. Suppose V is open in $\mathbf{R}^n$, $\boldsymbol{a} \in V$, $f : V \to \mathbf{R}$ is continuously differentiable at $\boldsymbol{a}$, and $w = f(\boldsymbol{x})$. If $\boldsymbol{dx} := (dx_1, \ldots, dx_n)$, where the dx_j's are small real numbers, then the first total differential $D^{(1)}f(\boldsymbol{a}, \boldsymbol{dx})$ is usually denoted by dw. In particular, if $y = f(x)$ then $dy = f'(x)\, dx$ and if $z = f(x, y)$, then $dz = f_x\, dx + f_y\, dy$.

Let $f : \mathbf{R} \to \mathbf{R}$ be differentiable. Recall from single-variable calculus that the change in $y = f(x)$ as x moves from a to $a + \Delta x$ is defined by $\Delta y = f(a + \Delta x) - f(a)$, and if $dx = \Delta x$ is a small real number, then the total differential dy approximates Δy (see Figure 6.2). If $z = f(x, y)$, does the total differential dz play an analogous geometric role in $\mathbf{R}^3$?

To answer this question, suppose f is differentiable at (a, b) and set $\Delta z = f(a + \Delta x, b + \Delta y) - f(a, b)$. Then the picture corresponding to Figure 6.2 involves a tangent plane and a wedge-shaped region (see Figure 6.3). Namely, let $z_0 = f(a, b)$ and consider the wedge-shaped region $\mathcal{W}$ with vertical sides parallel to the xz and yz planes whose base has vertices $\boldsymbol{c}_0 := (a, b, z_0)$, $\boldsymbol{c}_1 := (a + \Delta x, b, z_0)$, $\boldsymbol{c}_2 := (a, b + \Delta y, z_0)$, $\boldsymbol{c}_3 := (a + \Delta x, b + \Delta y, z_0)$, and whose top is tangent to $z = f(x, y)$ at $\boldsymbol{c}_0$. Let A represent the length of the vertical edge of $\mathcal{W}$ based at $\boldsymbol{c}_1$, B the length of the edge based at $\boldsymbol{c}_2$, and C the length of the edge based at $\boldsymbol{c}_3$. If dz is to play the same role in Figure 6.3 that dy plays in Figure 6.2, then it must be the case that $C = dz$. This is actually easy to verify. Since the diagonals of rectangles bisect one another, the line segment from the intersection of the diagonals in the base of $\mathcal{W}$ to the intersection of the diagonals in the top of $\mathcal{W}$ must be parallel to the z axis. Thus, the length D of this line segment can be computed two ways. On the one hand, $D = C/2$. On the other hand, $D = (A + B)/2$. Therefore, $C = A + B$. But from one-dimensional calculus, $A = f_x(a, b)\, dx$ and $B = f_y(a, b)\, dy$. Consequently,

$$C = A + B = \frac{\partial f}{\partial x}(a, b)\, dx + \frac{\partial f}{\partial y}(a, b)\, dy = dz.$$

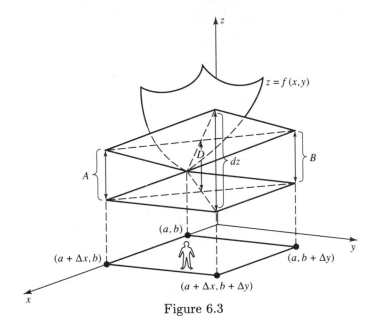

Figure 6.3

EXERCISES

1. Let $f : \mathbf{R}^n \to \mathbf{R}$. Suppose that for each unit vector $\boldsymbol{u} \in \mathbf{R}^n$, the directional derivative $D_{\boldsymbol{u}}f(\boldsymbol{a} + t\boldsymbol{u})$ exists for $t \in [0, 1]$ (see Exercise 3 in Section 6.2). Prove

$$f(\boldsymbol{a} + \boldsymbol{u}) - f(\boldsymbol{a}) = D_{\boldsymbol{u}}f(\boldsymbol{a} + t\boldsymbol{u})$$

   for some $t \in (0, 1)$.

2. Suppose r and α are positive numbers, E is a convex subset of $\mathbf{R}^n$, $\boldsymbol{0}$ is a cluster point of E, and $\overline{E} \subset B_r(\boldsymbol{0})$. If $f : B_r(\boldsymbol{0}) \to \mathbf{R}$ satisfies $|f(\boldsymbol{x})| \le \|\boldsymbol{x}\|^\alpha$ for all $\boldsymbol{x} \in E$ and f is continuously differentiable on $B_r(\boldsymbol{0})$, prove there is an $M > 0$ such that $|f(\boldsymbol{x})| \le M \|\boldsymbol{x}\|$ for all $\boldsymbol{x} \in E$.

3. a) Write out an expression in powers of $(x + 1)$ and $(y - 1)$ for $f(x, y) = x^2 + xy + y^2$.
   b) Write Taylor's Formula for $f(x, y) = \sqrt{x} + \sqrt{y}$, $\boldsymbol{a} = (1, 4)$, and $p = 3$.
   c) Write Taylor's Formula for $f(x, y) = e^{xy}$, $\boldsymbol{a} = (0, 0)$, and $p = 4$.

4. Suppose $f : \mathbf{R}^2 \to \mathbf{R}$ is $\mathcal{C}^p$ on $B_r(x_0, y_0)$ for some $r > 0$. Prove that given $(x, y) \in B_r(x_0, y_0)$, there is a point (c, d) on the line segment between (x_0, y_0) and (x, y) such that

$$f(x, y) = f(x_0, y_0)$$
$$+ \sum_{k=1}^{p-1} \frac{1}{k!} \left(\sum_{j=0}^{k} \binom{k}{j} (x - x_0)^j (y - y_0)^{k-j} \frac{\partial^k f}{\partial x^j \partial y^{k-j}} (x_0, y_0) \right)$$
$$+ \frac{1}{p!} \sum_{j=0}^{p} \binom{p}{j} (x - x_0)^j (y - y_0)^{p-j} \frac{\partial^p f}{\partial x^j \partial y^{p-j}} (c, d).$$

5. Let $r > 0$, $a, b \in \mathbf{R}$, $f : B_r(a, b) \to \mathbf{R}$ be differentiable, and $(x, y) \in B_r(a, b)$.

   a) Let $g(t) = f(tx + (1 - t)a, y) + f(a, ty + (1 - t)b)$ and compute the derivative of g.

   b) Prove there are numbers c between a and x and d between b and y such that

   $$f(x, y) - f(a, b) = (x - a)f_x(c, y) + (y - b)f_y(a, d).$$

   (This is Exercise 12.20 in Apostol [1].)

6. [INTEGRAL FORM OF TAYLOR'S FORMULA]. Let $p \in \mathbf{N}$, V be an open set in $\mathbf{R}^n$, $\boldsymbol{x}, \boldsymbol{a} \in V$, and $f : V \to \mathbf{R}$ be $\mathcal{C}^p$ on V. If $L(\boldsymbol{x}; \boldsymbol{a}) \subset V$ and $\boldsymbol{h} = \boldsymbol{x} - \boldsymbol{a}$, prove

   $$f(\boldsymbol{x}) - f(\boldsymbol{a}) = \sum_{k=1}^{p-1} \frac{1}{k!} D^{(k)} f(\boldsymbol{a}; \boldsymbol{h}) + \frac{1}{(p-1)!} \int_0^1 (1 - t)^{p-1} D^{(p)} f(\boldsymbol{a} + t\boldsymbol{h}; \boldsymbol{h}) \, dt.$$

7. Suppose V is open in $\mathbf{R}^n$, $f : V \to \mathbf{R}$ is $\mathcal{C}^2$ on V, and $f_{x_j}(\boldsymbol{a}) = 0$ for some $\boldsymbol{a} \in V$ and all $j = 1, \ldots, n$. Prove that if H is a compact convex subset of V, then there is a constant M such that

   $$|f(\boldsymbol{x}) - f(\boldsymbol{a})| \le M \|\boldsymbol{x} - \boldsymbol{a}\|^2$$

   for all $\boldsymbol{x} \in H$.

8. Suppose V is an open subset of $\mathbf{R}^2$, $(a, b) \in V$, and $f : V \to \mathbf{R}$ is $\mathcal{C}^3$ on V. Prove

   $$\lim_{r \to 0} \frac{4}{\pi r^2} \int_0^{2\pi} f(a + r \cos \theta, b + r \sin \theta) \cos(2\theta) \, d\theta = f_{xx}(a, b) - f_{yy}(a, b).$$

9. Suppose V is an open subset of $\mathbf{R}^2$, $H = [a, b] \times [0, c] \subset V$, $u : V \to \mathbf{R}$ is $\mathcal{C}^2$ on V, and $u(x_0, t_0) \ge 0$ for all $(x_0, t_0) \in \partial H$.

   a) Show that given $\varepsilon > 0$, there is a compact set $K \subset H^\circ$ such that $u(x, t) \ge -\varepsilon$ for all $(x, t) \in H \setminus K$.

   b) Suppose $u(x_1, t_1) = -\ell < 0$ for some $(x_1, t_1) \in H^\circ$ and choose $r > 0$ so small that $2rt_1 < \ell$. Apply part a) to $\varepsilon := \ell/2 - rt_1$ to choose the compact set K, and prove that the minimum of

   $$w(x, t) := u(x, t) + r(t - t_1)$$

   on H occurs at some $(x_2, t_2) \in K$.

   c) Prove that if u satisfies the *heat equation*, i.e., $u_{xx} - u_t = 0$ on V, and if $u(x_0, t_0) \ge 0$ for all $(x_0, t_0) \in \partial H$, then $u(x, t) \ge 0$ for all $(x, t) \in H$.

6.5 THE INVERSE FUNCTION THEOREM

By Theorem 2.22, if $g : \mathbf{R} \to \mathbf{R}$ is 1–1 and differentiable with $g'(x_0) \neq 0$, then g^{-1} is differentiable at $y_0 = g(x_0)$ and

$$(g^{-1})'(y_0) = \frac{1}{g'(x_0)}.$$

In this section we prove a multivariable analogue of this result, i.e., an Inverse Function Theorem for functions $f : \mathbf{R}^n \to \mathbf{R}^n$ which are differentiable at a point $\boldsymbol{a}$. Since, in the analogy between real numbers and matrices, reciprocals correspond to matrix inverses (see Appendix C), we shall replace the reciprocal $1/g'(x_0)$ by the matrix inverse $[Df(\boldsymbol{a})]^{-1}$. Recall that a matrix is invertible if and only if its determinant is nonzero. Thus, $[Df(\boldsymbol{a})]^{-1}$ exists if and only if the Jacobian $\Delta_f(\boldsymbol{a})$ is nonzero.

The proof of this theorem lies somewhat deeper than the previous results of this chapter and we precede it by three preliminary results which explore the consequences of a nonzero Jacobian. Our first preliminary result shows that if f is 1–1 and its Jacobian is nonzero, then f takes open sets to open sets.

Lemma 1. *Let V be open in $\mathbf{R}^n$, $f : V \to \mathbf{R}^n$, $\boldsymbol{a} \in V$, and $r > 0$ be so small that $\overline{B_r(\boldsymbol{a})} \subset V$. Suppose f is continuous and 1–1 on $\overline{B_r(\boldsymbol{a})}$, and its first-order partial derivatives exist at every point in $B_r(\boldsymbol{a})$. If $\Delta_f \neq 0$ on $B_r(\boldsymbol{a})$, then there is a $\rho > 0$ such that $B_\rho(f(\boldsymbol{a})) \subset f(B_r(\boldsymbol{a}))$.*

STRATEGY: The idea behind this proof is simple. Let $y \in B_\rho(f(\boldsymbol{a}))$ where ρ is to be determined later. To show $B_\rho(f(\boldsymbol{a})) \subset f(B_r(\boldsymbol{a}))$ we must show that $\boldsymbol{y} = f(\boldsymbol{b})$ for some $\boldsymbol{b} \in B_r(\boldsymbol{a})$, i.e., $\boldsymbol{y} - f(\boldsymbol{a}) = \boldsymbol{0}$. This leads us to believe that $\boldsymbol{b} \in B_r(\boldsymbol{a}))$ must be chosen so that $f(\boldsymbol{b})$ is closest to $\boldsymbol{y}$. The only problem with this strategy is that since $B_r(\boldsymbol{a})$ is not compact, we cannot assume such a closest $f(\boldsymbol{b})$ exists. No problem. Choose $\boldsymbol{b}$ in the compact set $\overline{B_r(\boldsymbol{a})}$ so that $f(\boldsymbol{b})$ is closest to $\boldsymbol{y}$ and let $\rho < m$, where m is the minimal distance from $f(\partial B_r(\boldsymbol{a}))$ to $f(\boldsymbol{a})$. Since $\rho < m$, the chosen $\boldsymbol{b}$ cannot belong to $\partial B_r(\boldsymbol{a})$. Thus, $\boldsymbol{b} \in B_r(\boldsymbol{a})$ as required. Here are the details.

PROOF. Let

$$g(\boldsymbol{x}) = \|f(\boldsymbol{x}) - f(\boldsymbol{a})\|, \qquad \boldsymbol{x} \in \overline{B_r(\boldsymbol{a})}.$$

By hypothesis $g : \overline{B_r(\boldsymbol{a})} \to \mathbf{R}$ is continuous. Since f is 1–1, $g(\boldsymbol{x}) > 0$ for all $\boldsymbol{x} \neq \boldsymbol{a}$. Since $\partial B_r(\boldsymbol{a})$ is compact, it follows that

$$m = \inf_{\boldsymbol{x} \in \partial B_r(\boldsymbol{a})} g(\boldsymbol{x}) > 0.$$

Set $\rho = m/2$ and fix $\boldsymbol{y} \in B_\rho(f(\boldsymbol{a}))$. Since the function

$$h(\boldsymbol{x}) := \|f(\boldsymbol{x}) - \boldsymbol{y}\|, \qquad \boldsymbol{x} \in \overline{B_r(\boldsymbol{a})}$$

is continuous on the compact set $\overline{B_r(\boldsymbol{a})}$, it attains its minimum there. Thus, there is a $\boldsymbol{b} \in \overline{B_r(\boldsymbol{a})}$ such that $h(\boldsymbol{b}) \leq h(\boldsymbol{x})$ for all $\boldsymbol{x} \in \overline{B_r(\boldsymbol{a})}$.

To show $\boldsymbol{b} \in B_r(\boldsymbol{a})$, suppose to the contrary that $\boldsymbol{b} \notin B_r(\boldsymbol{a})$. Then $\boldsymbol{b} \in \partial B_r(\boldsymbol{a})$. Since $h(\boldsymbol{a}) = \|f(\boldsymbol{a}) - \boldsymbol{y}\| < \rho$, the minimum $h(\boldsymbol{b})$ must also satisfy $h(\boldsymbol{b}) < \rho$. Since $\boldsymbol{b} \in \partial B_r(\boldsymbol{a})$, it follows from the Triangle Inequality and the choice of ρ that

$$\rho > h(\boldsymbol{b}) = \|f(\boldsymbol{b}) - \boldsymbol{y}\| \geq \|f(\boldsymbol{b}) - f(\boldsymbol{a})\| - \|f(\boldsymbol{a}) - \boldsymbol{y}\| = g(\boldsymbol{b}) - h(\boldsymbol{a}) > 2\rho - \rho = \rho,$$

a contradiction.

It remains to prove $\boldsymbol{y} = f(\boldsymbol{b})$. Notice that since $h(\boldsymbol{b}) \geq 0$, $h^2(\boldsymbol{b})$ is the minimum of h^2 on $\overline{B_r(\boldsymbol{a})}$. Thus, by one-dimensional calculus,

$$\frac{\partial h^2}{\partial x_k}(\boldsymbol{b}) = 0$$

for $k = 1, \ldots, n$. Since $h^2(\boldsymbol{x}) = \sum_{j=1}^{n}(f_j(\boldsymbol{x}) - y_j)^2$, it follows that

$$0 = \frac{\partial h^2}{\partial x_k}(\boldsymbol{b}) = \sum_{j=1}^{n} 2(f_j(\boldsymbol{b}) - y_j)\frac{\partial f_j}{\partial x_k}(\boldsymbol{b}).$$

This is a system of n linear equations in n unknowns $f_j(\boldsymbol{b}) - y_j$. Since the matrix of coefficients of this system has determinant $2^n \Delta_f(\boldsymbol{b}) \neq 0$, it follows from Cramer's Rule (see Appendix C) that this system has only the trivial solution, i.e., $f_j(\boldsymbol{b}) - y_j = 0$ for all $j = 1, \ldots, n$. In particular, $\boldsymbol{y} = f(\boldsymbol{b})$ and $\boldsymbol{y} \in f(B_r(\boldsymbol{a}))$. ∎

Next, we show that f^{-1} is continuous when f is 1–1 and Δ_f is nonzero.

THEOREM 6.19. *Let V be an open set in $\mathbf{R}^n$ and $f : V \to \mathbf{R}^n$ be continuous. If f is 1–1 and has first-order partial derivatives on V, and if $\Delta_f \neq 0$ on V, then f^{-1} is continuous on $f(V)$.*

PROOF. By Theorem 5.25 (applied to f^{-1}), we must show that $f(W)$ is open in $\mathbf{R}^n$ for every open $W \subseteq V$ in $\mathbf{R}^n$. Let $\boldsymbol{b} \in f(W)$, i.e., $\boldsymbol{b} = f(\boldsymbol{a})$ for some $\boldsymbol{a} \in W$. Since W is open, choose $q > 0$ such that $B_q(\boldsymbol{a}) \subset W$. Fix $0 < r < q$ and notice that $\overline{B_r(\boldsymbol{a})} \subset W$. Since f is 1–1 on $V \supseteq W$, apply Lemma 1 to choose $\rho > 0$ such that

$$B_\rho(\boldsymbol{b}) = B_\rho(f(\boldsymbol{a})) \subset f(B_r(\boldsymbol{a})).$$

Since $f(B_r(\boldsymbol{a})) \subset f(W)$, this proves $f(W)$ is open. ∎

Our final preliminary result shows that if the Jacobian of a continuously differentiable function f is nonzero at a point, then f must be 1–1 near that point. (This will provide a key step in the proof of the Inverse Function Theorem below.)

Lemma 2. *Let V be open in $\mathbf{R}^n$ and $f : V \to \mathbf{R}^n$ be continuously differentiable on V. If $\Delta_f(\boldsymbol{a}) \neq 0$ for some $\boldsymbol{a} \in V$, then there is an $r > 0$ such that $B_r(\boldsymbol{a}) \subset V$, f is 1–1 on $B_r(\boldsymbol{a})$, $\Delta_f(\boldsymbol{x}) \neq 0$ for all $\boldsymbol{x} \in B_r(\boldsymbol{a})$, and*

$$\det \left[\frac{\partial f_i}{\partial x_j}(\boldsymbol{c}_i) \right]_{n \times n} \neq 0$$

for all $c_1, \ldots c_n \in B_r(a)$.

STRATEGY: The idea behind the proof is simple. If f is not 1–1 on some $B_r(a)$, then there exist $x, y \in B_r(a)$ such that $x \neq y$ and $f(x) = f(y)$. Since $L(x;y) \subset B_r(a)$, we have by Corollary 6.15 (the Mean Value Theorem) that

$$(20) \qquad 0 = f_i(y) - f_i(x) = \sum_{k=1}^{n} \frac{\partial f_i}{\partial x_k}(c_i)(y_k - x_k)$$

for $x = (x_1, \ldots, x_n)$, $y = (y_1, \ldots, y_n)$, $c_i \in L(x;y)$, and $i = 1, \ldots, n$. Notice that (20) is a system of n linear equations in n unknowns ($y_k - x_k$). If we can show for sufficiently small r that the matrix of coefficients of (20) has nonzero determinant for any choice of $c_i \in B_r(a)$, then by Cramer's Rule the linear system (20) has only one solution: $y_k - x_k = 0$ for $k = 1, \ldots, n$. This would imply $x = y$, a contradiction. Here are the details.

PROOF. To show there is an $r > 0$ such that the matrix of coefficients of the linear system (20) is nonzero for all $c_i \in B_r(a)$, define $h : \mathbf{R}^{n^2} \to \mathbf{R}$ by

$$h(x_1, x_2, \ldots, x_n) = \det \left[\frac{\partial f_i}{\partial x_j}(x_i) \right]_{n \times n}.$$

Since the determinant of a matrix is defined using products and differences of its entries (see Appendix C), we have by hypothesis that h is continuous on $V^{(n)} = V \times \cdots \times V$, the n-fold Cartesian product of V with itself. Since $h(a, \ldots, a) = \Delta_f(a) \neq 0$, it follows that there is an $r > 0$ such that $B_r(a) \subset V$ and $h(c_1, \ldots, c_n) \neq 0$ for $c_i \in B_r(a)$. In particular, the matrix of coefficients of the linear system (20) is nonzero for all $c_i \in B_r(a)$, and $\Delta_f(x) = h(x, \ldots, x) \neq 0$ for all $x \in B_r(a)$. ∎

We now prove a multidimensional version of the Inverse Function Theorem.

THEOREM 6.20 [THE INVERSE FUNCTION THEOREM]. *Let V be open in $\mathbf{R}^n$, $f : V \to \mathbf{R}^n$ be continuously differentiable on V, and $W = f(V)$. If $\Delta_f(a) \neq 0$ for some $a \in V$, then there exist open sets $V_0 \subset V$ and $W_0 \subset W$ such that*

  i) *$a \in V_0$, $f(a) \in W_0$,*
  ii) *f is 1–1 from V_0 onto W_0 and f^{-1} is 1–1 from W_0 onto V_0,*
  iii) *f^{-1} is continuously differentiable on W_0, and*
  iv) *for each $y = f(x) \in W_0$,*

$$D(f^{-1})(y) = [Df(x)]^{-1},$$

where $[\ ]^{-1}$ represents matrix inversion.

PROOF. By Lemma 2, there is an open ball B centered at a such that f is 1–1 and $\Delta_f \neq 0$ on B, and

$$\Delta := \det \left[\frac{\partial f_i}{\partial x_j}(c_i) \right]_{n \times n} \neq 0$$

for all $\boldsymbol{c}_i \in B$. By Theorem 6.19, f^{-1} is continuous on $f(B)$. Let B_0 be an open ball centered at $\boldsymbol{a}$ which is smaller than B; i.e., the radius of B_0 is strictly less than the radius of B. Then $\overline{B}_0 \subset B$ and by Lemma 1, there is an open ball B_1 centered at $f(\boldsymbol{a})$ such that $B_1 \subset f(B_0)$.

Fix $\boldsymbol{y}_0 \in B_1$ and $1 \le i, k \le n$. Choose $t \in \mathbf{R}$ so small that $\boldsymbol{y}_0 + t\boldsymbol{e}_k \in B_1$ and set

$$\boldsymbol{x}_0 = f^{-1}(\boldsymbol{y}_0), \qquad \boldsymbol{x}_1 = f^{-1}(\boldsymbol{y}_0 + t\boldsymbol{e}_k).$$

By Corollary 6.15 (the Mean Value Theorem), there are points $\boldsymbol{c}_i \in L(\boldsymbol{x}_0; \boldsymbol{x}_1)$ such that

$$(21) \qquad \nabla f_i(\boldsymbol{c}_i) \cdot \frac{\boldsymbol{x}_1 - \boldsymbol{x}_0}{t} = \frac{f_i(\boldsymbol{x}_1) - f_i(\boldsymbol{x}_0)}{t}, \qquad i = 1, 2, \dots, n.$$

Let $x_0(j)$ (respectively, $x_1(j)$) denote the jth component of $\boldsymbol{x}_0$ (respectively, $\boldsymbol{x}_1$). Since (21) is a system of n linear equations in n variables $(x_1(j) - x_0(j))/t$ whose coefficient matrix has determinant Δ (which is nonzero by the choice of B), we see by Cramer's Rule that the solutions of (21) satisfy

$$(22) \qquad \frac{f_j^{-1}(\boldsymbol{y}_0 + t\boldsymbol{e}_k) - f_j^{-1}(\boldsymbol{y}_0)}{t} := \frac{x_1(j) - x_0(j)}{t} = \mathcal{Q}_j(t),$$

where $\mathcal{Q}_j(t)$ is a quotient of determinants whose entries are components of f evaluated at $\boldsymbol{x}_1$ and $\boldsymbol{x}_0$, or first-order partial derivatives of components of f evaluated at the $\boldsymbol{c}_i$'s. Since $t \to 0$ implies $\boldsymbol{x}_1 \to \boldsymbol{x}_0$, $\boldsymbol{c}_i \to \boldsymbol{x}_0$, and $\boldsymbol{y}_0 + t\boldsymbol{e}_k \to \boldsymbol{y}_0$, $\mathcal{Q}_j(t)$ converges to $\mathcal{Q}_j$, a quotient of determinants whose entries are components of f or first-order partial derivatives of components of f, all evaluated at $\boldsymbol{x}_0 = f^{-1}(\boldsymbol{y}_0)$. Since f^{-1} is continuous on $f(B)$, it follows that $\mathcal{Q}_j$ is continuous at each $\boldsymbol{y}_0 \in B_1$. Taking the limit of (22) as $t \to 0$, we see that the first-order partial derivatives of f_j^{-1} exist at $\boldsymbol{y}_0$ and equal $\mathcal{Q}_j$, i.e., f^{-1} is continuously differentiable on B_1.

Let W_0 be an open ball centered at $f(\boldsymbol{a})$ which is smaller than B_1 and set

$$V_0 = f^{-1}(W_0) \cap B_0.$$

Clearly, i), ii), and iii) hold. Moreover, for each $\boldsymbol{y} \in W_0$ we have by the Chain Rule and Theorem 6.11 that

$$I = DI(\boldsymbol{y}) = D(f \circ f^{-1})(\boldsymbol{y}) = Df(f^{-1}(\boldsymbol{y}))Df^{-1}(\boldsymbol{y}).$$

By the uniqueness of matrix inverses, we conclude that

$$D(f^{-1})(\boldsymbol{y}) = [Df(f^{-1}(\boldsymbol{y}))]^{-1}. \quad \blacksquare$$

Remark 1. *The hypothesis "$\Delta_f \ne 0$" in Theorem 6.19 can be relaxed.*

PROOF. If $f(x) = x^3$, then $f : \mathbf{R} \to \mathbf{R}$ and its inverse $f^{-1}(x) = \sqrt[3]{x}$ are continuous on $\mathbf{R}$, but $\Delta_f(0) = f'(0) = 0$. $\blacksquare$

Remark 2. *The hypothesis "$\Delta_f \neq 0$" in Theorem 6.20 cannot be relaxed. In fact, if $f : B_r(\mathbf{a}) \to \mathbf{R}^n$ is differentiable at $\mathbf{a}$ and its inverse f^{-1} exists and is differentiable at $f(\mathbf{a})$, then $\Delta_f(\mathbf{a}) \neq 0$.*

PROOF. If f were such a function and $\Delta_f(\mathbf{a}) = 0$, then by Theorem 6.11 and the Chain Rule,

$$I = D(f^{-1} \circ f)(\mathbf{a}) = D(f^{-1})(f(\mathbf{a}))Df(\mathbf{a}).$$

Taking the determinant of this identity, we have

$$1 = \Delta_{f^{-1}}(f(\mathbf{a}))\Delta_f(\mathbf{a}) = 0,$$

a contradiction. ∎

Remark 3. *The hypothesis "f is continuously differentiable on V" in Theorem 6.20 cannot be relaxed.*

PROOF. If $f(x) = x + 2x^2 \sin(1/x)$, $x \neq 0$, and $f(0) = 0$, then $f : \mathbf{R} \to \mathbf{R}$ is differentiable on $(-1, 1)$ and $f'(0) = 1 \neq 0$. However, since

$$f\left(\frac{2}{(4k-1)\pi}\right) < f\left(\frac{2}{(4k+1)\pi}\right) < f\left(\frac{2}{(4k-3)\pi}\right)$$

for $k \in \mathbf{N}$, f is not 1–1 on any open set V_0 which contains 0. ∎

Although Theorem 6.20 says f must be 1–1 on some subset V_0 of V, it does not say that f is 1–1 on V.

Remark 4. *The set V_0 in Theorem 6.20 is in general a proper subset of V, even when V is connected.*

PROOF. Set $f(x, y) = (x^2 - y^2, xy)$ and $V = \mathbf{R}^2 \setminus \{(0,0)\}$. Then $\Delta_f = 2(x^2 + y^2) \neq 0$ for $(x, y) \in V$ but $f(x, -y) = f(-x, y)$ for all $(x, y) \in \mathbf{R}^2$. Thus, f is not 1–1 on V. ∎

Sometimes functions from $\mathbf{R}^p$ to $\mathbf{R}^n$ are defined implicitly by relations on $\mathbf{R}^{n+p}$.

Example 1. *If $x_0^2 + s_0^2 + t_0^2 = 1$ and $x_0 \neq 0$, prove there exist an $r > 0$ and a function $g(s, t)$, continuously differentiable on $B_r(s_0, t_0)$, such that $x_0 = g(s_0, t_0)$ and*

$$x^2 + s^2 + t^2 = 1$$

for $x = g(s, t)$ and $(s, t) \in B_r(s_0, t_0)$.

PROOF. The identity $x^2 + s^2 + t^2 = 1$ is equivalent to

$$x = \pm\sqrt{1 - s^2 - t^2}.$$

If $x_0 > 0$, set $g(s, t) = \sqrt{1 - s^2 - t^2}$. By the Chain Rule,

$$\frac{\partial g}{\partial s} = \frac{-s}{\sqrt{1 - s^2 - t^2}} \quad \text{and} \quad \frac{\partial g}{\partial t} = \frac{-t}{\sqrt{1 - s^2 - t^2}}.$$

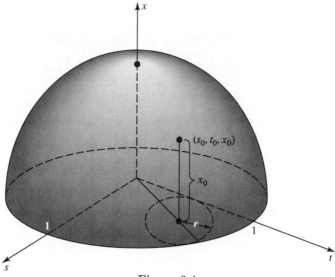

Figure 6.4

Thus, g is differentiable at any point (s, t) which lies inside the two-dimensional unit ball, i.e., which satisfies $s^2 + t^2 < 1$. Since $x_0^2 + s_0^2 + t_0^2 = 1$ and $x_0 > 0$, (s_0, t_0, x_0) lies on the boundary of the three-dimensional unit ball in stx space a distance x_0 units above the st plane (see Figure 6.4). In particular, if $r := 1 - \sqrt{1 - x_0^2}$ and $(s, t) \in B_r(s_0, t_0)$, then $s^2 + t^2 < 1$. Therefore, g is continuously differentiable on $B_r(s_0, t_0)$. If $x_0 < 0$, a similar argument works for $g(s, t) = -\sqrt{1 - s^2 - t^2}$. ∎

Not all relations can be solved explicitly. For situations like this, the following result can prove useful. (In this theorem we use the notation $(\boldsymbol{x}, \boldsymbol{t})$ to represent the vector $(x_1, \dots, x_n, t_1, \dots, t_p)$.)

THEOREM 6.21 [THE IMPLICIT FUNCTION THEOREM]. *Suppose V is open in $\mathbf{R}^{n+p}$ and $F = (F_1, \dots, F_n) : V \to \mathbf{R}^n$ is continuously differentiable on V. Suppose further that $F(\boldsymbol{x}_0, \boldsymbol{t}_0) = \mathbf{0}$ for some $(\boldsymbol{x}_0, \boldsymbol{t}_0) \in V$, where $\boldsymbol{x}_0 \in \mathbf{R}^n$ and $\boldsymbol{t}_0 \in \mathbf{R}^p$. If*

$$\frac{\partial(F_1, \dots, F_n)}{\partial(x_1, \dots, x_n)}(\boldsymbol{x}_0, \boldsymbol{t}_0) \neq 0,$$

then there is an open set $W \subset \mathbf{R}^p$ containing $\boldsymbol{t}_0$ and a unique continuously differentiable function $g : W \to \mathbf{R}^n$ such that $g(\boldsymbol{t}_0) = \boldsymbol{x}_0$, and $F(g(\boldsymbol{t}), \boldsymbol{t}) = \mathbf{0}$ for all $\boldsymbol{t} \in W$.

STRATEGY: The idea behind the proof is simple. If F took its range in $\mathbf{R}^{n+p}$ instead of $\mathbf{R}^n$ and had nonzero Jacobian, then by the Inverse Function Theorem, F^{-1} would exist and be differentiable on some open set. Presumably, the first n components of F^{-1} would solve F for the variables $x_1, \dots, x_n$. Thus, we should extend F (in the simplest possible way) to a function $\widetilde{F}$ which takes its range in $\mathbf{R}^{n+p}$ and has nonzero Jacobian, and apply the Inverse Function Theorem to $\widetilde{F}$. Here are the details.

PROOF. For each $(\boldsymbol{x}, \boldsymbol{t}) \in V$, set

$$(23) \qquad \widetilde{F}(\boldsymbol{x}, \boldsymbol{t}) = (F_1(\boldsymbol{x}, \boldsymbol{t}), \ldots, F_n(\boldsymbol{x}, \boldsymbol{t}), t_1, \ldots, t_p).$$

Clearly, $\widetilde{F} : V \to \mathbf{R}^{n+p}$ and

$$D\widetilde{F} = \begin{bmatrix} \left[\dfrac{\partial F_i}{\partial x_j} \right]_{n \times n} & B \\[2mm] O_{p \times n} & I_{p \times p} \end{bmatrix}$$

where $O_{p \times n}$ represents a zero matrix, $I_{p \times p}$ represents an identity matrix, and B represents a certain $n \times p$ matrix whose entries are first-order partial derivatives of F_j's with respect to t_k's. Expanding the determinant of $D\widetilde{F}$ along the bottom rows first, we see by hypothesis that

$$\Delta_{\widetilde{F}}(\boldsymbol{x}_0, \boldsymbol{t}_0) = 1 \cdot \frac{\partial(F_1, \ldots, F_n)}{\partial(x_1, \ldots, x_n)}(\boldsymbol{x}_0, \boldsymbol{t}_0) \neq 0.$$

Since $\widetilde{F}(\boldsymbol{x}_0, \boldsymbol{t}_0) = (\boldsymbol{0}, \boldsymbol{t}_0)$, it follows from the Inverse Function Theorem that there exist open sets Ω_1 containing $(\boldsymbol{x}_0, \boldsymbol{t}_0)$ and Ω_2 containing $(\boldsymbol{0}, \boldsymbol{t}_0)$ such that $\widetilde{F}$ is 1–1 from Ω_1 onto Ω_2, $G := \widetilde{F}^{-1}$ is 1–1 from Ω_2 onto Ω_1, and G is continuously differentiable on Ω_2.

Let $\phi = (G_1, \ldots, G_n)$. Since $G = \widetilde{F}^{-1}$ is 1–1 from Ω_2 onto Ω_1, it is evident by (23) that

$$(24) \qquad \phi(\widetilde{F}(\boldsymbol{x}, \boldsymbol{t})) = \boldsymbol{x}$$

for all $(\boldsymbol{x}, \boldsymbol{t}) \in \Omega_1$ and

$$(25) \qquad \widetilde{F}(\phi(\boldsymbol{x}, \boldsymbol{t}), \boldsymbol{t}) = (\boldsymbol{x}, \boldsymbol{t})$$

for all $(\boldsymbol{x}, \boldsymbol{t}) \in \Omega_2$. Define g on $W := \{\boldsymbol{t} \in \mathbf{R}^p : (\boldsymbol{0}, \boldsymbol{t}) \in \Omega_2\}$ by $g(\boldsymbol{t}) = \phi(\boldsymbol{0}, \boldsymbol{t})$. Since Ω_2 is open in $\mathbf{R}^{n+p}$, W is open in $\mathbf{R}^p$. Since G is continuously differentiable on Ω_2 and ϕ represents the first n components of G, g is continuously differentiable on W. By (23) and (24) we have

$$g(\boldsymbol{t}_0) = \phi(\boldsymbol{0}, \boldsymbol{t}_0) = \phi(\widetilde{F}(\boldsymbol{x}_0, \boldsymbol{t}_0)) = \boldsymbol{x}_0.$$

Moreover, by (23) and (25) we have $F(\phi(\boldsymbol{x}, \boldsymbol{t}), \boldsymbol{t}) = \boldsymbol{x}$ for all $(\boldsymbol{x}, \boldsymbol{t}) \in \Omega_2$. Specializing to the case $\boldsymbol{x} = \boldsymbol{0}$, we obtain $F(g(\boldsymbol{t}), \boldsymbol{t}) = \boldsymbol{0}$ for $\boldsymbol{t} \in W$.

It remains to show uniqueness. But if $h : W \to \mathbf{R}^n$ satisfies $F(h(\boldsymbol{t}), \boldsymbol{t}) = \boldsymbol{0} = F(g(\boldsymbol{t}), \boldsymbol{t})$, i.e., $\widetilde{F}(h(\boldsymbol{t}), \boldsymbol{t}) = (\boldsymbol{0}, \boldsymbol{t}) = \widetilde{F}(g(\boldsymbol{t}), \boldsymbol{t})$, then $g(\boldsymbol{t}) = h(\boldsymbol{t})$ for all $\boldsymbol{t} \in W$, since $\widetilde{F}$ is 1–1. ∎

Theorem 6.21 is an existence theorem. It states that a solution g exists without giving us any idea how to find it. Fortunately, for many applications it is not as important to be able to write an explicit formula for g as it is to know that g exists.

Here is an example for which an explicit solution is unobtainable.

Example 2. Prove that there is a function $g(s,t)$, continuously differentiable on some $B_r(1,0)$, such that $1 = g(1,0)$, and

$$sx^2 + tx^3 + 2\sqrt{t+s} + t^2 x^4 - x^5 \cos t - x^6 = 1$$

for $x = g(s,t)$ and $(s,t) \in B_r(1,0)$.

PROOF.   If $F(x,s,t) = sx^2 + tx^3 + 2\sqrt{t+s} + t^2 x^4 - x^5 \cos t - x^6 - 1$, then $F(1,1,0) = 0$ and $F_x = 2sx + 3tx^2 + 4t^2 x^3 - 5x^4 \cos t - 6x^5$ is nonzero at the point $(1,1,0)$. Applying the Implicit Function Theorem to F, with $n = 1$, $p = 2$, $x_0 = 1$, and $(s_0, t_0) = (1,0)$, we conclude that such a g exists.   ∎

Even when an explicit solution is obtainable, it is frequently easier to apply the Implicit Function Theorem than it is to solve a relation explicitly for one or more of its variables. Indeed, consider Example 1 again. Let $F(x,s,t) = 1 - x^2 - s^2 - t^2$ and notice that $F_x = -2x$. Thus, by the Implicit Function Theorem, a continuously differentiable solution $x = g(s,t)$ exists for each $x_0 \neq 0$.

The following example shows that the Implicit Function Theorem can be used to show several differentiable solutions exist simultaneously.

Example 3. Prove there exist functions $u, v : \mathbf{R}^4 \to \mathbf{R}$, continuously differentiable on some ball B centered at the point $(x, y, z, w) = (2, 1, -1, -2)$, such that $u(2, 1, -1, -2) = 4$, $v(2, 1, -1, -2) = 3$, and the equations

$$u^2 + v^2 + w^2 = 29, \qquad \frac{u^2}{x^2} + \frac{v^2}{y^2} + \frac{w^2}{z^2} = 17$$

both hold for all (x, y, z, w) in B.

PROOF.   Set $n = 2$, $p = 4$, and

$$F(u, v, x, y, z, w) = (u^2 + v^2 + w^2 - 29, u^2/x^2 + v^2/y^2 + w^2/z^2 - 17).$$

Then $F(4, 3, 2, 1, -1, -2) = (0, 0)$ and

$$\frac{\partial(F_1, F_2)}{\partial(u, v)} = \det \begin{bmatrix} 2u & 2v \\ 2u/x^2 & 2v/y^2 \end{bmatrix} = 4uv \left(\frac{1}{y^2} - \frac{1}{x^2} \right).$$

This determinant is nonzero when $u = 4$, $v = 3$, $x = 2$, and $y = 1$. Therefore, such functions u, v exist by the Implicit Function Theorem.   ∎

EXERCISES

1. Prove that Lemma 1 is true if "f is 1–1 on $\overline{B_r(\boldsymbol{a})}$" is replaced by "$f(\boldsymbol{x}) \neq f(\boldsymbol{a})$ for all $\boldsymbol{x} \in \partial B_r(\boldsymbol{a})$."

2. Let $E = \{(x, y) : 0 < y < x\}$ and set $f(x, y) = (x + y, xy)$ for $(x, y) \in E$.

   a) Prove f is 1–1 from E onto $\{(s, t) : s > 2\sqrt{t}, t > 0\}$ and find a formula for $f^{-1}(s, t)$.

   b) Use the Inverse Function Theorem to compute $D(f^{-1})(f(x, y))$ for $x \neq y$.

   c) Use the formula you obtained in part a) to compute $D(f^{-1})(s, t)$ directly. (Check to see that this answer agrees with the one you found in part b).)

3. For $u \in \mathbf{R}$ and $v > 0$, let $f(u, v) = (u^3 - v^2, \sin u - \log v)$. Prove that f^{-1} exists and is differentiable near $(-1, 0)$ and compute $D(f^{-1})(-1, 0)$.

4. Suppose $f : \mathbf{R}^2 \to \mathbf{R}^2$ has continuous first-order partial derivatives in some ball $B_r(x_0, y_0)$, $r > 0$. Prove that if $\Delta_f(x_0, y_0) \neq 0$, then

$$\frac{\partial f_1^{-1}}{\partial x}(f(x_0, y_0)) = \frac{\partial f_2/\partial y(x_0, y_0)}{\Delta_f(x_0, y_0)}, \qquad \frac{\partial f_1^{-1}}{\partial y}(f(x_0, y_0)) = \frac{-\partial f_1/\partial y(x_0, y_0)}{\Delta_f(x_0, y_0)},$$

and

$$\frac{\partial f_2^{-1}}{\partial x}(f(x_0, y_0)) = \frac{-\partial f_2/\partial x(x_0, y_0)}{\Delta_f(x_0, y_0)}, \qquad \frac{\partial f_2^{-1}}{\partial y}(f(x_0, y_0)) = \frac{\partial f_1/\partial x(x_0, y_0)}{\Delta_f(x_0, y_0)}.$$

5. Prove there exist functions $u(x, y)$, $v(x, y)$ and $w(x, y)$ and an $r > 0$ such that u, v, w are continuously differentiable and satisfy the equations

$$u^5 + xv^2 - y + w = 0$$
$$v^5 + yu^2 - x + w = 0$$
$$w^4 + y^5 - x^4 = 1$$

   on $B_r(1, 1)$, and $u(1, 1) = 1$, $v(1, 1) = 1$, $w(1, 1) = -1$.

6. Find conditions on a point (x_0, y_0, u_0, v_0) such that there exist real-valued functions $u(x, y)$ and $v(x, y)$ which are continuously differentiable near (x_0, y_0) and satisfy the simultaneous equations

$$xu^2 + yv^2 + xy = 9$$
$$xv^2 + yu^2 - xy = 7.$$

   Prove that the solutions satisfy $u^2 + v^2 = 16/(x + y)$.

7. Given nonzero numbers $x_0, y_0, u_0, v_0, s_0, t_0$ which satisfy the simultaneous equations

(*)
$$u^2 + sx + ty = 0$$
$$v^2 + tx + sy = 0$$
$$s^2 x + t^2 y = 0$$
$$s^2 x - t^2 y = 0,$$

   prove that there exist functions $u(x, y)$, $v(x, y)$, $s(x, y)$, $t(x, y)$, and an open ball B containing (x_0, y_0) such that u, v, s, t are continuously differentiable and satisfy (*) on B, and such that $u(x_0, y_0) = u_0$, $v(x_0, y_0) = v_0$, $s(x_0, y_0) = s_0$, and $t(x_0, y_0) = t_0$.

$\boxed{8}$. **This exercise is used in Section 6.6.** Let $F(x, y, z)$ be a relation in $\mathbf{R}^3$ and consider its graph defined by

$$\mathcal{G} := \{(x, y, z) \in \mathbf{R}^3 : F(x, y, z) = 0\}.$$

$\mathcal{G}$ is said to have a tangent plane at a point (a, b, c) if $F(x, y, z) = 0$ can be "solved" for one of the variables in a differentiable way, e.g., there is a function $x = f(y, z)$, differentiable at the point (b, c), such that $F(f(y, z), y, z) = 0$ for all (y, z) in an open ball centered at (b, c). Let $F : \mathbf{R}^3 \to \mathbf{R}$ be continuously differentiable at (a, b, c) with $\nabla F(a, b, c) \neq 0$.

a) Prove $\mathcal{G}$ has a tangent plane at (a, b, c).

b) Show that if Π is a plane passing through the point (a, b, c) with equation $\lambda(x, y, z) = d$ such that the line $\lambda(x, b, z) = d$ is tangent to the curve $F(x, b, z) = 0$ at $x = a$ and the line $\lambda(a, y, z) = d$ is tangent to the curve $F(a, y, z) = 0$ at $y = b$, then a normal vector to Π is given by $\nabla F(a, b, c)$.

$\boxed{9}$. **This exercise is used in Section 7.5.** If V is open in $\mathbf{R}^n$, $\phi : V \to \mathbf{R}^n$ is $\mathcal{C}^1$ and $\Delta_\phi \neq 0$ on V, prove that $\phi(V)$ is open.

^{e}6.6   EXTREMA   *This section uses no material from any other enrichment section.*

In this section we discuss optimization of functions of several variables.

DEFINITION 6.6.   Let V be an open subset of $\mathbf{R}^n$, $\boldsymbol{a} \in V$, and $f : V \to \mathbf{R}$.

i) $f(\boldsymbol{a})$ is called a *local minimum* of f if there is an $r > 0$ such that $f(\boldsymbol{a}) \leq f(\boldsymbol{x})$ for all $\boldsymbol{x} \in B_r(\boldsymbol{a})$.

ii) $f(\boldsymbol{a})$ is called a *local maximum* of f if there is an $r > 0$ such that $f(\boldsymbol{a}) \geq f(\boldsymbol{x})$ for all $\boldsymbol{x} \in B_r(\boldsymbol{a})$.

iii) $f(\boldsymbol{a})$ is called a *local extremum* of f if $f(\boldsymbol{a})$ is a local maximum or a local minimum of f.

The following result shows that as in the one-dimensional case, extrema of real-valued differentiable functions occur at points where the "derivative" is zero.

Remark 1.   *If the first-order partial derivatives of f exist at $\boldsymbol{a}$ and $f(\boldsymbol{a})$ is a local extremum of f, then*

$$\nabla f(\boldsymbol{a}) = \mathbf{0}.$$

PROOF.   The one-dimensional function $g(t) = f(a_1, \ldots, a_{j-1}, t, a_{j+1}, \ldots, a_n)$ has a local extremum at $t = a_j$ for each $j = 1, \ldots, n$. Hence, by the one-dimensional theory,

$$\frac{\partial f}{\partial x_j}(\boldsymbol{a}) = g'(a_j) = 0. \quad \blacksquare$$

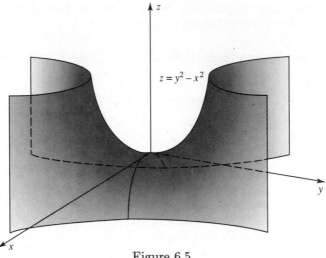

Figure 6.5

As in the one-dimensional case, $\nabla f(\boldsymbol{a}) = \boldsymbol{0}$ is necessary but not sufficient for $f(\boldsymbol{a})$ to be a local extremum.

Remark 2. *There exist continuously differentiable functions which satisfy* $\nabla f(\boldsymbol{a}) = \boldsymbol{0}$ *such that* $f(\boldsymbol{a})$ *is neither a local maximum nor a local minimum.*

PROOF. Consider

$$f(x, y) = y^2 - x^2.$$

Since the first-order partial derivatives of f exist and are continuous everywhere on $\mathbf{R}^2$, f is continuously differentiable on $\mathbf{R}^2$. Moreover, it is evident that $\nabla f(\boldsymbol{0}) = \boldsymbol{0}$ but $f(\boldsymbol{0})$ is not a local extremum (see Figure 6.5). ∎

The fact that the graph of this function resembles a saddle motivates the following terminology.

DEFINITION 6.7. Let V, $\boldsymbol{a} \in V$, and $f : V \to \mathbf{R}$ be differentiable at $\boldsymbol{a}$. Then $\boldsymbol{a}$ is called a *saddle point* of f if $\nabla f(\boldsymbol{a}) = \boldsymbol{0}$ and there is a $r_0 > 0$ such that given any $0 < \rho < r_0$ there are points $\boldsymbol{x}, \boldsymbol{y} \in B_\rho(\boldsymbol{a})$ which satisfy

$$f(\boldsymbol{x}) < f(\boldsymbol{a}) < f(\boldsymbol{y}).$$

By the Extreme Value Theorem, if f is continuous on a compact set H, then it *attains* its maximum and minimum on H; i.e., there exist points $\boldsymbol{a}, \boldsymbol{b} \in H$ such that

$$f(\boldsymbol{a}) = \sup_{\boldsymbol{x} \in H} f(\boldsymbol{x}) \quad \text{and} \quad f(\boldsymbol{b}) = \inf_{\boldsymbol{x} \in H} f(\boldsymbol{x}).$$

When f is a function of two variables, these points can be found by combining Remark 1 with one-dimensional techniques.

Example 1. Find the maximum and minimum of $f(x,y) = x^2 - x + y^2 - 2y$ on $H = \overline{B_1(0,0)}$.

SOLUTION. Since $\nabla f(x,y) = (0,0)$ implies $(x,y) = (1/2,1)$, f has no local extrema inside H. Thus, the extrema of f on H must occur on ∂H. Using polar coordinates, we can describe ∂H by $(x,y) = (\cos\theta, \sin\theta)$, where $0 \le \theta < 2\pi$. Set

$$h(\theta) := f(\cos\theta, \sin\theta) = 1 - \cos\theta - 2\sin\theta.$$

Notice that the derivative of h is zero when $\tan\theta = 2$, i.e., $\theta = \arctan 2 \approx 1.10715$ or $\theta = \arctan 2 + \pi \approx 4.24874$. Therefore, candidates for the extrema of f on ∂H are $(x,y) \approx (0.4472, 0.8944)$ and $(x,y) \approx (-0.4472, -0.8944)$. Checking the sign of $h''(\theta)$, we see that the first point corresponds to a minimum and the second point corresponds to a maximum. Therefore, the maximum of f on H is $f(-0.4472, -0.8944) \approx 3.236$ and the minimum of f on H is $f(0.4472, 0.8944) \approx -1.236$. ∎

Using the second-order total differential $D^{(2)}f$ introduced in Section 6.4, we can obtain a multidimensional analogue of the Second Derivative Test. First, we prove a technical result.

Lemma 1. *Let V be open in $\mathbf{R}^n$, $\boldsymbol{a} \in V$ and $f : V \to \mathbf{R}$. If all second-order partial derivatives of f exist at $\boldsymbol{a}$ and $D^{(2)}f(\boldsymbol{a};\boldsymbol{h}) > 0$ for all $\boldsymbol{h} \ne \boldsymbol{0}$, then there is an $m > 0$ such that*

$$(26) \qquad\qquad D^{(2)}f(\boldsymbol{a};\boldsymbol{x}) \ge m\|\boldsymbol{x}\|^2$$

for all $\boldsymbol{x} \in \mathbf{R}^n$.

PROOF. Set $H = \{\boldsymbol{x} \in \mathbf{R}^n : \|\boldsymbol{x}\| = 1\}$ and consider the function

$$g(\boldsymbol{x}) = D^{(2)}f(\boldsymbol{a};\boldsymbol{x}) = \sum_{j=1}^n \sum_{k=1}^n \frac{\partial^2 f}{\partial x_k \partial x_j}(\boldsymbol{a})x_j x_k, \qquad \boldsymbol{x} \in \mathbf{R}^n.$$

By hypothesis, g is continuous and positive on $\mathbf{R}^n \setminus \{\boldsymbol{0}\}$, hence on H. Since H is compact, it follows from the Extreme Value Theorem that g has a positive minimum m on H.

Clearly, (26) holds for $\boldsymbol{x} = \boldsymbol{0}$. If $\boldsymbol{x} \ne \boldsymbol{0}$, then $\boldsymbol{x}/\|\boldsymbol{x}\| \in H$, and it follows from the definition of g and m that

$$D^{(2)}f(\boldsymbol{a};\boldsymbol{x}) = \frac{g(\boldsymbol{x})}{\|\boldsymbol{x}\|^2}\|\boldsymbol{x}\|^2 = g\left(\frac{\boldsymbol{x}}{\|\boldsymbol{x}\|}\right)\|\boldsymbol{x}\|^2 \ge m\|\boldsymbol{x}\|^2.$$

We conclude that (26) holds for all $\boldsymbol{x} \in \mathbf{R}^n$. ∎

THEOREM 6.22 [THE SECOND DERIVATIVE TEST]. *Let V be open in $\mathbf{R}^n$, $\boldsymbol{a} \in V$ and $f : V \to \mathbf{R}$ satisfy $\nabla f(\boldsymbol{a}) = \boldsymbol{0}$. Suppose further that all second-order partial derivatives of f exist on V and are continuous at $\boldsymbol{a}$.*

   i) *If $D^{(2)}f(\boldsymbol{a};\boldsymbol{h}) > 0$ for all $\boldsymbol{h} \ne \boldsymbol{0}$, then $f(\boldsymbol{a})$ is a local minimum of f.*
   ii) *If $D^{(2)}f(\boldsymbol{a};\boldsymbol{h}) < 0$ for all $\boldsymbol{h} \ne \boldsymbol{0}$, then $f(\boldsymbol{a})$ is a local maximum of f.*
   iii) *If $D^{(2)}f(\boldsymbol{a};\boldsymbol{h})$ takes on both positive and negative values for $\boldsymbol{h} \in \mathbf{R}^n$, then $\boldsymbol{a}$ is a saddle point of f.*

PROOF. Choose $r > 0$ such that $B_r(\boldsymbol{a}) \subset V$ and suppose for a moment that there is a function $\varepsilon : B_r(\boldsymbol{0}) \to \mathbf{R}$ such that $\varepsilon(\boldsymbol{h}) \to 0$ as $\boldsymbol{h} \to \boldsymbol{0}$ and

$$(27) \qquad f(\boldsymbol{a}+\boldsymbol{h}) - f(\boldsymbol{a}) = \frac{1}{2} D^{(2)} f(\boldsymbol{a};\boldsymbol{h}) + \|\boldsymbol{h}\|^2 \varepsilon(\boldsymbol{h})$$

for $\boldsymbol{h}$ sufficiently small. If $D^{(2)} f(\boldsymbol{a};\boldsymbol{h}) > 0$ for $\boldsymbol{h} \neq \boldsymbol{0}$, then (26) and (27) imply

$$f(\boldsymbol{a}+\boldsymbol{h}) - f(\boldsymbol{a}) \geq \left(\frac{m}{2} + \varepsilon(\boldsymbol{h})\right) \|\boldsymbol{h}\|^2$$

for $\boldsymbol{h}$ sufficiently small. Since $m > 0$ and $\varepsilon(\boldsymbol{h}) \to 0$ as $\boldsymbol{h} \to \boldsymbol{0}$, it follows that $f(\boldsymbol{a}+\boldsymbol{h}) - f(\boldsymbol{a}) > 0$ for $\boldsymbol{h}$ sufficiently small, i.e., $f(\boldsymbol{a})$ is a local minimum. Similarly, if $D^{(2)} f(\boldsymbol{a};\boldsymbol{h}) < 0$ for $\boldsymbol{h} \neq \boldsymbol{0}$, then $f(\boldsymbol{a})$ is a local maximum. This proves parts i and ii.

To prove part iii, fix $\boldsymbol{h} \in \mathbf{R}^n$ and notice that (27) implies

$$f(\boldsymbol{a}+t\boldsymbol{h}) - f(\boldsymbol{a}) = t^2 \left(\frac{1}{2} D^{(2)} f(\boldsymbol{a};\boldsymbol{h}) + \|\boldsymbol{h}\|^2 \varepsilon(t\boldsymbol{h})\right)$$

for $t \in \mathbf{R}$. Since $\varepsilon(t\boldsymbol{h}) \to 0$ as $t \to 0$, it follows that $f(\boldsymbol{a}+t\boldsymbol{h}) - f(\boldsymbol{a})$ takes on the same sign as $D^{(2)} f(\boldsymbol{a};\boldsymbol{h})$ for t small. In particular, if $D^{(2)} f(\boldsymbol{a};\boldsymbol{h})$ takes on both positive and negative values as $\boldsymbol{h}$ varies, then $\boldsymbol{a}$ is a saddle point.

It remains to find a function $\varepsilon : B_r(\boldsymbol{0}) \to \mathbf{R}$ such that $\varepsilon(\boldsymbol{h}) \to 0$ as $\boldsymbol{h} \to \boldsymbol{0}$, and (27) holds for all $\boldsymbol{h}$ sufficiently small. Set $\varepsilon(\boldsymbol{0}) = 0$ and

$$\varepsilon(\boldsymbol{h}) = \frac{f(\boldsymbol{a}+\boldsymbol{h}) - f(\boldsymbol{a}) - \frac{1}{2} D^{(2)} f(\boldsymbol{a};\boldsymbol{h})}{\|\boldsymbol{h}\|^2}, \qquad \boldsymbol{h} \in B_r(\boldsymbol{0}), \ \boldsymbol{h} \neq \boldsymbol{0}.$$

By the definition of $\varepsilon(\boldsymbol{h})$, (27) holds for $\boldsymbol{h} \in B_r(\boldsymbol{0})$. If $\boldsymbol{h} = (h_1, h_2, \ldots, h_n) \in B_r(\boldsymbol{0})$ is fixed, then $\nabla f(\boldsymbol{a}) = \boldsymbol{0}$ and Taylor's Formula imply

$$f(\boldsymbol{a}+\boldsymbol{h}) - f(\boldsymbol{a}) = \frac{1}{2} D^{(2)} f(\boldsymbol{c};\boldsymbol{h})$$

for some $\boldsymbol{c} \in L(\boldsymbol{a}; \boldsymbol{a}+\boldsymbol{h})$, i.e.,

$$f(\boldsymbol{a}+\boldsymbol{h}) - f(\boldsymbol{a}) - \frac{1}{2} D^{(2)} f(\boldsymbol{a};\boldsymbol{h}) = \frac{1}{2} \left(D^{(2)} f(\boldsymbol{c};\boldsymbol{h}) - D^{(2)} f(\boldsymbol{a};\boldsymbol{h})\right)$$

$$= \frac{1}{2} \sum_{j=1}^{n} \sum_{k=1}^{n} \left(\frac{\partial^2 f}{\partial x_j \partial x_k}(\boldsymbol{c}) - \frac{\partial^2 f}{\partial x_j \partial x_k}(\boldsymbol{a})\right) h_j h_k.$$

Since $|h_j h_k| \leq \|\boldsymbol{h}\|^2$ and the second-order partial derivatives of f are continuous at $\boldsymbol{a}$, it follows that

$$0 \leq |\varepsilon(\boldsymbol{h})| \leq \frac{1}{2} \left(\sum_{j=1}^{n} \sum_{k=1}^{n} \left|\frac{\partial^2 f}{\partial x_j \partial x_k}(\boldsymbol{c}) - \frac{\partial^2 f}{\partial x_j \partial x_k}(\boldsymbol{a})\right|\right) \to 0$$

as $\boldsymbol{h} \to \boldsymbol{0}$. We conclude by the Squeeze Theorem that $\varepsilon(\boldsymbol{h}) \to 0$ as $\boldsymbol{h} \to \boldsymbol{0}$. ∎

The following result shows that the strict inequalities in Theorem 6.22 cannot be relaxed.

Remark 3. *If $D^{(2)}f(a;h) \geq 0$, then $f(a)$ can be a local minimum or a can be a saddle point.*

PROOF. $f(0,0)$ is a local minimum of $f(x,y) = x^4 + y^2$, and $(0,0)$ is a saddle point of $f(x,y) = x^3 + y^2$. ∎

In practice, it is not easy to determine the sign of $D^{(2)}f(a;h)$. For the case $n = 2$, the second total differential $D^{(2)}f(a;h)$ is a *quadratic form*, i.e., has the form $Ah^2 + 2Bhk + Ck^2$. The following result shows that the sign of a quadratic form is completely determined by the *discriminant $D = B^2 - AC$*.

Lemma 2. *Let $A, B, C \in \mathbf{R}$, $D = B^2 - AC$, and $\phi(h,k) = Ah^2 + 2Bhk + Ck^2$.*

   i) *If $D < 0$, then A and $\phi(h,k)$ have the same sign for all $(h,k) \neq (0,0)$.*
   ii) *If $D > 0$, then $\phi(h,k)$ takes on both positive and negative values as (h,k) varies over $\mathbf{R}^2$.*

PROOF. i) Suppose $D < 0$. Then $A \neq 0$ and $A\phi(h,k)$ is a sum of two squares:

$$A\phi(h,k) = A^2 h^2 + 2ABhk + ACk^2 = (Ah + Bk)^2 + |D|k^2.$$

Since $A \neq 0 \neq D$, at least one of these squares is positive for each $(h,k) \neq (0,0)$. It follows that A and $\phi(h,k)$ have the same sign for all $(h,k) \neq (0,0)$.

ii) Suppose $D > 0$. Then either $A \neq 0$ or $B \neq 0$.

If $A \neq 0$, then $A\phi(h,k)$ is a difference of two squares:

$$A\phi(h,k) = (Ah + Bk - \sqrt{D}k)(Ah + Bk + \sqrt{D}k).$$

The lines $Ah + Bk - \sqrt{D}k = 0$ and $Ah + Bk + \sqrt{D}k = 0$ divide the hk plane into four open regions (see Figure 6.6). Since $A\phi(h,k)$ is positive on two of these regions and negative on the other two, it follows that $\phi(h,k)$ takes on both positive and negative values as (h,k) varies over $\mathbf{R}^2$.

If $A = 0$ and $B \neq 0$, then

$$\phi(h,k) = 2Bhk + Ck^2 = (2Bh + Ck)k.$$

Since $B \neq 0$, the lines $2Bh + Ck = 0$ and $k = 0$ divide the hk plane into four open regions. As before, $\phi(h,k)$ takes on both positive and negative values as (h,k) varies over $\mathbf{R}^2$. ∎

This result leads us to the following simple test for extrema and saddle points.

THEOREM 6.23. *Let V be open in $\mathbf{R}^2$, $(a,b) \in V$, and suppose $f : V \to \mathbf{R}$ satisfies $\nabla f(a,b) = \mathbf{0}$. Suppose further that all second-order partial derivatives of f exist on V, are continuous at (a,b), and*

$$D = f_{xy}^2(a,b) - f_{xx}(a,b)f_{yy}(a,b).$$

   i) *If $D < 0$ and $f_{xx}(a,b) > 0$, then $f(a,b)$ is a local minimum.*
   ii) *If $D < 0$ and $f_{xx}(a,b) < 0$, then $f(a,b)$ is a local maximum.*
   iii) *If $D > 0$, then (a,b) is a saddle point.*

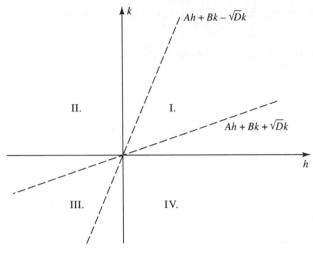

Figure 6.6

PROOF. Let $A = f_{xx}(a, b)$, $B = f_{xy}(a, b)$, and $C = f_{yy}(a, b)$. Apply Theorem 6.22 and Lemma 2. ∎

(For a discriminant which works for functions of three variables, see Widder [14], p. 134.)

Remark 4. *If the discriminant $D = 0$, $f(a, b)$ may be a local maximum, a local minimum, or (a, b) may be a saddle point.*

PROOF. The function $f(x, y) = x^2$ has zero discriminant at $(a, b) = (0, 0)$, and $0 = f(0, 0)$ is a local minimum for f. On the other hand, $f(x, y) = x^3$ has zero discriminant at $(a, b) = (0, 0)$, and $(0, 0)$ is a saddle point for f. ∎

In practice, one often searches for extrema subject to constraints.

DEFINITION 6.8. Let V be open in $\mathbf{R}^n$, $\boldsymbol{a} \in V$, and $f, g_j : V \to \mathbf{R}$ for $j = 1, 2, \ldots, m$.

   i) $f(\boldsymbol{a})$ is called a *local minimum of f subject to the constraints* $g_j(\boldsymbol{a}) = 0$, $j = 1, \ldots, m$, if there is a $\rho > 0$ such that $\boldsymbol{x} \in B_\rho(\boldsymbol{a})$ and $g_j(\boldsymbol{x}) = 0$ for all $j = 1, \ldots, m$ imply $f(\boldsymbol{x}) \geq f(\boldsymbol{a})$.

   ii) $f(\boldsymbol{a})$ is called a *local maximum of f subject to the constraints* $g_j(\boldsymbol{a}) = 0$, $j = 1, \ldots, m$, if there is a $\rho > 0$ such that $\boldsymbol{x} \in B_\rho(\boldsymbol{a})$ and $g_j(\boldsymbol{x}) = 0$ for all $j = 1, \ldots, m$ imply $f(\boldsymbol{x}) \leq f(\boldsymbol{a})$.

Example 1. Find all points on the ellipsoid $x^2 + 2y^2 + 3z^2 = 1$ (see Appendix D) which lie closest to or furthest from the origin.

SOLUTION. We must optimize the distance formula $\sqrt{x^2 + y^2 + z^2}$ (equivalently, the function $f(x, y, z) = x^2 + y^2 + z^2$) subject to the constraint $g(x, y, z) = x^2 + 2y^2 + 3z^2 - 1 = 0$. Using g to eliminate the variable x in f, we see that f takes on the form

$$\phi(y, z) = 1 - y^2 - 2z^2.$$

Solving $\nabla\phi(y, z) = (0, 0)$, we obtain $(y, z) = (0, 0)$, i.e., $x^2 = 1$. Thus, elimination of x leads to the points $(\pm 1, 0, 0)$. Similarly, elimination of y leads to $(0, \pm 1/\sqrt{2}, 0)$ and elimination of z leads to $(0, 0, \pm 1/\sqrt{3})$. Checking the distance formula, we see that the maximum distance is 1, which occurs at the points $(\pm 1, 0, 0)$, and the minimum distance is $1/\sqrt{3}$, which occurs at the points $(0, 0, \pm 1/\sqrt{3})$. (The points $(0, \pm 1/\sqrt{2}, 0)$ are saddle points, i.e., correspond neither to a maximum nor a minimum.) ∎

Solving an extremal problem with constraints by eliminating one or more of the variables is called the *direct method*. There is another, more geometric, method for solving Example 1. Notice that for the ellipsoid $g(x, y, z) = x^2 + 2y^2 + 3z^2 - 1 = 0$, the points closest to and furthest from the origin occur when the tangent planes of $g(x, y, z)$ and the sphere $f(x, y, z) = 1$ are parallel (see Figure 6.7). Recall that two nonzero vectors $\boldsymbol{a}$ and $\boldsymbol{b}$ are parallel if and only if $\boldsymbol{a} + \lambda\boldsymbol{b} = \boldsymbol{0}$ for some scalar $\lambda \neq 0$. Since normal vectors of the tangent planes of $f(x, y, z) = 1$ and $g(x, y, z) = 0$ are ∇f and ∇g (see Exercise 8 in Section 6.5), it follows that points (x, y, z) closest to and furthest from the origin must satisfy

$$(28) \qquad\qquad \nabla f(x, y, z) + \lambda\nabla g(x, y, z) = \boldsymbol{0}$$

for some $\lambda \neq 0$. For the case at hand, (28) implies $(2x, 2y, 2z) + \lambda(2x, 4y, 6z) = (0, 0, 0)$. Combining this equation with the constraint $g(x, y, z) = 0$, we have four equations in four unknowns:

$$x(\lambda + 1), \qquad y(2\lambda + 1) = 0, \qquad z(3\lambda + 1) = 0, \quad \text{and} \quad x^2 + 2y^2 + 3z^2 = 1.$$

Solving these equations, we obtain three pairs of solutions: $(\pm 1, 0, 0)$ (when $\lambda = -1$), $(0, \pm 1/\sqrt{2}, 0)$ (when $\lambda = -1/2$), and $(0, 0, \pm 1/\sqrt{3})$ (when $\lambda = -1/3$). Hence, we obtain the same solutions with the geometric method as we did with the direct method.

The following result shows that the geometric method is valid, even in the case when the functions have nothing to do with spheres and ellipsoids, and even when several constraints are used. This is fortunate since the direct method cannot be used unless the constraints are relatively simple.

THEOREM 6.24 [LAGRANGE MULTIPLIERS]. *Let $m < n$, V be an open set in $\mathbf{R}^n$, and $f, g_j : V \to \mathbf{R}$ be continuously differentiable on V for $j = 1, 2, \ldots, m$. Suppose there is an $\boldsymbol{a} \in V$ such that*

$$\frac{\partial(g_1, \ldots, g_m)}{\partial(x_1, \ldots, x_m)}(\boldsymbol{a}) \neq 0.$$

If $f(\boldsymbol{a})$ is a local extremum of f subject to the constraints $g_k(\boldsymbol{a}) = 0$, $k = 1, \ldots, m$, then there exist scalars $\lambda_1, \lambda_2, \ldots, \lambda_m$ such that

$$(29) \qquad\qquad \nabla f(\boldsymbol{a}) + \sum_{k=1}^{n} \lambda_k \nabla g_k(\boldsymbol{a}) = \boldsymbol{0}.$$

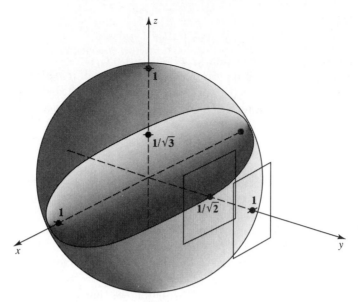

Figure 6.7

Proof. Equation (29) is a system of n equations in m unknowns $\lambda_1, \lambda_2, \ldots, \lambda_m$:

$$(30) \qquad \sum_{k=1}^{m} \lambda_k \frac{\partial g_k}{\partial x_j}(\boldsymbol{a}) = -\frac{\partial f}{\partial x_j}(\boldsymbol{a}), \qquad j = 1, 2, \ldots, n.$$

If we restrict our attention to the indices $j = 1, 2, \ldots, m$, then we see by hypothesis that the determinant of the matrix of coefficients of the system (30) is nonzero. Thus, the λ_k's are uniquely determined by the first m equations of (30). What remains to be seen is that these λ_k's also satisfy (30) for $j = m + 1, \ldots, n$. This is a question about implicit functions.

Let $p = n - m$. As in the proof of the Implicit Function Theorem, write vectors in $\mathbf{R}^{m+p}$ in the form

$$\boldsymbol{x} = (\boldsymbol{y}, \boldsymbol{t}) = (y_1, \ldots, y_m, t_1, \ldots, t_p).$$

We must show

$$(31) \qquad 0 = \frac{\partial f}{\partial t_\ell}(\boldsymbol{a}) + \sum_{k=1}^{m} \lambda_k \frac{\partial g_k}{\partial t_\ell}(\boldsymbol{t}_0)$$

for $\ell = 1, \ldots, p$.

Let $g = (g_1, \ldots, g_m)$, and choose $\boldsymbol{y}_0 \in \mathbf{R}^m$, $\boldsymbol{t}_0 \in \mathbf{R}^p$ such that $\boldsymbol{a} = (\boldsymbol{y}_0, \boldsymbol{t}_0)$. By hypothesis, $g(\boldsymbol{y}_0, \boldsymbol{t}_0) = \boldsymbol{0}$ and the Jacobian of g (with respect to the variables y_j) is nonzero at $(\boldsymbol{y}_0, \boldsymbol{t}_0)$. Hence, by the Implicit Function Theorem, there is an open set $W \subset \mathbf{R}^p$ which contains $\boldsymbol{t}_0$ and a function $h : W \to \mathbf{R}^m$ such that h is continuously differentiable on W, $h(\boldsymbol{t}_0) = \boldsymbol{y}_0$, and

$$(32) \qquad g(h(\boldsymbol{t}), \boldsymbol{t}) = \boldsymbol{0}, \qquad \boldsymbol{t} \in W.$$

For each $\boldsymbol{t} \in W$ and $k = 1, \ldots, m$, set

$$G_k(\boldsymbol{t}) = g_k(h(\boldsymbol{t}), \boldsymbol{t}) \quad \text{and} \quad F(\boldsymbol{t}) = f(h(\boldsymbol{t}), \boldsymbol{t}).$$

We shall use the functions $G_1, \ldots, G_m$ and F to verify (31) for $\ell = 1, \ldots, p$. Fix such an ℓ. By (32), each G_k is identically zero on W, hence has derivative zero there. Since $\boldsymbol{t}_0 \in W$ and $(h(\boldsymbol{t}_0), \boldsymbol{t}_0) = (\boldsymbol{y}_0, \boldsymbol{t}_0) = \boldsymbol{a}$, it follows from the Chain Rule that

$$O = DG_k(\boldsymbol{t}_0) = \begin{bmatrix} \dfrac{\partial g_k}{\partial x_1}(\boldsymbol{a}) & \cdots & \dfrac{\partial g_k}{\partial x_n}(\boldsymbol{a}) \end{bmatrix} \begin{bmatrix} \dfrac{\partial h_1}{\partial t_1}(\boldsymbol{t}_0) & \cdots & \dfrac{\partial h_1}{\partial t_p}(\boldsymbol{t}_0) \\ \vdots & \ddots & \vdots \\ \dfrac{\partial h_m}{\partial t_1}(\boldsymbol{t}_0) & \cdots & \dfrac{\partial h_m}{\partial t_p}(\boldsymbol{t}_0) \\ 1 & \cdots & 0 \\ \vdots & \ddots & \vdots \\ 0 & \cdots & 1 \end{bmatrix}.$$

Hence, the ℓth component of $DG_k(\boldsymbol{t}_0)$ is given by

$$(33) \qquad 0 = \sum_{j=1}^{m} \frac{\partial g_k}{\partial x_j}(\boldsymbol{a}) \frac{\partial h_j}{\partial t_\ell}(\boldsymbol{t}_0) + \frac{\partial g_k}{\partial t_\ell}(\boldsymbol{a})$$

for $k = 1, 2, \ldots, m$. Multiplying (33) by λ_k and adding, we obtain

$$0 = \sum_{k=1}^{m} \sum_{j=1}^{m} \lambda_k \frac{\partial g_k}{\partial x_j}(\boldsymbol{a}) \frac{\partial h_j}{\partial t_\ell}(\boldsymbol{t}_0) + \sum_{k=1}^{m} \lambda_k \frac{\partial g_k}{\partial t_\ell}(\boldsymbol{a})$$

$$= \sum_{j=1}^{m} \left(\sum_{k=1}^{m} \lambda_k \frac{\partial g_k}{\partial x_j}(\boldsymbol{a}) \right) \frac{\partial h_j}{\partial t_\ell}(\boldsymbol{t}_0) + \sum_{k=1}^{m} \lambda_k \frac{\partial g_k}{\partial t_\ell}(\boldsymbol{a}).$$

Hence, it follows from (30) that

$$(34) \qquad 0 = -\sum_{j=1}^{m} \frac{\partial f}{\partial x_j}(\boldsymbol{a}) \frac{\partial h_j}{\partial t_\ell}(\boldsymbol{t}_0) + \sum_{k=1}^{m} \lambda_k \frac{\partial g_k}{\partial t_\ell}(\boldsymbol{a}).$$

Suppose $f(\boldsymbol{a})$ is a local maximum subject to the constraints $g(\boldsymbol{a}) = \boldsymbol{0}$. Set $E_0 = \{\boldsymbol{x} \in V : g(\boldsymbol{x}) = \boldsymbol{0}\}$ and choose an n-dimensional open ball $B(\boldsymbol{a})$ such that

$$(35) \qquad \boldsymbol{x} \in B(\boldsymbol{a}) \cap E_0 \quad \text{implies} \quad f(\boldsymbol{x}) \le f(\boldsymbol{a}).$$

Since h is continuous, choose a p-dimensional open ball $B(\boldsymbol{t}_0)$ such that $\boldsymbol{t} \in B(\boldsymbol{t}_0)$ implies $(h(\boldsymbol{t}), \boldsymbol{t}) \in B(\boldsymbol{a})$. By (35), $F(\boldsymbol{t}_0)$ is a local maximum of F on $B(\boldsymbol{t}_0)$. Hence, $\nabla F(\boldsymbol{t}_0) = \boldsymbol{0}$. Applying the Chain Rule as above, we obtain

$$(36) \qquad 0 = \sum_{j=1}^{m} \frac{\partial f}{\partial x_j}(\boldsymbol{a}) \frac{\partial h_j}{\partial t_\ell}(\boldsymbol{t}_0) + \frac{\partial f}{\partial t_\ell}(\boldsymbol{a}).$$

(compare with (33)). Adding (36) and (34), we conclude that

$$0 = \frac{\partial f}{\partial t_\ell}(\boldsymbol{a}) + \sum_{k=1}^{m} \lambda_k \frac{\partial g_k}{\partial t_\ell}(\boldsymbol{a}). \quad \blacksquare$$

Example 2. Find all extrema of $x^2 + y^2 + z^2$ subject to the constraints $x - y = 1$ and $y^2 - z^2 = 1$.

SOLUTION. Let $f(x,y,z) = x^2 + y^2 + z^2$, $g(x,y,z) = x - y - 1$ and $h(x,y,z) = y^2 - z^2 - 1$. Then (29) takes on the form $\nabla f + \lambda \nabla g + \mu \nabla h = \boldsymbol{0}$, i.e.,

$$(2x, 2y, 2z) + \lambda(1, -1, 0) + \mu(0, 2y, -2z) = (0, 0, 0).$$

In particular, $2x + \lambda = 0$, $2y + 2\mu y - \lambda = 0$ and $2z - 2\mu z = 0$. From this last equation, either $\mu = 1$ or $z = 0$.

If $\mu = 1$, then $\lambda = 4y$. Since $2x + \lambda = 0$, we find that $x = -2y$. From $g = 0$ we obtain $-3y = 1$, i.e., $y = -1/3$. Substituting this into $h = 0$, we obtain $z^2 = -8/9$, a contradiction.

If $z = 0$, then from $h = 0$ we obtain $y = \pm 1$. Since $g = 0$, we obtain $x = 2$ when $y = 1$ and $x = 0$ when $y = -1$. Thus, the only candidates for extrema of f subject to the constraints $g = 0 = h$ are $f(2, 1, 0) = 5$ and $f(0, -1, 0) = 1$. To decide whether these are maxima, minima, or neither, look at the problem from a geometric point of view. The problem requires us to find points on the intersection of the plane $x - y = 1$ and the hyperbolic cylinder $y^2 - z^2 = 1$ which lie closest to the origin. Evidently, both of these points correspond to local minima and there is no maximum (see Figure 6.8). In particular, the minimum of $x^2 + y^2 + z^2$ subject to the given constraints is 1, attained at the point $(0, -1, 0)$. $\blacksquare$

EXERCISES

1. Find all local extrema of each of the following functions.
    a) $f(x, y) = x^2 - xy + y^3 - y$.
    b) $f(x, y) = \sin x + \cos y$.
    c) $f(x, y, z) = e^{x+y} \cos z$.
    d) $f(x, y) = ax^2 + bxy + cy^2$ where $a \neq 0$ and $b^2 - 4ac \neq 0$.
2. For each of the following, find the maximum and minimum of f on H.
    a) $f(x, y) = x^2 + 2x - y^2$ and $H = \{(x, y) : x^2 + 4y^2 \leq 4\}$.
    b) $f(x, y) = x^2 + 2xy + 3y^2$ and H is the region bounded by the triangle with vertices $(1, 0)$, $(1, 2)$, $(3, 0)$.
    c) $f(x, y) = x^3 + 3xy - y^3$ and $H = [-1, 1] \times [-1, 1]$.
3. For each of the following, use Lagrange Multipliers to find all extrema of f subject to the given constraints
    a) $f(x, y) = x + y^2$ and $x^2 + y^2 = 4$.

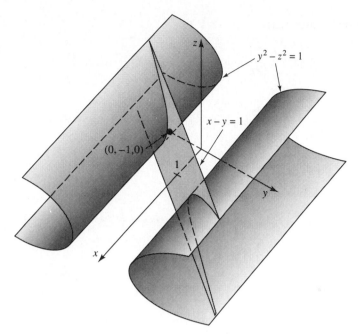

Figure 6.8

b) $f(x,y) = x^2 - 4xy + 4y^2$ and $x^2 + y^2 = 1$.

c) $f(x,y,z) = xy$, $x^2 + y^2 + z^2 = 1$ and $x + y + z = 0$.

d) $f(x,y,z,w) = 3x + y + w$, $3x^2 + y + 4z^3 = 1$ and $-x^3 + 3z^4 + w = 0$.

4. Suppose $f : \mathbf{R}^n \to \mathbf{R}^m$ is differentiable at $\boldsymbol{a}$ and $g : \mathbf{R}^m \to \mathbf{R}$ is differentiable at $\boldsymbol{b} = f(\boldsymbol{a})$. Prove that if $g(\boldsymbol{b})$ is a local extremum of g, then $\nabla(g \circ f)(\boldsymbol{a}) = \boldsymbol{0}$.

5. Suppose V is open in $\mathbf{R}^2$, $(a,b) \in V$, and $f : V \to \mathbf{R}$ has second-order partial derivatives on V with $f_x(a,b) = f_y(a,b) = 0$. If the second-order partial derivatives of f are continuous at (a,b) and exactly two of the three numbers $f_{xx}(a,b)$, $f_{xy}(a,b)$, and $f_{yy}(a,b)$ are zero, prove that (a,b) is a saddle point if and only if $f_{xy}(a,b) \neq 0$.

6. Suppose V is an open set in $\mathbf{R}^n$, $\boldsymbol{a} \in V$, and $f : V \to \mathbf{R}$ is $\mathcal{C}^2$ on V. If $f(\boldsymbol{a})$ is a local minimum of f, prove $D^{(2)}f(\boldsymbol{a})(\boldsymbol{h}) \geq 0$ for all $\boldsymbol{h} \in \mathbf{R}^n$.

7. Let a,b,c,D,E be real numbers with $c \neq 0$.

   a) If $DE > 0$, find all extrema of $ax + by + cz$ subject to the constraint $z = Dx^2 + Ey^2$. Prove that a maximum occurs when $cD < 0$ and a minimum when $cD > 0$.

   b) What can you say when $DE < 0$?

8. [IMPLICIT METHOD].

   a) Suppose $f, g : \mathbf{R}^3 \to \mathbf{R}$ are differentiable at a point (a,b,c) and $f(a,b,c)$ is an extremum of f subject to the constraint $g(x,y,z) = k$, where k is a constant. Prove that

$$\frac{\partial f}{\partial x}(a, b, c)\frac{\partial g}{\partial z}(a, b, c) - \frac{\partial f}{\partial z}(a, b, c)\frac{\partial g}{\partial x}(a, b, c) = 0$$

and

$$\frac{\partial f}{\partial y}(a, b, c)\frac{\partial g}{\partial z}(a, b, c) - \frac{\partial f}{\partial z}(a, b, c)\frac{\partial g}{\partial y}(a, b, c) = 0.$$

b) Use part a) to find all extrema of $f(x, y, z) = 4xy + 2xz + 2yz$ subject to the constraint $xyz = 16$.

9 . **This exercise is used in Section ^{e}9.4.**

a) Let $p > 1$. Find all extrema of $f(\boldsymbol{x}) = \sum_{k=1}^{n} x_k^2$ subject to the constraint $\sum_{k=1}^{n} |x_k|^p = 1$.

b) Prove that

$$\frac{1}{n^{(2-p)/(2p)}} \left(\sum_{k=1}^{n} |x_k|^p \right)^{1/p} \le \left(\sum_{k=1}^{n} x_k^2 \right)^{1/2} \le \left(\sum_{k=1}^{n} |x_k|^p \right)^{1/p}$$

for all $x_1, \ldots, x_n \in \mathbf{R}$, $n \in \mathbf{N}$, and $1 \le p \le 2$.

c) Prove that if $\{x_k\}$ is a sequence of real numbers which satisfies $\sum_{k=1}^{\infty} |x_k| < \infty$, then

$$\left(\sum_{k=1}^{\infty} x_k^2 \right)^{1/2} < \infty.$$

^{e}6.7   DIFFERENTIABILITY AND TANGENT PLANES   *This section requires no material from any other enrichment section.*

In Section 6.2 we examined differentiability from an algebraic-analytic point of view. In this section we examine differentiability from a geometric point of view. First, we show (Theorem 6.25 below) that a function f is differentiable at a point (a, b) if and only if the graph of $z = f(x, y)$ has a unique tangent plane at the point $(a, b, f(a, b))$. (The proof presented here is based on Taylor [13][1].) Next, we show that just as one-dimensional differentials can be used to approximate the change of a function along its tangent line, multivariable differentials can be used to approximate the change of a function along its tangent hyperplane (see Remark 1 below).

We begin with the following geometric definition of a tangent plane.

DEFINITION 6.9. Let V be open in $\mathbf{R}^2$, $(a, b) \in V$, and $f : V \to \mathbf{R}$. The graph of $z = f(x, y)$ has a *unique tangent plane* at the point $\boldsymbol{c} = (a, b, f(a, b))$ if there is a

---

[1]Angus E. Taylor, <u>Advanced Calculus</u> (Boston: Ginn and Company, 1955). Reprinted with permission of John Wiley & Sons, Inc.

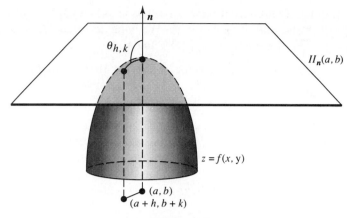

$$\text{Figure 6.9}$$

plane $\varPi$ which contains the point c such that the angle $\theta_{h,k}$ between n, any normal of $\varPi$, and

$$(37) \quad (h, k, f(a+h, b+k) - f(a,b)) = (a+h, b+k, f(a+h, b+k)) - (a, b, f(a,b)),$$

a vector from c to a general point on the surface $z = f(x,y)$, converges to $\pi/2$ as $(h,k) \to (0,0)$ (see Figure 6.9). In this case we say the $\varPi$ is *tangent* to $z = f(x,y)$ at c.

Notice that by (37), a plane $\varPi$ is the unique tangent of $z = f(x,y)$ if and only if the angle between the nearest vector which lies in $\varPi$ and the vector $(h, k, f(a+h, b+k) - f(a,b))$ converges to 0 as $(h, k) \to (0,0)$.

If the plane $\varPi$ is nonvertical, any normal (α, β, γ) of $\varPi$ has a nonzero third component γ. Dividing by γ, we may suppose that $\varPi$ has a normal of the form $n = (n_1, n_2, 1)$. The following result shows that in this case, the condition $\theta_{h,k} \to \pi/2$ looks like a differentiability condition.

Lemma. *Let V be open in $\mathbf{R}^2$, $(a,b) \in V$, $f : V \to \mathbf{R}$, $c = (a, b, f(a,b))$, and $n = (n_1, n_2, 1)$ for some $n_1, n_2 \in \mathbf{R}$. Then the plane $\varPi_n(c)$ is the unique tangent of the graph of $z = f(x,y)$ at c if and only if*

$$(38) \qquad \lim_{(h,k) \to (0,0)} \frac{f(a+h, b+k) - f(a,b) + n_1 h + n_2 k}{\|(h, k, f(a+h, b+k) - f(a,b))\|} = 0.$$

Proof. Since $\theta_{h,k}$ is the angle between n and $(h, k, f(a+h, b+k) - f(a,b))$, we have by (2) in Section 5.1 that

$$\cos \theta_{h,k} = \frac{n \cdot (h, k, f(a+h, b+k) - f(a,b))}{\|n\| \, \|(h, k, f(a+h, b+k) - f(a,b))\|}$$

$$= \frac{1}{\|n\|} \cdot \frac{f(a+h, b+k) - f(a,b) + n_1 h + n_2 k}{\|(h, k, f(a+h, b+k) - f(a,b))\|}.$$

Therefore, (38) holds if and only if $\cos\theta_{h,k} \to 0$ as $(h,k) \to (0,0)$. But $\cos\theta_{h,k} \to 0$ as $(h,k) \to (0,0)$ if and only if $\theta_{h,k} \to \pi/2$. We conclude that $\Pi_{\boldsymbol{n}}(\boldsymbol{c})$ is the unique tangent of $z = f(x,y)$ at $\boldsymbol{c}$ if and only if (38) holds. $\blacksquare$

Comparing (38) with (3) in Section 6.2, we guess that $\nabla f(a,b) = (-n_1, -n_2)$, i.e., that $n_1 = -f_x(a,b)$ and $n_2 = -f_y(a,b)$ (see also Remark 2 in Section 6.2). The following result shows that this guess is correct.

THEOREM 6.25. *Let V be open in $\mathbf{R}^2$, $(a,b) \in V$, $f : V \to \mathbf{R}$, and $\boldsymbol{c} = (a, b, f(a,b))$. Then the graph of $z = f(x,y)$ has a unique nonvertical tangent plane Π at $\boldsymbol{c}$ if and only if f is differentiable at (a,b), in which case $\Pi = \Pi_{\boldsymbol{n}}(\boldsymbol{c})$, where*

$$(39) \qquad \boldsymbol{n} = (-f_x(a,b), -f_y(a,b), 1).$$

PROOF. Choose $r > 0$ such that $B_r(a,b) \subset V$ and suppose $z = f(x,y)$ has a unique nonvertical tangent plane Π at $\boldsymbol{c}$ with normal $\boldsymbol{n} = (n_1, n_2, 1)$. For each $(h,k) \in B_r(\boldsymbol{0}) \setminus \{(0,0)\}$ set

$$\varepsilon = \frac{\Delta z + \delta}{\sqrt{h^2 + k^2 + (\Delta z)^2}},$$

where $\Delta z := f(a+h, b+k) - f(a,b)$ and $\delta := n_1 h + n_2 k$. The expression defining ε is a quadratic in Δz. Solving for Δz, we have

$$(40) \qquad \Delta z = \frac{\delta \pm \varepsilon\sqrt{\delta^2 + (1-\varepsilon^2)(h^2+k^2)}}{\varepsilon^2 - 1}.$$

Notice that $|\delta| = |(n_1, n_2) \cdot (h,k)| \leq \|\boldsymbol{n}\| \, \|(h,k)\|$. Hence, it follows from (40) that

$$(41) \qquad |f(a+h, b+k) - f(a,b) + n_1 h + n_2 k| := |\Delta z + \delta| \leq |G(\varepsilon)| \, \|(h,k)\|$$

where $G(\varepsilon) := (\varepsilon^2\|\boldsymbol{n}\| + \varepsilon\sqrt{\|\boldsymbol{n}\|^2 + (1-\varepsilon^2)})/|\varepsilon^2 - 1|$. By the lemma, $\varepsilon \to 0$ as $(h,k) \to (0,0)$. Consequently, (41) implies

$$0 \leq \frac{|f(a+h, b+k) - f(a,b) + (n_1, n_2) \cdot (h,k)|}{\|(h,k)\|} \leq |G(\varepsilon)| \to 0$$

as $(h,k) \to (0,0)$. Therefore, f is differentiable at $\boldsymbol{a}$ and $\nabla f(a,b) = (-n_1, -n_2)$. Conversely, suppose f is differentiable at (a,b) and define $\boldsymbol{n}$ by (39). Since

$$\|(h, k, f(a+h, b+k) - f(a,b))\| \geq \|(h,k)\|$$

we have

$$0 \leq \frac{|f(a+h, b+k) - f(a,b) + n_1 h + n_2 k|}{\|(h, k, f(a+h, b+k) - f(a,b))\|}$$
$$\leq \frac{|f(a+h, b+k) - f(a,b) - \nabla f(a,b) \cdot (h,k)|}{\|(h,k)\|}.$$

Since f is differentiable, this last quotient converges to zero as $(h, k) \to (0, 0)$. Hence, by the Squeeze Theorem, the middle quotient must also converge to zero as $(h, k) \to (0, 0)$. We conclude by the lemma that $\Pi_{\mathbf{n}}(\mathbf{c})$ is the unique tangent of $z = f(x, y)$ at $\mathbf{c}$. $\blacksquare$

Recall that the total differential of a real-valued differentiable function of n variables at $\mathbf{a}$ is defined by

$$(42) \qquad dz := \sum_{j=1}^{n} \frac{\partial f}{\partial x_j}(\mathbf{a}) \, dx_j.$$

We close this section by showing how to use such differentials to approximate functions.

Remark 1. *Let $f : \mathbf{R}^n \to \mathbf{R}$ be differentiable at $\mathbf{a}$ and $\Delta \mathbf{x} = (\Delta x_1, \ldots, \Delta x_n)$. If the change in $z = f(\mathbf{x})$ as $\mathbf{x}$ moves from $\mathbf{a}$ to $\mathbf{a} + \Delta \mathbf{x}$ is defined by $\Delta z = f(\mathbf{a} + \Delta \mathbf{x}) - f(\mathbf{a})$, then*

$$\frac{\Delta z - dz}{\|\Delta \mathbf{x}\|} \to 0 \qquad as \qquad \Delta \mathbf{x} \to \mathbf{0}.$$

In particular, the differential dz approximates Δz.

PROOF. By definition, if f is differentiable at $\mathbf{a}$, then given $\varepsilon > 0$ there is a $\delta > 0$ such that

$$|f(\mathbf{a} + \mathbf{h}) - f(\mathbf{a}) - \nabla f(\mathbf{a}) \cdot \mathbf{h}| \le \varepsilon \|\mathbf{h}\|$$

for all $\|\mathbf{h}\| < \delta$. If $\mathbf{h} = \Delta \mathbf{x} = (\Delta x_1, \ldots, \Delta x_n)$, then $\Delta z = f(\mathbf{a} + \mathbf{h}) - f(\mathbf{a})$ and $dz = \nabla f(\mathbf{a}) \cdot \mathbf{h}$. Thus $|\Delta z - dz| < \varepsilon \|\Delta \mathbf{x}\|$ for $\|\Delta \mathbf{x}\|$ sufficiently small. $\blacksquare$

Hence, if f is differentiable at $\mathbf{a}$, then the differential of f can be used to approximate the change of f as $\mathbf{x}$ moves from $\mathbf{a}$ to $\mathbf{a} + \mathbf{h}$ for $\mathbf{h}$ sufficiently small.

Example 1. Use differentials to approximate the change of $f(x, y) = x^2 y - y^3$ as (x, y) moves from $(0, 1)$ to $(.02, 1.01)$.

SOLUTION. Let $z = x^2 y - y^3$, $a = 0$, and $b = 1$. Then $dx = .02$ and $dy = .01$. Since $dz = 2xy \, dx + (x^2 - 3y^2) \, dy$, we have

$$\Delta z \approx 0(.02) - 3(.01) = -.03.$$

Note that $\Delta z = f(.02, 1.01) - f(0, 1) = -.029897 \ldots$ is very close to $-.03$. $\blacksquare$

Example 2. Use differentials to approximate $(5.97) \sqrt[4]{16.03}$.

SOLUTION. Let $z = y \sqrt[4]{x}$, $a = 16$, and $b = 6$. Then $dx = .03$ and $dy = -.03$. Since

$$dz = \frac{y}{4\sqrt[4]{x^3}} dx + \sqrt[4]{x} \, dy,$$

we have

$$\Delta z \approx \frac{6(.03)}{4\sqrt[4]{(16)^3}} dx + \sqrt[4]{16}(-.03) \approx -.054375.$$

Thus,

$$z \approx 6\sqrt[4]{16} - .054375 = 11.945625.$$

Note that the actual value of $5.97\sqrt[4]{16.03}$ is $11.945593\ldots$ Thus, our approximation is good to three decimal places. ∎

Example 3. Find the maximum percentage error for the calculated value of the volume of a right circular cylinder if the radius can be measured with a maximum error of 3% and the altitude can be measured with a maximum error of 2%.

SOLUTION. The volume of a right circular cylinder is $V = \pi r^2 h$, where r is the radius and h is the altitude. Hence, the differential of V is $dV = 2\pi r h\, dr + \pi r^2\, dh$. Thus,

$$\frac{dV}{V} = 2\frac{dr}{r} + \frac{dh}{h}.$$

Since the percentage error of a variable x is $\Delta x/x \approx dx/x$, it follows that the maximum percentage error in calculating the volume V is approximately 8%:

$$\frac{dV}{V} = 2(\pm 0.03) + (\pm 0.02) = \pm 0.08. \quad ∎$$

EXERCISES

1. Compute the differential of the each of the following functions.

   a) $z = x^2 + y^2$.   b) $z = \sin(xy)$.   c) $z = \dfrac{xy}{1 + x^2 + y^2}$.

2. Let $w = x^2 y + z$. Use differentials to approximate Δw as (x, y, z) moves from $(1, 2, 1)$ to $(1.01, 1.98, 1.03)$. Compare your approximation with the actual value of Δw.

3. Let $f, g : \mathbf{R}^n \to \mathbf{R}$ be differentiable functions and $u = f(\mathbf{x})$ and $v = g(\mathbf{x})$.

   a) Prove that the sum and product rules hold for differentials, i.e.,

   $$d(u + v) = du + dv, \qquad d(uv) = u\, dv + v\, du.$$

   b) If $v \neq 0$, prove that the quotient rule holds for differentials, i.e.,

   $$d\left(\frac{u}{v}\right) = \frac{v\, du - u\, dv}{v^2}.$$

4. For each of the following functions, find an equation of the tangent plane to $z = f(x, y)$ at $\mathbf{c}$.
   a) $f(x, y) = x^3 \sin y$, $\mathbf{c} = (0, 0, 0)$.
   b) $f(x, y) = x^3 y - xy^3$, $\mathbf{c} = (1, 1, 0)$.

5. The time T it takes for a pendulum to complete one full swing is given by

$$T = 2\pi\sqrt{\frac{L}{g}},$$

where g is the acceleration due to gravity and L is the length of the pendulum. If g can be measured with a maximum error of 1%, how accurately must L be measured (in terms of percentage error) so that the calculated value of T has a maximum error of 2%?

6. Suppose

$$\frac{1}{w} = \frac{1}{x} + \frac{1}{y} + \frac{1}{z}$$

where each variable x, y, z can be measured with a maximum error of $p\%$. Prove that the calculated value of w also has a maximum error of $p\%$.

7. Find all points on the paraboloid $z = x^2 + y^2$ (see Appendix D) where the tangent plane is parallel to the plane $x + y + z = 1$. Find equations of the corresponding tangent planes. Sketch the graphs of these functions to see that your answer agrees with your intuition.

8. Let $\mathcal{H}$ be the hyperboloid of one sheet given by $x^2 + y^2 - z^2 = 1$.

a) Prove that at every point $(a, b, c) \in \mathcal{H}$, $\mathcal{H}$ has a tangent plane whose normal is given by $(-a, -b, c)$.

b) Find an equation of each plane tangent to $\mathcal{H}$ which is perpendicular to the xy plane.

c) Find an equation of each plane tangent to $\mathcal{H}$ which is parallel to the plane $x + y - z = 1$.

Chapter 7

INTEGRATION ON $\mathbf{R}^n$

7.1 JORDAN REGIONS

In this section we define grids (a multidimensional analogue of partitions) and use them to identify special subsets of $\mathbf{R}^n$ called Jordan regions. As partitions were used to define integrals of one-variable functions on intervals, grids will be used to define integrals of multivariable functions on Jordan regions.

Let R be an n-dimensional rectangle, i.e.,

$$(1) \qquad R = [a_1, b_1] \times \cdots \times [a_n, b_n] := \{\boldsymbol{x} \in \mathbf{R}^n : x_j \in [a_j, b_j] \text{ for } j = 1, \ldots, n\}.$$

A *grid* on R is a collection of n-dimensional rectangles $\mathcal{G} = \{R_1, \ldots, R_p\}$ obtained by subdividing the sides of R, i.e., for each $j = 1, \ldots, n$ there are integers $\nu_j \in \mathbf{N}$ and partitions $\mathcal{P}_j = \mathcal{P}_j(\mathcal{G}) = \{x_k^{(j)} : k = 1, \ldots, \nu_j\}$ of $[a_j, b_j]$ such that $\mathcal{G}$ is the collection of rectangles of the form $I_1 \times \cdots \times I_n$, where each $I_j = [x_{k-1}^{(j)}, x_k^{(j)}]$ for some $k = 1, \ldots, \nu_j$ (see Figure 7.1). A grid $\mathcal{G}$ is said to be *finer* than a grid $\mathcal{H}$ if each partition $\mathcal{P}_j(\mathcal{G})$ is finer than the corresponding partition $\mathcal{P}_j(\mathcal{H})$, $j = 1, \ldots, n$.

If R is an n-dimensional rectangle of the form (1), then the *volume* of R is defined to be

$$|R| = (b_1 - a_1) \ldots (b_n - a_n).$$

(When $n = 1$, we shall call $|R|$ the *length* of R and when $n = 2$ we shall call $|R|$ the *area* of R.) Notice that given $\varepsilon > 0$ there exists a rectangle R^* such that $R \subset (R^*)^o$ and $|R^*| = |R| + \varepsilon$. Indeed, since $b_j - a_j + 2\delta \to b_j - a_j$ as $\delta \to 0$, we can choose $\delta > 0$ so small that $R^* := [a_1 - \delta, b_1 + \delta] \times \cdots \times [a_n - \delta, b_n + \delta]$ satisfies $|R^*| = |R| + \varepsilon$.

We want to define the integral of a multivariable function on a variety of sets, for example, the integral of a function of two variables on rectangles, discs, triangles, ellipses, and the integral of a function of three variables on balls, cones, ellipsoids,

336

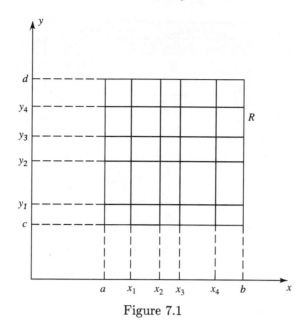

Figure 7.1

pyramids, etc. One property these regions all have in common is that they all have a well-defined "area" or "volume."

How shall we define the volume of a general set E? Not every set will have a well-defined volume, but for those which do, the situation is analogous to the definition of the integral using upper and lower sums. Given $E \subset \mathbf{R}^n$, a subset of some n-dimensional rectangle R, and $\mathcal{G} = \{R_j : j = 1, \ldots, p\}$, a grid on R, the *outer* and *inner sums* of E with respect to $\mathcal{G}$ are defined by

$$V(E;\mathcal{G}) := \sum_{R_j \cap \overline{E} \neq \emptyset} |R_j| \quad \text{and} \quad v(E;\mathcal{G}) := \sum_{R_j \subset E^\circ} |R_j|,$$

where the empty sum is interpreted to be zero. Thus, $V(\emptyset;\mathcal{G}) = v(\emptyset;\mathcal{G}) = 0$ and $v(E;\mathcal{G}) = 0$ for all grids $\mathcal{G}$ and all sets E satisfying $E^\circ = \emptyset$.

Inner and outer sums will be used to define inner and outer volume in the same way that upper and lower sums were used to define upper and lower integrals (see Definition 7.1 below). In this regard, notice that if $\mathcal{G}$ is fine enough, each inner and outer sum of a set E is a sum of volumes of rectangles which approximates the "volume" of E if it exists (see Figure 7.2); $V(E;\mathcal{G})$ overestimates the "volume" of E and $v(E;\mathcal{G})$ underestimates the "volume" of E. (In Figure 7.2, the lighter shading covers the underestimate $v(E;\mathcal{G})$; the lighter and heavier shading covers the overestimate $V(E;\mathcal{G})$.)

Before we define inner and outer volume, we prove three elementary results about inner and outer sums. Our first result reveals why the sum $V(E;\mathcal{G})$ is taken over rectangles which intersect the closure of E and the sum $v(E;\mathcal{G})$ is taken over rectangles which are contained in the interior of E.

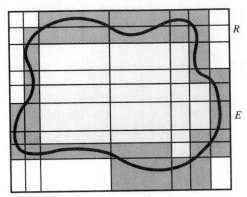

Figure 7.2

Remark 1. *Let R be an n-dimensional rectangle and E be a subset of R. For all grids $\mathcal{G}$ on R,*

$$V(E;\mathcal{G}) - v(E;\mathcal{G}) = V(\partial E;\mathcal{G}).$$

PROOF. Let $R_j \in \mathcal{G}$. Recall from Theorems 5.9 and 5.7 that $\partial E = \overline{E} \setminus E^o$ is closed. Thus, $R_j \cap \overline{\partial E} \neq \emptyset$ implies R_j intersects $\overline{E}$ but R_j is not a subset of E^o. On the other hand, if R_j intersects $\overline{E}$, R_j is not a subset of E^o, and $R_j \cap \overline{\partial E} = \emptyset$, then the pair E^o, $(\mathbf{R}^n \setminus E)^o$ separates R_j, a contradiction since all rectangles are connected. It follows that $R_j \cap \overline{\partial E} \neq \emptyset$ if and only if R_j intersects $\overline{E}$ and R_j is not a subset of E^o. Thus, by definition,

$$V(\partial E;\mathcal{G}) = V(E;\mathcal{G}) - v(E;\mathcal{G}). \quad \blacksquare$$

The next result shows that as the grid gets finer, the estimates $V(E;\mathcal{G})$ and $v(E;\mathcal{G})$ of the "volume" of E get better.

Remark 2. *Let R be an n-dimensional rectangle, E be a subset of R, and $\mathcal{G}$, $\mathcal{H}$ be grids on R. If $\mathcal{G}$ is finer than $\mathcal{H}$, then*

$$0 \le v(E;\mathcal{H}) \le v(E;\mathcal{G}) \le V(E;\mathcal{G}) \le V(E;\mathcal{H}).$$

PROOF. Since $v(E;\mathcal{H})$ is either zero or a sum of nonnegative terms, it is clear that $v(E;\mathcal{H}) \ge 0$ for all grids $\mathcal{H}$.

The third inequality is trivial if E^o is empty because in this case $v(E;\mathcal{G}) = 0$ for all grids $\mathcal{G}$. If E^o is nonempty, then the sum defining $V(E;\mathcal{G})$ contains every term from the sum defining $v(E;\mathcal{G})$ because $R_j \subset E^o$ implies $R_j \cap \overline{E} \neq \emptyset$. Thus, $v(E;\mathcal{G}) \le V(E;\mathcal{G})$ in either case.

It remains to verify the second and fourth inequalities. Since their proofs are similar, we supply the details only for the fourth inequality. Since $\mathcal{G}$ is finer than $\mathcal{H}$, each $Q \in \mathcal{H}$ is a finite union of R_j's in $\mathcal{G}$. If $Q \cap \overline{E} \neq \emptyset$, then some of the R_j's in Q intersect $\overline{E}$ and others might not (see Figure 7.3 where the darker lines

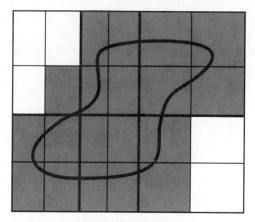

Figure 7.3

represent the grid $\mathcal{H}$, the lighter lines represent $\mathcal{G} \setminus \mathcal{H}$, and the R_j's which intersect $\overline{E}$ are shaded). Let $\mathcal{I}_1 = \{R \in \mathcal{G} : R \cap \overline{E} \neq \emptyset\}$ and $\mathcal{I}_2 = \{R \in \mathcal{G} \setminus \mathcal{I}_1 : R \subseteq Q$ for some $Q \in \mathcal{H}$ with $Q \cap \overline{E} \neq \emptyset\}$. Then

$$V(E;\mathcal{H}) = \sum_{R \in \mathcal{I}_1} |R| + \sum_{R \in \mathcal{I}_2} |R| \geq \sum_{R \in \mathcal{I}_1} |R| = V(E;\mathcal{G}). \quad \blacksquare$$

Remark 3. *Let R be an n-dimensional rectangle and E be a subset of R. If $\mathcal{G}$ and $\mathcal{H}$ are grids on R, then*

(2) $$0 \leq v(E;\mathcal{G}) \leq V(E;\mathcal{H}).$$

PROOF. Let $\mathcal{I}$ be a grid finer than both $\mathcal{G}$ and $\mathcal{H}$. (Such a grid can be constructed by taking $\mathcal{P}_j(\mathcal{I}) = \mathcal{P}_j(\mathcal{G}) \cup \mathcal{P}_j(\mathcal{H})$ for $j = 1, \dots, n$.) By Remark 2,

$$0 \leq v(E;\mathcal{G}) \leq v(E;\mathcal{I}) \leq V(E;\mathcal{I}) \leq V(E;\mathcal{H}). \quad \blacksquare$$

Using the sums $v(E;\mathcal{G})$ and $V(E;\mathcal{G})$, we define inner and outer volume.

DEFINITION 7.1. Let $E \subset \mathbf{R}^n$ and R be an n-dimensional rectangle which satisfies $E \subset R$. The *inner volume* of E is defined by

$$\underline{\text{Vol}}(E) := \sup\{v(E;\mathcal{G}) : \ \mathcal{G} \text{ ranges over all grids on } R\},$$

and the *outer volume* of E is defined by

$$\overline{\text{Vol}}(E) := \inf\{V(E;\mathcal{G}) : \ \mathcal{G} \text{ ranges over all grids on } R\}.$$

There is a logical difficulty with Definition 7.1 which we must address before continuing. Does the definition of inner and outer volume of a set E depend on the rectangle R used to generate the grids $\mathcal{G}$? To show that the answer to this question is no, let R and Q be rectangles which contain E. Since the intersection of two rectangles is a rectangle, we may suppose $E \subset Q \subset R$. To simplify matters, we first suppose that $\overline{E} \subset Q^o$. Let $\mathcal{G}$ be any grid on R and let $\mathcal{G}_0$ be the grid on R formed by "adding the endpoints of Q to $\mathcal{G}$," i.e., if $Q = [c_1, d_1] \times \cdots \times [c_n, d_n]$, then $\mathcal{P}_j(\mathcal{G}_0) = \mathcal{P}_j(\mathcal{G}) \cup \{c_j, d_j\}$. Clearly, $\mathcal{G}_0$ is finer than $\mathcal{G}$ and $\mathcal{H}_0 := \mathcal{G}_0 \cap Q$ is a grid on Q. Since $\overline{E} \subset Q^o \subset R^o$ implies $V(E; \mathcal{H}_0) = V(E; \mathcal{G}_0)$, it follows from Remark 2 that

$$\inf_{\mathcal{H} \text{ on } Q} V(E; \mathcal{H}) \leq V(E; \mathcal{H}_0) = V(E; \mathcal{G}_0) \leq V(E; \mathcal{G}).$$

Taking the infimum of this inequality over all grids $\mathcal{G}$ on R, we obtain

$$\inf_{\mathcal{H} \text{ on } Q} V(E; \mathcal{H}) \leq \inf_{\mathcal{G} \text{ on } R} V(E; \mathcal{G}).$$

To obtain the reverse inequality, let $\mathcal{H}$ be any grid on Q. Let $\mathcal{H}_0$ be the grid on R formed by "adding the endpoints of R to $\mathcal{H}$." Since $\overline{E} \subset Q^o \subset R^o$, we have

$$V(E; \mathcal{H}) = V(E; \mathcal{H}_0) \geq \inf_{\mathcal{G} \text{ on } R} V(E; \mathcal{G}).$$

Taking the infimum of this inequality over all grids $\mathcal{H}$ on Q, we obtain

$$\inf_{\mathcal{H} \text{ on } Q} V(E; \mathcal{H}) \geq \inf_{\mathcal{G} \text{ on } R} V(E; \mathcal{G}).$$

Therefore,

$$\inf_{\mathcal{H} \text{ on } Q} V(E; \mathcal{H}) = \inf_{\mathcal{G} \text{ on } R} V(E; \mathcal{G})$$

when $\overline{E} \subset Q^o \subset R^o$. To handle the general case $E \subset Q \subset R$, let $\varepsilon > 0$ and choose rectangles Q^*, R^* such that $Q \subset (Q^*)^o \subset (R^*)^o$, $|Q^*| = |Q| + \varepsilon$, and $|R^*| = |R| + \varepsilon$. Then $\overline{E} \subseteq Q \subset (Q^*)^o \subset (R^*)^o$, and it follows that

$$\inf_{\mathcal{H} \text{ on } Q} V(E; \mathcal{H}) \leq \inf_{\mathcal{H} \text{ on } Q^*} V(E; \mathcal{H}) = \inf_{\mathcal{G} \text{ on } R^*} V(E; \mathcal{G}) \leq \inf_{\mathcal{G} \text{ on } R} V(E; \mathcal{G}) + \varepsilon,$$

and

$$\inf_{\mathcal{H} \text{ on } Q} V(E; \mathcal{H}) + \varepsilon \geq \inf_{\mathcal{H} \text{ on } Q^*} V(E; \mathcal{H}) = \inf_{\mathcal{G} \text{ on } R^*} V(E; \mathcal{G}) \geq \inf_{\mathcal{G} \text{ on } R} V(E; \mathcal{G}).$$

We conclude that the definition of outer volume does not depend on the rectangle R. A similar argument shows that the same is true for inner volume.

Having established that inner and outer volume are well defined, we continue our exposition. The sums $V(E; \mathcal{G})$ (respectively, $v(E; \mathcal{G})$) overestimate (respectively, underestimate) the "volume" of E. If E has a well defined volume, we expect $V(E; \mathcal{G}) - v(E; \mathcal{G})$ to approach zero as $\mathcal{G}$ gets finer, i.e., $\overline{\text{Vol}}(E) = \underline{\text{Vol}}(E)$. (In Figure 7.2, the shaded rectangles represent the difference $V(E; \mathcal{G}) - v(E; \mathcal{G})$.) Thus, we make the following definition.

DEFINITION 7.2. A subset E of $\mathbf{R}^n$ is called a *Jordan region* if $\overline{\mathrm{Vol}}(E) = \underline{\mathrm{Vol}}(E)$. In this case we shall denote $\overline{\mathrm{Vol}}(E) = \underline{\mathrm{Vol}}(E)$ by $\mathrm{Vol}(E)$ and call it the *volume* (or *Jordan content*) of E.

If $n = 1$, we shall call $\mathrm{Vol}(E)$ the *length* of E and denote it by $\ell(E)$; if $n = 2$ we shall call $\mathrm{Vol}(E)$ the *area* of E and denote it by $\mathrm{Area}(E)$.

A set E is said to be *of volume zero* if $\mathrm{Vol}(E) = 0$. Notice that $\emptyset$ is of volume zero. Indeed, since the empty sum is interpreted to be zero, $\underline{\mathrm{Vol}}(\emptyset) = \overline{\mathrm{Vol}}(\emptyset) = 0$.

A collection of Jordan regions $\{E_\ell\}_{\ell \in \mathbf{N}}$ is said to be *nonoverlapping* (respectively, *pairwise disjoint*) if $E_j \cap E_k$ is of volume zero (respectively, $E_j \cap E_k = \emptyset$) for $j \neq k$. Notice that a collection of pairwise disjoint sets is nonoverlapping but not conversely (see Exercise 6).

We make several elementary observations about Jordan regions and volume. Our first observation is the following characterization of Jordan regions.

Remark 4. *For every bounded subset E of $\mathbf{R}^n$, $0 \leq \underline{\mathrm{Vol}}(E) \leq \overline{\mathrm{Vol}}(E)$. Moreover, E is a Jordan region if and only if $\overline{\mathrm{Vol}}(E) \leq \underline{\mathrm{Vol}}(E)$.*

PROOF. It suffices to prove the first string of inequalities. Let R be a rectangle containing E and fix a grid $\mathcal{H}$ on R. Taking the supremum of (2) over all grids $\mathcal{G}$ on R we obtain $0 \leq \underline{\mathrm{Vol}}(E) \leq V(E; \mathcal{H})$. Taking the infimum of this last inequality over all grids $\mathcal{H}$ on R finishes the proof. ∎

Our second observation is that when restricted to rectangles, $\mathrm{Vol}(R)$ agrees with the usual definition of volume.

Remark 5. *If R is a rectangle, then R is a Jordan region and $\mathrm{Vol}(R) = |R|$.*

PROOF. By Remark 4, it suffices to show that $\underline{\mathrm{Vol}}(R) \geq |R| \geq \overline{\mathrm{Vol}}(R)$. Since $\mathcal{G} = \{R\}$ is a grid on R, it is clear by definition that $|R| = V(R; \mathcal{G}) \geq \overline{\mathrm{Vol}}(R)$. On the other hand, let $\varepsilon > 0$ and suppose

$$R = [a_1, b_1] \times \cdots \times [a_n, b_n].$$

Since $b_j - a_j - 2\delta \to b_j - a_j$ as $\delta \to 0$, we can choose $\delta > 0$ so small that if

$$Q = [a_1 + \delta, b_1 - \delta] \times \cdots \times [a_n + \delta, b_n - \delta]$$

then $|R| - |Q| < \varepsilon$. Let $\mathcal{G}$ be the grid on R determined by

$$\mathcal{P}_j(\mathcal{G}) = \{a_j, a_j + \delta, b_j - \delta, b_j\}.$$

Then Q is the only rectangle in $\mathcal{G}$ which satisfies $Q \subset R^\circ$. Hence, by definition,

$$\underline{\mathrm{Vol}}(R) \geq v(R; \mathcal{G}) = |Q| > |R| - \varepsilon,$$

i.e., $\underline{\mathrm{Vol}}(R) > |R| - \varepsilon$. Taking the limit of this last inequality as $\varepsilon \to 0$, we conclude that $\underline{\mathrm{Vol}}(R) \geq |R|$. ∎

Our third observation concerns sets of volume zero.

Remark 6. *Let* $E \subset \mathbf{R}^n$. *Then* E *is a Jordan region of volume zero if and only if* $\overline{\text{Vol}}(E) = 0$. *Moreover, if* E *is a Jordan region of volume zero and* $E_0 \subseteq E$, *then* E_0 *is a Jordan region of volume zero.*

PROOF. If $\text{Vol}(E) = 0$, then by Definition 7.2, $\overline{\text{Vol}}(E) = 0$. Conversely, suppose $\overline{\text{Vol}}(E) = 0$. Then by Remark 4 and Definition 7.2, $\underline{\text{Vol}}(E) = \overline{\text{Vol}}(E) = \text{Vol}(E) = 0$.

To finish the proof, suppose E is a Jordan region of volume zero and $E_0 \subseteq E$. By hypothesis and the Monotone Property of Suprema, $0 \le \overline{\text{Vol}}(E_0) \le \overline{\text{Vol}}(E) = 0$. Hence, by what we just proved, E_0 is a Jordan region and $\text{Vol}(E_0) = 0$. ∎

According to Remark 6, *any* subset of a set of volume zero is a Jordan region. This statement is false if the hypothesis "of volume zero" is omitted.

Remark 7. *There exist subsets of the unit square* $R = [0,1] \times [0,1]$ *which are not Jordan regions.*

PROOF. Consider the set of ordered pairs

$$E = \{(x,y) : x, y \in \mathbf{Q} \cap [0,1]\}.$$

Since no rectangle satisfies $R_j \subseteq E$ (much less $R_j \subset E^o$), $v(E; \mathcal{G}) = 0$ for all grids $\mathcal{G}$. On the other hand, if R_j is any rectangle contained in R, then $R_j \cap E \ne \emptyset$; hence, $R_j \cap \overline{E} \ne \emptyset$. Thus, $V(E; \mathcal{G}) = 1$ for all grids $\mathcal{G}$ on R. It follows that

$$\underline{\text{Vol}}(E) = 0 < 1 = \overline{\text{Vol}}(E). \quad \blacksquare$$

Evidently, some sets are Jordan regions and others are not. How shall we determine whether or not a given set is a Jordan region? Since not even every open set is a Jordan region (see Spivak [12], p. 56), we cannot answer this question using topology alone.

Let us examine this question from another point of view. Notice that by definition, if E is a Jordan region, then given $\varepsilon > 0$ there is a grid $\mathcal{G}$ such that $V(E; \mathcal{G}) - v(E; \mathcal{G}) < \varepsilon$. Since by Remark 1, $V(E; \mathcal{G}) - v(E; \mathcal{G}) = V(\partial E; \mathcal{G})$ (see Figure 7.2), we guess that a set is a Jordan region if and only if its boundary is of volume zero. The following result shows this guess is correct when E is bounded.

THEOREM 7.1. *Let* $E \subset \mathbf{R}^n$ *be bounded. Then* E *is a Jordan region if and only if its boundary* ∂E *is of volume zero.*

PROOF. Let R be a rectangle which contains E and $\mathcal{G}$ be a grid on R. By Remark 1 and Definition 7.1, $V(\partial E; \mathcal{G}) = V(E; \mathcal{G}) - v(E; \mathcal{G}) \ge \overline{\text{Vol}}(E) - \underline{\text{Vol}}(E)$. Taking the infimum of this inequality over all grids $\mathcal{G}$, we have

$$(3) \qquad\qquad \overline{\text{Vol}}(\partial E) \ge \overline{\text{Vol}}(E) - \underline{\text{Vol}}(E).$$

On the other hand, given $\varepsilon > 0$ choose grids $\mathcal{H}_1$ and $\mathcal{H}_2$ such that

$$\overline{\text{Vol}}(E) + \varepsilon > V(E; \mathcal{H}_1) \quad \text{and} \quad \underline{\text{Vol}}(E) - \varepsilon < v(E; \mathcal{H}_2).$$

If $\mathcal{G}$ is a grid on R which is finer than both $\mathcal{H}_1$ and $\mathcal{H}_2$, it follows from Remark 2 that

$$\overline{\mathrm{Vol}}\,(E) + \varepsilon > V(E;\mathcal{G}) \quad \text{and} \quad \underline{\mathrm{Vol}}\,(E) - \varepsilon < v(E;\mathcal{G}).$$

Subtracting these inequalities, we see by Remark 1 that

$$V(\partial E;\mathcal{G}) = V(E;\mathcal{G}) - v(E;\mathcal{G}) < \overline{\mathrm{Vol}}\,(E) - \underline{\mathrm{Vol}}\,(E) + 2\varepsilon.$$

Therefore, $\overline{\mathrm{Vol}}\,(\partial E) < \overline{\mathrm{Vol}}\,(E) - \underline{\mathrm{Vol}}\,(E) + 2\varepsilon$. Taking the limit of this inequality as $\varepsilon \to 0$, we obtain $\overline{\mathrm{Vol}}\,(\partial E) \le \overline{\mathrm{Vol}}\,(E) - \underline{\mathrm{Vol}}\,(E)$. This inequality, together with (3), proves $\overline{\mathrm{Vol}}\,(\partial E) = \overline{\mathrm{Vol}}\,(E) - \underline{\mathrm{Vol}}\,(E)$. Hence, by Definition 7.2, E is a Jordan region if and only if $\overline{\mathrm{Vol}}\,(\partial E) = 0$. We conclude by Remark 6 that E is a Jordan region if and only if ∂E is a Jordan region of volume zero. ∎

This result can be used to show all balls are Jordan regions (see Exercise 9 below). We shall find a formula for the volume of a ball in Section 7.6.

In view of Theorem 7.1, we need a quantitative characterization of sets of volume zero. One such characterization is the following.

THEOREM 7.2. *Let $E \subset \mathbf{R}^n$. Then $\mathrm{Vol}\,(E) = 0$ if and only if given $\varepsilon > 0$ there is a finite collection of cubes Q_k of the same size, i.e., all with sides of length s, such that*

$$\overline{E} \subset \bigcup_{k=1}^{p} Q_k \quad \text{and} \quad \sum_{k=1}^{p} |Q_k| < \varepsilon.$$

PROOF. If $\mathrm{Vol}\,(E) = 0$, then by definition there exists a grid $\mathcal{G} = \{R_1, \ldots, R_q\}$ such that

$$\overline{E} \subset \bigcup_{j=1}^{q} R_j \quad \text{and} \quad \sum_{j=1}^{q} |R_j| < \frac{\varepsilon}{2}.$$

By increasing the size of the R_j's slightly, we may suppose that the sides of each R_j have rational lengths and $\sum_{j=1}^{q} |R_j| < \varepsilon$. (These rectangles may no longer be nonoverlapping.) The lengths of the sides of the R_j's have a common denominator, say d. By using a grid fine enough, we can divide each R_j into cubes $Q_k^{(j)}$, for $k = 1, 2, \ldots \nu_j$ and some choice of $\nu_j \in \mathbf{N}$, such that each $Q_k^{(j)}$ has sides of common length $s = 1/d$. Since $|R_j| = \sum_{k=1}^{\nu_j} |Q_k^{(j)}|$, it follows that

$$\sum_{j=1}^{q} \sum_{k=1}^{\nu_j} |Q_k^{(j)}| = \sum_{j=1}^{q} |R_j| < \varepsilon.$$

Conversely, if such cubes exist let R be a rectangle which contains the union of the Q_k's and suppose

$$Q_k = [a_1^{(k)}, b_1^{(k)}] \times \cdots \times [a_n^{(k)}, b_n^{(k)}].$$

For each $j = 1, 2, \ldots, n$, the endpoints $\{a_j^{(1)}, b_j^{(1)}, \ldots, a_j^{(p)}, b_j^{(p)}\}$ can be arranged in increasing order to form a partition of the jth side of R. Thus there is a grid

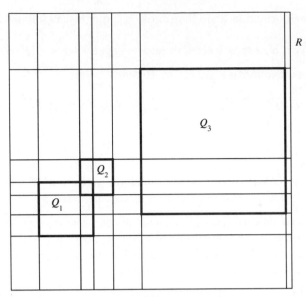

Figure 7.4

$\mathcal{G} = \{R_1, \ldots, R_q\}$ so fine that each Q_k is a union of the R_j's (see Figure 7.4). Since $V(E;\mathcal{G}) \le \sum_{k=1}^{p} |Q_k| < \varepsilon$, it follows that $\overline{\mathrm{Vol}}(E) < \varepsilon$ for all $\varepsilon > 0$. We conclude that $\mathrm{Vol}(E) = 0$. ∎

This result can be used to prove that the collection of Jordan regions is closed under finite unions (see also Exercise 4 below).

Remark 8. *If E_1 and E_2 are Jordan regions, then $E_1 \cup E_2$ is a Jordan region and*

$$\mathrm{Vol}(E_1 \cup E_2) \le \mathrm{Vol}(E_1) + \mathrm{Vol}(E_2).$$

PROOF. By Theorem 7.2, the union of two sets of volume zero is a set of volume zero. Since Theorem 5.10 implies $\partial(E_1 \cup E_2) \subseteq \partial E_1 \cup \partial E_2$, it follows from Theorem 7.1 that $E_1 \cup E_2$ is a Jordan region.

To estimate the volume of this set, let $\mathcal{G}$ be a grid on a rectangle which contains $E_1 \cup E_2$. If R_j intersects $\overline{E_1 \cup E_2}$, then R_j intersects $\overline{E}_1$ or $\overline{E}_2$ (or both). Hence, every term of the sum $V(E_1 \cup E_2;\mathcal{G})$ appears in $V(E_1;\mathcal{G})$ or $V(E_2;\mathcal{G})$ (or both). Consequently,

$$\mathrm{Vol}(E_1 \cup E_2) \le V(E_1 \cup E_2;\mathcal{G}) \le V(E_1;\mathcal{G}) + V(E_2;\mathcal{G})$$

for any grid $\mathcal{G}$. Taking the infimum of this last inequality over all grids $\mathcal{G}$, we obtain $\mathrm{Vol}(E_1 \cup E_2) \le \mathrm{Vol}(E_1) + \mathrm{Vol}(E_2)$. ∎

We close this section by showing that under suitable conditions, the image of a Jordan region is a Jordan region. (We shall use this result later to obtain a change of variables formula for integrals of multivariable functions.)

THEOREM 7.3. *Let V be a bounded open set in $\mathbf{R}^n$, E be a Jordan region, and $\phi : V \to \mathbf{R}^n$ be 1–1 and continuously differentiable on V. If $\overline{E} \subset V$ and $\Delta_\phi \neq 0$ on V, then $\phi(E)$ is a Jordan region.*

STRATEGY: The idea behind the proof is the following. By Theorem 7.2, a closed set is of volume zero if and only if it can be covered by cubes of small volume. If E is a Jordan region, then ∂E is a set of volume zero by Theorem 7.1, hence can be covered by cubes of small volume. Since $\|\phi(\boldsymbol{x}) - \phi(\boldsymbol{y})\|$ is small when $\|\boldsymbol{x} - \boldsymbol{y}\|$ is, ϕ takes a set of small diameter to a set of small diameter, i.e., a cube of small volume to a subset of a cube of small volume. Hence, it seems likely that $\phi(\partial E)$ can be covered by cubes of small volume, i.e., is also of volume zero. Since $\phi(E)$ is a Jordan region if and only if $\partial(\phi(E))$ is of volume zero, it would remain to prove that

$$(4) \qquad\qquad \partial(\phi(E)) \subseteq \phi(\partial E).$$

(Here alone the condition $\Delta_\phi \neq 0$ enters the picture.) We now make this precise.

PROOF. By Theorem 6.19, the set $\phi(E^o)$ is open and by Theorem 5.26, the set $\phi(\overline{E})$ is compact, hence closed. It follows from Theorem 5.9 that $\phi(E^o) \subseteq (\phi(E))^o$ and $\phi(\overline{E}) \supseteq \overline{\phi(E)}$. Therefore,

$$\partial(\phi(E)) = \overline{\phi(E)} \setminus (\phi(E))^o \subseteq \phi(\overline{E}) \setminus \phi(E^o) = \phi(\overline{E} \setminus E^o) = \phi(\partial E),$$

i.e., (4) holds. Hence, by Theorem 7.1, it suffices to show $\mathrm{Vol}\,(\phi(\partial E)) = 0$.

For each $\boldsymbol{x} \in \overline{E}$, choose $r = r(\boldsymbol{x}) > 0$ such that $\overline{B_r(\boldsymbol{x})} \subset V$. Since E is bounded, $\overline{E}$ is compact. Hence, we can choose points $\boldsymbol{x}_j \in \overline{E}$ and radii $r_j = r(\boldsymbol{x}_j)$ such that

$$\overline{E} \subset \bigcup_{j=1}^{N} B_{r_j}(\boldsymbol{x}_j) \subset H := \bigcup_{j=1}^{N} \overline{B_{r_j}(\boldsymbol{x}_j)}.$$

Evidently, H is compact and $\overline{E} \subset H^o \subset H \subset V$.

Since rectangles are compact and convex, apply Corollary 6.16 to choose a constant $M > 0$ (which depends only on ϕ and H) such that

$$(5) \qquad\qquad \|\phi(\boldsymbol{x}) - \phi(\boldsymbol{y})\| \leq M\|\boldsymbol{x} - \boldsymbol{y}\|$$

for all $\boldsymbol{x}, \boldsymbol{y} \in R$ and any rectangle $R \subseteq H$. Let $\varepsilon > 0$. Since ∂E is a closed subset of H^o of volume zero, use Theorem 7.2 to choose cubes $Q_1, \ldots, Q_p$ all with sides of length s such that $Q_j \subset H$,

$$(6) \qquad\qquad \partial E \subset \bigcup_{j=1}^{p} Q_j, \quad \text{and} \quad \sum_{j=1}^{p} |Q_j| < \frac{\varepsilon}{M^n n^{n/2}}.$$

Since each $|Q_j| = s^n$, (6) implies

$$s < \left(\frac{\varepsilon}{p M^n n^{n/2}} \right)^{1/n} = \left(\frac{\varepsilon}{p} \right)^{1/n} \cdot \frac{1}{M\sqrt{n}}.$$

Since $\|\mathbf{x} - \mathbf{y}\| \le s\sqrt{n}$ for any $\mathbf{x}, \mathbf{y} \in Q_j$, it follows from (5) that each $\phi(Q_j)$ is contained in a cube R_j with sides of length $< (\varepsilon/p)^{1/n}$. Therefore,

$$\phi(\partial E) \subset \bigcup_{j=1}^{p} R_j \quad \text{and} \quad \sum_{j=1}^{p} |R_j| < \varepsilon.$$

Finally, we note that since ∂E is compact, $\phi(\partial E)$ is compact, hence closed. Thus, we have shown that $\overline{\phi(\partial E)}$ can be covered by cubes of small volume. We conclude by Theorem 7.2 that $\phi(\partial E)$ is of volume zero. $\blacksquare$

EXERCISES

1. Prove that every finite subset of $\mathbf{R}^n$ is a Jordan region of volume zero. Show this is not true if "finite" is replaced by "countable."

2. Let $E \subset \mathbf{R}^n$ be a Jordan region.

   a) Prove E^o and $\overline{E}$ are Jordan regions.
   b) Prove $\text{Vol}(E^o) = \text{Vol}(\overline{E}) = \text{Vol}(E)$.
   c) Prove that $\text{Vol}(E) > 0$ if and only if $E^o \ne \emptyset$.

3. **This exercise is used in Section 7.6.** Let $E \subset \mathbf{R}^n$. The *translation* of E by an $\mathbf{x} \in \mathbf{R}^n$ is the set

$$\mathbf{x} + E = \{\mathbf{y} \in \mathbf{R}^n : \mathbf{y} = \mathbf{x} + \mathbf{z} \text{ for some } \mathbf{z} \in E\}$$

and the *dilation* of E by a scalar $\alpha > 0$ is the set

$$\alpha E = \{\mathbf{y} \in \mathbf{R}^n : \mathbf{y} = \alpha \mathbf{z} \text{ for some } \mathbf{z} \in E\}.$$

   a) Prove that E is a Jordan region if and only if $\mathbf{x} + E$ is a Jordan region, in which case $\text{Vol}(\mathbf{x} + E) = \text{Vol}(E)$.
   b) Prove that E is a Jordan region if and only if αE is a Jordan region, in which case $\text{Vol}(\alpha E) = \alpha^n \, \text{Vol}(E)$.

4. **This exercise is used in Section 7.5.** Suppose E_1, E_2 are Jordan regions in $\mathbf{R}^n$.

   a) Prove that if $E_1 \subseteq E_2$, then $\text{Vol}(E_1) \le \text{Vol}(E_2)$.
   b) Prove that $E_1 \cap E_2$ and $E_1 \setminus E_2$ are Jordan regions.
   c) Prove that if E_1, E_2 are nonoverlapping, then $\text{Vol}(E_1 \cup E_2) = \text{Vol}(E_1) + \text{Vol}(E_2)$.
   d) If $E_2 \subseteq E_1$, prove $\text{Vol}(E_1 \setminus E_2) = \text{Vol}(E_1) - \text{Vol}(E_2)$.
   e) Prove that $\text{Vol}(E_1 \cup E_2) = \text{Vol}(E_1) + \text{Vol}(E_2) - \text{Vol}(E_1 \cap E_2)$.

5. a) Let $f : [a, b] \to \mathbf{R}$ be continuous on $[a, b]$. Prove that the graph of $y = f(x)$, $x \in [a, b]$, is a Jordan region in $\mathbf{R}^2$ of area zero.
   b) Does part a) hold if "continuous" is replaced by "integrable"? How about "bounded"?

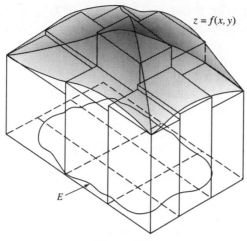

Figure 7.5

6. Prove that every grid is a nonoverlapping collection of Jordan regions.

$\boxed{7}$. **This exercise is used in Sections 7.4 and 7.5.** Suppose V is a bounded open set in $\mathbf{R}^n$ and $\phi : V \to \mathbf{R}^n$ is continuously differentiable on V.

   a) Show that if E is of volume zero and $\overline{E} \subset V$, then $\phi(E)$ is of volume zero.

   b) If ϕ is 1–1, $\Delta_\phi(\boldsymbol{x}) \neq 0$ for $\boldsymbol{x} \in V$, and $\{E_k\}_{k \in \mathbf{N}}$ is a nonoverlapping collection of Jordan regions with $\overline{E_k} \subset V$ for all $k \in \mathbf{N}$, prove that $\{\phi(E_k)\}_{k \in \mathbf{N}}$ is a nonoverlapping collection of Jordan regions.

8. Show that if $E \subset \mathbf{R}^n$ is bounded and has only finitely many cluster points, then E is a Jordan region.

9. a) Prove that the boundary of an open ball $B_r(\boldsymbol{a})$ is given by

$$\partial B_r(\boldsymbol{a}) = \{\boldsymbol{x} : \|\boldsymbol{x} - \boldsymbol{a}\| = r\}.$$

   b) Prove that $B_r(\boldsymbol{a})$ is a Jordan region for all $\boldsymbol{a} \in \mathbf{R}^n$ and all $r \geq 0$.

*10. A set $E \subset \mathbf{R}^n$ is said to be of *measure zero* if given $\varepsilon > 0$ there is a countable collection of rectangles $R_1, \ldots$ which covers E such that $\sum_{k=1}^{\infty} |R_k| < \varepsilon$.

   a) Prove that if $E \subset \mathbf{R}^n$ is of volume zero, then E is of measure zero.

   b) Prove that if $E \subset \mathbf{R}^n$ is countable, then E is of measure zero.

   c) Prove that there is a set $E \subset \mathbf{R}^2$ of measure zero which does not have zero area, in fact is not even a Jordan region.

7.2   RIEMANN INTEGRABILITY ON JORDAN REGIONS

By analogy with the one-variable case, the integral of a nonnegative function f over a Jordan region E should be the volume of the set $\{(\boldsymbol{x}, t) : \boldsymbol{x} \in E, 0 \leq t \leq f(\boldsymbol{x})\}$. We should be able to approximate this volume by using $(n + 1)$-dimensional rectangles whose heights approximate $t = f(\boldsymbol{x})$ and whose bases belong to some grid on E (see Figure 7.5). This leads us to the following definition (compare with Definition 3.4).

DEFINITION 7.3. Let E be a Jordan region in $\mathbf{R}^n$, $f : E \to \mathbf{R}$ be a bounded function, R be an n-dimensional rectangle such that $E \subseteq R$, and $\mathcal{G} = \{R_1, \ldots, R_p\}$ be a grid on R.

i) The *upper sum* of f on E with respect to $\mathcal{G}$ is

$$U(f, \mathcal{G}) := \sum_{R_j \cap E \neq \emptyset} M_j |R_j|,$$

where $M_j = \sup_{\boldsymbol{x} \in R_j \cap E} f(\boldsymbol{x})$.

ii) The *lower sum* of f on E with respect to $\mathcal{G}$ is

$$L(f, \mathcal{G}) := \sum_{R_j \cap E \neq \emptyset} m_j |R_j|,$$

where $m_j = \inf_{\boldsymbol{x} \in R_j \cap E} f(\boldsymbol{x})$.

iii) The *upper* and *lower integrals* of function f on E are defined by

$$(L) \int_E f(\boldsymbol{x}) \, d\boldsymbol{x} := (L) \int_E f \, dV := \sup_{\mathcal{G}} L(f, \mathcal{G})$$

and

$$(U) \int_E f(\boldsymbol{x}) \, d\boldsymbol{x} := (U) \int_E f \, dV := \inf_{\mathcal{G}} U(f, \mathcal{G}),$$

where the supremum and infimum are taken over all grids $\mathcal{G}$ on R.

iv) The function f is said to be (*Riemann*) *integrable* on E if

(7)
$$(L) \int_E f(\boldsymbol{x}) \, d\boldsymbol{x} = (U) \int_E f(\boldsymbol{x}) \, d\boldsymbol{x}.$$

When f is integrable on E, we denote the common value in (7) by

$$\int_E f(\boldsymbol{x}) \, d\boldsymbol{x} \quad \text{or} \quad \int_E f \, dV$$

and call it the *integral* of f over E. For $n = 2$ (respectively, $n = 3$) we shall frequently denote the integral $\int_E f \, dV$ by $\iint_E f \, dA$ (respectively, by $\iiint_E f \, dV$).

Modifying the proofs of Lemma 2, Remark 6, and Theorem 3.2 in Section 3.1, we can prove the following three results.

Remark 1. *Let E be a Jordan region in $\mathbf{R}^n$, $f : E \to \mathbf{R}$ be bounded, and E be a subset of an n-dimensional rectangle R. If $\mathcal{G}$ and $\mathcal{H}$ are grids on R, then*

$$L(f, \mathcal{G}) \leq U(f, \mathcal{H}).$$

Remark 2. *Let E be a Jordan region in $\mathbf{R}^n$ and $f : E \to \mathbf{R}$ be bounded. The upper and lower integrals of f over E exist, do not depend on the choice of R, and satisfy*

$$(L) \int_E f(\boldsymbol{x}) \, d\boldsymbol{x} \leq (U) \int_E f(\boldsymbol{x}) \, d\boldsymbol{x}.$$

Remark 3. *Let E be a Jordan region in $\mathbf{R}^n$ and $f : E \to \mathbf{R}$ be bounded. Then f is integrable on E if and only if given $\varepsilon > 0$ there is a grid $\mathcal{G}$ such that*

$$U(f,\mathcal{G}) - L(f,\mathcal{G}) \le \varepsilon.$$

The following result is a multidimensional analogue of Theorem 3.1.

THEOREM 7.4. *If E is a Jordan region in $\mathbf{R}^n$ and $f : E \to \mathbf{R}$ is uniformly continuous on E, then f is integrable on E.*

PROOF. Let $\varepsilon > 0$ and R be a rectangle containing E. Since f is uniformly continuous on E, choose $\delta > 0$ such that

$$\|\boldsymbol{x} - \boldsymbol{y}\| < \delta \quad \text{and} \quad \boldsymbol{x}, \boldsymbol{y} \in E \quad \text{imply} \quad |f(\boldsymbol{x}) - f(\boldsymbol{y})| < \frac{\varepsilon}{|R|}.$$

Let $\mathcal{G} = \{R_1, \dots, R_p\}$ be a grid on R so fine that $\boldsymbol{x}, \boldsymbol{y} \in R_j$ implies $\|\boldsymbol{x} - \boldsymbol{y}\| < \delta$. Then $M_j - m_j \le \varepsilon/|R|$ for each j which satisfies $R_j \cap E \ne \emptyset$. It follows that

$$U(f,\mathcal{G}) - L(f,\mathcal{G}) \le \frac{\varepsilon}{|R|} \sum_{R_j \cap E \ne \emptyset} |R_j| \le \varepsilon.$$

We conclude by Remark 3 that f is integrable on E. ∎

The volume of a Jordan region can be computed by integration.

COROLLARY 7.5. *If E is a Jordan region, then*

$$\mathrm{Vol}\,(E) = \int_E 1 \, d\boldsymbol{x}.$$

PROOF. By Theorem 7.4, $f(x) = 1$ is integrable on E. Suppose E^o is nonempty. Then $R_j \subset E^o$ implies $R_j \cap E \ne \emptyset$ and $R_j \cap E \ne \emptyset$ implies $R_j \cap \overline{E} \ne \emptyset$, hence

$$v(E;\mathcal{G}) \le L(1,\mathcal{G}) \le U(1,\mathcal{G}) \le V(E;\mathcal{G})$$

for all grids $\mathcal{G}$. Since the empty sum is defined to be zero, this string of inequalities also holds when E^o is empty. Therefore,

$$\mathrm{Vol}\,(E) = \sup_{\mathcal{G}} v(E;\mathcal{G}) \le \int_E 1 \, d\boldsymbol{x} \le \inf_{\mathcal{G}} V(E;\mathcal{G}) = \mathrm{Vol}\,(E). \quad ∎$$

As in the one-dimensional case, the integral of a sum of functions over a union of regions can be broken into simpler pieces.

THEOREM 7.6 [LINEAR PROPERTIES]. *Let E be a Jordan region in $\mathbf{R}^n$, f, g : $E \to \mathbf{R}$, and α be a scalar.*

   i) *If f, g are integrable on E, then so are αf and $f + g$. In fact,*

$$(8) \qquad\qquad \int_E \alpha f(\boldsymbol{x})\, d\boldsymbol{x} = \alpha \int_E f(\boldsymbol{x})\, d\boldsymbol{x}$$

   and

$$(9) \qquad\qquad \int_E (f(\boldsymbol{x}) + g(\boldsymbol{x}))\, d\boldsymbol{x} = \int_E f(\boldsymbol{x})\, d\boldsymbol{x} + \int_E g(\boldsymbol{x})\, d\boldsymbol{x}.$$

   ii) *If $E_1, E_2 \subseteq E$ are nonoverlapping Jordan regions and f is integrable on E_1 and E_2, then f is integrable on $E_1 \cup E_2$ and*

$$(10) \qquad\qquad \int_{E_1 \cup E_2} f(\boldsymbol{x})\, d\boldsymbol{x} = \int_{E_1} f(\boldsymbol{x})\, d\boldsymbol{x} + \int_{E_2} f(\boldsymbol{x})\, d\boldsymbol{x}.$$

PROOF. We suppose for simplicity that $\alpha > 0$. Let $\varepsilon > 0$ and choose a grid $\mathcal{G}$ such that

$$(11) \qquad\qquad U(f, \mathcal{G}) - \varepsilon < \int_E f(\boldsymbol{x})\, d\boldsymbol{x} < L(f, \mathcal{G}) + \varepsilon.$$

Notice that $U(\alpha f, \mathcal{G}) = \alpha U(f, \mathcal{G})$ and $L(\alpha f, \mathcal{G}) = \alpha L(f, \mathcal{G})$. Multiplying (11) by α we obtain

$$U(\alpha f, \mathcal{G}) - \alpha \varepsilon < \alpha \int_E f(\boldsymbol{x})\, d\boldsymbol{x} < L(\alpha f, \mathcal{G}) + \alpha \varepsilon.$$

In particular,

$$\inf_{\mathcal{G}} U(\alpha f, \mathcal{G}) < \alpha \int_E f(\boldsymbol{x})\, d\boldsymbol{x} + \alpha \varepsilon$$

and

$$\sup_{\mathcal{G}} L(\alpha f, \mathcal{G}) > \alpha \int_E f(\boldsymbol{x})\, d\boldsymbol{x} - \alpha \varepsilon.$$

Taking the limit of these inequalities as $\varepsilon \to 0$, we conclude that

$$\inf_{\mathcal{G}} U(\alpha f, \mathcal{G}) \leq \alpha \int_E f(\boldsymbol{x})\, d\boldsymbol{x} \leq \sup_{\mathcal{G}} L(\alpha f, \mathcal{G}).$$

This proves (8).

   To prove (9), choose a grid $\mathcal{G}$ such that

$$U(f, \mathcal{G}) - \varepsilon < \int_E f(\boldsymbol{x})\, d\boldsymbol{x} < L(f, \mathcal{G}) + \varepsilon$$

and

$$U(g, \mathcal{G}) - \varepsilon < \int_E g(\boldsymbol{x}) \, d\boldsymbol{x} < L(g, \mathcal{G}) + \varepsilon.$$

Adding these inequalities, we have

$$U(f, \mathcal{G}) + U(g, \mathcal{G}) - 2\varepsilon < \int_E f(\boldsymbol{x}) \, d\boldsymbol{x} + \int_E g(\boldsymbol{x}) \, d\boldsymbol{x} < L(f, \mathcal{G}) + L(g, \mathcal{G}) + 2\varepsilon.$$

By definition, $U(f + g, \mathcal{G}) \leq U(f, \mathcal{G}) + U(g, \mathcal{G})$ and $L(f + g, \mathcal{G}) \geq L(f, \mathcal{G}) + L(g, \mathcal{G})$. Therefore,

$$U(f + g, \mathcal{G}) - 2\varepsilon < \int_E f(\boldsymbol{x}) \, d\boldsymbol{x} + \int_E g(\boldsymbol{x}) \, d\boldsymbol{x} < L(f + g, \mathcal{G}) + 2\varepsilon,$$

i.e.,

$$\inf_{\mathcal{G}} U(f + g, \mathcal{G}) \leq \int_E f(\boldsymbol{x}) \, d\boldsymbol{x} + \int_E g(\boldsymbol{x}) \, d\boldsymbol{x} \leq \sup_{\mathcal{G}} L(f + g, \mathcal{G}).$$

This proves (9).

To prove (10), set $\Omega = \partial E_1 \cup \partial E_2 \cup (E_1 \cap E_2)$ and notice that $\mathrm{Vol}\,(\Omega) = 0$. Choose grids $\mathcal{G}_i$, $i = 1, 2, 3$, such that

$$(12) \qquad U(f, \mathcal{G}_i) - \varepsilon < \int_{E_i} f(\boldsymbol{x}) \, d\boldsymbol{x} < L(f, \mathcal{G}_i) + \varepsilon$$

for $i = 1, 2$, and

$$(13) \qquad V(\Omega; \mathcal{G}_3) < \varepsilon.$$

Let $\mathcal{G} = \{R_1, \ldots, R_p\}$ be a grid finer than $\mathcal{G}_1$, $\mathcal{G}_2$, and $\mathcal{G}_3$. Let

$$M = \max\{|M_1|, \ldots, |M_p|\}, \text{ where } M_j = \sup_{\boldsymbol{x} \in R_j \cap (E_1 \cup E_2)} f(\boldsymbol{x}), \text{ for } 1 \leq j \leq p,$$

$\mathcal{I}_1 = \{j : R_j \subseteq E_1\}$, $\mathcal{I}_2 = \{j : R_j \subseteq E_2\}$, and $\mathcal{I}_3 = \{j \notin \mathcal{I}_1 \cup \mathcal{I}_2 : R_j \cap (E_1 \cup E_2) \neq \emptyset\}$. Notice that if $j \in \mathcal{I}_3$ and $R_j \cap (E_1 \cap E_2) = \emptyset$, then $R_j \cap \partial E_1 \neq \emptyset$ or $R_j \cap \partial E_2 \neq \emptyset$. Consequently,

$$(14) \qquad \sum_{j \in \mathcal{I}_3} M_j |R_j| \leq M \, V(\Omega; \mathcal{G}).$$

Since $\mathcal{G}$ is finer than $\mathcal{G}_i$, $i = 1, 2, 3$, it follows from (12), (13), and (14) that

$$U(f, \mathcal{G}) = \sum_{j \in \mathcal{I}_1} M_j |R_j| + \sum_{j \in \mathcal{I}_2} M_j |R_j| + \sum_{j \in \mathcal{I}_3} M_j |R_j|$$
$$\leq U(f, \mathcal{G}_1) + U(f, \mathcal{G}_2) + M \, V(\Omega; \mathcal{G}_3)$$
$$< \int_{E_1} f(\boldsymbol{x}) \, d\boldsymbol{x} + \int_{E_2} f(\boldsymbol{x}) \, d\boldsymbol{x} + (2 + M)\varepsilon.$$

Consequently,

$$\inf_{\mathcal{G}} U(f, \mathcal{G}) \leq \int_{E_1} f(\boldsymbol{x})\, d\boldsymbol{x} + \int_{E_2} f(\boldsymbol{x})\, d\boldsymbol{x}.$$

A similar argument establishes

$$\sup_{\mathcal{G}} L(f, \mathcal{G}) \geq \int_{E_1} f(\boldsymbol{x})\, d\boldsymbol{x} + \int_{E_2} f(\boldsymbol{x})\, d\boldsymbol{x}.$$

Thus, (10) holds. ∎

The following result shows that the value of an integral remains the same when the integrand is changed on a set of volume zero (compare with Exercise 2 in Section 3.1).

COROLLARY 7.7. *Let $\ddot{E}$ be a Jordan region in $\mathbf{R}^n$ and $g : E \to \mathbf{R}$ be bounded.*

i) *If E_0 is a subset of E of volume zero, then g is integrable on E_0 and*

$$\int_{E_0} g(\boldsymbol{x})\, d\boldsymbol{x} = 0.$$

ii) *If $f : E \to \mathbf{R}$ is integrable on E and*

$$E_0 = \{\boldsymbol{x} \in E : f(\boldsymbol{x}) \neq g(\boldsymbol{x})\}$$

is a Jordan region of volume zero, then g is integrable on E and

$$\int_E g(\boldsymbol{x})\, d\boldsymbol{x} = \int_E f(\boldsymbol{x})\, d\boldsymbol{x}.$$

PROOF. i) Let $\varepsilon > 0$ and set $M = \sup_{\boldsymbol{x} \in E_0} |g(\boldsymbol{x})|$. Since E_0 is of volume zero, choose a grid $\mathcal{G}$ such that $V(E_0; \mathcal{G}) < \varepsilon/M$. Then

$$-\varepsilon < -M\, V(E_0; \mathcal{G}) \leq U(g, \mathcal{G}) \leq M\, V(E_0; \mathcal{G}) < \varepsilon,$$

i.e., $(U) \int_{E_0} g(\boldsymbol{x})\, d\boldsymbol{x} = 0$. A similar argument shows that $(L) \int_{E_0} g(\boldsymbol{x})\, d\boldsymbol{x} = 0$. Therefore, g is integrable on E_0 and $\int_{E_0} g(\boldsymbol{x})\, d\boldsymbol{x} = 0$.

ii) By part i), g is integrable on E_0 and

$$\int_{E_0} g(\boldsymbol{x})\, d\boldsymbol{x} = \int_{E_0} f(\boldsymbol{x})\, d\boldsymbol{x} = 0.$$

Since $g = f$ on $E \setminus E_0$, g is also integrable on $E \setminus E_0$. Hence, by Theorem 7.6ii, g is integrable on E and it follows that

$$\int_E g(\boldsymbol{x})\, d\boldsymbol{x} = \int_{E \setminus E_0} g(\boldsymbol{x})\, d\boldsymbol{x} + \int_{E_0} g(\boldsymbol{x})\, d\boldsymbol{x}$$

$$= \int_{E \setminus E_0} f(\boldsymbol{x})\, d\boldsymbol{x} + \int_{E_0} f(\boldsymbol{x})\, d\boldsymbol{x} = \int_E f(\boldsymbol{x})\, d\boldsymbol{x}. \ \blacksquare$$

This suggests a way to define the integral of f on E when f is not defined on all of E. Indeed, if f is defined on $E \setminus E_0$, where E is a Jordan region and E_0 is of volume zero, and the function

$$g(x) := \begin{cases} f(x) & x \in E \setminus E_0 \\ 0 & x \in E_0 \end{cases}$$

is integrable on E, then we shall define

$$\int_E f(x)\, dx := \int_E g(x)\, dx.$$

For example,

$$\int_0^2 \frac{x^2 - 1}{x - 1}\, dx = \int_0^2 (x + 1)\, dx = 4.$$

Henceforth, the phrase "$f : E \to \mathbf{R}$ is integrable" includes the possibility that f may not be defined on a subset of E of volume zero.

The following result is a multidimensional analogue of Theorem 3.7.

THEOREM 7.8 [COMPARISON THEOREM FOR MULTIPLE INTEGRALS]. *Let E be a Jordan region in $\mathbf{R}^n$ and $f, g : E \to \mathbf{R}$ be integrable on E.*

  i) *If $f(x) \le g(x)$ for $x \in E$, then*

$$\int_E f(x)\, dx \le \int_E g(x)\, dx.$$

 ii) *If m, M are scalars which satisfy $m \le f(x) \le M$ for $x \in E$, then*

$$m \operatorname{Vol}(E) \le \int_E f(x)\, dx \le M \operatorname{Vol}(E).$$

iii) *The function $|f|$ is integrable on E and*

(15)
$$\left| \int_E f(x)\, dx \right| \le \int_E |f(x)|\, dx.$$

PROOF. i) If $f \le g$ on E, then $L(f, \mathcal{G}) \le L(g, \mathcal{G})$ for any grid $\mathcal{G}$. Taking the supremum of this inequality over all grids $\mathcal{G}$ verifies part i).

ii) By Corollary 7.5, (8), and part i),

$$m \operatorname{Vol}(E) = \int_E m\, dx \le \int_E f(x)\, dx \le \int_E M\, dx = M \operatorname{Vol}(E).$$

iii) Let $\varepsilon > 0$ and choose a grid $\mathcal{G} = \{R_1, \dots, R_p\}$ such that

$$U(f, \mathcal{G}) - \varepsilon < \int_E f(x)\, dx < L(f, \mathcal{G}) + \varepsilon,$$

i.e.,

$$(16) \qquad\qquad U(f, \mathcal{G}) - L(f, \mathcal{G}) < 2\varepsilon.$$

It is easy to see that

$$\sup_{\boldsymbol{x} \in R_j \cap E} |f(\boldsymbol{x})| - \inf_{\boldsymbol{x} \in R_j \cap E} |f(\boldsymbol{x})| \le \sup_{\boldsymbol{x} \in R_j \cap E} f(\boldsymbol{x}) - \inf_{\boldsymbol{x} \in R_j \cap E} f(\boldsymbol{x})$$

(see (10) in Section 3.2). Hence, it follows from (16) that

$$U(|f|, \mathcal{G}) - L(|f|, \mathcal{G}) \le U(f, \mathcal{G}) - L(f, \mathcal{G}) < 2\varepsilon.$$

Thus, $|f|$ is integrable on E. Since $-|f| \le f \le |f|$, we conclude by part i) that

$$-\int_E |f(\boldsymbol{x})| \, d\boldsymbol{x} \le \int_E f(\boldsymbol{x}) \, d\boldsymbol{x} \le \int_E |f(\boldsymbol{x})| \, d\boldsymbol{x}. \quad \blacksquare$$

THEOREM 7.9 [MEAN VALUE THEOREM FOR MULTIPLE INTEGRALS]. *Let E be a Jordan region in* $\mathbf{R}^n$ *and $f, g : E \to \mathbf{R}$ be integrable on E with $g(\boldsymbol{x}) \ge 0$ for all $\boldsymbol{x} \in E$.*

   i) *There is a number c satisfying*

$$(17) \qquad\qquad \inf_{\boldsymbol{x} \in E} f(\boldsymbol{x}) \le c \le \sup_{\boldsymbol{x} \in E} f(\boldsymbol{x})$$

   such that

$$(18) \qquad\qquad c \int_E g(\boldsymbol{x}) \, d\boldsymbol{x} = \int_E f(\boldsymbol{x}) g(\boldsymbol{x}) \, d\boldsymbol{x}.$$

   ii) *There is a number c satisfying (17) such that*

$$c \, \mathrm{Vol}\,(E) = \int_E f(\boldsymbol{x}) \, d\boldsymbol{x}.$$

PROOF. i) By hypothesis, the product fg is integrable on E (see Exercise 3 below). Let $m = \inf_{\boldsymbol{x} \in E} f(\boldsymbol{x})$ and $M = \sup_{\boldsymbol{x} \in E} f(\boldsymbol{x})$. Since $g \ge 0$, Theorem 7.8 implies

$$(19) \qquad\qquad m \int_E g(\boldsymbol{x}) \, d\boldsymbol{x} \le \int_E f(\boldsymbol{x}) g(\boldsymbol{x}) \, d\boldsymbol{x} \le M \int_E g(\boldsymbol{x}) \, d\boldsymbol{x}.$$

If $\int_E g(\boldsymbol{x}) \, d\boldsymbol{x} = 0$, then (19) implies $\int_E f(\boldsymbol{x}) g(\boldsymbol{x}) \, d\boldsymbol{x} = 0$ and (18) holds for any c. If $\int_E g(\boldsymbol{x}) \, d\boldsymbol{x} \ne 0$, then (18) holds for

$$c = \frac{\int_E f(\boldsymbol{x}) g(\boldsymbol{x}) \, d\boldsymbol{x}}{\int_E g(\boldsymbol{x}) \, d\boldsymbol{x}}.$$

   ii) Apply part i) to $g(\boldsymbol{x}) = 1$. $\blacksquare$

We close this section with some optional material which generalizes a concept introduced in Section 5.7.

*DEFINITION 7.4. A set $E \subset \mathbf{R}^n$ is said to be of *measure zero* if given $\varepsilon > 0$ there is a countable collection of rectangles $\{R_j\}_{j \in \mathbf{N}}$ such that

$$E \subset \bigcup_{j=1}^{\infty} R_j \quad \text{and} \quad \sum_{j=1}^{\infty} |R_j| < \varepsilon.$$

*Remark 4. *If $E_1, E_2, \ldots$ is a sequence of subsets of $\mathbf{R}^n$ and each E_k is of measure zero, then*

$$E = \bigcup_{k=1}^{\infty} E_k$$

is also of measure zero.

PROOF. Let $\varepsilon > 0$. For each $k \in \mathbf{N}$, choose a collection of rectangles $\{R_j^{(k)}\}_{j \in \mathbf{N}}$ which covers E_k such that

$$\sum_{j=1}^{\infty} |R_j^{(k)}| < \frac{\varepsilon}{2^k}.$$

Clearly, the collection $\{R_j^{(k)}\}_{j,k \in \mathbf{N}}$ is countable, covers E, and

$$\sum_{k=1}^{\infty} \sum_{j=1}^{\infty} |R_j^{(k)}| \leq \sum_{k=1}^{\infty} \frac{\varepsilon}{2^k} = \varepsilon.$$

Consequently, E is of measure zero. ∎

Every singleton $E = \{a\}$ in $\mathbf{R}^n$ is of measure zero. In fact, by comparing Definition 7.4 with Theorem 7.2, it is clear that every set of volume zero is a set of measure zero. The converse of this statement is false. Indeed, for each $a \in \mathbf{R}$ the set $\{(a, y) : y \in [0, 1]\}$ is of volume zero, hence is of measure zero. Thus, by Remark 4, $E = \mathbf{Q} \times [0, 1]$ is a set of measure zero. On the other hand, it is clear that $\underline{\text{Vol}}(E) = 0 < 1 \leq \overline{\text{Vol}}(E)$, so E is not a set of volume zero; in fact, E is not even a Jordan region.

An analogue of Lebesgue's Theorem holds for multiple integrals.

*Remark 5. *A function f is Riemann integrable on a Jordan region E if and only if the set of points of discontinuity of f on E is of measure zero.*

The proof is similar to the proof of Theorem 5.33 (see Spivak [12], p. 53).

EXERCISES

1. Using Exercise 1 in Section 1.2, compute the upper and lower sums of $f(x, y) = xy$ on $[0, 1] \times [0, 1]$ generated by the grid $\mathcal{G}_m$ which satisfies

$$P_j(\mathcal{G}_m) = \{k/2^m : k = 0, 1, \ldots, 2^m\}$$

for $j = 1, 2$. Prove

$$\lim_{m\to\infty} U(f, \mathcal{G}_m) - L(f, \mathcal{G}_m) = 0.$$

2. a) Let $E_1 \subset E$ be Jordan regions in $\mathbf{R}^n$. Prove that if $f : E \to \mathbf{R}$ is integrable on E, then f is integrable on E_1.

   b) Prove that if $f : \mathbf{R}^n \to \mathbf{R}$ is continuous on $\mathbf{R}^n$, then f is integrable on any Jordan region in $\mathbf{R}^n$.

3. Let E be a Jordan region in $\mathbf{R}^n$ and $f, g : E \to \mathbf{R}$ be integrable on E.

   a) Modifying the proof of Corollary 3.9, prove that fg is integrable on E.

   b) Prove that $f \vee g$ and $f \wedge g$ are integrable on E (see Exercise 8 in Section 2.1).

4. a) Suppose E is a Jordan region in $\mathbf{R}^n$ and $f, f_k : E \to \mathbf{R}$ are integrable on E. If $f_k \to f$ uniformly on E as $k \to \infty$, prove

$$\lim_{k\to\infty} \int_E f_k(\boldsymbol{x}) \, d\boldsymbol{x} = \int_E f(\boldsymbol{x}) \, d\boldsymbol{x}.$$

   b) Prove that

$$\lim_{k\to\infty} \iint_E \cos(x/k) e^{y/k} \, dA$$

   exists and find its value for any Jordan region E in $\mathbf{R}^2$.

$\boxed{5}$. **This exercise is used in Sections 7.4, 8.5, and 8.6.**
   Let E be an open Jordan region in $\mathbf{R}^n$ and $\boldsymbol{x}_0 \in E$. If $f : E \to \mathbf{R}$ is integrable on E and continuous at $\boldsymbol{x}_0$, prove

$$\lim_{r\to 0} \frac{1}{\text{Vol}\,(B_r(\boldsymbol{x}_0))} \int_{B_r(\boldsymbol{x}_0)} f(\boldsymbol{x}) \, d\boldsymbol{x} = f(\boldsymbol{x}_0).$$

6. Let E be an open Jordan region in $\mathbf{R}^n$, $f, g : E \to \mathbf{R}$, and $\boldsymbol{x}_0 \in E$. Suppose both f and g are integrable on E and g is continuous at $\boldsymbol{x}_0$. Prove

$$I = \lim_{r\to 0} \frac{1}{\text{Vol}\,(B_r(\boldsymbol{x}_0))} \int_{B_r(\boldsymbol{x}_0)} f(\boldsymbol{x}) \, d\boldsymbol{x}$$

exists if and only if

$$J = \lim_{r\to 0} \frac{1}{\text{Vol}\,(B_r(\boldsymbol{x}_0))} \int_{B_r(\boldsymbol{x}_0)} f(\boldsymbol{x}) g(\boldsymbol{x}) \, d\boldsymbol{x}$$

exists, in which case $J = g(\boldsymbol{x}_0) I$.

7. Let H be a compact connected Jordan region and $f : H \to \mathbf{R}$ be continuous on H.

   a) If $g : H \to \mathbf{R}$ is integrable and nonnegative on H, prove there is an $\boldsymbol{x}_0 \in H$ such that

$$f(\boldsymbol{x}_0) \int_H g(\boldsymbol{x}) \, d\boldsymbol{x} = \int_H f(\boldsymbol{x}) g(\boldsymbol{x}) \, d\boldsymbol{x}.$$

   b) If $H^o \neq \emptyset$, prove there is a point $\boldsymbol{x}_0 \in H^o$ such that

$$\int_H f(\boldsymbol{x}) \, d\boldsymbol{x} = f(\boldsymbol{x}_0) \, \text{Vol}\,(H).$$

8. Prove the following special case of Remark 5. If E is a compact Jordan region in $\mathbf{R}^n$, E_0 is a Jordan region of volume zero, and $f : E \to \mathbf{R}$ is a bounded function which is continuous on $E \setminus E_0$, then f is integrable on E.

9. Suppose V is open in $\mathbf{R}^n$ and $f : V \to \mathbf{R}$ is continuous. Prove that if

$$\int_E f(\boldsymbol{x}) \, d\boldsymbol{x} = 0$$

for all Jordan regions $E \subset V$, then $f = 0$ on V.

7.3 ITERATED INTEGRALS

If $f(x_1, \ldots, x_k, \ldots, x_j, \ldots, x_n)$ is defined for $x_k \in [c, d]$ and $x_j \in [a, b]$, $j \neq k$, then we shall call

$$\int_c^d \int_a^b f(x_1, \ldots, x_n) \, dx_j \, dx_k := \int_c^d \left(\int_a^b f(x_1, \ldots, x_n) \, dx_j \right) dx_k$$

an *iterated integral*, when the integrals on the right side exist. In a similar way, we define higher-order iterated integrals.

In the preceding section we defined the integral of a multivariable function but developed no practical way to evaluate it. In this section we show that for a large collection of Jordan regions E, an integral over E can be evaluated using iterated integrals.

For simplicity, we begin with the two-dimensional case. Recall that for each $\phi : [a, b] \to \mathbf{R}$, $(U) \int_a^b \phi(x) \, dx$ represents the upper Riemann integral of ϕ and $(L) \int_a^b \phi(x) \, dx$ represents the lower Riemann integral of ϕ.

Lemma 1. *Let $R = [a, b] \times [c, d]$ be a two-dimensional rectangle and $f : R \to \mathbf{R}$ be bounded. If $f(x, \cdot)$ is integrable on $[c, d]$ for each $x \in [a, b]$, then*

$$(20) \qquad (L) \iint_R f \, dA \le (L) \int_a^b \left(\int_c^d f(x, y) dy \right) dx$$

$$\le (U) \int_a^b \left(\int_c^d f(x, y) dy \right) dx \le (U) \iint_R f \, dA.$$

PROOF. Let $R_{ij} = [x_{i-1}, x_i] \times [y_{j-1}, y_j]$, where $\{x_0, \ldots, x_k\}$ is a partition of $[a, b]$ and $\{y_0, \ldots, y_\ell\}$ is a partition of $[c, d]$. Then $\mathcal{G} = \{R_{ij} : i = 1, 2, \ldots, k, j = 1, 2 \ldots, \ell\}$ is a grid on R. Choose $\mathcal{G}$ so that

$$(21) \qquad\qquad U(f; \mathcal{G}) - \varepsilon < (U) \iint_R f \, dA.$$

and set

$$M_{ij} = \sup_{(x, y) \in R_{ij}} f(x, y).$$

Since $(U) \int_a^b \phi(x)\, dx = \sum_{i=1}^k (U) \int_{x_{i-1}}^{x_i} \phi(x)\, dx$ and

$$(U) \int_a^b (\phi(x) + \psi(x))\, dx \le (U) \int_a^b \phi(x)\, dx + (U) \int_a^b \psi(x)\, dx$$

for any bounded functions ϕ and ψ defined on $[a, b]$ (see Exercise 7 in Section 3.1), we can write

$$(U) \int_a^b \left(\int_c^d f(x,y) dy \right) dx = \sum_{i=1}^k (U) \int_{x_{i-1}}^{x_i} \left(\sum_{j=1}^\ell \int_{y_{j-1}}^{y_j} f(x,y) dy \right) dx$$

$$\le \sum_{i=1}^k \sum_{j=1}^\ell (U) \int_{x_{i-1}}^{x_i} \left(\int_{y_{j-1}}^{y_j} f(x,y) dy \right) dx$$

$$\le \sum_{i=1}^k \sum_{j=1}^\ell M_{ij} (x_i - x_{i-1})(y_j - x_{j-1}) = U(f, \mathcal{G}).$$

It follows from (21) that

$$(U) \int_a^b \left(\int_c^d f(x,y) dy \right) dx < (U) \iint_R f\, dA + \varepsilon.$$

Taking the limit of this inequality as $\varepsilon \to 0$, we obtain

$$(U) \int_a^b \left(\int_c^d f(x,y) dy \right) dx \le (U) \iint_R f\, dA.$$

Similarly,

$$(L) \int_a^b \left(\int_c^d f(x,y) dy \right) dx \ge (L) \iint_R f\, dA. \quad \blacksquare$$

We are now prepared to show that under reasonable conditions, a double integral over a rectangle reduces to an iterated integral.

THEOREM 7.10 [FUBINI'S THEOREM]. *Let* $R = [a, b] \times [c, d]$ *be a two-dimensional rectangle and* $f : R \to \mathbf{R}$. *Suppose that* $f(x, \cdot)$ *is integrable on* $[c, d]$ *for each* $x \in [a, b]$, $f(\cdot, y)$ *is integrable on* $[a, b]$ *for each* $y \in [c, d]$, *and* f *is integrable on* R. *Then*

$$(22) \qquad \iint_R f\, dA = \int_a^b \left(\int_c^d f(x,y) dy \right) dx = \int_c^d \left(\int_a^b f(x,y) dx \right) dy.$$

PROOF. For each $x \in [a, b]$, set $g(x) = \int_c^d f(x, y)\, dy$. Since f is integrable on R, Lemma 1 implies

$$\iint_R f\, dA = (U) \int_a^b g(x)\, dx = (L) \int_a^b g(x)\, dx.$$

Hence, g is integrable on $[a,b]$ and the first identity in (22) holds. Reversing the roles of x and y, we obtain

$$\iint_R f\, dA = \int_c^d \left(\int_a^b f(x,y)dx \right) dy.$$

Hence, the second identity in (22) holds. ∎

Notice that the hypotheses of Fubini's Theorem hold if f is continuous on the rectangle $[a,b] \times [c,d]$.

The second identity in Fubini's Theorem is as important as the first. It tells us that under certain conditions, the order of integration in an iterated integral can be reversed. Frequently one of these iterated integrals is easier to evaluate than the other.

Example 1. Find

$$\int_0^1 \int_0^1 y^3 e^{xy^2}\, dy\, dx.$$

SOLUTION. This iterated integral looks tough to integrate. However, if we change the order of integration using Fubini's Theorem, we obtain

$$\int_0^1 \int_0^1 y^3 e^{xy^2}\, dx\, dy = \int_0^1 y(e^{y^2}-1)\, dy = \frac{e-2}{2}. \quad ∎$$

The following two remarks show that the hypotheses of Fubini's Theorem cannot be relaxed.

Remark 1. *There exists a function $f : \mathbf{R}^2 \to \mathbf{R}$ such that $f(x,\cdot)$ and $f(\cdot,y)$ are both integrable on $[0,1]$, but the iterated integrals are not equal.*

PROOF. Set

$$\phi_k(t) = \begin{cases} 2^k & 2^{-k} \le t < 2^{-k+1} \\ 0 & \text{otherwise,} \end{cases}$$

for each $k \in \mathbf{N}$ and $t \in \mathbf{R}$, and observe that $\phi_k(t) = 0$ for $t \notin [0,1)$, and

(23) $$\int_0^1 \phi_k(t)\, dt = 1, \qquad k \in \mathbf{N}.$$

Consider the function $f : \mathbf{R}^2 \to \mathbf{R}$ defined by

$$f(x,y) = \sum_{k=1}^{\infty} (\phi_k(x) - \phi_{k+1}(x))\phi_k(y).$$

This series converges pointwise on $[0,1) \times [0,1)$. Indeed, since each fixed $y \in [0,1)$ belongs to exactly one of the intervals $[2^{-k}, 2^{-k+1})$, $k \in \mathbf{N}$, the series defining f collapses to one nonzero term for each $(x,y) \in [0,1) \times [0,1)$, namely,

$$f(x,y) = (\phi_k(x) - \phi_{k+1}(x))\phi_k(y), \qquad y \in [2^{-k}, 2^{-k+1}),\ k \in \mathbf{N}.$$

In particular, the function $f(\cdot, y)$ is integrable on $[0, 1]$ and

$$\int_0^1 \left(\int_0^1 f(x, y)\, dx \right) dy = \sum_{k=1}^{\infty} \int_{2^{-k}}^{2^{-k+1}} \left(\int_0^1 f(x, y)\, dx \right) dy$$

$$= \sum_{k=1}^{\infty} \int_{2^{-k}}^{2^{-k+1}} \phi_k(y) \left(\int_0^1 (\phi_k(x) - \phi_{k+1}(x))\, dx \right) dy = 0$$

by Exercise 6 in Section 3.2 and (23).

Similarly, for each fixed $x \in [2^{-k}, 2^{-k+1})$,

$$f(x, y) = \begin{cases} \phi_1(x)\phi_1(y) & k = 1 \\ \phi_k(x)(\phi_k(y) - \phi_{k-1}(y)) & k \geq 2. \end{cases}$$

Therefore,

$$\int_0^1 \left(\int_0^1 f(x, y)\, dy \right) dx = \sum_{k=1}^{\infty} \int_{2^{-k}}^{2^{-k+1}} \left(\int_0^1 f(x, y)\, dy \right) dx$$

$$= \int_{1/2}^1 \int_0^1 \phi_1(x)\phi_1(y)\, dy\, dx + \sum_{k=2}^{\infty} \int_{2^{-k}}^{2^{-k+1}} \int_0^1 \phi_k(x)(\phi_k(y) - \phi_{k-1}(y))\, dy\, dx$$

$$= 1 + 0 = 1.$$

In particular, the iterated integrals are not equal even though both exist and are finite. ∎

Thus, we cannot be sure a function of several variables is integrable just because its iterated integrals exist. (See also Exercises 5 and 9 below.)

The following remark is starred because it uses Lebesgue's characterization of Riemann integrability (Theorem 5.33 and Remark 5 in Section 7.2).

*Remark 2. *There exists a function $f : \mathbf{R}^2 \to \mathbf{R}$ such that f integrable on $[0, 1] \times [0, 1]$, $f(\cdot, y)$ is integrable on $[0, 1]$ for all $y \in [0, 1]$, but $f(x, \cdot)$ is not integrable on $[0, 1]$ for infinitely many $x \in [0, 1]$.*

PROOF. Let

$$f(x, y) = \begin{cases} 0 & \text{when } x = 0 \text{ or when } x \text{ or } y \text{ is irrational} \\ 1/q & \text{when } x, y \in \mathbf{Q} \text{ and } x = p/q \text{ is in reduced form.} \end{cases}$$

By the argument of Example 4 in Section 2.2, the function f is continuous and zero on the set $([0, 1] \setminus \mathbf{Q}) \times [0, 1]$. Hence, by Lebesgue's Theorem, f is integrable on the square $R = [0, 1] \times [0, 1]$. Moreover, since $\mathbf{Q} \times [0, 1]$ is of volume zero, $\iint_R f\, dA = 0$. Similarly, for each $y \in [0, 1]$, $f(\cdot, y)$ is integrable on $[0, 1]$ with $\int_0^1 f(x, y)\, dx = 0$. Thus,

$$\int_0^1 \left(\int_0^1 f(x, y)\, dx \right) dy = \iint_R f\, dA = 0.$$

On the other hand, since for each nonzero $x \in \mathbf{Q}$ the function $f(x, \cdot)$ is nowhere continuous, it cannot be integrable on $[0, 1]$. Therefore, the other iterated integral in Fubini's Theorem does not exist. ∎

Fubini's Theorem shows us how to evaluate a double integral over a rectangle using iterated integrals. The following result is a first step toward evaluating multiple integrals over nonrectangular Jordan regions.

THEOREM 7.11. *Let E be a Jordan region in $\mathbf{R}^n$, R be an n-dimensional rectangle which contains E, and $f : E \to \mathbf{R}$ be integrable on E. If*

$$g(\pmb{x}) = \begin{cases} f(\pmb{x}) & \pmb{x} \in E \\ 0 & \pmb{x} \notin E, \end{cases}$$

then g is integrable on R and

(24) $$\int_E f(\pmb{x}) \, d\pmb{x} = \int_R g(\pmb{x}) \, d\pmb{x}.$$

PROOF. Set $f^+ = f \vee 0$ and $f^- = (-f) \vee 0$ and observe that both f^+ and f^- are nonnegative, integrable on E, and $f = f^+ - f^-$ (see Exercise 2 in Section 3.2). Since $g^+ = f^+$ and $g^- = f^-$, we may suppose, by Theorem 7.6i, that $f \geq 0$.

Let $\varepsilon > 0$ and choose a grid $\mathcal{G} = \{R_1, \ldots, R_p\}$ on R such that

(25) $$U(f, \mathcal{G}) - \varepsilon < \int_E f(\pmb{x}) \, d\pmb{x} < L(f, \mathcal{G}) + \varepsilon$$

and

(26) $$V(\partial E; \mathcal{G}) = V(E; \mathcal{G}) - v(E; \mathcal{G}) < \varepsilon.$$

Let $M_j = \sup_{\pmb{x} \in R_j} g(\pmb{x})$ and split the Riemann sum of g over R into three pieces:

$$U(g, \mathcal{G}) = S_1 + S_2 + S_3 := \sum_{j \in \mathcal{I}_1} M_j |R_j| + \sum_{j \in \mathcal{I}_2} M_j |R_j| + \sum_{j \in \mathcal{I}_3} M_j |R_j|,$$

where $\mathcal{I}_1 = \{j : R_j \subseteq E\}$, $\mathcal{I}_2 = \{j \notin \mathcal{I}_1 : R_j \cap \partial E \neq \emptyset\}$, and $\mathcal{I}_3 = \{j \notin \mathcal{I}_2 : R_j \subseteq R \setminus E\}$. If $M = \max\{|M_1|, \ldots, |M_p|\}$, then (26) implies $S_2 \leq M\varepsilon$. Since $j \in \mathcal{I}_1$ implies $g(\pmb{x}) = f(\pmb{x})$ for $\pmb{x} \in R_j$, we have $S_1 \leq U(f, \mathcal{G})$ (because $f \geq 0$). And, since $j \in \mathcal{I}_3$ implies $g(\pmb{x}) = 0$ for $\pmb{x} \in R_j$, we have $S_3 = 0$. Consequently, it follows from (25) that

$$U(g, \mathcal{G}) = S_1 + S_2 + S_3 \leq U(f, \mathcal{G}) + M\varepsilon < \int_E f(\pmb{x}) \, d\pmb{x} + (M+1)\varepsilon.$$

In particular,

$$(U) \int_R g(\pmb{x}) \, d\pmb{x} \leq \int_E f(\pmb{x}) \, d\pmb{x}.$$

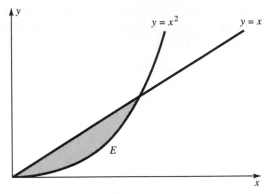

Figure 7.6

A similar argument proves

$$(L) \int_R g(\boldsymbol{x}) \, d\boldsymbol{x} \geq \int_E f(\boldsymbol{x}) \, d\boldsymbol{x}.$$

We conclude that g is integrable on R and (24) holds. ∎

A set $E \subset \mathbf{R}^n$ is said to be a *projectable region* if there is a compact Jordan region $H \subset \mathbf{R}^{n-1}$, an index $j \in \{1, \ldots, n\}$, and continuous functions $\phi, \psi : H \to \mathbf{R}$ such that

$$E = \{(x_1, \ldots, x_n) \in \mathbf{R}^n : (x_1, \ldots, \widehat{x}_j, \ldots, x_n) \in H$$
$$\text{and} \quad \phi(x_1, \ldots, \widehat{x}_j, \ldots, x_n) \leq x_j \leq \psi(x_1, \ldots, \widehat{x}_j, \ldots, x_n)\}.$$

(The notation $\widehat{x}_j$ means the variable x_j is missing; hence, $(x_1, \ldots, \widehat{x}_j, \ldots, x_n)$ is a point in $\mathbf{R}^{n-1}$.) In this case, we say that E is *generated* by j, H, ϕ, and ψ.

We are more specific for regions in $\mathbf{R}^2$ and $\mathbf{R}^3$. A set $E \subset \mathbf{R}^2$ is called a *region of type I* if $E = \{(x, y) : x \in [a, b], \phi(x) \leq y \leq \psi(x)\}$ and a *region of type II* if $E = \{(x, y) : y \in [a, b], \phi(y) \leq x \leq \psi(y)\}$, where $\phi, \psi : [a, b] \to \mathbf{R}$ are continuous functions. Similarly, a set $E \subset \mathbf{R}^3$ is called a *region of type I* if $E = \{(x, y, z) : (x, y) \in H, \phi(x, y) \leq z \leq \psi(x, y)\}$, a *region of type II* if $E = \{(x, y, z) : (x, z) \in H, \phi(x, z) \leq y \leq \psi(x, z)\}$, and a *region of type III* if $E = \{(x, y, z) : (y, z) \in H, \phi(y, z) \leq x \leq \psi(y, z)\}$, where $\phi, \psi : H \to \mathbf{R}$ are continuous functions and H is a compact Jordan region in $\mathbf{R}^2$.

Example 2. Prove that the set E in $\mathbf{R}^2$ bounded by $y = x$ and $y = x^2$ is a region of types I and II.

PROOF. The set E can be described by

$$\{(x, y) : x^2 \leq y \leq x, \ x \in [0, 1]\} \quad \text{or} \quad \{(x, y) : y \leq x \leq \sqrt{y}, \ y \in [0, 1]\}$$

(see Figure 7.6). ∎

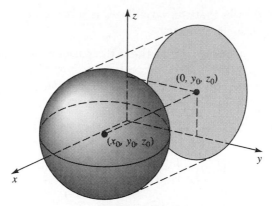

Figure 7.7

Example 3. Prove that the set E of points (x, y, z) which satisfy $4x^2 + y^2 + z^2 \leq 1$ is a region of types I, II, and III.

Proof. The set E, an ellipsoid, can be described by

$$E = \{(x, y, z) : -\sqrt{1 - 4x^2 - y^2} \leq z \leq \sqrt{1 - 4x^2 - y^2}, \ (x, y) \in H\}$$

where $H = \{(x, y) : 4x^2 + y^2 \leq 1\}$. A similar argument shows E is of types II and III. ∎

Before we show how to evaluate multiple integrals over projectable regions, we introduce additional terminology. For each $k = 1, \ldots, n$ the set

$$\Pi_k = \{\boldsymbol{x} \in \mathbf{R}^n : x_k = 0\}$$

will be called a *coordinate hyperplane*. Given a set $E \subseteq \mathbf{R}^n$, the *projection* of E onto the coordinate hyperplane Π_k is the set E_k of points $(x_1, \ldots, x_{k-1}, 0, x_{k+1}, \ldots, x_n)$ such that $(x_1, \ldots, x_k, \ldots, x_n) \in E$ for some $x_k \in \mathbf{R}$. For example, in $\mathbf{R}^3$ the coordinate hyperplane Π_1 corresponds to the yz plane and the projection of the three-dimensional ball $B_r(x_0, y_0, z_0)$ onto Π_1 is essentially the two-dimensional ball $B_r(y_0, z_0)$ (see Figure 7.7).

The following result shows that the integral of a continuous function over a rectangle in $\mathbf{R}^n$ can be evaluated using n partial integrals.

Lemma 2. *Let $1 \leq k \leq n$ be integers, $R = [a_1, b_1] \times \cdots \times [a_n, b_n]$ be an n-dimensional rectangle, and $f : R \to \mathbf{R}$ be integrable on R.   If $f(x_1, \ldots, x_{k-1}, \cdot, x_{k+1}, \ldots, x_n)$ is integrable on $[a_k, b_k]$ then*

$$\int_{a_k}^{b_k} f(x_1, \ldots, x_{k-1}, t, x_{k+1}, \ldots, x_n) \, dt$$

is integrable on the projection R_k and

$$(27) \qquad \int_R f(\boldsymbol{x}) \, d\boldsymbol{x} = \int_{R_k} \int_{a_k}^{b_k} f(x_1, \ldots, x_n) \, dx_k \, d(x_1 \ldots, \widehat{x_k}, \ldots, x_n).$$

(The notation $\widehat{x_k}$ indicates that the variable x_k is missing in the outer integral; i.e., it is an $(n-1)$-dimensional integral.)

PROOF. Since the projection R_k of the n-dimensional rectangle R onto the hyperplane Π_k is an $(n-1)$-dimensional rectangle, it is clear that R_k is a Jordan region. By repeating the argument of Lemma 1, we have

$$(L) \int_R f(\boldsymbol{x})\, d\boldsymbol{x} \le (L) \int_{R_k} \int_{a_k}^{b_k} f(x_1,\dots,x_n)\, dx_k\, d(x_1\dots,\widehat{x_k},\dots,x_n)$$

$$\le (U) \int_{R_k} \int_{a_k}^{b_k} f(x_1,\dots,x_n)\, dx_k\, d(x_1\dots,\widehat{x_k},\dots,x_n)$$

$$\le (U) \int_R f(\boldsymbol{x})\, d\boldsymbol{x}$$

for any bounded f. Since f is integrable on R, it follows that (27) holds. ∎

We will use the following result to show that multiple integrals over certain projectable regions can be evaluated using iterated integrals.

THEOREM 7.12. *Let E be a projectable region in $\mathbf{R}^n$ generated by j, H, ϕ, and ψ. Then E is a Jordan region in $\mathbf{R}^n$. Moreover, if $f : E \to \mathbf{R}$ is continuous on E, then*

$$(28) \quad \int_E f(\boldsymbol{x})\, d\boldsymbol{x} = \int_H \left(\int_{\phi(x_1,\dots\widehat{x_j},\dots x_n)}^{\psi(x_1,\dots,\widehat{x_j},\dots x_n)} f(x_1,\dots,x_n)\, dx_j \right) d(x_1,\dots,\widehat{x_j},\dots,x_n).$$

PROOF. By symmetry, we may suppose $j = n$. Thus,

$$E = \{(\boldsymbol{x},t) : \boldsymbol{x} = (x_1,\dots,x_{n-1}) \in H \quad \text{and} \quad \phi(\boldsymbol{x}) \le t \le \psi(\boldsymbol{x})\}.$$

To show E is a Jordan region, we must show that the volume of ∂E is zero. Now ∂E is made up of "lower dimensional pieces," a bottom $B = \{(\boldsymbol{x},t) : \boldsymbol{x} \in H \text{ and } t = \phi(\boldsymbol{x})\}$, a top $T = \{(\boldsymbol{x},t) : \boldsymbol{x} \in H \text{ and } t = \psi(\boldsymbol{x})\}$, and a side $S = \{(\boldsymbol{x},t) : \boldsymbol{x} \in \partial H \text{ and } \phi(\boldsymbol{x}) \le t \le \psi(\boldsymbol{x})\}$ (Figure 7.8 illustrates the situation for the case $n = 3$.) Hence, we must show that B, T, and S are of volume zero.

To estimate the volume of B, notice that since H is compact, ϕ is uniformly continuous on H. Thus, given $\varepsilon > 0$ there is a $\delta > 0$ such that

$$(29) \qquad \boldsymbol{x}, \boldsymbol{y} \in H \quad \text{and} \quad \|\boldsymbol{x} - \boldsymbol{y}\| < \delta \quad \text{imply} \quad \|\phi(\boldsymbol{x}) - \phi(\boldsymbol{y})\| < \varepsilon.$$

Since compact sets are bounded, H is contained in some $(n-1)$-dimensional cube Q. Divide Q into subcubes $Q_1,\dots,Q_p$ such that $\boldsymbol{x}, \boldsymbol{y} \in Q_k$ implies $\|\boldsymbol{x} - \boldsymbol{y}\| < \delta$, and let $R_k = Q_k \times [\phi(\boldsymbol{a}_k) - 2\varepsilon, \phi(\boldsymbol{a}_k) + 2\varepsilon]$ for some $\boldsymbol{a}_k \in Q_k$, $k = 1, 2, \dots, p$. Then each R_k is an n-dimensional rectangle and

$$\sum_{k=1}^p |R_k| = 4\varepsilon \sum_{k=1}^p |Q_k| = 4\varepsilon |Q|.$$

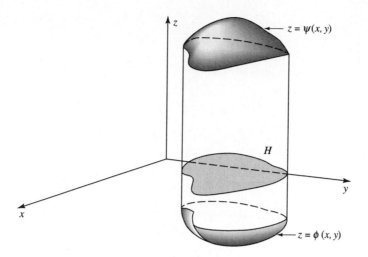

Figure 7.8

Moreover, by (29), B is covered by the rectangles R_k. Hence, $\overline{\mathrm{Vol}}(B) < 4\varepsilon|Q|$. Taking the limit of this inequality as $\varepsilon \to 0$, we see that B is of volume zero. A similar argument shows that T is of volume zero.

To estimate the volume of S, set

$$M = \sup_{\boldsymbol{x} \in H} \psi(\boldsymbol{x}) \quad \text{and} \quad m = \inf_{\boldsymbol{x} \in H} \phi(\boldsymbol{x}).$$

Since H is a Jordan region, choose $(n-1)$-dimensional cubes $Q_1, \ldots, Q_p$ which cover ∂H such that

$$\sum_{k=1}^{p} |Q_k| < \varepsilon.$$

Set $R_k = Q_k \times [m, M]$ and observe that

$$S \subseteq \bigcup_{k=1}^{p} R_k \quad \text{and} \quad \sum_{k=1}^{p} |R_k| < (M - m)\varepsilon.$$

Hence, $\overline{\mathrm{Vol}}(S) \le (M - m)\varepsilon$ and it follows that S is of volume zero. We conclude that ∂E is of volume zero; i.e., E is a Jordan region.

To prove (28), let $R = [a_1, b_1] \times \cdots \times [a_n, b_n]$ be an n-dimensional rectangle which contains E and define g on R by $g(\boldsymbol{x}, t) = f(\boldsymbol{x}, t)$ when $(\boldsymbol{x}, t) \in E$, and $g(\boldsymbol{x}, t) = 0$ otherwise. By Theorem 7.11 and Lemma 2,

$$\int_E f(\boldsymbol{x}, t)\, d(\boldsymbol{x}, t) = \int_{a_1}^{b_1} \cdots \int_{a_n}^{b_n} g(x_1, \ldots, x_n)\, dx_n \ldots dx_1$$

$$= \int_H \left(\int_{a_n}^{b_n} g(\boldsymbol{x}, t)\, dt \right) d\boldsymbol{x}.$$

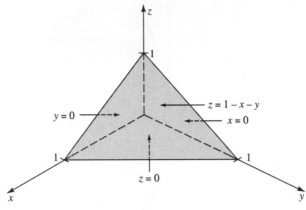

Figure 7.9

But for each $\boldsymbol{x} = (x_1, \ldots, x_{n-1}) \in H$, we have

$$g(\boldsymbol{x}, t) = \begin{cases} f(\boldsymbol{x}, t) & \phi(\boldsymbol{x}) \leq t \leq \psi(\boldsymbol{x}) \\ 0 & \text{otherwise.} \end{cases}$$

Therefore,

$$\int_{a_n}^{b_n} g(\boldsymbol{x}, t) \, dt = \int_{\phi(\boldsymbol{x})}^{\psi(\boldsymbol{x})} f(\boldsymbol{x}, t) \, dt. \quad \blacksquare$$

Although we have stated Theorem 7.12 for continuous f, the result is evidently true whenever Lemma 2 applies, e.g., if f is integrable on E and $f(\boldsymbol{x}, \cdot)$ is integrable on $[a_n, b_n]$ for each fixed $\boldsymbol{x} \in H$.

If the set H is itself projectable, then Theorem 7.12 can be applied again to H. Thus, if E is nice enough, an integral over E can be evaluated using n partial integrals. We close this section with several examples which illustrate this principle for the cases $n = 2$ and $n = 3$.

Example 4. Find the integral of $f(x, y, z) = x$ over the region E bounded by $z = 1 - x - y$, $x = 0$, $y = 0$, and $z = 0$.

SOLUTION. The surfaces $z = 0$ and $z = 1 - x - y$ intersect when $y = 1 - x$. The projection E_3 is bounded by the curves $x = 0$, $y = 0$, and $y = 1 - x$. These last two curves intersect when $x = 1$. Thus, E is a region of the type I:

$$E = \{(x, y, z) : 0 \leq x \leq 1, \ 0 \leq y \leq 1 - x, \ 0 \leq z \leq 1 - x - y\}$$

(see in Figure 7.9). It follows that

$$\iiint_E f \, dV = \int_0^1 \int_0^{1-x} \int_0^{1-x-y} x \, dz \, dy \, dx$$

$$= \int_0^1 \int_0^{1-x} (x - x^2 - xy) \, dy \, dx$$

$$= \frac{1}{2} \int_0^1 (x - 2x^2 + x^3) \, dx = \frac{1}{24}. \quad \blacksquare$$

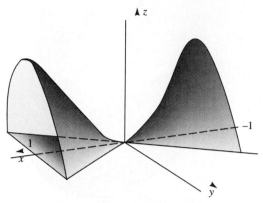

Figure 7.10

Example 5. Find the integral of $f(x, y, z) = x^2$ over the region E bounded by $|x| = 1$, $z = x^2 - y^2$, where $z \geq 0$.

SOLUTION. The surfaces $z = 0$ and $z = x^2 - y^2$ intersect when $x^2 - y^2 = 0$, i.e., $y = \pm x$. The curves $y = \pm x$ and $|x| = 1$ intersect when $x = \pm 1$. Thus, the region E is of type I:

$$E = \{(x, y, z) : -1 \leq x \leq 1, \ -|x| \leq y \leq |x|, \ 0 \leq z \leq x^2 - y^2\}$$

(see Figure 7.10). It follows that

$$\iiint_E f \, dV = \int_{-1}^{1} \int_{-|x|}^{|x|} \int_{0}^{x^2 - y^2} x^2 \, dz \, dy \, dx$$

$$= \int_{-1}^{1} \int_{-|x|}^{|x|} (x^2 - y^2) x^2 \, dy \, dx$$

$$= 4 \int_{0}^{1} \int_{0}^{x} (x^2 - y^2) x^2 dy \, dx = \frac{8}{3} \int_{0}^{1} x^5 \, dx = \frac{4}{9}. \ \blacksquare$$

Although Theorem 7.12 can be used in conjunction with Theorem 7.6 to handle the case when E is a finite union of projectable subregions, we can sometimes avoid breaking E into subregions by changing our point of view. Here is a typical example.

Example 6. Find the integral of $f(x, y, z) = x - z$ over the region bounded by $z = y^2$, $z = 1$, $z = x$, and $x = 0$.

SOLUTION. Since the region E is a union of two regions of type I (see Figure 7.11), we must use two integrals if we integrate dz first, the integral where z varies between y^2 and 1, and the integral where z varies from x to 1. It looks complicated to set up. The solution is simpler if we integrate dx first. Indeed, E is a single region of type III since

$$E = \{(x, y, z) : -1 \leq y \leq 1, \ y^2 \leq z \leq 1, \ 0 \leq x \leq z\}.$$

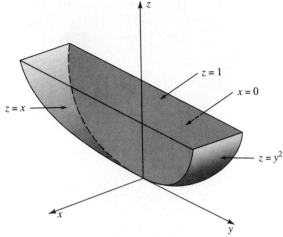

Figure 7.11

Thus,

$$\iiint_E f\, dV = \int_{-1}^{1}\int_{y^2}^{1}\int_{0}^{z}(x-z)\,dx\,dz\,dy$$

$$= -\frac{1}{2}\int_{-1}^{1}\int_{y^2}^{1} z^2\,dz\,dy = \frac{1}{6}\int_{-1}^{1}(y^6-1)\,dy = -\frac{2}{7}.\ \blacksquare$$

EXERCISES

1. Evaluate each of the following iterated integrals.

a) $\displaystyle\int_0^1\int_0^1 (x^2+y)\,dx\,dy.$ b) $\displaystyle\int_0^1\int_0^1 \sqrt{xy+x}\,dx\,dy.$ c) $\displaystyle\int_0^{\pi/2}\int_0^{\pi/2} y\cos(xy)\,dy\,dx.$

2. Evaluate each of the following iterated integrals. Write each as an integral over a region E and sketch E in each case.

a) $\displaystyle\int_0^1\int_x^{x^2+1}(x+y)\,dy\,dx.$ b) $\displaystyle\int_0^1\int_{\sqrt{y}}^1\int_0^{x^2+y^2} 3\,dz\,dx\,dy.$

c) $\displaystyle\int_0^1\int_y^1 \sin(x^2)\,dx\,dy.$ d) $\displaystyle\int_0^1\int_{\sqrt{y}}^1\int_{x^3}^1 \sqrt{x^3+z}\,dz\,dx\,dy.$

3. For each of the following, evaluate $\int_E f(\mathbf{x})\,d\mathbf{x}$.

a) $f(x,y) = x\sqrt{y}$ and E is bounded by $y=x$ and $y=x^2$.
b) $f(x,y) = x+y$ and E is the triangle with vertices $(0,0)$, $(0,1)$, $(2,0)$.
c) $f(x,y) = x$ and E is bounded by $y=\sqrt{x}$, $x=-\sqrt{y}$, and $y=4$.
d) $f(x,y,z) = x$ and E is the set of points (x,y,z) such that $0 \le z \le 1-x^2$, $0 \le y \le x^2+z^2$, and $x \ge 0$.

4. Compute the volume of each of the following regions.

   a) E is bounded by the surfaces $x + y + z = 3$, $z = 0$, and $x^2 + y^2 = 1$.

   b) E lies under the plane $z = x + y$ and over the region in the xy plane bounded by the curves $x = \sqrt{y/2}$, $x = 2\sqrt{y}$, $x + y = 3$.

   c) E is bounded by $z = y^2$, $x = y^2 + z^2$, $x = 0$, $z = 1$.

   d) E is bounded by $y = x^3$, $x = z^2$, $z = x^2$, and $y = 0$.

5. a) Verify that the hypotheses of Fubini's Theorem hold when f is continuous on R.

   b) Modify the proof of Remark 1 to show that Fubini's Theorem does not hold for nonintegrable f, even if $f(x, y)$ is continuous in each variable separately, i.e., if $f(x, \cdot)$ is continuous for each $x \in [a, b]$ and $f(\cdot, y)$ is continuous for each $y \in [c, d]$.

6. a) Suppose f_k is integrable on $[a_k, b_k]$ for $k = 1, \ldots, n$ and set $R = [a_1, b_1] \times \cdots \times [a_n, b_n]$. Prove that

$$\int_R f_1(x_1) \ldots f_n(x_n) \, d(x_1, \ldots, x_n) = \left(\int_{a_1}^{b_1} f_1(x_1) \, dx_1 \right) \cdots \left(\int_{a_n}^{b_n} f_n(x_n) \, dx_n \right).$$

   b) If $Q = [0, 1]^n$ and $\mathbf{y} := (1, 1, \ldots, 1)$, prove

$$\int_Q e^{-\mathbf{x} \cdot \mathbf{y}} \, d\mathbf{x} = \left(\frac{e - 1}{e} \right)^n.$$

7. Let R be a two-dimensional rectangle, $\phi : \mathbf{R} \to \mathbf{R}$ be defined by $\phi(x) = x - n - 1/2$ for $n \le x < n + 1$, $n = 0, \pm 1, \pm 2, \ldots$, and set $\psi(x, y) = \phi(x)\phi(y)$.

   a) Prove that $\iint_R \psi \, dA = 0$ if and only if at least one side of R has integer length.

   b) Suppose R is tiled by rectangles $R_1 \ldots, R_N$, i.e., the R_j's are nonoverlapping and $R = \cup_{j=1}^N R_j$. Prove that if each R_j has at least one side of integer length, then R has at least one side of integer length.

8. Let E be a Jordan region in $\mathbf{R}^2$ and $f : E \to [0, \infty)$ be integrable on E. Prove that the volume of the region $\Omega = \{(x, y, z) : (x, y) \in E, \ 0 \le z \le f(x, y)\}$ (as given by Definition 7.2) satisfies

$$\text{Vol}(\Omega) = \iint_E f \, dA.$$

9. Let $R = [a, b] \times [c, d]$ be a two-dimensional rectangle and $f : R \to \mathbf{R}$ be bounded.

   a) Prove

$$(L) \iint_R f \, dA \le (L) \int_a^b \left((X) \int_c^d f(x, y) \, dy \right) dx$$

$$\le (U) \int_a^b \left((X) \int_c^d f(x, y) \, dy \right) dx$$

$$\le (U) \iint_R f \, dA$$

for $X = U$ or $X = L$.

b) Prove that if f is integrable on R, then

$$\iint_R f \, dA = \int_a^b \left((L) \int_c^d f(x,y) \, dy \right) dx = \int_a^b \left((U) \int_c^d f(x,y) \, dy \right) dx.$$

c) Compute the two iterated integrals in part b) for

$$f(x,y) = \begin{cases} 1 & y \in \mathbf{Q} \\ x & y \notin \mathbf{Q} \end{cases}$$

and $R = [0,1] \times [0,1]$. Prove that f is not integrable on R.

10. [FUBINI'S THEOREM FOR IMPROPER INTEGRALS]. Suppose $a < b$ are extended real numbers, $c < d$ are finite real numbers, and $f : (a,b) \times [c,d] \to \mathbf{R}$ is continuous. If

$$F(y) = \int_a^b f(x,y) \, dx$$

converges uniformly on $[c,d]$, prove

$$\int_c^d f(x,y) \, dy$$

is improperly integrable on (a,b) and

$$\int_c^d \int_a^b f(x,y) \, dx \, dy = \int_a^b \int_c^d f(x,y) \, dy \, dx.$$

7.4 CHANGE OF VARIABLES

Recall (Exercise 11 in Section 3.3) that if $\phi : [a,b] \to \mathbf{R}$ is continuously differentiable and $\phi' \neq 0$ on $[a,b]$, then

$$\int_{\phi([a,b])} f(t) \, dt = \int_{[a,b]} f(\phi(x)) \, |\phi'(x)| \, dx$$

for all f integrable on $\phi([a,b])$. We shall generalize this result to functions of several variables, namely, we shall identify conditions under which

(30) $$\int_{\phi(E)} f(\mathbf{u}) \, d\mathbf{u} = \int_E f(\phi(\mathbf{x})) |\Delta_\phi(\mathbf{x})| \, d\mathbf{x},$$

holds. (At this point you may wish to read the discussion preceding Example 1 below to see that Δ_ϕ takes on a familiar form when ϕ is the change from polar to rectangular coordinates.) The proof of (30), arguably the most difficult in this text, will be presented in several steps spread over two sections. (The proofs of two key steps, Theorems 7.13 and 7.20, are based on Spivak [12].[1])

A function f is said to satisfy a certain property P "locally" on a set E if given $\boldsymbol{a} \in E$ there is an open ball B containing $\boldsymbol{a}$ such that f satisfies P on $B \cap E$. f is said to satisfy the property P "globally" on E if f satisfies P for all points in E. In this section we obtain several preliminary results which culminate in a "local" change of variables formula (see Theorem 7.13) and a "global" change of variables formula for functions whose Jacobians are never zero (see Theorem 7.14). We shall also state a global change of variables formula for functions whose Jacobians are zero on a set of volume zero (see Theorem 7.15). The proof of this deeper result will be given in Section 7.5.

Our first preliminary result shows that if (30) holds for two functions, then it holds for their composition.

Lemma 1. *Let V be open in $\mathbf{R}^n$, $\psi : V \to \mathbf{R}^n$ be continuously differentiable on V, $\psi(V)$ be open, and $\sigma : \psi(V) \to \mathbf{R}^n$ be continuously differentiable on $\psi(V)$. Suppose further that E is a subset of V such that E, $\psi(E)$, and $\sigma(\psi(E))$ are all Jordan regions. If*

$$(31) \qquad \int_{\psi(E)} g(\boldsymbol{u})\, d\boldsymbol{u} = \int_E (g \circ \psi)(\boldsymbol{x}) |\Delta_\psi(\boldsymbol{x})|\, d\boldsymbol{x}$$

and

$$(32) \qquad \int_{\sigma(\psi(E))} f(\boldsymbol{u})\, d\boldsymbol{u} = \int_{\psi(E)} (f \circ \sigma)(\boldsymbol{x}) |\Delta_\sigma(\boldsymbol{x})|\, d\boldsymbol{x}$$

for all f and g such that f is integrable on $\sigma(\psi(E))$, $f \circ \sigma |\Delta_\sigma|$ is integrable on $\psi(E)$, g is integrable on $\psi(E)$, and $g \circ \psi |\Delta_\psi|$ is integrable on E, then

$$\int_{\phi(E)} f(\boldsymbol{u})\, d\boldsymbol{u} = \int_E f(\phi(\boldsymbol{x})) |\Delta_\phi(\boldsymbol{x})|\, d\boldsymbol{x}$$

for $\phi = \sigma \circ \psi$, provided f is integrable on $\phi(E)$, and $f \circ \phi |\Delta_\phi|$ is integrable on E.

PROOF. By the Chain Rule,

$$(33) \qquad \Delta_\phi(\boldsymbol{x}) = \Delta_\sigma(\psi(\boldsymbol{x})) \Delta_\psi(\boldsymbol{x})$$

for all $\boldsymbol{x} \in V$. Hence, by (32), (31) and (33), we have

$$\int_{\phi(E)} f(\boldsymbol{u})\, d\boldsymbol{u} = \int_{\sigma(\psi(E))} f(\boldsymbol{u})\, d\boldsymbol{u}$$

$$= \int_{\psi(E)} (f \circ \sigma)(\boldsymbol{y}) |\Delta_\sigma(\boldsymbol{y})|\, d\boldsymbol{y}$$

$$= \int_E (f \circ \sigma \circ \psi)(\boldsymbol{x}) |\Delta_\sigma(\psi(\boldsymbol{x}))|\, |\Delta_\psi(\boldsymbol{x})|\, d\boldsymbol{x}$$

$$= \int_E (f \circ \phi)(\boldsymbol{x}) |\Delta_\phi(\boldsymbol{x})|\, d\boldsymbol{x}. \qquad \blacksquare$$

[1]M. Spivak, <u>Calculus on Manifolds</u>, (New York: W. A. Benjamin, Inc., 1965). Reprinted with permission of Addison-Wesley Publishing Company.

Next, we show that if ϕ is nice enough, it suffices to prove (30) for the special cases $f = 1$, $E = \phi^{-1}(R)$, and $E = \phi(R)$, where R is a rectangle. This should come as no surprise since every integrable function is nearly continuous, hence, is essentially constant on small sets, and every Jordan region has volume which can be "approximated" by the sums $V(E; \mathcal{G})$, hence, is essentially a finite union of rectangles.

Lemma 2. *Let V be open in $\mathbf{R}^n$, $\phi : V \to \mathbf{R}^n$ be 1–1 and continuously differentiable on V, $\phi^{-1} : \phi(V) \to \mathbf{R}^n$ be continuously differentiable on $\phi(V)$, and $\Delta_\phi \neq 0$ on V. If*

$$(34) \qquad |R| = \int_{\phi^{-1}(R)} |\Delta_\phi(\boldsymbol{x})| \, d\boldsymbol{x}$$

for every n-dimensional rectangle $R \subset \phi(V)$ and

$$(35) \qquad |Q| = \int_{\phi(Q)} |\Delta_{\phi^{-1}}(\boldsymbol{x})| \, d\boldsymbol{x}$$

for every n-dimensional rectangle $Q \subset V$, then

$$\int_{\phi(E)} f(\boldsymbol{u}) \, d\boldsymbol{u} = \int_E (f \circ \phi)(\boldsymbol{x}) |\Delta_\phi(\boldsymbol{x})| \, d\boldsymbol{x}$$

holds for every Jordan region E which satisfies $\overline{E} \subset V$, provided f is integrable on $\phi(E)$, and $f \circ \phi |\Delta_\phi|$ is integrable on E.

PROOF. Let E be a fixed Jordan region which satisfies $\overline{E} \subset V$ and let $f : \phi(E) \to \mathbf{R}$ be integrable on $\phi(E)$. (Notice that by Theorem 7.3, $\phi(E)$ is a Jordan region. Hence, it makes sense to talk about the integral of f on $\phi(E)$.) By the argument opening the proof of Lemma 2 in Section 7.3, we may suppose $f \geq 0$ on $\phi(E)$.

Let H be a rectangle containing $\phi(E)$ and $\mathcal{G}$ be a grid on H. Given $\varepsilon > 0$, choose a grid $\mathcal{H} = \{R_1, \ldots, R_p\}$ finer than $\mathcal{G}$ such that $V(\partial(\phi(E)); \mathcal{H}) < \varepsilon$. Set $\mathcal{I}_1 = \{j : R_j \subseteq \phi(E)\}$, $\mathcal{I}_2 = \{j \notin \mathcal{I}_1 : R_j \cap \phi(E) \neq \emptyset\}$,

$$m_j = \inf_{\boldsymbol{u} \in R_j \cap \phi(E)} f(\boldsymbol{u}) = \inf_{\boldsymbol{x} \in \phi^{-1}(R_j) \cap E} f(\phi(\boldsymbol{x}))$$

for each $j \in \mathcal{I}_1 \cup \mathcal{I}_2$, and $M = \sup\{m_1, \ldots, m_p\}$. Notice that $\{\phi^{-1}(R_j)\}_{j \in \mathcal{I}_1}$ is a nonoverlapping collection of Jordan regions (see Exercise 7 in Section 7.1) and

$$\Omega := \bigcup_{j \in \mathcal{I}_1} \phi^{-1}(R_j) \subseteq E.$$

It follows from (34), Theorem 7.8, and Theorem 7.6 that

$$
\begin{aligned}
L(f, \mathcal{G}) \leq L(f, \mathcal{H}) &= \sum_{j \in \mathcal{I}_1} m_j |R_j| + \sum_{j \in \mathcal{I}_2} m_j |R_j| \\
&\leq \sum_{j \in \mathcal{I}_1} m_j |R_j| + MV(\partial(\phi(E)); \mathcal{H}) \\
&< \sum_{j \in \mathcal{I}_1} m_j |R_j| + M\varepsilon \\
&= \sum_{j \in \mathcal{I}_1} m_j \int_{\phi^{-1}(R_j)} |\Delta_\phi(\boldsymbol{x})| \, d\boldsymbol{x} + M\varepsilon \\
&\leq \sum_{j \in \mathcal{I}_1} \int_{\phi^{-1}(R_j)} f(\phi(\boldsymbol{x}))|\Delta_\phi(\boldsymbol{x})| \, d\boldsymbol{x} + M\varepsilon \\
&= \int_\Omega f(\phi(\boldsymbol{x}))|\Delta_\phi(\boldsymbol{x})| \, d\boldsymbol{x} + M\varepsilon \\
&\leq \int_E f(\phi(\boldsymbol{x}))|\Delta_\phi(\boldsymbol{x})| \, d\boldsymbol{x} + M\varepsilon.
\end{aligned}
$$

(This last step is the only place $f \geq 0$ is used.) Taking the supremum of this inequality over all grids $\mathcal{G}$ and the limit as $\varepsilon \to 0$, we obtain

$$
(36) \qquad \int_{\phi(E)} f(\boldsymbol{u}) \, d\boldsymbol{u} \leq \int_E f(\phi(\boldsymbol{x}))|\Delta_\phi(\boldsymbol{x})| \, d\boldsymbol{x}.
$$

To obtain the reverse inequality, notice that by Theorem 6.19, $\phi(V)$ is open. Also notice that by the Chain Rule, $\Delta_{\phi^{-1}} \neq 0$ on $\phi(V)$ since $\Delta_\phi \neq 0$ on V. Hence, we can apply (36) with ϕ^{-1} in place of ϕ, $f \circ \phi |\Delta_\phi|$ in place of f, and $E = \phi^{-1}(\phi(E))$ in place of $\phi(E)$. We obtain

$$
(37) \qquad \int_E f(\phi(\boldsymbol{x}))|\Delta_\phi(\boldsymbol{x})| \, d\boldsymbol{x} \leq \int_{\phi(E)} f(\phi(\phi^{-1}(\boldsymbol{u})))|\Delta_\phi(\phi^{-1}(\boldsymbol{u}))| \, |\Delta_{\phi^{-1}}(\boldsymbol{u})| \, d\boldsymbol{u}.
$$

But $\phi \circ \phi^{-1} = I$ on $\mathbf{R}^n$, hence, by the Chain Rule,

$$
|\Delta_\phi(\phi^{-1}(\boldsymbol{u}))| \, |\Delta_{\phi^{-1}}(\boldsymbol{u})| = |\Delta_I(\boldsymbol{u})| = 1
$$

for all $\boldsymbol{u} \in \phi(V) \supset \phi(E)$. It follows from (37) that

$$
\int_E f(\phi(\boldsymbol{x}))|\Delta_\phi(\boldsymbol{x})| \, d\boldsymbol{x} \leq \int_{\phi(E)} f(\boldsymbol{u}) \, d\boldsymbol{u}.
$$

We conclude by (36) that equality holds. ∎

Here is a local version of the change of variables formula we want.

THEOREM 7.13. *Suppose V is open in $\mathbf{R}^n$, $\boldsymbol{a} \in V$, and $\phi : V \to \mathbf{R}^n$ is 1–1 and continuously differentiable on V. If $\Delta_\phi(\boldsymbol{a}) \neq 0$, then there exists an open rectangle $W \subset V$ containing $\boldsymbol{a}$ such that*

$$\int_{\phi(E)} f(\boldsymbol{u})\, d\boldsymbol{u} = \int_E f(\phi(\boldsymbol{x})) |\Delta_\phi(\boldsymbol{x})|\, d\boldsymbol{x}$$

holds for every Jordan region E which satisfies $\overline{E} \subset W$ provided f is integrable on $\phi(E)$ and $f \circ \phi |\Delta_\phi|$ is integrable on E.

STRATEGY: The idea behind the proof is relatively simple. The proof is by induction on n. By Lemma 2 and Exercise 11 in Section 3.3, Theorem 7.13 holds when $n = 1$. Assuming the result holds for $n - 1$, we write $\phi = \sigma \circ \psi$ where σ is essentially a one-dimensional function and ψ is essentially an $(n - 1)$-dimensional function. By Lemma 1, it suffices to show that Theorem 7.13 holds for ψ and σ, i.e., it suffices to show that the hypotheses of Lemma 2 hold for ψ and σ in place of ϕ. Now the crux of the matter is that by the Inverse Function Theorem and the inductive hypothesis, the hypotheses of Lemma 2 do hold for ψ and σ provided V is sufficiently small. Here are the details.

PROOF. Suppose the theorem holds on $\mathbf{R}^{n-1}$ for some $n > 1$. Let $\phi : V \to \mathbf{R}^n$ satisfy $\Delta_\phi(\boldsymbol{a}) \neq 0$. Expanding this determinant along the first column, we see that $\partial \phi_i / \partial x_1 \cdot \Delta_i$ must be nonzero at $\boldsymbol{a}$ for some i, where Δ_i represents the cofactor of $\partial \phi_i / \partial x_1$ in $D\phi(\boldsymbol{a})$. Interchange ϕ_i and ϕ_1, and for each $\boldsymbol{x} = (x_1, \dots, x_n) \in V$ set

$$\psi(\boldsymbol{x}) = (x_1, \phi_2(\boldsymbol{x}), \dots, \phi_n(\boldsymbol{x}))$$

and

$$\sigma(\boldsymbol{x}) = (\phi_1(\psi^{-1}(\boldsymbol{x})), x_2, \dots, x_n).$$

Then $\phi = \psi \circ \sigma$ and, since $(\phi_1)_{x_1} \cdot \Delta_\psi = \partial \phi_1 / \partial x_1 \cdot \Delta_1$ is nonzero at $\boldsymbol{a}$, $\Delta_\psi(\boldsymbol{a}) \neq 0$.

By Lemma 1, it suffices to show that there is an open set W such that $\boldsymbol{a} \in W \subset V$ and (31), (32) both hold for W in place of V. Hence, by Lemma 2, it suffices to show that there is an open set $W \subset V$ containing $\boldsymbol{a}$ such that (34) and (35) hold for ψ (respectively, σ) in place of ϕ and all rectangles $R \subset \psi(W)$ (respectively, $R \subset \sigma(W)$) and $Q \subset W$. The arguments are similar and (35) is simpler to prove than (34). Hence, we provide the details only for (34).

Since $\Delta_\psi(\boldsymbol{a}) \neq 0$, use Theorem 6.20 (the Inverse Function Theorem) to choose an open set $W \subset V$ containing $\boldsymbol{a}$ such that $\Delta_\psi(\boldsymbol{x}) \neq 0$ for $\boldsymbol{x} \in W$, ψ^{-1} is continuously differentiable on $\psi(W)$, and $\Delta_{\psi^{-1}} \neq 0$ on $\psi(W)$.

Making W smaller if necessary, we may suppose that W is an open rectangle; i.e., there exist open intervals I_j, $j = 1, \dots, n$ such that $W = I_1 \times \cdots \times I_n$. Set $W_0 = I_2 \times \cdots \times I_n$, fix $x_1 \in I_1$, and for each $\boldsymbol{y} = (x_2, \dots, x_n)$ which satisfies $(x_1, \boldsymbol{y}) \in W$ set

$$\phi_0(\boldsymbol{y}) = (\phi_2(x_1, \boldsymbol{y}), \dots, \phi_n(x_1, \boldsymbol{y})).$$

Clearly, $\phi_0 : W_0 \to \mathbf{R}^{n-1}$ satisfies $\psi(x_1, \boldsymbol{y}) = (x_1, \phi_0(\boldsymbol{y}))$. Hence, ϕ_0 is continuously differentiable and by the Chain Rule satisfies $\Delta_{\phi_0} = \Delta_\psi \neq 0$ on W_0. It follows from the inductive hypothesis that

$$(38) \qquad |R_0| = \int_{\phi_0^{-1}(R_0)} |\Delta_{\phi_0}(\boldsymbol{x})| \, d\boldsymbol{x}$$

for every $(n-1)$-dimensional rectangle $R_0 \subset \phi_0(W_0)$, and

$$|Q_0| = \int_{\phi_0(Q_0)} |\Delta_{\phi_0^{-1}}(\boldsymbol{x})| \, d\boldsymbol{x}$$

for every $(n-1)$-dimensional rectangle $Q_0 \subset W_0$. (W_0, hence, W may have gotten smaller again.)

Suppose R is an n-dimensional rectangle which satisfies $R \subset \psi(W)$. Then $R = [a, b] \times R_0$ where R_0 is an $(n-1)$-dimensional rectangle contained in $\phi_0(W_0)$ and

$$\psi^{-1}(\{x_1\} \times R_0) = \{x_1\} \times \phi_0^{-1}(R_0)$$

for all $x_1 \in [a, b]$. By (38) and Theorem 7.12,

$$|R| = (b - a)|R_0| = \int_{\phi_0^{-1}(R_0)} \int_a^b |\Delta_{\phi_0}(\boldsymbol{y})| \, dx_1 \, d\boldsymbol{y} = \int_{\psi^{-1}(R)} |\Delta_\psi(\boldsymbol{x})| \, d\boldsymbol{x}.$$

This proves $|R| = \int_{\psi^{-1}(R)} |\Delta_\psi(\boldsymbol{x})| \, d\boldsymbol{x}$ for all n-dimensional rectangles $R \subset \psi(W)$; i.e., (34) holds for ψ in place of ϕ. A similar argument shows that (34) holds for σ in place of ϕ. ∎

This local change of variables formula contains the following global result.

THEOREM 7.14. *Suppose V is open in $\mathbf{R}^n$ and $\phi : V \to \mathbf{R}^n$ is 1–1 and continuously differentiable on V. If $\Delta_\phi \neq 0$ on V and E is a Jordan region which satisfies $\overline{E} \subset V$, then*

$$\int_{\phi(E)} f(\boldsymbol{u}) \, d\boldsymbol{u} = \int_E f(\phi(\boldsymbol{x})) |\Delta_\phi(\boldsymbol{x})| \, d\boldsymbol{x},$$

provided f is integrable on $\phi(E)$ and $f \circ \phi |\Delta_\phi|$ is integrable on E.

PROOF. Let $f : \phi(E) \to \mathbf{R}$, $H = \overline{E}$, suppose f is integrable on $\phi(E)$, and $(f \circ \phi) |\Delta_\phi|$ is integrable on E. By Theorem 7.13, given $\boldsymbol{a} \in H$ there is an open rectangle $V_{\boldsymbol{a}}$ such that $\boldsymbol{a} \in V_{\boldsymbol{a}} \subset V$ and

$$(39) \qquad \int_{\phi(E_i)} f(\boldsymbol{u}) \, d\boldsymbol{u} = \int_{E_i} f(\phi(\boldsymbol{x})) |\Delta_\phi(\boldsymbol{x})| \, d\boldsymbol{x}$$

for every Jordan region E_i which satisfies $\overline{E_i} \subset V_{\boldsymbol{a}}$. Let $W_{\boldsymbol{a}}$ be an open rectangle which satisfies $\boldsymbol{a} \in W_{\boldsymbol{a}} \subset \overline{W_{\boldsymbol{a}}} \subset V_{\boldsymbol{a}}$. Since H is compact and

$$H \subset \bigcup_{\boldsymbol{a} \in H} W_{\boldsymbol{a}},$$

we can choose open rectangles $W_j := W_{\mathbf{a}_j}$ such that

$$H \subset \bigcup_{j=1}^{k} W_j.$$

Let R be a huge rectangle which contains H and $\mathcal{G} = \{R_i : i = 1, \dots, N\}$ be a grid on R so fine that each rectangle in $\mathcal{G}$ which intersects H is a subset of some $\overline{W_j}$. (This is possible since there are only finitely many W_j's.) Let $E_i = R_i \cap E$. Then $\overline{E_i} \subseteq R_i \cap H \subseteq \overline{W_j} \subset V_{\mathbf{a}_j}$ for some $j \in \{1, \dots, k\}$, i.e., (39) holds. Moreover, the collection $\{E_i : i = 1, \dots, N\}$ is a nonoverlapping family of Jordan regions whose union is E hence, by Theorem 1.30 and Exercise 7 in Section 7.1, the collection $\{\phi(E_i) : i = 1, \dots, N\}$ is a nonoverlapping family of Jordan regions whose union is $\phi(E)$. It follows from Theorem 7.6 and (39) that

$$\int_{\phi(E)} f(\mathbf{u}) \, d\mathbf{u} = \sum_{i=1}^{N} \int_{\phi(E_i)} f(\mathbf{u}) \, d\mathbf{u}$$

$$= \sum_{i=1}^{N} \int_{E_i} f(\phi(\mathbf{x})) |\Delta_\phi(\mathbf{x})| \, d\mathbf{x} = \int_{E} f(\phi(\mathbf{x})) |\Delta_\phi(\mathbf{x})| \, d\mathbf{x}. \quad \blacksquare$$

Although this change of variables formula is global, it is not general enough. The condition that Δ_ϕ be nonzero on *all* of V is too restrictive, even for simple changes of variables like $\phi(x) = x^3$ on $(-1, 1)$. This obstacle can usually be surmounted, when the Jacobian vanishes on a set of volume zero, by using the following result.

THEOREM 7.15. *Suppose W is open in $\mathbf{R}^n$, $\phi : W \to \mathbf{R}^n$ is continuously differentiable, and E is a Jordan region such that $\overline{E} \subset W$ and $\phi(E)$ is a Jordan region. If there exists a closed set Z such that $E \cap Z$ is of volume zero, and ϕ is 1–1 and $\Delta_\phi(\mathbf{x}) \neq 0$ for all $\mathbf{x} \in E^\circ \setminus Z$, then*

$$(40) \qquad \int_{\phi(E)} f(\mathbf{u}) \, d\mathbf{u} = \int_{E} f(\phi(\mathbf{x})) |\Delta_\phi(\mathbf{x})| \, d\mathbf{x}$$

provided f is integrable on $\phi(E)$ and $f \circ \phi |\Delta_\phi|$ is integrable on E.

The proof of Theorem 7.15, which requires additional machinery, is postponed until the next section. In the meantime, we shall use this result on examples and exercises as the need arises. Although it looks complicated to apply, with more than half a dozen hypotheses, in practice most of these hypotheses can be verified by inspection. Hypotheses to be careful about are ϕ is 1–1 and Δ_ϕ is nonzero off Z, and $E \cap Z$ is of volume zero.

Our first application of Theorem 7.15 involves a familiar change of variables in $\mathbf{R}^2$. Recall that polar coordinates in $\mathbf{R}^2$ have the form

$$x = r \cos\theta, \qquad y = r \sin\theta,$$

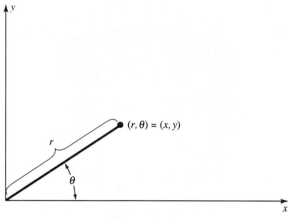

Figure 7.12

where $r = \|(x, y)\|$ and θ is the angle measured counterclockwise from the positive x axis to the line segment $L((0,0); (x, y))$ (see Figure 7.12). Set $\phi(r, \theta) = (r\cos\theta, r\sin\theta)$. Although ϕ is not 1–1 (for example, $\phi(0, \theta) = (0, 0)$ for all $\theta \in \mathbf{R}$), ϕ is 1–1 from $\Omega := \{(r, \theta) : r > 0, 0 \le \theta < 2\pi\}$ onto $\mathbf{R}^2 \setminus \{(0, 0)\}$. Moreover,

$$\Delta_\phi = \det \begin{bmatrix} \cos\theta & \sin\theta \\ -r\sin\theta & r\cos\theta \end{bmatrix} = r(\cos^2\theta + \sin^2\theta) = r$$

is nonzero on Ω. Thus, ϕ is 1–1 and its Jacobian is nonzero off the set $Z := \{(r, \theta) : r = 0\}$. Since $Z \cap E$ is a set of area zero, Theorem 7.15 implies that (40) holds for any Jordan region E, provided $\phi(E)$ is a Jordan region, f is integrable on $\phi(E)$, and $f \circ \phi |\Delta_\phi|$ is integrable on E. We shall abbreviate the change of variables formula from polar coordinates to rectangular coordinates by $dx\, dy = r\, dr\, d\theta$.

The next two examples show that polar coordinates can be used to evaluate integrals which cannot be easily evaluated using rectangular coordinates.

Example 1. Find the volume of the region E bounded by $z = x^2 + y^2$, $x^2 + y^2 = 4$ and $z = 0$.

SOLUTION. Clearly, E lies under the function $f(x, y) = x^2 + y^2$ over the region $B = B_2(0, 0)$ (see Figure 7.13). Using polar coordinates, we obtain

$$\text{Vol}\,(E) = \iint_B (x^2 + y^2)\, dA = \int_0^{2\pi} \int_0^2 r^3\, dr\, d\theta = 8\pi. \quad \blacksquare$$

Example 2. Evaluate

$$\iint_E \frac{\sqrt{x^2 + y^2}}{x}\, dA,$$

where $E = \{(x, y) : a^2 \le y \le x,\ x^2 + y^2 \le 1\}$ for some $0 < a < 1$.

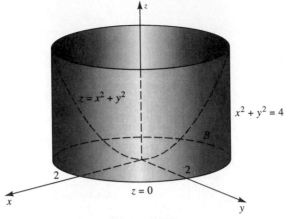

Figure 7.13

SOLUTION. Changing to polar coordinates, we see that

$$\iint_E \frac{\sqrt{x^2+y^2}}{x}\, dA = \int_0^{\pi/4} \int_a^1 \frac{r^2}{r\cos\theta}\, dr\, d\theta = \frac{1-a^3}{3}\int_0^{\pi/4} \sec\theta\, d\theta.$$

To integrate $\sec\theta$, multiply and divide by $\sec\theta + \tan\theta$. Using the change of variables $u = \sec\theta + \tan\theta$, we obtain

$$\int_0^{\pi/4} \sec\theta\, d\theta = \int_0^{\pi/4} \frac{\sec\theta\tan\theta + \sec^2\theta}{\sec\theta + \tan\theta}\, d\theta$$

$$= \int_1^{1+\sqrt{2}} \frac{du}{u} = \log(1+\sqrt{2}).$$

Consequently,

$$\iint_E \frac{\sqrt{x^2+y^2}}{x}\, dA = \frac{(1-a^3)\log(1+\sqrt{2})}{3}. \quad \blacksquare$$

Recall that *cylindrical coordinates* in $\mathbf{R}^3$ have the form

$$x = r\cos\theta, \qquad y = r\sin\theta, \qquad z = z,$$

where $r = \|(x,y,z)\|$ and θ is the angle measured counterclockwise from the positive x axis to the line segment $L((0,0,0);(x,y,0))$. It is easy to see that this change of variables is 1–1 on $\Omega := \{(r,\theta,z) : r > 0, 0 \le \theta < 2\pi, z \in \mathbf{R}\}$ and its Jacobian, r, is nonzero off $Z := \{(r,\theta,z) : r = 0\}$. Thus, by Theorem 7.15, (40) holds for any Jordan region E, provided $\phi(E)$ is a Jordan region, f is integrable on $\phi(E)$, and $f \circ \phi |\Delta_\phi|$ is integrable on E. We shall abbreviate the change of variables formula from cylindrical coordinates to rectangular coordinates by $dx\, dy\, dz = r\, dz\, dr\, d\theta$.

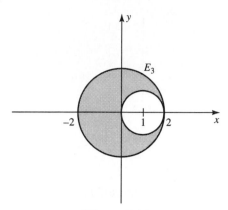

Figure 7.14

Example 3. Find the volume of the region E which lies inside the paraboloid $x^2 + y^2 + z = 4$, outside the cylinder $x^2 - 2x + y^2 = 0$, and above the plane $z = 0$.

SOLUTION. The paraboloid $z = 4 - x^2 - y^2$ has vertex $(0, 0, 4)$ and opens downward about the z axis. The cylinder $x^2 - 2x + y^2 = (x-1)^2 + y^2 - 1 = 0$ has base centered at $(1, 0)$ with radius 1. Hence, the projection E_3 lies inside the circle $x^2 + y^2 = 4$ and outside the circle $x^2 + y^2 = 2x$ (see Figure 7.14). This last circle can be described in polar coordinates by $r^2 = 2r \cos\theta$, i.e., $r = 2\cos\theta$. Thus,

$$\text{Vol}(E) = \iiint_E 1 \, dV = \iint_{E_3} \int_0^{4-r^2} dz \, dA$$

$$= \int_{-\pi/2}^{\pi/2} \int_{2\cos\theta}^{2} (4 - r^2) r \, dr \, d\theta + \int_{\pi/2}^{3\pi/2} \int_0^2 (4 - r^2) r \, dr \, d\theta = 3\pi. \quad \blacksquare$$

Recall that spherical coordinates in $\mathbf{R}^3$ have the form

$$x = \rho \sin\varphi \cos\theta, \qquad y = \rho \sin\varphi \sin\theta, \qquad z = \rho \cos\varphi,$$

where $\rho = \|(x, y, z)\|$, θ is the angle measured counterclockwise from the positive x axis to the line segment $L((0,0,0); (x, y, 0))$, and ϕ is the angle measured from the positive z axis to the vector (x, y, z) (see Figure 7.15). Notice that this change of variables is 1–1 on $\{(\rho, \varphi, \theta) : \rho > 0, 0 < \varphi < \pi, 0 \le \theta < 2\pi\}$ and its Jacobian, $\rho^2 \sin\varphi$ (see Exercise 1), is nonzero off $Z := \{(\rho, \varphi, \theta) : \varphi = 0, \pi, \rho = 0\}$. Thus, (40) holds for any Jordan region E, provided $\phi(E)$ is a Jordan region, f is integrable on $\phi(E)$, and $f \circ \phi|\Delta_\phi|$ is integrable on E. We shall abbreviate the change of variables formula from spherical coordinates to rectangular coordinates by $dx \, dy \, dz = \rho^2 \sin\varphi \, d\rho \, d\varphi \, d\theta$. (For spherical coordinates in $\mathbf{R}^n$, see the proof of Theorem 7.22.)

Example 4. Find

$$\iiint_Q x \, dV,$$

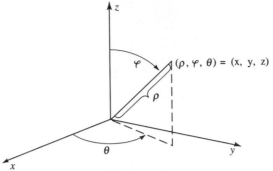

Figure 7.15

where $Q = B_3(0,0,0) \setminus B_2(0,0,0)$.

SOLUTION. Using spherical coordinates, we have

$$\iiint_Q x\,dV = \int_0^{2\pi} \int_0^{\pi} \int_2^3 \rho \sin\varphi \cos\theta(\rho^2 \sin\varphi)\,d\rho\,d\varphi\,d\theta = 0. \quad \blacksquare$$

Theorems 7.14 and 7.15 can be used for other changes of variables besides polar, cylindrical, and spherical coordinates.

Example 5. Find

$$\iint_E \sin(x+y)\cos(2x-y)\,dA,$$

where E is the region bounded by $y = 2x - 1$, $y = 2x + 3$, $y = -x$, and $y = -x + 1$.

SOLUTION. Let $u = 2x - y$, $v = x + y$. By Cramer's Rule, for each fixed $u, v \in \mathbf{R}$, this system of two linear equations in two unknowns has a unique solution in x, y. Hence, the change of variables $\phi(x,y) = (2x - y, x + y)$ is 1–1 and its Jacobian can be obtained by using the Inverse Function Theorem:

$$\frac{\partial(u,v)}{\partial(x,y)} = \det \begin{bmatrix} 2 & -1 \\ 1 & 1 \end{bmatrix} = 3 \quad \text{implies} \quad \frac{\partial(x,y)}{\partial(u,v)} = 3^{-1} = \frac{1}{3}.$$

To find what happens to E under ϕ, notice that $y = 2x - 1$ implies $u = 1$, $y = 2x + 3$ implies $u = -3$, $y = -x$ implies $v = 0$, and $y = -x + 1$ implies $v = 1$. Thus, $\phi(E) = [-3, 1] \times [0, 1]$, and it follows from Theorem 7.14 that

$$\iint_E \sin(x+y)\cos(2x-y)\,dA = \frac{1}{3} \int_0^1 \int_{-3}^1 \sin v \cos u \, du\, dv$$

$$= \frac{1}{3}(\sin(1) + \sin(3))(1 - \cos(1)). \quad \blacksquare$$

EXERCISES

1. a) Compute the Jacobian of the change of variables from spherical coordinates to rectangular coordinates.
   b) Assuming Vol is translation and rotation invariant (see Exercise 3 in Section 7.1 and Exercise 8 below), verify the following classical formulas: the volume of a sphere of radius r is $\frac{4}{3}\pi r^3$ and the volume of a right circular cone of altitude h and radius r is $\pi r^2 h/3$.

2. Evaluate each of the following integrals.

   a)
   $$\int_0^2 \int_0^{\sqrt{4-x^2}} \sin(x^2 + y^2)\, dy\, dx.$$

   b)
   $$\int_0^1 \int_0^x \sqrt[3]{(2y - y^2)^2}\, dy\, dx.$$

   c)
   $$\int_a^b \int_0^x \sqrt{x^2 + y^2}\, dy\, dx, \qquad 0 \le a < b.$$

3. For each of the following, find $\iint_E f\, dA$.
   a) $f(x, y) = \cos(3x^2 + y^2)$ and E is the set of points satisfying $x^2 + y^2/3 \le 1$.
   b) $f(x, y) = y\sqrt{x - 2y}$ and E is bounded by the triangle with vertices $(0,0)$, $(4,0)$, and $(4,2)$.

4. For each of the following, find $\iiint_E f\, dV$.
   a) $f(x, y, z) = z^2$ and E is the set of points satisfying $x^2 + y^2 + z^2 \le 6$ and $z \ge x^2 + y^2$.
   b) $f(x, y, z) = e^z$ and E is the set of points satisfying $x^2 + y^2 + z^2 \le 9$, $x^2 + y^2 \le 1$, and $z \ge 0$.
   c) $f(x, y, z) = (x - y)z$ and E is the set of points satisfying $x^2 + y^2 + z^2 \le 4$, $z \ge \sqrt{x^2 + y^2}$, and $x \ge 0$.

5. a) Prove that the volume bounded by the ellipsoid
   $$\frac{x^2}{a^2} + \frac{y^2}{b^2} + \frac{z^2}{c^2} = 1$$
   is $4\pi abc/3$.
   b) Let a, b, c, d be positive numbers and $r^2 < d^2/(b^2 + c^2)$. Find the volume of the region bounded by $y^2 + z^2 = r^2$ and $ax + by + cz = d$.
   c) Show that for any $a \ge 0$, the volume of the region bounded by the cylinders $x^2 + z^2 = a^2$ and $y^2 + z^2 = a^2$ is $16a^3/3$.

6. a) Compute $\iint_E \sqrt{x-y}\sqrt{x+2y}\, dA$, where E is the parallelogram with vertices $(0,0)$, $(2/3,-1/3)$, $(1,0)$, $(1/3,1/3)$.

   b) Compute $\iint_E \sqrt[3]{2x^2 - 5xy - 3y^2}\, dA$, where E is the parallelogram bounded by the lines $y = x/3$, $y = (x-1)/3$, $y = -2x$, $y = 1 - 2x$.

   c) Find

   $$\iint_E e^{(y-x)/(y+x)}\, dA,$$

   where E is the trapezoid with vertices $(1,1)$, $(2,2)$, $(2,0)$, $(4,0)$.

   d) Given $\int_0^1 (1-x)f(x)\, dx = 5$, find

   $$\int_0^1 \int_0^x f(x-y)\, dy\, dx.$$

7. Suppose V is open in **R**^{n} and $f : V \to$ **R** is continuously differentiable with $\Delta_f \neq 0$ on V. Prove that

   $$\lim_{r \to 0} \frac{\operatorname{Vol}\left(f(B_r(\mathbf{x}_0))\right)}{\operatorname{Vol}\left(B_r(\mathbf{x}_0)\right)} = |\Delta_f(\mathbf{x}_0)|$$

   for every $\mathbf{x}_0 \in V$.

8. Show that Vol is *rotation invariant* in **R**^{2}, i.e., if ϕ is a rotation on **R**^{2} (see Exercise 9 in Section 5.1) and E is a Jordan region in **R**^{2}, then

   $$\operatorname{Vol}(\phi(E)) = \operatorname{Vol}(E).$$

9. Let $\mathbf{v}_j = (v_{j1}, \ldots, v_{jn}) \in$ **R**^{n}, $j = 1, \ldots, n$, be fixed. The parallelepiped determined by the vectors $\mathbf{v}_j$ is the set

   $$\mathcal{P}(\mathbf{v}_1, \ldots, \mathbf{v}_n) := \{t_1\mathbf{v}_1 + \cdots + t_n\mathbf{v}_n : t_j \in [0,1]\}$$

   and the determinant of the $\mathbf{v}_j$'s is the number

   $$\det(\mathbf{v}_1, \ldots, \mathbf{v}_n) := \det [v_{jk}]_{n \times n}.$$

   Prove that
   $$\operatorname{Vol}\left(\mathcal{P}(\mathbf{v}_1, \ldots, \mathbf{v}_n)\right) = |\det(\mathbf{v}_1, \ldots, \mathbf{v}_n)|.$$

   Check this formula for $n = 2$ and $n = 3$ to see that it agrees with the classical formulas for the area of a parallelogram and the volume of a parallelepiped.

$\boxed{10}$. **This exercise is used in Section 7.6.**

   a) Prove that the improper integral $\int_0^\infty e^{-x^2}\, dx$ converges to a finite real number I.

   b) Prove that

   $$I^2 = \lim_{N \to \infty} \int_0^{\pi/2} \int_0^N e^{-r^2} r\, dr\, d\theta.$$

c) Show

$$\int_0^\infty e^{-x^2}\, dx = \frac{\sqrt{\pi}}{2}.$$

d) Let Q_k represent the n-dimensional cube $[-k, k] \times \cdots \times [-k, k]$. Find

$$\lim_{k \to \infty} \int_{Q_k} e^{-\|\boldsymbol{x}\|^2}\, d\boldsymbol{x}.$$

^{e}**7.5   PARTITIONS OF UNITY**   *This section uses no material from any other enrichment section.*

In this section we show that a smooth function can be broken into a sum of smooth functions, each of which is zero except on a small set, and use this to prove the global change of variables formula stated in Section 7.4. Later, this same technique will be used to prove the Fundamental Theorem of Calculus on manifolds (see Theorem 10.8).

DEFINITION 7.5. Let $f : \mathbf{R}^n \to \mathbf{R}$.

   i) The *support* of f is the closure of the set of points where f is nonzero, i.e.,

$$\operatorname{spt} f := \overline{\{\boldsymbol{x} \in \mathbf{R}^n : f(\boldsymbol{x}) \neq 0\}}.$$

   ii) A function f is said to have *compact support* if $\operatorname{spt} f$ is a compact set.

Example 1. If

$$f(x) = \begin{cases} 1 & x \in \mathbf{Q} \\ 0 & x \notin \mathbf{Q}, \end{cases}$$

then $\operatorname{spt} f = \mathbf{R}$.

Example 2. If

$$f(x) = \begin{cases} 1 & x \in (0, 1) \\ 2 & x \in (1, 2) \\ 0 & \text{otherwise,} \end{cases}$$

then $\operatorname{spt} f = [0, 2]$.

Since the support of a function is always closed, f has compact support if and only if $\operatorname{spt} f$ is bounded.

The following result shows that if two functions have compact support, then so does their sum (see also Exercises 1 and 2).

Remark 1. *If $f, g : \mathbf{R}^n \to \mathbf{R}$, then*

$$\operatorname{spt} (f + g) \subseteq \operatorname{spt} f \cup \operatorname{spt} g.$$

PROOF. If $(f+g)(\boldsymbol{x}) \neq 0$, then $f(\boldsymbol{x}) \neq 0$ or $g(\boldsymbol{x}) \neq 0$. Thus,

$$\{\boldsymbol{x} \in \mathbf{R}^n : (f+g)(\boldsymbol{x}) \neq 0\} \subseteq \{\boldsymbol{x} \in \mathbf{R}^n : f(\boldsymbol{x}) \neq 0\} \cup \{\boldsymbol{x} \in \mathbf{R}^n : g(\boldsymbol{x}) \neq 0\}.$$

Since the closure of a union equals the union of its closures (see Theorem 5.10), it follows that $\operatorname{spt}(f+g) \subseteq \operatorname{spt} f \cup \operatorname{spt} g$. ∎

Let $p \in \mathbf{N}$ or $p = \infty$. The symbol $\mathcal{C}_c^p(\mathbf{R}^n)$ will denote the collection of functions $f : \mathbf{R}^n \to \mathbf{R}$ which are $\mathcal{C}^p$ on $\mathbf{R}^n$ and have compact support. In particular, it follows from Remark 1 that if $f_j \in \mathcal{C}_c^p(\mathbf{R}^n)$ for $j = 1, \ldots, N$, then

$$\sum_{j=1}^{N} f_j \in \mathcal{C}_c^p(\mathbf{R}^n).$$

We will use this observation several times below.

If f is analytic (a condition stronger than $\mathcal{C}^\infty$) and has compact support, then f is identically zero (see Exercise 3 below). Thus, it is not at all obvious that $\mathcal{C}_c^\infty(\mathbf{R}^n)$ contains anything but the zero function. Nevertheless, we shall show that $\mathcal{C}_c^\infty(\mathbf{R}^n)$ not only contains nonzero functions, but has enough functions to "approximate" any compact set (see Theorem 7.16 and Exercise 6 below).

First, we deal with the one-dimensional case.

Lemma 1. *For every $a < b$ there is a function $\phi \in \mathcal{C}_c^\infty(\mathbf{R})$ such that $\phi(t) > 0$ for $t \in (a, b)$ and $\phi(t) = 0$ for $t \notin (a, b)$.*

PROOF. The function

$$f(t) = \begin{cases} e^{-1/t^2} & t \neq 0 \\ 0 & t = 0 \end{cases}$$

belongs to $\mathcal{C}^\infty(\mathbf{R})$ and $f^{(j)}(0) = 0$ for all $j \in \mathbf{N}$ (see Exercise 3 in Section 2.5). Hence,

$$\phi(t) = \begin{cases} e^{-1/(t-a)^2} e^{-1/(t-b)^2} & t \in (a, b) \\ 0 & \text{otherwise} \end{cases}$$

belongs to $\mathcal{C}^\infty(\mathbf{R})$, satisfies $\phi(t) > 0$ for $t \in (a, b)$, and $\operatorname{spt} \phi = [a, b]$. ∎

Next, we show that there exists a nonzero $\mathcal{C}^\infty$ function which is constant everywhere except on a small interval.

Lemma 2. *For each $\delta > 0$ there is a function ψ, $\mathcal{C}^\infty$ on $\mathbf{R}$, such that $0 \leq \psi \leq 1$ on $\mathbf{R}$, $\psi(t) = 0$ for $t \leq 0$, and $\psi(t) = 1$ for $t > \delta$.*

PROOF. By Lemma 1, choose $\phi \in \mathcal{C}_c^\infty(\mathbf{R})$ such that $\phi(t) > 0$ for $t \in (0, \delta)$ and $\phi(t) = 0$ for $t \notin (0, \delta)$. Set

$$\psi(t) = \frac{\int_0^t \phi(u)\, du}{\int_0^\delta \phi(u)\, du}.$$

By the Fundamental Theorem of Calculus, $\psi \in \mathcal{C}^\infty(\mathbf{R})$, by construction $0 \le \psi \le 1$, and

$$\psi(t) = \begin{cases} 0 & t \le 0 \\ 1 & t > \delta. \end{cases} \blacksquare$$

Finally, we use these one-dimensional $\mathcal{C}^\infty$ functions to construct nonzero functions in $\mathcal{C}_c^\infty(\mathbf{R}^n)$.

THEOREM 7.16 [$\mathcal{C}^\infty$ VERSION OF URYSOHN'S LEMMA]. *Let H be compact, V be open in $\mathbf{R}^n$, and $H \subset V$. Then there is an $h \in \mathcal{C}_c^\infty(\mathbf{R}^n)$ such that $0 \le h(\boldsymbol{x}) \le 1$ for all $\boldsymbol{x} \in \mathbf{R}^n$, $h(\boldsymbol{x}) = 1$ for all $\boldsymbol{x} \in H$, and $\mathrm{spt}\, h \subset V$.*

PROOF. Let $\phi \in \mathcal{C}_c^\infty(\mathbf{R})$ satisfy $\phi(t) > 0$ for $t \in (-1,1)$ and $\phi(t) = 0$ for $t \notin (-1,1)$. For each $\varepsilon > 0$ and each $\boldsymbol{x} \in \mathbf{R}^n$, let $Q_\varepsilon(\boldsymbol{x})$ represent the n-dimensional cube

$$Q_\varepsilon(\boldsymbol{x}) = \{\boldsymbol{y} \in \mathbf{R}^n : |y_j - x_j| \le \varepsilon \text{ for all } j = 1, \ldots, n\}.$$

Set

(41)
$$g_\varepsilon(\boldsymbol{x}) = \phi\left(\frac{x_1}{\varepsilon}\right) \cdots \phi\left(\frac{x_n}{\varepsilon}\right),$$

and observe by Theorem 2.13 (the Product Rule) that g_ε is $\mathcal{C}^\infty$ on $\mathbf{R}^n$. By construction, $g_\varepsilon(\boldsymbol{y}) \ge 0$ on $\mathbf{R}^n$, $g_\varepsilon(\boldsymbol{y}) > 0$ for all $\boldsymbol{y}$ in the open ball $B_\varepsilon(\mathbf{0})$ and the support of g_ε is a subset of the cube $Q_\varepsilon(\mathbf{0})$. In particular, $g_\varepsilon \in \mathcal{C}_c^\infty(\mathbf{R}^n)$.

We will use sums of translates of these g_ε's to construct a $\mathcal{C}^\infty$ function, supported on V, which is strictly positive on H. It is here that the compactness of H enters in a crucial way.

For each $\boldsymbol{x} \in H$, choose $\varepsilon := \varepsilon(\boldsymbol{x}) > 0$ such that $Q_\varepsilon(\boldsymbol{x}) \subset V$. Set

$$h_{\boldsymbol{x}}(\boldsymbol{y}) = g_\varepsilon(\boldsymbol{y} - \boldsymbol{x}), \qquad \boldsymbol{y} \in \mathbf{R}^n,$$

and notice that $h_{\boldsymbol{x}} \ge 0$ on $\mathbf{R}^n$, $h_{\boldsymbol{x}}(\boldsymbol{y}) > 0$ for all $\boldsymbol{y} \in B_\varepsilon(\boldsymbol{x})$, $h_{\boldsymbol{x}}(\boldsymbol{y}) = 0$ for all $\boldsymbol{y} \notin Q_\varepsilon(\boldsymbol{x})$, and $h_{\boldsymbol{x}} \in \mathcal{C}_c^\infty(\mathbf{R}^n)$. Since H is compact and

$$H \subset \bigcup_{\boldsymbol{x} \in H} B_\varepsilon(\boldsymbol{x}),$$

choose points $\boldsymbol{x}_j \in H$ and positive numbers $\varepsilon_j = \varepsilon(\boldsymbol{x}_j)$, $j = 1, \ldots, N$, such that

$$H \subset B_{\varepsilon_1}(\boldsymbol{x}_1) \cup \cdots \cup B_{\varepsilon_N}(\boldsymbol{x}_N).$$

Set $Q = Q_{\varepsilon_1}(\boldsymbol{x}_1) \cup \cdots \cup Q_{\varepsilon_N}(\boldsymbol{x}_N)$ and $f = h_{\boldsymbol{x}_1} + \cdots + h_{\boldsymbol{x}_N}$. Clearly, Q is compact, $Q \subset V$, and f is $\mathcal{C}^\infty$ on $\mathbf{R}^n$. If $\boldsymbol{x} \notin Q$, then $\boldsymbol{x} \notin Q_{\varepsilon_j}(\boldsymbol{x}_j)$ for all j, hence $f(\boldsymbol{x}) = 0$. Thus, $\mathrm{spt}\, f \subseteq Q$. If $\boldsymbol{x} \in H$, then $\boldsymbol{x} \in B_{\varepsilon_j}(\boldsymbol{x}_j)$ for some j, hence $f(\boldsymbol{x}) > 0$. It remains to flatten f so that it is identically 1 on H. This is where Lemma 2 comes in.

Since $f > 0$ on the compact set H, f has a nonzero minimum on H. Thus, there is a $\delta > 0$ such that $f(\boldsymbol{x}) > \delta$ for $\boldsymbol{x} \in H$. By Lemma 2, choose $\psi \in \mathcal{C}^\infty(\mathbf{R})$ such that $\psi(t) = 0$ when $t \le 0$, and $\psi(t) = 1$ when $t > \delta$. Set $h = \psi \circ f$. Clearly, $h \in \mathcal{C}_c^\infty(\mathbf{R}^n)$, $\mathrm{spt}\, h \subseteq Q \subset V$ and, since $f > \delta$ on H, $h = 1$ on H. Finally, since $0 \le \psi \le 1$, the same is true of h. $\blacksquare$

This result leads directly to a decomposition theorem for $\mathcal{C}^\infty$ functions.

THEOREM 7.17 [$\mathcal{C}^\infty$ PARTITIONS OF UNITY]. *Let* $\Omega \subset \mathbf{R}^n$ *and* $\{V_\alpha\}_{\alpha \in A}$ *be an open covering of* Ω. *Then there exist functions* $\phi_j \in \mathcal{C}_c^\infty(\mathbf{R}^n)$ *and indices* $\alpha_j \in A$, $j \in \mathbf{N}$, *such that the following properties hold.*

i) $$\phi_j \geq 0 \text{ for all } j \in \mathbf{N}.$$

ii) $$\operatorname{spt} \phi_j \subset V_{\alpha_j} \text{ for all } j \in \mathbf{N}.$$

iii) $$\sum_{j=1}^\infty \phi_j(\boldsymbol{x}) = 1 \text{ for all } \boldsymbol{x} \in \Omega.$$

iv) *If* H *is a compact subset of* Ω, *then there is an open set* $W \supset H$ *and an integer* N *such that* $\phi_j(\boldsymbol{x}) = 0$ *for all* $j \geq N$ *and* $\boldsymbol{x} \in W$. *In particular,*

$$\sum_{j=1}^N \phi_j(\boldsymbol{x}) = 1 \quad \text{for all } \boldsymbol{x} \in W.$$

PROOF. For each $\boldsymbol{x} \in \Omega$, choose a bounded open set $W(\boldsymbol{x})$ and an index $\alpha \in A$ such that

$$\boldsymbol{x} \in W(\boldsymbol{x}) \subset \overline{W}(\boldsymbol{x}) \subset V_\alpha.$$

Then $\mathcal{W} = \{W(\boldsymbol{x}) : \boldsymbol{x} \in \Omega\}$ is an open covering of Ω and by Lindelöf's Theorem, we may suppose $\mathcal{W}$ is countable, i.e., $\mathcal{W} = \{W_j\}_{j \in \mathbf{N}}$.

By construction, given $j \in \mathbf{N}$, there is an index $\alpha_j \in A$ such that

$$W_j \subset \overline{W}_j \subset V_{\alpha_j}.$$

Choose by Theorem 7.16 functions $h_j \in \mathcal{C}_c^\infty(\mathbf{R}^n)$ such that $0 \leq h_j \leq 1$ on $\mathbf{R}^n$, $h_j = 1$ on $\overline{W}_j$, and $\operatorname{spt} h_j \subset V_{\alpha_j}$ for $j \in \mathbf{N}$. Set $\phi_1 = h_1$ and for $j > 1$, set

$$\phi_j = (1 - h_1) \dots (1 - h_{j-1}) h_j.$$

Then $\phi_j \geq 0$ on $\mathbf{R}^n$, and $\phi_j \in \mathcal{C}_c^\infty(\mathbf{R}^n)$ with $\operatorname{spt} \phi_j \subseteq \operatorname{spt} h_j \subset V_{\alpha_j}$ for $j \in \mathbf{N}$. This proves parts i) and ii).

An easy induction argument establishes

$$\sum_{j=1}^k \phi_j = 1 - (1 - h_1) \dots (1 - h_k)$$

for $k \in \mathbf{N}$. If $\boldsymbol{x} \in \Omega$, then $\boldsymbol{x} \in W_{j_0}$ for some j_0 so $1 - h_{j_0}(\boldsymbol{x}) = 0$. Thus,

$$\sum_{j=1}^{k} \phi_j = 1 - 0 = 1$$

for $k \geq j_0$. If H is a compact subset of Ω, then $H \subset W_1 \cup \cdots \cup W_N$ for some $N \in \mathbf{N}$. If $W = W_1 \cup \cdots \cup W_N$, then $\boldsymbol{x} \in W$ implies $h_k(\boldsymbol{x}) = 1$ for some $1 \leq k \leq N$, i.e., $\phi_j(\boldsymbol{x}) = 0$ for all $j > N$. Hence,

$$\sum_{j=1}^{N} \phi_j(\boldsymbol{x}) = \sum_{j=1}^{\infty} \phi_j(\boldsymbol{x}) = 1$$

for all $\boldsymbol{x} \in W$. ∎

A sequence of functions $\{\phi_j\}_{j \in \mathbf{N}}$ is called a (C^0) *partition of unity on Ω subordinate to a covering* $\{V_\alpha\}_{\alpha \in A}$ if the ϕ_j's are continuous with compact support on Ω and satisfy statements i) through iv) of Theorem 7.17. By a C^p *partition of unity on Ω* we shall mean a partition of unity on Ω whose functions ϕ_j are also C^p on Ω. By Theorem 7.17, given any open covering $\mathcal{V}$ of any open set $\Omega \subseteq \mathbf{R}^n$ and any extended real number $p \geq 0$, there exists a C^p partition of unity on Ω subordinate to $\mathcal{V}$.

C^p partitions of unity can be used to decompose a function f into a sum of functions f_j which have small support and are as smooth as f. For example, let f be defined on a set Ω, $\{\phi_j\}_{j \in \mathbf{N}}$ be a C^p partition of unity on Ω subordinate to a covering $\{V_j\}_{j \in \mathbf{N}}$, and $f_j = f\phi_j$. Then

$$f(\boldsymbol{x}) = f(\boldsymbol{x}) \sum_{j=1}^{\infty} \phi_j(\boldsymbol{x}) = \sum_{j=1}^{\infty} f(\boldsymbol{x})\phi_j(\boldsymbol{x}) = \sum_{j=1}^{\infty} f_j(\boldsymbol{x})$$

for all $\boldsymbol{x} \in \Omega$. If f is continuous on Ω and $p \geq 0$, then each f_j is continuous on Ω; if f is continuously differentiable on Ω and $p \geq 1$, then each f_j is continuously differentiable on Ω. Thus, f can be written as a sum of functions f_j which are as smooth as f. This allows us to pass from local results to global ones; e.g., if we know that a certain property holds on small open sets in Ω, then we can show that a similar property holds on all of Ω by using a partition of unity subordinate to a covering of Ω which consists of small open sets.

To illustrate the power of this point of view, we now show that the integral can be extended from Jordan regions to open bounded sets, even though such sets are not always Jordan regions.

STRATEGY: The idea behind this extension is fairly simple. Let V be a bounded open set and let f be *locally integrable* on V, i.e., $f : V \to \mathbf{R}$ is integrable on every closed Jordan region $H \subset V$. For each $\boldsymbol{x} \in V$, choose an open Jordan region $V(\boldsymbol{x})$ so small that $\boldsymbol{x} \in V(\boldsymbol{x}) \subset V$. (For example, $V(\boldsymbol{x})$ could be an open ball.) Then $\{V(\boldsymbol{x})\}_{\boldsymbol{x} \in V}$ is an open covering of V and by Lindelöf's Theorem it has a countable subcover, say $\mathcal{V} = \{V_j\}_{j \in \mathbf{N}}$. Let $\{\phi_j\}_{j \in \mathbf{N}}$ be a partition of unity on V subordinate to $\mathcal{V}$. Since f is locally integrable on V, each $f\phi_j$ is integrable. Since $f = \sum_{j=1}^{\infty} f\phi_j$ it seems reasonable to define

$$\int_V f(\boldsymbol{x})\,d\boldsymbol{x} = \sum_{j=1}^{\infty} \int_{V_j} f(\boldsymbol{x})\phi_j(\boldsymbol{x})\,d\boldsymbol{x}.$$

Before we can proceed, we must answer two questions. Does this series converge? And if it does, will its value change when the partition of unity changes? The next two results answer these questions.

Lemma 3. *Let V be a bounded open set in $\mathbf{R}^n$ and $\mathcal{V} = \{V_j\}_{j \in \mathbf{N}}$ be a sequence of open Jordan regions in V which satisfies*

$$V = \bigcup_{j=1}^{\infty} V_j.$$

Suppose $f : V \to \mathbf{R}$ is bounded on V and integrable on each V_j. If ϕ_j is any partition of unity on V subordinate to the covering $\mathcal{V}$, then

(42)
$$\sum_{j=1}^{\infty} \int_{V_j} \phi_j(\boldsymbol{x})f(\boldsymbol{x})\,d\boldsymbol{x}$$

converges absolutely.

PROOF. Let R be an n-dimensional rectangle containing V and $M = \sup_{\boldsymbol{x} \in V} |f(\boldsymbol{x})|$. Since ϕ_j is supported on V_j, the function $\phi_j f$ is integrable on V_j. Moreover, if $E = \cup_{j=1}^{N} V_j$ we have

$$\sum_{j=1}^{N} \left| \int_{V_j} \phi_j(\boldsymbol{x})f(\boldsymbol{x})\,d\boldsymbol{x} \right| \leq \sum_{j=1}^{N} \int_E |\phi_j(\boldsymbol{x})f(\boldsymbol{x})|\,d\boldsymbol{x}$$

$$= \int_E \sum_{j=1}^{N} |\phi_j(\boldsymbol{x})f(\boldsymbol{x})|\,d\boldsymbol{x}$$

$$\leq M \int_E \sum_{j=1}^{N} |\phi_j(\boldsymbol{x})|\,d\boldsymbol{x} \leq M \operatorname{Vol}(E) \leq M|R| < \infty.$$

Therefore, the series in (42) converges absolutely. ∎

The value of the series in (42) depends neither on the partition of unity chosen nor the covering $\mathcal{V}$.

Lemma 4. *Let V be a bounded open set in $\mathbf{R}^n$. Suppose that $\mathcal{V} = \{V_j\}_{j \in \mathbf{N}}$ and $\mathcal{W} = \{W_k\}_{k \in \mathbf{N}}$ are sequences of open Jordan regions in $\mathbf{R}^n$ such that*

$$V = \bigcup_{j=1}^{\infty} V_j = \bigcup_{k=1}^{\infty} W_k.$$

Suppose further that $f : V \to \mathbf{R}$ *is bounded and locally integrable on* V. *If* $\{\phi_j\}_{j \in \mathbf{N}}$ *is a partition of unity on* $\mathcal{V}$ *subordinate to* $\mathcal{V}$ *and* $\{\psi_k\}_{k \in \mathbf{N}}$ *is a partition of unity on* V *subordinate to* $\mathcal{W}$, *then*

$$(43) \qquad \sum_{j=1}^{\infty} \int_{V_j} \phi_j(\boldsymbol{x}) f(\boldsymbol{x})\, d\boldsymbol{x} = \sum_{k=1}^{\infty} \int_{W_k} \psi_k(\boldsymbol{x}) f(\boldsymbol{x})\, d\boldsymbol{x}.$$

PROOF. By Lemma 3, both sums in (43) converge absolutely. By Exercise 5, $\{\phi_j \psi_k\}_{j,k \in \mathbf{N}}$ is a partition of unity on V subordinate to the covering $\{V_j \cap W_k\}_{j,k \in \mathbf{N}}$. Thus,

$$\sum_{j=1}^{\infty} \sum_{k=1}^{\infty} \int_V \phi_j(\boldsymbol{x}) \psi_k(\boldsymbol{x}) f(\boldsymbol{x})\, d\boldsymbol{x}$$

also converges absolutely. Fix $j \in \mathbf{N}$. Since spt ϕ_j is compact, choose $N \in \mathbf{N}$ so large that $\psi_k(\boldsymbol{x}) = 0$ for $k > N$ and $\boldsymbol{x} \in$ spt ϕ_j. Hence,

$$\int_{V_j} \phi_j(\boldsymbol{x}) f(\boldsymbol{x})\, d\boldsymbol{x} = \int_{V_j} \phi_j(\boldsymbol{x}) \sum_{k=1}^{N} \psi_k(\boldsymbol{x}) f(\boldsymbol{x})\, d\boldsymbol{x}$$

$$= \sum_{k=1}^{N} \int_{V_j \cap W_k} \phi_j(\boldsymbol{x}) \psi_k(\boldsymbol{x}) f(\boldsymbol{x})\, d\boldsymbol{x}$$

$$= \sum_{k=1}^{\infty} \int_{V_j \cap W_k} \phi_j(\boldsymbol{x}) \psi_k(\boldsymbol{x}) f(\boldsymbol{x})\, d\boldsymbol{x}.$$

Thus,

$$\sum_{j=1}^{\infty} \int_{V_j} \phi_j(\boldsymbol{x}) f(\boldsymbol{x})\, d\boldsymbol{x} = \sum_{j=1}^{\infty} \sum_{k=1}^{\infty} \int_{V_j \cap W_k} \phi_j(\boldsymbol{x}) \psi_k(\boldsymbol{x}) f(\boldsymbol{x})\, d\boldsymbol{x}.$$

Reversing the roles of j and k we also have

$$\sum_{k=1}^{\infty} \int_{W_k} \psi_k(\boldsymbol{x}) f(\boldsymbol{x})\, d\boldsymbol{x} = \sum_{k=1}^{\infty} \sum_{j=1}^{\infty} \int_{V_j \cap W_k} \phi_j(\boldsymbol{x}) \psi_k(\boldsymbol{x}) f(\boldsymbol{x})\, d\boldsymbol{x}.$$

Since these series are absolutely convergent, we may reverse the order of summation in the last double series. ∎

Using Lemma 4, we define the integral of a locally integrable function f over a bounded open set V as follows.

DEFINITION 7.6. Let V be a bounded open set in $\mathbf{R}^n$ and $f : V \to \mathbf{R}$ be bounded and locally integrable on V. The *integral* of f on V is defined to be

$$I_V(f) := \sum_{j=1}^{\infty} \int_{V_j} \phi_j(\boldsymbol{x}) f(\boldsymbol{x})\, d\boldsymbol{x},$$

where $\{\phi_j\}_{j\in\mathbf{N}}$ is any partition of unity on V subordinate to an open covering $\mathcal{V} = \{V_j\}_{j\in\mathbf{N}}$ such that each V_j is a Jordan region and

$$V = \bigcup_{j=1}^{\infty} V_j.$$

The following result shows that this definition agrees with the old one when V is a Jordan region. Thus, we shall use the notation $\int_V f(\boldsymbol{x})\,d\boldsymbol{x}$ for $I_V(f)$.

THEOREM 7.18. *If E is an open Jordan region in $\mathbf{R}^n$ and $f : E \to \mathbf{R}$ is integrable on E, then*

$$\int_E f(\boldsymbol{x})\,d\boldsymbol{x} = I_E(f).$$

PROOF. Let $\varepsilon > 0$. Since E is a Jordan region, choose a grid $\mathcal{G} = \{Q_1, \ldots, Q_p\}$ of some n-dimensional rectangle $R \supset E$ such that

$$(44) \qquad\qquad \sum_{Q_\ell \cap \partial E \neq \emptyset} |Q_\ell| < \varepsilon.$$

Let

$$H = \bigcup_{Q_\ell \subset E} Q_\ell.$$

Clearly, H is compact and by (44), $\mathrm{Vol}\,(E \setminus H) < \varepsilon$ (see Exercise 4d in Section 7.1).

Set $M = \sup_{\boldsymbol{x}\in E} |f(\boldsymbol{x})|$. Let $\{R_j\}_{j\in\mathbf{N}}$ be a sequence of rectangles such that $R_j \subset E$ and $E = \bigcup_{j=1}^{\infty} R_j^o$, and let $\{\phi_j\}_{j\in\mathbf{N}}$ be a partition of unity on E subordinate to $\mathcal{V} = \{R_j^o\}_{j\in\mathbf{N}}$. Since H is compact, choose $N_1 \in \mathbf{N}$ such that $\phi_j(\boldsymbol{x}) = 0$ for $j > N_1$ and $\boldsymbol{x} \in H$. Then for any $N \geq N_1$ we have

$$\left| \int_E f(\boldsymbol{x})\,d\boldsymbol{x} - \sum_{j=1}^{N} \int_{R_j} \phi_j(\boldsymbol{x}) f(\boldsymbol{x})\,d\boldsymbol{x} \right| = \left| \int_E f(\boldsymbol{x})\,d\boldsymbol{x} - \sum_{j=1}^{N} \int_E \phi_j(\boldsymbol{x}) f(\boldsymbol{x})\,d\boldsymbol{x} \right|$$

$$\leq \int_E \left| f(\boldsymbol{x}) - \sum_{j=1}^{N} \phi_j(\boldsymbol{x}) f(\boldsymbol{x}) \right| d\boldsymbol{x}$$

$$\leq M \int_E \left| 1 - \sum_{j=1}^{N} \phi_j(\boldsymbol{x}) \right| d\boldsymbol{x}$$

$$\leq M\,\mathrm{Vol}\,(E \setminus H) < M\varepsilon.$$

We conclude that $I_E(f)$ exists and equals $\int_E f(\boldsymbol{x})\,d\boldsymbol{x}$. ∎

We now prove a change of variables formula valid for all open bounded sets.

THEOREM 7.19 [CHANGE OF VARIABLES FOR MULTIPLE INTEGRALS]. *Suppose V is a bounded open set in $\mathbf{R}^n$, $\phi : V \to \mathbf{R}^n$ is 1–1 and continuously differentiable on V, and $\phi(V)$ is bounded. If $\Delta_\phi \neq 0$ on V, then*

$$\int_{\phi(V)} f(\boldsymbol{u}) \, d\boldsymbol{u} = \int_V f(\phi(\boldsymbol{x}))|\Delta_\phi(\boldsymbol{x})| \, d\boldsymbol{x}$$

for all bounded $f : \phi(V) \to \mathbf{R}$, provided f is locally integrable on $\phi(V)$ and $(f \circ \phi)|\Delta_\phi|$ is locally integrable on V.

PROOF. For each $\boldsymbol{a} \in V$, choose by Theorem 7.13 an open rectangle $W(\boldsymbol{a})$ such that $\overline{W}(\boldsymbol{a}) \subset V$ and

$$(45) \qquad \int_{\phi(W(\boldsymbol{a}))} f(\boldsymbol{u}) \, d\boldsymbol{u} = \int_{W(\boldsymbol{a})} f(\phi(\boldsymbol{x}))|\Delta_\phi(\boldsymbol{x})| \, d\boldsymbol{x}.$$

Set $\mathcal{W} = \{W(\boldsymbol{a})\}_{\boldsymbol{a} \in V}$. Then $\mathcal{W}$ is an open covering of V. By Lindelöf's Theorem, we may assume that $\mathcal{W} = \{W_j\}_{j \in \mathbf{N}}$. Let $\{\phi_j\}_{j \in \mathbf{N}}$ be a partition of unity on V subordinate to $\mathcal{W}$, i.e., a sequence of $\mathcal{C}^\infty$ functions such that

$$\operatorname{spt} \phi_j \subset W_j \subset V, \quad j \in \mathbf{N}, \quad \text{and} \quad \sum_{j=1}^\infty \phi_j(\boldsymbol{x}) = 1$$

for all $\boldsymbol{x} \in V$. By Theorem 7.3, each $\phi(W_j)$ is a Jordan region. By Exercise 9 in Section 6.5, each $\phi(W_j)$ is open. And by Exercise 4 below, $\{\phi_j \circ \phi^{-1}\}_{j \in \mathbf{N}}$ is a partition of unity on $\phi(V)$ subordinate to the open covering $\{\phi(W_j)\}_{j \in \mathbf{N}}$. Hence, by Definition 7.6 and (45),

$$\int_{\phi(V)} f(\boldsymbol{u}) \, d\boldsymbol{u} = \sum_{j=1}^\infty \int_{\phi(W_j)} (\phi_j \circ \phi^{-1})(\boldsymbol{u}) f(\boldsymbol{u}) \, d\boldsymbol{u}$$

$$= \sum_{j=1}^\infty \int_{W_j} \phi_j(\boldsymbol{x}) f(\phi(\boldsymbol{x}))|\Delta_\phi(\boldsymbol{x})| \, d\boldsymbol{x}$$

$$= \int_V f(\phi(\boldsymbol{x}))|\Delta_\phi(\boldsymbol{x})| \, d\boldsymbol{x}. \quad \blacksquare$$

Finally, we are prepared to prove a change of variables formula for functions whose Jacobians are zero on a set of volume zero.

PROOF OF THEOREM 7.15. Set $V := E^o \setminus Z$ and observe that V is open and bounded. Since $\phi(E) \supseteq \phi(V)$, $\phi(V)$ is also bounded. By hypothesis, $E \setminus E^o \subseteq \partial E$ and $E \cap Z$ are of volume zero. Moreover, by Exercise 7a in Section 7.1, $\phi(E \setminus E^o)$ and $\phi(E \cap Z)$ are of volume zero. Since $E = V \cup (E \cap Z) \cup (E \setminus E^o)$ and $\phi(E) = \phi(V) \cup \phi(E \cap Z) \cup \phi(E \setminus E^o)$, it follows from Theorem 7.6, Corollary 7.7, and Theorem 7.19 that

$$\int_{\phi(E)} f(\boldsymbol{u}) \, d\boldsymbol{u} = \int_{\phi(V)} f(\boldsymbol{u}) \, d\boldsymbol{u}$$

$$= \int_V f(\phi(\boldsymbol{x}))|\Delta_\phi(\boldsymbol{x})| \, d\boldsymbol{x} = \int_E f(\phi(\boldsymbol{x}))|\Delta_\phi(\boldsymbol{x})| \, d\boldsymbol{x}. \quad \blacksquare$$

We close by noting that even this result, general as it is, can be improved. If something called the Lebesgue integral is used instead of the Riemann integral, the integrability condition concerning $(f \circ \phi)|\Delta_\phi|$ and the condition $\Delta_\phi \neq 0$ can be dropped altogether (see Apostol [1], p. 421, and Spivak [12], p. 72, respectively.)

EXERCISES

1. If $f, g : \mathbf{R}^n \to \mathbf{R}$, prove spt $(fg) \subseteq$ spt $f \cap$ spt g.
2. Prove that if $f, g \in C_c^\infty(\mathbf{R}^n)$, then so are fg and αf for any scalar α.
*3. Prove that if f is analytic on $\mathbf{R}$ and $f(x_0) \neq 0$ for some $x_0 \in \mathbf{R}$, then $f \notin C_c^\infty(\mathbf{R})$.
4. Suppose V is a bounded open set in $\mathbf{R}^n$ and $\phi : V \to \mathbf{R}^n$ is 1–1 and continuously differentiable on V with $\Delta_\phi \neq 0$ on V. Let $\mathcal{W} = \{W_j\}_{j \in \mathbf{N}}$ be an open covering of V and $\{\phi_j\}_{j \in \mathbf{N}}$ be a C^p partition of unity on V subordinate to $\mathcal{W}$, where $p \geq 1$. Prove $\{\phi_j \circ \phi^{-1}\}_{j \in \mathbf{N}}$ is a C^1 partition of unity on $\phi(V)$ subordinate to the open covering $\{\phi(W_j)\}_{j \in \mathbf{N}}$.
5. Let V be open in $\mathbf{R}^n$ and $\mathcal{V} = \{V_j\}_{j \in \mathbf{N}}$, $\mathcal{W} = \{W_k\}_{k \in \mathbf{N}}$ be coverings of V. If $\{\phi_j\}_{j \in \mathbf{N}}$ is a C^p partition of unity on V subordinate to $\mathcal{V}$ and $\{\psi_k\}_{k \in \mathbf{N}}$ is a C^p partition of unity on V subordinate to $\mathcal{W}$, prove that $\{\phi_j \psi_k\}_{j,k \in \mathbf{N}}$ is a C^p partition of unity on V subordinate to the covering $\{V_j \cap W_k\}_{j,k \in \mathbf{N}}$.
6. Show that given any compact Jordan region $H \subset \mathbf{R}^n$, there is a sequence of C^∞ functions ϕ_j such that

$$\lim_{j \to \infty} \int_{\mathbf{R}^n} \phi_j(\boldsymbol{x}) \, d\boldsymbol{x} = \text{Vol}(H)$$

^{e}**7.6   THE GAMMA FUNCTION AND VOLUME**   *The last result of this section uses Dini's Theorem from Section 5.7.*

In this section we introduce the gamma function and use it to find a formula for the volume of any n-dimensional ball and an asymptotic estimate of $n!$.

Recall that if $f : (0, \infty) \to \mathbf{R}$ is locally integrable on $(0, \infty)$, then

$$\int_0^\infty f(t) \, dt = \lim_{x \to 0+, y \to \infty} \int_x^y f(t) \, dt.$$

In particular, it is easy to check that $\int_0^\infty e^{-\alpha t} \, dt$ is finite for all $\alpha > 0$.

DEFINITION 7.7. The *gamma function* is defined by

$$\Gamma(x) = \int_0^\infty t^{x-1} e^{-t} \, dt, \qquad x \in (0, \infty),$$

when this integral converges.

By definition,

$$\Gamma(1) = \int_0^\infty e^{-t}\, dt = 1,$$

and

$$\Gamma(1/2) = \int_0^\infty t^{-1/2} e^{-t}\, dt = 2 \int_0^\infty e^{-u^2}\, du = \sqrt{\pi}.$$

(We used the change of variables $t = u^2$ and Exercise 10 in Section 7.4.)

It turns out that $\Gamma(x)$ is defined for all $x \in (0, \infty)$.

THEOREM 7.20. *For each $x \in (0, \infty)$, $\Gamma(x)$ exists and is finite, $\Gamma(x+1) = x\Gamma(x)$, and $\Gamma(n) = (n-1)!$ for $n \in \mathbf{N}$.*

PROOF. Write

$$\Gamma(x) = \int_0^1 t^{x-1} e^{-t}\, dt + \int_1^\infty t^{x-1} e^{-t}\, dt = I_1 + I_2.$$

By L'Hôpital's Rule,

$$\lim_{t \to \infty} e^{-\beta t} t^y = 0$$

for all $\beta > 0$, $t > 0$ and $y \in \mathbf{R}$. Hence, $e^{-t} t^{x-1} \leq e^{-t/2}$ for t large and it follows from Theorem 3.17 (the Comparison Theorem) that I_2 is finite for all $x \in \mathbf{R}$.

To show I_1 is finite for $x > 0$, suppose first that $x \geq 1$. Then $t^{x-1} \leq 1$ for all $t \in [0, 1]$ and

$$I_1 = \int_0^1 t^{x-1} e^{-t}\, dt \leq \int_0^1 e^{-t}\, dt = 1 - \frac{1}{e} < \infty.$$

Therefore, $\Gamma(x)$ is finite for all $x \geq 1$. Next suppose $0 < x < 1$. Then $x + 1 \geq 1$ so $\Gamma(x+1)$ is finite. Integration by parts yields

$$\Gamma(x) = \int_0^\infty t^{x-1} e^{-t}\, dt = \frac{t^x e^{-t}}{x} \Big|_{t=0}^\infty + \frac{1}{x} \int_0^\infty t^x e^{-t}\, dt = \frac{1}{x} \Gamma(x+1).$$

Therefore, $\Gamma(x)$ is finite when $0 < x < 1$.

This argument also verifies $x\Gamma(x) = \Gamma(x+1)$ for $x \in (0, \infty)$. Since $\Gamma(0) = 1$, it follows that $\Gamma(n) = (n-1)!$ for all $n \in \mathbf{N}$. ∎

The gamma function can be used to evaluate certain integrals which cannot be evaluated using elementary techniques of integration.

THEOREM 7.21. *If $x, y \in (0, \infty)$, then*

i)
$$\int_0^1 v^{y-1} (1-v)^{x-1}\, dv = \frac{\Gamma(x)\Gamma(y)}{\Gamma(x+y)},$$

and

ii)
$$\int_0^{\pi/2} \cos^{2x-1} \varphi \sin^{2y-1} \varphi\, d\varphi = \frac{\Gamma(x)\Gamma(y)}{2\Gamma(x+y)}.$$

In particular,

iii)
$$\int_0^\pi \sin^{k-2}\varphi\,d\varphi = \frac{\Gamma((k-1)/2)\Gamma(1/2)}{\Gamma(k/2)}$$

holds for all integers $k > 2$.

PROOF. To prove part i), make the change of variables $v = u/(1+u)$ and write

$$\int_0^1 v^{y-1}(1-v)^{x-1}\,dv = \int_0^\infty \left(\frac{u}{1+u}\right)^{y-1}\left(1-\frac{u}{1+u}\right)^{x-1}\frac{du}{(1+u)^2}$$
$$= \int_0^\infty u^{y-1}\left(\frac{1}{1+u}\right)^{x+y}du.$$

It follows from two more changes of variables ($s = t/(1+u)$ and $w = su$) and Fubini's Theorem, that

$$\Gamma(x+y)\int_0^1 v^{y-1}(1-v)^{x-1}\,dv$$

$$= \int_0^\infty \int_0^\infty u^{y-1}\left(\frac{1}{1+u}\right)^{x+y}t^{x+y-1}e^{-t}\,dt\,du$$

$$= \int_0^\infty \int_0^\infty u^{y-1}s^{x+y-1}e^{-s(u+1)}\,ds\,du$$

$$= \int_0^\infty s^{x-1}e^{-s}\left(\int_0^\infty u^{y-1}s^y e^{-su}\,du\right)ds$$

$$= \int_0^\infty s^{x-1}e^{-s}\left(\int_0^\infty w^{y-1}e^{-w}\,dw\right)ds = \Gamma(x)\Gamma(y).$$

To prove part ii) use the change of variables $v = \sin^2\varphi$ and part i) to verify

$$\int_0^{\pi/2}\cos^{2x-1}\varphi\sin^{2y-1}\varphi\,d\varphi = \frac{1}{2}\int_0^1 v^{y-1}(1-v)^{x-1}\,dv = \frac{\Gamma(x)\Gamma(y)}{2\Gamma(x+y)}.$$

Specializing to the case $y = (k-1)/2$ and $x = 1/2$, we obtain part iii). ∎

The connection between the gamma function and volume is contained in the following result.

THEOREM 7.22. *If $r > 0$ and $\mathbf{a} \in \mathbf{R}^n$, then*

$$\mathrm{Vol}\,(B_r(\mathbf{a})) = \frac{2r^n \pi^{n/2}}{n\Gamma(n/2)}.$$

PROOF. By translation invariance (see Exercise 3 in Section 7.1) and Corollary 7.5, $\mathrm{Vol}\,(B_r(\mathbf{a})) = \int_B 1\,d\mathbf{x}$ for $B = B_r(\mathbf{0})$. We suppose for simplicity that $n \geq 2$, and introduce a change of variables in $\mathbf{R}^n$ analogous to spherical coordinates. Namely, let

$$x_1 = \rho \cos \varphi_1, \quad x_2 = \rho \sin \varphi_1 \cos \varphi_2, \quad x_3 = \rho \sin \varphi_1 \sin \varphi_2 \cos \varphi_3, \quad \ldots,$$

$$x_{n-1} = \rho \sin \varphi_1 \ldots \sin \varphi_{n-2} \cos \theta, \quad \text{and} \quad x_n = \rho \sin \varphi_1 \ldots \sin \varphi_{n-2} \sin \theta,$$

where $0 \le \rho \le r$, $0 \le \theta \le 2\pi$, and $0 \le \varphi_j \le \pi$ for $j = 1, \ldots, n - 2$. An easy induction argument shows this change of variables has Jacobian

(47) $$\Delta := \rho^{n-1} \sin^{n-2} \varphi_1 \sin^{n-3} \varphi_2 \ldots \sin^2 \varphi_{n-3} \sin \varphi_{n-2}.$$

Hence, by Theorems 7.15 and 7.21iii,

$$\begin{aligned}
\operatorname{Vol}(B_r(\mathbf{a})) &= \int_B 1 \, d\mathbf{x} \\
&= \int_0^r \int_0^\pi \cdots \int_0^\pi \int_0^{2\pi} \rho^{n-1} \sin^{n-2} \varphi_1 \ldots \sin \varphi_{n-2} \, d\theta \, d\varphi_1 \ldots d\varphi_{n-2} \, d\rho \\
&= \frac{2\pi r^n}{n} \left(\int_0^\pi \sin^{n-2} \varphi \, d\varphi \right) \cdots \left(\int_0^\pi \sin \varphi \, d\varphi \right) \\
&= \frac{2\pi r^n}{n} \cdot \frac{\Gamma((n-1)/2)\Gamma(1/2)}{\Gamma(n/2)} \cdot \frac{\Gamma((n-2)/2)\Gamma(1/2)}{\Gamma((n-1)/2)} \cdots \frac{\Gamma(1)\Gamma(1/2)}{\Gamma(3/2)}.
\end{aligned}$$

Canceling all superfluous factors and substituting the value $\sqrt{\pi}$ for $\Gamma(1/2)$, we conclude that

$$\operatorname{Vol}(B_r(\mathbf{a})) = \frac{2\pi r^n}{n} \left(\frac{\Gamma^{n-2}(1/2)}{\Gamma(n/2)} \right) = \frac{2r^n \pi^{n/2}}{n\Gamma(n/2)}. \quad \blacksquare$$

This formula agrees with what we already know. For $n = 1$ we have

$$\operatorname{Vol}(B_r(0)) = \frac{2r\pi^{1/2}}{\Gamma(1/2)} = 2r,$$

for $n = 2$ we have

$$\operatorname{Vol}(B_r(0,0)) = \frac{2r^2\pi}{2\Gamma(1)} = \pi r^2,$$

and for $n = 3$ we have

$$\operatorname{Vol}(B_r(0,0,0)) = \frac{2r^3\pi^{3/2}}{3\Gamma(3/2)} = \frac{2r^3\pi^{3/2}}{(3/2)\Gamma(1/2)} = \frac{4}{3}\pi r^3.$$

We close this section with an asymptotic estimate of $n!$. First we obtain an integral representation for $n!/(n^{n+1/2}e^{-n})$.

Lemma 1. *If* $\phi(x) = x - \log x - 1$, $x > 0$, *then*

$$\frac{n!}{n^{n+1/2}e^{-n}} = \int_{-\sqrt{n}}^{\infty} e^{-n\phi(1+t/\sqrt{n})} \, dt.$$

PROOF. By Definition 7.7 and Theorem 7.20, we can write

$$n! = \Gamma(n+1) = \int_0^{\infty} x^n e^{-x} \, dx.$$

Making two changes of variables (first $x = ny$, then $y = 1 + t/\sqrt{n}$), we conclude that

$$\frac{n!}{n^{n+1/2}e^{-n}} = \frac{1}{\sqrt{n}} \int_0^{\infty} \left(\frac{x}{n}\right)^n e^{-x+n} \, dx$$

$$= \sqrt{n} \int_0^{\infty} y^n e^{-n(y-1)} \, dy$$

$$= \sqrt{n} \int_0^{\infty} e^{-n\phi(y)} \, dy = \int_{-\sqrt{n}}^{\infty} e^{-n\phi(1+t/\sqrt{n})} \, dt. \ \blacksquare$$

Next, we derive some inequalities which will be used, in conjunction with Dini's Theorem, to evaluate the limit of the integral which appears in Lemma 1.

Lemma 2. *If* $\phi(x) = x - \log x - 1$, $x > 0$, *then*

$$(x-1)\phi'(x) - 2\phi(x) > 0 \quad \text{for } 0 < x < 1,$$

and

$$(x-1)\phi'(x) - 2\phi(x) < 0 \quad \text{for } x > 1.$$

Moreover, there is an absolute constant $M > 0$ *such that*

(48) $$\phi(x) \geq M(x-1)^2 \quad \text{for } 0 < x < 2,$$

and

(49) $$\phi(x) \geq M(x-1) \quad \text{for } x \geq 2.$$

PROOF. Let $\psi(x) = 2\log x - x + 1/x$ and observe that $(x-1)\phi'(x) - 2\phi(x) = \psi(x)$. Since $\psi'(x) = -(x-1)^2/x^2 < 0$ for all $x \neq 1$, ψ is decreasing on $(0, \infty)$. Since $\psi(1) = 0$, it follows that $\psi > 0$ on $(0, 1)$ and $\psi < 0$ on $(1, \infty)$. This proves the first pair of inequalities.

To prove the second pair of inequalities, observe first that by Taylor's Formula,

$$\phi(x) = \phi(1) + \phi'(1)(x-1) + \phi''(c)\frac{(x-1)^2}{2!} = \frac{(x-1)^2}{2c^2}$$

for some c between x and 1. Thus, $\phi(x) \geq (x-1)^2/8$ for all $0 < x < 2$. Next, observe since $\phi(x) > 0$ for $x > 1$ and $\phi(x)/(x-1) \to 1$ as $x \to \infty$, that $\phi(x)/(x-1)$ has a positive minimum, say m, on $[2, \infty)$. Thus, (48) and (49) hold for $M := \min\{m, 1/8\}$. $\blacksquare$

Our final preliminary result evaluates the limit of the integral which appears in Lemma 1.

Lemma 3. If $\phi(x) = x - \log x - 1$, $x > 0$, and $F_n(t) = e^{-n\phi(1+t/\sqrt{n})}$, $n \in \mathbf{N}$, $t > -\sqrt{n}$, then

$$\lim_{n\to\infty} \int_{-\sqrt{n}}^{\infty} F_n(t)\, dt = \int_{-\infty}^{\infty} e^{-t^2/2}\, dt.$$

STRATEGY: The idea behind the proof is simple. By L'Hôpital's Rule,

$$\lim_{n\to\infty} n\phi\left(1 + \frac{t}{\sqrt{n}}\right) = \lim_{n\to\infty} \frac{t}{2}\frac{\phi'(1+t/\sqrt{n})}{1/\sqrt{n}} = \frac{t^2}{2}\lim_{n\to\infty}\phi''\left(1+\frac{t}{\sqrt{n}}\right) = \frac{t^2}{2},$$

so $F_n(t) \to e^{-t^2/2}$, as $n \to \infty$, for every $t \in \mathbf{R}$. Thus, $\int_{-\sqrt{n}}^{\infty} F_n(t)\, dt$ should converge to $\int_{-\infty}^{\infty} e^{-t^2/2}\, dt$ as $n \to \infty$. To prove this, we break the integral over $(-\sqrt{n}, \infty)$ into three pieces: one over $(-\sqrt{n}, -\sqrt{a})$, one over $(\sqrt{a}, \infty)$, and one over $(-\sqrt{a}, \sqrt{a})$. Since $e^{-t^2/2}$ is integrable on $(-\infty, \infty)$, the first two integrals should be small for a sufficiently large. Once a is fixed, we shall use Dini's Theorem on the third integral. Here are the details.

PROOF. Let M be the constant given in Lemma 2, and choose $a > 0$ so large that

$$(50) \qquad \int_{|t| \geq \sqrt{a}} e^{-Mt^2}\, dt < \frac{\varepsilon}{4}, \qquad \int_{\sqrt{a}}^{\infty} e^{-Mt}\, dt < \frac{\varepsilon}{4},$$

and

$$(51) \qquad \int_{|t| \geq \sqrt{a}} e^{-t^2/2}\, dt < \frac{\varepsilon}{4}.$$

Fix $t > -\sqrt{a}$ and consider the function $G(x) = e^{-x\phi(1+t/\sqrt{x})}$, $x > 0$. By the Product Rule,

$$G'(x) = e^{-x\phi(1+t/\sqrt{x})}\left(\frac{t}{2\sqrt{x}}\phi'\left(1+\frac{t}{\sqrt{x}}\right) - \phi\left(1+\frac{t}{\sqrt{x}}\right)\right)$$

$$= \frac{e^{-x\phi(y)}}{2}((y-1)\phi'(y) - 2\phi(y)),$$

where $y = 1 + t/\sqrt{x}$. Thus, by Lemma 2, $G'(x) > 0$ for $x > a$, $-\sqrt{a} < t < 0$, and $G'(x) < 0$ for $x > 0$, $t > 0$. It follows that for each $t \in (-\sqrt{a}, 0)$, $F_n(t) \uparrow e^{-t^2/2}$ as $n \to \infty$, and for each $t \in (0, \infty)$, $F_n(t) \downarrow e^{-t^2/2}$ as $n \to \infty$. Hence, by Dini's Theorem (Theorem 5.32), given $\varepsilon > 0$ there is an $N \in \mathbf{N}$ such that $n \geq N$ implies

$$(52) \qquad \left| \int_{-\sqrt{a}}^{\sqrt{a}} \left(F_n(t) - e^{-t^2/2}\right) dt \right| < \frac{\varepsilon}{4}.$$

Let $n > \max\{N, a\}$. By (48) and (49),

$$n\phi\left(1 + \frac{t}{\sqrt{n}}\right) \geq nM\frac{t^2}{n} = Mt^2 \quad \text{for } -\sqrt{n} < t < \sqrt{n},$$

and

$$n\phi\left(1 + \frac{t}{\sqrt{n}}\right) \geq nM\frac{t}{\sqrt{n}} \geq Mt \quad \text{for } t \geq \sqrt{n}.$$

Since $n > a$, it follows that

$$\int_{\sqrt{a}}^{\infty} |F_n(t)|\, dt + \int_{-\sqrt{n}}^{-\sqrt{a}} |F_n(t)|\, dt \leq \int_{\sqrt{a} \leq |t| \leq \sqrt{n}} e^{-Mt^2}\, dt + \int_{\sqrt{n}}^{\infty} e^{-Mt}\, dt$$

$$< \int_{|t| \geq \sqrt{a}} e^{-Mt^2}\, dt + \int_{\sqrt{a}}^{\infty} e^{-Mt}\, dt.$$

Combining this inequality with (50), we obtain

$$(53) \qquad \int_{\sqrt{a}}^{\infty} |F_n(t)|\, dt + \int_{-\sqrt{n}}^{-\sqrt{a}} |F_n(t)|\, dt < \frac{\varepsilon}{2}.$$

Consequently, we conclude by (51), (52), and (53) that

$$\left| \int_{-\sqrt{n}}^{\infty} F_n(t)\, dt - \int_{-\sqrt{n}}^{\infty} e^{-t^2/2}\, dt \right|$$

$$\leq \int_{\sqrt{a}}^{\infty} |F_n(t)|\, dt + \int_{-\sqrt{n}}^{-\sqrt{a}} |F_n(t)|\, dt + \int_{|t| \geq \sqrt{a}} e^{-t^2/2}\, dt$$

$$+ \left| \int_{-\sqrt{a}}^{\sqrt{a}} \left(F_n(t) - e^{-t^2/2} \right)\, dt \right|$$

$$< \frac{\varepsilon}{2} + \frac{\varepsilon}{4} + \frac{\varepsilon}{4} = \varepsilon. \ \blacksquare$$

THEOREM 7.23 [STIRLING'S FORMULA]. *For $n \in \mathbf{N}$ sufficiently large, $n! \approx \sqrt{2\pi}(n^{n+1/2})e^{-n}$, i.e.,*

$$\lim_{n \to \infty} \frac{n!}{\sqrt{2\pi}(n^{n+1/2})e^{-n}} = 1.$$

PROOF. By Exercise 10 in Section 7.4 and the change of variables $t = \sqrt{2}u$, we have

$$\int_{-\infty}^{\infty} e^{-t^2/2}\, dt = \sqrt{2} \int_{-\infty}^{\infty} e^{-u^2}\, du = 2\sqrt{2} \int_{0}^{\infty} e^{-u^2}\, du = \sqrt{2\pi}.$$

Therefore, it follows from Lemmas 1 and 3 that

$$\lim_{n \to \infty} \frac{n!}{\sqrt{2\pi}n^{n+1/2}e^{-n}} = \lim_{n \to \infty} \frac{1}{\sqrt{2\pi}} \int_{-\sqrt{n}}^{\infty} e^{-n\phi(1+t/\sqrt{n})}\, dt$$

$$= \frac{1}{\sqrt{2\pi}} \int_{-\infty}^{\infty} e^{-t^2/2}\, dt = 1. \ \blacksquare$$

EXERCISES

1. Show
$$\int_0^\infty t^2 e^{-t^2}\, dt = \frac{\sqrt{\pi}}{4}.$$

2. Show
$$\int_0^1 \frac{dx}{\sqrt{-\log x}} = \sqrt{\pi}$$

3. Show
$$\int_{-\infty}^\infty e^{\pi t - e^t}\, dt = \Gamma(\pi).$$

4. Show that the volume of a four-dimensional ball of radius r is $\pi^2 r^4/2$ and the volume of a five-dimensional ball of radius r is $8\pi^2 r^5/15$.

5. Verify (47).

6. For $n > 2$, prove that the volume of the n-dimensional ellipsoid
$$E = \left\{ (x_1,\ldots,x_n) : \frac{x_1^2}{a_1^2} + \frac{x_2^2}{a_2^2} + \cdots + \frac{x_n^2}{a_n^2} \le 1 \right\}$$
is
$$\mathrm{Vol}\,(E) = \frac{2a_1 \ldots a_n \pi^{n/2}}{n\Gamma(n/2)}.$$

7. For $n > 2$, prove that the volume of the n-dimensional cone
$$C = \{ (x_1,\ldots,x_n) : (h/r)\sqrt{x_2^2 + \ldots x_n^2} \le x_1 \le h \}$$
is
$$\mathrm{Vol}\,(C) = \frac{2hr^{n-1}\pi^{(n-1)/2}}{n(n-1)\Gamma((n-1)/2)}.$$

8. Find the value of
$$\int_{B_r(0)} x_k^2\, d(x_1,\ldots,x_n)$$
for each $k \in \mathbf{N}$.

9. If $f : B_1(0) \to \mathbf{R}$ is differentiable with
$$f(0) = 0 \quad \text{and} \quad \|\nabla f(x)\| \le 1$$
for $x \in B_1(0)$, prove that the following exists and equals 0.
$$\lim_{k\to\infty} \int_{B_1(0)} |f(x)|^k\, dx.$$

10. a) Prove that Γ is differentiable on $(0,\infty)$ with
$$\Gamma'(x) = \int_0^\infty e^{-t} t^{x-1} \log t\, dt.$$

*b) Prove that Γ is C^∞ and convex on $(0,\infty)$.

Chapter 8

FUNDAMENTAL THEOREMS OF MULTIVARIABLE CALCULUS

This chapter is more descriptive and less rigorous than previous chapters. Our goal is to lay some practical foundations for a very abstract Chapter 10.

8.1 CURVES

According to the dictionary, a curve is a smooth line which bends without corners, a one-dimensional object with length but no breadth. Of course, this definition is too imprecise. It is also too restrictive. Our concept of curve will include not only "smooth" objects like the graphs of the function $y = x^2$ and the relation $x^2 + y^2 = 1$, but also objects with "corners" like the graph of $y = |x|$.

Recall that if $I \subseteq \mathbf{R}$ and $\phi : I \to \mathbf{R}^m$, then the image of I under ϕ is the set

$$\phi(I) = \{\boldsymbol{x} \in \mathbf{R}^m : \boldsymbol{x} = \phi(t) \quad \text{for some } t \in I\}.$$

Let $\boldsymbol{x}_0, \boldsymbol{a} \in \mathbf{R}^m$. It is easy to see that the image of $\mathbf{R}$ under $\phi(t) := t\boldsymbol{a} + \boldsymbol{x}_0$ is a straight line (see Exercise 2). We shall call it the *straight line* through $\boldsymbol{x}_0$ *in the direction* of $\boldsymbol{a}$. This is the simplest type of curve in $\mathbf{R}^m$.

A naive attempt to define a general curve in $\mathbf{R}^m$ is to insist that it be the image of an interval under some continuous function $\phi : \mathbf{R} \to \mathbf{R}^m$. It turns out this definition is too broad. There are continuous functions (called "space filling curves") which take the unit interval $[0, 1]$ onto the unit square $[0, 1] \times [0, 1]$ (see Boas [2]). One way to fix this definition is to use homeomorphisms, i.e., continuous functions whose inverses are also continuous. Since we are primarily interested in the differential structure of curves, we take a different approach, using differentiable functions to define curves (see Definition 8.1 below).

We begin by extending the definition of partial differentiation to include functions defined on nonopen domains. Let $m, n, p \in \mathbf{N}$ and E be a subset of $\mathbf{R}^n$. A function $f : E \to \mathbf{R}^m$ is said to be $\mathcal{C}^p$ (on E) if there is an open set $V \supseteq E$ and a function

$g : V \to \mathbf{R}$ whose partial derivatives of orders $j \le p$ exist on V such that $f(\boldsymbol{x}) = g(\boldsymbol{x})$ for all $\boldsymbol{x} \in E$. In this case we define the *partial derivatives* of f to be equal to the partial derivatives of g, e.g., $\partial f_j / \partial x_k(\boldsymbol{x}) = \partial g_j / \partial x_k(\boldsymbol{x})$ for $k = 1, 2, \ldots, n$, $j = 1, 2, \ldots, m$, and $\boldsymbol{x} \in E$. A function $f : E \to \mathbf{R}^m$ is said to be C^∞ (on E) if f is C^p on E for all $p \in \mathbf{N}$.

Henceforth, p will denote an element of $\mathbf{N}$ or the extended real number ∞.

For technical reasons, we want a curve to be more than just a set of points. Hence, we define a general curve to be a set of ordered pairs as follows.

DEFINITION 8.1.

    i) A (C^p) *curve* in $\mathbf{R}^m$ is a pair $C = (\phi, I)$, where I is an interval (bounded or unbounded) and $\phi : I \to \mathbf{R}^m$ is C^p on I. The set $\phi(I)$ will be called the *trace* of C. A point $\boldsymbol{x} \in \mathbf{R}^m$ is said to *lie on the curve* C if it belongs to the trace of C.

    ii) A curve $C = (\phi, I)$ is called an *arc* if $I = [a, b]$ is closed and bounded, in which case $\phi(a), \phi(b)$ are called the *endpoints* of C.

    iii) An arc $(\phi, [a, b])$ is called *closed* if $\phi(a) = \phi(b)$.

    iv) A nonclosed curve (ϕ, I) is called *simple* if ϕ is 1–1 on I. A closed curve $(\phi, [a, b])$ is called *simple* if ϕ is 1–1 on $[a, b)$ and $\phi(a) = \phi(b)$.

Simple closed curves are also called *Jordan curves* because of the Jordan Curve Theorem. This theorem states that the trace of every simple closed curve (ϕ, I) in $\mathbf{R}^2$ separates $\mathbf{R}^2$ into two pieces, a bounded connected set E and an unbounded connected set Ω, where $\partial E = \partial \Omega = \phi(I)$ (see Griffiths [3]). (It is interesting to note that the set E is not necessarily a Jordan region. This fact was discovered by W.F. Osgood[1].)

It is important to realize from the beginning that a given set of points can be the trace of many different curves. For example, the line segment $\{(x, y) \in \mathbf{R}^2 : y = x, 0 < x \le 1\}$ is the trace of $\phi(t) = (t, t)$ on $(0, 1]$, of $\psi(t) = (t/2, t/2)$ on $(0, 2]$, and of $\sigma(t) = (1/t, 1/t)$ on $[1, \infty)$. Although these functions trace the same line segment, each of them traces it differently. The function ψ traces the line "twice as slowly" as ϕ and σ traces the line "backwards" from ϕ. Therefore, a curve (ϕ, I) is not merely a set of points, but a set of points $\phi(I)$ together with a specific way $\phi : I \to \phi(I)$ of tracing those points.

To discuss different ways of tracing out the same set of points, we introduce the following terminology. By a C^p *parametrization* of a curve (ϕ, I) we mean a simple C^p curve (ψ, J) which satisfies $\psi(J) = \phi(I)$. The equations

$$x_j = \psi_j(t), \qquad t \in J, \quad j = 1, \ldots, m$$

are called the *parametric equations* of (ϕ, I) induced by the parametrization (ψ, J). (For a physical interpretation of parametrization, see Remark 1 below.)

The following result shows that the graph of a C^p real-valued function on an interval is the trace of a simple C^p curve.

---

[1]"A Jordan Curve of Positive Area," *Transactions of the American Mathematical Society*, vol. 4 (1903), pp. 107–112.

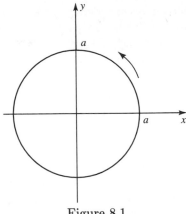

Figure 8.1

Example 1. Let I be an interval and $f : I \to \mathbf{R}$ be a $\mathcal{C}^p$ function. Prove that there is a 1–1 $\mathcal{C}^p$ function $\phi : I \to \mathbf{R}^2$ whose trace coincides with the graph of $y = f(x)$, $x \in I$.

PROOF. If $\phi(t) = (t, f(t))$, then ϕ is $\mathcal{C}^p$ and 1–1 on I, and $\phi(I)$ is the graph of $y = f(x)$ on I. (We shall call (ϕ, I) the *trivial parametrization* of $y = f(x)$.) ∎

By an *explicit curve* we mean a curve of the form (ϕ, I) where $\phi(t) = (t, f(t))$ (respectively, $\phi(t) = (f(t), t)$) for some $\mathcal{C}^p$ function $f : I \to \mathbf{R}$. Notice that the trace of an explicit curve is the set of points (x, y) which satisfies $y = f(x)$ (respectively, $x = f(y)$).

By the proof of Example 1, every explicit curve is a simple curve in $\mathbf{R}^2$. The following result shows that the converse of this statement is false.

Example 2. Prove that the circle $x^2 + y^2 = a^2$ is the trace of a simple closed $\mathcal{C}^\infty$ curve in $\mathbf{R}^2$ (see Figure 8.1).

PROOF. This circle can be described in polar coordinates by $r = a$, i.e., in rectangular coordinates by $x = a\cos\theta$, $y = a\sin\theta$. Thus, the set of points $(x, y) \in \mathbf{R}^2$ such that $x^2 + y^2 = a^2$ is the trace of the simple closed $\mathcal{C}^\infty$ curve (ϕ, I) defined by $\phi(t) = (a\cos t, a\sin t)$ and $I = [0, 2\pi]$. ∎

The length of a curve can be approximated by lengths of straight line segments (see Figure 8.2). This observation leads us to the following definition.

DEFINITION 8.2. The *arc length* of a simple $\mathcal{C}^p$ curve $C = (\phi, I)$ is the extended real number

$$L(C) := \sup\left\{ \sum_{j=1}^{k} \|\phi(t_j) - \phi(t_{j-1})\| : \{t_0, t_1, \ldots, t_k\} \text{ is a partition of } I \right\}.$$

The curve C is called *rectifiable* if $L(C)$ is finite.

(Note: If the curve were not simple, some part of it could be traced more than once, giving an inflated value of its arc length.)

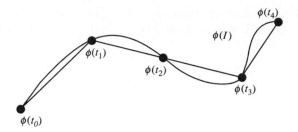

$\phi(t_4)$

$\phi(I)$

$\phi(t_1)$

$\phi(t_2)$

$\phi(t_3)$

$\phi(t_0)$

Figure 8.2

The following result shows that every C^p curve with endpoints is rectifiable for $p \geq$ 1. (The space filling curve shows this result can be false if ϕ is merely continuous, and Exercise 4 shows that it can be false if the curve has no endpoints.)

THEOREM 8.1. *If $C = (\phi, I)$ is a simple C^1 arc, then the arc length of C is*

$$L(C) = \int_I \|\phi'(t)\| \, dt.$$

STRATEGY: The idea behind the proof is simple. By the Mean Value Theorem, each term $\|\phi(t_j) - \phi(t_{j-1})\|$ which appears in the definition of arc length is approximately $\|\phi'(t_j)\|(t_j - t_{j-1})$, a term of a Riemann sum of the integral of $\|\phi'(t)\|$. Thus, we begin by controlling the size of $\|\phi'(t_j)\|$.

PROOF. Let $\varepsilon > 0$, write $\phi = (\phi_1, \phi_2, \ldots, \phi_m)$, and set

$$F(x_1, \ldots, x_m) = \left(\sum_{\ell=1}^m |\phi'_\ell(x_\ell)|^2 \right)^{1/2}$$

for $(x_1, \ldots, x_m)$ in the cube $I^m := I \times \cdots \times I$. By hypothesis, F is continuous on I^m and I^m is evidently compact. Thus, F is uniformly continuous on I^m; i.e., there is a $\delta > 0$ such that

$$\boldsymbol{x}, \boldsymbol{y} \in I^m \quad \text{and} \quad \|\boldsymbol{x} - \boldsymbol{y}\| < \delta \quad \text{imply} \quad |F(\boldsymbol{x}) - F(\boldsymbol{y})| < \frac{\varepsilon}{2|I|}.$$

Let $\mathcal{P} = \{u_0, \ldots, u_N\}$ be any partition of I. By Theorem 3.4, choose a partition $\mathcal{P}_0 = \{t_0, t_1, \ldots, t_k\}$ of I finer than $\mathcal{P}$ such that $\|\mathcal{P}_0\| < \delta/\sqrt{m}$ and

$$\int_I \|\phi'(t)\| \, dt - \frac{\varepsilon}{2} < \sum_{j=1}^k \|\phi'(t_j)\|(t_j - t_{j-1}) < \int_I \|\phi'(t)\| \, dt + \frac{\varepsilon}{2}.$$

Fix $\ell \in \{1, \ldots, m\}$ and $j \in \{1, \ldots, k\}$. By Theorem 2.16 (the one-dimensional Mean Value Theorem), choose a point $c_j(\ell) \in [t_{j-1}, t_j]$ such that

$$\phi_\ell(t_j) - \phi_\ell(t_{j-1}) = \phi'_\ell(c_j(\ell))(t_j - t_{j-1}).$$

Since $\|\mathcal{P}_0\| < \delta/\sqrt{m}$, we have $\|F(t_j,\ldots,t_j)-F(c_j(1),\ldots,c_j(m))\| < \varepsilon/(2|I|)$. Since $\phi'(t) = (\phi'_1(t),\ldots,\phi'_m(t))$, we also have $F(t_j,\ldots,t_j) = \|\phi'(t_j)\|$ and

$$F(c_j(1),\ldots,c_j(m))(t_j - t_{j-1}) = \left(\sum_{\ell=1}^{m} |\phi'_\ell(c_j(\ell))|^2\right)^{1/2} (t_j - t_{j-1})$$

$$= \|\phi(t_j) - \phi(t_{j-1})\|.$$

It follows that

$$\sum_{j=1}^{k} \|\phi'(t_j)\|(t_j - t_{j-1}) - \frac{\varepsilon}{2} < \sum_{j=1}^{k} \|\phi(t_j) - \phi(t_{j-1})\| < \sum_{j=1}^{k} \|\phi'(t_j)\|(t_j - t_{j-1}) + \frac{\varepsilon}{2}.$$

Combining this double inequality with the previous one, we obtain

$$\int_I \|\phi'(t)\|\, dt - \varepsilon < \sum_{j=1}^{k} \|\phi(t_j) - \phi(t_{j-1})\| < \int_I \|\phi'(t)\|\, dt + \varepsilon.$$

Using the left-hand inequality and Definition 8.2, we have

$$L(C) \geq \sum_{j=1}^{k} \|\phi(t_j) - \phi(t_{j-1})\| > \int_I \|\phi'(t)\|\, dt - \varepsilon,$$

i.e., $L(C) \geq \int_I \|\phi'(t)\|\, dt$. On the other hand, since $\mathcal{P}_0 = \{t_0, t_1, \ldots, t_k\}$ is finer than $\mathcal{P}$, it follows from the right-hand inequality that

$$\sum_{i=1}^{N} \|\phi(u_i) - \phi(u_{i-1})\| \leq \sum_{j=1}^{k} \|\phi(t_j) - \phi(t_{j-1})\| < \int_I \|\phi'(t)\|\, dt + \varepsilon.$$

Taking the supremum over all partitions $\{u_0, \ldots, u_N\}$ of I, we have

$$L(C) \leq \int_I \|\phi'(t)\|\, dt + \varepsilon,$$

i.e., $L(C) \leq \int_I \|\phi'(t)\|\, dt$. ∎

It is useful to interpret the trace of a curve as the path of a particle p moving through space and each of its parametrizations (ϕ, I) as a record of a particular flight plan. (The independent variable t is used to suggest "time.")

Remark 1. *Under this interpretation, if $t_0 \in I^\circ$ and $\boldsymbol{x}_0 = \phi(t_0)$, then $\|\phi'(t_0)\|$ is the speed of the particle p at $\boldsymbol{x}_0$ and $\phi'(t_0)$, when nonzero, is a vector "tangent" to the curve C at $\boldsymbol{x}_0$ which points in the direction of flight.*

PROOF. Let $t_0 \in I^\circ$ and notice that for each sufficiently small $h > 0$, the quotient

$$\frac{\phi(t_0 + h) - \phi(t_0)}{h}$$

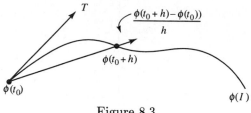

Figure 8.3

is a vector which points in the direction of flight along the curve C and approximates the direction of the line tangent to C at x_0 (see Figure 8.3). To calculate the speed of p, define the *natural parameter* of the curve C by

$$(1) \qquad\qquad s := \ell(t) := \int_a^t \|\phi'(u)\| \, du, \qquad t \in [a, b].$$

By the Fundamental Theorem of Calculus, $ds/dt = \ell'(t) = \|\phi'(t)\|$. Thus, the change of arc length s with respect to time t, i.e., the speed of the particle p at x_0, is precisely $\|\phi'(t_0)\|$. ∎

Recall that the graph of a differentiable function on an open interval is "smooth," i.e., has a unique tangent line at each of its points. The following example shows that this is not the case for the trace of a C^p curve.

Example 3. Let $\phi(t) = (\cos^3 t, \sin^3 t)$ and $I = [0, 2\pi]$. Prove that (ϕ, I) is a simple closed C^∞ curve in $\mathbf{R}^2$ which has "corners." (This curve is called the *astroid.*)

PROOF. Clearly, ϕ is C^∞ on I and 1–1 on $[0, 2\pi)$. Let $x = \cos^3 t$ and $y = \sin^3 t$ and observe by a double angle formula that

$$x^2 + y^2 = \frac{3}{4} \cos^2(2t) + \frac{1}{4}.$$

Hence, $\sqrt{x^2 + y^2}$ varies from a maximum of 1 (attained when $t = 0, \pi/2, \pi, 3\pi/2, 2\pi$) to a minimum of $1/2$ (attained when $t = \pi/4, 3\pi/4, 5\pi/4, 7\pi/4$). Since I is connected and ϕ is differentiable, hence, continuous, the set $\phi(I)$ must also be connected. Plotting a few points, we see that $\phi(I)$ is a four-cornered star, starting at $(1, 0)$ and moving in a counterclockwise direction from $\partial B_1(0, 0)$ to $\partial B_{1/2}(0, 0)$ and back again (see Figure 8.4). As t runs from 0 to 2π, this curve makes one complete circuit. ∎

We shall say that a curve (ϕ, I) in $\mathbf{R}^2$ has a *unique tangent line* at a point $(x_0, y_0) = \phi(t_0)$ if the trace $\phi(I)$ looks like the graph of a differentiable function near (x_0, y_0). When $t_0 \in I^\circ$, this means that there exist open intervals I_0 and J and a differentiable function $f : J \to \mathbf{R}$ such that $t_0 \in I_0 \subset I$, and either $x_0 \in J$ and the graph of $y = f(x)$ on J coincides with $\phi(I_0)$, or $y_0 \in J$ and the graph of $x = f(y)$ on J coincides with $\phi(I_0)$.

How shall we identify curves which have unique tangent lines at each of their points? To answer this question, let (ϕ, I) be the astroid (see Example 3), and

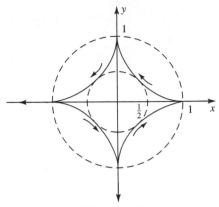

Figure 8.4

notice that $\phi'(t) = (0,0)$ when $t = 0, \pi/2, \pi, 3\pi/2, 2\pi$, i.e., exactly at the points where the curve (ϕ, I) fails to have a unique tangent line. This agrees with Remark 1, since it is impossible to draw the trace of the astroid without pausing at the corners. Thus, we guess that a curve (ϕ, I) has a unique tangent line at some point $(x_0, y_0) = \phi(t_0)$ when $\phi'(t_0) \neq \mathbf{0}$. The following result shows that this guess is correct except possibly at the endpoints.

Remark 2. *If (ϕ, I) is a $\mathcal{C}^1$ curve in $\mathbf{R}^2$ which has no tangent line at $(x_0, y_0) = \phi(t_0)$, where $t_0 \in I^\circ$, then $\phi'(t_0) = \mathbf{0}$.*

PROOF. Let (ϕ_1, ϕ_2) represent the components of ϕ. Suppose $\phi'(t_0) \neq \mathbf{0}$. We may suppose that $\phi_1'(t_0) \neq 0$. By the Implicit Function Theorem, applied to $F(x, t) = \phi_1(t) - x$, there is an open interval J containing x_0 and a continuously differentiable function $g : J \to I$ such that $\phi_1(g(x)) = x$ for all $x \in J$ and $g(x_0) = t_0$. Set $f = \phi_2 \circ g$. Then the graph of $y = f(x)$, $x \in J$, coincides with the trace of ϕ on $g(J)$, i.e., near (x_0, y_0). Since f is differentiable at x_0, it follows from elementary calculus that the graph of $y = f(x)$ has a tangent line at x_0, a contradiction. ∎

Accordingly, we make the following definition.

DEFINITION 8.3. Let (ϕ, I) be a $\mathcal{C}^1$ curve or parametrization.

   i) (ϕ, I) is said to be *smooth* at $t_0 \in I$ if $\phi'(t_0) \neq \mathbf{0}$.
  ii) A nonclosed (ϕ, I) is called *smooth* if it is smooth at each point of I. A closed $(\phi, [a, b])$ is called *smooth* if $(\phi, [a, b))$ is smooth and $\phi'(a) = \phi'(b)$.
 iii) If $C = (\phi, I)$ is simple and smooth, then the *unit tangent vector* of C at $\mathbf{x}_0 = \phi(t_0)$ is defined by $T(\mathbf{x}_0) := \phi'(t_0)/\|\phi'(t_0)\|$.

By Remark 2, a curve which has no tangent line at some interior point cannot be smoothly parametrized. The converse of this statement is false.

Remark 3. *A curve which has a unique tangent line at each of its points can have a nonsmooth parametrization.*

PROOF. Let $\psi(t) = (t^3, 3t^3)$. The trace of ψ is the line $y = 3x$ which has a unique tangent line at each of its points. Nevertheless, $\psi'(0) = \mathbf{0}$. (Of course, there are other parametrizations of this line which satisfy $\phi'(t) \neq \mathbf{0}$, for example, the trivial parametrization $\phi(t) = (t, 3t)$.) ∎

In view of this, we must be careful to distinguish between those properties which are the same for all parametrizations of a given curve, and those which are not. For example, is the arc length of a given curve, as computed in Theorem 8.1, independent of parametrization?

To answer this question, we begin by showing that any two parametrizations of the same arc are related by a one-dimensional change of variables τ.

Remark 4. *Let* I, J *be closed bounded intervals and* $\phi : I \rightarrow \mathbf{R}^m$ *be 1–1 and continuous. Then* $\phi(I) = \psi(J)$ *for some continuous* $\psi : J \rightarrow \mathbf{R}^m$ *if and only if there is a continuous function* τ *from* J *onto* I *such that* $\psi = \phi \circ \tau$.

PROOF. Since I is compact and ϕ is 1–1 and continuous on I, ϕ^{-1} is continuous from $\phi(I)$ onto I (see Exercise 6 in Section 5.6). Since $\psi(J) = \phi(I)$, it follows that $\tau = \phi^{-1} \circ \psi$ is continuous from J onto I.

Conversely, if τ is continuous from J onto I, then $\psi = \phi \circ \tau$ is continuous from J onto $\phi(I)$, i.e., $\psi(J) = \phi(I)$. ∎

Notice that if (ψ, J), (ϕ, I) are parametrizations of the same C^1 curve and $\tau = \phi^{-1} \circ \psi$ is differentiable, then by the Chain Rule,

$$(2) \qquad \psi'(u) = \phi'(\tau(u))\tau'(u), \qquad u \in J.$$

In particular, if (ϕ, I) is smooth, then (ψ, J) is smooth if and only if $\tau'(u) \neq 0$ for all $u \in J$. This leads us to the following definition.

DEFINITION 8.4. Let $p \geq 1$. Two smooth C^p curves (or parametrizations) (ϕ, I), (ψ, J) are said to be *smoothly equivalent* if $\phi(I) = \psi(J)$ and there is a C^p function $\tau : J \rightarrow \mathbf{R}$, which takes J onto I, such that $\psi = \phi \circ \tau$ and $\tau'(u) \neq 0$ for all $u \in J$. The function τ is called the *transition* from $\psi(J)$ to $\phi(I)$.

Notice that since τ' is continuous and nonzero, either τ' is positive on J or τ' is negative on J. Hence, by Theorem 2.19, a transition τ is always 1–1. Also notice that if two curves are smoothly equivalent, then one is simple if and only if the other one is simple (see Exercise 5 in Section 8.4).

We now answer the question posed above concerning the computation of arc length.

Remark 5. *If* (ϕ, I) *and* (ψ, J) *are smoothly equivalent, then*

$$\int_I \|\phi'(t)\| \, dt = \int_J \|\psi'(u)\| \, du.$$

PROOF. If τ is the transition from $\psi(J)$ to $\phi(I)$, then $\tau(J) = I$. Hence, it follows from (2) and the Change of Variables Formula (Theorem 7.14) that

$$\int_I \|\phi'(t)\| \, dt = \int_{\tau(J)} \|\phi'(t)\| \, dt = \int_J \|\phi'(\tau(u))\| \, |\tau'(u)| \, du = \int_J \|\psi'(u)\| \, du. \quad \blacksquare$$

We note that the condition $\tau'(u) \neq 0$ can be relaxed at finitely many points in J (see Exercise 8).

The following integral can be interpreted as the mass of a wire on a curve with density g (see Appendix E).

DEFINITION 8.5. Let $C = (\phi, I)$ be a smooth simple arc in $\mathbf{R}^m$ and $g : \phi(I) \to \mathbf{R}$ be continuous. Then the *line integral* of g on C is

$$(3) \qquad \int_C g\,ds := \int_{\phi(I)} g\,ds := \int_I g(\phi(t))\|\phi'(t)\|\,dt.$$

For an explicit curve C given by $y = f(x)$, $x \in [a, b]$, this integral becomes

$$\int_C g\,ds = \int_a^b g(x, f(x))\sqrt{1 + |f'(x)|^2}\,dx.$$

We note that by Theorem 8.1, the line integral (3) equals the arc length of C when $g = 1$. This explains the notation ds. Indeed, the parameter s represents arc length (see (1) above) and by the Fundamental Theorem of Calculus, $ds/dt = \|\phi'(t)\|$. Hence, the differential of s is defined by $ds = \|\phi'(t)\|\,dt$. We also note that the line integral of a function g on a curve is the same under smoothly equivalent parametrizations (see Exercise 8).

By Definition 8.5, a line integral is a one-dimensional integral which can be evaluated by the techniques discussed in Chapter 3.

Example 4. Find $\int_{\phi(I)} g\,ds$, where $g(x, y) = 2x + y$, $\phi(t) = (\cos t, \sin t)$, and $I = [0, \pi/2]$.

SOLUTION. Since $\|\phi'(t)\| = \|(-\sin t, \cos t)\| = 1$ we have

$$\int_{\phi(I)} g\,ds = \int_0^{\pi/2} (2\cos t + \sin t)\,dt = 3. \quad \blacksquare$$

For some applications, it is necessary to extend these ideas to a finite union of smooth curves. Let $C = \cup_{j=1}^N C_j$, where each C_j is a smooth simple curve with endpoints. We shall call C a *piecewise smooth curve* if either of the following conditions holds. The first condition is that the traces of C_j and C_k are disjoint for each $j \neq k$. (This allows for nonconnected smooth curves, e.g., the boundary of an annulus $0 < a^2 < x^2 + y^2 < b^2$.) The second condition is that the C_j's are joined end to end, i.e., if $C_j = (\phi_j, [a_j, b_j])$ then $\phi_{j-1}(b_{j-1}) = \phi_j(a_j)$ for $j = 2, \ldots, N$ (see Figure 8.5). (This allows for curves with finitely many corners, for example, the perimeter of a triangle, or the boundary of a rectangle.) If C is a piecewise smooth curve C which satisfies the second condition, then the points $\phi(a_1)$ and $\phi(b_N)$ are called the *endpoints* of C.

We extend the concepts defined above to piecewise smooth curves as follows. Let $C = \cup_{j=1}^N C_j$. The *trace* of C is defined to be the union of the traces of the C_j's. By a *parametrization* of C we mean a collection of smooth parametrizations (ϕ_j, I_j) of

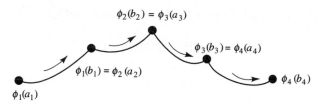

$$\phi_2(b_2) = \phi_3(a_3)$$

$$\phi_3(b_3) = \phi_4(a_4)$$

$$\phi_1(b_1) = \phi_2(a_2)$$

$$\phi_4(b_4)$$

$$\phi_1(a_1)$$

Figure 8.5

C_j. By changing parameters, we may suppose that $\cup_{j=1}^N I_j$ is itself an interval (see Exercise 4 in Section 8.2). Two parametrizations $\cup_{j=1}^N(\phi_j, I_j)$ and $\cup_{j=1}^N(\psi_j, J_j)$ of C are said to be smoothly equivalent if (ϕ_j, I_j) and (ψ_j, J_j) are smoothly equivalent for each $j \in \{1, \ldots, N\}$. C is said to be *rectifiable* if each C_j is rectifiable, in which case we set

$$L(C) = \sum_{j=1}^N L(C_j).$$

If g is continuous and real-valued on the trace of C, then the *line integral* of g on C is defined by

$$\int_C g\,ds = \sum_{j=1}^N \int_{C_j} g\,ds.$$

A piecewise smooth curve C is said to be *simple* if its trace does not intersect itself except possibly at the endpoints, and *closed* if it is connected and its endpoints are identical.

Example 5. Parametrize the boundary C of the unit square $[0,1] \times [0,1]$ and compute $\int_C g\,ds$, where $g(x,y) = x^2 + y^3$.

Solution. C has four smooth pieces which can be parametrized by

$$\phi_1(t) = (t,0), \quad \phi_2(t) = (1,t), \quad \phi_3(t) = (1-t, 1), \quad \phi_4(t) = (0, 1-t),$$

for $t \in [0,1]$. Since $\|\phi_j'(t)\| = 1$, we have by definition,

$$\int_C g\,ds = \int_0^1 t^2\,dt + \int_0^1 (1 + t^3)\,dt + \int_0^1 (t^2 + 1)\,dt + \int_0^1 t^3\,dt = \frac{19}{6}. \quad \blacksquare$$

EXERCISES

1. Let $\psi(t) = (a\sin t, a\cos t)$, $\sigma(t) = (a\cos 2t, a\sin 2t)$, $I = [0, 2\pi)$, and $J = [0, \pi)$. Sketch the traces of (ψ, I) and (σ, J). Note the "direction of flight" and the "speed" of each parametrization. Compare these parametrizations with the one given in Example 2.

2. Let $\boldsymbol{x}_0, \boldsymbol{a} \in \mathbf{R}^m$, $\boldsymbol{a} \neq \mathbf{0}$, and set $\phi(t) = t\boldsymbol{a} + \boldsymbol{x}_0$. Show that $C = (\phi, \mathbf{R})$ is a smooth simple unbounded curve which contains $\boldsymbol{x}_0$ and $\boldsymbol{x}_0 + \boldsymbol{a}$. Prove that the angle between $\phi(t_1) - \phi(0)$ and $\phi(t_2) - \phi(0)$ for any $t_1, t_2 \neq 0$ is 0 or π.

3. Let I be an interval and $f : I \to \mathbf{R}$ be continuously differentiable with

$$|f(\theta)|^2 + |f'(\theta)|^2 \neq 0$$

for all $\theta \in I$. Prove that the graph of $r = f(\theta)$ (in polar coordinates) is the trace of a smooth curve in $\mathbf{R}^2$.

4. Show that the curve $y = \sin(1/x)$, $0 < x \leq 1$, is not rectifiable. Thus, show that Theorem 8.1 can be false if C is not an arc.

5. Sketch the trace and compute the arc length of each of the following curves.

   a) $\phi(t) = (e^t \sin t, e^t \cos t, e^t)$, $t \in [0, 2\pi]$.
   b) $y^3 = x^2$ from $(-1, 1)$ to $(1, 1)$.
   c) $\phi(t) = (t^2, t^2, t^2)$, $t \in [0, 2]$.
   d) The astroid of Example 3.

6. For each of the following, find a (piecewise) smooth parametrization of C and compute $\int_C g \, ds$.

   a) C is the curve $y = \sqrt{9 - x^2}$, $x \geq 0$, and $g(x, y) = xy^2$.
   b) C is the portion of the ellipse $x^2/a^2 + y^2/b^2 = 1$, $a, b > 0$, which lies in the first quadrant and $g(x, y) = xy$.
   c) C is the intersection of the surfaces $x^2 + z^2 = 4$ and $y = x^2$, and $g(x, y, z) = \sqrt{1 + yz^2}$.
   d) C is the triangle with vertices $(0, 0, 0)$, $(1, 0, 0)$ and $(0, 2, 0)$, and $g(x, y, z) = x + y + z^3$.

7. Let (ϕ, I) be a smooth arc and $g_k : \phi(I) \to \mathbf{R}$ be continuous for $n \in \mathbf{N}$.

   a) If $g_k \to g$ uniformly on $\phi(I)$, prove $\int_C g_k \, ds \to \int_C g \, ds$ as $k \to \infty$.
   *b) Suppose $\{g_k\}$ is pointwise monotone and $g_k \to g$ pointwise on $\phi(I)$ as $k \to \infty$. If g is continuous on $\phi(I)$, prove $\int_C g_k \, ds \to \int_C g \, ds$ as $k \to \infty$.

8. Suppose (ϕ, I) is a smooth simple arc in $\mathbf{R}^m$ and $\tau : J \to \mathbf{R}$ is a C^1 function, 1–1 from J onto I. If $\tau'(u) \neq 0$ for all but finitely many $u \in J$, $\psi = \phi \circ \tau$, and $g : \phi(I) \to \mathbf{R}$ is continuous, prove

$$\int_I g(\phi(t)) \|\phi'(t)\| \, dt = \int_J g(\psi(u)) \|\psi'(u)\| \, du.$$

9. [THE FOLIUM OF DESCARTES]. Let C be the piecewise smooth curve $(\phi, I_1) \cup (\phi, I_2)$, where $I_1 = (-\infty, -1)$, $I_2 = (-1, \infty)$, and

$$\phi(t) = \left(\frac{3t}{1 + t^3}, \frac{3t^2}{1 + t^3} \right).$$

Show that if $(x, y) = \phi(t)$, then $x^3 + y^3 = 3xy$. Sketch the trace of C.

10. The *absolute curvature* of a smooth curve (ψ, I) at a point $\mathbf{x}_0 = \psi(t_0)$ is the number

$$\kappa(\mathbf{x}_0) = \lim_{t \to t_0} \frac{\theta(t)}{\ell(t)},$$

when this limit exists, where $\theta(t)$ is the angle between $\psi'(t)$ and $\psi'(t_0)$, and $\ell(t)$ is the arc length of (ψ, I) from $\psi(t)$ to $\psi(t_0)$. (Thus, κ measures how rapidly $\theta(t)$ changes with respect to arc length.)

a) Given $\boldsymbol{a}, \boldsymbol{b} \in \mathbf{R}^n$, prove that the absolute curvature of the line $\Lambda = (\psi, I)$, where $\psi(t) := t\boldsymbol{a} + \boldsymbol{b}$ and $I := (-\infty, \infty)$, is zero at each point $\boldsymbol{x}_0$ on Λ.

b) Given $(a, b) \in \mathbf{R}^2$, prove that the absolute curvature of the circle C of radius r given by $\psi(t) = (r\cos t, r\sin t)$ is $1/r$ at each point $\boldsymbol{x}_0$ on C.

11. Let $C = (\phi, [a, b])$ be a smooth simple arc in $\mathbf{R}^m$ and $s = \ell(t)$ be given by (1). The *natural parametrization* of C is the pair $(\nu, [0, L])$, where

$$\nu(s) = (\phi \circ \ell^{-1})(s) \quad \text{and} \quad L = L(C).$$

a) Prove that $\|\nu'(s)\| = 1$ for all $s \in [0, L]$ and the arc length of a subcurve $(\nu, [c, d])$ of C is $d - c$. (This is why $(\nu, [0, L])$ is called the natural parametrization.)

b) Show $\nu'(s)$ and $\nu''(s)$ are orthogonal for each $s \in [0, L]$.

c) Prove that the absolute curvature (see Exercise 10 above) of $(\nu, [0, L])$ at $\boldsymbol{x}_0 = \nu(s_0)$ is $\kappa(\boldsymbol{x}_0) = \|\nu''(s_0)\|$.

d) Show if $\boldsymbol{x}_0 = \phi(t_0) = \nu(s_0)$, then

$$\kappa(\boldsymbol{x}_0) = \|\nu'(s_0) \times \nu''(s_0)\| = \frac{\|\phi'(t_0) \times \phi''(t_0)\|}{\|\phi'(t_0)\|^3}.$$

e) Let $p \geq 1$. Prove that the absolute curvature of an explicit $\mathcal{C}^p$ curve $y = f(x)$ at (x_0, y_0) is

$$\kappa = \frac{|y''(x_0)|}{(1 + (y'(x_0))^2)^{3/2}}.$$

8.2 ORIENTED CURVES

Every parametrization (ϕ, I) of a smooth curve C implicitly determines a "direction of flight" along C; i.e., the direction $\phi(t)$ moves as t increases on I, equivalently, the direction $\phi'(t)$ points. This direction is called the *orientation of C induced by (ϕ, I)*. (The arrows in Figures 8.1, 8.4, and 8.5 above represent the orientation of the given parametrization.) Suppose (ϕ, I) and (ψ, J) are smoothly equivalent parametrizations of the same curve with transition τ. Since τ is continuously differentiable and nonzero, either $\tau'(u) > 0$ for all $u \in J$ or $\tau'(u) < 0$ for all $u \in J$. In the first case, the vectors $\phi'(\tau(u))$ and $\psi'(u)$ point in the same direction (see (2) in Section 8.1); hence, these parametrizations determine the same orientation. In the second case, the vectors $\phi'(\tau(u))$ and $\psi'(u)$ point in opposite directions, hence, determine different orientations. Accordingly, we make the following definition.

DEFINITION 8.6. Two smooth curves (or parametrizations) (ϕ, I) and (ψ, J) are said to be *orientation equivalent* if they are smoothly equivalent and the transition τ from $\psi(J)$ to $\phi(I)$ satisfies $\tau'(u) > 0$ for all $u \in J$.

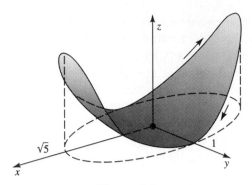

Figure 8.6

In practice, a curve and its orientation are often described geometrically. The reader must provide a parametrization which traces the curve in the prescribed orientation. Here are two typical examples.

Example 1. Find a smooth parametrization of the curve C in $\mathbf{R}^3$, oriented in the clockwise direction when viewed from high up the positive z axis, formed by intersecting the surfaces $x^2 + 5y^2 = 5$ and $z = x^2$.

SOLUTION. The elliptical cylinder $x^2 + 5y^2 = 5$ intersects the parabolic cylinder $z = x^2$ to form a "sagging ellipse" (the shaded region in Figure 8.6 represents that part of $z = x^2$ which lies inside the cylinder $x^2 + 5y^2 = 5$). Using $x = \sqrt{5}\sin t$, $y = \cos t$ to incorporate clockwise motion around the ellipse $x^2 + 5y^2 = 5$, we see that $z = x^2 = 5\sin^2 t$. Thus, a smooth parametrization of C is $\phi(t) = (\sqrt{5}\sin t, \cos t, 5\sin^2 t)$ on $I = [0, 2\pi]$. ∎

Example 2. Find a smooth parametrization of the curve C in $\mathbf{R}^3$, oriented from right to left when viewed from far out the line $y = x$ in the xy plane, formed by intersecting the surfaces $z = x^2 - y^2$ and $x + y = 1$.

SOLUTION. The saddle surface $z = x^2 - y^2$ intersects the plane $x + y = 1$ to form a curve which cuts across the surface. Using $x = t$ as a parameter to incorporate right to left orientation, we see that $y = 1 - t$ and $z = t^2 - (1 - t)^2 = 2t - 1$. Thus, a smooth parametrization of C is $\phi(t) = (t, 1 - t, 2t - 1)$ on $I = \mathbf{R}$. In particular, C is a line in the direction $(1, -1, 2)$ passing though the point $(0, 1, -1)$. ∎

The following integral arises naturally in the study of fluids, electricity, and magnetism.

DEFINITION 8.7. Let $C = (\phi, I)$ be a smooth simple arc in $\mathbf{R}^m$ and $F : \phi(I) \to \mathbf{R}^m$ be continuous. Then the *oriented line integral* of F *along* C is

$$(4) \qquad \int_C F \cdot T \, ds := \int_{\phi(I)} F \cdot T \, ds := \int_{\phi(I)} F \cdot d\phi := \int_I F(\phi(t)) \cdot \phi'(t) \, dt.$$

The notation $F \cdot d\phi$ is self-explanatory. The notation $F \cdot T \, ds$ is consistent with (3) in Section 8.1. Indeed, $T = \phi'(t)/\|\phi'(t)\|$ is the unit tangent vector of C and $ds =$

$\|\phi'(t)\|\,dt$ is the arc length differential associated with C. Thus, when evaluating $F\cdot T\,ds$, the scalars $\|\phi'(t)\|$ cancel each other out.

What does this number represent? If F represents the flow of a fluid, then $F\cdot T$ is the tangential component of F, i.e., a measure of fluid flow in the direction which the tangent T points (see Appendix E). For example, suppose C is the unit circle oriented in the counterclockwise direction and $F(x,y)=(-y,x)$. The unit tangent to C at a point (x,y) is $(-y,x)$ so F points in the same direction that T does. Hence, $F\cdot T=1$ is an indication that the fluid is flowing "with the tangent" rather than against it. On the other hand, if $G(x,y)=(y,-x)$ or $H(x,y)=(x,y)$, then $G\cdot T=-1$ because the fluid is flowing against the tangent, and $H\cdot T=0$ because the fluid is flowing orthogonally to T (e.g., neither with nor against it). Therefore, the integral of $F\cdot T\,ds$ over C is a measure of the circulation of F around C in the direction of the tangent vector. If this integral is positive, it means that the net flow of the fluid is with T rather than against T.

By Definition 8.7, an oriented line integral is an integral of a real-valued function, hence, can be evaluated by techniques introduced in Chapter 3. Here is a typical example.

Example 3. Describe the trace of $\phi(t)=(\cos t,\sin t,t)$, $t\in I=[0,4\pi]$, and compute

$$\int_C F\cdot T\,ds,$$

where $F(x,y,z)=(1,\cos z,xy)$ and $C=(\phi,I)$.

SOLUTION. Let $(x,y,z)=\phi(t)$. Since $x^2+y^2=1$, the trace of C lies on the cylinder $x^2+y^2=1$, $0\le z\le 4\pi$. As t increases, the point (x,y) winds around the unit circle $x^2+y^2=1$ in a counterclockwise direction. Thus, the trace of ϕ is a spiral (called the *circular helix*) which winds around the cylinder $x^2+y^2=1$ (see Figure 8.7). As t runs from 0 to 4π this spiral winds around the cylinder twice and z runs from 0 to 4π. Since $\phi'(t)=(-\sin t,\cos t,1)$, we have

$$\int_C F\cdot T\,ds=\int_0^{4\pi}(1,\cos t,\cos t\sin t)\cdot(-\sin t,\cos t,1)\,dt$$

$$=\int_0^{4\pi}(-\sin t+\cos^2 t+\sin t\cos t)\,dt=2\pi.\ \blacksquare$$

The following result shows that unlike the line integral $\int_C g\,ds$, the oriented line integral $\int_C F\cdot T\,ds$ can give different values for different smoothly equivalent parametrizations of the same curve.

Remark 1. *If (ϕ,I) and (ψ,J) are simple arcs which are smoothly equivalent but not orientation equivalent, then*

$$\int_I F(\phi(t))\cdot\phi'(t)\,dt=-\int_J F(\psi(u))\cdot\psi'(u)\,du.$$

PROOF. Let τ be the transition from $\psi(J)$ to $\phi(I)$. Since τ' is continuous and nonzero, it is either positive on J or negative on J. Since (ϕ,I) and (ψ,J) are not

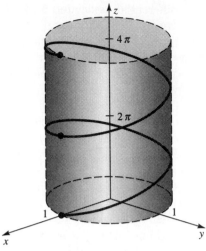

Figure 8.7

orientation equivalent, it follows that τ' is negative on J, i.e., $|\tau'(u)| = -\tau'(u)$ for $u \in J$. Combining this observation with the Change of Variables Formula (Theorem 7.14) and (2) in Section 8.1, we conclude that

$$\int_I F(\phi(t)) \cdot \phi'(t)\, dt = \int_J F(\phi(\tau(u)) \cdot \phi'(\tau(u))|\tau'(u)|\, du$$
$$= -\int_J F(\psi(u)) \cdot \psi'(u)\, du. \quad \blacksquare$$

By the same method, one can show that the oriented integral (4) gives identical values for orientation equivalent parametrizations of the same curve (see Exercise 5). Therefore, to evaluate an oriented integral over a curve C whose orientation has been described geometrically, we can use any smoothly equivalent parametrization of C and adjust the sign of the integral to reflect the prescribed orientation. Here is a typical example.

Example 4. Find

$$\int_C F \cdot T\, ds,$$

where $F(x, y) = (y, xy)$ and C is the unit circle $x^2 + y^2 = 1$ oriented in the clockwise direction.

SOLUTION. The parametrization $\phi(t) = (\cos t, \sin t)$, $t \in [0, 2\pi]$, of C has counterclockwise orientation (see Example 2 in Section 8.1). Thus, by Remark 1,

$$\int_C F \cdot T\, ds = -\int_0^{2\pi} (\sin t, \sin t \cos t) \cdot (-\sin t, \cos t)\, dt$$
$$= \int_0^{2\pi} (\sin^2 t - \sin t \cos^2 t)\, dt = \pi. \quad \blacksquare$$

There is another way to represent the oriented integral (4) which uses differential notation. Recall that if $x_j = \phi_j(t)$, then $dx_j = \phi'_j(t)dt$. Hence, formally, $F(\phi(t)) \cdot \phi'(t)\,dt$ looks like

$$(F_1(\phi(t))\phi'_1(t) + \cdots + F_m(\phi(t))\phi'_m(t))\,dt = F_1\,dx_1 + \ldots F_m\,dx_m.$$

This last expression is called a *1–form* on $\mathbf{R}^m$ and the functions F_j are called its *coefficients*. A 1–form is said to be *continuous* on a set E if each of its coefficients is continuous on E. The *oriented integral* of a continuous 1–form on a smooth simple arc $C = (\phi, I)$ in $\mathbf{R}^m$ is defined by

$$\int_C F_1\,dx_1 + \cdots + F_m\,dx_m := \int_{\phi(I)} F \cdot T\,ds,$$

where $F = (F_1, \ldots, F_m)$.

The following example illustrates the fact that differential forms provide a shorthand for the way an oriented line integral is computed.

Example 5. Find

$$\int_C y\,dx + \cos x\,dy,$$

where C is the explicit curve $y = x^2 + \sin x$ oriented from $(0,0)$ to (π, π^2).

SOLUTION. Since $y = x^2 + \sin x$ and $dy = (2x + \cos x)\,dx$, we have

$$\int_C y\,dx + \cos x\,dy = \int_0^\pi (x^2 + \sin x)\,dx + \int_0^\pi \cos x\,(2x + \cos x)\,dx$$

$$= \frac{\pi^3}{3} + \frac{\pi}{2} - 2. \ \blacksquare$$

Let $C = \bigcup_{j=1}^N C_j$ be a piecewise smooth curve in $\mathbf{R}^m$ (see the discussion preceding Example 5 in Section 8.1) with trace E. If $F : E \to \mathbf{R}^m$ is continuous, then the *oriented line integral* of F *along* C is defined to be

$$\int_C F \cdot T\,ds = \sum_{j=1}^N \int_{C_j} F \cdot T\,ds.$$

If ω is a 1–form continuous on E, then the *oriented integral* of ω *along* C is defined to be

$$\int_C \omega = \sum_{j=1}^N \int_{C_j} \omega.$$

Example 6. Find

$$\int_C xy\,dx + (x^2 + y^2)\,dy,$$

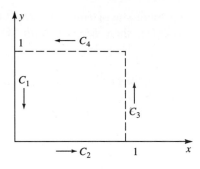

Figure 8.8

where C is the boundary of $Q = [0,1] \times [0,1]$ oriented in the counterclockwise direction.

SOLUTION. The boundary $C = \partial Q$ consists of four smooth pieces (see Figure 8.8): C_1 (which lies in the line $x = 0$), C_2 (in $y = 0$), C_3 (in $x = 1$), and C_4 (in $y = 1$). For C_1, let $x = 0$ and y run from 1 to 0 (to maintain counterclockwise orientation on C). Then

$$\int_{C_1} xy\, dx + (x^2 + y^2)\, dy = \int_1^0 y^2\, dy = -\frac{1}{3}.$$

Similarly, the integrals over C_2, C_3, and C_4 are 0, 4/3, and −1/2. Hence,

$$\int_C F \cdot T\, ds = -\frac{1}{3} + 0 + \frac{4}{3} - \frac{1}{2} = \frac{1}{2}. \quad \blacksquare$$

EXERCISES

1. For each of the following, sketch the trace of $(\phi, \mathbf{R})$, describe its orientation, and verify that it is a subset of the set S.

   a) $\phi(t) = (3t, 3\sin t, \cos t)$, $S = \{(x, y, z) : y^2 + 9z^2 = 9\}$.
   b) $\phi(t) = (t^2, t^3, t^2)$, $S = \{(x, y, z) : z = x\}$.
   c) $\phi(t) = (t, t^2, \sin t)$, $S = \{(x, y, z) : y = x^2\}$.
   d) $\phi(t) = (\cos t, \sin t, \cos t)$, $S = \{(x, y, z) : y^2 + z^2 = 1\}$.
   e) $\phi(t) = (\sin t, \sin t, t)$, $S = \{(x, y, z) : y = x\}$.

2. For each of the following, find a (piecewise) smooth parametrization of C and compute $\int_C F \cdot T\, ds$.

   a) C is the curve $y = x^2$ from $(1,1)$ to $(3,9)$, and $F(x, y) = (xy, y - x)$.
   b) C is the intersection of the elliptical cylinder $y^2 + 2z^2 = 1$ with the plane $x = -1$, oriented in the counterclockwise direction when viewed from far out the positive x axis, and $F(x, y, z) = (\sqrt{x^3 + y^3 + 5}, z, x^2)$.
   c) C is the intersection of the bent plane $y = |x|$ with the elliptical cylinder $x^2 + 3z^2 = 1$, oriented in the clockwise direction when viewed from far out the positive y axis, and $F(x, y, z) = (z, -z, x + y)$.

3. For each of the following, compute $\int_C \omega$.

   a) C is the polygonal path consisting of the line segment from $(1,1)$ to $(2,1)$ followed by the line segment from $(2,1)$ to $(2,3)$, and $\omega = y\,dx + x\,dy$.

   b) C is the intersection of $z = x^2 + y^2$ and $x^2 + y^2 + z^2 = 1$, oriented in the counterclockwise direction when viewed from high up the positive z axis, and $\omega = dx + (x+y)\,dy + (x^2 + xy + y^2)\,dz$.

   c) C is the boundary of the rectangle $R = [a,b] \times [c,d]$, oriented in the counterclockwise direction, and $\omega = xy\,dx + (x+y)\,dy$.

   d) C is the intersection of $y = x$ and $y = z^2$, $0 \le z \le 1$, oriented from left to right when viewed from far out the y axis, and $\omega = \sqrt{x}\,dx + \cos y\,dy - dz$.

4. a) Let $c \in \mathbf{R}$, $\delta > 0$, and set $\tau(u) = \delta u + c$ for $u \in \mathbf{R}$. Prove that if (ϕ, I) is a smooth curve, $J = \tau^{-1}(I)$, and $\psi = \phi \circ \tau$ for $\tau(u) = \delta u + c$, then (ψ, J) is orientation equivalent to (ϕ, I).

   b) Prove that if (ϕ, I) is a smooth arc, then it has an orientation equivalent parametrization of the form $(\psi, [0,1])$.

   c) Obtain an analogue of b) for piecewise smooth curves.

   d) Let C be the boundary of the unit square $[0,1] \times [0,1]$ oriented in the counterclockwise direction. Parametrize C on $[0,1]$, e.g., find a parametrization (ϕ_1, I_1), (ϕ_2, I_2), (ϕ_3, I_3), (ϕ_4, I_4) such that $I_j = [(j-1)/4, j/4]$.

5. Suppose (ϕ, I) is a smooth simple arc and τ is a C^1 function, 1–1 from J onto I, which satisfies $\tau'(u) > 0$ for all but finitely many $u \in J$. If $\psi = \phi \circ \tau$ prove that

$$\int_I F(\phi(t)) \cdot \phi'(t)\,dt = \int_J F(\psi(u)) \cdot \psi'(u)\,du$$

   for any continuous $F : \phi(I) \to \mathbf{R}^m$.

$\boxed{6}$. **This exercise is used in Section 8.5.** Let $f : [a,b] \to \mathbf{R}$ be C^1 on $[a,b]$ with $f'(t) \ne 0$ for $t \in [a,b]$. Prove that the explicit curve $x = f^{-1}(y)$, as y runs from $f(a)$ to $f(b)$, is orientation equivalent to the explicit curve $y = f(x)$, as x runs from a to b.

7. Let V be open in $\mathbf{R}^2$. A function $F : V \to \mathbf{R}^2$ is said to be *conservative* on V if there is a function $f : V \to \mathbf{R}$ such that $F = \nabla f$ on V. Let $(x,y) \in V$ and $F = (P,Q) : V \to \mathbf{R}^2$ be continuous on V.

   a) Suppose $C(x)$ is a horizontal line segment in V terminating at (x,y), i.e., a line segment of the form $L((x_1,y);(x,y))$, oriented from (x_1,y) to (x,y), whose trace is a subset of V. Prove

$$\frac{\partial}{\partial x} \int_{C(x)} F \cdot T\,ds = P(x,y).$$

   Make and prove a similar statement for $\partial/\partial y$ and vertical line segments in V terminating at (x,y).

   b) Let $(x_0, y_0) \in V$. Prove that

   (*) $$\int_C F \cdot T\,ds = 0$$

for all closed piecewise smooth curves $C \subset V$ if and only if for all $(x, y) \in V$, the integrals

$$f(x, y) := \int_{C(x,y)} F \cdot T \, ds$$

give the same value for all piecewise smooth curves $C(x, y)$ whose traces are subsets of V which start at (x_0, y_0) and end at (x, y).

c) Prove that F is conservative on V if and only if (*) holds for all closed piecewise smooth curves C whose traces are subsets of V.

d) Prove that if F satisfies (*) for all closed piecewise smooth curves C whose traces are subsets of V, then

$$\frac{\partial P}{\partial y} = \frac{\partial Q}{\partial x}.$$

Note: If V is nice enough, the converse of this statement also holds (see Exercise 7 in Section 8.6 or Theorem 10.9).

*8. Suppose $f : [0, 1] \to \mathbf{R}$ is increasing and continuously differentiable on $[0, 1]$. Let T be the right triangle whose vertices are $(0, f(0))$, $(1, f(0))$, and $(1, f(1))$. If c represents the hypotenuse of T, a and b represent the legs of T, and L represents the arc length of the explicit curve $y = f(x)$, $x \in [0, 1]$, prove that $c \leq L \leq a + b$.

8.3 SURFACES

In this section we define surfaces and surface integrals, concepts which are two-dimensional analogues of curves and the curve integrals discussed in Section 8.1. Recall that a smooth arc is parametrized on a closed bounded interval. On what shall we parametrize a smooth surface? Evidently, we need to use some type of closed bounded set in $\mathbf{R}^2$. Although we could use rectangles, we find it more convenient to use *two-dimensional regions*. A description of this type of set is contained in the following definition.

DEFINITION 8.8. An *m-dimensional region* is a set $E \subset \mathbf{R}^m$ such that $E = \overline{V}$ for some nonempty, open, connected Jordan region V in $\mathbf{R}^m$.

Notice that a one-dimensional region is a closed bounded interval. Also notice that every two-dimensional rectangle and the closure of every two-dimensional ball is a two-dimensional region.

DEFINITION 8.9.

i) A (C^p) *surface* (in $\mathbf{R}^3$) is a pair $S = (\phi, E)$, where E is a two-dimensional region and $\phi : E \to \mathbf{R}^3$ is C^p on E. The *trace* of S is the set $\phi(E)$. A point $(x, y, z) \in \mathbf{R}^3$ is said to *lie on* (or *belong to*) S if it belongs to the trace of S.

ii) A surface (ϕ, E) is said to be *simple* if ϕ is 1–1 on E.

iii) A pair (ψ, B) is called a C^p *parametrization* of a surface (ϕ, E) if (ψ, B) is a simple C^p surface and $\psi(B) = \phi(E)$. The equations

$$x = \psi_1(u, v), \quad y = \psi_2(u, v), \quad z = \psi_3(u, v), \qquad (u, v) \in B,$$

are called *parametric equations* of (ϕ, E) induced by the parametrization (ψ, B).

Earlier, we called the graph of a function $z = f(x, y)$ a surface. The following result shows that this designation is compatible with Definition 8.9 when f is C^p.

Example 1. Let E be a two-dimensional region and $f : E \to \mathbf{R}$ be a C^p function. Prove that the graph of $z = f(x, y)$ is the trace of a simple C^p surface.

PROOF. If $\phi(u, v) = (u, v, f(u, v))$, then ϕ is C^p and 1–1 on E, and $\phi(E)$ is the graph of $z = f(x, y)$. (This is called the *trivial parametrization* of $z = f(x, y)$.) ∎

In a similar way we define trivial parametrizations of surfaces of the form $x = f(y, z)$ and $y = f(x, z)$. For example, the trivial parametrization of the surface $x = f(y, z)$, $(y, z) \in E$, is given by (ϕ, E), where $\phi(u, v) = (f(u, v), u, v)$. By an *explicit surface over* E we shall mean the trivial parametrization of a surface of the form $x = f(y, z)$, $y = f(x, z)$, or $z = f(x, y)$, where $f : E \to \mathbf{R}$ is a C^p function and E is a two-dimensional region. By the proof of Example 1, every explicit surface is a simple surface.

The next four examples, which show that not every surface is an explicit surface, provide model parametrizations for certain kinds of surfaces.

Example 2. Show that the truncated cylinder $x^2 + y^2 = 1$, $0 \le z \le 2$, is the trace of a simple C^∞ surface.

PROOF. Let $\phi(u, v) = (\cos u, \sin u, v)$ and $E = [0, 2\pi] \times [0, 2]$, and notice that ϕ is C^∞ on E. The corresponding parametric equations are $x = \cos u$, $y = \sin u$, $z = v$. Clearly, $x^2 + y^2 = 1$. Thus, $\phi(E)$ is a subset of the cylinder $x^2 + y^2 = 1$, $0 \le z \le 2$. Since E is connected, so is $\phi(E)$. To see that $\phi(E)$ is the entire cylinder, look at the images of horizontal line segments in E. The image of the line segment $v = v_0$ is a circle lying in the plane $z = v_0$, centered at $(0, 0, v_0)$, of radius 1 (see Figure 8.9). Thus, as v_0 ranges from 0 to 2, the images of horizontal lines $v = v_0$ cover the entire cylinder $x^2 + y^2 = 1$, $0 \le z \le 2$. ∎

Example 3. Show that the sphere $x^2 + y^2 + z^2 = a^2$ is the trace of a C^∞ surface.

PROOF. Let $\phi(u, v) = (a \cos u \cos v, a \sin u \cos v, a \sin v)$ and $E = [0, 2\pi] \times [-\pi/2, \pi/2]$. Clearly, ϕ is C^∞ on E. The corresponding parametric equations are $x = a \cos u \cos v$, $y = a \sin u \cos v$, $z = a \sin v$. Since $x^2 + y^2 = a^2 \cos^2 v$, we have $x^2 + y^2 + z^2 = a^2$. Thus, $\phi(E)$ is a subset of the sphere centered at the origin of radius a. The image of the horizontal line segment $v = v_0$ is a circle, lying in the plane $z = a \sin v_0$, centered at $(0, 0, a \sin v_0)$ of radius $a \cos v_0$ (see Figure 8.10). The image of the top edge (respectively, bottom edge) of E, i.e., of the horizontal line $v = \pi/2$ (respectively, $v = -\pi/2$), is the north pole $(0, 0, a)$ (respectively, the south

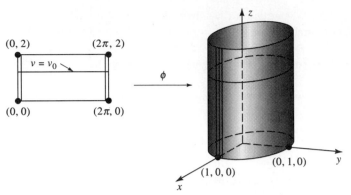

Figure 8.9

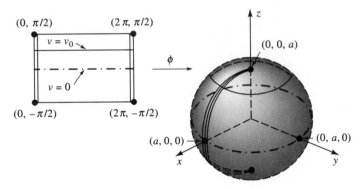

Figure 8.10

pole $(0,0,-a)$). Thus, as v_0 ranges from $-\pi/2$ to $\pi/2$, the images of horizontal lines $v = v_0$ cover the entire sphere $x^2 + y^2 + z^2 = a^2$. ∎

Let C represent the circle in the xz plane centered at $(a, 0, 0)$ of radius b, where $a > b$. The *torus* centered at the origin with radii $a > b$ is the donut-shaped surface obtained by revolving C about the z axis (see Figure 8.11).

Example 4. Show that the torus centered at the origin with radii $a > b$ is the trace of a C^∞ surface.

PROOF. Let $\phi(u, v) = ((a + b\cos v)\cos u, (a + b\cos v)\sin u, b\sin v)$ and $E = [-\pi, \pi] \times [-\pi, \pi]$, and notice that ϕ is C^∞ on E. The image of $u = 0$ is a circle in the xz plane centered at $(a, 0, 0)$ of radius b. The images of horizontal lines $v = v_0$ are circles, parallel to the xy plane, centered at $(0, 0, b\sin v_0)$ of radius $(a + b\cos v_0)$. The image of the lines $v = \pm\pi$ is a circle in the xy plane centered at $(0, 0, 0)$ of radius $a - b$. Thus, $\phi(E)$ covers the entire torus. ∎

Example 5. Show that the truncated cone $z = \sqrt{x^2 + y^2}$, $0 \le z \le b$, is the trace of a C^∞ surface.

PROOF. Let $(x, y, z) = \phi(u, v) = (v\cos u, v\sin u, v)$ and $E = [0, 2\pi] \times [0, b]$, and

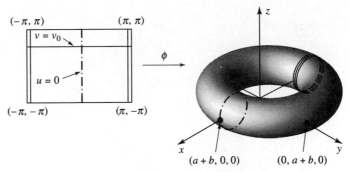

Figure 8.11

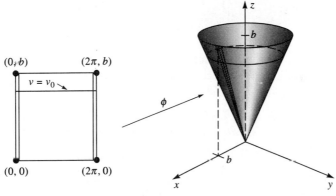

Figure 8.12

notice that ϕ is C^∞ on E. Clearly, $x^2 + y^2 = z^2$ and $0 \le z \le b$. Thus, $\phi(E)$ is a subset of the given cone. The image of a horizontal line $v = v_0$, $0 < v_0 \le b$, is a circle in the plane $z = v_0$ centered at $(0, 0, v_0)$ of radius v_0 (see Figure 8.12). Thus, $\phi(E)$ is the cone $z = \sqrt{x^2 + y^2}$, $0 \le z \le b$. Notice that the image of the line $v = 0$ is the vertex $(0, 0, 0)$. ∎

   We shall say that a surface $S = (\phi, E)$ has a unique tangent plane at a point $(x_0, y_0, z_0) = \phi(u_0, v_0)$ if the trace of S near (x_0, y_0, z_0) looks like the graph of a differentiable function. When $(u_0, v_0) \in E^o$, this means that there exist open sets E_0 and V, and a differentiable function $f : V \to \mathbf{R}$, such that $(u_0, v_0) \in E_0 \subset E$, and one of the following conditions holds: $(x_0, y_0) \in V$ and the graph of $z = f(x, y)$ coincides with $\phi(E_0)$; $(x_0, z_0) \in V$ and the graph of $y = f(x, z)$ coincides with $\phi(E_0)$; or $(y_0, z_0) \in V$ and the graph of $x = f(y, z)$ coincides with $\phi(E_0)$. In this case, the tangent plane of the explicit surface generated by f is called the *tangent plane* of S at (x_0, y_0, z_0), and its *normal (vector)* can be computed by applying Remark 2 in Section 6.2 or Theorem 6.25. For example, if the trace of S looks like the graph of $x = f(y, z)$, then a normal to the tangent plane is given by $(1, -f_y(y_0, z_0), -f_z(y_0, z_0))$.

   Normal vectors play the same role for surfaces that tangent vectors played for

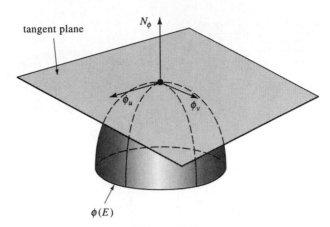

Figure 8.13

curves. For example, we shall use normal vectors to define area of surfaces, smooth surfaces, and orientation of surfaces. The following result shows how to compute a normal vector to a surface given in parametric form.

Remark 1. *Let $S = (\phi, E)$ be a $\mathcal{C}^p$ surface for some $p \geq 1$ and $\phi = (\phi_1, \phi_2, \phi_3)$. If S has a unique tangent plane at a point $(x_0, y_0, z_0) = \phi(u_0, v_0)$, where $(u_0, v_0) \in E^\circ$, then a normal to S at (x_0, y_0, z_0) is given by*

$$N_\phi := \phi_u(u_0, v_0) \times \phi_v(u_0, v_0).$$

PROOF. By Remark 1 in Section 8.1, $\phi_u(u_0, v_0)$ is tangent, at (x_0, y_0, z_0), to the curve $\phi(u, v_0)$ and $\phi_v(u_0, v_0)$ is tangent, at (x_0, y_0, z_0), to the curve $\phi(u_0, v)$. Hence, a normal to the tangent plane at (x_0, y_0, z_0) is given by the cross product of these two vectors (see Figure 8.13). Therefore, $\phi_u(u_0, v_0) \times \phi_v(u_0, v_0)$ is normal to S at (x_0, y_0, z_0). ∎

If $z = f(x, y)$ is an explicit surface and ϕ is its trivial parametrization, an easy calculation verifies $N_\phi = (-f_x, -f_y, 1)$. Notice that this agrees with Theorem 6.25 and with Remark 2 in Section 6.2.

By replacing ϕ' by N_ϕ, we develop a theory of surfaces which is parallel to the theory of curves presented in the previous two sections. For example, compare the following definition with Definition 8.3.

DEFINITION 8.10. Let (ϕ, E) be a $\mathcal{C}^p$ surface or a parametrization of a $\mathcal{C}^p$ surface, where $p \geq 1$.

   i) (ϕ, E) is said to be *smooth* at a point $(u_0, v_0) \in E$ if $N_\phi(u_0, v_0) \neq \mathbf{0}$ (equivalently, if $\|N_\phi(u_0, v_0)\| > 0$).

   ii) (ϕ, E) is said to be *smooth* if it is smooth at each point in E.

   iii) (ϕ, E) is said to be smooth off a set $E_0 \subset E$ if (ϕ, E) is smooth at each point in $E \setminus E_0$.

Notice that an explicit surface, under the trivial parametrization, is always smooth.

Analogous to the situation for curves, a surface (ϕ, E) which has no tangent plane at $\phi(u_0, v_0)$, where $(u_0, v_0) \in E^o$, cannot be smooth at (u_0, v_0) (see Exercise 7). On the other hand, there are surfaces with tangent planes at each point which have non-smooth parametrizations. (This is consistent with our position that a surface is not just a set of points, but a set of points together with a particular parametrization.) For example, the parametrization ϕ of the sphere given in Example 3 satisfies

$$\|\phi_u \times \phi_v\| = \|(a^2 \cos u \cos^2 v, a^2 \sin u \cos^2 v, a^2 \sin v \cos v)\| = a^2 |\cos v|.$$

Thus, although the sphere has a tangent plane at each point in its trace, the parametrization (ϕ, E) is not smooth at $v = \pm\pi/2$. (This happens because this parametrization takes the lines $v = \pm\pi/2$ to the north and south pole, hence, is not 1–1 there.) In particular, smoothness, as defined here, depends on the parametrization, not just the shape of the surface.

The following result shows what happens to the normal vector N_ϕ under a change of parameter.

THEOREM 8.2. *Let E, B be two-dimensional regions in $\mathbf{R}^2$, (ϕ, E) and (ψ, B) be C^p surfaces in $\mathbf{R}^3$, and $p \geq 1$. If τ is a C^p function which takes B into E such that $\psi = \phi \circ \tau$, then*

$$N_\psi(u, v) = \Delta_\tau(u, v) N_\phi(\tau(u, v))$$

for each $u, v \in B$.

PROOF. Let $\phi = (\phi_1, \phi_2, \phi_3)$ and $\psi = (\psi_1, \psi_2, \psi_3)$. By Remark 1,

$$N_\psi = (\Delta_{(\psi_2, \psi_3)}, \Delta_{(\psi_3, \psi_1)}, \Delta_{(\psi_1, \psi_2)}).$$

Since by hypothesis, $\psi = \phi \circ \tau$, i.e., $(\psi_i, \psi_j) = (\phi_i, \phi_j) \circ \tau$ for $i, j = 1, 2, 3$, it follows from the Chain Rule that

$$\Delta_{(\psi_i, \psi_j)}(u, v) = \Delta_\tau(u, v) \Delta_{(\phi_i, \phi_j)}(\tau(u, v))$$

for any $u, v \in B$. Therefore, $N_\psi = \Delta_\tau(N_\phi \circ \tau)$ on B. ∎

This leads us to the following definition (compare with Definition 8.4).

DEFINITION 8.11. *Let $p \geq 1$. Two smooth C^p surfaces (or parametrizations) (ϕ, E), (ψ, B) are said to be* smoothly equivalent *if $\psi(B) = \phi(E)$ and there is a 1–1 C^p function τ, which takes B onto E, such that $\psi = \phi \circ \tau$ and $\Delta_\tau(u, v) \neq 0$ for all $(u, v) \in B$. The function τ is called the* transition *from $\psi(B)$ to $\phi(E)$.*

Notice that if two surfaces are smoothly equivalent, then one is simple if and only if the other one is simple (see Exercise 5 in Section 8.4).

Analogous to Theorem 8.1 and Definition 8.5, we define surface area and surface integrals as follows.

DEFINITION 8.12. Let $S = (\phi, E)$ be a smooth simple surface with normal $N_\phi = \phi_u \times \phi_v$.

  i) The *surface area* of S is defined to be

$$\sigma(S) := \int_E \|N_\phi(u, v)\| \, d(u, v).$$

  ii) If $g : \phi(E) \to \mathbf{R}$ is continuous, then the *surface integral* of g on S is defined to be

(5) $$\iint_S g \, d\sigma := \iint_{\phi(E)} g \, d\sigma := \int_E g(\phi(u, v)) \|N_\phi(u, v)\| \, d(u, v).$$

Notice that by Corollary 7.7, (5) makes sense when ϕ is 1–1 off a set of area zero or $N_\phi(u, v)$ is undefined on a set of area zero, in particular, if (ϕ, E) is smooth and simple off some set Z of area zero. Thus, these concepts can be extended to certain nonsmooth nonsimple surfaces including all cones and spheres.

For an explicit $\mathcal{C}^p$ surface S given by $z = f(x, y)$, $(x, y) \in E$, integral (5) takes on the form

(6) $$\iint_S g \, d\sigma = \int_E g(x, y) \sqrt{f_x^2(x, y) + f_y^2(x, y) + 1} \, d(x, y).$$

It can be argued on heuristic grounds that this is the right definition for surface area (see Appendix E). Also, the surface integral (5) can be interpreted as the mass of a membrane with shape $\phi(E)$ and density g (see Appendix E).

We could have defined the surface area of S by approximating it with planar regions as we defined arc length by approximating with line segments (see Price [10], p. 360). This approach, however, works only under suitable restrictions. Indeed, even when using triangular regions to approximate a bounded cylinder, the total area of the approximating regions may become infinite (see Spivak [12], p. 130).

It is easy to see that surface area and the surface integral are invariant under smoothly equivalent parametrizations, even when the condition $\Delta_\tau \neq 0$ is relaxed on a closed set of area zero (see Exercise 5). It is also easy to see that if the trace of S is a subset of $\mathbf{R}^2$, then its surface area, as defined by Definition 8.12, is the same as the area of the trace of S, as defined by Definition 7.2 (see Exercise 4).

To compute a surface integral, one must find a suitable parametrization of the given surface and apply Definition 8.12.

Example 6. Find $\iint_S g \, d\sigma$, where S is the hemisphere $z = \sqrt{a^2 - x^2 - y^2}$ and $g(x, y, z) = \sqrt[4]{z}$.

SOLUTION. Let ϕ be the function defined in Example 3 and $E_0 = [0, 2\pi] \times [0, \pi/2]$. Then (ϕ, E_0) is a parametrization of the hemisphere S and $\|N_\phi\| = a^2 \cos v$. Since $\cos v \neq 0$ for $v \in [0, \pi/2)$, $\phi(E_0)$ is smooth off $[0, 2\pi] \times \{\pi/2\}$, a set of area zero. Hence,

$$\iint_S g \, d\sigma = \iint_{E_0} a^2 \cos v \sqrt[4]{a \sin v} \, du \, dv = 2\pi a^{9/4} \int_0^{\pi/2} \cos v \sqrt[4]{\sin v} \, dv = \frac{8\pi}{5} a^{9/4}. \quad \blacksquare$$

Continuity of g is assumed in Definition 8.12 only so the integral on the right-hand side of (5) makes sense. If one of the iterated integrals is a convergent improper integral, we can extend the definition of the surface integral in the obvious way. Using this observation, we now offer a second solution to Example 6 using the trivial parametrization.

ALTERNATE SOLUTION. The explicit surface $z = \sqrt{a^2 - x^2 - y^2}$ has normal $N = (-z_x, -z_y, 1) = (x/z, y/z, 1)$. (This normal does not exist on $\partial B_a(0,0)$ but since $\partial B_a(0,0)$ is of area zero, we can ignore it when integrating over $B_a(0,0)$.) On S, $\|N\| = a/z$. Thus, by (6) and polar coordinates,

$$\iint_S g \, d\sigma = \int_{B_a(0,0)} \frac{a\sqrt[4]{z}}{z} \, d(x,y) = a \int_0^{2\pi} \int_0^a r(a^2 - r^2)^{-3/8} \, dr \, d\theta = \frac{8\pi}{5} a^{9/4}.$$

(The inner integral (with respect to r) is an improper integral.) ∎

The theory of surfaces is more complicated than the theory of curves and we will use informal geometric descriptions, rather than formal statements, to define several concepts including what it means to be a piecewise smooth surface. For now these vague descriptions will suffice because the concrete surfaces we use are easy to visualize. (Chapter 10 contains a rigorous and more mathematically satisfying treatment of these ideas.)

Before describing how to patch several smooth surfaces together to form a piecewise smooth surface, we must distinguish between interior points, i.e., points which lie "inside" a surface and boundary points, i.e., points which lie on the "edge" of a surface. To illustrate the difference, consider the cylinder S parametrized by (ϕ, E) in Example 2. A point $(x, y, z) \in S$ lies inside S if $0 < z < 2$ and on its edge if $z = 0$ or $z = 2$. (Look at Figure 8.9 and see why this terminology is appropriate.) Naively, we might guess that (x, y, z) lies inside (ϕ, E) if $(x, y, z) \in \phi(E^o)$. This guess is incorrect, even for the cylinder; for example, $(-1, 0, 1) = \phi(\pi, 1) \in \phi(E^o)$ lies on a "seam" of S but not an edge. Evidently, to define the interior and boundary of a general surface S, we must use the trace of S and retreat from our earlier position that a surface is fully described by a particular parametrization (ϕ, E).

Accordingly, let S be (the trace of) a $\mathcal{C}^p$ surface in $\mathbf{R}^3$. Imagine yourself standing on a point $(x, y, z) \in S$. We shall say that (x, y, z) is *interior* to S if you are surrounded on all sides by points in S; i.e., if you take a sufficiently small step in any direction you remain on S. We shall denote the set of interior points of a surface S by $\text{Int}(S)$ and shall define the (*manifold*) *boundary* of a surface S by $\partial S := S \setminus \text{Int}(S)$.

We have used the same notation to denote the boundary of a surface as we did to denote the boundary of a set (see Definition 5.7) even though these concepts are not the same. We made this choice because it homogenizes the statements of all the fundamental theorems of multidimensional calculus. To avoid ambiguity, we shall henceforth refer to the boundary of a set E (i.e., to $\overline{E} \setminus E^o$) as the *topological boundary* of E. No confusion will arise because the only boundary we use in connection with surfaces is the manifold boundary, and the only boundary we use in connection with m-dimensional regions is the topological boundary.

A surface S is said to be *closed* if $\partial S = \emptyset$. For example, if $a > 0$ then the sphere $x^2 + y^2 + z^2 = a^2$ is closed, but the hemisphere $z = \sqrt{a^2 - x^2 - y^2}$ (respectively, the truncated paraboloid $z = x^2 + y^2$, $0 \le z \le 1$) is not closed, since its boundary is $x^2 + y^2 = a^2$, $z = 0$ (respectively, $x^2 + y^2 = 1$, $z = 1$).

By the Jordan Curve Theorem, a closed smooth curve C divides $\mathbf{R}^2$ into two disjoint sets, the set of points "surrounded" by C and the set of points which lie "outside" C. This is not the case for surfaces. Indeed, there are closed smooth surfaces (the Klein bottle is one example) which surround no points, hence, do not divide $\mathbf{R}^3$ into two disjoint sets (see Griffiths [3], p. 22, or Hocking and Young [4], p. 237).

Let $S = \cup_{j=1}^{N} S_j$, where each $S_j = (\phi_j, E_j)$ is a smooth simple surface. S will be called *piecewise smooth* if either of the following conditions holds. The first condition is that the traces of S_j and S_k are disjoint for each $j \ne k$. (For example, the topological boundary of the *corona* $0 < a \le \|(x,y,z)\| \le b$ is a smooth surface.) The second condition is that the S_j's are nonoverlapping, i.e., if $\phi_j(E_j^o) \cap \phi_k(E_k^o) = \emptyset$ and a portion of the boundary of each S_j is matched to a portion of the boundary of some S_k. (For example, the topological boundary of a three-dimensional rectangle is a piecewise smooth surface.) We make the further restriction that the intersection of any three S_j's is empty or a finite set. This prevents a piecewise smooth surface from doubling back on itself more than once along any given edge. We shall distinguish between these two types of surfaces by calling the first type a nonconnected piecewise smooth surface and the second type a connected piecewise smooth surface.

The *trace* of a piecewise smooth surface $S = \cup_{j=1}^{N}$ is defined to be the union of the traces of the S_j's. If S is a nonconnected piecewise smooth surface, the *boundary* ∂S of S is defined to be the union of the boundaries ∂S_j. If S is a connected piecewise smooth surface, the *boundary* ∂S of S is defined to be the union of all points which belong to the closure of an unmatched portion of ∂S_j. (For example, the boundary of the box formed by removing the face $z = 1$ from the unit cube $[0,1] \times [0,1] \times [0,1]$ is the unit square in the plane $z = 1$, and the boundary of the union of $x^2 + y^2 = 1$, $-3 \le z \le 0$, and $z = \sqrt{1 - x^2 - y^2}$ is the unit circle in the plane $z = -3$.) A *piecewise smooth parametrization* of $S = \cup_{j=1}^{N} S_j$ is a collection of smooth parametrizations (ϕ_j, E_j), where $S_j = (\phi_j, E_j)$. The surface area of S is defined by

$$\sigma(S) = \sum_{j=1}^{N} \sigma(S_j)$$

and the *surface integral* of a real-valued function g continuous on the trace of S is defined by

$$\iint_S g\, d\sigma = \sum_{j=1}^{N} \iint_{S_j} g\, d\sigma.$$

Example 7. Find a piecewise smooth parametrization of the tetrahedron S bounded by $x = 0$, $y = 0$, $z = 0$, and $x + y + z = 1$, and compute $\iint_S g\, d\sigma$, where $g(x,y,z) = x + y^2 + z^3$.

SOLUTION. The tetrahedron S has four faces. They can be parametrized by $\phi_1(u, v) = (u, v, 0)$, $\phi_2(u, v) = (0, u, v)$, $\phi_3(u, v) = (u, 0, v)$, $\phi_4(u, v) = (u, v, 1 - u - v)$, where (u, v) belongs to E, the triangular region with vertices $(0, 0)$, $(1, 0)$, and $(0, 1)$. Since $\|N_{\phi_j}\| = 1$ for $j = 1, 2, 3$ and $\|N_{\phi_4}\| = \sqrt{3}$, we have

$$
\iint_S g \, d\sigma = \int_0^1 \int_0^{1-u} (u + v^2) \, dv \, du + \int_0^1 \int_0^{1-u} (u^2 + v^3) \, dv \, du
$$
$$
+ \int_0^1 \int_0^{1-u} (u + v^3) \, dv \, du
$$
$$
+ \sqrt{3} \int_0^1 \int_0^{1-u} (u + v^2 + (1 - u - v)^3)) \, dv \, du
$$
$$
= \int_0^1 \int_0^{1-u} ((2 + \sqrt{3})u + u^2 + (1 + \sqrt{3})v^2 + 2v^3
$$
$$
+ \sqrt{3}(1 - u - v)^3) \, dv \, du
$$
$$
= \int_0^1 ((2 + \sqrt{3})u - (1 + \sqrt{3})u^2 - u^3 + \frac{1 + \sqrt{3}}{3}(1 - u)^3
$$
$$
+ \frac{2 + \sqrt{3}}{4}(1 - u)^4) \, du
$$
$$
= \frac{3}{10}(2 + \sqrt{3}). \quad \blacksquare
$$

EXERCISES

1. For each of the following, find the surface area of S.

   a) S is the conical shell given by $z = \sqrt{x^2 + y^2}$, where $a \le z \le b$.
   b) S is the sphere given in Example 3.
   c) S is the torus given in Example 4.

2. For each of the following, find a (piecewise) smooth parametrization of S and of ∂S, and compute $\iint_S g \, d\sigma$.

   a) S is the portion of the surface $z = x^2 - y^2$ which lies above the xy plane and between the planes $x = 1$ and $x = -1$, and $g(x, y, z) = \sqrt{1 + 4x^2 + 4y^2}$.
   b) S is the surface $y = x^3$, $0 \le y \le 8$, $0 \le z \le 4$, and $g(x, y, z) = x^3 z$.
   c) S is the portion of the hemisphere $z = \sqrt{9 - x^2 - y^2}$ which lies outside the cylinder $2x^2 + 2y^2 = 9$, and $g(x, y, z) = x + y + z$.

3. Find a parametrization (ϕ, E) of the ellipsoid

$$
\frac{x^2}{a^2} + \frac{y^2}{b^2} + \frac{z^2}{c^2} = 1
$$

which is smooth off the topological boundary ∂E.

4. a) Suppose E is a two-dimensional region and $S = \{(x, y, z) \in \mathbf{R}^3 : (x, y) \in E$ and $z = 0\}$. Prove

$$\text{Area}\,(E) = \iint_S d\sigma$$

and

$$\iint_S g\,d\sigma = \int_E g(x, y, 0)\,d(x, y)$$

for each continuous $g : E \to \mathbf{R}$.

b) Let $f : [a, b] \to \mathbf{R}$ be a $\mathcal{C}^p$ function, $p \geq 1$, C be the curve in $\mathbf{R}^2$ determined by $z = f(x)$, $a \leq x \leq b$, and S be the surface in $\mathbf{R}^3$ determined by $z = f(x)$, $a \leq x \leq b$, $c \leq y \leq d$. Show $\sigma(S) = (d - c)L(C)$.

c) Let $f : [a, b] \to \mathbf{R}$ be a $\mathcal{C}^p$ function, $p \geq 1$, and S be the surface obtained by revolving the curve $y = f(x)$, $a \leq x \leq b$, about the y axis. Prove that the surface area of S is

$$\sigma(S) = 2\pi \int_a^b |f(x)|\sqrt{1 + |f'(x)|^2}\,dx.$$

5. Suppose (ψ, B) and (ϕ, E) are simple $\mathcal{C}^p$ surfaces, Z is a closed subset of B of area zero, (ψ, B) is smooth off Z, and $\tau : B \to \mathbf{R}^2$ is a $\mathcal{C}^1$ function from B onto E. If τ is 1–1 and $\Delta_\tau \neq 0$ on $B^\circ \setminus Z$, and $\psi = \phi \circ \tau$, prove

$$\iint_E g(\phi(u, v))\|N_\phi(u, v)\|\,du\,dv = \iint_B g(\psi(s, t))\|N_\psi(s, t)\|\,ds\,dt$$

for all continuous $g : \phi(E) \to \mathbf{R}$.

6. Suppose $f : B_3(0, 0) \to \mathbf{R}$ is differentiable with $\|\nabla f(x, y)\| \leq 1$ for all $(x, y) \in B_3(0, 0)$. Prove that if S is the paraboloid $2z = x^2 + y^2$, $0 \leq z \leq 4$, then

$$\iint_S |f(x, y) - f(0, 0)|\,d\sigma \leq 40\pi.$$

7. Suppose (ϕ, E) is a $\mathcal{C}^p$ surface and $(x_0, y_0, z_0) = \phi(u_0, v_0)$, where $(u_0, v_0) \in E^\circ$. If (ϕ, E) has no tangent plane at (x_0, y_0, z_0), prove $N_\phi(u_0, v_0) = \mathbf{0}$.

8. Let $S = (\psi, B)$ be a smooth simple surface. Set $E = \|\psi_u\|$, $F = \psi_u \cdot \psi_v$, and $G = \|\psi_v\|$. Prove that the surface area of S is

$$\int_B \sqrt{E^2 G^2 - F^2}\,d(u, v).$$

9. Suppose $S = (\phi, E)$ is a $\mathcal{C}^1$ surface smooth at $(x_0, y_0, z_0) = \phi(u_0, v_0)$. Let $C = (\psi, I)$ be a $\mathcal{C}^1$ curve whose trace is a subset of E which passes through the point (u_0, v_0) (i.e., there is a $t_0 \in I$ such that $\psi(t_0) = (u_0, v_0)$). Prove $(\phi \circ \psi)'(t_0) \cdot (\phi_u \times \phi_v)(u_0, v_0) = 0$.

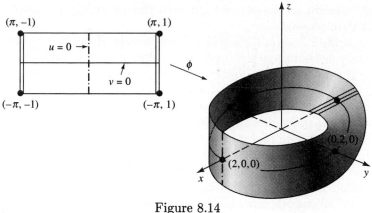

Figure 8.14

8.4   ORIENTED SURFACES

Recall that a smooth simple curve (ϕ, I) is oriented by using the tangent vector $\phi'(t)$ to choose a "positive direction." This works because a connected curve can be traversed in two and only two directions. Analogously, a smooth simple surface (ϕ, E) is oriented by using the normal vector N_ϕ to choose a "positive side" of S. For a connected surface S, such a choice is possible if S has two and only two sides.

A new complication arises here. There are smooth connected surfaces which have only one side. (The following example of such a surface can be made out of paper by taking a long narrow strip by the narrow edges, twisting it once, and gluing the narrow edges together.)

Example 1. [THE MÖBIUS STRIP]. Sketch the trace of the surface (ϕ, E), where $\phi(u, v) = ((2 + v\sin(u/2))\cos u, (2 + v\sin(u/2))\sin u, v\cos(u/2))$ and $E = [-\pi, \pi] \times [-1, 1]$.

SOLUTION. The image of the horizontal line $v = 0$ under ϕ is $(2\cos u, 2\sin u, 0)$, i.e., the circle in the xy plane centered at the origin of radius 2. The image of each vertical line $u = u_0$ is a line segment in $\mathbf{R}^3$. For example, the image of $u = 0$ is $(2, 0, v)$, $-1 \le v \le 1$, and the image of $u = \pm\pi$ is $(-2 \mp v, 0, 0)$, $-1 \le v \le 1$, i.e., the set of points $S_0 := \{(x, 0, 0) : -3 \le x \le -1\}$. Thus the trace of (ϕ, E) is given in Figure 8.14. ∎

To avoid such anomalies, we introduce the following concepts. The *unit normal* of a smooth surface $S = (\phi, E)$ at a point (x_0, y_0, z_0) on S is the vector $\boldsymbol{n}(x_0, y_0, z_0) = N_\phi(u_0, v_0)/\|N_\phi(u_0, v_0)\|$, where $\phi(u_0, v_0) = (x_0, y_0, z_0)$. Evidently, the unit normal $\boldsymbol{n}$ is well defined only when

$$\frac{N_\phi(u_0, v_0)}{\|N_\phi(u_0, v_0)\|} = \frac{N_\phi(u_1, v_1)}{\|N_\phi(u_1, v_1)\|} \ne \boldsymbol{0}$$

for all $(u_j, v_j) \in E$ which satisfy $\phi(u_j, v_j) = (x_0, y_0, z_0)$ for $j = 0, 1$. This will surely be the case if ϕ is 1–1 and smooth on E. If ϕ fails to be 1–1 on E, however, the unit normal $\boldsymbol{n}$ might not be well defined even though (ϕ, E) is smooth on E (see the Möbius strip above).

A smooth surface $S = (\phi, E)$ is said to be *orientable* if the unit normal $\boldsymbol{n}$ can be unambiguously defined at each point on the surface S so that it varies continuously over S, i.e., if $\phi(u_0, v_0) = \phi(u_1, v_1)$, then $N_\phi(u_0, v_0)$ points in the same direction as $N_\phi(u_1, v_1)$ and if (u_2, v_2) is near (u_0, v_0), then $N_\phi(u_2, v_2)$ points in approximately the same direction as $N_\phi(u_0, v_0)$. (A formal definition of orientable will be given in Section 10.2.) The problem with the Möbius strip is that near the "seam" S_0, N_ϕ points both upward and downward.

Notice that if S is connected, orientable, and smooth, then S has two and only two sides, a "positive" side (the side from which $\boldsymbol{n}$ points) and a "negative" side (the other side). Clearly, an explicit surface is always orientable. The Möbius strip shows that not all smooth surfaces are orientable.

DEFINITION 8.13. Two smooth simple surfaces (or parametrizations) (ϕ, E) and (ψ, B) are said to be *orientation equivalent* if they are orientable, smoothly equivalent with transition τ, and $\Delta_\tau(u, v) > 0$ for all $(u, v) \in B$.

By Theorem 8.2, if (ϕ, E) and (ψ, B) are orientation equivalent, then the normal vectors they determine point in the same direction. Thus, the positive side chosen by (ϕ, E) is the same as the positive side chosen by (ψ, B).

Oriented surface integrals can be defined using the unit normal in the same way that oriented line integrals were defined using the unit tangent (compare the following definition with Definition 8.7).

DEFINITION 8.14. Let $S = (\phi, E)$ be a smooth simple orientable surface with unit normal $\boldsymbol{n}$, and $F : \phi(E) \to \mathbf{R}^3$ be continuous. The *oriented surface integral* of F on S is

$$\iint_S F \cdot \boldsymbol{n} \, d\sigma := \iint_{\phi(E)} F \cdot \boldsymbol{n} \, d\sigma := \int_E (F \circ \phi)(u, v) \cdot N_\phi(u, v) \, d(u, v).$$

This notation is consistent with the notation in (5) since $\boldsymbol{n} = N_\phi/\|N_\phi\|$ and $d\sigma = \|N_\phi\| \, d(u, v)$.

For an explicit surface S given by $z = f(x, y)$, $(x, y) \in E$, Definition 8.14 takes the form

$$(7) \qquad \iint_S F \cdot \boldsymbol{n} \, d\sigma = \int_E F(x, y, f(x, y)) \cdot (-f_x, -f_y, 1) \, d(x, y).$$

What does an oriented surface integral represent? If F represents the flow of an incompressible fluid at points in the trace of $S = (\phi, E)$, then $F \cdot \boldsymbol{n}$ represents the normal component of F, i.e., the amount of fluid which flows in the direction of $\boldsymbol{n}$ (see Appendix E). Thus, the integral of $F \cdot \boldsymbol{n} \, d\sigma$ on S, a measure of the flow of the fluid across the trace of S in the direction of $\boldsymbol{n}$, is sometimes called the *flux* of F across S.

As was the case for the oriented line integral, the oriented surface integral can be defined on some nonsmooth surfaces (see Exercise 4). One needs to be careful, however, with the definition of orientable. If the collection of nonsmooth points cuts across the entire surface (like the peak of a pup tent or the edge of a pyramid), one has difficulty defining what it means to have a "continuously varying" normal. We shall address this problem for piecewise smooth surfaces at the end of this section. In the meantime, notice that one can define what it means for a surface $S = (\phi, E)$ to be orientable if the set of *singularities* (i.e., the set of $(x, y, z) \in \mathbf{R}^3$ such that $(x, y, z) = \phi(u, v)$ for some $(u, v) \in E$ which satisfies $N_\phi(u, v) = \mathbf{0}$) is finite. In particular, all spheres and cones are orientable.

It is easy to see that the integral of $F \cdot \boldsymbol{n} \, d\sigma$ on a surface S which is smooth off a set of area zero does not change when orientation equivalent parametrizations are used (see Exercise 4). The following result shows that a change of orientation changes the value of the oriented surface integral by a minus sign.

Remark 2. *If (ϕ, E) and (ψ, B) are smoothly equivalent but not orientation equivalent, then*

$$\int_E F(\phi(u, v)) \cdot N_\phi(u, v) \, d(u, v) = -\int_B F(\psi(s, t)) \cdot N_\psi(s, t) \, d(s, t).$$

PROOF. Let τ be the transition from $\psi(B)$ to $\phi(E)$. Since Δ_τ is continuous on the connected set B and (ϕ, E) and (ψ, B) are not orientation equivalent, we have $\Delta_\tau < 0$ on B. Hence, it follows from Theorem 8.2 and Theorem 7.14 (the Change of Variables Formula) that

$$\int_B F(\psi(s, t)) \cdot N_\psi(s, t) \, d(s, t) = -\int_B |\Delta_\tau(s, t)| (F \circ \phi \circ \tau)(s, t) \cdot (N_\phi \circ \tau)(s, t)$$

$$= -\int_{\tau(B)} F(\phi(u, v)) \cdot N_\phi(u, v) \, d(u, v)$$

$$= -\int_E F(\phi(u, v)) \cdot N_\phi(u, v) \, d(u, v). \quad \blacksquare$$

Therefore, when evaluating an oriented integral on a surface S whose orientation has been described geometrically, we can use any smoothly equivalent parametrization of S and adjust the sign of the integral to reflect the prescribed orientation. Here is a typical example.

Example 2. Find the value of $\iint_S F \cdot \boldsymbol{n} \, d\sigma$, where $F(x, y, z) = (xy, x - y, z)$, S is the planar region $x + y + z = 1$, $(x, y) \in [0, 1] \times [0, 1]$, and $\boldsymbol{n}$ is the downward pointing normal.

SOLUTION. The usual normal $(1, 1, 1)$ of the plane $x + y + z = 1$ points upward rather than downward. Thus, by Remark 2,

$$\iint_S F \cdot \boldsymbol{n} \, d\sigma = -\int_0^1 \int_0^1 (xy, x - y, 1 - x - y) \cdot (1, 1, 1) \, dx \, dy = -\frac{1}{4}. \quad \blacksquare$$

It is convenient to have a "differential" version of oriented surface integrals. To see how to define differentials of degree two, let $S = (\phi, E)$ be a smooth orientable surface and $x = \phi_1(u,v)$, $y = \phi_2(u,v)$, $z = \phi_3(u,v)$. By Remark 1 in Section 8.3,

$$N_\phi = \left(\frac{\partial(y,z)}{\partial(u,v)}, \frac{\partial(z,x)}{\partial(u,v)}, \frac{\partial(x,y)}{\partial(u,v)} \right).$$

Therefore, the oriented surface integral of a function $F = (P,Q,R) : \phi(E) \to \mathbf{R}^3$ has the form

$$\int_E \left(P \frac{\partial(y,z)}{\partial(u,v)} + Q \frac{\partial(z,x)}{\partial(u,v)} + R \frac{\partial(x,y)}{\partial(u,v)} \right) d(u,v)$$

$$= \iint_S P \, dy \, dz + Q \, dz \, dx + R \, dx \, dy,$$

where

$$dy \, dz = \frac{\partial(y,z)}{\partial(u,v)} d(u,v), \quad dz \, dx = \frac{\partial(z,x)}{\partial(u,v)} d(u,v), \quad \text{and} \quad dx \, dy = \frac{\partial(x,y)}{\partial(u,v)} d(u,v).$$

(These are two-dimensional analogues of the differential $dy = f'(x) \, dx$.) By a 2–form (or a *differential form of degree* 2) on a set $\Omega \subset \mathbf{R}^3$ we mean an expression of the form

$$P \, dy \, dz + Q \, dz \, dx + R \, dx \, dy,$$

where $P,Q,R : \Omega \to \mathbf{R}$. A 2–form is said to be continuous on Ω if its *coefficients* P,Q,R are continuous on Ω. The *oriented integral* of a continuous 2–form on a smooth simple surface S oriented with a unit normal $\boldsymbol{n}$ is defined by

$$\iint_S P \, dy \, dz + Q \, dz \, dx + R \, dx \, dy = \iint_S (P,Q,R) \cdot \boldsymbol{n} \, d\sigma.$$

Differential forms of degree one were formal devices used in certain computations, e.g., to compute an oriented line integral or to estimate the increment of a function. Similarly, differential forms of degree two are formal devices which will be used in certain computations, e.g., to compute an oriented surface integral. They can also be used to unify the three fundamental theorems of vector calculus presented in the next two sections (see Exercise 4 in Section 10.1). (There is a less formal but time-consuming way to introduce differentials in which the differential dx can be interpreted as the derivative of the projection operator $(x,y,z) \longmapsto x$ (see Spivak [12], p. 89).)

In general, the boundary of a surface is a curve. Since the boundary of the Möbius strip is a simple closed curve, the boundary of a surface may be orientable even when the surface is not.

Suppose S is an oriented surface and ∂S is a piecewise smooth curve. The orientation of S can be used to induce an orientation on ∂S in the following way. Imagine yourself standing on the positive side of S close to ∂S. The direction of positive flow on ∂S moves from right to left; i.e., as you walk around the boundary on the positive side of S in the direction of positive flow, the surface lies on your left. This orientation of ∂S is called the *positive orientation*, the *right-hand orientation*, or the orientation on ∂S *induced* by the orientation of S. When S is a subset of $\mathbf{R}^2$, i.e., of the xy plane, we shall say that ∂S is oriented *positively* if it carries the orientation induced by the upward pointing normal on S, i.e., the normal which points toward upper half space $z \geq 0$. Thus, if S is a bounded subset of $\mathbf{R}^2$ whose boundary is a connected piecewise smooth closed curve, then the usual orientation on S induces a counterclockwise orientation on ∂S when viewed from high up the positive z axis. This is not the case, however, when E has interior "bubbles." For example, if $E = \{(x,y) : a^2 < x^2 + y^2 < b^2\}$ for some $a > 0$, then the positive orientation is counterclockwise on $\{(x,y) : x^2 + y^2 = b^2\}$, but clockwise on $\{(x,y) : x^2 + y^2 = a^2\}$.

A formal definition of the positive or induced orientation will be given in Section 10.2. In the meantime, the informal geometric description given above is sufficient to identify the induced orientation in most concrete situations. Here is a typical example.

Example 3. Let S be the truncated paraboloid $z = x^2 + y^2$, $0 \leq z \leq 4$, with outward pointing normal. Parametrize ∂S with positive orientation.

SOLUTION. The boundary of S is the circle $x^2 + y^2 = 4$ which lies in the $z = 4$ plane. The positive orientation is clockwise when viewed from high up the z axis. Therefore, a parametrization of ∂S is given by $\phi(t) = (2\sin t, 2\cos t, 4)$, $t \in [0, 2\pi]$. ∎

If $S = \cup S_j = \cup(\phi_j, E_j)$ is a nonconnected piecewise smooth surface, we shall call S *orientable* if each S_j is orientable. This definition must be modified for connected surfaces, since the Möbius strip is the union of two orientable surfaces, namely (ϕ, E_1) and (ϕ, E_2) where ϕ is given by Example 1 and $E_k = [\pi(k - 2), \pi(k - 1)] \times [-1, 1]$, $k = 1, 2$. We shall say that a connected piecewise smooth surface $S = \cup S_j = \cup(\phi_j, E_j)$ is *orientable* if it has two distinct sides. In this case, an orientation on S can be given using the normals $\pm N_{\phi_j}$ to generate a unit normal $\boldsymbol{n}_j$ on each piece S_j which identifies the "positive side" in a consistent way, e.g., all normals point outward.

Oriented surface integrals can be defined for orientable piecewise smooth surfaces in the natural way. Indeed, if $S = \cup_{j=1}^N S_j$ is orientable, then the *oriented surface integral* of a vector-valued function F continuous on the trace of S is defined to be

$$\iint_S F \cdot \boldsymbol{n}\, d\sigma = \sum_{j=1}^{N} \iint_{S_j} F \cdot \boldsymbol{n}_j\, d\sigma.$$

The following three examples provide further explanation of these ideas.

Example 4. Evaluate

$$\iint_S F \cdot \boldsymbol{n}\, d\sigma,$$

where S is the topological boundary of the solid bounded by the cylinder $x^2 + y^2 = 1$ and the planes $z = 0$, $z = 2$, $\boldsymbol{n}$ is the outward pointing normal, and $F(x, y, z) = (x, 0, y)$.

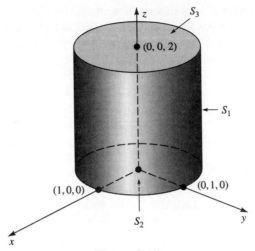

Figure 8.15

SOLUTION. This surface has three smooth pieces: a vertical side S_1, a bottom S_2, and a top S_3 (see Figure 8.15). Parametrize S_1 by $\phi(u, v) = (\cos u, \sin u, v)$, where $E = [0, 2\pi] \times [0, 2]$. Thus, $N_\phi = (\cos u, \sin u, 0)$ and

$$\iint_{S_1} F \cdot n \, d\sigma = \int_0^2 \int_0^{2\pi} \cos^2 u \, du \, dv = 2\pi.$$

Since the outward pointing unit normal to S_2 is $n = (0, 0, -1)$, we see by Exercise 4a in Section 8.3 that

$$\iint_{S_2} F \cdot n \, d\sigma = -\int_{B_1(0,0)} y \, d(x, y) = -\int_0^{2\pi} \int_0^1 r^2 \sin \theta \, dr \, d\theta = 0.$$

Similarly, the integral on S_3 is also zero. Therefore,

$$\iint_S F \cdot n \, d\sigma = 2\pi + 0 + 0 = 2\pi. \quad \blacksquare$$

Example 5. Find $\iint_S F \cdot n \, d\sigma$, where $F(x, y, z) = (x + z^2, x, z)$, S is the topological boundary of the solid bounded by the paraboloid $z = x^2 + y^2$ and the plane $z = 1$, and n is the outward pointing normal.

SOLUTION. The surface S has two smooth pieces: the paraboloid S_1 given by $z = x^2 + y^2$, $0 \le z \le 1$, and the disc S_2 given by $x^2 + y^2 \le 1$, $z = 1$. The trivial parametrization of S_1 is $\phi(u, v) = (u, v, u^2 + v^2)$, $(u, v) \in B_1(0, 0)$. Note that $N_\phi = (-2u, -2v, 1)$ points inward (the wrong way). Thus, by Remark 2 and polar coordinates,

$$\iint_{S_1} F \cdot n \, d\sigma = -\int_{B_1(0,0)} (-2u^2 - 2u(u^2 + v^2)^2 - 2uv + (u^2 + v^2)) \, d(u, v)$$

$$= \int_0^1 \int_0^{2\pi} (2r^2 \cos^2 \theta + 2r^5 \cos \theta + 2r^2 \cos \theta \sin \theta - r^2) r \, d\theta \, dr = 0.$$

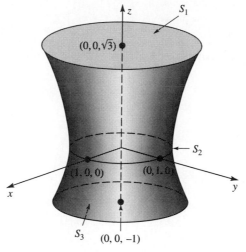

Figure 8.16

Since the unit outward pointing normal of S_2 is $\boldsymbol{n} = (0,0,1)$ and on S_2, $F \cdot \boldsymbol{n} = z = 1$, we see by Exercise 4a in Section 8.3 that

$$\iint_{S_2} F \cdot \boldsymbol{n} \, d\sigma = \int_{B_1(0,0)} d(x,y) = \text{Area}\,(B_1(0,0)) = \pi.$$

Therefore,

$$\iint_S F \cdot \boldsymbol{n} \, d\sigma = 0 + \pi = \pi. \quad \blacksquare$$

Example 6. Compute $\iint_S F \cdot \boldsymbol{n} \, d\sigma$, where $F(x,y,z) = (x,y,z)$, S is the topological boundary of the solid bounded by the hyperboloid of one sheet $x^2 + y^2 - z^2 = 1$ and the planes $z = -1$, $z = \sqrt{3}$, and $\boldsymbol{n}$ is the outward pointing normal to S.

SOLUTION. The surface S has three smooth pieces: a top S_1, a side S_2, and a bottom S_3 (see Figure 8.16). Using $\boldsymbol{n} = (0,0,1)$ for S_1, we have

$$\iint_{S_1} F \cdot \boldsymbol{n} \, d\sigma = \int_{B_2(0,0)} \sqrt{3} \, d(x,y) = 4\sqrt{3}\pi.$$

Similarly,

$$\iint_{S_3} F \cdot \boldsymbol{n} \, d\sigma = 2\pi.$$

To integrate $F \cdot \boldsymbol{n}$ on S_2, let $z = u$ and note that $x^2 + y^2 = 1 + u^2$. Thus, $\phi(u,v) = ((1+u^2)\cos v, (1+u^2)\sin v, u)$, $(u,v) \in [-1, \sqrt{3}] \times [0, 2\pi]$, is a parametrization of S_2. Since $N_\phi = (-(1+u^2)\cos v, -(1+u^2)\sin v, 2u(1+u^2))$ points inward and

$$F \cdot N_\phi = ((1+u^2)\cos v, (1+u^2)\sin v, u) \cdot$$
$$\cdot (-(1+u^2)\cos v, -(1+u^2)\sin v, 2u(1+u^2))$$
$$= -(1+u^2)^2 + 2u^2(1+u^2) = u^4 - 1$$

we have

$$\iint_{S_2} F \cdot n \, d\sigma = -\int_{-1}^{\sqrt{3}} \int_0^{2\pi} (u^4 - 1) \, dv \, du$$

$$= 2\pi \int_{-1}^{\sqrt{3}} (1 - u^4) \, du = \frac{8\pi}{5}(1 - \sqrt{3}).$$

Therefore,

$$\iint_S F \cdot n \, d\sigma = 4\sqrt{3}\pi + 2\pi + \frac{8\pi}{5}(1 - \sqrt{3}) = \frac{6\pi}{5}(3 + 2\sqrt{3}). \quad \blacksquare$$

EXERCISES

1. For each of the following, find a (piecewise) smooth parametrization of ∂S which agrees with the induced orientation and compute $\int_{\partial S} F \cdot T \, ds$.

   a) S is the truncated paraboloid $y = 9 - x^2 - z^2$, $y \geq 0$, with outward pointing normal, and $F(x, y, z) = (x^2 y, y^2 x, x + y + z)$.

   b) S is the portion of the plane $x + 2y + z = 1$ which lies in the first octant, with normal which points away from the origin, and $F(x, y, z) = (x - y, y - x, xz^2)$.

   c) S is the truncated paraboloid $z = x^2 + y^2$, $1 \leq z \leq 4$, with outward pointing normal, and $F(x, y, z) = (5y + \cos z, 4x - \sin z, 3x \cos z + 2y \sin z)$.

2. For each of the following, compute $\iint_S F \cdot n \, d\sigma$.

   a) S is the truncated paraboloid $z = x^2 + y^2$, $0 \leq z \leq 1$, n is the outward pointing normal, and $F(x, y, z) = (x, y, z)$.

   b) S is the truncated half cylinder $z = \sqrt{4 - y^2}$, $0 \leq x \leq 1$, n is outward pointing normal, and $F(x, y, z) = (x^2 + y^2, yz, z^2)$.

   c) S is the torus in Example 4 in Section 8.3, n is the outward pointing normal, and $F(x, y, z) = (y, -x, z)$.

   d) S is the portion of $z = x^2$ which lies inside the cylinder $x^2 + y^2 = 1$, n is the upward pointing normal, and $F(x, y, z) = (y^2 z, \cos(2 + \log(2 - x^2 - y^2)), x^2 z)$.

3. For each of the following, compute $\iint_S \omega$.

   a) S is the portion of the surface $z = x^4 + y^2$ which lies over the unit square $[0, 1] \times [0, 1]$, with upward pointing normal, and $\omega = x \, dy \, dz + y \, dz \, dx + z \, dx \, dy$.

   b) S is the upper hemisphere $z = \sqrt{a^2 - x^2 - y^2}$, with outward pointing normal, and $\omega = x \, dy \, dz + y \, dz \, dx$.

   c) S is the spherical cap $z = \sqrt{a^2 - x^2 - y^2}$ which lies inside the cylinder $x^2 + y^2 = b^2$, $0 < b < a$, with upward pointing normal, and $\omega = xz \, dy \, dz + dz \, dx + z \, dx \, dy$.

   d) S is the truncated cone $z = 2\sqrt{x^2 + y^2}$, $0 \leq z \leq 2$, with normal which points away from the z axis, and $\omega = x \, dy \, dz + y \, dz \, dx + z^2 \, dx \, dy$.

4. Suppose (ψ, B) and (ϕ, E) are simple C^p surfaces, Z is a closed subset of B of area zero, (ψ, B) is smooth off Z, and $\tau : B \to \mathbf{R}^2$ is a C^1 function from B onto E. If τ is 1–1 and $\Delta_\tau > 0$ on $B^o \setminus Z$, and $\psi = \phi \circ \tau$, prove

$$\int_E F(\phi(u,v)) \cdot N_\phi(u,v)\, d(u,v) = \int_B F(\psi(s,t)) \cdot N_\psi(s,t)\, d(s,t)$$

for all continuous $F : \phi(E) \to \mathbf{R}^3$.

5. Suppose M_1 and M_2 are both smooth curves or both smooth surfaces. Prove that if M_1 and M_2 are smoothly equivalent or orientation equivalent, then M_1 is simple if and only if M_2 is simple.

6. Let E be the solid tetrahedron bounded by $x = 0$, $y = 0$, $z = 0$, and $x+y+z = 1$, and suppose its topological boundary, $T = \partial E$, is oriented with outward pointing normal. Prove

$$\iint_{\partial E} P\, dy\, dz + Q\, dz\, dx + R\, dx\, dy = \iiint_E (P_x + Q_y + R_z)\, dV$$

for all C^1 functions $P, Q, R : E \to \mathbf{R}$.

7. Let T be the topological boundary of the tetrahedron in Exercise 6, with outward pointing normal, and S be the surface obtained by taking away the slanted face from T, i.e., S has three triangular faces, one each in the planes $x = 0$, $y = 0$, $z = 0$. If ∂S is oriented positively, prove

$$\int_{\partial S} P\, dx + Q\, dy + R\, dz = \iint_S (R_y - Q_z)\, dy\, dz + (P_z - R_x)\, dz\, dx + (Q_x - P_y)\, dx\, dy$$

for all C^1 functions $P, Q, R : S \to \mathbf{R}$.

8.5 THEOREMS OF GREEN AND GAUSS

Recall by the Fundamental Theorem of Calculus that if f is a C^1 function, then

$$f(b) - f(a) = \int_a^b f'(t)\, dt.$$

Thus, the integral of the derivative f' on $[a, b]$ is completely determined by the values f takes on the topological boundary $\{a, b\}$ of $[a, b]$.

In the next two sections we shall obtain analogues of this theorem where $f : [a, b] \to \mathbf{R}$ is replaced by $F : \Omega \to \mathbf{R}^m$, and Ω is a surface or an m-dimensional region, $m = 2$ or 3. Namely, we shall show that the integral of a "derivative" of F on Ω is completely determined by the values F takes on the "boundary" of Ω. Which "derivative" and "boundary" we use depends on whether Ω is a surface or an m-dimensional region and whether $m = 2$ or 3.

Our first fundamental theorem applies to two-dimensional regions in the plane.

THEOREM 8.3 [GREEN'S THEOREM]. *Let E be a two-dimensional region whose topological boundary ∂E is a piecewise smooth curve oriented positively. If $P, Q : E \to \mathbf{R}$ are C^1 and $F = (P, Q)$, then*

$$\int_{\partial E} F \cdot T \, ds = \iint_E \left(\frac{\partial Q}{\partial x} - \frac{\partial P}{\partial y} \right) dA.$$

PROOF FOR SPECIAL REGIONS. Suppose for simplicity that E is of types I and II. Write the integral on the left in differential notation

$$\int_{\partial E} P \, dx + Q \, dy = \int_{\partial E} P \, dx + \int_{\partial E} Q \, dy = I_1 + I_2.$$

We evaluate I_1 first. Since E is of type I, choose continuous functions $f, g : [a, b] \to \mathbf{R}$ such that

$$E = \{(x, y) \in \mathbf{R}^2 : a \le x \le b, \ f(x) \le y \le g(x)\}.$$

Thus, ∂E has a top $y = g(x)$, a bottom $y = f(x)$, and (possibly) one or two vertical sides (see Figure 8.17).

Since the positive orientation is counterclockwise, the trivial parametrization of the top is $y = g(x)$ where x runs from b to a and of the bottom is $y = f(x)$ where x runs from a to b. Since $dx = 0$ on any vertical curve, the contribution of the vertical sides to I_1 is zero. Thus, it follows from Definition 8.7 and the one-dimensional Fundamental Theorem of Calculus that

$$I_1 = \int_{\partial E} P \, dx = \int_a^b P(x, f(x)) \, dx + \int_b^a P(x, g(x)) \, dx$$

$$= -\int_a^b (P(x, g(x)) - P(x, f(x))) \, dx$$

$$= -\int_a^b \int_{f(x)}^{g(x)} \frac{\partial P}{\partial y}(x, y) \, dy \, dx = -\iint_E \frac{\partial P}{\partial y} \, dA.$$

Since E is of type II, a similar argument establishes

$$I_2 = \int_{\partial E} Q \, dy = \iint_E \frac{\partial Q}{\partial x} \, dA.$$

(Here, we have changed parametrizations of ∂E, e.g., replaced $y = f(x)$ by $x = f^{-1}(y)$. The value of the oriented integral does not change because these parametrizations are orientation equivalent—see Exercise 6 in Section 8.2.) Adding I_1 and I_2 completes the proof. ∎

The hypothesis that E be of types I and II was added to make the proof simple. For a proof of Green's Theorem as stated, see Theorem 10.8 and the reference which follows it. In the meantime, it is easy to check that Green's Theorem holds for any two-dimensional region which can be divided into a finite number of regions each of

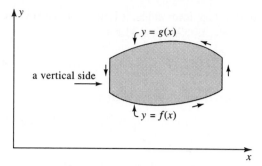

Figure 8.17

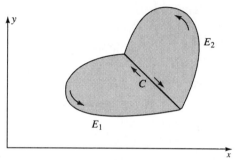

Figure 8.18

which is of types I and II. For example, consider the region E illustrated in Figure 8.18. Notice that E is neither of type I nor of type II, but can be divided into E_1, E_2, both of which are of types I and II. Applying Theorem 8.3 to each piece, we find

$$\iint_E \left(\frac{\partial Q}{\partial x} - \frac{\partial P}{\partial y} \right) dA = \iint_{E_1} \left(\frac{\partial Q}{\partial x} - \frac{\partial P}{\partial y} \right) dA + \iint_{E_2} \left(\frac{\partial Q}{\partial x} - \frac{\partial P}{\partial y} \right) dA$$

$$= \int_{\partial E_1} F \cdot T \, ds + \int_{\partial E_2} F \cdot T \, ds$$

$$= \int_{\partial E} F \cdot T \, ds + \int_{C \cap \partial E_1} F \cdot T \, ds + \int_{C \cap \partial E_2} F \cdot T \, ds,$$

where C is the common border between E_1 and E_2. Since ∂E_1 and ∂E_2 are oriented in the counterclockwise direction, the orientation of $C \cap \partial E_1$ is different from the orientation of $C \cap \partial E_2$. Since a change of orientation changes the sign of the integral, the integrals along C drop out. The end result is the integral of $F \cdot T \, ds$ on ∂E, as promised.

Green's Theorem is often used to avoid tedious parametrizations.

Example 1. Find $\int_{\partial E} F \cdot T \, ds$, where $E = [0,2] \times [1,3]$, ∂E has the counterclockwise orientation, and $F(x,y) = (xy, x^2 + y^2)$.

SOLUTION. Since ∂E has four sides, direct evaluation requires four separate parametrizations. However, by Green's Theorem,

$$\int_{\partial E} F \cdot T \, ds = \int_0^2 \int_1^3 (2x - x) \, dy \, dx = 4. \quad \blacksquare$$

Green's Theorem is also used to avoid difficult integrals.

Example 2. Find $\int_{\partial E} F \cdot T \, ds$, where $E = B_1(0,0)$, ∂E has the clockwise orientation, and $F = (xy^2, \arctan(\log(y+3)) - x))$.

SOLUTION. The second component of F looks tough to integrate. However, by Green's Theorem,

$$\int_{\partial E} F \cdot T \, ds = -\iint_{B_1(0,0)} (-1 - 2xy) \, dx \, dy$$

$$= \int_0^{2\pi} \int_0^1 (1 + 2r^2 \cos\theta \sin\theta) r \, dr \, d\theta = \pi.$$

(Note: The minus sign appears because ∂E is oriented in the clockwise direction.) $\blacksquare$

By Green's Theorem, the "derivative" used to obtain a fundamental theorem of calculus for two-dimensional regions in $\mathbf{R}^2$ is $Q_x - P_y$. Here are the "derivatives" which will be used when Ω is a surface in $\mathbf{R}^3$ or a three-dimensional region.

DEFINITION 8.15. Let E be a subset of $\mathbf{R}^3$ and $F = (P, Q, R) : E \to \mathbf{R}^3$ be $\mathcal{C}^1$ on E. The *curl* of F is

$$\operatorname{curl} F = \left(\frac{\partial R}{\partial y} - \frac{\partial Q}{\partial z}, \frac{\partial P}{\partial z} - \frac{\partial R}{\partial x}, \frac{\partial Q}{\partial x} - \frac{\partial P}{\partial y} \right),$$

and the *divergence* of F is

$$\operatorname{div} F = \frac{\partial P}{\partial x} + \frac{\partial Q}{\partial y} + \frac{\partial R}{\partial z}.$$

Notice that if $F = (P, Q, 0)$, then $\operatorname{curl} F = (0, 0, Q_x - P_y)$ is the derivative used for Green's Theorem.

These derivatives take on a more easily remembered form using the notation

$$\nabla = (\frac{\partial}{\partial x}, \frac{\partial}{\partial y}, \frac{\partial}{\partial z}).$$

Indeed, $\operatorname{curl} F = \nabla \times F$ and $\operatorname{div} F = \nabla \cdot F$.

If E is a three-dimensional region whose topological boundary is a piecewise smooth orientable surface, then the *positive orientation* on ∂E is determined by the unit normal which points away from E^o. If E is convex, this means that $\boldsymbol{n}$ points outward. This is not the case, however, when E has interior "bubbles." For example, if $E = \{\boldsymbol{x} : a < \|\boldsymbol{x}\| < b\}$ for some $a > 0$, then $\boldsymbol{n}$ points away from the origin on $\{\boldsymbol{x} : \|\boldsymbol{x}\| = b\}$ but toward the origin on $\{\boldsymbol{x} : \|\boldsymbol{x}\| = a\}$.

Our next fundamental theorem applies to the case when Ω is a three-dimensional region. This result is also called the *Divergence Theorem*.

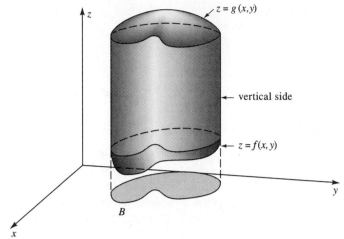

Figure 8.19

THEOREM 8.4 [GAUSS'S THEOREM]. *Let E be a three-dimensional region whose topological boundary ∂E is a piecewise smooth surface oriented positively. If $F : E \to \mathbf{R}^3$ is C^1 on E, then*

$$\iint_{\partial E} F \cdot \boldsymbol{n} \, d\sigma = \iiint_E \operatorname{div} F \, dV.$$

PROOF FOR SPECIAL REGIONS. Suppose for simplicity that E is a region of types I, II, and III. Let $F = (P, Q, R)$ and write the surface integral in differential form:

$$\iint_{\partial E} F \cdot \boldsymbol{n} \, d\sigma = \iint_{\partial E} P \, dy \, dz + \iint_{\partial E} Q \, dz \, dx + \iint_{\partial E} R \, dx \, dy = I_1 + I_2 + I_3.$$

We evaluate I_3 first.

Since E is of type I, there exist a two-dimensional region $B \subset \mathbf{R}^2$ and continuous functions $f, g : B \to \mathbf{R}$ such that

$$E = \{(x, y, z) \in \mathbf{R}^3 : (x, y) \in B, \ f(x, y) \le z \le g(x, y)\}.$$

Thus, ∂E has a top $z = g(x, y)$, a bottom $z = f(x, y)$, and (possibly) a vertical side (see Figure 8.19). Any normal to ∂E on the vertical side is parallel to the xy plane. Since $dx \, dy$ is the third component of a normal to ∂E, it must be zero on the vertical portion. Therefore, I_3 can be evaluated by integrating over the top and bottom of ∂E. Notice that by hypothesis, the unit normal on the bottom portion points downward and the unit normal on the top portion points upward. By using trivial parametrizations and Theorem 3.13 (the Fundamental Theorem of Calculus), we obtain

$$I_3 = \iint_{\partial E} R \, dx \, dy = \int_B \left(R(x, y, g(x, y)) - R(x, y, f(x, y)) \right) d(x, y)$$

$$= \int_B \int_{f(x,y)}^{g(x,y)} \frac{\partial R}{\partial z}(x, y, z) \, dz \, d(x, y) = \iiint_E \frac{\partial R}{\partial z} \, dV.$$

Similarly, since E is of type II,

$$I_2 = \iiint_E \frac{\partial Q}{\partial y}\, dV,$$

and since E is of type III,

$$I_1 = \iiint_E \frac{\partial P}{\partial x}\, dV.$$

Adding $I_1 + I_2 + I_3$ verifies the theorem. ∎

The hypothesis that E be of types I, II, and III was added to make the proof simple. For a proof of Gauss's Theorem as stated, see Theorem 10.8 and the reference which follows it. In the meantime, it is easy to check that Gauss's Theorem holds for any three-dimensional region E which can be divided into a finite number of regions E_j, each of which is of types I, II, and III. For example, if $E = E_1 \cup E_2$, then

$$\iiint_E \operatorname{div} F\, dV = \iiint_{E_1} \operatorname{div} F\, dV + \iiint_{E_2} \operatorname{div} F\, dV$$

$$= \iint_{\partial E} F \cdot \boldsymbol{n}\, d\sigma + \iint_{S \cap \partial E_1} F \cdot \boldsymbol{n}\, d\sigma + \iint_{S \cap \partial E_2} F \cdot \boldsymbol{n}\, d\sigma,$$

where S is the common surface between E_1 and E_2. Since E_1 and E_2 have outward pointing normals, the orientation of $S \cap \partial E_1$ is different from the orientation of $S \cap \partial E_2$ and the integrals over S cancel each other out.

The next two examples show that, like Green's Theorem, Gauss's Theorem can be used to avoid difficult integrals and tedious parametrizations.

Example 3. Use Theorem 8.4 to evaluate $\iint_S F \cdot \boldsymbol{n}\, d\sigma$, where S is the topological boundary of the solid

$$E = \{(x, y, z) : x^2 + y^2 \le z \le 1\},$$

$\boldsymbol{n}$ is the outward pointing normal, and $F(x, y, z) = (2x + z^2, x^5 + z^7, \cos(x^2) + \sin(y^3) - z^2)$.

SOLUTION. Since $\operatorname{div} F = 2 - 2z$, it follows from Gauss's Theorem that

$$\iint_S F \cdot \boldsymbol{n}\, d\sigma = \iiint_E (2 - 2z)\, dV = 2 \int_0^{2\pi} \int_0^1 \int_{r^2}^1 (1 - z) r\, dz\, dr\, d\theta = \frac{\pi}{3}. \quad \blacksquare$$

Example 4. Evaluate $\iint_{\partial Q} F \cdot \boldsymbol{n}\, d\sigma$, where Q is the unit cube $[0, 1] \times [0, 1] \times [0, 1]$, $\boldsymbol{n}$ is the outward pointing normal, and $F(x, y, z) = (2x - z, x^2 y, -xz^2)$.

SOLUTION. Since ∂Q has six sides, direct evaluation of this integral requires six separate integrals. However by Gauss's Theorem,

$$\iint_{\partial Q} F \cdot \boldsymbol{n}\, d\sigma = \int_0^1 \int_0^1 \int_0^1 (2 + x^2 - 2xz)\, dx\, dy\, dz = \frac{11}{6}. \quad \blacksquare$$

These definitions and results take on new meaning when examined in the context of fluid flow. When F represents the flow of an incompressible fluid near a point $\boldsymbol{a}$, $\operatorname{curl} F(\boldsymbol{a})$ measures the tendency of the fluid to swirl in a counterclockwise direction about $\boldsymbol{a}$ (see Exercise 6 in Section 8.6) and $\operatorname{div} F(\boldsymbol{a})$ measures the tendency of the fluid to spread out from $\boldsymbol{a}$ (see Exercise 7 below). (This explains the etymology of the words *curl* and *divergence*.) For example, if $F(x, y, z) = (x, y, z)$, then the fluid is not swirling at all but spreading straight out from the origin. Accordingly, $\operatorname{curl} F = \boldsymbol{0}$ and $\operatorname{div} F = 3$. On the other hand, if $G(x, y, z) = (y, z, x)$, then the fluid is swirling around in a circular motion about the origin. Accordingly, $\operatorname{curl} G = (-1, -1, -1)$ but $\operatorname{div} G = 0$. Note the minus signs in the components of $\operatorname{curl} G$. This fluid swirls about the origin in a clockwise direction so runs against counterclockwise motion.

When the fluid flows over a two-dimensional region $E \subset \mathbf{R}^2$, the integral of $F \cdot T \, ds$ over C represents the circulation of the fluid around C in the direction of T (see the comments following Example 5 in Section 8.2). Thus, Green's Theorem tells us that the circulation of a fluid around ∂E in the direction of the tangent is determined by how strongly the fluid swirls inside E. When F represents the flow of an incompressible fluid through a three-dimensional region $E \subset \mathbf{R}^3$ and $S = \partial E$, the integral $\iint_S F \cdot \boldsymbol{n}$ represents the flux of the fluid across the surface S (see the comments following Definition 8.14). Thus, Gauss's Theorem tells us that the flux of the fluid across $S = \partial E$ is determined by how strongly the fluid is spreading out inside E.

We close this section by admitting that the interpretations of curl and divergence given above are imperfect at best. For example, the vector field $F(x, y, z) = (0, z, 0)$ has curl $(-1, 0, 0)$. Here the fluid is shearing in layers with flow parallel to the xy plane in the direction of the positive y axis when $z > 0$. Although the fluid is not swirling, it does tend to rotate a stick placed in the fluid parallel to the z axis (for example, the line segment $\{(0, 1, z) : 0 \le z \le 1\}$) because more force is applied to the top than the bottom. This tendency toward rotation is reflected by the value of the curl. (Notice that the rotation is clockwise and the curl has a negative first component.)

EXERCISES

1. For each of the following, evaluate $\int_C F \cdot T \, ds$.
   a) C is the topological boundary of the two-dimensional region in the first quadrant bounded by $x = 0$, $y = 0$, and $y = \sqrt{4 - x^2}$, oriented in the counterclockwise direction, and $F(x, y) = (\sin(\sqrt{x^3 - x^2}), xy)$.
   b) C is the perimeter of the rectangle with vertices $(0, 0)$, $(2, 0)$, $(0, 3)$, $(2, 3)$, oriented in the counterclockwise direction, and $F(x, y) = (e^y, \log(x + 1))$.
   c) $C = C_1 \cup C_2$, where $C_1 = \partial B_1(0, 0)$ oriented in the counterclockwise direction, $C_2 = \partial B_2(0, 0)$ oriented in the clockwise direction, and $F(x, y) = (f(x^2 + y^2), xy^2)$, where f is a C^1 function on $[1, 2]$.

2. For each of the following, evaluate $\int_C \omega$.
   a) C is the topological boundary of the rectangle $[a, b] \times [c, d]$, oriented in the

counterclockwise direction, and $\omega = (f(x) + y)\, dx + xy\, dy$, where $f : [0, 1] \to$ **R** is any continuous function.

b) C is the topological boundary of the two-dimensional region bounded by $y = x^2$ and $y = x$, oriented in the clockwise direction, and $\omega = yf(x)\, dx + (x^2 + y^2)\, dy$, where $f : [0, 1] \to$ **R** is continuous and satisfies $\int_0^1 xf(x)\, dx = \int_0^1 x^2 f(x)\, dx$.

c) C is the topological boundary of a two-dimensional region E which satisfies the conclusion of Green's Theorem, oriented positively, and $\omega = e^x \sin y\, dy - e^x \cos y\, dx$.

3. For each of the following, evaluate $\iint_S F \cdot \boldsymbol{n}\, d\sigma$, where $\boldsymbol{n}$ is the outward pointing normal.

   a) S is the topological boundary of the rectangle $[0, 1] \times [0, 2] \times [0, 3]$ and $F(x, y, z) = (x + e^z, y + e^z, e^z)$.

   b) S is the truncated cylinder $x^2 + y^2 = 1$, $0 \le z \le 1$ together with the discs $x^2 + y^2 \le 1$, $z = 0, 1$, and $F(x, y, z) = (x^2, y^2, z^2)$.

   c) S is the topological boundary of E, where $E \subset$ **R**^{3} is bounded by $z = 2 - x^2$, $z = x^2$, $y = 0$, $z = y$, and $F(x, y, z) = (x + f(y, z), y + g(x, z), z + h(x, y))$ and $f, g, h :$ **R**$^2 \to$ **R** are continuous.

   d) S is the ellipsoid $x^2/a^2 + y^2/b^2 + z^2/c^2 = 1$ and $F(x, y, z) = (x|y|, y|z|, z|x|)$.

4. For each of the following, find $\int_S \omega$, where $\boldsymbol{n}$ is the outward pointing normal.

   a) S is the topological boundary of the three-dimensional region enclosed by $y = x^2$, $z = 0$, $z = 1$, $y = 4$, and $\omega = xyz\, dy\, dz + (x^2 + y^2 + z^2)\, dz\, dx + (x + y + z)\, dx\, dy$.

   b) S is the truncated hyperboloid of one sheet $x^2 - y^2 + z^2 = 1$, $0 \le y \le 1$, together with the discs $x^2 + z^2 \le 1$, $y = 0$, and $x^2 + z^2 \le 2$, $y = 1$, and $\omega = xy|z|\, dy\, dz + x^2|z|\, dz\, dx + (x^3 + y^3)\, dx\, dy$.

   c) S is the topological boundary of E, where $E \subset$ **R**^{3} is bounded by the surfaces $x^2 + y + z^2 = 4$ and $4x + y + 2z = 5$, and $\omega = (x + y^2 + z^2)\, dy\, dz + (x^2 + y + z^2)\, dz\, dx + (x^2 + y^2 + z)\, dx\, dy$.

5. a) Prove that if E is a Jordan region whose topological boundary is a smooth curve oriented in the counterclockwise direction, then

$$\text{Area}\, (E) = \frac{1}{2} \int_{\partial E} x\, dy - y\, dx.$$

   b) Find the area enclosed by the loop in the Folium of Descartes, i.e., by

$$\phi(t) = \left(\frac{3t}{1 + t^3}, \frac{3t^2}{1 + t^3} \right), \qquad t \in [0, \infty).$$

   c) Find an analogue of part a) for the volume of a Jordan region E in **R**^{3}.

   d) Compute the volume of the torus with radii $a > b$ (see Example 4 in Section 8.3).

6. a) Show Green's Theorem does not hold if continuity of P, Q is relaxed at one point in E. Hint: Consider $P = y/(x^2 + y^2)$, $Q = -x/(x^2 + y^2)$, and $E = B_1(0,0)$.

   b) Show that Gauss's Theorem does not hold if continuity of F is relaxed at one point in E.

7. **This exercise is used in Section 8.6.** Suppose V is an open set in $\mathbf{R}^3$ and $F : V \to \mathbf{R}^3$ is C^1. Prove that

$$\operatorname{div} F(\boldsymbol{x}_0) = \lim_{r \to 0} \frac{1}{\operatorname{Vol}(B_r(\boldsymbol{x}_0))} \iint_{\partial B_r(\boldsymbol{x}_0)} F \cdot \boldsymbol{n} \, d\sigma$$

for each $\boldsymbol{x}_0 \in V$, where $\boldsymbol{n}$ is the outward pointing normal of $B_r(\boldsymbol{x}_0)$.

8. Let $F, G : \mathbf{R}^3 \to \mathbf{R}^3$ and $f : \mathbf{R}^3 \to \mathbf{R}$ be differentiable. Prove the following analogues of the Sum and Product Rules for the "derivatives" curl and divergence.

   a) $$\nabla \times (F + G) = (\nabla \times F) + (\nabla \times G)$$

   b) $$\nabla \times (fF) = f(\nabla \times F) + (\nabla f \times F)$$

   c) $$\nabla \cdot (fF) = \nabla f \cdot F + f \cdot (\nabla \cdot F)$$

   d) $$\nabla \cdot (F + G) = \nabla \cdot F + \nabla \cdot G$$

   e) $$\nabla \cdot (F \times G) = (\nabla \times F) \cdot G - (\nabla \times G) \cdot F$$

9. **This exercise is used in Section 8.6.** Let $E \subset \mathbf{R}^2$. Recall that the *gradient* of a C^1 function $f : E \to \mathbf{R}$ is defined by

$$\operatorname{grad} f := \nabla f := (f_x, f_y).$$

   a) Suppose E and F satisfy the conclusion of Green's Theorem and $f : E \to \mathbf{R}$ is a C^2 function. If $F = \operatorname{grad} f$ on E, prove

   $$\int_{\partial E} F \cdot T \, ds = 0.$$

   b) Prove that if f and F are C^2 at $\boldsymbol{x}_0$, then $\operatorname{curl} \operatorname{grad} f(\boldsymbol{x}_0) = \mathbf{0}$ and $\operatorname{div} \operatorname{curl} F(\boldsymbol{x}_0) = 0$.

   c) Suppose E and F satisfy the conclusion of Gauss's Theorem and $f : E \to \mathbf{R}$ is a C^2 function. If $F = \operatorname{grad} f$ on E, prove

   $$\iint_{\partial E} fF \cdot \boldsymbol{n} \, d\sigma = \iiint_E F \cdot F \, dV.$$

10. Let E be a set in $\mathbf{R}^m$. For each $u : E \to \mathbf{R}$ which has second-order partial derivatives on E, *Laplace's equation* is defined by

$$\Delta u := \sum_{j=1}^{m} \frac{\partial^2 u}{\partial x_j^2}.$$

a) Show that if u is $\mathcal{C}^2$ on E, then $\Delta u = \nabla \cdot (\nabla u)$ on E.

b) [GREEN'S FIRST IDENTITY]. Show that if $E \subset \mathbf{R}^3$ satisfies the conclusion of Gauss's Theorem for all $\mathcal{C}^1$ functions F, then

$$\iiint_E (u\Delta v + \nabla u \cdot \nabla v)\, dV = \iint_{\partial E} u \nabla v \cdot \boldsymbol{n}\, d\sigma$$

for all $\mathcal{C}^2$ functions $u, v : E \to \mathbf{R}$.

c) [GREEN'S SECOND IDENTITY]. Show that if $E \subset \mathbf{R}^3$ satisfies the conclusion of Gauss's Theorem for all $\mathcal{C}^1$ functions F, then

$$\iiint_E (u\Delta v - v\Delta u)\, dV = \iint_{\partial E} (u\nabla v - v\nabla u) \cdot \boldsymbol{n}\, d\sigma$$

for all $\mathcal{C}^2$ functions $u, v : E \to \mathbf{R}$.

d) A function $u : E \to \mathbf{R}$ is said to be *harmonic* on E if u is $\mathcal{C}^2$ on E and $\Delta u(\boldsymbol{x}) = 0$ for all $\boldsymbol{x} \in E$. Suppose E is an open region in $\mathbf{R}^3$ which satisfies the conclusion of Gauss's Theorem for all $\mathcal{C}^1$ functions F. If u is harmonic on E, u is continuous on $\overline{E}$, and $u = 0$ on ∂E, prove $u = 0$ on $\overline{E}$.

e) Suppose V is open in $\mathbf{R}^2$, u is $\mathcal{C}^2$ on V, and u is continuous on $\overline{V}$. Prove that u is harmonic on V if and only if

$$\int_{\partial E} (u_x\, dy - u_y\, dx) = 0$$

for all two-dimensional regions $E \subset V$ which satisfy the conclusion of Green's Theorem for all $\mathcal{C}^1$ functions $F = (P, Q)$.

8.6 STOKES'S THEOREM

Our final fundamental theorem applies to surfaces in $\mathbf{R}^3$ whose boundaries are curves.

THEOREM 8.5 [STOKES'S THEOREM]. *Let S be an oriented $\mathcal{C}^2$ surface in $\mathbf{R}^3$ with unit normal $\boldsymbol{n}$. If the boundary ∂S is a piecewise smooth curve oriented positively and $F : S \to \mathbf{R}^3$ is $\mathcal{C}^1$, then*

$$\int_{\partial S} F \cdot T\, ds = \iint_S \operatorname{curl} F \cdot \boldsymbol{n}\, d\sigma.$$

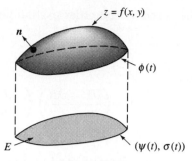

$z = f(x, y)$

n

$\phi(t)$

E

$(\psi(t), \sigma(t))$

Figure 8.20

PROOF FOR CERTAIN EXPLICIT SURFACES. Suppose for simplicity that S is an explicit $\mathcal{C}^2$ surface which lies over E, a two-dimensional region which satisfies the conclusion of Green's Theorem for all $\mathcal{C}^1$ functions F. Let $F = (P, Q, R)$ and write the line integral in differential notation:

$$\int_{\partial S} F \cdot T \, ds = \int_{\partial S} P \, dx + Q \, dy + R \, dz.$$

Without loss of generality, suppose S is determined by $z = f(x, y)$, $(x, y) \in E$, where $f : E \to \mathbf{R}$ is a $\mathcal{C}^2$ function and S is oriented with the upward pointing normal. Thus, $\boldsymbol{n} = N/\|N\|$, where $N = (-f_x, -f_y, 1)$.

Let $(\psi(t), \sigma(t))$, $t \in [a, b]$, be a piecewise smooth parametrization of ∂E oriented in the counterclockwise direction. Then

$$\phi(t) = (\psi(t), \sigma(t), f(\psi(t), \sigma(t))), \qquad t \in [a, b],$$

is a piecewise smooth parametrization of ∂S which is oriented positively (see Figure 8.20). If $x = \psi(t)$, $y = \sigma(t)$, and $z = f(\psi(t), \sigma(t))$, then $dx = \psi'(t) \, dt$, $dy = \sigma'(t) \, dt$, and

$$dz = \frac{\partial z}{\partial x} \, dx + \frac{\partial z}{\partial y} \, dy.$$

Thus, by definition,

(8) $$\int_{\partial S} P \, dx + Q \, dy + R \, dz = \int_{\partial E} (P + R\frac{\partial z}{\partial x}) \, dx + (Q + R\frac{\partial z}{\partial y}) \, dy.$$

We shall apply Green's Theorem to this last integral. By the Chain Rule and the Product Rule,

$$\frac{\partial}{\partial x}(Q + R\frac{\partial z}{\partial y}) = \frac{\partial Q}{\partial x} + \frac{\partial Q}{\partial z}\frac{\partial z}{\partial x} + \frac{\partial R}{\partial x}\frac{\partial z}{\partial y} + \frac{\partial R}{\partial z}\frac{\partial z}{\partial x}\frac{\partial z}{\partial y} + R\frac{\partial^2 z}{\partial x \partial y}$$

and

$$\frac{\partial}{\partial y}(P + R\frac{\partial z}{\partial x}) = \frac{\partial P}{\partial y} + \frac{\partial P}{\partial z}\frac{\partial z}{\partial y} + \frac{\partial R}{\partial y}\frac{\partial z}{\partial x} + \frac{\partial R}{\partial z}\frac{\partial z}{\partial y}\frac{\partial z}{\partial x} + R\frac{\partial^2 z}{\partial y \partial x}.$$

Since $z = f(x, y)$ is C^2, the mixed second-order partial derivatives above are equal. Therefore,

$$\frac{\partial}{\partial x}(Q + R\frac{\partial z}{\partial y}) - \frac{\partial}{\partial y}(P + R\frac{\partial z}{\partial x})$$

$$= \left(\frac{\partial R}{\partial y} - \frac{\partial Q}{\partial z}\right)\left(-\frac{\partial z}{\partial x}\right) + \left(\frac{\partial P}{\partial z} - \frac{\partial R}{\partial x}\right)\left(-\frac{\partial z}{\partial y}\right) + \left(\frac{\partial Q}{\partial x} - \frac{\partial P}{\partial y}\right)$$

$$= \operatorname{curl} F \cdot N.$$

Hence, it follows from (8), Green's Theorem, and (7) that

$$\int_{\partial S} F \cdot T \, ds = \int_E \operatorname{curl} F \cdot N \, d(x, y) = \iint_S \operatorname{curl} F \cdot \boldsymbol{n} \, d\sigma. \quad \blacksquare$$

The hypothesis that S be an explicit surface over a "Green's region" was added to make the proof simple. For a proof of Stokes's Theorem as stated, see Theorem 10.8 and the reference which follows it. In the meantime, it is easy to check that Stokes's Theorem holds for any surface which can be divided into a finite number of such explicit surfaces. As before, the common boundaries appear twice, each time in a different orientation, hence, cancel each other out.

Stokes's Theorem can be used to replace complicated line integrals by simple surface integrals.

Example 1. Compute $\int_C F \cdot T \, ds$, where C is the circle $x^2 + z^2 = 1$, $y = 0$, oriented in the counterclockwise direction when viewed from far out the y axis, and $F(x, y, z) = (x^2 z + \sqrt{x^3 + x^2 + 2}, xy, xy + \sqrt{z^3 + z^2 + 2})$.

SOLUTION. Since $\operatorname{curl} F = (x, x^2 - y, y)$, using Stokes's Theorem is considerably easier than trying to integrate $F \cdot T \, ds$ directly. Let S be the disc $x^2 + z^2 \leq 1$, $y = 0$, and notice that $\partial S = C$. Since C is oriented in the counterclockwise direction, the normal to S must point toward the positive y axis, i.e., $\boldsymbol{n} = (0, 1, 0)$. Thus, $\operatorname{curl} F \cdot \boldsymbol{n} = x^2 - y = x^2$ on S and Stokes's Theorem implies

$$\int_C F \cdot T \, ds = \iint_S x^2 \, dA = \int_0^{2\pi} \int_0^1 r^3 \cos^2 \theta \, dr \, d\theta = \frac{\pi}{4}. \quad \blacksquare$$

In Example 1, we could have chosen any surface S whose boundary is C. Thus, Stokes's Theorem can also be used to replace complicated surface integrals by simpler ones.

Example 2. Find $\iint_S \operatorname{curl} F \cdot \boldsymbol{n} \, d\sigma$, where S is the semi-ellipsoid $9x^2 + 4y^2 + 36z^2 = 36$, $z \geq 0$, $\boldsymbol{n}$ is the upward pointing normal, and

$$F(x, y, z) = (\cos x \sin z + xy, x^3, e^{x^2+z^2} - e^{y^2+z^2} + \tan(xy)).$$

SOLUTION. Let $C = \partial S$. The integral of $\operatorname{curl} F \cdot \boldsymbol{n} \, d\sigma$ over S and the integral of $F \cdot T \, ds$ over C are both complicated. But by Stokes's Theorem, the integral of $F \cdot T\,ds$ over C is the same as the integral of curl $F \cdot \boldsymbol{n}\,d\sigma$ over any surface E, provided $\partial E = C$. Let E be the two-dimensional region $9x^2 + 4y^2 \leq 36$. On E, $\boldsymbol{n} = (0,0,1)$. Thus, we need only compute the third component of curl F, which is

$$\frac{\partial}{\partial x}(x^3) - \frac{\partial}{\partial y}(\cos x \sin z + xy) = 3x^2 - x.$$

Therefore,

$$\iint_S \text{curl } F \cdot \boldsymbol{n}\,d\sigma = \int_E (3x^2 - x)\,d(x,y).$$

Let $x = 2r\cos\theta$ and $y = 3r\sin\theta$. By a change of variables,

$$\int_E (3x^2 - x)\,d(x,y) = \int_0^{2\pi}\int_0^1 (12r^2\cos^2\theta - 2r\cos\theta)6r\,dr\,d\theta = 18\pi. \quad \blacksquare$$

Stokes's Theorem can also be used to replace complicated surface integrals by simple line integrals.

Example 3. Let S be the union of the truncated paraboloid $z = x^2 + y^2$, $0 \leq z \leq 1$, and the truncated cylinder $x^2 + y^2 = 1$, $1 \leq z \leq 3$. Compute

$$\iint_S F \cdot \boldsymbol{n}\,d\sigma,$$

where $\boldsymbol{n}$ is the outward pointing normal and $F(x,y,z) = (x + z^2, 0, -z - 3)$.

SOLUTION. The boundary of S is $x^2 + y^2 = 1$, $z = 3$. To use Stokes's Theorem, we must find a function $G = (P,Q,R) : S \to \mathbf{R}^3$ such that

$$\text{curl } G = F,$$

i.e., such that

(9)
$$\frac{\partial R}{\partial y} - \frac{\partial Q}{\partial z} = x + z^2,$$

(10)
$$\frac{\partial P}{\partial z} - \frac{\partial R}{\partial x} = 0,$$

and

(11)
$$\frac{\partial Q}{\partial x} - \frac{\partial P}{\partial y} = -z - 3.$$

Starting with (9), set

(12)
$$\frac{\partial Q}{\partial z} = -x \quad \text{and} \quad \frac{\partial R}{\partial y} = z^2.$$

The left side of (12) implies $Q = -xz + g(x, y)$ for some $g : \mathbf{R}^2 \to \mathbf{R}$. Similarly, the right side of (12) leads to $R = z^2 y + h(x, z)$ for some $h : \mathbf{R}^2 \to \mathbf{R}$. Thus, $Q_x = -z + g_x$ will solve (11) if we set $g = 0$ and $P_y = 3$, i.e., $P = 3y + \sigma(x, z)$ for some $\sigma : \mathbf{R}^2 \to \mathbf{R}$. Hence, $P_z - R_x = \sigma_z - h_x$ will satisfy (10) if $\sigma = h = 0$. Therefore, $P = 3y$, $Q = -xz$ and $R = yz^2$, i.e., $G = (3y, -xz, yz^2)$.

Parametrize ∂S by $\phi(t) = (\sin t, \cos t, 3)$, $t \in [0, 2\pi]$, and observe that

$$(G \circ \phi) \cdot \phi' = (3 \cos t, -3 \sin t, 9 \cos t) \cdot (\cos t, -\sin t, 0) = 3 \cos^2 t + 3 \sin^2 t = 3.$$

Consequently, Stokes's Theorem implies

$$\iint_S F \cdot \mathbf{n} \, d\sigma = \iint_S \operatorname{curl} G \cdot \mathbf{n} \, d\sigma = \int_{\partial S} G \cdot T \, ds = \int_0^{2\pi} 3 \, dt = 6\pi. \quad \blacksquare$$

The solution to Example 3 involved finding a function G which satisfied $\operatorname{curl} G = F$. This function is not unique. Indeed, we could have begun with

$$\frac{\partial Q}{\partial z} = -z^2 \quad \text{and} \quad \frac{\partial R}{\partial y} = x$$

instead of (12). This leads to a different solution:

$$\widetilde{G}(x, y, z) = (zy, -(3x + z^3/3), xy).$$

The technique used to solve Example 3, however, is perfectly valid. Indeed, by Stokes's Theorem the value of the oriented line integral of $G \cdot T$ will be the same for all G which satisfy $\operatorname{curl} G = F$.

This technique will work only when the system of partial differential equations $\operatorname{curl} G = F$ has a solution G. To avoid searching for something which does not exist, we must be able to discern beforehand whether such a solution exists. To discover how this can be done, suppose G is a C^2 function which satisfies $\operatorname{curl} G = F$ on some set E. Then $\operatorname{div} F = 0$ on E by Exercise 9b in Section 8.5. Thus, the condition $\operatorname{div} F = 0$ is necessary for existence of a solution G to $\operatorname{curl} G = F$. The following result shows that if E is nice enough, this condition is also sufficient (see also Theorem 10.9).

THEOREM 8.6. *Let Ω be a ball centered at $(0, 0, 0)$ or a rectangle containing $(0, 0, 0)$, and $F : \Omega \to \mathbf{R}^3$ be C^1 on Ω. Then the following three statements are equivalent.*

    i) *There is a C^2 function $G : \Omega \to \mathbf{R}^3$ such that $\operatorname{curl} G = F$ on Ω.*

    ii) *If F, E, and $S = \partial E$ satisfy the conclusion of Gauss's Theorem and $E \subset \Omega$, then*

(13) $$\iint_S F \cdot \mathbf{n} \, d\sigma = 0,$$

    iii) *The identity $\operatorname{div} F = 0$ holds everywhere on Ω.*

PROOF. If i) holds, then $\operatorname{div} F = \operatorname{div}(\operatorname{curl} G) = 0$ since the first-order partial derivatives of G commute. Thus, (13) holds by Gauss's Theorem. (This works for any set Ω.)

If ii) holds, then by Gauss's Theorem and Exercise 7 in Section 8.5,

$$\operatorname{div} F(\boldsymbol{x}_0) = \lim_{r \to 0} \frac{1}{\operatorname{Vol}(B_r(\boldsymbol{x}_0))} \iiint_{B_r(\boldsymbol{x}_0)} \operatorname{div} F \, dV$$

$$= \lim_{r \to 0} \frac{1}{\operatorname{Vol}(B_r(\boldsymbol{x}_0))} \iint_{\partial B_r(\boldsymbol{x}_0)} F \cdot \boldsymbol{n} \, d\sigma = 0$$

for each $\boldsymbol{x}_0 \in \Omega^o$. (Recall by definition that any n-dimensional region has nonempty interior.) Since $\operatorname{div} F$ is continuous on Ω, it follows that $\operatorname{div} F = 0$ everywhere on Ω. (This works for any three-dimensional region Ω.)

Finally, suppose iii) holds. Let $F = (p, q, r)$ and suppose for simplicity that $G = (0, Q, R)$. If $\operatorname{curl} G = F$, then

$$(14) \qquad R_y - Q_z = p, \quad -R_x = q, \quad Q_x = r.$$

Integrating the last two identities, we have

$$R = -\int_0^x q(u, y, z) \, du + g(y, z) \quad \text{and} \quad Q = \int_0^x r(u, y, z) \, du + h(y, z)$$

for some $g, h : \mathbf{R}^2 \to \mathbf{R}$. (Note: By hypothesis, the line segment from $(0, y, z)$ to (x, y, z) is a subset of Ω so these integrals make sense.) Differentiating under the integral sign (Theorem 6.3), and using condition iii), the first identity becomes

$$p = R_y - Q_z = -\int_0^x (q_y(u, y, z) + r_z(u, y, z)) \, du + g_y - h_z$$

$$= \int_0^x p_x(u, y, z) \, du + g_y - h_z = p(x, y, z) - p(0, y, z) + g_y - h_z.$$

Thus, (14) can be solved by $g_y = p(0, y, z)$ and $h = 0$, i.e.,

$$Q = \int_0^x r(u, y, z) \, du \quad \text{and} \quad R = \int_0^y p(0, v, z) \, dv - \int_0^x q(u, y, z) \, du. \quad \blacksquare$$

We notice that Theorem 8.6 holds for any three-dimensional region Ω which satisfies the following property: given $(x, y, z) \in \Omega$, the line segments $L((0, y, z); (x, y, z))$ and $L((0, 0, z); (0, y, z))$ are both subsets of Ω.

The following result shows that without some restriction on Ω, Theorem 8.6 is false.

Remark 2. *Let* $\Omega = B_1(0, 0, 0) \setminus \{(0, 0, 0)\}$ *and*

$$F(x, y, z) = \left(\frac{x}{w^{3/2}}, \frac{y}{w^{3/2}}, \frac{z}{w^{3/2}} \right),$$

where $w = w(x, y, z) = x^2 + y^2 + z^2$. Then $\operatorname{div} F = 0$ on Ω but there is no G which satisfies $\operatorname{curl} G = F$.

PROOF. By definition

$$\operatorname{div} F = \frac{-2x^2 + y^2 + z^2}{w^{5/2}} + \frac{x^2 - 2y^2 + z^2}{w^{5/2}} + \frac{x^2 + y^2 - 2z^2}{w^{5/2}} = 0.$$

Let S represent the unit sphere $\partial B_1(0, 0, 0)$ oriented with the outward pointing normal, and suppose there is a G such that $\operatorname{curl} G = F$. On the one hand, since $F = (x, y, z) = \boldsymbol{n}$ on S implies $F \cdot \boldsymbol{n} = x^2 + y^2 + z^2 = 1$, we have

(15) $$\iint_S F \cdot \boldsymbol{n} \, d\sigma = \iint_S 1 \, dA = \sigma(S) = 4\pi.$$

On the other hand, dividing S into the upper hemisphere S_1 and the lower hemisphere S_2, we have by Stokes's Theorem that

(16) $$\iint_S F \cdot \boldsymbol{n} \, d\sigma = \iint_{S_1} F \cdot \boldsymbol{n} \, d\sigma + \iint_{S_2} F \cdot \boldsymbol{n} \, d\sigma$$
$$= \int_{\partial S_1} G \cdot T_1 \, ds + \int_{\partial S_2} G \cdot T_2 \, ds = 0.$$

This last step follows from the fact that $\partial S_1 = \partial S_2$ and $T_1 = -T_2$. Since (15) and (16) are incompatible, we conclude that there is no G which satisfies $\operatorname{curl} G = F$. ∎

EXERCISES

1. For each of the following, evaluate $\int_C F \cdot T \, ds$.

   a) C is the curve formed by intersecting the cylinder $x^2 + y^2 = 1$ with $z = -x$, oriented in the counterclockwise direction when viewed from high on the positive z axis and $F(x, y, z) = (xy^2, 0, xyz)$.

   b) C is the intersection of the cubic cylinder $z = y^3$ and the circular cylinder $x^2 + y^2 = 3$, oriented in the clockwise direction when viewed from high up the positive z axis, and $F(x, y, z) = (e^x + z, xy, ze^y)$.

2. For each of the following, evaluate $\iint_S \operatorname{curl} F \cdot \boldsymbol{n} \, d\sigma$.

   a) S is the "bottomless" surface in upper half space $z \geq 0$ bounded by $y = x^2$, $z = 1 - y$, $\boldsymbol{n}$ is the outward pointing normal, and $F(x, y, z) = (x \sin z^3, y \cos z^3, x^3 + y^3 + z^3)$.

   b) S is the truncated paraboloid $z = 3 - x^2 - y^2$, $z \geq 0$, $\boldsymbol{n}$ is the outward pointing normal, and $F(x, y, z) = (y, xyz, y)$.

   c) S is the hemisphere $z = \sqrt{10 - x^2 - y^2}$, $\boldsymbol{n}$ is the inward pointing normal, and $F(x, y, z) = (x, x, x^2 y^3 \log(z + 1))$.

   d) S is the "bottomless" tetrahedron in upper half space $z \geq 0$ bounded by $x = 0$, $y = 0$, $x + 2y + 3z = 1$, $z \geq 0$, $\boldsymbol{n}$ is the outward pointing normal, and $F(x, y, z) = (xy, yz, xz)$.

3. For each of the following, evaluate $\iint_S F \cdot \boldsymbol{n} \, d\sigma$ using Stokes's Theorem or Gauss's Theorem.

   a) S is the sphere $x^2 + y^2 + z^2 = 1$, $\boldsymbol{n}$ is the outward pointing normal, and $F(x, y, z) = (xz^2, x^2 y - z^3, 2xy + y^2 z)$.

   b) S is the portion of the plane $z = y$ which lies inside the ball $B_1(\boldsymbol{0})$, $\boldsymbol{n}$ is the upward pointing normal and $F(x, y, z) = (xy, xz, -yz)$.

   c) S is the truncated cone $y = 2\sqrt{x^2 + z^2}$, $2 \le y \le 4$, $\boldsymbol{n}$ is the outward pointing normal, and $F(x, y, z) = (x, -2y, z)$.

   d) S is a union of truncated paraboloids $z = 4 - x^2 - y^2$, $0 \le z \le 4$, and $z = x^2 + y^2 - 4$, $-4 \le z \le 0$, $\boldsymbol{n}$ is the outward pointing normal, and

   $$F(x, y, z) = (x + y^2 + \sin z, x + y^2 + \cos z, \cos x + \sin y + z).$$

   e) S is the union of three surfaces $z = x^2 + y^2$ $(0 \le z \le 2)$, $2 = x^2 + y^2$ $(2 \le z \le 5)$, and $z = 7 - x^2 - y^2$ $(5 \le z \le 6)$, $\boldsymbol{n}$ is the outward pointing normal, and $F(x, y, z) = (2y, 2z, 1)$.

4. For each of the following, evaluate $\int_S \omega$ using Stokes's Theorem or Gauss's Theorem.

   a) S is topological boundary of cylindrical solid $y^2 + z^2 \le 9$, $0 \le x \le 2$, with outward pointing normal, and $\omega = xy \, dy \, dz + (x^2 - z^2) \, dz \, dx + xz \, dx \, dy$.

   b) S is the truncated cylinder $x^2 + z^2 = 8$, $0 \le y \le 1$, with outward pointing normal, and $\omega = (x - 2z) \, dy \, dz - y \, dz \, dx$.

   c) S is the topological boundary of $R = [0, \pi/2] \times [0, 1] \times [0, 3]$, with outward pointing normal, and $\omega = e^y \cos x \, dy \, dz + x^2 z \, dz \, dx + (x + y + z) \, dx \, dy$.

   d) S is the intersection of the elliptic cylindrical solid $2x^2 + z^2 \le 1$ and the plane $x = y$, with normal which points toward the positive x axis, and $\omega = x \, dy \, dz - y \, dz \, dx + \sin y \, dx \, dy$.

5. Prove Green's Theorem is a corollary of Stokes's Theorem.

6. Let Π be a plane in $\mathbf{R}^3$ with unit normal $\boldsymbol{n}$ and $\boldsymbol{x}_0 \in \Pi$. For each $r > 0$ let S_r be the disc in Π centered at $\boldsymbol{x}_0$ of radius r, i.e., $S_r = B_r(\boldsymbol{x}_0) \cap \Pi$. Prove that if $F : B_1(\boldsymbol{x}_0) \to \mathbf{R}$ is C^1 and ∂S_r carries the orientation induced by $\boldsymbol{n}$, then

   $$\operatorname{curl} F(\boldsymbol{x}_0) \cdot \boldsymbol{n} = \lim_{r \to 0} \frac{1}{\sigma(S_r)} \int_{\partial S_r} F \cdot T \, ds.$$

7. Let S be an orientable surface with unit normal $\boldsymbol{n}$ and nonempty boundary ∂S which satisfies the conclusion of Stokes's Theorem for all C^1 functions F.

   a) Suppose $F : S \to \mathbf{R}^3 \setminus \{\boldsymbol{0}\}$ is C^1, ∂S is smooth, and T is the unit tangent vector on ∂S induced by $\boldsymbol{n}$. If the angle between $T(\boldsymbol{x}_0)$ and $F(\boldsymbol{x}_0)$ is never obtuse for any $\boldsymbol{x}_0 \in \partial S$, and $\iint_S \operatorname{curl} F \cdot \boldsymbol{n} \, d\sigma = 0$, prove that $T(\boldsymbol{x}_0)$ and $F(\boldsymbol{x}_0)$ are orthogonal for all $\boldsymbol{x}_0 \in \partial S$.

   b) If $F, F_k : S \to \mathbf{R}^3$ are C^1 and $F_k \to F$ uniformly on ∂S, prove that

   $$\lim_{k \to \infty} \iint_S \operatorname{curl} F_k \cdot \boldsymbol{n} \, d\sigma = \iint_S \operatorname{curl} F \cdot \boldsymbol{n} \, d\sigma.$$

8. Suppose E is a two-dimensional region such that if $(x, y) \in E$, then the line segments from $(0,0)$ to $(x,0)$ and from $(x,0)$ to (x,y) are both subsets of E. If $F : E \to \mathbf{R}^2$ is C^1, prove that the following three statements are equivalent.

a) $F = \nabla f$ on E for some $f : E \to \mathbf{R}$.

b) $F = (P, Q)$ is exact, i.e., $Q_x = P_y$ on E.

c) $\int_C F \cdot T \, ds = 0$ for all piecewise smooth curves $C = \partial\Omega$ oriented counterclockwise, where Ω is a two-dimensional region which satisfies the conclusion of Green's Theorem, and $\Omega \subset E$.

9. Let Ω be a three-dimensional region and $F : \Omega \to \mathbf{R}^3$ be C^1 on Ω. Suppose further that for each $(x, y, z) \in \Omega$, both the line segments $L((x, y, 0); (x, y, z))$ and $L((x, 0, 0); (x, y, 0))$ are subsets of Ω. Prove the following statements are equivalent.

a) There is a C^2 function $G : \Omega \to \mathbf{R}^3$ such that $\operatorname{curl} G = F$ on Ω.

b) If F, E, and $S = \partial E$ satisfy the conclusion of Gauss's Theorem and $E \subset \Omega$, then

$$\iint_S F \cdot n \, d\sigma = 0,$$

c) The identity $\operatorname{div} F = 0$ holds everywhere on Ω.

10. Let F be C^1 and exact on $\mathbf{R}^2 \setminus \{(0,0)\}$ (see Exercise 8b above).

a) Suppose C_1 and C_2 are disjoint smooth simple curves, oriented in the counterclockwise direction, and E is a two-dimensional region whose topological boundary ∂E is the union of the traces of C_1 and C_2. (Note: This means that E has a hole with one of the C_j's as the outer boundary and the other as the inner boundary.) If $(0,0) \notin E$, prove that

$$\int_{C_1} F \cdot T \, ds = \int_{C_2} F \cdot T \, ds.$$

b) Suppose E is a two-dimensional region which satisfies $(0,0) \in E^o$. If ∂E is a smooth simple curve oriented in the counterclockwise direction, and

$$F(x, y) = \left(\frac{-y}{x^2 + y^2}, \frac{x}{x^2 + y^2} \right),$$

compute $\int_{\partial E} F \cdot T \, ds$.

c) State and prove an analogue of part a) for functions $F : \mathbf{R}^3 \setminus \{(0,0,0)\}$, three-dimensional regions, and smooth simple surfaces.

Chapter 9

FOURIER SERIES

*e***9.1 INTRODUCTION** *This section uses no material from any other enrich-ment section.*

In Chapter 4 we studied power series and their partial sums, *classical* polynomials. In this chapter, we shall study the following objects.

DEFINITION 9.1. Let $a_k, b_k \in \mathbf{R}$ and N be a nonnegative integer.

i) A *trigonometric series* is a series of the form

$$\frac{a_0}{2} + \sum_{k=1}^{\infty}(a_k \cos kx + b_k \sin kx).$$

ii) A *trigonometric polynomial* of order N is a function $P : \mathbf{R} \to \mathbf{R}$ of the form

$$P(x) = \frac{a_0}{2} + \sum_{k=1}^{N}(a_k \cos kx + b_k \sin kx).$$

(Here, $\cos kx$ is shorthand for $\cos(kx)$ and $\sin kx$ is shorthand for $\sin(kx)$.)

Calculus was invented with the tacit assumption that power series provided a unified function theory; i.e., every function has a power series expansion (see Klein [5]). When Cauchy showed that this assumption was false (see Remark 2 in Section 4.7), mathematicians began to wonder whether some other type of series would provide a unified function theory. Euler (respectively, Fourier) had shown that the position of a vibrating string (respectively, the temperature along a metal rod) can be represented by trigonometric series. Thus, it was natural to ask: Does every function have a trigonometric series expansion? In this chapter we shall examine this question.

The following calculation will help answer this question.

Lemma [ORTHOGONALITY]. *Let k, j be nonnegative integers. Then*

i)
$$\int_{-\pi}^{\pi} \cos kx \cos jx \, dx = \begin{cases} 2\pi & k = j = 0 \\ \pi & k = j \neq 0 \\ 0 & k \neq j \end{cases}$$

ii)
$$\int_{-\pi}^{\pi} \sin kx \sin jx \, dx = \begin{cases} \pi & k = j \neq 0 \\ 0 & k \neq j \end{cases}$$

and

iii)
$$\int_{-\pi}^{\pi} \sin kx \cos jx \, dx = 0.$$

PROOF. Let

$$I = \int_{-\pi}^{\pi} \cos kx \cos jx \, dx.$$

If $k = j = 0$, then $I = \int_{-\pi}^{\pi} dx = 2\pi$. If $k = j \neq 0$, then by a half angle formula and elementary integration, we have

$$I = \int_{-\pi}^{\pi} \cos^2 kx \, dx = \frac{1}{2} \int_{-\pi}^{\pi} (1 + \cos 2kx) \, dx = \pi.$$

And if $k \neq j$, then by a sum angle formula and elementary integration, we have

$$I = \frac{1}{2} \int_{-\pi}^{\pi} (\cos(k+j)x + \cos(k-j)x) \, dx = 0.$$

This proves part i). Similar arguments prove parts ii) and iii). ∎

Notice once and for all that the question concerning representation of functions by trigonometric series has a built-in limitation. A function $f : \mathbf{R} \to \mathbf{R}$ is said to be *periodic* (of period 2π) if $f(x + 2\pi) = f(x)$ for all $x \in \mathbf{R}$. Since $\cos kx$ and $\sin kx$ are periodic, it is clear that every trigonometric polynomial is periodic. Therefore, any function which is the pointwise or uniform limit of a trigonometric series must also be periodic. For this reason, we will usually restrict our attention to the interval $[-\pi, \pi]$ and assume that $f(-\pi) = f(\pi)$.

The following definition, which introduces a special type of trigonometric series, plays a crucial role in the representation of periodic functions by trigonometric series.

DEFINITION 9.2. *Let f be integrable on $[-\pi, \pi]$ and N be a nonnegative integer.*

i) *The Fourier coefficients of f are the numbers*

$$a_k(f) = \frac{1}{\pi} \int_{-\pi}^{\pi} f(x) \cos kx \, dx, \qquad k = 0, 1, \ldots,$$

and

$$b_k(f) = \frac{1}{\pi} \int_{-\pi}^{\pi} f(x) \sin kx \, dx, \qquad k = 1, 2, \dots$$

ii) The *Fourier series* of f is the trigonometric series

$$(Sf)(x) = \frac{a_0(f)}{2} + \sum_{k=1}^{\infty} (a_k(f) \cos kx + b_k(f) \sin kx)$$

iii) The *partial sum* of Sf of order N is the trigonometric polynomial defined, for each $x \in \mathbf{R}$, by $(S_0 f)(x) = a_0(f)/2$ if $N = 0$, and

$$(S_N f)(x) = \frac{a_0(f)}{2} + \sum_{k=1}^{N} (a_k(f) \cos kx + b_k(f) \sin kx)$$

if $N \in \mathbf{N}$.

The following result shows why Fourier series play such an important role in the representation of periodic functions by trigonometric series.

THEOREM 9.1 [FOURIER]. *If a trigonometric series*

$$\frac{a_0}{2} + \sum_{k=1}^{\infty} (a_k \cos kx + b_k \sin kx)$$

converges uniformly on $\mathbf{R}$ *to a function* f, *then* S *is the Fourier series of* f, *i.e.,* $a_k = a_k(f)$ *for* $k = 0, 1, \dots$, *and* $b_k = b_k(f)$ *for* $k = 1, 2, \dots$

PROOF. Fix an integer $k \geq 0$. Since

$$f(x) = \frac{a_0}{2} + \sum_{j=1}^{\infty} (a_j \cos jx + b_j \sin jx)$$

converges uniformly and $\cos kx$ is bounded,

$$(1) \qquad f(x) \cos kx = \frac{a_0}{2} \cos kx + \sum_{j=1}^{\infty} (a_j \cos jx \cos kx + b_j \sin jx \cos kx)$$

also converges uniformly on $\mathbf{R}$. Since f is the uniform limit of continuous functions, f is continuous, hence, integrable on $[-\pi, \pi]$. Integrating (1) term by term and using orthogonality, we obtain

$$a_k(f) = \frac{1}{\pi} \int_{-\pi}^{\pi} f(x) \cos kx \, dx$$

$$= \frac{a_0}{2\pi} \int_{-\pi}^{\pi} \cos kx \, dx$$

$$+ \sum_{j=1}^{\infty} \left(\frac{a_j}{\pi} \int_{-\pi}^{\pi} \cos kx \cos jx \, dx + \frac{b_j}{\pi} \int_{-\pi}^{\pi} \cos kx \sin jx \, dx \right)$$

$$= a_k.$$

This verifies (5). A similar argument establishes (6). ∎

There are two central questions in the study of trigonometric series.

THE CONVERGENCE QUESTION. Given a function $f : \mathbf{R} \to \mathbf{R}$, periodic on $\mathbf{R}$ and integrable on $[-\pi, \pi]$, does the Fourier series of f converge to f?

THE UNIQUENESS QUESTION. If a trigonometric series S converges to some function f integrable on $[-\pi, \pi]$, is S the Fourier series of f?

We shall answer these questions for pointwise and uniform convergence when f is continuous and of bounded variation. We notice in passing that by Theorem 9.1, the answer to the Uniqueness Question is yes if uniform convergence is used.

The following special trigonometric polynomials come up naturally in connection with the Convergence Question (see Exercise 2).

DEFINITION 9.3. Let N be a nonnegative integer.

i) The *Dirichlet kernel* of order N is the function defined, for each $x \in \mathbf{R}$, by $D_0(x) = 1/2$ if $N = 0$, and

$$D_N(x) = \frac{1}{2} + \sum_{k=1}^{N} \cos kx$$

if $N \in \mathbf{N}$.

ii) The *Fejér kernel* of order N is the function defined, for each $x \in \mathbf{R}$, by $K_0(x) = 1/2$ if $N = 0$, and

$$(2) \qquad K_N(x) = \frac{1}{2} + \sum_{k=1}^{N} \left(1 - \frac{k}{N+1}\right) \cos kx$$

if $N \in \mathbf{N}$.

The following result shows that there is a simple relationship between Fejér kernels and Dirichlet kernels.

Remark 1. *If N is a nonnegative integer, then*

$$K_N(x) = \frac{D_0(x) + \cdots + D_N(x)}{N+1}$$

for all $x \in \mathbf{R}$.

PROOF. The identity is trivial if $N = 0$. To prove the identity for $N \in \mathbf{N}$, fix $x \in \mathbf{R}$. By definition,

$$K_N(x) = \frac{1}{N+1} \left(\frac{1}{2} + \frac{N}{2} + \sum_{k=1}^{N} (N - k + 1) \cos kx \right)$$

$$= \frac{1}{N+1} \left(\frac{1}{2} + \sum_{j=1}^{N} \left(\frac{1}{2} + \sum_{k=1}^{j} \cos kx \right) \right)$$

$$= \frac{D_0(x) + \cdots + D_N(x)}{N+1}. \quad \blacksquare$$

The next result shows that Dirichlet and Fejér kernels can be represented by quotients of trigonometric functions.

THEOREM 9.2. *If $x \in \mathbf{R}$ cannot be written in the form $2k\pi$ for any $k \in \mathbf{Z}$, then*

(3)
$$D_N(x) = \frac{\sin\left(N + \frac{1}{2}\right)x}{2\sin\frac{x}{2}}$$

and

(4)
$$K_N(x) = \frac{2}{N+1}\left(\frac{\sin\left(\frac{N+1}{2}\right)x}{2\sin\frac{x}{2}}\right)^2$$

for $N = 0, 1, \ldots$

PROOF. The formulas are trivial for $N = 0$. Fix $N \in \mathbf{N}$. Applying a sum angle formula and telescoping, we have

$$(D_N(x) - \frac{1}{2})\sin\frac{x}{2} = \sum_{k=1}^{N}\cos kx \sin\frac{x}{2}$$

$$= \frac{1}{2}\sum_{k=1}^{N}(\sin(k + \frac{1}{2})x - \sin(k - \frac{1}{2})x)$$

$$= \frac{1}{2}(\sin(N + \frac{1}{2})x - \sin\frac{x}{2}).$$

Solving for $D_N(x)$, we verify (3).

Let $k \in \mathbf{N}$. By (3) and another sum angle formula,

$$D_k(x)\sin^2\frac{x}{2} = \frac{1}{2}\sin\frac{x}{2}\sin(k + \frac{1}{2})x = \frac{1}{4}(\cos kx - \cos(k+1)x).$$

This identity also holds for $k = 0$. Applying Remark 1 and telescoping, we have

$$(N+1)K_N(x)\sin^2\frac{x}{2} = \sum_{k=0}^{N}D_k(x)\sin^2\frac{x}{2}$$

$$= \frac{1}{4}\sum_{k=0}^{N}(\cos kx - \cos(k+1)x)$$

$$= \frac{1}{4}(1 - \cos(N+1)x) = \frac{1}{2}\sin^2\left(\frac{N+1}{2}\right)x.$$

Solving for $K_N(x)$, we verify (4). ∎

These identities will be used in the next section to obtain a partial answer to the Convergence Question.

The next two examples illustrate the general principle that the Fourier coefficients of many common functions can be computed using integration by parts.

Example 1. Prove that the Fourier series of $f(x) = x$ is

$$2 \sum_{k=1}^{\infty} \frac{(-1)^{k+1}}{k} \sin kx.$$

PROOF. Since $x \cos kx$ is odd and $x \sin kx$ is even, it is easy to see that $a_k(f) = 0$ for $k = 0, 1, \ldots$, and

$$b_k(f) = \frac{2}{\pi} \int_0^{\pi} x \sin kx \, dx$$

for $k = 1, 2, \ldots$ Integrating by parts, we conclude that

$$b_k(f) = \frac{2}{\pi} \left(-\frac{x \cos kx}{k} \Big|_0^{\pi} + \frac{1}{k} \int_0^{\pi} \cos kx \, dx \right) = \frac{2(-1)^{k+1}}{k}. \quad \blacksquare$$

Example 2. Prove that the Fourier series of $f(x) = |x|$ is

$$\frac{\pi}{2} - \frac{4}{\pi} \sum_{k=1}^{\infty} \frac{\cos(2k-1)x}{(2k-1)^2}.$$

PROOF. Since $|x| \cos kx$ is even and $|x| \sin kx$ is odd, it is easy to see that $b_k(f) = 0$ for $k = 1, 2 \ldots$, and

$$a_k(f) = \frac{2}{\pi} \int_0^{\pi} x \cos kx \, dx$$

for $k = 0, 1, \ldots$ If $k = 0$, then

$$a_k(f) = \frac{2}{\pi} \left(\frac{\pi^2}{2} \right) = \pi,$$

i.e., $a_0(f)/2 = \pi/2$. If $k > 0$, then integration by parts yields

$$a_k(f) = \frac{2}{\pi k^2} (\cos k\pi - 1) = \begin{cases} 0 & \text{if } k \text{ is even,} \\ -\dfrac{4}{\pi k^2} & \text{if } k \text{ is odd.} \end{cases} \quad \blacksquare$$

EXERCISES

1. Compute the Fourier series of x^2 and of $\sin^2 x$.
2. Prove that if $f : \mathbf{R} \to \mathbf{R}$ is integrable on $[-\pi, \pi]$, then

$$(S_N f)(x) = \frac{1}{\pi} \int_{-\pi}^{\pi} f(t) D_N(x - t) \, dt$$

for all $x \in [-\pi, \pi]$ and $N \in \mathbf{N}$.

3. Show that if f, g are integrable on $[-\pi, \pi]$ and $\alpha \in \mathbf{R}$, then

$$a_k(f + g) = a_k(f) + a_k(g), \quad a_k(\alpha f) = \alpha a_k(f), \qquad k = 0, 1, \ldots$$

and

$$b_k(f + g) = b_k(f) + b_k(g), \quad b_k(\alpha f) = \alpha b_k(f), \qquad k = 1, 2, \ldots$$

4. Suppose $f : \mathbf{R} \to \mathbf{R}$ is differentiable and periodic and f' is integrable on $[-\pi, \pi]$. Prove that

$$a_k(f') = k b_k(f) \quad \text{and} \quad b_k(f') = -k a_k(f), \qquad k \in \mathbf{N}.$$

5. Suppose $f_N : [-\pi, \pi] \to \mathbf{R}$ are integrable and $f_N \to f$ uniformly on $[-\pi, \pi]$ as $N \to \infty$.

   a) Prove $a_k(f_N) \to a_k(f)$ and $b_k(f_N) \to b_k(f)$, as $N \to \infty$, uniformly in k.
   b) Show that a) holds under the weaker hypothesis

$$\lim_{N \to \infty} \int_{-\pi}^{\pi} |f(x) - f_N(x)| \, dx = 0.$$

6. Let

$$f(x) = \begin{cases} \dfrac{x}{|x|} & x \neq 0 \\ 0 & x = 0. \end{cases}$$

   a) Compute the Fourier coefficients of f.
   b) Prove that
$$(S_{2N} f)(x) = \frac{2}{\pi} \int_0^x \frac{\sin 2Nt}{\sin t} \, dt$$

   for $x \in [-\pi, \pi]$ and $N \in \mathbf{N}$.
   *c) [GIBBS'S PHENOMENON]. Prove that

$$\lim_{N \to \infty} (S_{2N} f)(\frac{\pi}{2N}) = \frac{2}{\pi} \int_0^\pi \frac{\sin t}{t} \, dt \approx 1.179$$

---

^{e}**9.2   SUMMABILITY OF FOURIER SERIES**   *This section uses material from Section 9.1.*

The Convergence Question posed in Section 9.1 is very difficult to answer, even for continuous functions. In this section we replace it with an easier question and show that the answer to this question is yes. Namely, we shall show that the Fourier series of any continuous periodic function f is uniformly summable to f. By summable, we mean the following concept.

DEFINITION 9.4. A series $\sum_{k=0}^{\infty} a_k$ with partial sums $s_N = \sum_{k=0}^{N} a_k$ is said to be *Cesàro summable* to L if its *Cesàro means*

$$\sigma_N := \frac{s_0 + \cdots + s_N}{N+1}$$

converge to L as $N \to \infty$.

The following result shows that summability is a generalization of convergence.

Remark 1. *If $\sum_{k=0}^{\infty} a_k$ converges to L, then it is Cesàro summable to L.*

PROOF. Let $\varepsilon > 0$. Choose $N_1 \in \mathbf{N}$ such that $k \geq N_1$ implies $|s_k - L| < \varepsilon/2$. Use the Archimedean Principle to choose $N_2 \in \mathbf{N}$ such that $N_2 > N_1$ and

$$\sum_{k=0}^{N_1} |s_k - L| < \frac{\varepsilon N_2}{2}.$$

If $N > N_2$, then

$$|\sigma_N - L| \leq \frac{1}{N+1} \sum_{k=0}^{N_1} |s_k - L| + \frac{1}{N+1} \sum_{k=N_1+1}^{N} |s_k - L|$$

$$\leq \frac{\varepsilon N_2}{2(N+1)} + \frac{\varepsilon}{2} \left(\frac{N - N_1}{N+1} \right) < \frac{\varepsilon}{2} + \frac{\varepsilon}{2} = \varepsilon. \ \blacksquare$$

The converse of Remark 1 is false. Indeed, although the series $\sum_{k=0}^{\infty} (-1)^k$ does not converge, its Cesàro means satisfy

$$\sigma_N = \begin{cases} \dfrac{N+2}{2(N+1)} & N \text{ is even} \\[2mm] \dfrac{1}{2} & N \text{ is odd,} \end{cases}$$

whence $\sigma_N \to 1/2$ as $N \to \infty$.

It is easier to show that a series is Cesàro summable than to show it converges. Thus, the following question is easier to answer than the Convergence Question.

THE SUMMABILITY QUESTION. Given a function $f : \mathbf{R} \to \mathbf{R}$, periodic on $\mathbf{R}$ and integrable on $[-\pi, \pi]$, is Sf Cesàro summable to f?

The Cesàro means of a Fourier series Sf are denoted by

$$(\sigma_N f)(x) := \frac{(S_0 f)(x) + \cdots + (S_N f)(x)}{N+1},$$

$N = 0, 1, \ldots$. The following result shows that the Cesàro means of a Fourier series can always be represented by an integral equation. This is important because it allows us to estimate the remainder $\sigma_N f - f$ using techniques of integration.

Lemma 1. *Let $f : \mathbf{R} \to \mathbf{R}$ be periodic on $\mathbf{R}$ and integrable on $[-\pi, \pi]$. Then*

$$(\sigma_N f)(x) = \frac{1}{\pi} \int_{-\pi}^{\pi} f(x - t) K_N(t)\, dt$$

for all $N = 0, 1, \ldots$ and all $x \in \mathbf{R}$.

PROOF. Fix $j, N \in \mathbf{N}$ and $x \in \mathbf{R}$. By definition and a sum angle formula,

$$a_j(f) \cos jx + b_j(f) \sin jx$$

$$= \frac{1}{\pi} \int_{-\pi}^{\pi} f(u) \cos ju \cos jx \, du + \frac{1}{\pi} \int_{-\pi}^{\pi} f(u) \sin ju \sin jx \, du$$

$$= \frac{1}{\pi} \int_{-\pi}^{\pi} f(u)(\cos ju \cos jx + \sin ju \sin jx) \, du$$

$$= \frac{1}{\pi} \int_{-\pi}^{\pi} f(u) \cos j(x - u) \, du.$$

Summing this identity over integers $j = 1, 2, \ldots, k$ and adding $a_0(f)/2$, we have

$$(S_k f)(x) = \frac{a_0(f)}{2} + \sum_{j=1}^{k} (a_j(f) \cos jx + b_j(f) \sin jx)$$

$$= \frac{1}{\pi} \int_{-\pi}^{\pi} f(u) \left(\frac{1}{2} + \sum_{j=1}^{k} \cos j(x - u) \right) du$$

$$= \frac{1}{\pi} \int_{-\pi}^{\pi} f(u) D_k(x - u) \, du$$

for $k = 0, 1, \ldots$. Making the change of variables $t = x - u$ and using the fact that both f and D_k are periodic, we obtain

$$(S_k f)(x) = \frac{1}{\pi} \int_{-\pi}^{\pi} f(x - t) D_k(t) \, dt, \qquad k = 0, 1, \ldots$$

We conclude by Remark 1 that

$$(\sigma_N f)(x) = \frac{1}{N+1} \sum_{k=0}^{N} (S_k f)(x)$$

$$= \frac{1}{N+1} \sum_{k=0}^{N} \frac{1}{\pi} \int_{-\pi}^{\pi} f(x - t) D_k(t) \, dt = \frac{1}{\pi} \int_{-\pi}^{\pi} f(x - t) K_N(t) \, dt. \ \blacksquare$$

To answer the Summability Question we need to know more about Fejér kernels. The following result shows that Fejér kernels satisfy some very nice properties.

Lemma 2. For each nonnegative integer N,

(5) $$K_N(t) \geq 0, \qquad t \in \mathbf{R},$$

and

(6) $$\frac{1}{\pi} \int_{-\pi}^{\pi} K_N(t)\, dt = 1.$$

Moreover, for each $0 < \delta < \pi$,

(7) $$\lim_{N \to \infty} \int_{\delta}^{\pi} |K_N(t)|\, dt = 0.$$

PROOF. Fix $N \geq 0$. If $t = 2j\pi$ for some $j \in \mathbf{Z}$, then $D_k(t) = k + 1/2 \geq 0$ for all $k \geq 0$, whence $K_N(t) \geq 0$. If $t \neq 2j\pi$ for any $j \in \mathbf{Z}$, then by Theorem 9.2,

$$K_N(t) = \frac{2}{N+1} \left(\frac{\sin(\frac{N+1}{2})t}{2\sin\frac{t}{2}} \right)^2 \geq 0.$$

This proves (5). By Definition 9.3 and orthogonality,

$$\int_{-\pi}^{\pi} K_N(t)\, dt = \int_{-\pi}^{\pi} \left(\frac{1}{2} + \sum_{k=1}^{N} \left(1 - \frac{k}{N+1} \right) \cos kt \right) dt = \pi.$$

This proves (6).

To prove (7), fix $0 < \delta < \pi$ and observe that $\sin\frac{t}{2} \geq \sin\frac{\delta}{2}$ for $t \in [\delta, \pi]$. Hence, it follows from Theorem 9.2 that

$$\int_{\delta}^{\pi} |K_N(t)|\, dt \leq \frac{2}{N+1} \int_{\delta}^{\pi} \left(\frac{\sin(\frac{N+1}{2})t}{2\sin\frac{\delta}{2}} \right)^2 dt \leq \frac{\pi}{2(N+1)\sin^2\frac{\delta}{2}}.$$

Since δ is fixed, this last expression tends to 0 as $N \to \infty$. ∎

Using these properties, we can answer the Summability Question for continuous functions (see also Exercise 7).

THEOREM 9.3 [FEJÉR]. *Suppose $f : \mathbf{R} \to \mathbf{R}$ is periodic on $\mathbf{R}$ and integrable on $[-\pi, \pi]$.*

    i) *If*

$$L = \lim_{h \to 0} \frac{f(x_0 + h) + f(x_0 - h)}{2}$$

    exists for some $x_0 \in \mathbf{R}$, then $(\sigma_N f)(x_0) \to L$ as $N \to \infty$.

    ii) *If f is continuous on some closed interval I, then $\sigma_N f \to f$ uniformly on I as $N \to \infty$.*

PROOF. Since f is periodic, we may suppose that $x_0 \in [-\pi, \pi]$. Fix $N \in \mathbf{N}$. By (6), Lemma 1, and a change of variables,

$$(8) \qquad (\sigma_N f)(x_0) - L = \frac{1}{\pi} \int_{-\pi}^{\pi} K_N(t)(f(x_0 - t) - L)\, dt$$

$$= \frac{2}{\pi} \int_0^{\pi} K_N(t) \left(\frac{f(x_0 + t) + f(x_0 - t)}{2} - L \right) dt$$

$$=: \frac{2}{\pi} \int_0^{\pi} K_N(t) F(x_0, t)\, dt.$$

Let $\varepsilon > 0$ and choose $0 < \delta < \pi$ such that $|t| < \delta$ implies $|F(x_0, t)| < \varepsilon/3$. By (5) and (6) we have

$$(9) \qquad \frac{2}{\pi} \int_0^{\delta} K_N(t)|F(x_0, t)|\, dt < \frac{2\varepsilon}{3\pi} \int_0^{\delta} K_N(t)\, dt \le \frac{2\varepsilon}{3}.$$

On the other hand, choose by (7) an $N_1 \in \mathbf{N}$ such that $N \ge N_1$ implies $\int_\delta^\pi K_N(t)\, dt < \varepsilon/3M$, where $M := \sup_{x \in \mathbf{R}} |f(x)|$. Then

$$\frac{2}{\pi} \int_\delta^\pi K_N(t)|F(x_0, t)|\, dt \le M \int_\delta^\pi K_N(t)\, dt < \frac{\varepsilon}{3},$$

and it follows from (8) and (9) that

$$(10) \quad |(\sigma_N f)(x_0) - L| \le \frac{2}{\pi} \int_0^{\delta} K_N(t)|F(x_0, t)|\, dt + \frac{2}{\pi} \int_\delta^\pi K_N(t)|F(x_0, t)|\, dt < \varepsilon$$

for all $N \ge N_1$. This proves part i).

To prove part ii), suppose f is continuous on some closed interval I. Since f is periodic, we may suppose that $I \subseteq [-\pi, \pi]$. Thus, I is closed and bounded, and f is uniformly continuous on I. Repeating the estimates above, we see that (10) holds uniformly for all $x_0 \in I$. ∎

COROLLARY 9.4. *If $f : \mathbf{R} \to \mathbf{R}$ is continuous and periodic, then $\sigma_N f$ converges to f uniformly on $\mathbf{R}$ as $N \to \infty$.*

COROLLARY 9.5 [COMPLETENESS]. *If $f : \mathbf{R} \to \mathbf{R}$ is continuous and periodic, and $a_{k-1}(f) = b_k(f) = 0$ for $k \in \mathbf{N}$, then $f(x) = 0$ for all $x \in \mathbf{R}$.*

PROOF. By hypothesis, $(\sigma_N f)(x) = 0$ for all $N \in \mathbf{N}$ and $x \in \mathbf{R}$. Hence, by Corollary 9.4, $f(x) = \lim_{N \to \infty} (\sigma_N f)(x) = 0$ for all $x \in \mathbf{R}$. ∎

COROLLARY 9.6. *Let $f : \mathbf{R} \to \mathbf{R}$ be continuous and periodic. Then there is a sequence of trigonometric polynomials $T_1, T_2, \ldots$ such that $T_N \to f$ uniformly on $\mathbf{R}$.*

PROOF. Set $T_N = \sigma_N f$ for $N \in \mathbf{N}$ and apply Corollary 9.4. ∎

This result can be used to prove the following density result for classical polynomials, i.e., polynomials of the form $P(x) = \sum_{k=0}^{n} c_k x^k$.

THEOREM 9.7 [WEIERSTRASS APPROXIMATION THEOREM]. *Let $[a, b]$ be a closed bounded interval and suppose $f : [a, b] \to \mathbf{R}$ is continuous on $[a, b]$. Given $\varepsilon > 0$ there is a (classical) polynomial P on $\mathbf{R}$ such that*

$$|f(x) - P(x)| < \varepsilon$$

for all $x \in [a, b]$.

PROOF. By considering $g(t) := f(a + (b - a)t/\pi)$, which is continuous on $[0, \pi]$, we may suppose that $a = 0$ and $b = \pi$.

Let $\varepsilon > 0$. Extend f from $[0, \pi]$ to $\mathbf{R}$ so that f is continuous and periodic. (For example, we could insist that the graph of $y = f(x)$ on $[\pi, 2\pi]$ is the chord from $(\pi, f(\pi))$ to $(2\pi, f(0))$ and then define $f(x + 2k\pi) := f(x)$ for $k \in \mathbf{Z}$.) By Corollary 9.6, there is a trigonometric polynomial T such that

$$|T(x) - f(x)| < \frac{\varepsilon}{2}$$

for $x \in \mathbf{R}$. Since each $\cos kx$ and $\sin kx$ is analytic on $\mathbf{R}$, so is T. Since analytic functions are uniform limits of their Taylor series, it follows that there is a polynomial P on $\mathbf{R}$ such that

$$|T(x) - P(x)| < \frac{\varepsilon}{2}$$

for $x \in [-\pi, \pi] \supseteq [a, b]$. We conclude that

$$|f(x) - P(x)| \le |f(x) - T(x)| + |T(x) - P(x)| < \varepsilon$$

for all $x \in [a, b]$. ∎

EXERCISES

1. Let $E \subseteq \mathbf{R}$ and suppose $f, f_k : \mathbf{R} \to \mathbf{R}$ are bounded functions. Prove that if $\sum_{k=0}^{\infty} f_k(x)$ converges to $f(x)$ uniformly on E, then

$$\sigma_N(x) := \sum_{k=0}^{N} \left(1 - \frac{k}{N+1}\right) f_k(x)$$

converges to $f(x)$ uniformly on E as $N \to \infty$.

2. If $f : \mathbf{R} \to \mathbf{R}$ is periodic on $\mathbf{R}$ and integrable on $[-\pi, \pi]$, prove that the Cesàro means of Sf are uniformly bounded, i.e., there is an $M > 0$ such that

$$|(\sigma_N f)(x)| \le M$$

for all $x \in \mathbf{R}$ and $N \in \mathbf{N}$.

3. Let

$$S = \frac{a_0}{2} + \sum_{k=1}^{\infty}(a_k \cos kx + b_k \sin kx)$$

be a trigonometric series and set

$$\sigma_N(x) = \frac{a_0}{2} + \sum_{k=1}^{N}\left(1 - \frac{k}{N+1}\right)(a_k \cos kx + b_k \sin kx)$$

for $x \in \mathbf{R}$ and $N \in \mathbf{N}$. Prove that S is the Fourier series of some continuous, periodic function $f : \mathbf{R} \to \mathbf{R}$ if and only if σ_N converges uniformly on $\mathbf{R}$, as $N \to \infty$.

4. Let f be integrable on $[-\pi, \pi]$ and $L \in \mathbf{R}$.

    a) Prove that if $(\sigma_N f)(x_0) \to L$ as $N \to \infty$ and if $(Sf)(x_0)$ converges, then $(S_N f)(x_0) \to L$.

    b) Prove that

$$\sin \sqrt{2}\pi + \sum_{k=1}^{\infty}\frac{4(-1)^k \sin \sqrt{2}\pi}{2 - k^2}\cos kx$$

    converges to $\sqrt{2}\pi \cos \sqrt{2}x$ uniformly on $\mathbf{R}$.

5. Suppose $f : [a, b] \to \mathbf{R}$ is continuous and

$$\int_a^b x^n f(x)\,dx = 0$$

for all integers $n \geq 0$.

    a) Evaluate $\int_a^b P(x)f(x)\,dx$ for any polynomial P on $\mathbf{R}$.

    b) Prove $\int_a^b |f(x)|^2\,dx = 0$.

    c) Show that $f(x) = 0$ for all $x \in [a, b]$.

6. [SUMMABILITY KERNELS]. Let $\phi_N : \mathbf{R} \to \mathbf{R}$ be a sequence of continuous, periodic functions on $\mathbf{R}$ which satisfy

$$\int_0^{2\pi} \phi_N(t)\,dt = 1 \quad \text{and} \quad \int_0^{2\pi} |\phi_N(t)|\,dt \leq M < \infty$$

for all $N \in \mathbf{N}$, and

$$\lim_{N\to\infty} \int_\delta^{2\pi - \delta} |\phi_N(t)|\,dt = 0$$

for each $0 < \delta < 2\pi$. Suppose $f : \mathbf{R} \to \mathbf{R}$ is continuous and periodic. Prove that

$$\lim_{N\to\infty} \int_0^{2\pi} f(x - t)\phi_N(t)\,dt = f(x)$$

uniformly for $x \in \mathbf{R}$.

*7. A sequence of functions $f_N : \mathbf{R} \to \mathbf{R}$ is said to converge *almost everywhere* to a function f if there is a set E of measure zero such that $f_N(x) \to f(x)$, as $N \to \infty$, for every $x \in \mathbf{R} \setminus E$. Suppose $f : \mathbf{R} \to \mathbf{R}$ is also periodic. Prove that if f is Riemann integrable on $[-\pi, \pi]$, then $\sigma_N f \to f$ almost everywhere as $N \to \infty$.

ᵉ9.3 GROWTH OF FOURIER COEFFICIENTS *This section uses material from Sections 3.5 and 9.2.*

By Theorem 9.3, a continuous periodic function f is completely determined by its Fourier coefficients. In this section we ask to what extent does smoothness of f affect the growth of these coefficients.

We begin with a computational result.

Lemma 1. *If $f : \mathbf{R} \to \mathbf{R}$ is integrable on $[-\pi, \pi]$ and N is a nonnegative integer, then*

$$(11) \qquad \frac{1}{\pi} \int_{-\pi}^{\pi} f(x)(S_N f)(x)\, dx = \frac{|a_0(f)|^2}{2} + \sum_{k=1}^{N} \left(|a_k(f)|^2 + |b_k(f)|^2 \right)$$

$$= \frac{1}{\pi} \int_{-\pi}^{\pi} |(S_N f)(x)|^2\, dx.$$

PROOF. Fix $N \geq 0$. Since f and $S_N f$ are integrable on $[-\pi, \pi]$, both integrals in (11) exist. By definition and orthogonality,

$$\frac{1}{\pi} \int_{-\pi}^{\pi} f(x) \frac{a_0(f)}{2}\, dx = \frac{|a_0(f)|^2}{2} = \frac{1}{\pi} \int_{-\pi}^{\pi} (S_N f)(x) \frac{a_0(f)}{2}\, dx.$$

Similarly,

$$\frac{1}{\pi} \int_{-\pi}^{\pi} f(x) a_k(f) \cos kx\, dx = |a_k(f)|^2 = \frac{1}{\pi} \int_{-\pi}^{\pi} (S_N f)(x) a_k(f) \cos kx\, dx$$

and

$$\frac{1}{\pi} \int_{-\pi}^{\pi} f(x) b_k(f) \sin kx\, dx = |b_k(f)|^2 = \frac{1}{\pi} \int_{-\pi}^{\pi} (S_N f)(x) b_k(f) \sin kx\, dx$$

for $k \in \mathbf{N}$. Adding these identities for $k = 0, \ldots, N$ verifies (11). ∎

Next, we use this result to identify a growth condition satisfied by the Fourier coefficients of any Riemann integrable function.

THEOREM 9.8 [BESSEL'S INEQUALITY]. *If $f : \mathbf{R} \to \mathbf{R}$ is (Riemann) integrable on $[-\pi, \pi]$, then $\sum_{k=1}^{\infty} |a_k(f)|^2$ and $\sum_{k=1}^{\infty} |b_k(f)|^2$ are convergent series. In fact,*

$$(12) \qquad \frac{|a_0(f)|^2}{2} + \sum_{k=1}^{\infty} \left(|a_k(f)|^2 + |b_k(f)|^2 \right) \leq \frac{1}{\pi} \int_{-\pi}^{\pi} |f(x)|^2\, dx.$$

PROOF. Fix $N \in \mathbf{N}$. By Lemma 1,

$$0 \le \frac{1}{\pi} \int_{-\pi}^{\pi} |f(x) - (S_N f)(x)|^2 \, dx$$

$$= \frac{1}{\pi} \int_{-\pi}^{\pi} |f(x)|^2 \, dx - \frac{2}{\pi} \int_{-\pi}^{\pi} f(x)(S_N f)(x) \, dx + \frac{1}{\pi} \int_{-\pi}^{\pi} |(S_N f)(x)|^2 \, dx$$

$$= \frac{1}{\pi} \int_{-\pi}^{\pi} |f(x)|^2 \, dx - \left(\frac{|a_0(f)|^2}{2} + \sum_{k=1}^{N} \left(|a_k(f)|^2 + |b_k(f)|^2 \right) \right).$$

Therefore,

$$\frac{|a_0(f)|^2}{2} + \sum_{k=1}^{N} \left(|a_k(f)|^2 + |b_k(f)|^2 \right) \le \frac{1}{\pi} \int_{-\pi}^{\pi} |f(x)|^2 \, dx$$

for all $N \in \mathbf{N}$. Taking the limit of this inequality as $N \to \infty$ verifies (12). Since $|f|^2$ is Riemann integrable when f is, it follows that both $\sum_{k=1}^{\infty} |a_k(f)|^2$ and $\sum_{k=1}^{\infty} |b_k(f)|^2$ are convergent series. ∎

COROLLARY 9.9 [RIEMANN–LEBESGUE LEMMA]. *If f is integrable on $[-\pi, \pi]$, then*

$$\lim_{k \to \infty} a_k(f) = \lim_{k \to \infty} b_k(f) = 0.$$

PROOF. Since the terms of a convergent series converge to zero, it follows from Bessel's inequality that $a_k(f)$ and $b_k(f)$ converge to zero as $k \to \infty$. ∎

Our next major result shows that Bessel's inequality is actually an identity when f is continuous and periodic. First, we show that the partial sums of the Fourier series of a function f are the best approximations to f in the following sense.

Lemma 2. *Let $N \in \mathbf{N}$. If f is (Riemann) integrable on $[-\pi, \pi]$ and*

$$T_N = \frac{c_0}{2} + \sum_{k=1}^{N} (c_k \cos kx + d_k \sin kx)$$

is any trigonometric polynomial of degree N, then

$$\int_{-\pi}^{\pi} |f(x) - (S_N f)(x)|^2 \, dx \le \int_{-\pi}^{\pi} |f(x) - T_N(x)|^2 \, dx.$$

PROOF. Notice by (11) that

$$\int_{-\pi}^{\pi} |f(x) - T_N(x)|^2 \, dx$$

$$= \int_{-\pi}^{\pi} |f(x) - (S_N f)(x) + (S_N f)(x) - T_N(x)|^2 \, dx$$

$$= \int_{-\pi}^{\pi} |f(x) - (S_N f)(x)|^2 \, dx$$

$$\quad + 2 \int_{-\pi}^{\pi} (f(x) - (S_N f)(x))((S_N f)(x) - T_N(x)) \, dx$$

$$\quad + \int_{-\pi}^{\pi} |(S_N f)(x) - T_N(x)|^2 \, dx$$

$$\geq \int_{-\pi}^{\pi} |f(x) - (S_N f)(x)|^2 \, dx + 2 \int_{-\pi}^{\pi} ((S_N f)(x) T_N(x) - f(x) T_N(x)) \, dx.$$

This last term is zero, since by orthogonality,

$$\frac{1}{\pi} \int_{-\pi}^{\pi} ((S_N f)(x) T_N(x) - f(x) T_N(x)) \, dx$$

$$= \frac{a_0(f) c_0}{4} + \sum_{k=1}^{N} (a_k(f) c_k + b_k(f) d_k)$$

$$\quad - \frac{c_0}{2\pi} \int_{-\pi}^{\pi} f(x) \, dx - \sum_{j=1}^{N} \frac{c_j}{\pi} \int_{-\pi}^{\pi} f(x) \cos jx \, dx$$

$$\quad - \sum_{j=1}^{N} \frac{d_j}{\pi} \int_{-\pi}^{\pi} f(x) \sin jx \, dx$$

$$= \frac{a_0(f) c_0}{4} + \sum_{k=1}^{N} (a_k(f) c_k + b_k(f) d_k)$$

$$\quad - \left(\frac{a_0(f) c_0}{4} + \sum_{k=1}^{N} a_k(f) c_k + b_k(f) d_k \right)$$

$$= 0.$$

Consequently,

$$\int_{-\pi}^{\pi} |f(x) - T_N(x)|^2 \, dx \geq \int_{-\pi}^{\pi} |f(x) - (S_N f)(x)|^2 \, dx. \quad \blacksquare$$

THEOREM 9.10 [PARSEVAL'S IDENTITY]. *If* $f : \mathbf{R} \to \mathbf{R}$ *is periodic and contin-uous, then*

$$(13) \qquad \frac{|a_0(f)|^2}{2} + \sum_{k=1}^{\infty} \left(|a_k(f)|^2 + |b_k(f)|^2 \right) = \frac{1}{\pi} \int_{-\pi}^{\pi} |f(x)|^2 \, dx.$$

PROOF. By Bessel's inequality, we need only show that the left side of (13) is greater than or equal to the right side of (13). Since f is continuous and periodic, $\sigma_N f \to f$ uniformly on $\mathbf{R}$ as $N \to \infty$ by Fejér's Theorem. Hence, it follows from Lemmas 1 and 2 that

$$\frac{1}{\pi} \int_{-\pi}^{\pi} |f(x)|^2 \, dx - \frac{|a_0(f)|^2}{2} - \sum_{k=1}^{N} \left(|a_k(f)|^2 + |b_k(f)|^2 \right)$$

$$= \frac{1}{\pi} \int_{-\pi}^{\pi} |f(x) - (S_N f)(x)|^2 \, dx \le \frac{1}{\pi} \int_{-\pi}^{\pi} |f(x) - (\sigma_N f)(x)|^2 \, dx \to 0$$

as $N \to \infty$. In particular,

$$\frac{1}{\pi} \int_{-\pi}^{\pi} |f(x)|^2 \, dx \le \frac{|a_0(f)|^2}{2} + \sum_{k=1}^{\infty} \left(|a_k(f)|^2 + |b_k(f)|^2 \right). \quad \blacksquare$$

The Riemann–Lebesgue Lemma can be improved if f is smooth and periodic. In fact, the following result shows that the smoother f is, the more rapidly its Fourier coefficients converge to zero.

THEOREM 9.11. *Let $f : \mathbf{R} \to \mathbf{R}$ and $j \in \mathbf{N}$. If $f^{(j)}$ exists and is integrable on $[-\pi, \pi]$ and $f^{(\ell)}$ is periodic for each $0 \le \ell < j$, then*

(14) $$\lim_{k \to \infty} k^j a_k(f) = \lim_{k \to \infty} k^j b_k(f) = 0.$$

PROOF. Fix $k \in \mathbf{N}$. Since f is periodic, integration by parts yields

$$a_k(f') = \frac{1}{\pi} \int_{-\pi}^{\pi} f'(x) \cos kx \, dx = \frac{k}{\pi} \int_{-\pi}^{\pi} f'(x) \sin kx \, dx = k b_k(f).$$

Similarly, $b_k(f') = -k a_k(f)$, hence $a_k(f'') = k b_k(f') = -k^2 a_k(f)$. Iterating, we obtain

$$|a_k(f^{(j)})| = \begin{cases} |k^j a_k(f)| & \text{when } j \text{ is even,} \\ |k^j b_k(f)| & \text{when } j \text{ is odd.} \end{cases}$$

A similar identity holds for $|b_k(f^{(j)})|$. Since the Riemann–Lebesgue Lemma implies $a_k(f^{(j)})$ and $b_k(f^{(j)}) \to 0$ as $k \to \infty$, it follows that $k^j a_k(f) \to 0$ and $k^j b_k(f) \to 0$ as $k \to \infty$. $\blacksquare$

This result shows that if f is continuously differentiable and periodic, then $k a_k(f)$ and $k b_k(f)$ both converge to zero as $k \to \infty$. Recall that if f is continuously differentiable on $[-\pi, \pi]$ then f is of bounded variation (see Remark 1 in Section 3.5). Thus, it is natural to ask: How rapidly do $k a_k(f)$ and $k b_k(f)$ grow when f is a function of bounded variation? To answer this question, let $\{x_0, x_1, \ldots, x_n\}$

be a partition of $[-\pi, \pi]$. Using Riemann sums, the Mean Value Theorem, Abel's Formula, and $\sin kx_0 = \sin kx_n = 0$, we can convince ourselves that

$$\pi a_k(f) = \int_{-\pi}^{\pi} f(x) \cos kx \, dx \approx \sum_{j=1}^{n} f(x_j) \cos kx_j (x_j - x_{j-1})$$

$$\approx \frac{1}{k} \sum_{j=1}^{n} f(x_j)(\sin kx_j - \sin kx_{j-1})$$

$$= \frac{1}{k} \sum_{j=1}^{n-1} (f(x_j) - f(x_{j+1})) \sin kx_j.$$

Since the absolute value of this last sum is bounded by $\mathrm{Var}\, f$, we guess that $k|a_k(f)| \le \mathrm{Var}\, f/\pi$.

To prove our guess is correct, suppose for a moment that f is increasing, periodic, and differentiable on $[-\pi, \pi]$, and $\phi(x) = \sin kx$. Then by Definition 9.2, periodicity, integration by parts, and the Fundamental Theorem of Calculus, we can estimate the Fourier coefficients of f as follows:

$$\pi k |a_k(f)| = \left| \int_{-\pi}^{\pi} f(x)\phi'(x) \, dx \right|$$

$$= \left| f(x)\phi(x) \, \Big|_{-\pi}^{\pi} - \int_{-\pi}^{\pi} f'(x)\phi(x) \, dx \right|$$

$$= \left| \int_{-\pi}^{\pi} f'(x)\phi(x) \, dx \right| \le \int_{-\pi}^{\pi} f'(x) \, dx$$

$$= \sum_{j=1}^{n} \int_{x_{j-1}}^{x_j} f'(x) \, dx = \sum_{j=1}^{n} f(x_j) - f(x_{j-1}) \le \mathrm{Var}\, f.$$

The following result shows that this estimate is valid even when f is neither differentiable nor increasing.

Lemma 3. *Suppose that f and ϕ are periodic, where f is of bounded variation on $[-\pi, \pi]$ and ϕ is continuously differentiable on $[-\pi, \pi]$. If $M := \sup_{x \in [-\pi, \pi]} |\phi(x)|$, then*

$$(15) \qquad \left| \int_{-\pi}^{\pi} f(x)\phi'(x) \, dx \right| \le M \mathrm{Var}\, f.$$

PROOF. Since f is of bounded variation and ϕ' is continuous on $[-\pi, \pi]$, the product $f\phi'$ is integrable on $[-\pi, \pi]$ (see Corollary 3.9 and the comments following Corollary 3.21).

Let $\varepsilon > 0$ and set $C = \sup_{x \in [-\pi, \pi]} |f(x)|$. Since ϕ' is uniformly continuous and $f\phi'$ is integrable on $[-\pi, \pi]$, choose a partition $P = \{x_0, x_1, \ldots, x_{2n}\}$ of $[-\pi, \pi]$ such that

$$(16) \qquad w, c \in [x_{j-1}, x_j] \quad \text{implies} \quad |\phi'(w) - \phi'(c)| < \frac{\varepsilon}{4\pi C}$$

and

(17)
$$\left| \sum_{j=1}^{2n} f(w_j)\phi'(w_j)(x_j - x_{j-1}) - \int_{-\pi}^{\pi} f(x)\phi'(x)\,dx \right| < \frac{\varepsilon}{2}$$

for any choice of $w_j \in [x_{j-1}, x_j]$.

Set

$$A := \sum_{j=1}^{2n} f(w_j)(\phi(x_j) - \phi(x_{j-1})),$$

where $w_j = x_j$ when j is even, $w_j = x_{j-1}$ when j is odd. By the Mean Value Theorem, choose $c_j \in [x_{j-1}, x_j]$ such that $\phi(x_j) - \phi(x_{j-1}) = \phi'(c_j)(x_j - x_{j-1})$. Then

$$A = \sum_{j=1}^{2n} f(w_j)\phi'(c_j)(x_j - x_{j-1}).$$

Hence, it follows from (17) and (16) that

$$\left| A - \int_{-\pi}^{\pi} f(x)\phi'(x)\,dx \right|$$

$$\leq \left| \sum_{j=1}^{2n} f(w_j)\phi'(c_j)(x_j - x_{j-1}) - \sum_{j=1}^{2n} f(w_j)\phi'(w_j)(x_j - x_{j-1}) \right|$$

$$+ \left| \sum_{j=1}^{2n} f(w_j)\phi'(w_j)(x_j - x_{j-1}) - \int_{-\pi}^{\pi} f(x)\phi'(x)\,dx \right|$$

$$< \sum_{j=1}^{2n} |f(w_j)|\,|\phi'(c_j) - \phi'(w_j)|(x_j - x_{j-1}) + \frac{\varepsilon}{2}$$

$$\leq \frac{\varepsilon}{4\pi} \sum_{j=1}^{2n} (x_j - x_{j-1}) + \frac{\varepsilon}{2} = \frac{\varepsilon}{2} + \frac{\varepsilon}{2} = \varepsilon.$$

Combining this observation with the Triangle Inequality, we obtain

(18)
$$\left| \int_{-\pi}^{\pi} f(x)\phi'(x)\,dx \right| \leq |A| + \varepsilon.$$

On the other hand, by the choice of the w_j's,

$$A = \sum_{j=1}^{n} f(x_{2j-2})(\phi(x_{2j-1}) - \phi(x_{2j-2})) + \sum_{j=1}^{n} f(x_{2j})(\phi(x_{2j}) - \phi(x_{2j-1}))$$

$$= \sum_{j=1}^{n} \phi(x_{2j-1})(f(x_{2j-2}) - f(x_{2j}))$$

$$+ \sum_{j=1}^{n} (f(x_{2j})\phi(x_{2j}) - f(x_{2j-2})\phi(x_{2j-2})).$$

Since f and ϕ are periodic, this last sum telescopes to 0. Therefore,

$$|A| = \left| \sum_{j=1}^{n} \phi(x_{2j-1})(f(x_{2j-2}) - f(x_{2j})) \right|$$

$$\leq \sum_{j=1}^{n} |\phi(x_{2j-1})| \, |f(x_{2j-2}) - f(x_{2j})| \leq M \operatorname{Var} f.$$

This, together with (18), proves

$$\left| \int_{-\pi}^{\pi} f(x)\phi'(x)\,dx \right| \leq M \operatorname{Var} f + \varepsilon.$$

Taking the limit of this inequality as $\varepsilon \to 0$, we conclude that (15) holds. ∎

We now estimate the rate of growth of Fourier coefficients of functions of bounded variation.

THEOREM 9.12. *If $f : \mathbf{R} \to \mathbf{R}$ is periodic and of bounded variation on $[-\pi, \pi]$, then*

$$|k a_k(f)| \leq \frac{\operatorname{Var} f}{\pi} \quad \text{and} \quad |k b_k(f)| \leq \frac{\operatorname{Var} f}{\pi}$$

for $k \in \mathbf{N}$.

PROOF. Fix $k \in \mathbf{N}$ and set $\phi(x) = \sin kx$. Then ϕ is periodic and $\phi'(x) = k \cos kx$ is continuously differentiable on $[0, 2\pi]$. Hence, it follows from Lemma 3 that

$$|k a_k(f)| = \left| \frac{1}{\pi} \int_{-\pi}^{\pi} f(x) k \cos kx \, dx \right| = \left| \frac{1}{\pi} \int_{-\pi}^{\pi} f(x)\phi'(x)\,dx \right| \leq \frac{\operatorname{Var} f}{\pi}.$$

A similar argument proves $|k b_k(f)| \leq \operatorname{Var} f / \pi$. ∎

EXERCISES

1. Prove that

$$\lim_{k \to \infty} \int_{-\pi}^{\pi} f(x) \sin(k + \alpha) x \, dx = 0$$

   for all f integrable on $[-\pi, \pi]$ and all $\alpha \in \mathbf{R}$.

2. Prove that there is no continuous function whose Fourier coefficients satisfy $|a_k(f)| \geq 1/\sqrt{k}$ for $k \in \mathbf{N}$.

3. Prove that if $f : \mathbf{R} \to \mathbf{R}$ belongs to $\mathcal{C}^2(\mathbf{R})$ and f, f' are both periodic, then Sf converges to f uniformly and absolutely on $\mathbf{R}$. (See also Exercise 5 in Section 9.4.)

4. If $f : \mathbf{R} \to \mathbf{R}$ belongs to $\mathcal{C}^\infty(\mathbf{R})$ and $f^{(j)}$ is periodic for all $j \geq 0$, prove that Sf is term-by-term differentiable on $\mathbf{R}$. In fact, show that

$$\frac{d^j f}{dx^j}(x) = \sum_{k=1}^{\infty} \frac{d^j}{dx^j}(a_k(f) \cos kx + b_k(f) \sin kx)$$

uniformly for all $j \in \mathbf{N}$.

5. Suppose $f : \mathbf{R} \to \mathbf{R}$ is periodic on $\mathbf{R}$, integrable on $[-\pi, \pi]$, and $a_k(f) \geq 0$ for $k = 0, 1, \ldots$

   a) Prove $(S_k f)(0) \geq (S_j f)(0)$ for all $k \geq j \geq 0$.
   b) Prove $S_N f(0) \leq 2\sigma_{2N} f(0)$ for $N \in \mathbf{N}$.
   c) Prove that $\sum_{k=1}^{\infty} |a_k(f)| < \infty$.
   d) Suppose f is also even. Prove that f must be continuous and Sf converges uniformly and absolutely on $\mathbf{R}$.

6. Suppose $f : \mathbf{R} \to \mathbf{R}$ is continuous and periodic. The *modulus of continuity* of f is defined by

$$\omega(f, \delta) = \sup_{\substack{t \in [0, 2\pi] \\ |h| \leq \delta}} |f(t + h) - f(t)|.$$

   a) Show that

$$a_k(f) = \frac{1}{2\pi} \int_{-\pi}^{\pi} \left(f(u) - f(u + \frac{\pi}{k}) \right) \cos ku \, du$$

   for $k \in \mathbf{N}$.
   b) Prove

$$|a_k(f)| \leq \omega(f, \frac{\pi}{k}) \quad \text{and} \quad |b_k(f)| \leq \omega(f, \frac{\pi}{k})$$

   for $k \in \mathbf{N}$.
   c) Use part b) to give a different proof the Riemann–Lebesgue Lemma in the special case when f is periodic and continuous.

^{e}9.4   CONVERGENCE OF FOURIER SERIES   *This section uses material from Sections 3.5, 9.2, and 9.3.*

We shall prove that under certain conditions, a summable series must also be convergent. Such results, called *Tauberian theorems*, will be used to obtain a partial answer to the Convergence Question posed in Section 9.1 and further results concerning the growth of Fourier coefficients.

The following result was the first Tauberian theorem discovered.

THEOREM 9.13 [TAUBER]. *Let $a_k \geq 0$ and $L \in \mathbf{R}$. If $\sum_{k=0}^{\infty} a_k$ is Cesàro summable to L, then*

$$\sum_{k=0}^{\infty} a_k = L.$$

PROOF. By Remark 1 in Section 9.2, it suffices to prove $\sum_{k=0}^{\infty} a_k < \infty$. Suppose to the contrary that $\sum_{k=0}^{\infty} a_k = \infty$. Then given $M > 0$ there is an $n_0 \in \mathbf{N}$ such that $n \geq n_0$ implies $s_n := \sum_{k=0}^{n} a_k \geq M$. Let $N > n_0$. Then

$$\sigma_N := \frac{s_0 + s_1 + \cdots + s_{n_0}}{N + 1} + \frac{s_{n_0+1} + \cdots + s_N}{N + 1} \geq 0 + \frac{N - n_0}{N + 1} M.$$

Taking the limit of this last inequality as $N \to \infty$, we obtain $L \geq M$ for all $M > 0$. We conclude that $L = \infty$, a contradiction. ∎

This result can be used to improve the Riemann–Lebesgue Lemma for certain types of functions.

COROLLARY 9.14. *Let $f : \mathbf{R} \to \mathbf{R}$ be periodic on $\mathbf{R}$ and integrable on $[-\pi, \pi]$. If $a_k(f) = 0$ and $b_k(f) \geq 0$ for $k \in \mathbf{N}$, then*

$$\sum_{k=1}^{\infty} \frac{b_k(f)}{k} < \infty.$$

PROOF. By considering $g = f - a_0(f)$ we may suppose that $a_0(f) = 0$. Let

$$F(x) = \int_0^x f(t)\, dt.$$

By Theorem 3.11, F is continuous on $\mathbf{R}$. Since $a_0(f) = 0$, F is also periodic. Hence, by Fejér's Theorem, $(\sigma_N F)(0) \to F(0) = 0$ as $N \to \infty$. Integrating by parts, we obtain

$$a_k(F) = \frac{b_k(f)}{k} \geq 0 \quad \text{and} \quad b_k(F) = -\frac{a_k(f)}{k} = 0.$$

It follows that $\sum_{k=1}^{\infty} b_k(f)/k$ is Cesàro summable (to $-a_0(F)/2$) and has nonnegative terms. We conclude by Tauber's Theorem that $\sum_{k=1}^{\infty} b_k(f)/k$ converges. ∎

It is now easy to see that the converse of the Riemann–Lebesgue Lemma is false. Indeed, if

$$\sum_{k=2}^{\infty} \frac{\sin kx}{\log k}$$

were the Fourier series of some integrable function, then by Corollary 9.14,

$$\sum_{k=2}^{\infty} \frac{1}{k \log k}$$

would converge, a contradiction of the Integral Test.

The following result is one of the deepest Tauberian theorems.

THEOREM 9.15 [HARDY]. *Let $E \subseteq \mathbf{R}$ and suppose $f_k : E \to \mathbf{R}$ is a sequence of functions which satisfies*

$$(19) \qquad\qquad |k f_k(x)| \leq M$$

for all $x \in E$, all $k \in \mathbf{N}$, and some $M > 0$. If $\sum_{k=0}^{\infty} f_k$ is uniformly Cesàro summable to f on E, then $\sum_{k=0}^{\infty} f_k$ converges uniformly to f on E.

PROOF. Fix $x \in E$ and suppose without loss of generality that $M \geq 1$. For each $n = 0, 1, \ldots,$ set

$$s_n(x) = \sum_{k=0}^{n} f_k(x), \qquad \sigma_n(x) = \frac{s_0(x) + \cdots + s_n(x)}{n+1},$$

and consider the delayed averages

$$\sigma_{n,k}(x) := \frac{s_n(x) + \cdots + s_{n+k}(x)}{k+1}$$

defined for $n, k \geq 0$.

Let $0 < \varepsilon < 1$. For each $n \in \mathbf{N}$ choose $k = k(n) \in \mathbf{N}$ such that $k \leq n\varepsilon/(2M) < k + 1$. Then

(20)
$$\frac{n-1}{k+1} < \frac{n}{k+1} < \frac{2M}{\varepsilon} < \infty.$$

Moreover, since

$$\sigma_{n,k}(x) - s_n(x) = \frac{(s_n(x) - s_n(x)) + \cdots + (s_{n+k}(x) - s_n(x))}{k+1}$$

$$= \sum_{j=n+1}^{n+k} \left(1 - \frac{j-n}{k+1}\right) f_j(x),$$

it follows from (19) and the choice of $k = k(n)$ that

(21)
$$|\sigma_{n,k}(x) - s_n(x)| \leq \sum_{j=n+1}^{n+k} |f_j(x)| \leq M \sum_{j=n+1}^{n+k} \frac{1}{j} < \frac{Mk}{n+1} < \frac{\varepsilon}{2}.$$

Since $\sigma_n \to f$ uniformly on E, choose $N \in \mathbf{N}$ such that

(22)
$$n \geq N \quad \text{and} \quad x \in E \quad \text{imply} \quad |\sigma_n(x) - f(x)| < \frac{\varepsilon^2}{12M}.$$

Since

$$\sigma_{n,k}(x) = \left(1 + \frac{n-1}{k+1}\right)\sigma_{n+k} - \left(\frac{n-1}{k+1}\right)\sigma_{n-1},$$

it follows (20), (21), and (22) that

$$|s_n(x) - f(x)| \leq |s_n(x) - \sigma_{n,k}(x)| + |\sigma_{n,k} - f(x)|$$

$$< \frac{\varepsilon}{2} + \left(1 + \frac{n-1}{k+1}\right)|\sigma_{n+k}(x) - f(x)|$$

$$+ \left(\frac{n-1}{k+1}\right)|\sigma_{n-1}(x) - f(x)|$$

$$< \frac{\varepsilon}{2} + \left(1 + \frac{2M}{\varepsilon}\right)\left(\frac{\varepsilon^2}{12M}\right) + \frac{2M}{\varepsilon}\left(\frac{\varepsilon^2}{12M}\right)$$

$$= \frac{\varepsilon}{2} + \frac{\varepsilon^2}{12M} + \frac{\varepsilon}{3} < \frac{\varepsilon}{2} + \frac{\varepsilon}{12} + \frac{\varepsilon}{3} < \varepsilon$$

for any $n > N$ and $x \in E$. We conclude that $s_n \to f$ uniformly on E as $n \to \infty$. ∎

We are prepared to answer the Convergence Question posed in Section 9.1 for piecewise continuous functions of bounded variation.

THEOREM 9.16 [DIRICHLET–JORDAN]. *If $f : \mathbf{R} \to \mathbf{R}$ is periodic on $\mathbf{R}$ and of bounded variation on $[-\pi, \pi]$, then*

$$\lim_{N \to \infty} (S_N f)(x) = \frac{f(x+) + f(x-)}{2}$$

for every $x \in \mathbf{R}$. If f is also continuous on some closed interval I, then

$$\lim_{N \to \infty} S_N f = f$$

uniformly on I.

PROOF. Since f is periodic and of bounded variation, the one-sided limits $f(x+)$ and $f(x-)$ exist for each $x \in \mathbf{R}$, and f is Riemann integrable on $[-\pi, \pi]$ (see the comments which follow the proof of Corollary 3.21). Hence, by Fejér's Theorem, both conclusions hold if S_N is replaced by σ_N. Since Theorem 9.12 implies

$$|ka_k(f) \cos kx| \quad \text{and} \quad |kb_k(f) \cos kx| \le \frac{\operatorname{Var} f}{\pi}$$

for $k \in \mathbf{N}$, it follows from Hardy's Theorem that both conclusions hold as stated. ∎

We close this section with an application of Fourier series to an extremal problem. We will show that among all smooth simple closed curves in $\mathbf{R}^2$ with a given arc length, the largest area is enclosed by a circle. (The proof presented here comes from Marsden [7].)

THEOREM 9.17 [THE ISOPERIMETRIC PROBLEM]. *Let E be a region in $\mathbf{R}^2$ whose topological boundary $C = \partial E$ is a smooth closed simple curve of length 2π. If $A = \operatorname{Area}(E)$, then $A \le \pi$. Moreover, $A = \pi$ if and only if $E = B_1(a, b)$ for some $a, b \in \mathbf{R}$.*

PROOF. Let $(\nu, [0, 2\pi])$ be the natural parametrization of C, i.e., $\|\nu'(s)\| = 1$ for all $s \in [0, 2\pi]$. Let

$$a = \frac{1}{2\pi} \int_0^{2\pi} \nu_1(s) \, ds, \qquad b = \frac{1}{2\pi} \int_0^{2\pi} \nu_2(s) \, ds,$$

$$P(s) = \nu_1(s) - a, \quad Q(s) = \nu_2(s) - b, \quad \text{and} \quad \phi(s) = (P(s), Q(s))$$

for $s \in [0, 2\pi]$. Clearly, $(\phi, [0, 2\pi])$ is a smooth parametrization of $\partial E - (a, b)$ whose trace is a smooth closed simple curve with arc length 2π which encloses a region whose area is A. Moreover,

$$(23) \qquad\qquad |P'(s)|^2 + |Q'(s)|^2 = 1,$$

$$(24) \qquad \frac{1}{2\pi} \int_0^{2\pi} P(s) \, ds = 0, \qquad \frac{1}{2\pi} \int_0^{2\pi} Q(s) \, ds = 0,$$

and by Green's Theorem,

(25)
$$A = \iint_E dA = \int_{\partial E} x \, dy = \int_0^{2\pi} P(s)Q'(s) \, ds.$$

Let a_k, b_k (respectively, c_k, d_k) represent the Fourier coefficients of P (respectively, Q). Since $(\phi, [0, 2\pi])$ is smooth and closed, P and Q are continuously differentiable and periodic. By (24) and the Dirichlet–Jordan Theorem,

(26) $$P(s) = \sum_{k=1}^\infty (a_k \cos ks + b_k \sin ks), \qquad Q(s) = \sum_{k=1}^\infty (c_k \cos ks + d_k \sin ks),$$

(27)
$$P'(s) = \sum_{k=1}^\infty (kb_k \cos ks - ka_k \sin ks), \quad \text{and} \quad Q'(s) = \sum_{k=1}^\infty (kd_k \cos ks - kc_k \sin ks)$$

uniformly on $[0, 2\pi]$. Hence, by (23) and Parseval's Identity,

$$2\pi = \int_0^{2\pi} (|P'(s)|^2 + |Q'(s)|^2) \, ds = \pi \sum_{k=1}^\infty k^2 (a_k^2 + b_k^2 + c_k^2 + d_k^2).$$

Moreover, by (25) and orthogonality

$$A = \int_0^{2\pi} P(s)Q'(s) \, ds = \pi \sum_{k=1}^\infty k(a_k d_k - b_k c_k).$$

It follows that

$$\pi - A = \frac{\pi}{2} \sum_{k=2}^\infty (k^2 - k)(a_k^2 + b_k^2 + c_k^2 + d_k^2) + \frac{\pi}{2} \sum_{k=1}^\infty k((a_k - d_k)^2 + (c_k + b_k)^2) \geq 0.$$

In particular, $A \leq \pi$ and $A = \pi$ if and only if $a_1 = d_1$, $c_1 = -b_1$, and $a_k = b_k = c_k = d_k = 0$ for $k \geq 2$.

Suppose $A = \pi$. Then $P(s) = a_1 \cos s + b_1 \sin s$ and $Q(s) = -b_1 \cos s + a_1 \sin s = -P(s + \frac{\pi}{2})$. Thus, $P'(s) = -Q(s)$ and $Q'(s) = -P''(s) = P(s)$ for all $s \in [0, 2\pi]$. It follows from (23) that $\phi([0, 2\pi])$ is a subset of $\partial B_1(0, 0)$. Since $\phi(0) = \phi(2\pi)$, we must have $\phi([0, 2\pi]) = \partial B_1(0, 0)$. Therefore, C is the boundary of the disc $E = B_1(a, b)$. ∎

EXERCISES

1. Suppose f is continuous and of bounded variation on $[-\pi, \pi]$. Prove that $S_N f \to f$ pointwise on $(-\pi, \pi)$ and uniformly on any $[a, b] \subset (-\pi, \pi)$.

2. a) Prove that

$$x = 2 \sum_{k=1}^{\infty} \frac{(-1)^{k+1}}{k} \sin kx$$

pointwise on $(-\pi, \pi)$ and uniformly on any $[a, b] \subset (-\pi, \pi)$.

b) Prove that

$$|x| = \frac{\pi}{2} - \frac{4}{\pi} \sum_{k=1}^{\infty} \frac{\cos(2k-1)x}{(2k-1)^2}$$

uniformly on $[-\pi, \pi]$.

c) Find a value for

$$\sum_{k=1}^{\infty} \frac{1}{(2k-1)^2}.$$

3. Prove that if f is continuous, odd, and periodic, then $\sum_{k=1}^{\infty} b_k(f)/k$ converges.

4. Let $L \in \mathbf{R}$. A series $\sum_{k=0}^{\infty} a_k$ is said to be *Abel summable* to L if

$$\lim_{r \to 1-} \sum_{k=0}^{\infty} a_k r^k = L.$$

a) Let $S_k = \sum_{j=0}^{k} a_k$. Prove that

$$\sum_{k=0}^{\infty} a_k r^k = (1-r) \sum_{k=0}^{\infty} S_k r^k = (1-r)^2 \sum_{k=0}^{\infty} (k+1) \sigma_k r^k,$$

provided any one of these series converges for all $0 < r < 1$.

b) Prove that if $\sum_{k=0}^{\infty} a_k$ is Cesàro summable to L, then it is Abel summable to L.

c) Prove that if f is continuous, periodic, and of bounded variation on $\mathbf{R}$, then Sf is Abel summable to f uniformly on $\mathbf{R}$.

d) Show that if $a_k \geq 0$ and $\sum_{k=0}^{\infty} a_k$ is Abel summable to L, then $\sum_{k=0}^{\infty} a_k$ converges to L.

5. [BERNSTEIN]. Let $f : \mathbf{R} \to \mathbf{R}$ be periodic and $\alpha > 0$. Suppose f is Lipschitz of order α i.e., there is a constant $M > 0$ such that

$$|f(x+h) - f(x)| \leq M|h|^{\alpha}$$

for all $x, h \in \mathbf{R}$.

a) Prove

$$\frac{1}{\pi} \int_{-\pi}^{\pi} |f(x+h) - f(x-h)|^2 \, dx = 4 \sum_{k=1}^{\infty} (a_k^2(f) + b_k^2(f)) \sin^2 kh$$

holds for each $h \in \mathbf{R}$.

b) If $h = \pi/2^{n+1}$, prove $\sin^2 kh \geq 1/2$ for all $k \in [2^{n-1}, 2^n]$.

c) Combine a) and b) to prove

$$\left\{ \sum_{k=2^{n-1}}^{2^n-1} (a_k^2(f) + b_k^2(f)) \right\}^{1/2} \leq M^2 \left(\frac{\pi}{2^{n+1}} \right)^{2\alpha}$$

for $n = 1, 2, 3, \ldots$.

d) Assuming

$$\sum_{k=2^{n-1}}^{2^n-1} (|a_k(f)| + |b_k(f)|) \leq 2^{n/2} \left(\sum_{k=2^{n-1}}^{2^n-1} (a_k^2(f) + b_k^2(f)) \right)^{1/2}$$

(see Exercise 9 in Section 6.6), prove that if f is Lipschitz of order α for some $\alpha > 1/2$, then Sf converges absolutely and uniformly on $\mathbf{R}$.

e) Prove that if $f : \mathbf{R} \to \mathbf{R}$ is periodic and continuously differentiable, then Sf converges absolutely and uniformly on $\mathbf{R}$.

*6. Suppose $f : \mathbf{R} \to \mathbf{R}$ is periodic and of bounded variation on $[-\pi, \pi]$. Prove that $S_N f \to f$ almost everywhere as $N \to \infty$ (see Exercise 7 in Section 9.2).

^e9.5 UNIQUENESS *This section uses material from Section 9.4.*

In this section we examine the Uniqueness Question posed in Section 9.1. We begin with the following generalization of the second derivative.

DEFINITION 9.5. Let $x_0 \in \mathbf{R}$ and I be an open interval containing x_0. A function $F : I \to \mathbf{R}$ is said to have a *second symmetric derivative* at x_0 if

$$D_2 F(x_0) = \lim_{h \to 0+} \frac{F(x_0 + 2h) + F(x_0 - 2h) - 2F(x_0)}{4h^2}$$

exists.

Remark 1. *Let $x_0 \in \mathbf{R}$ and I be an open interval containing x_0. If F is differentiable on I and $F''(x_0)$ exists, then F has a second symmetric derivative at x_0 and $D_2 F(x_0) = F''(x_0)$.*

PROOF. Set $G(t) = F(x_0 + 2t) + F(x_0 - 2t)$ for $t \in I$ and $H(t) = 4t^2$ and fix $t \in I$. By Theorem 2.15 (the Generalized Mean Value Theorem),

$$\frac{F(x_0 + 2t) + F(x_0 - 2t) - 2F(x_0)}{4t^2} = \frac{G(t) - G(0)}{H(t) - H(0)} = \frac{G'(c)}{H'(c)}$$

$$= \frac{F'(x_0 + 2c) - F'(x_0 - 2c)}{4c}$$

for some c between 0 and t. Since $c \to 0$ as $t \to 0$, it follows that

$$D_2 F(x_0) = \lim_{c \to 0} \frac{F'(x_0 + 2c) - F'(x_0 - 2c)}{4c}$$

$$= \frac{1}{2} \lim_{c \to 0} \left(\frac{F'(x_0 + 2c) - F(x_0)}{2c} + \frac{F(x_0) - F'(x_0 - 2c)}{2c} \right)$$

$$= \frac{1}{2} (F''(x_0) + F''(x_0)) = F''(x_0). \blacksquare$$

The converse of Remark 1 is false. Indeed, if

$$F(x) = \begin{cases} 1 & x > 0 \\ 0 & x = 0 \\ -1 & x < 0, \end{cases}$$

then $D_2 F(0) = 0$ but $F''(0)$ does not exist.

The following result reinforces further the analogy between the second derivative and the second symmetric derivative (see also Exercises 1 and 5).

Lemma 1. *Let $[a, b]$ be a closed bounded interval. If $F : [a, b] \to \mathbf{R}$ is continuous on $[a, b]$ and $D_2 F(x) = 0$ for all $x \in (a, b)$, then F is linear on $[a, b]$, i.e., there exist constants m, γ such that $F(x) = mx + \gamma$ for all $x \in [a, b]$.*

PROOF. Let $\varepsilon > 0$. By hypothesis,

$$\phi(x) := F(x) - F(a) - \left(\frac{F(b) - F(a)}{b - a} \right) (x - a) + \varepsilon(x - a)(x - b)$$

is continuous on $[a, b]$ and by Remark 1,

(29) $$D_2 \phi(x) = D_2 F(x) + 2\varepsilon = 2\varepsilon$$

for $x \in (a, b)$.

We claim that $\phi(x) \le 0$ for $x \in [a, b]$. Clearly, $\phi(a) = \phi(b) = 0$. If $\phi(x) > 0$ for some $x \in (a, b)$, then ϕ attains its maximum at some $x_0 \in (a, b)$. By Exercise 1, $D_2 \phi(x_0) \le 0$, hence, by (29), $2\varepsilon \le 0$, a contradiction. This proves the claim.

Fix $x \in [a, b]$. We have shown that

$$F(x) - F(a) - \left(\frac{F(b) - F(a)}{b - a} \right) (x - a) \le \varepsilon(x - a)(b - x).$$

A similar argument establishes

$$F(x) - F(a) - \left(\frac{F(b) - F(a)}{b - a} \right) (x - a) \ge -\varepsilon(x - a)(b - x).$$

Therefore,

$$\left| F(x) - F(a) - \left(\frac{F(b) - F(a)}{b - a} \right) (x - a) \right| \le \varepsilon(x - a)(b - x) \le \varepsilon(b - a)^2.$$

Taking the limit of this inequality as $\varepsilon \to 0$, we conclude that

$$F(x) = F(a) + \left(\frac{F(b) - F(a)}{b - a} \right) (x - a)$$

for all $x \in [a, b]$, i.e., F is linear on $[a, b]$. $\blacksquare$

DEFINITION 9.6. The *second formal integral* of a trigonometric series

$$S = \frac{a_0}{2} + \sum_{k=1}^{\infty}(a_k \cos kx + b_k \sin kx)$$

is the function

$$F(x) = \frac{a_0}{4}x^2 - \sum_{k=1}^{\infty}\frac{1}{k^2}(a_k \cos kx + b_k \sin kx).$$

By the Weierstrass–M Test, if the coefficients of S are bounded, then the second formal integral of S converges uniformly on $\mathbf{R}$. In particular, the second formal integral always exists when the coefficients of S converge to zero.

Notice that the second formal integral of a trigonometric series S is the result of integrating S twice term by term. Hence, it is not unreasonable to expect that two derivatives of the second formal integral F might recapture the original series S. Although this statement is not quite correct, the following result shows there is a simple connection between the limit of the series S and the second *symmetric* derivative of F.

THEOREM 9.18 [RIEMANN]. *Suppose*

$$S = \frac{a_0}{2} + \sum_{k=1}^{\infty}(a_k \cos kx + b_k \sin kx)$$

is a trigonometric series whose coefficients $a_k,\ b_k \to 0$ *as* $k \to \infty$ *and let* F *be the second formal integral of* S. *If* $S(x_0)$ *converges to* L *for some* $x_0 \in \mathbf{R}$, *then* $D_2 F(x_0) = L$.

PROOF. Let F_N denote the partial sums of F. After several applications of Theorem B.3, we observe that

$$\lim_{h \to 0}\frac{F_N(x_0 + 2h) + F_N(x_0 - 2h) - 2F_N(x_0)}{4h^2}$$

$$= \lim_{h \to 0}\left(\frac{a_0}{2} + \sum_{k=1}^{N}(a_k \cos kx_0 + b_k \sin kx_0)\left(\frac{\sin kh}{kh}\right)^2\right)$$

$$= \frac{a_0}{2} + \sum_{k=1}^{N}(a_k \cos kx_0 + b_k \sin kx_0)$$

holds for any $N \in \mathbf{N}$. Therefore, it suffices to show that given $\varepsilon > 0$ there is an $N \in \mathbf{N}$ such that

(30) $$|R_N| := \left|\sum_{k=N+1}^{\infty}(a_k \cos kx_0 + b_k \sin kx_0)\left(\frac{\sin kh}{kh}\right)^2\right| < \varepsilon$$

for all $|h| \le 1$.

Let

$$A_k = \sum_{j=k+1}^{\infty} (a_j \cos jx_0 + b_j \sin jx_0) \quad \text{and} \quad B_k = \left(\frac{\sin kh}{kh}\right)^2$$

for $k \in \mathbf{N}$. Since $A_n \to 0$ as $n \to \infty$, we have by Abel's Formula that

$$(31) \qquad R_N := \lim_{n\to\infty} \sum_{k=N+1}^{n} (A_{k-1} - A_k)B_k$$

$$= \lim_{n\to\infty} \left((A_N - A_n)B_n - \sum_{k=N+1}^{n-1} (A_N - A_k)(B_{k+1} - B_k) \right)$$

$$= A_N B_{N+1} + \sum_{k=N+1}^{\infty} A_k(B_{k+1} - B_k).$$

Moreover, by the Fundamental Theorem of Calculus,

$$(32) \qquad |B_{k+1} - B_k| = \left| \int_{kh}^{(k+1)h} \frac{d}{dt} \left(\frac{\sin t}{t}\right)^2 dt \right|.$$

Since

$$\frac{d}{dt} \left(\frac{\sin t}{t}\right)^2 = \frac{2\sin t}{t} \left(\frac{t\cos t - \sin t}{t^2}\right)$$

is bounded near $t = 0$ and is bounded by $2(t+1)/t^3 < 2/t^2$ for $t \geq 2$, it is clear that the improper integral

$$C = \int_0^{\infty} \left| \frac{d}{dt} \left(\frac{\sin t}{t}\right)^2 \right| dt$$

converges. Since $\{B_k\}$ is bounded and $A_N \to 0$ as $N \to \infty$, we can choose an $N \in \mathbf{N}$ such that

$$(33) \qquad |A_N B_{N+1}| < \frac{\varepsilon}{2} \quad \text{and} \quad k \geq N \quad \text{implies} \quad |A_k| < \frac{\varepsilon}{2C}.$$

It follows from (32) that

$$\sum_{k=N+1}^{\infty} A_k(B_{k+1} - B_k) \leq \frac{\varepsilon}{2C} \sum_{k=N+1}^{\infty} \left| \int_{kh}^{(k+1)h} \frac{d}{dt} \left(\frac{\sin t}{t}\right)^2 dt \right|$$

$$\leq \frac{\varepsilon}{2C} \int_0^{\infty} \left| \frac{d}{dt} \left(\frac{\sin t}{t}\right)^2 \right| dt = \frac{\varepsilon}{2}.$$

Combining this inequality with (31) and (33), we conclude that $|R_N| < \varepsilon$. ∎

The following result shows that the hypotheses of Riemann's Theorem are satisfied by any trigonometric series which converges pointwise on a nondegenerate interval.

THEOREM 9.19 [THE CANTOR–LEBESGUE LEMMA]. *If*

$$S = \frac{a_0}{2} + \sum_{k=1}^{\infty}(a_k \cos kx + b_k \sin kx)$$

is a trigonometric series which converges pointwise on a nondegenerate interval $[a, b]$, *then its coefficients satisfy* $a_k, b_k \to 0$ *as* $k \to \infty$.

PROOF. Set $\rho_0 = a_0/2$ and $\rho_k^2 = a_k^2 + b_k^2$ for $k \in \mathbf{N}$. If the result is false, then there is a $\delta > 0$ such that $\rho_k > \delta$ for infinitely many $k \in \mathbf{N}$.

Set $\theta_0 = 0$ and for each $k \in \mathbf{N}$ define $\theta_k \in \mathbf{R}$ so that $a_k = \rho_k \cos k\theta_k$, $b_k = \rho_k \sin k\theta_k$. By a sum angle formula,

$$\frac{a_0}{2} + \sum_{k=1}^{n}(a_k \cos kx + b_k \sin kx) = \sum_{k=0}^{n} \rho_k \cos k(x - \theta_k)$$

for each $x \in \mathbf{R}$ and $n \in \mathbf{N}$. Since S converges on $[a, b]$, it follows that

$$(34) \qquad \lim_{k \to \infty} \rho_k \cos k(x - \theta_k) = 0$$

for all $x \in [a, b]$.

Set $I_0 = [a, b]$ and $k_0 = 1$. Fix $j \geq 0$ and suppose a closed interval $I_j \subseteq I_0$ and an integer $k_j > k_0$ have been chosen. Choose $k_{j+1} > k_j$ such that $k_{j+1}|I_j| > 2\pi$ and $\rho_{k_{j+1}} > \delta$. Clearly, $k_{j+1}(x - \theta_{k_{j+1}})$ runs over an interval of length $> 2\pi$ as x runs over I_j. Hence, we can choose a closed interval $I_{j+1} \subseteq I_j$ such that

$$x \in I_{j+1} \quad \text{implies} \quad \cos k_{j+1}(x - \theta_{k_{j+1}}) \geq \frac{1}{2}.$$

By induction, then, there exist integers $1 < k_1 < k_2 < \ldots$ and a nested sequence of closed intervals $I_0 \supseteq I_1 \supseteq \ldots$ such that

$$(35) \qquad \rho_{k_j} \cos k_j(x - \theta_{k_j}) \geq \frac{\delta}{2}$$

for $x \in I_j$, $j \in \mathbf{N}$. By the Nested Interval Property, there is an $x \in I_j$ for all $j \in \mathbf{N}$. This x must satisfy (35) for all $j \in \mathbf{N}$ and must belong to $[a, b]$ by construction. Since this contradicts (34), we conclude that $\rho_k \to 0$ as $k \to \infty$. ∎

We are now prepared to answer the Uniqueness Question for continuous functions of bounded variation.

THEOREM 9.20 [CANTOR]. *Suppose*

$$S = \frac{a_0}{2} + \sum_{k=1}^{\infty}(a_k \cos kx + b_k \sin kx)$$

converges pointwise on $[-\pi, \pi]$ to a function f which is periodic and continuous on **R**, *and of bounded variation on $[-\pi, \pi]$. Then S is the Fourier series of f; i.e.,* $a_k = a_k(f)$ *for $k = 0, 1, \ldots$ and $b_k = b_k(f)$ for $k = 1, 2, \ldots$*

PROOF. Suppose first that $f(x) = 0$ for all $x \in \mathbf{R}$. By the Cantor–Lebesgue Lemma, the coefficients a_k, b_k tend to zero as $k \to \infty$. Thus, the second formal integral F of S is continuous on $\mathbf{R}$ and by Riemann's Theorem has a second symmetric derivative which satisfies $D_2 F(x) = 0$ for $x \in \mathbf{R}$. It follows that F is linear on $\mathbf{R}$; i.e., there exist numbers m and γ such that

$$mx + \gamma = \frac{a_0}{4}x^2 - \sum_{k=1}^{\infty} \frac{1}{k^2}(a_k \cos kx + b_k \sin kx)$$

for $x \in \mathbf{R}$. Since the series in this expression is periodic, it must be the case that $m = a_0 = 0$, i.e.,

$$\gamma + \sum_{k=1}^{\infty} \frac{1}{k^2}(a_k \cos kx + b_k \sin kx) = 0$$

for all $x \in \mathbf{R}$. Since this series converges uniformly, it follows from Theorem 9.1 that $\gamma = 0$ and $a_k = b_k = 0$ for $k \in \mathbf{N}$. This proves the theorem when $f = 0$.

If f is periodic, continuous, and of bounded variation on $[-\pi, \pi]$, then $S_N f \to f$ uniformly on $\mathbf{R}$ by Theorem 9.16. Hence, the series $S - Sf$ converges pointwise on $\mathbf{R}$ to zero. It follows from the case already considered that $a_k - a_k(f) = 0$ for $k = 0, 1, \ldots$, and $b_k - b_k(f) = 0$ for $k = 1, 2, \ldots$ ∎

EXERCISES

1. Suppose $F : \mathbf{R} \to \mathbf{R}$ has a second symmetric derivative at some x_0. Prove that if $F(x_0)$ is a local maximum, then $D_2 F(x_0) \leq 0$, and if $F(x_0)$ is a local minimum, then $D_2 F(x_0) \geq 0$.

2. Prove that if the coefficients of a trigonometric series are bounded, then its second formal integral converges uniformly on $\mathbf{R}$.

3. Prove that if $f : \mathbf{R} \to \mathbf{R}$ is periodic, then there exists at most one trigonometric series which converges to f pointwise on $\mathbf{R}$.

4. Suppose $f : \mathbf{R} \to \mathbf{R}$ is periodic, piecewise continuous, and of bounded variation on $\mathbf{R}$. Prove that if S is a trigonometric series which converges to $(f(x+) + f(x-))/2$ for all $x \in \mathbf{R}$, then S is the Fourier series of f.

*5. Suppose $F : (a, b) \to \mathbf{R}$ is continuous and $D_2 F(x) > 0$ for all $x \in (a, b)$. Prove that F is convex on (a, b).

Chapter 10

STOKE'S THEOREM
ON MANIFOLDS

This chapter is considerably more abstract than previous chapters. Our aim is to show that the theorems of Green, Gauss, and Stokes are special cases of a more general theory in which differential forms of degree one and two are replaced by differential forms of degree n, and curves and surfaces are replaced by n-dimensional manifolds. Differential forms of degree n are introduced in Section 10.1, n-dimensional manifolds are introduced in Section 10.2, and an n-dimensional version of Stokes's Theorem is proved in Section 10.3.

^{e}**10.1 DIFFERENTIAL FORMS ON R^n** *This section uses no material from any other enrichment section.*

We introduced differential forms of degree one and two in Sections 8.2 and 8.4. In this section we introduce differential forms of degree r. It turns out that, as far as calculus is concerned, the actual definition of differential forms is not as important as their algebraic structure. For this reason, we begin with the following formal definition. (For a more constructive approach to differential forms which interprets dx_i as a derivative of the projection operator $(x_1, \ldots, x_n) \longmapsto x_i$, see Spivak [12], p. 89.)

DEFINITION 10.1. Let $0 \le r \le n$ and V be open in $\mathbf{R}^n$.

   i) A 0–*form* (or *differential form of degree $r = 0$*) on V is a function $f : V \to \mathbf{R}$.

   ii) Let $r > 0$. An r–*form* (or *differential form of degree r*) on V is an expression of the form

(1)
$$\omega = \sum f_{i_1, \ldots, i_r} \, dx_{i_1} \ldots dx_{i_r},$$

where the sum is taken over all integers i_j which satisfy $1 \le i_1 < i_2 < \cdots < i_r \le n$, each *coefficient function* $f_{i_1, \ldots, i_r}$ is a 0–form on V, and the dx_{i_j}'s are symbols which (for us) will take on meaning only in the context of integration (see Definition 10.14 below). If all the coefficient functions are zero, then ω is called the *zero r–form* and is denoted by 0.

iii) Two r–forms $\omega = \sum f_{i_1,\ldots,i_r}\, dx_{i_1} \ldots dx_{i_r}$ and $\eta = \sum g_{i_1,\ldots,i_r}\, dx_{i_1} \ldots dx_{i_r}$ are said to be *equal* on V if $f_{i_1,\ldots,i_r}(\boldsymbol{x}) = g_{i_1,\ldots,i_r}(\boldsymbol{x})$ for all $1 \le i_1 < i_2 < \cdots < i_r \le n$ and all $\boldsymbol{x} \in V$.

iv) An r–form is said to be *decomposable* on V if there exist integers $1 \le i_1 < \cdots < i_r \le n$ and a 0–form f such that

$$\omega = f\, dx_{i_1} \ldots dx_{i_r}$$

on V.

v) An r–form is said to be continuous (respectively, $\mathcal{C}^p$) on V if all of its coefficient functions are continuous (respectively, $\mathcal{C}^p$) on V.

vi) The *support* of an r–form (notation: $\operatorname{spt} \omega$) is the union of the supports of its coefficient functions, i.e., if ω is given by (1), then

$$\operatorname{spt} \omega = \bigcup_{1 \le i_1 < \cdots < i_r \le n} \operatorname{spt}(f_{i_1,\ldots,i_r}).$$

If $\operatorname{spt} \omega \subseteq E$, then ω is said to be *supported on E*.

Let V be open in $\mathbf{R}^n$. Since there is only one collection of indices which satisfies $1 \le i_1 < \cdots < i_r \le n$ for $r = n$, an n–form on V is an expression of the form

$$\omega = f\, dx_1 \ldots dx_n$$

for some 0–form f (i.e., a function) on V. Thus, every n–form on $V \subseteq \mathbf{R}^n$ is decomposable. At the other extreme, a general 1–form on V is an expression of the form

$$\omega = \sum_{j=1}^n f_j\, dx_j,$$

where each f_j is a 0–form on V. An example of a 1–form is the total differential of a differentiable function $z = f(x,y)$, i.e., $dz = f_x\, dx + f_y\, dy$.

An $(n-1)$–form on $\mathbf{R}^n$ is an expression of the form

$$\omega = \sum_{j=1}^n f_j\, dx_1 \ldots \widehat{dx_j} \ldots dx_n.$$

The notation $\widehat{dx_j}$ indicates that the differential dx_j is missing. Thus, a 2–form on $\mathbf{R}^3$ is an expression of the form

(2) $$\omega = f_1\, dy\, dz + f_2\, dx\, dz + f_3\, dx\, dy.$$

In Chapter 8, we used Jacobians to define differential forms of degree two on a smooth orientable surface $S = (\phi, E)$ and to associate with each 2–form an oriented integral on S (see also Exercise 5 below). In the same way, we shall associate n–forms on $\mathbf{R}^n$ with oriented integrals over certain geometric objects called n-dimensional manifolds. First we introduce an algebraic structure on the collection of differential forms which is compatible with this identification.

Addition of differential forms can be realized by grouping like terms and simplifying coefficients. For example, the sum of $x^2 \, dy \, dz + y \, dx \, dy$ and $(1 - x^2) \, dy \, dz$ is

$$x^2 \, dy \, dz + y \, dx \, dy + (1 - x^2) \, dy \, dz = dy \, dz + y \, dx \, dy.$$

In particular, if V is open in $\mathbf{R}^n$ and

$$\omega = \sum f_{i_1,\ldots,i_r} \, dx_{i_1} \ldots dx_{i_r}, \qquad \eta = \sum g_{i_1,\ldots,i_r} \, dx_{i_1} \ldots dx_{i_r}$$

are r–forms on V, then

$$\omega + \eta = \sum (f_{i_1,\ldots,i_r} + g_{i_1,\ldots,i_r}) \, dx_{i_1} \ldots dx_{i_r}.$$

It is clear that addition of differential forms satisfies the usual laws of algebra, e.g., the Commutative Law and the Associative Law.

The product of a 0–form (this includes scalars) and an r–form can be defined by

$$g \left(\sum f_{i_1,\ldots,i_r} \, dx_{i_1} \ldots dx_{i_r} \right) = \sum g f_{i_1,\ldots,i_r} \, dx_{i_1} \ldots dx_{i_r}.$$

It is clear that if ω, η are r–forms and f, g are 0–forms, then

$$f(\omega + \eta) = f\omega + f\eta, \quad \text{and} \quad (f + g)\omega = f\omega + g\omega.$$

Multiplication of differential forms of degrees $r, s > 0$ is somewhat more complicated to describe. To explain what happens, recall that if $S = (\phi, E)$ is a smooth orientable surface in $\mathbf{R}^3$, then differential forms of degree two were defined by

$$dy \, dz = \frac{\partial(y, z)}{\partial(u, v)} \, d(u, v), \quad dz \, dx = \frac{\partial(z, x)}{\partial(u, v)} \, d(u, v), \quad \text{and} \quad dx \, dy = \frac{\partial(x, y)}{\partial(u, v)} \, d(u, v),$$

where $x = \phi_1(u, v)$, $y = \phi_2(u, v)$, $z = \phi_3(u, v)$. The notation $dx \, dy$ looks like a product of 1–forms. Does this product satisfy the usual algebraic laws? Certainly not. Since interchanging two rows of a determinant changes its sign (see Appendix C), it is clear that multiplication of 1–forms must satisfy the *Anticommutative Property*, e.g., $dx \, dy = -dy \, dx$. Since the determinant of any matrix with two identical rows is zero (see Appendix C), it is also clear that multiplication of 1–forms must satisfy the *Nilpotent Property*, e.g., $dx \, dx = 0$.

Based on these observations, we define multiplication of differential forms in the following way. First, we assume the *Anticommutative Property* and the *Nilpotent Property* hold for all decomposable 1–forms, i.e.,

$$dx_j \, dx_k = -dx_k \, dx_j \quad \text{and} \quad dx_j \, dx_j = 0,$$

for $k, j = 1, \ldots, n$. Next, we multiply two differential forms by assuming the Distributive Law holds, grouping like terms, and simplifying the resulting expression using the Nilpotent Property and the Anticommutative Property. For example, the product of $x^2\, dx$ and $y\, dy + z\, dz$ is

$$x^2\, dx(y\, dy + z\, dz) = x^2 y\, dx\, dy + x^2 z\, dx\, dz$$

and the product of $\sin x\, dz$ and $x^2\, dx + xy\, dy + \log z\, dz$ is

$$\sin x\, dz(x^2\, dx + xy\, dy + \log z\, dz) = -xy \sin x\, dy\, dz - x^2 \sin x\, dx\, dz.$$

In particular, if $\omega = \sum_{j=1}^{N} \omega_j$ and $\eta = \sum_{k=1}^{L} \eta_k$ is a sum of differential forms, then

$$\omega\eta = \sum_{j=1}^{N} \sum_{k=1}^{L} \omega_j \eta_k.$$

Although the Anticommutative Property and the Nilpotent Property may seem strange, they are natural consequences of the fact that $dx\, dy$ comes not from an iterated integral but an oriented integral. For example, the Anticommutative Property reflects the fact that when orientation is changed, the sign of the integral changes.

Example 1. Find $\omega + \eta$, $\omega - \eta$, and $\omega\eta$ if $\omega = x^2\, dx\, dz + xy\, dy\, dz$ and $\eta = 2y\, dx\, dz$.

SOLUTION. By definition,

$$\omega + \eta = (x^2 + 2y)\, dx\, dz + xy\, dy\, dz = xy\, dy\, dz - (x^2 + 2y)\, dz\, dx,$$

$$\omega - \eta = (x^2 - 2y)\, dx\, dz + xy\, dy\, dz = xy\, dy\, dz + (2y - x^2)\, dz\, dx,$$

and

$$\begin{aligned}
\omega\eta &= (x^2\, dx\, dz + xy\, dy\, dz)(2y\, dx\, dz) \\
&= 2x^2 y\, dx\, dz\, dx\, dz + 2xy^2\, dy\, dz\, dx\, dz \\
&= -2x^2 y\, dx\, dx\, dz\, dz - 2xy^2\, dy\, dx\, dz\, dz = 0. \quad \blacksquare
\end{aligned}$$

Using products of 1–forms, we see that an r–form is an expression of the form

$$\sum f_{i_1, \ldots, i_r}\, dx_{i_1} \cdots dx_{i_r},$$

where the sum is taken over all integers $i_j \in \{1, \ldots, n\}$, i.e., it is no longer necessary that the i_j's increase in j. Because of the connection between 2–forms and oriented surface integrals, we will frequently use

$$\omega = P\, dy\, dz + Q\, dz\, dx + R\, dx\, dy$$

to represent a generic 2–form on $\mathbf{R}^3$ rather than (2).

Here is a summary of the algebraic laws satisfied by addition and multiplication of differential forms.

THEOREM 10.1. *Let V be open in* **R**^{n}, *f be a 0–form on V, ω be an r–form on V, η be an s–form on V, and θ be a t–form on V.*

   i) *If $r = s$, then $\omega + \eta$ is an r–form, $\omega + \eta = \eta + \omega$ and $(\omega + \eta)\theta = \omega\theta + \eta\theta$. If $r = s = t$, then $(\omega + \eta) + \theta = \omega + (\eta + \theta)$.*

   ii) *For any r and s, $\omega\eta = (-1)^{rs}\eta\omega$.*

   iii) *For any r, s, and t, $(\omega\eta)\theta = \omega(\eta\theta)$ and $f(\omega\eta) = (f\omega)\eta = \omega(f\eta)$.*

PROOF. Properties i) and iii) hold by definition.

To prove ii), we may suppose that ω and η are decomposable, i.e., suppose that $\omega = f\,dx_{i_1} \dots dx_{i_r}$ and $\eta = g\,dx_{j_1} \dots dx_{j_s}$. By definition, the product of ω and η is the $(r + s)$–form

$$\omega\eta = fg\,dx_{i_1} \dots dx_{i_r}\,dx_{j_1} \dots dx_{j_s}.$$

Successive applications of the Anticommutative Property yield

$$\begin{aligned}
\omega\eta &= fg\,dx_{i_1} \dots dx_{i_r}\,dx_{j_1} \dots dx_{j_s}\\
&= (-1)^r fg\,dx_{j_1}\,dx_{i_1}\,dx_{i_2} \dots dx_{i_r}\,dx_{j_2} \dots dx_{j_s}\\
&= \cdots = (-1)^{rs} gf\,dx_{j_1} \dots dx_{j_s}\,dx_{i_1} \dots dx_{i_r} = (-1)^{rs}\eta\omega.
\end{aligned}$$

This completes the proof of part ii). ∎

In Section 6.4 we introduced the total differential of a function $z = f(x,y)$ of two variables as $dz = f_x\,dx + f_y\,dy$. This gives us two definitions for $dz\,dx$ and $dy\,dz$, one using Jacobians and one "multiplying" the total differential dz by the 1–forms dx and dy. These two definitions are compatible. Indeed, using the trivial parametrization of the surface $z = f(x,y)$, the Jacobian definition yields

(3) $$\qquad\qquad dy\,dz = -f_x\,d(x,y) \quad\text{and}\quad dz\,dx = -f_y\,d(x,y).$$

On the other hand, multiplying the 1–form $dz = f_x\,dx + f_y\,dy$ on the left by dy we have by the Nilpotent Property and the Anticommutative Property that

$$dy\,dz = dy(f_x\,dx + f_y\,dy) = f_x\,dy\,dx + f_y\,dy\,dy = -f_x\,dx\,dy.$$

A similar computation leads to $dz\,dx = -f_y\,dx\,dy$. Thus, if we identify $d(x,y)$ with $dx\,dy$, (3) holds no matter which definition we use. (Identification of $d(x,y)$ with $dx\,dy$ is justified by using the "identity chart"—see Remark 3 in Section 10.3.)

The following result contains an important computation which relates the n–fold product of n–forms on **R**^{n} to the determinant operator.

THEOREM 10.2. *Let V be open in* **R**^{n} *and $\omega_1, \omega_2, \dots, \omega_n$ be 1–forms on V. If $A = [a_{ij}]_{n \times n}$ is a real matrix, then*

$$\left(\sum_{j=1}^{n} a_{1j}\omega_j\right) \cdots \left(\sum_{j=1}^{n} a_{nj}\omega_j\right) = (\det A)\omega_1 \dots \omega_n.$$

PROOF. The proof is by induction on n. If $n = 1$ there is nothing to prove. Suppose the theorem holds for some integer $n \geq 1$. By Theorem 10.1 and the Nilpotent Property, we have

$$\left(\sum_{j=1}^{n} a_{1j}\omega_j \right) \cdots \left(\sum_{j=1}^{n} a_{nj}\omega_j \right)$$

$$= \left(a_{11}\omega_1 + \sum_{j=2}^{n} a_{1j}\omega_j \right) \cdots \left(a_{n1}\omega_1 + \sum_{j=2}^{n} a_{nj}\omega_j \right)$$

$$= a_{11}\omega_1 \left(\sum_{j=2}^{n} a_{2j}\omega_j \right) \cdots \left(\sum_{j=2}^{n} a_{nj}\omega_j \right)$$

$$+ \left(\sum_{j=2}^{n} a_{1j}\omega_j \right) a_{21}\omega_1 \left(\sum_{j=2}^{n} a_{3j}\omega_j \right) \cdots \left(\sum_{j=2}^{n} a_{nj}\omega_j \right)$$

$$+ \cdots + \left(\sum_{j=2}^{n} a_{1j}\omega_j \right) \cdots \left(\sum_{j=2}^{n} a_{(n-1)j}\omega_j \right) a_{n1}\omega_1$$

$$+ \left(\sum_{j=2}^{n} a_{1j}\omega_j \right) \cdots \left(\sum_{j=2}^{n} a_{nj}\omega_j \right).$$

Continue this string of identities using the Anticommutative Property, the inductive hypothesis, the definition of $\det A$ in terms of cofactors of A, and the Nilpotent Property. We obtain

$$\left(\sum_{j=1}^{n} a_{1j}\omega_j \right) \cdots \left(\sum_{j=1}^{n} a_{nj}\omega_j \right)$$

$$= a_{11} \det A_{11}(\omega_1\omega_2\ldots\omega_n) + (-1)^1 a_{21} \det A_{21}(\omega_1\omega_2\ldots\omega_n)$$

$$+ \cdots + (-1)^{n-1} a_{n1} \det A_{n1}(\omega_1\omega_2\ldots\omega_n)$$

$$+ \left(\sum_{j=2}^{n} a_{1j}\omega_j \right) \det A_{11}(\omega_2\ldots\omega_n)$$

$$= (\det A)\omega_1\ldots\omega_n + 0 = (\det A)\omega_1\ldots\omega_n. \quad \blacksquare$$

The derivative of a differential form is defined as follows. (This derivative can be used to unify the three operators *grad*, *curl*, and *div*—see Exercise 4 below.)

DEFINITION 10.2. Let V be open in $\mathbf{R}^n$ and $\omega = \sum f_{i_1,\ldots,i_r} \, dx_{i_1} \ldots dx_{i_r}$ be a $\mathcal{C}^1$ r–form on V.

i) If $\omega = f$ is a 0–form, then the *exterior derivative* of ω is the 1–form

$$d\omega := \sum_{j=1}^{n} \frac{\partial f}{\partial x_j} \, dx_j.$$

ii) If $\omega = f \, dx_{i_1} \ldots dx_{i_r}$ is a decomposable r–form, $r > 0$, then the *exterior derivative* of ω is the $r + 1$ form

$$d\omega := df \, dx_{i_1} \ldots dx_{i_r}.$$

iii) If ω is a differential form of degree $r > 0$, i.e., $\omega = \sum_{j=1}^{N} \omega_j$, where each ω_j is a decomposable r–form, then the *exterior derivative* of ω is the $r + 1$ form

$$d\omega := \sum_{j=1}^{N} d\omega_j.$$

iv) If ω is $\mathcal{C}^2$ r–form on V, then the *second exterior derivative* of ω is $d^2\omega := d(d\omega)$.

Example 2. Find $d\omega$ and $d^2\omega$ if $\omega(x, y, z, t) = xy \, dx \, dy + (x + z + t) \, dz \, dt$.

SOLUTION. By definition,

$$d\omega = (y \, dx + x \, dy) \, dx \, dy + (dx + dz + dt) \, dz \, dt = dx \, dz \, dt,$$

hence $d^2\omega = (d1) \, dx \, dy \, dz = 0.$ ∎

It is clear that for 0–forms, the exterior derivative satisfies the following rules.

Remark 1. *Let* ω *and* η *be* $\mathcal{C}^1$ *0–forms on some open set* $V \subset \mathbf{R}^n$, *and* α *be a scalar. Then* $d(\alpha\omega)$, $d(\omega + \eta)$, *and* $d(\omega\eta)$ *are continuous 1–forms on* V *with*

$$d(\alpha\omega) = \alpha d\omega$$

$$d(\omega + \eta) = d\omega + d\eta,$$

and

$$d(\omega\eta) = \eta d\omega + \omega d\eta.$$

Analogues of these rules hold for arbitrary r–forms.

THEOREM 10.3. *Let* V *be open in* $\mathbf{R}^n$ *and* α *be a scalar. If* ω *is a* $\mathcal{C}^1$ *r–form on* V *and* η *is a* $\mathcal{C}^1$ *s–form on* V, *then* $d(\alpha\omega)$ *and* $d(\omega + \eta)$ *(when* $r = s$*) are continuous* $(r + 1)$*–forms on* V, *and* $d(\omega\eta)$ *is a continuous* $(r + s + 1)$*–form on* V. *Moreover,*

$$d(\alpha\omega) = \alpha d\omega,$$

$$d(\omega + \eta) = d\omega + d\eta$$

(when $r = s$*), and*

(4)
$$d(\omega\eta) = d\omega \, \eta + (-1)^r \omega \, d\eta.$$

PROOF. By Definition 10.2iii, we may suppose ω and η are decomposable, i.e.,

$$\omega = f\,dx_{i_1}\ldots dx_{i_r} \quad \text{and} \quad \eta = g\,dx_{j_1}\ldots dx_{j_s}.$$

By Definition 10.2 and Remark 1,

$$d(\alpha\omega) = d(\alpha f)dx_{i_1}\ldots dx_{i_r} = \alpha df\,dx_{i_1}\ldots dx_{i_r} = \alpha d\omega.$$

Similarly, if $r = s$ and $i_\nu = j_\nu$, $\nu = 1,\ldots,r$, then

$$d(\omega + \eta) = d(f + g)\,dx_{i_1}\ldots dx_{i_r} = df\,dx_{i_1}\ldots dx_{i_r} + dg\,dx_{i_1}\ldots dx_{i_r} = d\omega + d\eta.$$

To prove (4), consider first the case $r = 0$, i.e., $\omega = f$ is a 0–form. By Remark 1 and Theorem 10.1ii,

$$\begin{aligned}
d(\omega\eta) &= d(fg)dx_{j_1}\ldots dx_{j_s} = (g\,df + f\,dg)dx_{j_1}\ldots dx_{j_s}\\
&= df\,g\,dx_{j_1}\ldots dx_{j_s} + f\,dg\,dx_{j_1}\ldots dx_{j_s} = d\omega\,\eta + \omega\,d\eta.
\end{aligned}$$

Next, suppose $r > 0$. If $i_\nu = j_\mu$ for some indices ν and μ, then the Nilpotent Property implies

$$\omega\eta = 0 = d\omega\,\eta = \omega\,d\eta.$$

On the other hand, if all the indices are distinct, then since g is a 0–form and dg is a 1–form, we have by Theorem 10.1ii that

$$\begin{aligned}
d(\omega\eta) &= d(fg\,dx_{i_1}\ldots dx_{i_r}dx_{j_1}\ldots dx_{j_s})\\
&= (g\,df + f\,dg)dx_{i_1}\ldots dx_{i_r}dx_{j_1}\ldots dx_{j_s}\\
&= df\,dx_{i_1}\ldots dx_{i_r}\,g\,dx_{j_1}\ldots dx_{j_s}\\
&\quad + (-1)^{r\cdot 1}f\,dx_{i_1}\ldots dx_{i_r}\,dg\,dx_{j_1}\ldots dx_{j_s}\\
&= d\omega\,\eta + (-1)^r\omega\,d\eta. \quad\blacksquare
\end{aligned}$$

Equation (4) is called the *Product Rule*.

The following result shows that the second exterior derivative of a C^2 r–form is always zero. (By Exercise 4 below, this result generalizes Exercise 9b in Section 8.5.)

THEOREM 10.4. *If ω is a C^2 r–form on an open set $V \subseteq \mathbf{R}^n$, then $d^2\omega = 0$.*

PROOF. We may suppose that ω is decomposable, i.e., $\omega = f\,dx_{i_1}\ldots dx_{i_r}$. The proof is by induction on r. Suppose $r = 0$, i.e., $\omega = f$. By the Nilpotent Property, the Anticommutative Property, and the fact that the first-order partial derivatives of f commute, we have

$$\begin{aligned}
d^2\omega &= d\left(\sum_{j=1}^n \frac{\partial f}{\partial x_j}\,dx_j\right) = \sum_{j=1}^n\sum_{k=1}^n \frac{\partial^2 f}{\partial x_k\partial x_j}\,dx_k\,dx_j\\
&= \sum_{j\neq k}\left(\frac{\partial^2 f}{\partial x_k\partial x_j} - \frac{\partial^2 f}{\partial x_j\partial x_k}\right)dx_k\,dx_j = 0.
\end{aligned}$$

Suppose $r = 1$, i.e., $\omega = f\,dx_k$. Since all first-order partial derivatives of the function 1 are zero, we have by definition that $d^2 x_k = d(1\,dx_k) = 0$. Thus, by the Product Rule and the case $r = 0$,

$$d^2 \omega = d(df\,dx_k) = d^2 f\,dx_k - df\,d^2 x_k = 0.$$

Finally, suppose there is an $r > 1$ such that the theorem holds for all s–forms, $0 \le s < r$. By definition,

$$d\omega = d(f\,dx_{i_1}\dots dx_{i_{r-1}})\,dx_{i_r}.$$

Hence, by the Product Rule and the inductive hypothesis (for $s = 1$ and $s = r-1$), we have

$$d^2 \omega = d^2(f\,dx_{i_1}\dots dx_{i_{r-1}})\,dx_{i_r} + d(f\,dx_{i_1}\dots dx_{i_{r-1}})\,d^2 x_{i_r} = 0. \quad \blacksquare$$

The following definition shows how to use a continuously differentiable function $\phi : \mathbf{R}^n \to \mathbf{R}^m$ to transform differentials from $\mathbf{R}^m$ to $\mathbf{R}^n$. (This concept will be used later to define integration of r–forms over manifolds.)

DEFINITION 10.3. Let U be open in $\mathbf{R}^n$, V be open in $\mathbf{R}^m$, $\phi : U \to V$ be $\mathcal{C}^1$ on U, and

$$\omega = \sum f_{i_1,\dots,i_r}\,dx_{i_1}\dots dx_{i_r}$$

be an r–form on V. Then the *differential transform* (induced by ϕ) of ω is the r–form on U defined by

$$\phi^*(\omega) = \sum \phi^*(f_{i_1,\dots,i_r})\phi^*(dx_{i_1})\dots\phi^*(dx_{i_r}),$$

where $\phi^*(f) = f \circ \phi$ for every 0–form f and

$$\phi^*(dx_i) = d\phi_i = \sum_{j=1}^{n} \frac{\partial \phi_i}{\partial u_j}\,du_j$$

for every $i = 1, 2, \dots, m$.

For the next several remarks, let U be open in $\mathbf{R}^n$, V be open in $\mathbf{R}^m$, and $\phi : U \to V$.

Remark 2. *If ω is a $\mathcal{C}^1$ r–form on V and ϕ is $\mathcal{C}^2$ on U, then $\phi^*(\omega)$ is a $\mathcal{C}^1$ r–form on U.*

PROOF. By definition, $(\phi^* \circ f)(\mathbf{u}) = f(\phi(\mathbf{u}))$ is a $\mathcal{C}^1$ 0–form on U for every $\mathcal{C}^1$ 0–form f on V and

$$\phi^*(dx_i) = \sum_{j=1}^{n} \frac{\partial \phi_i}{\partial u_j}\,du_j$$

is a C^1 1–form on U for $i = 1, 2, \ldots, m$. Hence, it is clear that $\phi^*(\omega)$ is a C^1 r–form on U when ω is a C^1 r–form on V. ∎

Remark 3. *The differential transform ϕ^* is linear, i.e., if ω and η are r–forms on V and ϕ is C^1 on U, then*

$$\phi^*(\omega + \eta) = \phi^*(\omega) + \phi^*(\eta).$$

PROOF. We may suppose that $\omega = f \, dx_{i_1} \ldots dx_{i_r}$ and $\eta = g \, dx_{i_1} \ldots dx_{i_r}$. Thus,

$$\phi^*(\omega + \eta) = \phi^*((f + g) \, dx_{i_1} \ldots dx_{i_r})$$
$$= (f \circ \phi + g \circ \phi)\phi^*(dx_{i_1}) \ldots \phi^*(dx_{i_r}) = \phi^*(\omega) + \phi^*(\eta). \ \blacksquare$$

Remark 4. *The differential transform ϕ^* is multiplicative, i.e., if ω is an r–form on V and η is an s–form on V and ϕ is C^1 on U, then*

$$\phi^*(\omega\eta) = \phi^*(\omega)\phi^*(\eta).$$

PROOF. We may suppose that $\omega = f \, dx_{i_1} \ldots dx_{i_r}$ and $\eta = g \, dx_{j_1} \ldots dx_{j_s}$. Thus,

$$\phi^*(\omega\eta) = \phi^*((fg) \, dx_{i_1} \ldots dx_{i_r} dx_{j_1} \ldots dx_{j_s})$$
$$= (f \circ \phi)(g \circ \phi)\phi^*(dx_{i_1}) \ldots \phi^*(dx_{i_r})\phi^*(dx_{j_1}) \ldots \phi^*(dx_{j_s})$$
$$= \phi^*(\omega)\phi^*(\eta). \ \blacksquare$$

Remark 5. *The differential transform ϕ^* and the exterior derivative d commute, i.e., if ω is a C^1 r–form on V and ϕ is C^2 on U, then*

(5) $$\phi^*(d\omega) = d(\phi^*(\omega)).$$

PROOF. We may suppose ω is decomposable. The proof is by induction on r. Suppose $r = 0$, i.e., $\omega = f$. Then by definition and the Chain Rule,

$$\phi^*(d\omega) = \phi^*\left(\sum_{k=1}^m \frac{\partial f}{\partial x_k} dx_k\right) = \sum_{k=1}^m \phi^*\left(\frac{\partial f}{\partial x_k}\right)\phi^*(dx_k)$$
$$= \sum_{k=1}^m \left(\frac{\partial f}{\partial x_k} \circ \phi\right)\sum_{j=1}^n \frac{\partial \phi_k}{\partial u_j} du_j$$
$$= \sum_{j=1}^n \frac{\partial (f \circ \phi)}{\partial u_j} du_j = d(f \circ \phi) = d(\phi^*(\omega)).$$

Suppose $r = 1$, i.e., $\omega = f \, dx_k$. Then by definition, the multiplicative property of ϕ^*, and the case $r = 0$, we have

$$\phi^*(d\omega) = \phi^*(df \, dx_k) = \phi^*(df)\phi^*(dx_k) = d(f \circ \phi) \, d\phi_k.$$

On the other hand, since $\phi^*(\omega) = (f \circ \phi)\phi^*(dx_k) = (f \circ \phi)\,d\phi_k$, it follows from the Product Rule, the Nilpotent Property, and Remark 1, that

$$d(\phi^*(\omega)) = d(f \circ \phi)\,d\phi_k + (f \circ \phi)\,d^2\phi_k = d(f \circ \phi)\,d\phi_k.$$

Thus, (5) holds when ω is a 1–form.

Finally, suppose there is an $r > 1$ such that (5) holds for all s–forms, $0 \le s < r$. Let ω be a decomposable r–form and write $\omega = \theta\eta$, where θ is a 1–form and η is an $(r - 1)$–form. By the Product Rule,

$$d\omega = (d\theta)\eta - \theta d\eta.$$

Hence, it follows from the inductive hypothesis, the Product Rule, and the multiplicative property of ϕ^* that

$$
\begin{aligned}
\phi^*(d\omega) &= \phi^*(d\theta)\phi^*(\eta) - \phi^*(\theta)\phi^*(d\eta) \\
&= d(\phi^*\theta)\phi^*(\eta) - \phi^*(\theta)d(\phi^*\eta) \\
&= d((\phi^*\theta)(\phi^*\eta)) = d(\phi^*(\theta\eta)) = d(\phi^*(\omega)). \quad \blacksquare
\end{aligned}
$$

The following result shows that differential transforms can be used to define the oriented line and surface integrals introduced in Sections 8.2 and 8.4 (see Exercise 5 below).

THEOREM 10.5 [FUNDAMENTAL THEOREM OF DIFFERENTIAL TRANSFORMS]. *Let $m \ge n$, U be open in $\mathbf{R}^n$, V be open in $\mathbf{R}^m$, and $\phi : U \to V$ be $\mathcal{C}^1$ on U. If*

$$\omega = \sum f_{i_1,\dots,i_n}\,dx_{i_1}\dots dx_{i_n}$$

is an n–form on V, then

$$\phi^*(\omega) = \sum (f_{i_1,\dots,i_n} \circ \phi)\,\frac{\partial(\phi_{i_1},\dots,\phi_{i_n})}{\partial(u_1,\dots,u_n)}\,du_1\dots du_n.$$

PROOF. We may suppose that ω is decomposable. If $n = 1$, i.e., $\omega = f\,dx_j$, then by definition,

$$\phi^*(\omega) = \phi^*(f)\,\phi^*(dx_j) = (f \circ \phi)\phi'du.$$

If $n > 1$, i.e., $\omega = f\,dx_{i_1}\dots dx_{i_n}$, then by Definition 10.3 and Theorem 10.2,

$$
\begin{aligned}
\phi^*(\omega) &= \phi^*(f)\,\phi^*(dx_{i_1})\dots\phi^*(dx_{i_n}) \\
&= (f \circ \phi)\left(\sum_{k=1}^n \frac{\partial\phi_{i_1}}{\partial u_k}\,du_k\right)\dots\left(\sum_{k=1}^n \frac{\partial\phi_{i_n}}{\partial u_k}\,du_k\right) \\
&= (f \circ \phi)\,\frac{\partial(\phi_{i_1},\dots,\phi_{i_n})}{\partial(u_1,\dots,u_n)}\,du_1\dots du_n. \quad \blacksquare
\end{aligned}
$$

EXERCISES

1. Algebraically simplify the following differential forms.

   a) $3(dx + dy) dz + 2(dx + dz) dy$.

   b) $(x\, dy - y\, dx)(x\, dz - z\, dy)$.

   c) $(x^2\, dx\, dy - \cos x\, dy\, dz)(y^2\, dy + \cos x\, dw) - (x^3\, dy\, dz - \sin x\, dy\, dw)(y^3\, dy + \sin x\, dz)$.

2. Compute the exterior derivatives of the following differential forms.

   a) $x^2\, dy - y^2\, dx$.

   b) $\sin(xy)\, dz\, dw + \cos(zw)\, dx\, dy$.

   c) $\sqrt{x^2 + y^2}\, dy\, dz - \sqrt{x^2 + y^2}\, dx\, dz$.

   d) $(e^{xy}\, dz + e^{yz}\, dx)(\sin x\, dy + \cos y\, dx)$.

3. a) Prove that if ω is an r–form, r odd, then $\omega^2 = 0$.

   b) Prove that if ω_j are decomposable r–forms, r even, and $\omega = \sum_{j=1}^{N} \omega_j$, then

$$\omega^2 = 2 \sum_{\substack{k,j=1 \\ j<k}}^{N} \omega_j \omega_k.$$

$\boxed{4}$. **This exercise is used in Section 10.3.** If f, g are 0–forms and ω, η are r–forms, define

$$(f, g) \cdot (\omega, \eta) = f\omega + g\eta.$$

a) Prove that if $f : \mathbf{R}^3 \to \mathbf{R}$ is C^1 and $\operatorname{grad} f := (f_x, f_y, f_z)$, then the exterior derivative of the 0–form $\omega = f$ can be written in the form

$$d\omega = (\operatorname{grad} f) \cdot (dx, dy, dz).$$

b) Prove that if $F = (P, Q, R) : \mathbf{R}^3 \to \mathbf{R}^3$ is C^1, then the exterior derivative of the 1–form $\omega = P\, dx + Q\, dy + R\, dz$ can be written in the form

$$d\omega = (\operatorname{curl} F) \cdot (dy\, dz, dz\, dx, dx\, dy).$$

and the exterior derivative of the 2–form $\eta = P\, dy\, dz + Q\, dz\, dx + R\, dx\, dy$ can be written in the form

$$d\eta = (\operatorname{div} F)\, dx\, dy\, dz.$$

$\boxed{5}$. **This exercise is used in Section 10.3.** Let I be an interval and E be a Jordan region. Define the integral of a continuous 1–form $\omega = f\, dt$ on I and a continuous 2–form $\eta = g\, du\, dv$ on E by

$$\int_I \omega = \int_I f(t)\, dt \quad \text{and} \quad \iint_E \eta = \iint_E g(u, v)\, d(u, v).$$

a) Let $C = (\phi, I)$ be a smooth simple C^1 curve in $\mathbf{R}^2$, $F = (P, Q) : \phi(I) \to \mathbf{R}^2$ be continuous, and $\omega = P\,dx + Q\,dy$. Prove that

$$\int_C F \cdot T\,ds = \int_I \phi^*(\omega).$$

b) Let $S = (\phi, E)$ be a smooth simple orientable C^1 surface in $\mathbf{R}^3$, $F = (P, Q, R) : \phi(E) \to \mathbf{R}^3$ be continuous, and

$$\eta = P\,dy\,dz + Q\,dz\,dx + R\,dx\,dy.$$

Prove that

$$\iint_S F \cdot \boldsymbol{n}\,d\sigma = \iint_E \phi^*(\eta).$$

10.2 DIFFERENTIABLE MANIFOLDS   *This section uses no material from any other enrichment section.*

In Chapter 8 we introduced one-dimensional objects (curves), two-dimensional objects (surfaces), and corresponding oriented integrals. We shall extend these ideas to higher dimensions.

A problem which surfaced several times in Chapter 8 is that one parametrization by itself does not fully describe a surface. For example, the boundary of a surface could not be defined using one parametrization alone. To avoid this problem, we adopt a different point of view here. Instead of thinking of a surface as a particular parametrization $\phi : E \to S$, we will think of a surface as a set of points S together with a class of functions $h_\alpha : S \to E$ related to each other in a natural way. (The h_α's can be thought of as inverses of parametrizations of pieces of S.) This point of view has been used by map makers for centuries. The earth (a particular surface) can be described by an atlas, which is itself a collection of two-dimensional maps (or charts) which represent overlapping portions of its surface. If we know how the individual charts fit together (see (7) below), we can study the whole surface using this atlas.

DEFINITION 10.4. Let M be a set.

i) An *n-dimensional chart* of M at a point $x \in M$ is a pair (V, h), where $x \in V$, $V \subseteq M$, $h : V \to \mathbf{R}^n$ is 1–1, and $h(V)$ is open in $\mathbf{R}^n$. (We shall drop the adjective "*n*-dimensional" when no confusion arises.)

ii) An *n-dimensional C^p atlas* of M is a collection

$$(6) \qquad \mathcal{A} = \{(V_\alpha, h_\alpha) : \alpha \in A\}$$

of *n*-dimensional charts of M such that $h_\beta(V_\alpha \cap V_\beta)$ is open in $\mathbf{R}^n$,

$$(7) \qquad h_\alpha \circ h_\beta^{-1} \text{ is } C^p \text{ on } h_\beta(V_\alpha \cap V_\beta) \text{ for all } \alpha, \beta \in A$$

and

$$M = \bigcup_{\alpha \in A} V_\alpha.$$

The functions $h_\alpha \circ h_\beta^{-1}$ are called the *transition maps* of the atlas $\mathcal{A}$.

Notice, then, that if (V, h) is a chart of M, then $(h^{-1}, h(V))$ can be thought of as a parametrization of a portion V of M.

Example 1. If $C = \phi(I)$, where (ϕ, I) is a simple curve and I is an open interval, prove $\{(C, \phi^{-1})\}$ is a one-dimensional $\mathcal{C}^\infty$ atlas of C.

PROOF. Since (ϕ, I) is simple, ϕ^{-1} exists on $\phi(I)$. (C, ϕ^{-1}) is evidently a one-dimensional chart of C, so $\{(C, \phi^{-1})\}$ is an atlas of C. Since the transition map $(\phi^{-1} \circ \phi)(\boldsymbol{x}) = \boldsymbol{x}$ is the identity function on C, this atlas is $\mathcal{C}^\infty$. ∎

A similar argument establishes the following two remarks.

Remark 1. *If V is open in $\mathbf{R}^2$, $\phi : V \to \mathbf{R}^m$ is 1-1 on V, and $S = \phi(V)$, then $\{(S, \phi^{-1})\}$ is a two-dimensional $\mathcal{C}^\infty$ atlas of S.*

Remark 2. *If V is open in $\mathbf{R}^n$ and $I(\boldsymbol{x}) = \boldsymbol{x}$ is the identity function on $\mathbf{R}^n$, then $\{(V, I)\}$ is an n-dimensional $\mathcal{C}^\infty$ atlas of V.* (We shall call (V, I) the *identity chart*.)

Not all atlases consist of one chart.

Example 2. For each $t \in \mathbf{R}$, set $\phi(t) = (\cos t, \sin t)$, $\psi(t) = (\cos(t + \pi), \sin(t + \pi))$, $V = \phi(I)$, and $U = \psi(I)$, where $I = (0, 2\pi)$. If $h = \phi^{-1}$ on V and $g = \psi^{-1}$ on U, prove

$$\mathcal{A} = \{(V, h), (U, g)\}$$

is a one-dimensional $\mathcal{C}^\infty$ atlas of the unit circle $x^2 + y^2 = 1$.

PROOF. Let M represent the set of points (x, y) such that $x^2 + y^2 = 1$. Since ϕ (respectively ψ) is 1-1 from I onto V (respectively, I onto U) and $V \cup U = M$, (V, h) and (U, g) are charts which cover M. It is easy to see that the transition maps are $\mathcal{C}^\infty$. For example, $g(V \cap U) = (0, \pi) \cup (\pi, 2\pi)$, and on the interval $(0, \pi)$, $(h \circ g^{-1})(t) = t + \pi$. Thus, $\mathcal{A}$ is a $\mathcal{C}^\infty$ atlas of M. ∎

The following concept is a replacement for smooth equivalence of parametrizations.

DEFINITION 10.5. Two n-dimensional atlases $\mathcal{A}$, $\mathcal{B}$ of M are said to be $\mathcal{C}^p$ *compatible* (notation: $\mathcal{A} \sim \mathcal{B}$) if $\mathcal{A} \cup \mathcal{B}$ is an n-dimensional $\mathcal{C}^p$ atlas on M.

It is easy to see that $\mathcal{C}^p$ compatibility is an equivalence relation (see Exercise 2 below), i.e., any atlas $\mathcal{A}$ is $\mathcal{C}^p$ compatible with itself; if $\mathcal{A}$ is $\mathcal{C}^p$ compatible with $\mathcal{B}$, then $\mathcal{B}$ is $\mathcal{C}^p$ compatible with $\mathcal{A}$; and if $\mathcal{A}$ is $\mathcal{C}^p$ compatible with $\mathcal{B}$, and $\mathcal{B}$ is $\mathcal{C}^p$ compatible with $\mathcal{D}$, then $\mathcal{A}$ is $\mathcal{C}^p$ compatible with $\mathcal{D}$. (For some elementary remarks about equivalence relations and equivalence classes, see Appendix F.)

DEFINITION 10.6. An *n-dimensional C^p manifold* is a set M together with an equivalence class $\overline{A}$ of n-dimensional C^p atlases on M. By an *atlas* of M we mean an atlas in $\overline{A}$. By a *chart* of M we mean a chart in some atlas of M.

Atlases of an n-dimensional manifold M can be used to "pull-back" concepts from $\mathbf{R}^n$ to M.

DEFINITION 10.7. Let M be an n-dimensional C^p manifold and $\mathcal{A}$ be an atlas of M. A set $W \subseteq M$ is said to be *open* if $h(V \cap W)$ is open in $\mathbf{R}^n$ for all charts $(V, h) \in \mathcal{A}$.

The following result shows that this definition does not depend on the atlas chosen from the manifold structure of M.

Remark 3. *Let $\mathcal{A}$ and $\mathcal{B}$ be C^p compatible atlases of M and suppose $W \subseteq M$. Then $h(V \cap W)$ is open in $\mathbf{R}^n$ for all $(V, h) \in \mathcal{A}$ if and only if $g(U \cap W)$ is open in $\mathbf{R}^n$ for all $(U, g) \in \mathcal{B}$.*

PROOF. Let $W \subseteq M$ such that $h(V \cap W)$ is open in $\mathbf{R}^n$ for all $(V, h) \in \mathcal{A}$, and suppose $(U, g) \in \mathcal{B}$. If $W \cap U = \emptyset$, then $g(W \cap U) = \emptyset$ is open in $\mathbf{R}^n$ by definition. If $W \cap U \neq \emptyset$, choose $(V, h) \in \mathcal{A}$ such that $W \cap V \cap U \neq \emptyset$. Since $h(W \cap V)$ and $g(U)$ are open in $\mathbf{R}^n$ and the transition map $h \circ g^{-1}$ is C^p, hence continuous, it follows that

$$g(W \cap V \cap U) = (g \circ h^{-1})(h(W \cap V)) \cap g(U) = (h \circ g^{-1})^{-1}(h(W \cap V)) \cap g(U)$$

is open in $\mathbf{R}^n$. Since

$$g(W \cap U) = \bigcup_{(V,h) \in \mathcal{A}} g(W \cap V \cap U),$$

we conclude that $g(W \cap U)$ is open in $\mathbf{R}^n$. Reversing the roles of $\mathcal{A}$ and $\mathcal{B}$ proves the converse. ∎

Using open sets, we can define what we mean by continuity of a function on a manifold (compare with Theorem 5.25.)

DEFINITION 10.8. Let M be an n-dimensional C^p manifold and $\mathcal{A}$ be an atlas of M.

   i) A function $f : M \to \mathbf{R}^k$ is said to be *continuous* on M if $f^{-1}(U)$ is open in M for every open set $U \subset \mathbf{R}^k$.
   ii) A function $f : \mathbf{R}^k \to M$ is said to be *continuous* on a set $E \subset \mathbf{R}^k$ if $f^{-1}(W) \cap E$ is relatively open in E for every open set W in M.

Remark 4. *If (V, h) is a chart from an atlas $\mathcal{A}$ of M, then h is a homeomorphism, i.e., h is continuous on V and h^{-1} is continuous on $h(V)$.*

PROOF. If $W \subseteq V$ is open in M, then $(h^{-1})^{-1}(W) = h(W)$ is open in $\mathbf{R}^n$. Hence, h^{-1} is continuous on $h(V)$ by Definition 10.8. On the other hand, suppose $\Omega \subset h(V)$ is open in $\mathbf{R}^n$ and $W = h^{-1}(\Omega)$. Let (U, g) be any chart in $\mathcal{A}$. Then

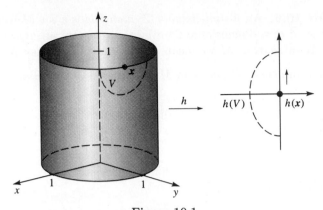

Figure 10.1

$g(U \cap W) = g(U) \cap g \circ h^{-1}(\Omega)$ is open in $\mathbf{R}^n$, i.e., W is open in M by Definition 10.7. Hence, h is continuous on V. ∎

To define the boundary of a manifold, we introduce the following terminology. By a *half space* of $\mathbf{R}^n$ we mean a set of the form

$$\{(x_1, \ldots, x_n) : x_j \geq \alpha\} \quad \text{or} \quad \{(x_1, \ldots, x_n) : x_j \leq \alpha\},$$

where $\alpha \in \mathbf{R}$ and $j \in \{1, 2, \ldots, n\}$. We shall refer to the special case

$$\mathcal{H}_1 := \{(x_1, \ldots, x_n) : x_1 \leq 0\}$$

as *left half space*. If $n = 2$, we shall refer to half spaces as *half planes*.

A simple curve parametrized on an open interval is a one-dimensional manifold (see Example 1 above). What about surfaces? A smooth surface with empty boundary is a two-dimensional manifold, but the restriction in Definition 10.4i that $h(V)$ be open prevents any surface whose boundary is nonempty from being a manifold. For example, the cylinder $1 = x^2 + y^2$, $0 \leq z \leq 1$, does not satisfy Definition 10.4 because there is no way to construct an "open" chart at points on its boundary (see Remark 5 below). Loosely speaking, this is because at a point on the boundary, the surface does not look like an open set, but rather like a relatively open set in a half plane (see Figure 10.1).

Accordingly, we make the following definition.

DEFINITION 10.9. Let M be a set.

  i) An *n-dimensional chart-with-smooth-boundary* of M at a point $x \in M$ is a pair (V, h), where $x \in V$, $V \subseteq M$, $h : V \to \mathbf{R}^n$ is 1-1 on V, and $h(V)$ is relatively open in some half space $\mathcal{H}$ of $\mathbf{R}^n$. If $h(V) \cap \partial \mathcal{H} = \emptyset$, then (V, h) is called an *interior chart*. If $h(V) \cap \partial \mathcal{H} \neq \emptyset$, then (V, h) is called a *boundary chart*.

  ii) An *n-dimensional C^p atlas-with-smooth-boundary* of M is a collection

(8) $$\mathcal{A} = \{(V_\alpha, h_\alpha) : \alpha \in A\}$$

of n-dimensional charts-with-smooth-boundary of M such that $h_\beta(V_\alpha \cap V_\beta)$ is relatively open in some half space $\mathcal{H}$,

(9) $$h_\alpha \circ h_\beta^{-1} \text{ is } \mathcal{C}^p \text{ on } h_\beta(V_\alpha \cap V_\beta) \text{ for all } \alpha, \beta \in A,$$

and

$$M = \bigcup_{\alpha \in A} V_\alpha.$$

The functions $h_\alpha \circ h_\beta^{-1}$ are called the *transition maps* of the atlas $\mathcal{A}$.

iii) Two n-dimensional atlases-with-smooth-boundary $\mathcal{A}, \mathcal{B}$ of M are said to be $\mathcal{C}^p$ *compatible* (notation: $\mathcal{A} \sim \mathcal{B}$) if $\mathcal{A} \cup \mathcal{B}$ is an n-dimensional $\mathcal{C}^p$ atlas-with-smooth-boundary on M.

It is easy to check that $\mathcal{C}^p$ compatibility of atlases-with-smooth-boundary is an equivalence relation. We also note that since any open subset of $\mathcal{H}$ is relatively open in $\mathcal{H}$, every atlas is an atlas-with-smooth-boundary.

We now expand the definition of manifold using atlases-with-smooth-boundary.

DEFINITION 10.10.

i) An *n-dimensional $\mathcal{C}^p$ manifold-with-smooth-boundary* is a set M together with an equivalence class of n-dimensional atlases-with-smooth-boundary.

ii) A point $x \in M$ is said to be a *boundary point* if it belongs to V for some boundary chart (V, h) of M, with $h(V)$ relatively open in some half space $\mathcal{H}$, and $h(x) \in \partial \mathcal{H}$. The collection of all boundary points, called the *boundary* of M, is denoted by ∂M.

The following result shows that if the transition maps of a manifold have nonzero Jacobian, then the definition of boundary point is unambiguous.

Remark 5. *Suppose M is an n-dimensional $\mathcal{C}^p$ manifold-with-smooth-boundary, $x \in M$, and (U, g), (V, h) are charts of M at x whose transition map $\phi = g \circ h^{-1}$ satisfies $\Delta_\phi \neq 0$ on $h(U \cap V)$. If $g(U)$ (respectively, $h(V)$) is relatively open in some half space $\mathcal{H}$ (respectively, $\mathcal{K}$), and $g(x) \in \partial \mathcal{H}$, then $h(x) \in \partial \mathcal{K}$.*

PROOF. Set $\Omega = (h(U \cap V))^o$. If $h(x) \notin \partial \mathcal{K}$, then $h(x) \in \Omega$. Since the transition map ϕ is $\mathcal{C}^p$ and has nonzero Jacobian, it follows from Theorem 6.19 (the Inverse Function Theorem) that $\phi(\Omega)$ is open in $\mathbf{R}^n$. But

$$g(x) \in \phi(\Omega) \subseteq (g \circ h^{-1})(h(U \cap V)) = g(U \cap V) \subseteq g(U).$$

Hence $g(x)$ belongs to the interior of $g(U)$, i.e., cannot belong to ∂H. ∎

Is Definition 10.10 general enough to include every smooth surface whose boundary is made up of smooth curves? At first glance, the answer to this question seems to be no because of the restriction that $h(V)$ be relatively open in some half plane, i.e., part of its boundary be a straight line. Nevertheless, if S is a smooth surface with smooth boundary, one can always find a smoothly equivalent parametrization

(ψ, B) of S such that ∂B is made up of straight lines (see Munkres [8], p. 51). In particular, every smooth surface with smooth boundary is a two-dimensional manifold-with-smooth-boundary.

Our goal is to prove Stokes's Theorem for manifolds-with-smooth-boundary. We shall deal exclusively with manifolds M which are subsets of $\mathbf{R}^m$ for some $m \geq n$. In this case we have two competing concepts: open sets defined by the manifold structure (Definition 10.7) and open sets defined by the relative topology (Definition 5.15). The purpose of the following definition is to make sure that these concepts coincide.

DEFINITION 10.11. An n-dimensional manifold M is said to be *continuously embedded* in $\mathbf{R}^m$ if $m \geq n$, M is a closed subset of $\mathbf{R}^m$, and if (V, h) is a chart of M, then V is relatively open in M and h is a *homeomorphism* from the relative topology on V to the usual topology on $\mathbf{R}^n$, i.e., if U is relatively open in V, then $h(U)$ is open in $\mathbf{R}^n$ and if Ω is open in $\mathbf{R}^n$, then $h^{-1}(\Omega) \cap V$ is relatively open in V.

From now on, by a manifold we mean a manifold-with-smooth-boundary (whose boundary may or may not be empty) which is continuously embedded in some $\mathbf{R}^m$.

We now define what it means for a manifold to be orientable.

DEFINITION 10.12. Let $m \geq n$ and $M \subset \mathbf{R}^m$.

i) A $\mathcal{C}^p$ atlas (8) is said to be *oriented* if

$$(10) \qquad\qquad \Delta_{h_\alpha \circ h_\beta^{-1}}(\boldsymbol{u}) > 0$$

    for all $\boldsymbol{u} \in h_\beta(V_\alpha \cap V_\beta)$ and $\alpha, \beta \in A$ (compare with Definition 8.13).

ii) An n-dimensional $\mathcal{C}^p$ manifold M is said to be *orientable* if it has an oriented atlas.

iii) Two oriented $\mathcal{C}^p$ atlases $\mathcal{A}$, $\mathcal{B}$ of an orientable manifold M are said to be *orientation compatible* if $\mathcal{A} \cup \mathcal{B}$ is an oriented atlas. (Note that orientation compatibility is an equivalence relation.)

iv) An *orientation* of an n-dimensional orientable $\mathcal{C}^p$ manifold M is an equivalence class of oriented $\mathcal{C}^p$ atlases. If $\mathcal{A}$ is an oriented $\mathcal{C}^p$ atlas of M, then the *orientation* generated by $\mathcal{A}$ is the orientation of M which contains $\mathcal{A}$.

An orientation of a manifold M can be used to induce an orientation of ∂M in the following way.

DEFINITION 10.13. Let M be a manifold with orientation $\mathcal{O}$ and $\mathcal{A}$ be an atlas of M consisting of charts (V, h) from $\mathcal{O}$ which satisfy $h_1(\boldsymbol{x}) \leq 0$ for all $\boldsymbol{x} \in V$, and $h_1(\boldsymbol{x}) = 0$ if and only if $\boldsymbol{x} \in \partial M \cap V$. The *orientation induced* on ∂M by $\mathcal{A}$ is the orientation of ∂M generated by the atlas

$$\widetilde{\mathcal{A}} = \{(\widetilde{V}, \widetilde{h}) : (V, h) \in \mathcal{A}\},$$

where $\widetilde{V} = V \cap \partial M$ and $\widetilde{h}(\boldsymbol{x}) = (h_2(\boldsymbol{x}), \ldots, h_n(\boldsymbol{x}))$.

The following result shows that $\widetilde{\mathcal{A}}$ is an oriented atlas of ∂M when $\mathcal{O}$ is an orientation of M.

Remark 6. *Suppose (V, h) and (U, g) are charts from an orientation $\mathcal{O}$ of an oriented manifold M which satisfy $h_1(\boldsymbol{x}) \leq 0$ (respectively, $g_1(\boldsymbol{x}) \leq 0$) for $\boldsymbol{x} \in V$ (respectively, $\boldsymbol{x} \in U$), and $h_1(\boldsymbol{x}) = 0$ (respectively, $g_1(\boldsymbol{x}) = 0$) if and only if $\boldsymbol{x} \in \partial M \cap V$ (respectively, $\boldsymbol{x} \in \partial M \cap U$). If $\widetilde{h} = (h_2, \ldots, h_n)$ and $\widetilde{g} = (g_2, \ldots, g_n)$, then*

$$\Delta_{\widetilde{h} \circ \widetilde{g}^{-1}}(\boldsymbol{u}) > 0 \ \textit{for all } \boldsymbol{u} \in \widetilde{g}(\widetilde{V} \cap \widetilde{U}).$$

Proof. Let $(t, \boldsymbol{u}) = (t, u_2, \ldots, u_n)$ represent a general point in $\mathbf{R}^n$, $\phi = h \circ g^{-1}$ be the transition from $g(U)$ to $h(V)$ and ϕ_1 be the first component of ϕ. By Remark 5, ϕ takes boundary points to boundary points. Since $h_1(\boldsymbol{x}) = g_1(\boldsymbol{x}) = 0$ for $\boldsymbol{x} \in \partial M \cap V \cap U$, it follows that $\phi_1(0, \boldsymbol{u}) = 0$ for all $\boldsymbol{u} \in \widetilde{g}(\widetilde{V} \cap \widetilde{U})$. Consequently, the first row of the Jacobian matrix $D(h \circ g^{-1})(0, \boldsymbol{u})$ is given by

$$\frac{\partial \phi_1}{\partial t}(0, \boldsymbol{u}) \quad 0 \quad \ldots \quad 0.$$

It follows that

$$\Delta_{h \circ g^{-1}}(0, \boldsymbol{u}) = \frac{\partial \phi_1}{\partial t}(0, \boldsymbol{u}) \cdot \Delta_{\widetilde{h} \circ \widetilde{g}^{-1}}(\boldsymbol{u}).$$

Moreover, the conditions $h_1 \leq 0$ on V and $g_1 \leq 0$ on U imply

$$\frac{\partial \phi_1}{\partial t}(0, \boldsymbol{u}) = \lim_{t \to 0-} \frac{\phi_1(t, \boldsymbol{u}) - \phi_1(0, \boldsymbol{u})}{t} = \lim_{t \to 0-} \frac{\phi_1(t, \boldsymbol{u})}{t} \geq 0.$$

Since $\Delta_{h \circ g^{-1}} > 0$ on $g(V \cap U)$, we conclude that $\Delta_{\widetilde{h} \circ \widetilde{g}^{-1}}(\boldsymbol{u}) > 0$ for each $\boldsymbol{u} \in \widetilde{g}(\widetilde{V} \cap \widetilde{U})$. ∎

For two-dimensional manifolds, the condition $h_1(\boldsymbol{x}) \leq 0$ makes the induced orientation as defined above agree with the right-handed orientation introduced in Section 8.4 (see Figure 10.1)

Our definition of n-dimensional manifolds is quite general, but not general enough. It does not include n-dimensional rectangles. (There is no way to parametrize a corner of a two-dimensional rectangle by using relatively open sets in a half plane).

One way to fix this is to extend the definition of charts to include "corner" charts. This extension is still not general enough to include all piecewise smooth curves; for example, it does not include curves with cusps (e.g., $y = x^{2/3}$). The theory can be extended once again by taking limits of "manifolds with corners." For details see Loomis and Sternberg [6].

We will take a less ambitious approach by treating the rectangular case separately. By a chart of an n-dimensional region R in $\mathbf{R}^n$ (this includes all n-dimensional rectangles) we mean a pair (E, h) where $R^o \subseteq E \subseteq R$ and $h : V \to \mathbf{R}^n$ is 1–1 and continuously differentiable on some open set V which contains R with $\Delta_h \neq 0$ on V. (Notice that by the Inverse Function Theorem, h is a homeomorphism on V and that by Theorem 7.3, $h(E)$ is a Jordan region.) Using such charts, we can define atlases and manifolds in the same way as above. The end result is that we

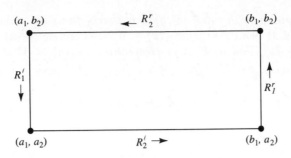

Figure 10.2

can consider any n-dimensional region in $\mathbf{R}^n$ to be a manifold. Notice that if I represents the identity function on $\mathbf{R}^n$, i.e., $I(\boldsymbol{x}) = \boldsymbol{x}$ for all $\boldsymbol{x} \in \mathbf{R}^n$, and if R is an n-dimensional rectangle, then $\{(R, I)\}$ is an atlas of R. The orientation generated by the identity chart is called the *usual orientation* on R. Also notice that since R is closed, the "manifold" boundary of R is precisely its topological boundary.

When R is a rectangle, what orientation is induced on ∂R by the usual orientation of R? The following result answers this question by showing how to find an atlas of the induced orientation for arbitrary $n \in \mathbf{N}$. Notice that for the special case $n = 2$, this orientation on ∂R is counterclockwise orientation (see Figure 10.2).

THEOREM 10.6. *Let $R = [a_1, b_1] \times \cdots \times [a_n, b_n]$ be an n-dimensional rectangle. For each $j = 1, \ldots, n$, set*

$$R_j^\ell = [a_1, b_1] \times \cdots \times \{a_j\} \times \cdots \times [a_n, b_n],$$

$$R_j^r = [a_1, b_1] \times \cdots \times \{b_j\} \times \cdots \times [a_n, b_n],$$

$$h_j(x_1, \ldots, x_n) = \begin{cases} (a_1 - x_1, -x_2, x_3, \ldots, x_n) & j = 1 \\ (a_j - x_j, (-1)^j x_1, x_2, \ldots, \widehat{x}_j, \ldots, x_n) & j \neq 1, \end{cases}$$

and

$$g_j(x_1, \ldots, x_n) = \begin{cases} (x_1 - b_1, x_2, x_3, \ldots, x_n) & j = 1 \\ (x_j - b_j, (-1)^{j+1} x_1, x_2, \ldots, \widehat{x}_j, \ldots, x_n) & j \neq 1. \end{cases}$$

(The notation $\widehat{x}_j$ indicates that this variable is missing.) If $V_j = R^o \cup R_j^\ell$ and $U_j = R^o \cup R_j^r$, then

$$\mathcal{A} = \{(V_j, h_j), (U_j, g_j) : j = 1, \ldots, n\}$$

is an oriented atlas of R which is compatible with the usual orientation. In particular, if $\widetilde{V}_j = V_j \cap R_j^\ell$, $\widetilde{U}_j = U_j \cap R_j^r$, $\widetilde{h}_j$, and $\widetilde{g}_j$ are defined as in Definition 10.13, then

$$\widetilde{\mathcal{A}} = \{(\widetilde{V}_j, \widetilde{h}_j), (\widetilde{U}_j, \widetilde{g}_j) : j = 1, \ldots, n\}$$

is an oriented atlas of ∂R which belongs to the orientation induced by the usual orientation.

PROOF. Fix $1 \le j \le n$ and let $I(\boldsymbol{x}) = \boldsymbol{x}$ represent the identity function on $\mathbf{R}^n$. By definition, if $j = 1$, then

$$\Delta_{h_j \circ I} = \det \begin{bmatrix} -1 & 0 & 0 & \cdots & 0 \\ 0 & -1 & 0 & \cdots & 0 \\ 0 & 0 & 1 & \cdots & 0 \\ \vdots & \vdots & \vdots & \ddots & \vdots \\ 0 & 0 & 0 & \cdots & 1 \end{bmatrix} = 1 > 0.$$

If $j \ne 1$, then by factoring -1 out of the first row and interchanging $j - 1$ rows we have

$$\Delta_{h_j \circ I} = (-1)^j \det \begin{bmatrix} (-1)^j & 0 & 0 & \cdots & 0 \\ 0 & 1 & 0 & \cdots & 0 \\ 0 & 0 & 1 & \cdots & 0 \\ \vdots & \vdots & \vdots & \ddots & \vdots \\ 0 & 0 & 0 & \cdots & 1 \end{bmatrix} = (-1)^{2j} = 1 > 0.$$

Thus, the chart (V_j, h_j) is compatible with the usual orientation on R.

Let h_{j1} represent the first component of the function h_j, $j = 1, \ldots, n$. Clearly if $\boldsymbol{x} \in R$, then $h_{j1}(\boldsymbol{x}) \le 0$; and $h_{j1}(\boldsymbol{x}) = 0$ if and only if $x_j = a_j$, i.e., if and only if $\boldsymbol{x} \in R_j^\ell$. Thus, $(\widetilde{V}_j, \widetilde{h}_j)$ belongs the orientation induced on ∂R by the usual orientation. A similar argument works for the "right-hand" boundaries R_j^r. ∎

We mentioned in Chapter 8 that a connected smooth curve or connected smooth orientable surface has only two orientations. This is a general principle shared by all connected orientable $\mathcal{C}^1$ manifolds (see Theorem 10.7 below). First we prove the following result.

Lemma. *If M is a connected orientable $\mathcal{C}^1$ manifold and $\mathcal{A}, \mathcal{B}$ are oriented atlases of M, then either*

$$\Delta_{h \circ g^{-1}}(\boldsymbol{u}) > 0$$

for all $(V, h) \in \mathcal{A}$, $(U, g) \in \mathcal{B}$, and $\boldsymbol{u} \in g(V \cap U)$, or

$$\Delta_{h \circ g^{-1}}(\boldsymbol{u}) < 0$$

for all $(V, h) \in \mathcal{A}$, $(U, g) \in \mathcal{B}$, and $\boldsymbol{u} \in g(V \cap U)$.

PROOF. Set

$$A = \{\boldsymbol{x} \in M : \Delta_{h \circ g^{-1}}(g(\boldsymbol{x})) > 0 \text{ for some } (V, h) \in \mathcal{A} \text{ and } (U, g) \in \mathcal{B}\},$$

and

$$B = \{\boldsymbol{x} \in M : \Delta_{h \circ g^{-1}}(g(\boldsymbol{x})) < 0 \text{ for some } (V, h) \in \mathcal{A} \text{ and } (U, g) \in \mathcal{B}\}.$$

We must show that $M = A$ or $M = B$. Since M is connected, it suffices to show that A and B are (relatively) open in M, $M = A \cup B$, and $A \cap B = \emptyset$ (see Theorem 5.19).

To show A is open in M, let $\boldsymbol{x}_0 \in A$ and choose $(V, h) \in \mathcal{A}$, $(U, g) \in \mathcal{B}$ such that $\Delta_{h \circ g^{-1}}(g(\boldsymbol{x}_0)) > 0$. Then

$$\boldsymbol{x}_0 \in \Omega := g^{-1}(\Delta_{h \circ g^{-1}}^{-1}((0, \infty))).$$

Since $h \circ g^{-1}$ is continuously differentiable on $g(U)$, its Jacobian is continuous on $g(U)$. Hence Ω, the inverse image of the open set $(0, \infty)$ under the continuous function $\Delta_{h \circ g^{-1}} \circ g$, must be open in M (Theorem 5.25). It follows that A is open in M. A similar argument proves B is open in M.

To show $M = A \cup B$ we must show that $\Delta_{h \circ g^{-1}}(g(\boldsymbol{x})) \neq 0$ for all $\boldsymbol{x} \in M$. Suppose to the contrary that $\Delta_{h \circ g^{-1}}(g(\boldsymbol{x})) = 0$ for some $\boldsymbol{x} \in M$. Since $\mathcal{A}$ is an atlas of M, choose $(W, \sigma) \in \mathcal{A}$ such that $\boldsymbol{x} \in W$ and set $\boldsymbol{u} = \sigma(\boldsymbol{x})$. By (10) and the Chain Rule,

$$0 < \Delta_{h \circ \sigma^{-1}}(\boldsymbol{u}) = \Delta_{h \circ g^{-1}}(g \circ \sigma^{-1}(\boldsymbol{u}))\Delta_{g \circ \sigma^{-1}}(\boldsymbol{u}) = 0 \cdot \Delta_{g \circ \sigma^{-1}}(\boldsymbol{u}) = 0,$$

a contradiction. Thus, $M = A \cup B$.

Finally, to show $A \cap B$ is empty, suppose to the contrary that there is an $\boldsymbol{x} \in A \cap B$. By definition, this means there exist charts $(V_i, h_i) \in \mathcal{A}$ and $(U_i, g_i) \in \mathcal{B}$ such that

$$(11) \qquad\qquad (-1)^{i+1}\Delta_{h_i \circ g_i^{-1}}(g_i(\boldsymbol{x})) > 0$$

for $i = 1, 2$. Since $\mathcal{A}$ is an orientation, we have by (10) and the Chain Rule that

$$0 < \Delta_{h_1 \circ h_2^{-1}}(h_2(\boldsymbol{x})) = \Delta_{h_1 \circ g_1^{-1}}(g_1 \circ g_2^{-1} \circ g_2 \circ h_2^{-1} \circ h_2(\boldsymbol{x})) \cdot$$
$$\cdot \Delta_{g_1 \circ g_2^{-1}}(g_2 \circ h_2^{-1} \circ h_2(\boldsymbol{x}))\Delta_{g_2 \circ h_2^{-1}}(h_2(\boldsymbol{x}))$$
$$= \Delta_{h_1 \circ g_1^{-1}}(g_1(\boldsymbol{x}))\Delta_{g_1 \circ g_2^{-1}}(g_2(\boldsymbol{x}))\Delta_{g_2 \circ h_2^{-1}}(h_2(\boldsymbol{x})).$$

By (11), the first (respectively, third) of these factors is positive (respectively, negative). Hence, the second factor must be negative. But the second factor is positive since both g_1 and g_2 come from the same oriented atlas $\mathcal{B}$. This contradiction proves the lemma. ∎

THEOREM 10.7. *Let M be a connected orientable $\mathcal{C}^p$ manifold. Then M has exactly two orientations.*

PROOF. We first show that M has at least two orientations. Let $\mathcal{A} = \{(V_\alpha, h_\alpha) : \alpha \in A\}$ be an oriented atlas of M with $h_\alpha = (h_{\alpha 1}, \ldots, h_{\alpha n})$, and consider

$$\mathcal{B} = \{(V_\alpha, g_\alpha) : \alpha \in A\},$$

where $g_\alpha = (-h_{\alpha 1}, h_{\alpha 2}, \ldots, h_{\alpha n})$. Clearly, $\mathcal{B}$ is a $\mathcal{C}^p$ atlas of M. Since

$$\Delta_{g_\alpha \circ g_\beta^{-1}} \circ g_\beta = \Delta_{h_\alpha \circ h_\beta^{-1}} \circ h_\beta > 0,$$

$\mathcal{B}$ is orientable. Since

$$\Delta_{g_\alpha \circ h_\beta^{-1}} \circ h_\beta = -\Delta_{h_\alpha \circ h_\beta^{-1}} \circ h_\beta < 0,$$

$\mathcal{B}$ is not orientation compatible with $\mathcal{A}$. Thus, M has at least two orientations.

To show that M has no more than two orientations, suppose to the contrary that M has three distinct orientations. Let $\mathcal{A}$, $\mathcal{B}$, and $\mathcal{O}$ be atlases from each of these orientations and choose $(V, h) \in \mathcal{A}$, $(U, g) \in \mathcal{B}$ such that $V \cap U \neq \emptyset$. Since these orientations are distinct, there exist $(W_i, \sigma_i) \in \mathcal{O}$, $i = 1, 2$, such that

$$\Delta_{h \circ \sigma_1^{-1}}(\sigma_1(\boldsymbol{x})) < 0 \quad \text{and} \quad \Delta_{\sigma_2 \circ g^{-1}}(g(\boldsymbol{y})) < 0$$

for some $\boldsymbol{x} \in V \cap W_1$ and $\boldsymbol{y} \in U \cap W_2$. By the lemma, $\Delta_{h \circ \sigma^{-1}} \circ \sigma < 0$ and $\Delta_{\sigma \circ g^{-1}} \circ g < 0$ on M for all $(W, \sigma) \in \mathcal{O}$. Let $\boldsymbol{x} \in V \cap U$ and choose $(W, \sigma) \in \mathcal{O}$ such that $\boldsymbol{x} \in W$. By the Chain Rule,

$$\Delta_{h \circ g^{-1}}(g(\boldsymbol{x})) = \Delta_{h \circ \sigma^{-1}}(\sigma(\boldsymbol{x}))\Delta_{\sigma \circ g^{-1}}(g(\boldsymbol{x})).$$

This is a product of two negative numbers, hence positive. It follows from the lemma that $\Delta_{h \circ g^{-1}} \circ g > 0$ on U for all $(V, h) \in \mathcal{A}$ and $(U, g) \in \mathcal{B}$. Therefore, $\mathcal{A}$ is orientation compatible with $\mathcal{B}$, a contradiction. ∎

EXERCISES

1. Let M be a $\mathcal{C}^p$ manifold (not necessarily continuously embedded in some $\mathbf{R}^m$).

   a) If $\{V_\alpha\}_{\alpha \in A}$ is a collection of open sets in M, prove that $\cup_{\alpha \in A} V_\alpha$ is open in M.

   b) If $V_1, \ldots, V_N$ are open in M, prove $\cap_{j=1}^{N} V_j$ is open in M.

2. Prove that $\mathcal{C}^p$ compatibility and orientation compatibility are equivalence relations.

3. Prove that the boundary of an n-dimensional $\mathcal{C}^p$ manifold-with-smooth-boundary is an $(n-1)$-dimensional manifold.

4. **This exercise is used in Section 10.3.** *Translation* on $\mathbf{R}^n$ by an $\boldsymbol{a} \in \mathbf{R}^n$ is defined by $\sigma(\boldsymbol{x}) = \boldsymbol{x} + \boldsymbol{a}$ for $\boldsymbol{x} \in \mathbf{R}^n$. *Dilation* on $\mathbf{R}^n$ by a $\delta > 0$ is defined by $\sigma(\boldsymbol{x}) = \delta \boldsymbol{x}$ for $\boldsymbol{x} \in \mathbf{R}^n$.

   a) Prove that if $\mathcal{A}$ is an oriented atlas of a manifold M and σ is a translation or a dilation, then $\mathcal{B} = \{(V, \sigma \circ h) : (V, h) \in \mathcal{A}\}$ is an atlas of M which is orientation compatible with $\mathcal{A}$.

   b) Let $\mathcal{A}$ be an orientation of a manifold M and $\boldsymbol{x} \in M$ be an interior point. Prove there is a chart (V, h) at $\boldsymbol{x}$ such that $h(V) = B_1(\mathbf{0})$ and $h(\boldsymbol{x}) = \mathbf{0}$.

5. Let $\mathcal{A}$ be an n-dimensional $\mathcal{C}^\infty$ atlas of a manifold M. A function $f : M \to \mathbf{R}^k$ is said to be $\mathcal{C}^p$ on M if $f \circ h^{-1} : h(V) \to \mathbf{R}^k$ is $\mathcal{C}^p$ for all charts $(V, h) \in \mathcal{A}$.

   a) Prove that this definition is independent of the atlas $\mathcal{A}$.

   b) Prove that the composition of $\mathcal{C}^p$ functions is a $\mathcal{C}^p$ function.

   c) Prove if (V, h) is a chart of M, then h is a $\mathcal{C}^\infty$ function on V.

6. Prove that the sphere $x_1^2 + \cdots + x_n^2 = a^2$ is an $(n-1)$-dimensional manifold in $\mathbf{R}^n$.

10.3 STOKES'S THEOREM ON MANIFOLDS   *This section uses material from Sections 7.5, 10.1, and 10.2.*

We shall define oriented integrals of n–forms on n-dimensional manifolds and obtain a fundamental theorem of calculus for these integrals. Recall that for us, a manifold M is closed and continuously embedded in $\mathbf{R}^m$ (see Definition 10.11). In particular, given W open in M, there is an open set $\Omega \subset \mathbf{R}^m$ such that $\Omega \cap M = W$. This assumption is not essential and all results stated in this section are valid without it. We make it to simplify the proof that partitions of unity exist on a manifold.

Lemma 1 $[\mathcal{C}^\infty$ PARTITIONS OF UNITY ON A COMPACT MANIFOLD]. *Let M be a compact n-dimensional $\mathcal{C}^p$ manifold in $\mathbf{R}^m$ with orientation $\mathcal{O}$ and $\{U_\alpha\}_{\alpha \in A}$ be an open covering of M. Then there exist $\mathcal{C}^\infty$ functions $\phi_j : \mathbf{R}^m \to \mathbf{R}$, $j = 1, 2, \ldots, N$, and an atlas $\{(V_j, h_j) \in \mathcal{O} : j = 1, \ldots, N\}$ of M such that*

   i) *given $j \in \{1, \ldots, N\}$ there is an $\alpha \in A$ such that $V_j \subset U_\alpha$,*
   ii) *$0 \le \phi_j(\boldsymbol{x}) \le 1$ for $\boldsymbol{x} \in \mathbf{R}^m$, $j = 1, \ldots, N$,*
   iii) *$\mathrm{spt}\,\phi_j \cap M \subset V_j$, for $j = 1, \ldots, N$, and*
   iv) *$\sum_{j=1}^N \phi_j(\boldsymbol{x}) = 1$ for $\boldsymbol{x} \in M$.*

PROOF. For each $\boldsymbol{x} \in M$ choose a chart $(V_{\boldsymbol{x}}, h_{\boldsymbol{x}}) \in \mathcal{O}$ such that $\boldsymbol{x} \in V_{\boldsymbol{x}} \subseteq U_\alpha$ for some $\alpha \in A$. Choose bounded open sets $\Omega_{\boldsymbol{x}}$ and $B_{\boldsymbol{x}}$ in $\mathbf{R}^m$ such that $\boldsymbol{x} \in \Omega_{\boldsymbol{x}} \subset \overline{\Omega}_{\boldsymbol{x}} \subset B_{\boldsymbol{x}}$ and $B_{\boldsymbol{x}} \cap M = V_{\boldsymbol{x}}$. By Theorem 7.16, there is a $\mathcal{C}^\infty$ function $\psi_{\boldsymbol{x}}$ on $\mathbf{R}^m$ such that $\psi_{\boldsymbol{x}} = 1$ on $\Omega_{\boldsymbol{x}}$ and $\mathrm{spt}\,\psi_{\boldsymbol{x}} \subset B_{\boldsymbol{x}}$. Clearly, $\{\Omega_{\boldsymbol{x}}\}_{\boldsymbol{x} \in M}$ is an open covering of M. Since M is compact, choose a finite subcover $\{\Omega_1, \ldots, \Omega_N\}$. If $\Omega_j = \Omega_{\boldsymbol{x}}$, set $\psi_j = \psi_{\boldsymbol{x}}$ and $(V_j, h_j) = (V_{\boldsymbol{x}}, h_{\boldsymbol{x}})$. By construction, $\psi_j = 1$ on Ω_j and $\mathrm{spt}\,\psi_j \cap M \subset V_j$.

Set $\phi_1 = \psi_1$, $\phi_2 = (1 - \psi_1)\psi_2$, $\ldots$, $\phi_N = (1 - \psi_1)\ldots(1 - \psi_{N-1})\psi_N$. Then ϕ_j is $\mathcal{C}^\infty$ on $\mathbf{R}^m$ and $\mathrm{spt}\,\phi_j \cap M \subset \mathrm{spt}\,\psi_j \cap M \subset V_j$. This verifies i), ii), and iii). It is easy to see by induction that

$$\sum_{j=1}^N \phi_j = 1 - (1 - \psi_1)\ldots(1 - \psi_N).$$

Since $\{\Omega_j\}$ covers M, this verifies iv). ∎

We shall call the functions $\phi_1, \ldots, \phi_N$ given in Lemma 1 a $\mathcal{C}^\infty$ *partition of unity on M subordinate* to the covering $\{U_\alpha\}_{\alpha \in A}$.

By a differential form on a manifold M we mean an r–form on some open set $\Omega \subset \mathbf{R}^m$ such that $M \subset \Omega$. Thus, a decomposable r–form on M has the form

$$\omega = f\, dx_{i_1} \ldots dx_{i_r},$$

where f is a 0–form on some open set Ω which contains M.

We are now prepared to define the integral of a differential form on an oriented manifold. (This definition includes oriented line integrals and oriented surface integrals—see Exercise 5 in Section 10.1.)

DEFINITION 10.14. Let $m \geq n$, M be an n-dimensional oriented C^p manifold in $\mathbf{R}^m$, ω be a continuous n-form on M, and $\mathcal{O}$ be an orientation of M.

i) If $\phi : V \to \mathbf{R}$ is continuous on V for some chart $(V, h) \in \mathcal{O}$ and spt $\phi \cap M \subseteq V$, then the *oriented integral* of $\phi\omega$ on M is defined by

$$\int_M \phi\omega = \int_{h(V)} (h^{-1})^*(\phi\omega)(\mathbf{u}) \, d\mathbf{u}.$$

ii) If M is compact, then the *oriented integral* of ω on M is defined by

$$\int_M \omega = \sum_{j=1}^{N} \int_M \phi_j \omega,$$

where $\phi_1, \ldots, \phi_N$ is any C^∞ partition of unity on M subordinate to the orientation $\mathcal{O}$.

The following two remarks show that these definitions make sense.

Remark 1. *The value of $\int_M \phi\omega$ does not depend on the chart chosen.*

Proof. Let $(U, g) \in \mathcal{O}$ be another chart which satisfies $U \supseteq$ spt $\phi \cap M$. We may suppose that ω is decomposable, i.e., $\omega = f \, dx_{i_1} \ldots dx_{i_n}$. Let $H = h^{-1}$ and $G = g^{-1}$. Since $\mathcal{O}$ is an orientation, $\Delta_{h \circ g^{-1}} \geq 0$ on $g(U)$. Moreover, by the Chain Rule,

$$\Delta_{(G_{i_1}, \ldots, G_{i_n})} = \Delta_{(H_{i_1}, \ldots, H_{i_n})} \circ h \circ g^{-1} \Delta_{h \circ g^{-1}}.$$

Since $h(V \cap U) = (h \circ g^{-1}) \circ g(V \cap U)$ and ϕ is supported in $V \cap U$, it follows from the Fundamental Theorem of Differential Transforms and a change of variables in $\mathbf{R}^n$ that

$$\int_{h(V)} (h^{-1})^*(\phi\omega)(\mathbf{u}) \, d\mathbf{u}$$

$$= \int_{h(V)} (\phi \circ h^{-1})(\mathbf{u}) \, (f \circ h^{-1})(\mathbf{u}) \Delta_{(H_{i_1}, \ldots, H_{i_n})}(\mathbf{u}) \, d\mathbf{u}$$

$$= \int_{g(U)} (\phi \circ g^{-1})(\mathbf{y}) \, (f \circ g^{-1})(\mathbf{y}) \cdot$$
$$\cdot \Delta_{(H_{i_1}, \ldots, H_{i_n})}(h \circ g^{-1}(\mathbf{y})) |\Delta_{h \circ g^{-1}}(\mathbf{y})| \, d\mathbf{y}$$

$$= \int_{g(U)} (\phi \circ g^{-1})(\mathbf{y}) \, (f \circ g^{-1})(\mathbf{y}) \Delta_{(G_{i_1}, \ldots, G_{i_n})}(\mathbf{y}) \, d\mathbf{y}$$

$$= \int_{g(U)} (g^{-1})^*(\phi\omega)(\mathbf{y}) \, d\mathbf{y}. \quad \blacksquare$$

Remark 2. *The value of $\int_M \omega$ does not depend on the C^∞ partition of unity $\phi_1, \ldots, \phi_N$ chosen.*

PROOF. If $\psi_1, \ldots, \psi_L$ is another $\mathcal{C}^\infty$ partition of unity subordinate to $\mathcal{O}$, then

$$\sum_{k=1}^{L} \int_M \psi_k \omega = \sum_{k=1}^{L} \int_M \psi_k \left(\sum_{j=1}^{N} \phi_j\right)\omega = \sum_{k=1}^{L} \sum_{j=1}^{N} \int_M \psi_k \phi_j \omega$$

$$= \sum_{j=1}^{N} \int_M \left(\sum_{k=1}^{L} \psi_k\right)\phi_j \omega = \sum_{j=1}^{N} \int_M \phi_j \omega. \ \blacksquare$$

The following result justifies the identification of $d(x, y)$ with $dx\, dy$ made below (3) in Section 10.1. (See also Exercise 5 in Section 10.1.)

Remark 3. *If R is an n-dimensional rectangle with the usual orientation and $\omega = f\, dx_1 \ldots dx_n$ is an n–form on some open $\Omega \supset R$, then*

$$\int_R \omega = \int_R f(\boldsymbol{x})\, d\boldsymbol{x}.$$

PROOF. By hypothesis, (R, I) is a chart of R. Hence, by Definition 10.14i,

$$\int_R \omega = \int_{I(R)} I^*(\omega)(\boldsymbol{x})\, d\boldsymbol{x} = \int_R f(\boldsymbol{x})\, d\boldsymbol{x}. \ \blacksquare$$

We shall prove that the oriented integral of the exterior derivative $d\omega$ of a differential form on a manifold M is determined by the behavior of ω on ∂M.

STRATEGY: The idea behind the proof is straightforward. First we prove the result when M is a rectangle (Lemma 2). (This case follows directly from the one dimensional Fundamental Theorem of Calculus because the boundary of a rectangle only moves in one dimension at a time.) Next, we pull-back this result to sufficiently small charts on M (Lemma 3). Finally, by using a $\mathcal{C}^\infty$ partition of unity subordinate to a covering by sufficiently small charts, we establish the general result. (The proofs of Lemma 3 and Theorem 10.9 presented here come from Spivak [12].[1])

Lemma 2. *Let $R = [a_1, b_1] \times \cdots \times [a_n, b_n]$ be an n-dimensional rectangle, ω be a $\mathcal{C}^1$ $(n-1)$–form on an open set U which contains R, and suppose R has the usual orientation. If ∂R carries the induced orientation, then*

$$\int_R d\omega = \int_{\partial R} \omega.$$

PROOF. Let $\widetilde{\mathcal{A}} = \{(\widetilde{V}_j, \widetilde{h}_j), (\widetilde{U}_j, \widetilde{g}_j) : j = 1, \ldots, n\}$ be the atlas of ∂R introduced in Theorem 10.6 and set $H_j = \widetilde{h}_j^{-1}$, $G_j = \widetilde{g}_j^{-1}$. We claim that

$$(12) \qquad (H_j)^*\omega = (-1)^j (f_j \circ \widetilde{h}_j^{-1}) du_1 \ldots du_{n-1}$$

---

[1]M. Spivak, Calculus on Manifolds, New York: W. A. Benjamin, Inc., 1965). Reprinted with permission of Addison-Wesley Publishing Company

and

(13)
$$(G_j)^* \omega = (-1)^{j+1}(f_j \circ \widetilde{g}_j^{-1}) du_1 \ldots du_{n-1}$$

for any $(n-1)$–form $\omega = \sum_{i=1}^n f_i dx_1 \ldots \widehat{dx_i} \ldots dx_n$ on U.

To prove (12), fix j and notice by construction that

$$H_j(u_1, \ldots, u_{n-1}) = \begin{cases} (a_1, -u_1, u_2, \ldots, u_{n-1}) & j = 1 \\ ((-1)^j u_1, u_2, \ldots, a_j, \ldots, u_{n-1}) & j > 1. \end{cases}$$

Representing the ith component of H_j by H_{ji}, we have

$$\Delta_i := \frac{\partial(H_{j1}, \ldots, \widehat{H_{ji}}, \ldots, H_{jn})}{\partial(u_1, \ldots, u_{n-1})} = \begin{cases} (-1)^j & i = j \\ 0 & i \neq j. \end{cases}$$

Hence, by the Fundamental Theorem of Differential Transforms,

$$(H_j)^* \omega = \sum_{i=1}^n (f_i \circ \widetilde{h}_j^{-1}) \Delta_i du_1 \ldots du_{n-1} = (-1)^j (f_j \circ \widetilde{h}_j^{-1}) du_1 \ldots du_{n-1}.$$

This proves (12). A similar argument proves (13).

Using (12) and (13), a change of variables in the second variable when $j = 1$ (respectively, in the first variable when $j > 1$), the Fundamental Theorem of Calculus in the jth variable, and Fubini's Theorem, we see that

$$\int_{\partial R} \omega = \sum_{j=1}^n \left(\int_{R_j^\ell} \omega + \int_{R_j^r} \omega \right)$$

$$= \sum_{j=1}^n (-1)^j \left(\int_{\widetilde{h}_j(\widetilde{V}_j)} (f_j \circ \widetilde{h}_j^{-1})(\mathbf{u}) \, d\mathbf{u} - \int_{\widetilde{g}_j(\widetilde{U}_j)} (f_j \circ \widetilde{g}_j^{-1})(\mathbf{u}) \, d\mathbf{u} \right)$$

$$= \sum_{j=1}^n (-1)^j \int_{a_1}^{b_1} \cdots \widehat{\int_{a_j}^{b_j}} \cdots \int_{a_n}^{b_n}$$

$$(f_j(x_1, \ldots, a_j, \ldots, x_n) - f_j(x_1, \ldots, b_j, \ldots, x_n)) \cdot$$

$$\cdot dx_n \ldots \widehat{dx_j} \ldots dx_1$$

$$= \sum_{j=1}^n (-1)^{j+1} \int_{a_1}^{b_1} \cdots \int_{a_n}^{b_n} \frac{\partial f_j}{\partial x_j}(x_1, \ldots, x_n) \, dx_n \ldots dx_1$$

$$= \int_R \sum_{j=1}^n (-1)^{j+1} \frac{\partial f_j}{\partial x_j}(\mathbf{x}) \, d\mathbf{x},$$

i.e.,

(14)
$$\int_{\partial R} \omega = \int_R \sum_{j=1}^n (-1)^{j+1} \frac{\partial f_j}{\partial x_j}(\mathbf{x}) \, d\mathbf{x}.$$

On the other hand, it is clear that

$$d(g dx_1 \ldots \widehat{dx_j} \ldots dx_n) = (-1)^{j+1} \frac{\partial g}{\partial x_j} dx_1 \ldots dx_n$$

for any differentiable function g. Thus

$$d\omega = \sum_{j=1}^{n} (-1)^{j+1} \frac{\partial f_j}{\partial x_j} dx_1 \ldots dx_n.$$

We conclude by Remark 3 and (14) that

$$\int_R d\omega = \int_R \sum_{j=1}^{n} (-1)^{j+1} \frac{\partial f_j}{\partial x_j}(\boldsymbol{x}) \, d\boldsymbol{x} = \int_{\partial R} \omega. \quad \blacksquare$$

Lemma 3. *Let M be an n-dimensional orientable C^2 manifold-with-smooth-boundary and $\mathcal{O}$ be an orientation of M. For each $\boldsymbol{x} \in M$ there is a chart $(V, h) \in \mathcal{O}$ at $\boldsymbol{x}$ such that if η is any C^1 $(n-1)$–form supported in V, then*

$$\int_V d\eta = \int_{V \cap \partial M} \eta.$$

PROOF. Suppose first that $\boldsymbol{x} \notin \partial M$. Then there is a chart (U, h) at $\boldsymbol{x}$ such that $h(U)$ is open in $\mathbf{R}^n$. Let R be an n-dimensional rectangle such that

$$h(\boldsymbol{x}) \in R^o \subset R \subset h(U)$$

and set $V = h^{-1}(R^o)$ (see Figure 10.3). Let η be an $(n-1)$–form supported in V. We may suppose that η is decomposable, i.e.,

$$(h^{-1})^*(\eta) = f dx_1 \ldots \widehat{dx_j} \ldots dx_n.$$

By definition and (5),

$$(h^{-1})^*(d\eta)) = d((h^{-1})^*(\eta)) = (-1)^{j-1} \frac{\partial f}{\partial x_j} dx_1 \ldots dx_n.$$

Since spt $f \subset h(V) = R^o$, it follows from Definition 10.14 and Lemma 2 (using the identity chart on R) that

$$\int_V d\eta = \int_R (-1)^{j-1} \frac{\partial f}{\partial x_j}(\boldsymbol{x}) \, d\boldsymbol{x} = \int_R d((h^{-1})^*(\eta)) = \int_{\partial R} (h^{-1})^*(\eta).$$

Since spt $((h^{-1})^*\eta) \subset R^o$ and $R^o \cap \partial R = \emptyset$, this last integral is zero, i.e., $\int_V d\eta = 0$. On the other hand, $h(U)$ is open in $\mathbf{R}^n$ so $h(V) \subseteq h(U)$ contains no boundary points of M. Therefore, $V \cap \partial M = \emptyset$ and

$$\int_{V \cap \partial M} \eta = 0 = \int_V d\eta.$$

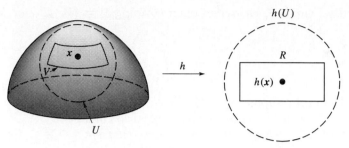

Figure 10.3

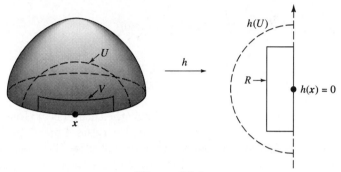

Figure 10.4

Next, suppose $\boldsymbol{x} \in \partial M$. Let (U, h) be a chart at $\boldsymbol{x}$ such that $h(U)$ is relatively open in the left half space $\mathcal{H}_1$ and $h(\boldsymbol{x}) = \boldsymbol{0}$. Let R be a n-dimensional rectangle such that $R^o \subset R \subset U$ and $R \cap \partial \mathcal{H}_1 = R_1^r$ (see Figure 10.4), and set $V = h^{-1}(R^o \cup R_1^r)$. Let η be a decomposable $\mathcal{C}^1$ $(n-1)$-form supported on V, with

$$(h^{-1})^*(\eta) = f \, dx_1 \ldots \widehat{dx_j} \ldots dx_n,$$

and $\boldsymbol{u} = (u_1, \ldots, u_{n-1}) \in \partial R$. Then f is identically zero on $\partial R \setminus R_1^r$, and it follows from Definition 10.14 and Lemma 2 that

$$\int_{V \cap \partial M} \eta = \int_{R_1^r} f(\boldsymbol{u}) \, d\boldsymbol{u} = \sum_{j=1}^n (-1)^j \left(\int_{R_j^\ell} f(\boldsymbol{u}) \, d\boldsymbol{u} - \int_{R_j^r} f(\boldsymbol{u}) \, d\boldsymbol{u} \right)$$

$$= \int_{\partial R} (h^{-1})^*(\eta)(\boldsymbol{u}) \, d\boldsymbol{u} = \int_R (h^{-1})^*(d\eta)(\boldsymbol{x}) \, d\boldsymbol{x}$$

$$= \int_{h(V)} (h^{-1})^*(d\eta)(\boldsymbol{x}) \, d\boldsymbol{x} = \int_V d\eta. \quad \blacksquare$$

We are now prepared to prove the general result.

THEOREM 10.8 [Stokes's Theorem on Manifolds]. *Let M be a compact n-dimensional oriented $\mathcal{C}^2$ manifold-with-smooth-boundary. If ω is a $\mathcal{C}^1$ $(n-1)$-form on M, then*

$$\int_M d\omega = \int_{\partial M} \omega.$$

PROOF. By Lemma 3, choose an open covering $\mathcal{V} = \{V_{\boldsymbol{x}}\}_{\boldsymbol{x} \in M}$ of M such that $\boldsymbol{x} \in V_{\boldsymbol{x}}$ and

$$
(15) \qquad \int_{V_{\boldsymbol{x}}} d\eta = \int_{V_{\boldsymbol{x}} \cap \partial M} \eta
$$

for all $(n-1)$–forms η supported in $V_{\boldsymbol{x}}$. Since M is compact, choose open sets $V_j = V_{\boldsymbol{x}_j}$, $j = 1, \ldots, N$, which cover M and a $\mathcal{C}^\infty$ partition of unity $\phi_1, \ldots, \phi_N$ on M such that $\operatorname{spt} \phi_j \cap M \subseteq V_j$. Set $\eta_j = \phi_j \omega$ and observe that $\operatorname{spt} \eta_j \cap M \subseteq \operatorname{spt} \phi_j \cap M \subseteq V_j$ for each j and $\omega = \sum_{j=1}^{N} \eta_j$. Hence, by (15),

$$
\int_M d\omega = \sum_{j=1}^{N} \int_M d\eta_j = \sum_{j=1}^{N} \int_{V_j} d\eta_j
$$

$$
= \sum_{j=1}^{N} \int_{V_j \cap \partial M} \eta_j = \sum_{j=1}^{N} \int_{\partial M} \eta_j = \int_{\partial M} \omega. \qquad \blacksquare
$$

This result extends Theorems 8.3, 8.4, and 8.5 (the theorems of Green, Gauss, and Stokes) to regions with smooth boundaries (see Exercises 4 and 5 in Section 10.1). Theorem 10.8 also holds for manifolds with singularities, i.e., piecewise smooth boundaries. (For a treatment of manifolds with singularities, see Loomis and Sternberg [6].)

We close this section with an n-dimensional analogue of Theorem 8.6. A set $V \subset \mathbf{R}^n$ is said to be *star-shaped* (centered at $\mathbf{0}$) if for each $\boldsymbol{x} \in V$ the line segment between $\boldsymbol{x}$ and $\mathbf{0}$ lies in V, i.e., $t\boldsymbol{x} \in V$ for all $0 \leq t \leq 1$. An r–form ω is said to be *exact* on V if there is an $(r-1)$–form η on V such that $d\eta = \omega$.

THEOREM 10.9 [THE POINCARÉ LEMMA]. *Let V be an open star-shaped set in $\mathbf{R}^n$ and ω be a $\mathcal{C}^1$ r–form on V. Then ω is exact on V if and only if $d\omega = 0$ on V.*

PROOF. For each r–form

$$
\omega = \sum_{1 \leq i_1 < \cdots < i_r \leq n} f_{i_1, \ldots, i_r}(\boldsymbol{x}) \, dx_{i_1} \ldots dx_{i_r}
$$

on V, define an $(r-1)$–form $\Lambda(\omega)$ on V by

$$
(16) \qquad \Lambda(\omega) = \sum_{1 \leq i_1 < \cdots < i_r \leq n} \sum_{k=1}^{r} (-1)^{k-1} \left(\int_0^1 t^{r-1} f_{i_1, \ldots, i_r}(t\boldsymbol{x}) \, dt \right) x_{i_k} \cdot
$$
$$
\cdot dx_{i_1} \ldots \widehat{dx_{i_k}} \ldots dx_{i_r}.
$$

Since V is star-shaped and f is defined on V, the integrals in (16) make sense for each $\boldsymbol{x} \in V$. Thus, $\Lambda(\omega)$ is an $(r-1)$–form on V. We claim that

$$
(17) \qquad \Lambda(d\omega) + d(\Lambda(\omega)) = \omega
$$

for every r–form ω on V.

To prove (17) we may suppose that ω is decomposable, i.e., $\omega = f\, dx_{i_1} \ldots dx_{i_r}$. By definition,

$$d\omega = \sum_{j=1}^{n} \frac{\partial f}{\partial x_j}\, dx_j\, dx_{i_1} \ldots dx_{i_r}$$

is an $(r+1)$–form on V. Letting $i_0 = j$, we have by (16) that

(18)

$$\Lambda(d\omega) = \sum_{k=0}^{r}(-1)^k \sum_{j=1}^{n}\left(\int_0^1 t^r \frac{\partial f}{\partial x_j}(t\boldsymbol{x})\, dt \right) x_{i_k}\, dx_{i_0}\, dx_{i_1} \ldots \widehat{dx_{i_k}} \ldots dx_{i_r}$$

$$= \sum_{j=1}^{n}\left(\int_0^1 t^r \frac{\partial f}{\partial x_j}(t\boldsymbol{x})\, dt \right) x_j\, dx_{i_1} \ldots dx_{i_r}$$

$$- \sum_{j=1}^{n}\sum_{k=1}^{r}(-1)^{k-1}\left(\int_0^1 t^r \frac{\partial f}{\partial x_j}(t\boldsymbol{x})\, dt \right) x_{i_k} \cdot$$

$$\cdot\, dx_j\, dx_{i_1} \ldots \widehat{dx_{i_k}} \ldots dx_{i_r}$$

On the other hand, by the Product Rule, differentiating under the integral sign (see Theorem 6.3), and the Chain Rule, we have

$$d\left(\left(\int_0^1 t^{r-1} f(t\boldsymbol{x})\, dt \right) x_{i_k} \right)$$

$$= d\left(\int_0^1 t^{r-1} f(t\boldsymbol{x})\, dt \right) x_{i_k} + \left(\int_0^1 t^{r-1} f(t\boldsymbol{x})\, dt \right) dx_{i_k}$$

$$= \sum_{j=1}^{n}\left(\int_0^1 t^r \frac{\partial f}{\partial x_j}(t\boldsymbol{x})\, dt \right) x_{i_k}\, dx_j + \left(\int_0^1 t^{r-1} f(t\boldsymbol{x})\, dt \right) dx_{i_k}.$$

Thus, by the Anticommutative Property, the exterior derivative of $\Lambda(\omega)$ is

(19)

$$d(\Lambda(\omega)) = \sum_{j=1}^{n}\sum_{k=1}^{r}(-1)^{k-1}\left(\int_0^1 t^r \frac{\partial f}{\partial x_j}(t\boldsymbol{x})\, dt \right) x_{i_k}\, dx_j\, dx_{i_1} \ldots \widehat{dx_{i_k}} \ldots dx_{i_r}$$

$$+ r\left(\int_0^1 t^{r-1} f(t\boldsymbol{x})\, dt \right) dx_{i_1} \ldots dx_{i_r}.$$

Adding (18) and (19), we obtain by the Product Rule and the one-dimensional

Fundamental Theorem of Calculus that

$$\Lambda(d\omega) + d(\Lambda(\omega)) = \sum_{j=1}^{n} \left(\int_0^1 t^r \frac{\partial f}{\partial x_j}(t\boldsymbol{x})\, dt \right) x_j\, dx_{i_1} \ldots dx_{i_r}$$

$$+ r \left(\int_0^1 t^{r-1} f(t\boldsymbol{x})\, dt \right) dx_{i_1} \ldots dx_{i_r}$$

$$= \left(\int_0^1 \frac{d}{dt}\left(t^r f(t\boldsymbol{x}) \right) dt \right) dx_{i_1} \ldots dx_{i_r}$$

$$= f(\boldsymbol{x})\, dx_{i_1} \ldots dx_{i_r} = \omega.$$

This proves (17).

Theorem 10.9 is now easy to prove. If ω is exact and $\mathcal{C}^1$, then there is a $\mathcal{C}^2$ form η such that $d\eta = \omega$. Thus $d\omega = d^2\eta = 0$ by Theorem 10.4. (This part works whether V is star-shaped or not.)

Conversely, if $d\omega = 0$, then by (17), $d(\Lambda(\omega)) = \omega$. Thus set $\eta = \Lambda(\omega)$. ∎

EXERCISES

1. Compute $\int_{\partial B_a(0,0,0,0)} x^3\, dy\, dz\, dw + y^2\, dx\, dz\, dw$.
2. Compute $\int_M \sum_{j=1}^{n} x_j^2 dx_1 \ldots \widehat{dx_j} \ldots dx_n$, where M is the boundary of the unit n-dimensional cube $Q = [0,1] \times \cdots \times [0,1]$.
3. Let E be a compact n-dimensional Jordan region in $\mathbf{R}^n$. If ∂E is an $(n-1)$-dimensional manifold, prove

$$\int_{\partial E} \sum_{j=1}^{n} dx_1 \ldots \widehat{dx_j} \ldots dx_n = \begin{cases} \mathrm{Vol}\,(E) & \text{if } n \text{ is odd} \\ 0 & \text{if } n \text{ is even.} \end{cases}$$

4. Let $r \in \mathbf{N}$, $m > n = 2r + 2$, V be a star-shaped open set in $\mathbf{R}^m$, M be an n-dimensional $\mathcal{C}^2$ manifold-with-smooth-boundary in $\mathbf{R}^m$, and $M \subset V$. Suppose ω is an exact $\mathcal{C}^1$ $r+1$–form on V with $\omega = d\eta$. Prove

$$\int_{\partial M} \eta\omega = \int_M \omega^2.$$

APPENDICES

A. ALGEBRAIC LAWS

In this section we derive several consequences of Postulate 1 in Section 1.1.

THEOREM A.1. *Let $x, a \in \mathbf{R}$.*

  i) *If $a = x + a$, then $x = 0$.*
  ii) *If $a = x \cdot a$ and $a \neq 0$, then $x = 1$.*

PROOF. i) Since the additive inverse of a exists, we can add $-a$ to the identity $a = x + a$. Using the Associative Property, and the fact that 0 is the additive identity, we obtain

$$0 = a + (-a) = (x + a) + (-a) = x + (a + (-a)) = x + 0 = x.$$

ii) Since the multiplicative inverse of a exists, we can multiply $a = x \cdot a$ by a^{-1}. Using the Associative Property, and the fact that 1 is the multiplicative identity, we obtain

$$1 = a \cdot a^{-1} = (x \cdot a) \cdot a^{-1} = x \cdot (a \cdot a^{-1}) = x \cdot 1 = x. \quad \blacksquare$$

Theorem A.1 shows that the additive and multiplicative identities are unique. The following result shows that additive and multiplicative inverses are also unique. Thus, "unique" can be dropped from the statements in Postulate 1.

THEOREM A.2.

  i) *If $a, b \in \mathbf{R}$ and $a + b = 0$ then $b = -a$.*
  ii) *If $a, b \in \mathbf{R}$ and $ab = 1$ then $b = a^{-1}$.*

PROOF. i) By hypothesis and the Associative Property,

$$-a = -a + (a + b) = (-a + a) + b = 0 + b = b.$$

ii) Since $1 \neq 0$, $a \neq 0$. Thus, it follows from hypothesis and the Associative Property that

$$a^{-1} = a^{-1}(ab) = (a^{-1}a)b = 1 \cdot b = b. \quad \blacksquare$$

THEOREM A.3. *For all* $a, b \in \mathbf{R}$, $0 \cdot a = 0$, $-a = (-1) \cdot a$, $-(-a) = a$, $(-1)^2 = 1$, *and* $-(a - b) = b - a$.

PROOF. Since 1 is the multiplicative identity and 0 is the additive identity, it follows from the Distributive Property that

$$a + 0 \cdot a = 1 \cdot a + 0 \cdot a = (1 + 0) \cdot a = 1 \cdot a = a.$$

Hence, by Theorem A.1, $0 \cdot a = 0$. Similarly,

$$a + (-1) \cdot a = (1 + (-1)) \cdot a = 0 \cdot a = 0.$$

Since additive inverses are unique, it follows that $(-1) \cdot a = -a$. Since $-a + a = a + (-a) = 0$, a similar argument proves $-(-a) = a$. Combining the last two statements, we have

$$(-1)(-1) = -(-1) = 1.$$

Finally,

$$-(a - b) = (-1)(a - b) = (-1)a + (-1)(-b) = -a + b = b - a. \quad \blacksquare$$

THEOREM A.4. *Let* $a, b, c \in \mathbf{R}$.

   i) *If* $a \cdot b = 0$, *then* $a = 0$ *or* $b = 0$.
   ii) *If* $a \cdot b = a \cdot c$ *and* $a \neq 0$, *then* $b = c$.

PROOF. i) If $a = 0$, we are done. If $a \neq 0$, then multiplying the identity $0 = a \cdot b$ by a^{-1}, we have
$$0 = a^{-1} \cdot 0 = a^{-1} \cdot (a \cdot b) = (a^{-1} \cdot a) \cdot b = 1 \cdot b = b.$$

ii) If $a \cdot b = a \cdot c$, then by Theorem A.3 we have

$$a \cdot (b - c) = a \cdot (b + (-1)c) = a \cdot b + (-1)a \cdot c = a \cdot b - a \cdot c = 0.$$

Since $a \neq 0$, it follows from part i) that $b - c = 0$, i.e., $b = c$. $\quad \blacksquare$

B. TRIGONOMETRY

In this section we derive some trigonometric identities using elementary geometry and algebra.

Let (x, y) be a point on the unit circle $x^2 + y^2 = 1$ and θ be the angle measured counterclockwise from the positive x axis to the line segment from $(0, 0)$ to (x, y) (see Figure B.1a). (We shall refer to (x, y) as the point *determined* by the angle θ.) Define

$$\sin \theta = y, \qquad \cos \theta = x, \qquad \text{and} \quad \tan \theta = \frac{y}{x}.$$

By the Law of Similar Triangles, given a right triangle with base angle θ, altitude a, base b, and hypotenuse h (see Figure B.1b), $\sin \theta = a/h$, $\cos \theta = b/h$, and $\tan \theta = a/b = \sin \theta / \cos \theta$.

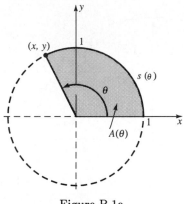

Figure B.1a   Figure B.1b

THEOREM B.1. *Given a circle $C : x^2 + y^2 = r^2$ of radius r, let $s(\theta)$ represent the length of the arc on C swept out by θ and $A(\theta)$ represent the area of the angular sector swept out by θ (see Figure B.1a). If the angle θ is measured in radians (not degrees), then*

$$s(\theta) = r\theta \quad and \quad A(\theta) = \frac{r^2\theta}{2}.$$

PROOF. Since there are 2π radians in a complete circle and the circumference of a circle of radius r is $2\pi r$, we have

$$\frac{s(\theta)}{2\pi r} = \frac{\theta}{2\pi},$$

i.e., $s(\theta) = r\theta$. Similarly, since the area of a circle is πr^2, we have

$$\frac{A(\theta)}{\pi r^2} = \frac{\theta}{2\pi},$$

i.e., $A(\theta) = r^2\theta/2$. ∎

THEOREM B.2.

   i) $\sin(0) = 0$ and $\cos(0) = 1$.
   ii) *For any $\theta \in \mathbf{R}$*, $|\sin\theta| \le 1$, $|\cos\theta| \le 1$, $\sin(-\theta) = -\sin\theta$, $\cos(-\theta) = \cos\theta$, *and* $\sin^2\theta + \cos^2\theta = 1$.
   iii) *If θ is measured in radians, then* $\sin(\pi/2) = 1$, $\cos(\pi/2) = 0$, $\sin(\theta + 2\pi) = \sin\theta$, *and* $\cos(\theta + 2\pi) = \cos\theta$. *Moreover, if $0 < \theta < \pi/2$, then $0 < \theta\cos\theta < \sin\theta < \theta$.*
   iv) *If $\theta \in \mathbf{R}$ is measured in radians, then* $|\sin\theta| \le |\theta|$.

PROOF. Let $\theta \in \mathbf{R}$ and (x, y) be the point on the unit circle determined by θ.
i) If $\theta = 0$, then $(x, y) = (1, 0)$ (see Figure B.1a). Hence, $\sin(0) = 0$ and $\cos(0) = 1$.
ii) Clearly, $|\sin\theta| = |y| = \sqrt{y^2} \le \sqrt{x^2 + y^2} = 1$, and similarly, $|\cos\theta| \le 1$. By definition (see Figure B.2a), $\sin(-\theta) = -y = -\sin\theta$ and $\cos(-\theta) = x = \cos\theta$. Moreover,

$$\sin^2\theta + \cos^2\theta = x^2 + y^2 = 1.$$

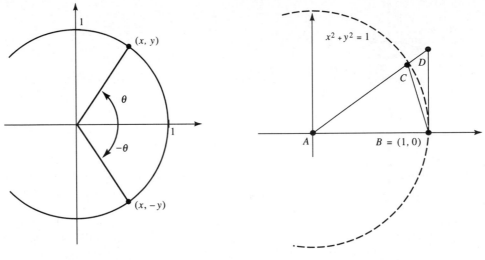

Figure B.2a   Figure B.2b

iii) If $\theta = \pi/2$, then $(x, y) = (0, 1)$, so $\sin(\pi/2) = 1$ and $\cos(\pi/2) = 0$. Fix $\theta \in (0, \pi/2)$ and consider Figure B.2b. Since $\sin\theta$ is the altitude of triangle ABC and the shortest distance between two points is a straight line, we have by Theorem B.1 that

$$\sin\theta < s(\theta) = \theta.$$

On the other hand, the triangle ABC is a proper subset of the angular sector swept out by θ which is a proper subset of the triangle ABD. Hence,

$$\text{Area}\,(ABC) < A(\theta) < \text{Area}\,(ABD).$$

Since the area of a triangle is one-half the product of its base and its altitude, it follows from Theorem B.1 that

(1) $$\frac{\sin\theta}{2} < \frac{\theta}{2} < \frac{\tan\theta}{2}.$$

But $0 < \cos\theta < 1$ for all $\theta \in (0, \pi/2)$. Multiplying (1) by $2\cos\theta$, we conclude that

(2) $$\sin\theta \cos\theta < \theta \cos\theta < \sin\theta.$$

iv) By part iii), $|\sin\theta| = \sin\theta \le \theta = |\theta|$ for all $0 \le \theta \le \pi/2$. Since $\sin(-\theta) = -\sin\theta$, it follows that $|\sin\theta| \le |\theta|$ for all $\theta \in [-\pi/2, \pi/2]$. But if $\theta \notin [-\pi/2, \pi/2]$, then $|\sin\theta| \le 1 < \pi/2 < |\theta|$. Therefore, $|\sin\theta| \le |\theta|$ for all $\theta \in \mathbf{R}$. ∎

The next result shows how to compute the sine and cosine of a sum of angles.

THEOREM B.3.

i) [SUM ANGLE FORMULAS]. If $\theta, \varphi \in \mathbf{R}$, then

$$\cos(\theta \pm \varphi) = \cos\theta \cos\varphi \mp \sin\theta \sin\varphi,$$

and

$$\sin(\theta \pm \varphi) = \sin\theta\cos\varphi \pm \cos\theta\sin\varphi.$$

ii) [DOUBLE ANGLE FORMULAS]. *If $\theta \in \mathbf{R}$, then*

$$\cos^2\theta = \frac{1 + \cos(2\theta)}{2},$$

$$\sin^2\theta = \frac{1 - \cos(2\theta)}{2},$$

and

$$\cos\theta = 1 - 2\sin^2(\theta/2).$$

iii) [SHIFT FORMULAS]. *If φ is measured in radians, then*

$$\sin\varphi = \cos\left(\frac{\pi}{2} - \varphi\right)$$

and

$$\cos\varphi = \sin\left(\frac{\pi}{2} - \varphi\right)$$

for all $\varphi \in \mathbf{R}$.

PROOF. Suppose first that $\theta > \varphi$. Consider the chord A cut from the unit circle by a central angle $\theta - \varphi$ and the chord B cut from the unit circle by a central angle $\varphi - \theta$ (see Figure B.3). Since $\sin^2\theta + \cos^2\theta = 1$, we have

(3) $$A^2 = (\cos\theta - \cos\varphi)^2 + (\sin\theta - \sin\varphi)^2 = 2 - 2(\cos\theta\cos\varphi + \sin\theta\sin\varphi)$$

and

$$B^2 = (\cos(\theta - \varphi) - 1)^2 + (\sin(\theta - \varphi))^2 = 2 - 2\cos(\theta - \varphi).$$

Since $|\theta - \varphi| = |\varphi - \theta|$, the lengths of these chords must be equal. Thus

(4) $$\cos(\theta - \varphi) = \cos\theta\cos\varphi + \sin\theta\sin\varphi$$

for $\theta < \varphi$. A similar argument establishes (4) for $\varphi < \theta$. Since (4) is trivial when $\theta = \varphi$, we have proved that (4) holds for all θ and φ. Combining the identities $\sin(-\theta) = -\sin\theta$ and $\cos(-\theta) = \cos\theta$ with (4), we obtain

$$\cos(\theta + \varphi) = \cos(\theta - (-\varphi)) = \cos\theta\cos\varphi - \sin\theta\sin\varphi.$$

This and (4) verifies the first identity in part i).

Applying this identity to $\theta = \pi/2$, we see by Theorem B.2ii that

$$\cos\left(\frac{\pi}{2} - \varphi\right) = \cos\left(\frac{\pi}{2}\right)\cos\varphi + \sin\left(\frac{\pi}{2}\right)\sin\varphi = \sin\varphi,$$

i.e., the first identity in part iii) holds. Combining the first identities in parts i) and iii), we obtain

$$\sin(\theta \pm \varphi) = \cos\left(\left(\frac{\pi}{2} - \theta\right) \mp \varphi\right)$$

$$= \cos\left(\frac{\pi}{2} - \theta\right)\cos(-\varphi) \mp \sin\left(\frac{\pi}{2} - \theta\right)\sin(-\varphi)$$

$$= \sin\theta\cos\varphi \pm \cos\theta\sin\varphi.$$

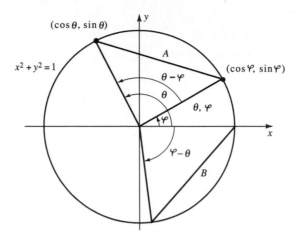

Figure B.3

This proves the second identity in part i). Specializing to the case $\theta = \pi/2$, we obtain $\sin(\pi/2 - \varphi) = \cos\varphi$. Thus, parts i) and iii) have been proved.

To establish part ii), notice by part i) and Theorem B.2 that

$$\cos(2\theta) = \cos(\theta + \theta) = \cos^2\theta - \sin^2\theta = 2\cos^2\theta - 1.$$

Hence, $\cos^2\theta = (1 + \cos(2\theta))/2$. Similar arguments establish the rest of part ii). ∎

We close this section with the Law of Cosines, a generalization of the Pythagorean Theorem.

THEOREM B.4 [LAW OF COSINES]. *If T is a triangle with sides of length a, b, c, and θ is the angle opposite the side of length c, then*

$$c^2 = a^2 + b^2 - 2ab\cos\theta.$$

PROOF. Suppose without loss of generality that θ is acute, and rotate T so b is its base. Let h be the altitude of T and notice that h cuts a right triangle out of T whose sides are a and h and the angle opposite h is θ. By the definition of $\sin\theta$ and $\cos\theta$, $h = a\sin\theta$ and the length d of the base of this right triangle is $d = b - a\cos\theta$. Substituting these values into the equation $c^2 = h^2 + d^2$ (which follows directly from the Pythagorean Theorem), we obtain

$$c^2 = (a\sin\theta)^2 + (b - a\cos\theta)^2$$
$$= a^2\sin^2\theta + a^2\cos^2\theta + b^2 - 2ab\cos\theta = a^2 + b^2 - 2ab\cos\theta. ∎$$

C. MATRICES AND DETERMINANTS

In this section we prove several elementary results about matrices and determinants. We assume the student is familiar with the concept of row and column reduction to canonical form.

Recall that an $m \times n$ matrix B is a rectangular array which has m rows and n columns:

$$B = [b_{ij}]_{m \times n} = \begin{bmatrix} b_{11} & b_{12} & \cdots & b_{1n} \\ b_{21} & b_{22} & \cdots & b_{2n} \\ \vdots & \vdots & \ddots & \vdots \\ b_{m1} & b_{m2} & \cdots & b_{mn} \end{bmatrix}.$$

The notation b_{ij} indicates the *entry* in the ith row and jth column. We shall call B *real* if all its entries b_{ij} belong to $\mathbf{R}$.

DEFINITION C.1. Let $B = [b_{ij}]_{m \times n}$ and $C = [c_{k\ell}]_{p \times q}$ be real matrices.

i) B and C are said to be *equal* if $m = p$, $n = q$, and $b_{ij} = c_{ij}$ for $i = 1, 2, \ldots, m$ and $j = 1, 2, \ldots, n$.

ii) The $m \times n$ *zero matrix* is the matrix $O = O_{m \times n} = [b_{ij}]_{m \times n}$ where $b_{ij} = 0$ for $i = 1, \ldots, m$ and $j = 1, \ldots, n$.

iii) The $n \times n$ *identity matrix* is the matrix $I = I_{n \times n} = [b_{ij}]_{n \times n}$ where $b_{ii} = 1$ for $i = 1, \ldots, n$ and $b_{ij} = 0$ for $i \neq j$, $i, j = 1, \ldots, n$.

iv) The *product* of a matrix B and a scalar α is defined by

$$\alpha B = [\alpha b_{ij}]_{m \times n}.$$

v) The *negative* of a matrix B is defined by $-B = (-1)B$.

vi) When $m = p$ and $n = q$, the *sum* of B and C is defined by

$$B + C = [b_{ij} + c_{ij}]_{m \times n}.$$

vii) When $n = p$, the *product* of B and C is defined by

$$BC = \left[\sum_{\nu=1}^{n} b_{i\nu} c_{\nu j} \right]_{m \times q}.$$

Example 1. Compute $B + C$, $3B$, $-C$, BC, and CB, where

$$B = \begin{bmatrix} 1 & 0 \\ 2 & 3 \end{bmatrix} \quad \text{and} \quad C = \begin{bmatrix} -1 & 1 \\ -2 & 0 \end{bmatrix}.$$

SOLUTION. By definition,

$$B + C = \begin{bmatrix} 0 & 1 \\ 0 & 3 \end{bmatrix}, \qquad 3B = \begin{bmatrix} 3 & 0 \\ 6 & 9 \end{bmatrix},$$

$$-C = \begin{bmatrix} 1 & -1 \\ 2 & 0 \end{bmatrix}, \qquad BC = \begin{bmatrix} -1 & 1 \\ -8 & 2 \end{bmatrix} \quad \text{and} \quad CB = \begin{bmatrix} 1 & 3 \\ -2 & 0 \end{bmatrix}. \ \blacksquare$$

These operations do not satisfy all the usual laws of algebra. (For example, the last two computations show matrix multiplication is not commutative.) Here is a list of algebraic laws satisfied by real matrices.

THEOREM C.1. *Let* $A = [a_{ij}]$, $B = [b_{ij}]$, *and* $C = [c_{ij}]$ *be real matrices and* α, β *be scalars.*

i) $(\alpha + \beta)C = \alpha C + \beta C$.

ii) *If* $B + C$ *is defined, then* $\alpha(B + C) = \alpha B + \alpha C$, *and* $B + C = C + B$.

iii) *If* BC *is defined, then* $\alpha(BC) = (\alpha B)C = B(\alpha C)$.

iv) *If* AB *and* AC *are defined, then* $A(B + C) = AB + AC$. *If* BA *and* CA *are defined then* $(B + C)A = BA + CA$.

v) *If* $A + B$ *and* $B + C$ *are defined, then* $(A + B) + C = A + (B + C)$. *If* AB *and* BC *are defined then* $(AB)C = A(BC)$.

vi) *If* B *is an* $m \times n$ *matrix, then*

$$B + O_{m \times n} = B, \qquad B - B = O_{m \times n},$$

$$BO_{n \times q} = O_{m \times q}, \qquad O_{p \times m}B = O_{p \times n} \quad \text{and} \quad 0B = O_{m \times n}.$$

vii) *If* B *is an* $n \times n$ *matrix, then*

$$I_{n \times n}B = BI_{n \times n} = B.$$

PROOF. By definition,

$$(\alpha + \beta)C = [(\alpha + \beta)c_{ij}] = [\alpha c_{ij} + \beta c_{ij}] = \alpha C + \beta C$$

and

$$\alpha(B + C) = \alpha[b_{ij} + c_{ij}] = [\alpha(b_{ij} + c_{ij})] = [\alpha b_{ij}] + [\alpha c_{ij}] = \alpha B + \alpha C.$$

A similar argument establishes $B + C = C + B$.

Let B be an $m \times n$ matrix and C be an $n \times q$ matrix. By definition,

$$\alpha(BC) = \left[\alpha \sum_{\nu=1}^{n} b_{i\nu}c_{\nu j}\right] = \left[\sum_{\nu=1}^{n} (\alpha b_{i\nu})c_{\nu j}\right] = (\alpha B)C.$$

A similar argument establishes $\alpha(BC) = B(\alpha C)$.

Let A be an $m \times n$ matrix and B, C be $n \times q$ matrices. By definition,

$$A(B + C) = \left[\sum_{\nu=1}^{n} a_{i\nu}(b_{\nu j} + c_{\nu j})\right] = \left[\sum_{\nu=1}^{n} a_{i\nu}b_{\nu j} + \sum_{\nu=1}^{n} a_{i\nu}c_{\nu j}\right] = AB + AC.$$

A similar argument establishes $(B + C)A = BA + CA$.

Let A be an $m \times n$ matrix, B be an $n \times p$ matrix, and C be a $p \times q$ matrix. By definition,

$$(AB)C = \left[\sum_{\nu=1}^{n} a_{i\nu}b_{\nu j}\right][c_{jk}]$$

$$= \left[\sum_{j=1}^{p} \left(\sum_{\nu=1}^{n} a_{i\nu}b_{\nu j}\right)c_{jk}\right]$$

$$= \left[\sum_{\nu=1}^{n} \left(a_{i\nu} \sum_{j=1}^{p} b_{\nu j}c_{jk}\right)\right]$$

$$= A\left[\sum_{j=1}^{p} b_{\nu j}c_{jk}\right] = A(BC).$$

A similar argument establishes $(A + B) + C = A + (B + C)$.

By definition,

$$B + O_{m \times n} = [b_{ij} + 0] = [b_{ij}] = B, \qquad B - B = [b_{ij} - b_{ij}] = O_{m \times n},$$

$$BO_{n \times q} = [\sum_{\nu=1}^{n} b_{i\nu} \cdot 0] = O_{m \times q}, \qquad O_{p \times m} B = [\sum_{\nu=1}^{m} 0 \cdot b_{\nu j}] = O_{p \times n},$$

and $0 \cdot B = [0 \cdot b_{ij}] = O_{m \times n}$. And, since $I = [\delta_{ij}]$, where

$$\delta_{ij} = \begin{cases} 1 & i = j, \\ 0 & i \neq j, \end{cases}$$

we have $I_{n \times n} B = [\sum_{\nu=1}^{n} \delta_{i\nu} b_{\nu j}] = [b_{ij}] = B = BI_{n \times n}$. ∎

A square matrix is a matrix with as many rows as columns. Clearly, if B and C are square real matrices of the same size, then both $B + C$ and BC are defined. This gives room for more algebraic structure. An $n \times n$ real matrix B is said to be *invertible* if there is an $n \times n$ matrix B^{-1}, called the *inverse* of B, which satisfies

$$BB^{-1} = B^{-1}B = I.$$

The following result shows that matrix inverses are unique.

THEOREM C.2. *Let A, B be $n \times n$ real matrices. If B is invertible and $BA = I$, then $B^{-1} = A$.*

PROOF. By Theorem C.1 and definition,

$$B^{-1} = B^{-1}I = B^{-1}(BA) = (B^{-1}B)A = IA = A. \quad ∎$$

If $B = [b_{ij}]_{n \times n}$ is square, recall that the *minor matrix* B_{ij} of B is the $(n-1) \times (n-1)$ matrix obtained by removing the ith row and the jth column from B. For example, if

$$B = \begin{bmatrix} 1 & 2 & 3 \\ -1 & -2 & -3 \\ 4 & 5 & 6 \end{bmatrix}$$

then

$$B_{21} = \begin{bmatrix} 2 & 3 \\ 5 & 6 \end{bmatrix}.$$

Minor matrices can be used to define an operation on square real matrices (the determinant) which makes invertible matrices easy to identify (see Theorem C.4 below).

The *determinant* can be defined recursively as follows. Let B be an $n \times n$ real matrix.

   i) If $n = 1$, then the determinant of B is defined by $\det[b] = b$.

  ii) If $n = 2$, then the determinant of B is defined by

$$\det \begin{bmatrix} a & b \\ c & d \end{bmatrix} = ad - bc.$$

 iii) If $n > 2$ then the determinant of B is defined recursively by

$$\det[b_{ij}]_{n \times n} = b_{11} \det B_{11} - b_{12} \det B_{12} + \cdots + (-1)^{n-1} b_{1n} \det B_{1n},$$

where B_{1j} are minor matrices of B.

The following result shows what an elementary column operation does to the determinant of a matrix.

THEOREM C.3. *Let* $B = [b_{ij}]$ *and* $C = [c_{ij}]$ *be* $n \times n$ *real matrices,* $n \geq 2$.

   i) *If* C *is obtained from* B *by interchanging two columns, then* $\det C = -\det B$.

   ii) *If* C *is obtained from* B *by multiplying one column of* B *by a scalar* α, *then* $\det C = \alpha \det B$.

   iii) *If* C *is obtained from* B *by multiplying one column of* B *by a scalar and adding it to another column of* B, *then* $\det C = \det B$.

PROOF. Since

$$\det \begin{bmatrix} a & b \\ c & d \end{bmatrix} = ad - bc = -(bc - ad) = -\det \begin{bmatrix} b & a \\ d & c \end{bmatrix},$$

part i) holds for 2×2 matrices. Suppose part i) holds for $(n-1) \times (n-1)$ matrices. Suppose further that there are indices $j_0 < j_1$ such that $b_{ij_0} = c_{ij_1}$ and $b_{ij_1} = c_{ij_0}$ for $i = 1, \ldots, n$. By the inductive hypothesis, $\det C_{1j} = -\det B_{1j}$ for $j \neq j_0$ and $j \neq j_1$,

$$\det C_{1j_0} = (-1)^{j_1 - j_0} \det B_{1j_1}, \quad \text{and} \quad \det C_{1j_1} = (-1)^{j_0 - j_1} \det B_{1j_0}.$$

Hence, by definition,

$$\det C = c_{11} \det C_{11} - c_{12} \det C_{12} + \cdots + (-1)^{n-1} c_{1n} \det C_{1n}$$
$$= -b_{11} \det B_{11} + b_{12} \det B_{12} + \cdots - (-1)^{n-1} b_{1n} \det B_{1n} = -\det B.$$

Thus, i) holds for all $n \in \mathbf{N}$. Similar arguments establish parts ii) and iii). ∎

In the same way we can show that Theorem C.3 holds if "column" is replaced by "row." It follows that we can compute the determinant of a real matrix by expanding along any row or any column with an appropriate adjustment of signs. For example, to expand along the ith row, interchange the ith row with the first row, expand along the new first row, and use Theorem C.3 to relate everything back to B. In particular, we see that

$$\det[b_{ij}]_{n \times n} = (-1)^{i+1} b_{i1} \det B_{i1} + (-1)^{i+2} b_{i2} \det B_{i2} + \cdots + (-1)^{i+n} b_{in} \det B_{in}.$$

The numbers $(-1)^{i+j} \det B_{ij}$ are called the *cofactors* of b_{ij} in $\det B$

The operations in Theorem C.3 are called *elementary column operations*. They can be simulated by matrix multiplication. Indeed, an *elementary matrix* is a matrix obtained from the identity matrix by a single elementary column operation. Thus, elementary matrices fall into three categories: $E(i \leftrightarrow j)$, the matrix obtained by interchanging the ith and jth columns of I; $E(\alpha i)$, the matrix obtained by multiplying the ith column of I by $\alpha \neq 0$; and $E(\alpha i + j)$, the matrix obtained by multiplying the ith column of I by $\alpha \neq 0$ and adding it to the jth column. Notice that an elementary column operation on B can be obtained by multiplying B by an elementary matrix, e.g., $E(i \leftrightarrow j)B$ is the matrix obtained by interchanging the ith and jth columns of B.

These observations can be used to show that the determinant is multiplicative.

THEOREM C.4. *If* B, C *are* $n \times n$ *real matrices, then*

$$\det(BC) = \det B \det C.$$

Moreover, *B is invertible if and only if* $\det(B) \neq 0$.

PROOF. It is easy to check that

$$\det(E(i \leftrightarrow j)) = -1, \quad \det(E(\alpha i)) = \alpha, \quad \text{and} \quad \det(E(\alpha i + j)) = 1.$$

Hence, by Theorem C.3,

(5) $$\det(EA) = \det E \det A$$

holds for any $n \times n$ matrix A and any $n \times n$ elementary matrix E.

The matrix B can be reduced, by a sequence of elementary column operations, to a matrix V where $V = I$ if B is invertible, and V has at least one zero column if B is not invertible (see Noble and Daniel [9], p. 85). It follows that there exist elementary matrices $E_1, \ldots, E_p$ such that $A = E_1 \ldots E_p V$. Hence, by (5),

$$\det(B) = \det(E_1 \ldots E_p V)$$
$$= \det(E_1) \det(E_2 \ldots E_p V) = \cdots = \det(E_1 \ldots E_p) \det(V).$$

In particular, B is invertible if and only if $\det B \neq 0$.

Suppose B is invertible. Then $V = I$ and by (5),

$$\det(BC) = \det(E_1 \ldots E_p) \det(VC) = \det B \det C.$$

If B is not invertible, then BC is not invertible either (see Noble and Daniel [9], p. 204). Hence, $\det(BC) = 0$ and we have

$$\det(BC) = 0 = \det B \det C. \quad \blacksquare$$

The *transpose* of a matrix $B = [b_{ij}]$ is the matrix B^T obtained from B by making the ith row of B the ith column of B^T, i.e., the $(i \times j)$th entry of B^T is b_{ji}. The *adjoint* of an $n \times n$ matrix B is the transpose of the matrix of cofactors of B, i.e.,

$$adj\,(B) = [(-1)^{i+j} \det B_{ij}]^T.$$

The adjoint can be used to give an explicit formula for the inverse of an invertible matrix.

THEOREM C.5. *Suppose B is a square real matrix. If B is invertible, then*

(6) $$B^{-1} = \frac{1}{\det B} adj\,(B).$$

PROOF. Set $[c_{ij}] = B\,adj\,(B)$. By definition,

$$c_{ij} = (-1)^{1+j} b_{i1} B_{j1} + \cdots + (-1)^{n+j} b_{in} B_{jn}.$$

If $i = j$, then c_{ij} is an expansion of the determinant of B along the ith row of B, i.e., $c_{ii} = \det B$. If $i \neq j$, then c_{ij} is a determinant of a matrix with two identical rows so c_{ij} is zero. It follows that

$$B\,adj\,(B) = \det B \cdot I.$$

We conclude by Theorem C.2 that (6) holds. $\blacksquare$

The following result shows how the determinant can be used to solve systems of linear equations. (This result is of great theoretical interest but of little practical use because it requires lots of storage to use on a computer. Most packaged routines which solve systems of linear equations use methods more efficient than Cramer's Rule, e.g., Gaussian elimination.)

THEOREM C.6 [CRAMER'S RULE]. *Let $c_1, c_2, \ldots, c_n \in \mathbf{R}$ and $B = [b_{ij}]_{n \times n}$ be a square real matrix. The system*

(7)
$$b_{11}x_1 + b_{12}x_2 + \cdots + b_{1n}x_n = c_1$$
$$b_{21}x_1 + b_{22}x_2 + \cdots + b_{2n}x_n = c_2$$
$$\vdots \qquad\qquad \vdots$$
$$b_{n1}x_1 + b_{n2}x_2 + \cdots + b_{nn}x_n = c_n$$

of n linear equations in n unknowns has a unique solution if and only if the matrix B has a nonzero determinant, in which case

$$x_j = \frac{\det C(j)}{\det B},$$

where $C(j)$ is obtained from B by replacing the jth column of B by the column matrix $[c_1 \ldots c_n]^T$. In particular, if $c_j = 0$ for all j and $\det B \neq 0$ then the system (7) has only the trivial solution $x_j = 0$ for $j = 1, 2, \ldots, n$.

PROOF. The system (7) is equivalent to the matrix equation

$$BX = C,$$

where $B = [b_{ij}]$, $X = [x_1 \ldots x_n]^T$, and $C = [c_1 \ldots c_n]^T$. If $\det B \neq 0$, then by Theorem C.5,

$$X = B^{-1}C = \frac{1}{\det B} adj\,(B)C.$$

By definition, $adj\,(B)C$ is a column matrix whose jth "row" is the number

$$(-1)^{1+j}c_1 \det B_{1j} + (-1)^{2+j}c_2 \det B_{2j} + \cdots + (-1)^{n+j}c_n \det B_{nj} = \det C(j).$$

(We expanded the determinant of $C(j)$ along the jth column.) Thus, $x_j = \det C(j)$.

Conversely, if $BX = C$ has a unique solution, B can be row reduced to I. Thus, B is invertible, i.e., $\det B \neq 0$. ∎

D. QUADRIC SURFACES

A *quadric surface* is a surface which is the graph of a relation in $\mathbf{R}^3$ of the form

$$Ax^2 + By^2 + Cz^2 + Dx + Ey + Fz + Gxy + Hyz + Izx = J,$$

where $A, B, \ldots, J \in \mathbf{R}$ and not all A, B, C, G, H, I are zero. We shall only consider the cases when $G = H = I = 0$. These include the following special types.

1. The *ellipsoid*, the graph of

$$Ax^2 + By^2 + Cz^2 = 1,$$

where A, B, C are all positive.

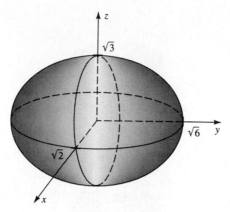

Figure D.1

2. The *hyperboloid of one sheet*, the graph of

$$Ax^2 + By^2 + Cz^2 = 1,$$

where two of A, B, C are positive and the other is negative.

3. The *hyperboloid of two sheets*, the graph of

$$Ax^2 + By^2 + Cz^2 = 1,$$

where two of A, B, C are negative and the other is positive.

4. The *cone*, the graph of

$$Ax^2 + By^2 + Cz^2 = 0,$$

where two of A, B, C are positive and the other is negative.

5. The *paraboloid*, the graph of

$$z = Ax^2 + By^2,$$

where A, B are both positive or both negative.

6. The *hyperbolic paraboloid*, the graph of

$$z = Ax^2 + By^2,$$

where one of A, B is positive and the other is negative.

The *trace* of a surface S in a plane Π is defined to be the intersection of S with Π. Graphs of many surfaces, including all quadrics, can be visualized by looking at their traces in various planes. We illustrate this technique with a typical example of each type of quadric.

Example 1. The ellipsoid $3x^2 + y^2 + 2z^2 = 6$.

SOLUTION. The trace of this surface in the xy plane is the ellipse $3x^2 + y^2 = 6$. The trace of this surface in the yz plane is the ellipse $y^2 + 2z^2 = 6$. And the trace of this surface in the xz plane is the ellipse $3x^2 + 2z^2 = 6$. This surface is sketched in Figure D.1. ∎

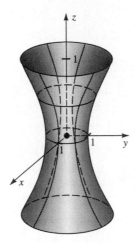

Figure D.2

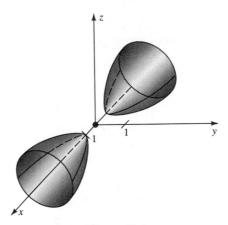

Figure D.3

Example 2. The hyperboloid of one sheet $x^2 + y^2 - z^2 = 1$.

SOLUTION. The trace of this surface in the plane $z = a$ is the circle $x^2 + y^2 = 1 + a^2$. The trace of this surface in $x = 0$ is the hyperbola $y^2 - z^2 = 1$. This surface is sketched in Figure D.2. ∎

Example 3. The hyperboloid of two sheets $x^2 - y^2 - z^2 = 1$.

SOLUTION. The trace of this surface in the plane $z = 0$ is the hyperbola $x^2 - y^2 = 1$. The trace of this surface in $y = 0$ is the hyperbola $x^2 - z^2 = 1$. This surface has no trace in $x = 0$. This surface is sketched in Figure D.3. ∎

Example 4. The cone $z^2 = x^2 + y^2$.

SOLUTION. The trace of this surface in the plane $z = a$ is the circle $x^2 + y^2 = a^2$. The trace of this surface in $y = 0$ is a pair of lines $z = \pm x$. This surface is sketched in

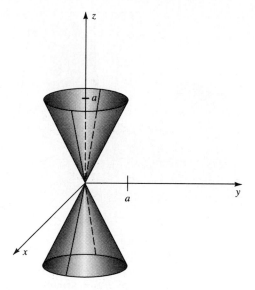

Figure D.4

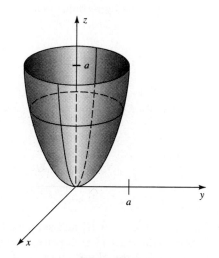

Figure D.5

Figure D.4. ∎

Example 5. The paraboloid $z = x^2 + y^2$.

SOLUTION. If $a > 0$, the trace of this surface in the plane $z = a$ is the circle $x^2 + y^2 = a$. The trace of this surface in $y = 0$ is the parabola $z = x^2$. This surface is sketched in Figure D.5. ∎

Example 6. The hyperbolic paraboloid $z = x^2 - y^2$.

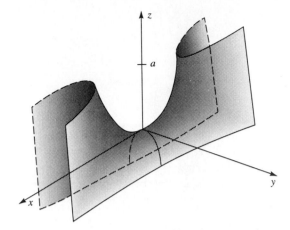

Figure D.6

SOLUTION. The trace of this surface in the plane $z = a$ is the hyperbola $a = x^2 - y^2$. (It opens up around the xz plane when $a > 0$ and around the yz plane when $a < 0$.) The trace of this surface in the plane $y = 0$ is the parabola $z = x^2$. This surface is sketched in Figure D.6. (Note: The scale along the x axis has been exaggerated to enhance perspective. Because of this, the hyperbolas below the $z = 0$ plane are barely discernible.) ∎

E. VECTOR CALCULUS AND PHYSICS

Throughout this section $C = (\varphi, I)$ is a smooth arc in $\mathbf{R}^2$, $S = (\psi, E)$ is a smooth surface in $\mathbf{R}^3$, $\{t_0, \dots, t_N\}$ is a partition of I, and $\{R_1, \dots, R_N\}$ is a grid on E.

Remark 1. *The integral*

$$(8) \qquad \iint_S d\sigma = \int_E \|N_\psi(u, v)\| \, d(u, v)$$

can be interpreted as the surface area of S.

Let (u_j, v_j) be the lower left-hand corner of R_j and suppose R_j has sides Δu, Δv (see Figure E.1). If R_j is small enough, the trace of each piece $S_j = (\psi, R_j)$ is approximately equal to the parallelogram determined by the vectors $\Delta u\, \psi_u$ and $\Delta v\, \psi_v$. Hence, by Exercise 7 in Section 5.1,

$$A(S_j) \approx \|(\Delta u\, \psi_u(u_j, v_j)) \times (\Delta v\, \psi_v(u_j, v_j))\|$$
$$= \|N_\psi(u_j, v_j)\| \Delta u \Delta v = \|N_\psi(u_j, v_j)\| |R_j|.$$

Summing over j, we obtain

$$A(S) \approx \sum_{j=1}^{N} \|N_\psi(u_j, v_j)\| |R_j|$$

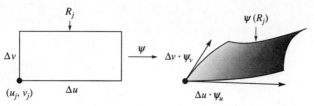

Figure E.1

which is a Riemann sum of the integral (8).

Remark 2. *If w is a thin wire lying along C whose density (mass per unit length) at a point (x, y) is given by $g(x, y)$, then*

$$\int_C g\, ds$$

can be interpreted as the mass of w.

Since mass is the product of density and length, an approximation to the mass of the piece of w lying along $C_k = (\phi, [t_{k-1}, t_k])$ is given by

$$g(t_k) \cdot L\,(C_k) = \int_{t_{k-1}}^{t_k} g(t_k) \|\phi'(t)\|\, dt$$

(see Theorem 8.1). Summing over k, an approximation to the mass of w is

$$\sum_{j=1}^{N} \int_{t_{k-1}}^{t_k} g(t_k) \|\phi'(t)\|\, dt,$$

which is nearly a Riemann sum of the integral

$$\int_I g(\phi(t)) \|\phi'(t)\|\, dt = \int_C g\, ds.$$

The following remark has a similar justification.

Remark 3. *If S is a thin sheet of metal whose density at a point (x, y, z) is given by $g(x, y, z)$, then*

$$\iint_S g\, d\sigma$$

can be interpreted as the mass of S.

Work done by a force F acting on an object as it moves a distance d is defined to be $W = Fd$. There are many situations where the force changes from point to point. Examples include the force of gravity (which is weaker at higher altitudes), the velocity of a fluid flowing through a constricted tube (which gets faster at places where the tube narrows), the force on an electron moving through an electric field, and the force on a copper coil moving through a magnetic field.

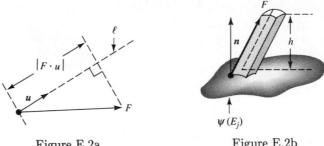

Figure E.2a   Figure E.2b

Remark 4. *If an object acted on by a force* $F : \mathbf{R}^3 \to \mathbf{R}$ *moves along the curve* $C = (\phi, I)$, *then the unoriented line integral*

$$\int_C F \, ds$$

can be interpreted as the work done by F *along* C.

An approximation to the work done along $C_k = (\phi, [t_{k-1}, t_k])$ is

$$W_k \approx F(\phi(t_k))\|\phi(t_k) - \phi(t_{k-1})\| = \int_{t_{k-1}}^{t_k} F(\phi(t_k))\|\phi'(t)\| \, dt.$$

Summing over k, we find that an approximation to the total work along C is given by

$$\sum_{k=1}^{N} \int_{t_{k-1}}^{t_k} F(\phi(t_k))\|\phi'(t)\| \, dt,$$

which is nearly a Riemann sum of the integral

$$\int_I F(\phi(t))\|\phi'(t)\| \, dt = \int_C F \, ds.$$

The following remark explains why $F \cdot T$ is called the tangential component of F and $F \cdot \boldsymbol{n}$ is called the normal component of F.

Remark 5. *Let* $\boldsymbol{u}$ *be a unit vector in* $\mathbf{R}^2$ *(respectively,* $\mathbf{R}^3$*) and* F *be a function whose range is a subset of* $\mathbf{R}^2$ *(respectively,* $\mathbf{R}^3$*). If* ℓ *is the line in the direction* $\boldsymbol{u}$ *passing through the origin, then* $|F \cdot \boldsymbol{u}|$ *is the length of the projection of* F *onto* ℓ *(see Figure E.2a).*

Let θ represent the angle between $\boldsymbol{u}$ and F. By (2) in Section 5.1,

$$|F \cdot \boldsymbol{u}| = \cos\theta\|F\| \, \|\boldsymbol{u}\| = \cos\theta\|F\|.$$

Hence, by trigonometry, $|F \cdot \boldsymbol{u}|$ is the length of the projection of F onto ℓ. Notice that $F \cdot \boldsymbol{u}$ is positive when θ is acute and negative when θ is obtuse.

Combining Remarks 4 and 5, we see that $\int_C F \cdot T \, ds$ represents the work done by the tangential component of a force field $F : \mathbf{R}^3 \to \mathbf{R}^3$ along C.

Remark 6. *If $S = (\psi, E)$ is a thin membrane submerged in an incompressible fluid which passes through S and $F(x, y, z)$ represents the velocity vector of the flow of that fluid at the point (x, y, z), then the oriented integral of $F \cdot \boldsymbol{n}$ can be interpreted as the volume of fluid flowing through S in unit time.*

Let $\{E_j\}$ be a grid which covers E and let h be the length of the line segment obtained by projecting F onto the normal line to S at a point $(x_j, y_j, z_j) \in \psi(E_j)$ (see Figure E.2b). If E_j is so small that F is essentially constant on the trace of $S_j = (\psi, E_j)$, then an approximation to the volume of fluid passing through S_j per unit time is given by

$$V_j = A(S_j) \cdot h = A(S_j) \cdot F(x_j, y_j, z_j) \cdot \boldsymbol{n}$$
$$= A(S_j) F(\psi(u_j, v_j)) \cdot N_\psi(u_j, v_j) / \|N_\psi(u_j, v_j)\|.$$

Summing over j and replacing $A(S_j)$ by $\|N_\psi\| |E_j|$ (see Remark 1 above), we see that an approximation to the volume V of fluid passing through S per unit time is given by

$$\sum_{j=1}^{N} F(\psi(u_j, v_j)) \cdot N_\psi(u_j, v_j) |E_j|.$$

This is a Riemann sum of the oriented integral

$$\iint_E F(\psi(u, v)) \cdot N_\psi(u, v) \, dA = \iint_S F \cdot \boldsymbol{n} \, d\sigma.$$

F.   EQUIVALENCE RELATIONS

A *partition* of a set X is a family of nonempty sets $\{E_\alpha\}_{\alpha \in A}$ such that

$$X = \bigcup_{\alpha \in A} E_\alpha \quad \text{and} \quad E_\alpha \cap E_\beta = \emptyset$$

for $\alpha \neq \beta$. A *binary relation* $\sim$ on X is a subset of $X \times X$. If (x, y) belongs to $\sim$, we shall write $x \sim y$. Examples of binary relations include $=$ on $\mathbf{R}$, $\leq$ on $\mathbf{R}$, and "parallel to" on the class of straight lines in $\mathbf{R}^2$.

A binary relation is called an *equivalence relation* if it satisfies three additional properties.

[THE REFLEXIVE PROPERTY]   For every $x \in X$, $x \sim x$.

[THE SYMMETRIC PROPERTY]   If $x \sim y$, then $y \sim x$.

[THE TRANSITIVE PROPERTY]   If $x \sim y$ and $y \sim z$, then $x \sim z$.

Notice that $=$ is an equivalence relation on $\mathbf{R}$, "parallel to" is an equivalence relation on the class of straight lines in $\mathbf{R}^2$, but $\leq$ is not an equivalence relation on $\mathbf{R}$ (it fails to satisfy the Symmetric Property).

If $\sim$ is an equivalence relation on a set X, then

$$\overline{x} := \{y \in X : y \sim x\}$$

is called the *equivalence class* of X which contains x.

THEOREM F.1. *If $\sim$ is an equivalence relation on a set X, then the set of equivalence classes $\{\bar{x} : x \in X\}$ forms a partition of X.*

PROOF. Since $\sim$ is reflexive, each equivalence class $\bar{x}$ contains x, i.e., $\bar{x}$ is nonempty. Suppose $\bar{x} \cap \bar{y} \neq \emptyset$, i.e., some $z \in X$ belongs to both these equivalence classes. Then $z \sim x$ and $z \sim y$. By the Symmetric Property and the Transitive Property, we have $x \sim y$, i.e., $y \in \bar{x}$. By the Transitive Property, it follows that $\bar{y} \subseteq \bar{x}$. Reversing the roles of x and y, we also have $\bar{x} \subseteq \bar{y}$. Thus, $\bar{x} = \bar{y}$. ∎

REFERENCES

1. APOSTOL, TOM M., *Mathematical Analysis*. Reading, Massachusetts: Addison-Wesley Publishing Co., 1974.

2. BOAS, RALPH P., JR., *Primer of Real Functions*, Carus Monograph 13. New York: Mathematical Association of America and John Wiley and Sons, Inc., 1960.

3. GRIFFITHS, HUBERT B., *Surfaces*. London and New York: Cambridge University Press, 1976.

4. HOCKING, JOHN G. AND GAIL S. YOUNG, *Topology*. Reading, Massachusetts: Addison-Wesley Publishing Co., 1961.

5. KLINE, MORRIS, *Mathematical Thought from Ancient to Modern Times*. New York: Oxford University Press, 1972.

6. LOOMIS, LYNN H. AND SHLOMO STERNBERG, *Advanced Calculus*. Reading, Massachusetts: Addison-Wesley Publishing Co., 1968.

7. MARSDEN, JERROLD E., *Elementary Classical Analysis*. New York, N.Y.: W. H. Freeman & Co. Publishers, 1990.

8. MUNKRES, JAMES R., *Elementary Differential Topology*, Annals of Mathematical Studies. Princeton: Princeton University Press, 1963.

9. NOBLE, BEN AND JAMES W. DANIEL, *Applied Linear Algebra*, 2nd ed. Englewood Cliffs: Prentice-Hall, Inc., 1977.

10. PRICE, G. BALEY, *Multivariable Analysis*. New York: Springer-Verlag New York, Inc., 1984.

11. RUDIN, WALTER, *Principles of Mathematical Analysis*, 3rd ed. New York: McGraw-Hill Book Co., 1976.

12. SPIVAK, MICHAEL, *Calculus on Manifolds*. New York: W. A. Benjamin, Inc., 1965.

13. TAYLOR, ANGUS E., *Advanced Calculus*. Boston: Ginn and Company, 1955.

14. WIDDER, DAVID V. *Advanced Calculus*, 2nd ed. Englewood Cliffs: Prentice-Hall, Inc., 1961.

15. ZYGMUND, ANTONI, *Trigonometric Series*, Vol. I, 2nd ed. London and New York: Cambridge University Press, 1968.

ANSWERS AND HINTS TO SELECTED EXERCISES

CHAPTER 1

1.1 Ordered Field Axioms.

2. a) $(-3,7)$. b) $(-3,5)$. c) $(-1,-1/2) \cup (1,\infty)$. d) $(-2,1)$.

3. Consider the cases $x = 0$ and $x \neq 0$.

4. To prove (7), multiply the first inequality in (7) by c and the second inequality in (7) by b. Prove (8) and (9) by contradiction.

5. a) Apply (6) to $1 - a$. b) Apply (6) to $a - 1$. c) Prove $2\sqrt{ab} \leq a + b$ first.

6. a) Use uniqueness of multiplicative inverses to prove $(nq)^{-1} = n^{-1}q^{-1}$. b) Use a). c) Use proof by contradiction for the sum. Use a similar argument for the product and identify all rationals q such that $xq \in \mathbf{Q}$ for a given $x \in \mathbf{R} \setminus \mathbf{Q}$. d) Use the Multiplicative Properties.

7. a) Prove $|x| \leq 1$ implies $|x + 1| \leq 2$. b) Prove $-1 \leq x \leq 2$ implies $|x + 2| \leq 4$.

8. a) $n > 99$. b) $n \geq 20$. c) $n \geq 23$.

9. Show first that the given inequality implies $2a_1 b_1 a_2 b_2 \leq a_2^2 b_1^2 + a_1^2 b_2^2$.

10. a) Observe that $|xy - ab| = |xy - xb + xb - ab|$ and $|x| < |a| + \varepsilon$.

11. a) The Trichotomy Property implies i), the Additive and Multiplicative Properties imply ii).

1.2 The Well-Ordering Principle.

3. $\sum_{k=1}^{n} \binom{n}{k} x^{n-k} h^{k-1}$.

4. See Exercise 5 in Section 1.1.

5. Observe that $x^2 - x - 2 < 0$ for all $0 < x < 2$.

6. b) First prove that $2n + 1 < 2^n$ for $n = 3, 4, \ldots$.

8. a) Show that $n^2 + 3n$ cannot be the square of an integer when $n > 1$. b) The expression is rational if and only if $n = 9$.

1.3 The Completeness Axiom.

1. a) $\inf E = 1$, $\sup E = 8$. b) $\inf E = (3 - \sqrt{29})/2$, $\sup E = (3 + \sqrt{29})/2$. c) $\inf E = a$, $\sup E = b$. d) $\inf E = 0$, $\sup E = \sqrt{2}$. e) $\inf E = 0$, $\sup E = 2$. f) $\inf E = -1$, $\sup E = 2$. g) $\inf E = 0$, $\sup E = 3/2$.

2. $t_1 \le t_2 \le \dots$.

3. Notice that $a - \sqrt{2} < b - \sqrt{2}$ and use Exercise 6c in Section 1.1.

5. b) Apply Theorem 1.6 to $-E$.

6. b) Apply the Completeness Axiom to $-E$.

8. Prove $\sup E$ must be an integer.

9. Use the proof of Theorem 1.8 as a model.

11. First prove that $\max\{\sup A, \sup B\} \le \sup E$.

1.4 Sequences.

1. Use a proof by contradiction.

4. You may wish to prove that $\sqrt{x_n} - \sqrt{x} = (x_n - x)/(\sqrt{x_n} + \sqrt{x})$.

6. a) -3. b) $1/5$. c) $\sqrt{2}$. d) 0.

7. b) Use the Comparison Theorem.

8. If $x = \lim_{n\to\infty} x_n$ exists, what is $\lim_{n\to\infty} x_{n+1}$?

1.5 The Bolzano–Weierstrass Theorem.

1. You only need to prove that $\{x_n\}$ *has* a convergent subsequence, not actually find it.

4. See Exercise 4a in Section 1.2

5. Prove $x \le \sqrt{2x + 3}$ for $-3/2 \le x \le 3$.

6. See Exercise 4b in Section 1.2.

8. Prove that $\{x_n\}$ is monotone.

9. a) See Exercise 5c in Section 1.1.

1.6 The Extended Real Number System.

1. a) $2, 4$. b) $-1, 1$. c) $-1, 1$. d) $1/2, 1/2$. e) $0, 0$. f) $0, \infty$. g) ∞, ∞.

5. a) First prove $\inf_{k \ge n} x_k + \inf_{k \ge n} y_k \le \inf_{k \ge n}(x_k + y_k)$. c) By b), the first and final inequalities can only be strict if neither $\{x_n\}$ nor $\{y_n\}$ converges.

8. Let $s = \inf_{n \in \mathbf{N}}(\sup_{k \ge n} x_k)$ and consider the cases $s = \infty$, $s = -\infty$, and $s \in \mathbf{R}$.

9. Let $s = \liminf_{n\to\infty} x_n$ and consider the cases $s = \infty$, $s = 0$, and $0 < s < \infty$.

1.7 Functions, Countability, and the Algebra of Sets.

1. a) $f^{-1}(x) = (x + 7)/3$. b) $f^{-1}(x) = 1/\log x$. c) $f^{-1}(x) = \arctan x$. d) $f^{-1}(x) = (-3 + \sqrt{33 + 4x})/2$. e) $f^{-1}(x) = (x - 2)/3$ when $x \le 2$, $f^{-1}(x) = x - 2$ when $2 < x \le 4$, and $f^{-1}(x) = (x + 2)/3$ when $x > 4$.

2. a) $f(x) = \sqrt[3]{1 - x^3}$, $\mathrm{Dom}\,(f) = \mathbf{R}$. b) $f(x) = \log(x - 1)/x$, $\mathrm{Dom}\,(f) = (1, \infty)$. c) $f_1(x) = (-x + \sqrt{8 - 3x^2})/2$, $f_2(x) = (-x - \sqrt{8 - 3x^2})/2$, $\mathrm{Dom}\,(f_1) = \mathrm{Dom}\,(f_2) = [-\sqrt{8/3}, \sqrt{8/3}]$. d) $f_1(x) = \sqrt{(2 + x - 2x^2 - x^3)/2}$, $f_2(x) = -f_1(x)$, $f_3(x) = f_1(x)$ when $x \in (-\infty, -2]$, $f_3(x) = f_2(x)$ when $x \in [-1, 1]$, $f_4(x) = -f_3(x)$, and $\mathrm{Dom}\,(f_j) = (-\infty, -2] \cup [-1, 1]$ for $j = 1, 2, 3, 4$.

3. a) $f(E) = (-4, 16)$, $f^{-1}(E) = (0, 4/5)$. b) $f(E) = [0, 16]$, $f^{-1}(E) = [-2, 2]$. c) $f(E) = [-1/4, 2]$, $f^{-1}(E) = ((-1 - \sqrt{5})/2), (-1 + \sqrt{5})/2)$.
d) $f(E) = (\log(7/4), \log(31)]$, $f^{-1}(E) = [-1 - \sqrt{4e^5 - 3})/2, (-1 - \sqrt{4\sqrt{e} - 3})/2) \cup ((-1 + \sqrt{4\sqrt{e} - 3})/2, -1 + \sqrt{4e^5 - 3})/2]$.
e) $f(E) = [-1, 1]$, $f^{-1}(E) = \cup_{k \in \mathbf{Z}}[2k\pi, (2k + 1)\pi]$.

5. a) $[-1, 2]$. b) $[0, 1]$. c) $[0, 1]$. d) $\{0\}$.

6. First prove that $f(A) \setminus f(B) \subseteq f(A \setminus B)$ and $A \subseteq f^{-1}(f(A))$ hold whether f is 1–1 or not.

CHAPTER 2

2.1 Limits.

1. a) $\mathbf{R}$. b) $[a, b]$. c) $\emptyset$. d) $\{x\}$ if E is infinite, $\emptyset$ if E is finite. e) $\emptyset$.
2. a) 0. b) 1. c) -1. d) -1. e) 0.
3. a) 3/2. b) -3. c) $-\infty$. d) Does not exist. e) 0. f) Does not exist. g) n.
4. b) Use Theorem 2.2.
7. b) Use Exercise 6.
9. Prove that if $f(x)$ does not converge to L as $x \to \infty$, then there is a sequence $\{x_n\}$ such that $x_n \to \infty$ but $f(x_n)$ does not converge to L as $n \to \infty$.
10. See Exercise 9 in Section 1.4.

2.2 Continuity.

8. d) If the statement is true, then m must equal $f(1)$.

2.3 Uniform Continuity.

6. c) and e) Prove $f(x) = x$ and $g(x) = x^2$ are both uniformly continuous on $(0, 1)$ but only one of them is uniformly continuous on $[0, \infty)$.
9. You may wish to prove that if $P(x) = a_n x^n + \cdots + a_0$ is a polynomial of degree $n \geq 1$ whose leading coefficient satisfies $a_n > 0$, then $P(x) \to \infty$ as $x \to \infty$.

2.4 Differentiability.

1. a) $(5x^2 - 6x + 3)/(2\sqrt{x})$, $x > 0$. b) $-(2x+1)/(x^2 + x - 1)^2$, $x \neq (-1 \pm \sqrt{5})/2$. c) $(1 + \log x)x^x$, $x > 0$. d) $f'(x) = (3x^2 + 4x - 1)(x^3 + 2x^2 - x - 2)/|x^3 + 2x^2 - x - 2|$, $x \neq 1, -1, -2$.
2. a) $3a + c$. b) $(2b - d)/8$. c) bc. d) bc.
6. b) Observe that $y^n = x^m$ and use part a) together with the Chain Rule. c) To handle the case $q < 0$, first prove that $(x^{-1})' = -x^{-2}$ for all $x \neq 0$.
7. a) Use (ii) and (vi) to prove $\sin x \to 0$ as $x \to 0$. Use (iii) to prove $\cos x \to 1$ as $x \to 0$. b) First prove $\sin x = \sin(x - x_0)\cos x_0 + \cos(x - x_0)\sin x_0$ for any $x, x_0 \in \mathbf{R}$. c) Inequality (vi) and $0 \leq 1 - \cos x \leq 1 - \cos^2 x$ play a prominent role here. d) Use (iv) and part c).

2.5 The Mean Value Theorem.

1. a) 3. b) $-\infty$. c) $e^{1/6}$. d) 1. e) $-1/\pi$. f) -1.
2. No, f is not differentiable at 0.
3. b) First prove that if $g(x) = e^{-1/x^2}/x^k$ for some $k \in \mathbf{N}$, then $g(x) \to 0$ as $x \to 0$. Next, prove that given $n \in \mathbf{N}$, there are integers $N = N(n) \in \mathbf{N}$ and $a_k = a_k^{(n)} \in \mathbf{Z}$ such that

$$f^{(n)}(x) = \begin{cases} \sum_{k=0}^{N}(a_k/x^k)e^{-1/x^2} & x \neq 0 \\ 0 & x = 0. \end{cases}$$

(Note: Although for each $n \in \mathbf{N}$ many of the a_k's are zero, this fact is not needed in this exercise.)
4. b) Find the maximum of $f(x) = \log x/x^\alpha$ for $x \in [1, \infty)$.
7. a) Compare with Exercise 5 in Section 2.4.
9. This is the only exercise in this section which has nothing to do with the Mean Value Theorem or L'Hôpital's Rule.

2.6 Monotone Functions and the Inverse Function Theorem.

1. a) $a > -3$. b) $a \geq -3/4$. c) f is strictly decreasing on $(-\infty, 1]$ and strictly increasing on $[1, \infty)$.
5. $f(x) = \pm\sqrt{\alpha}x + c$ for some $c \in \mathbf{R}$.
7. Use Duhamel's Theorem.
8. Observe that if $x = \sin y$, then $\cos y = \sqrt{1 - x^2}$.

CHAPTER 3

3.1 The Riemann Integral.

4. a) Use the Sign Preserving Property.
5. First show that $\int_I f(x)\, dx = 0$ for all subintervals I of $[a, b]$.

3.2 Riemann Sums.

1. a) $1/4$. b) $\pi a^2/4$. c) 9. d) $(3/2)(b^2 - a^2) + (b - a)$. (Note: If $a \geq -1/3$ or $b \leq -1/3$, the integral represents the area of a trapezoid; if $a < -1/3 < b$, the integral represents the difference of the areas of two triangles, one above the x axis and the other below the x axis.)
5. b) You may use the fact that $\int x^n\, dx = x^{n+1}/(n + 1)$.
8. a) If $|f(x_0)| > M - \varepsilon/2$ for some $x_0 \in [a, b]$, can you choose a nondegenerate interval I such that $|f(x)| > M - \varepsilon$ for all $x \in I$? b) See Example 1 in Section 1.5.

3.3 The Fundamental Theorem of Calculus.

1. a) 15. b) $-5/3$. c) $(4^{100} - 1)/300$. d) $(e^2 + 1)/4$. e) $(e^{\pi/2} + 1)/2$. f) $4\sqrt{3} - 2\sqrt{11}$.
3. a) $g(-t)$.
7. a) See Exercise 4 in Section 3.1. b) See Exercise 3 in Section 2.5.

3.4 Improper Riemann Integration.

1. a) $3/2$. b) π. c) $3/2$. d) 4.
2. a) $p > 1$. b) $p < 1$. c) $p > 1$. b) $p > 1$.
3. Compare with Example 3.
4. a) diverges. b) diverges. c) converges. d) diverges. e) converges.
10. You might begin by verifying $\sin x \geq \sqrt{2}/2$ for $x \in [\pi/4, \pi/2]$ and $\sin x \geq 2x/\pi$ for $x \in [0, \pi/4]$.

3.5 Functions of Bounded Variation.

9. For the bounded case, prove $(L) \int_a^b |f'(x)|\, dx \leq \operatorname{Var} f \leq (U) \int_a^b |f'(x)|\, dx$.

3.6 Convex Functions.

5. Use Remark 2.

CHAPTER 4

4.1 Introduction.

2. a) $1/(1 + \pi)$. b) $5/6$. c) $21/4$. d) $e/(e - 2)$.

3. a) 1. b) $\log(2/3)$. c) $-1 + \pi/4$.

6. b) Consider the Geometric Series.

7. c) Notice that if the partial sums of $\sum_{k=1}^{\infty} b_k$ are bounded, then $b = 0$.

8. b) See Exercise 7b. d) First prove that if $a_k \geq 0$ and $\sum_{k=0}^{\infty} a_k$ diverges, then $\sum_{k=0}^{\infty} a_k = \infty$.

4.2 Tests for Convergence.

2. a) at most 100 terms. b) at most 15 terms. e) at most 10 terms. (To prove $\{a_k\}$ is monotone, show $a_{k+1}/a_k < 1$.)

3. a) It converges when $p > 1$ and diverges when $0 \leq p \leq 1$.

5. See Exercise 8 in Section 2.6.

11. It diverges when $0 < q \leq 1$ and converges when $q > 1$.

4.3 Absolute Convergence.

1. a) $n = 5$. b) $n = 7$. c) $n = 10$. d) $n = 7$.

2. a) Absolutely convergent. b) Absolutely convergent. c) Absolutely convergent. d) Conditionally convergent. e) Absolutely convergent. f) Absolutely convergent.

3. a) $[-1, 1)$. b) $(-\sqrt[3]{2}, \sqrt[3]{2})$. c) $[-1, 1]$. (Use Raabe's Test when $x = \pm 1$.) d) $[-3, -1]$.

4. a) $(1, \infty)$. b) $(0, \infty)$. c) $\emptyset$. d) $(1, \infty)$. e) $(-\infty, -1) \cup (1, \infty)$.

11. See Exercise 8b) in Section 4.1.

4.4 Uniform Convergence of Sequences.

2. Be sure your argument works for $a = 0$ as well as $a > 0$.

7. Modify the proof of Example 3 in Section 2.5 to show $(1 + x/n)^n \uparrow e^x$ as $n \to \infty$. To prove this is a uniform limit, choose N so large that $[a, b] \subset [-N, N]$ and find the maximum of $e^x - (1 + x/N)^N$ on $[a, b]$.

4.5 Uniform Convergence of Series.

3. See Example 6 in Section 4.2.

6. See Exercise 3a in Section 4.1.

4.6 Power Series.

1. a) $(-2, 2)$. b) $(3/4, 5/4)$. c) $[-1, 1)$. d) $[-1/\sqrt{2}, 1/\sqrt{2}]$. (Use Raabe's Test for the endpoints.)

2. a) $f(x) = 3x^2/(1 - x^3)$ for $x \in (-1, 1)$. b) $f(x) = (2 - x)/(1 - x)^2$ for $x \in (-1, 1)$. c) $f(x) = 2(\log x + 1/x - 1)/(1 - x)$ for $x \in (0, 2)$, $x \neq 1$, and $f(1) = 0$. d) $f(x) = \log(1/(1 - x^3))/x^3$ for $x \in [-1, 1)$, $x \neq 0$, and $f(0) = 1$.

4. Use Exercise 9 in Section 1.6 to prove that if $\limsup |a_k/a_{k+1}| < R$ then there is an $r < R$ such that $\{|a_k r^k|\}$ is increasing for k large, i.e., that $\sum_{k=1}^{\infty} a_k r^k$ diverges for some $r < R$.

8. Use the method of Example 2 to estimate $|f'(x)|$.

10. a) Use Theorem 4.8i to estimate $\log(n!) = \sum_{k=1}^{n} \log k$. b) $x \in (-1/e, 1/e)$.

4.7 Analytic Functions.

1. a) $\cos(3x) = \sum_{k=0}^{\infty} (-9)^k x^{2k}/(2k)!$. b) $2^x = \sum_{k=0}^{\infty} x^k \log^k 2/k!$. c) $\cos^2 x = 1 + \sum_{k=1}^{\infty} (-1)^k 2^{2k-1} x^{2k}/(2k)!$. d) $\sin^2 x + \cos^2 x = 1$. e) $x^3 e^{x^2} = \sum_{k=0}^{\infty} x^{2k+3}/k!$.

2. a) $\log(1-x) = -\sum_{k=1}^{\infty} x^k/k$. b) $x^2/(1-x^3) = \sum_{k=0}^{\infty} x^{3k+2}$. c) $e^x/(1-x) = \sum_{k=0}^{\infty}(\sum_{j=0}^{k} 1/j!)x^k$. d) $x^3/(1-x)^2 = \sum_{k=1}^{\infty} kx^{k+2}$.

e) $\arcsin x = \sum_{k=0}^{\infty} \binom{-1/2}{k}(-1)^k x^{2k+1}/(2k+1)$. (Use Theorem 4.45.)

3. a) $(x^2-1)e^x = -1 - x\sum_{k=2}^{\infty}(k^2 - k - 1)x^k/k!$.

b) $e^x \cos x = \sum_{k=0}^{\infty}(\sum_{j \in A_k}(-1)^j/((2j)!(k-2j)!)) \cdot x^k$, where $A_k := \{j \in \mathbf{N} : 0 \leq j \leq k/2\}$.

c) $\sin x/e^x = \sum_{k=1}^{\infty}(\sum_{j \in A_k}(-1)^{k-j+1}/((2j+1)!(k-2j-1)!)) \cdot x^k$, where $A_k := \{j \in \mathbf{N} : 0 \leq j \leq (k-1)/2\}$.

d) $f(x) = \sum_{k=0}^{\infty} x^k \log^{k+1} a/(k+1)!$.

4. a) $\log_{10} x = \sum_{k=1}^{\infty}(-1)^{k+1}(x-1)^k/(k \log 10)$, valid for $x \in (0,2]$. b) $x^2+2x-1 = 2 + 4(x-1) + (x-1)^2$, valid for $x \in \mathbf{R}$. c) $e^x = \sum_{k=0}^{\infty} e(x-1)^k/k!$, valid for $x \in \mathbf{R}$.

9. See Exercise 4 in Section 3.1 and use analytic continuation.

10. First use the Binomial Series to verify that $(1+x)^\beta \geq 1 + x^\beta$ for any $0 < x < 1$.

4.8 Applications.

1. The first seven places of the only real root are given by -0.3176721.

CHAPTER 5

5.1 Algebraic Structure of $\mathbf{R}^n$.

1. a) (a,a,a), $a \neq 0$. b) $(a,(20-8a)/7,(8+a)/7)$, $a \neq 0$. c) $x - 2y - z = 0$. d) $x - 4y - z = -6$.

4. vi) Write $\|\mathbf{x} \times \mathbf{y}\|^2 = (\mathbf{x} \times \mathbf{y}) \cdot (\mathbf{x} \times \mathbf{y})$ and use parts iv) and v).

10. If (x_0, y_0, z_0) does not lie on Π, let (x_2, y_2, z_2) be a point on Π different from (x_1, y_1, z_1), let θ represent the angle between $\mathbf{w} := (x_0 - x_2, y_0 - y_2, z_0 - z_2)$ and the normal (a, b, c), and compute $\cos\theta$ two different ways, once in terms of $\mathbf{w}$ and a second time in terms of the distance from (x_0, y_0, z_0) to Π.

5.2 Open and Closed Sets in $\mathbf{R}^n$.

2. a) Closed. $E^\circ = \{(x,y) : x^2 + 4y^2 < 1\}$ and $\partial E = \{(x,y) : x^2 + 4y^2 = 1\}$. b) Closed. $E^\circ = \emptyset$ and $\partial E = E$. c) Neither open nor closed. $E^\circ = \{(x,y) : y > x^2, 0 < y < 1\}$, $\overline{E} = \{(x,y) : y \geq x^2, 0 \leq y \leq 1\}$, and $\partial E = \{(x,y) : y = x^2, 0 \leq y \leq 1\} \cup \{(x,1) : -1 \leq x \leq 1\}$. d) Open. $\overline{E} = \{x^2 - y^2 \leq 1, -1 \leq y \leq 1\}$ and $\partial E = \{(x,y) : x^2 - y^2 = 1, -\sqrt{2} \leq x \leq \sqrt{2}\} \cup \{(x,1) : -\sqrt{2} \leq x \leq \sqrt{2}\} \cup \{(x,-1) : -\sqrt{2} \leq x \leq \sqrt{2}\}$.

9. These examples can be provided using subsets of $\mathbf{R}$.

5.3 Sequences and Compact Sets in $\mathbf{R}^n$.

1. a) $(0,-3)$. b) $(1,0,1)$. c) $(-1/2,1,0)$.

4. a) Compact. b) Compact. c) Not compact. $H = E \cup \{(0,y) : -1 \leq y \leq 1\}$. d) Not compact. There is no compact set H which contains E.

5.4 Convex Sets and Connected Sets in $\mathbf{R}^n$.

2. b) Use Theorem 5.15.

9. Convex functions are defined in Section 3.6.

5.5 Limits of Functions on $\mathbf{R}^n$.

1. a) $\lim_{y\to 0}\lim_{x\to 0}f(x,y) = \lim_{x\to 0}\lim_{y\to 0}f(x,y) = 0$, but $f(x,y)$ has no limit as $(x,y) \to (0,0)$. b) $\lim_{y\to 0}\lim_{x\to 0}f(x,y) = \lim_{x\to 0}\lim_{y\to 0}f(x,y) = 0$, and $f(x,y) \to 0$ as $(x,y) \to (0,0)$. c) $\lim_{y\to 0}\lim_{x\to 0}f(x,y) = \lim_{x\to 0}\lim_{y\to 0}f(x,y)$ $= 0$, and $f(x,y) \to 0$ as $(x,y) \to (0,0)$. d) $\lim_{y\to 0}\lim_{x\to 0}f(x,y) = 1/2$, $\lim_{x\to 0}\lim_{y\to 0}f(x,y) = 1$, so $f(x,y)$ has no limit as $(x,y) \to (0,0)$. e) $\lim_{y\to 0}\lim_{x\to 0}f(x,y) = \lim_{x\to 0}\lim_{y\to 0}f(x,y) = 0$, and $f(x,y) \to 0$ as $(x,y) \to (0,0)$.

2. a) $\mathrm{Dom}\,f = \{(x,y) : x \neq 1, y \neq 1\}$ and the limit is $(0,3)$. b) $\mathrm{Dom}\,f = \{(x,y) : x \neq 0, y \neq 0\}$ and the limit is $(1,0,1)$. c) $\mathrm{Dom}\,f = \{(x,y) : (x,y) \neq (0,0)\}$ and the limit is $(0,0)$. d) $\mathrm{Dom}\,f = \{(x,y) : (x,y) \neq (1,1)\}$ and the limit is $(0,0)$.

5.6 Continuous Functions on $\mathbf{R}^n$.

2. First prove A is relatively closed in E if and only if $E \setminus A$ is relatively open in E. Then combine Theorem 5.25 with Theorem 1.30ii.

6. b) See Exercise 6 in Section 1.7.

8. a) A polygonal path in E can be described as the image of a continuous function $f : [0,1] \to E$. Use this to prove every polygonal path is connected. c) Prove that if E is not polygonally connected, then there are nonempty open sets $U,V \subset E$ such that $U \cap V = \emptyset$ and $U \cup V = E$.

5.7 Applications.

2. See Exercise 8 in Section 4.5.

4. a) $\omega_f(t) = 1$ for all t. b) $\omega_f(t) = 0$ if $t \neq 0$ and $\omega_f(0) = 1$. c) $\omega_f(t) = 0$ if $t \neq 0$ and $\omega_f(0) = 2$.

7. a) $\sqrt{1/2}$. b) $f(0)/3$. c) $1/4$. d) $(e^4 - 1)/(2e^2)$.

9. d) You may wish to use the lemma in Section 2.6.

5.8 Metric Spaces.

9. Use Exercise 8.

11. Use Exercise 10.

CHAPTER 6

6.1 Partial Derivatives and Partial Integrals.

1. a) $f_{xy} = f_{yx} = e^y$. b) $f_{xy} = f_{yx} = -\sin(xy) - xy\cos(xy)$. c) $f_{xy} = f_{yx} = -2x/(x^2+1)^2$.

2. a) $f_x = 2x + y\cos(xy)$ and $f_y = x\cos(xy)$ are continuous everywhere on $\mathbf{R}^2$. b) $f_x = y/(1+z)$, $f_y = x/(1+z)$, and $f_z = -xy/(1+z)^2$ are continuous except when $z = -1$. c) $f_x = x/\sqrt{x^2+y^2}$ and $f_y = y/\sqrt{x^2+y^2}$ are continuous everywhere except at the origin.

3. a) $f_x = (2x^5 + 4x^3y^2 - 2xy^4)/(x^2+y^2)^2$ for $(x,y) \neq (0,0)$ and $f_x(0,0) = 0$. f_x is continuous on $\mathbf{R}^2$. b) $f_x = (2x/3) \cdot (2x^2 + 4y^2)/(x^2+y^2)^{4/3}$ for $(x,y) \neq (0,0)$ and $f_x(0,0) = 0$. f_x is continuous on $\mathbf{R}^2$.

7. a) 1. b) $\sqrt{2}/2$. c) $9/10$. d) $e^{-\pi}/2$.

9. c) Choose $\delta > 0$ such that $|\phi(t)| < \varepsilon$ for $0 \le t < \delta$ and break the integral in part b) into two pieces, one corresponding to $0 \le t \le \delta$ and the other to $\delta \le t < \infty$.

   d) Combine part b) with Theorem 6.6.

10. a) $\mathcal{L}\{te^t\} = 1/(s-1)^2$. b) $\mathcal{L}\{t\sin \pi t\} = 2s\pi/(s^2 + \pi^2)^2$. c) $\mathcal{L}\{t^2 \cos t\} = 2(s^3 - 3s)/(s^2 + 1)^3$.

6.2 The Definition of Differentiability.

1. a)
$$Df(x,y) = \begin{pmatrix} \cos x & 0 \\ y/(2\sqrt{xy}) & x/(2\sqrt{xy}) \\ 0 & -\sin y \end{pmatrix}.$$

b)
$$Df(s,t,u,v) = \begin{pmatrix} t & s & 2u & 0 \\ -2s & 0 & v & u \end{pmatrix}.$$

c)
$$Df(t) = \begin{pmatrix} 1/t \\ -1/(1+t)^2 \end{pmatrix}.$$

d)
$$Df(r,\theta) = \begin{pmatrix} \cos\theta & -r\sin\theta \\ \sin\theta & r\cos\theta \end{pmatrix}.$$

9. b) The function f might not be differentiable when $\alpha = 1$.

6.3 Differentiability Theorems.

2.
$$\frac{\partial w}{\partial p} = \frac{\partial F}{\partial x}\frac{\partial x}{\partial p} + \frac{\partial F}{\partial y}\frac{\partial y}{\partial p} + \frac{\partial F}{\partial z}\frac{\partial z}{\partial p}, \quad \frac{\partial w}{\partial q} = \frac{\partial F}{\partial x}\frac{\partial x}{\partial q} + \frac{\partial F}{\partial y}\frac{\partial y}{\partial q} + \frac{\partial F}{\partial z}\frac{\partial z}{\partial q},$$

$$\frac{\partial^2 w}{\partial p^2} = \frac{\partial}{\partial p}\left(\frac{\partial F}{\partial x}\right)\frac{\partial x}{\partial p} + \frac{\partial F}{\partial x}\frac{\partial^2 x}{\partial p^2} + \frac{\partial}{\partial p}\left(\frac{\partial F}{\partial y}\right)\frac{\partial y}{\partial p}$$

$$+ \frac{\partial F}{\partial y}\frac{\partial^2 y}{\partial p^2} + \frac{\partial}{\partial p}\left(\frac{\partial F}{\partial z}\right)\frac{\partial z}{\partial p} + \frac{\partial F}{\partial z}\frac{\partial^2 z}{\partial p^2}$$

$$= \frac{\partial F}{\partial x}\frac{\partial^2 x}{\partial p^2} + \frac{\partial F}{\partial y}\frac{\partial^2 y}{\partial p^2} + \frac{\partial F}{\partial z}\frac{\partial^2 z}{\partial p^2}$$

$$+ \frac{\partial^2 F}{\partial x^2}\left(\frac{\partial x}{\partial p}\right)^2 + \frac{\partial^2 F}{\partial y^2}\left(\frac{\partial y}{\partial p}\right)^2 + \frac{\partial^2 F}{\partial z^2}\left(\frac{\partial z}{\partial p}\right)^2$$

$$+ 2\frac{\partial^2 F}{\partial x \partial y}\left(\frac{\partial x}{\partial p}\right)\left(\frac{\partial y}{\partial p}\right) + 2\frac{\partial^2 F}{\partial x \partial z}\left(\frac{\partial x}{\partial p}\right)\left(\frac{\partial z}{\partial p}\right) + 2\frac{\partial^2 F}{\partial y \partial z}\left(\frac{\partial y}{\partial p}\right)\left(\frac{\partial z}{\partial p}\right).$$

7. Compute the derivative of $f \cdot f$ using the Dot Product Rule.

11. Notice that by Exercise 9, this result still holds if "f is in $\mathcal{C}^2$" is replaced by "the first-order partial derivatives of f are differentiable."

6.4 The Mean Value Theorem and Taylor's Formula.

3. a) $f(x,y) = 1-(x+1)+(y-1)+(x+1)^2+(x+1)(y-1)+(y-1)^2$. b) $\sqrt{x}+\sqrt{y} = 3+(x-1)/2+(y-4)/4-(x-1)^2/8-(y-4)^2/64 +(x-1)^3/16\sqrt{c^5}+(y-1)^3/16\sqrt{d^5}$ for some $(c,d) \in L((x,y);(1,4))$. c) $e^{xy} = 1+xy+((dx+cy)^4+12(dx+cy)^2xy+ 12x^2y^2)e^{cd}/4!$ for some $(c,d) \in L((x,y);(0,0))$.

4. Notice that by Exercise 9 in Section 6.3, this result still holds if "f is in C^p" is replaced by "the $(p-1)$-st order partial derivatives of f are differentiable."

8. Apply Taylor's Formula to $f(a+x,b+y)$ for $p = 3$, $x = r\cos\theta$, and $y = r\sin\theta$, and prove that

$$\frac{4}{\pi r^2}\int_0^{2\pi} f(a+r\cos\theta, b+r\sin\theta)\cos(2\theta)\, d\theta = f_{xx}(a,b) - f_{yy}(a,b) + F(r),$$

where $F(r)$ is a function which converges to 0 as $r \to 0$.

9. c) Let (x_2,t_2) be the point identified in part b), and observe by one-dimensional theory that $u_t(x_2,t_2) = 0$. Use this observation and Taylor's Formula to obtain the contradiction $w_{xx}(x_2,t_2) - w_t(x_2,t_2) \geq 0$.

6.5 The Inverse Function Theorem.

2. a) $f^{-1}(s,t) = ((s+\sqrt{s^2-4t})/2, (s-\sqrt{s^2-4t})/2)$.

   b) $D(f^{-1})(f(x,y)) = \begin{pmatrix} x/(x-y) & 1/(y-x) \\ y/(y-x) & 1/(x-y) \end{pmatrix}$ (see Theorem C.5).

3. $D(f^{-1})(-1,0) = \begin{pmatrix} -1/2 & 1 \\ -1/2 & 0 \end{pmatrix}$.

6. $F(x_0,y_0,u_0,v_0) = (0,0)$, $x_0^2 \neq y_0^2$, and $u_0 \neq 0 \neq v_0$, where $F(x,y,u,v) = (xu^2+yv^2+xy-9, xv^2+yu^2-xy-7)$.

8. See Remark 2 in Section 6.2.

6.6 Extrema.

1. a) $f(1/3,2/3) = -13/27$ is a local minimum and $(-1/4,-1/2)$ is a saddle point. b) Let $k,j \in \mathbf{Z}$. $f((2k+1)\pi/2, j\pi) = 2$ is a local maximum if k and j are even, $f((2k+1)\pi/2, j\pi) = -2$ is a local minimum if k and j are odd, and $((2k+1)\pi/2, j\pi)$ is a saddle point if $k+j$ is odd. c) This function has no local extrema. d) $f(0,0) = 0$ is a local minimum if $a > 0$ and $b^2-4ac < 0$, a local maximum if $a < 0$ and $b^2-4ac < 0$, and $(0,0)$ is a saddle point if $b^2-4ac > 0$.

2. a) $f(2,0) = 8$ is the maximum and $f(-4/5, \sqrt{21}/5) = -9/5$ is the minimum. b) $f(1,2) = 17$ is the maximum and $f(1,0) = 1$ is the minimum. c) $f(1,1) = f(-1,-1) = 3$ is the maximum and $f(-1,1) = -5$ is the minimum.

3. a) $f(2,0) = 2$ is the minimum and $f(1/2,\pm\sqrt{15}/2) = 17/4$ is the maximum. b) $f(\pm 2/\sqrt{5},\pm 1/\sqrt{5}) = 0$ is the minimum and $f(\pm 1/\sqrt{5},\mp 2/\sqrt{5}) = 5$ is the maximum. c) $\lambda = xy$, $3\mu = x+y$, $f(\pm 1/\sqrt{2},\mp 1/\sqrt{2},0) = -1/2$ is the minimum and $f(\pm 1/\sqrt{6},\pm 1/\sqrt{6},\mp 2/\sqrt{6}) = 1/6$ is the maximum. d) $f(1,-2,0,1) = 2$ is the minimum and $f(1,2,-1,-2) = 3$ is the maximum.

7. b) If $DE < 0$, then $ax+by+cz$ has no extremum subject to the constraint $z = Dx^2 + Ey^2$.

8. b) $f(2,2,4) = 48$ is the minimum. There is no maximum.

6.7 Differentiability and Tangent Planes.

1. a) $dz = 2x\,dx + 2y\,dy$. b) $dz = y\cos(xy)\,dx + x\cos(xy)\,dy$. c) $dz = (1 - x^2 + y^2)y/(1 + x^2 + y^2)^2\,dx + (1 + x^2 - y^2)x/(1 + x^2 + y^2)^2\,dy$.
2. $dw = .05$ and $\Delta w \approx 0.049798$.
4. a) $z = 0$. b) $2x - 2y - z = 0$.
5. L must be measured with no more than 5% error.
7. $(-1/2, -1/2, 1/2)$, $2x + 2y + 2z = -1$.
8. b) $ax + by = 1$, where $a^2 + b^2 = 1$. c) $x + y - z = \pm 1$.

CHAPTER 7

7.1 Integration on $\mathbf{R}^n$.

2. a) See Theorem 5.9. b) Use Exercise 4a below.
4. d) Apply part c) to $E_1 = (E_1 \setminus E_2) \cup E_2$. e) Apply parts c) and d) to $(E_1 \cup E_2) = (E_1 \setminus (E_1 \cap E_2)) \cup (E_2 \setminus (E_1 \cap E_2)) \cup (E_1 \cap E_2)$.
7. a) Consider the proof of Theorem 7.3. b) See Exercise 6 in Section 1.7.

7.2 Riemann Integration of Functions on $\mathbf{R}^n$.

4. b) Area (E).
7. b) The restriction $H^o \neq \emptyset$ is not needed for the case $\inf_{x \in H} f(x) < c < \sup_{x \in H} f(x)$.

7.3 Iterated Integrals.

1. a) $5/6$. b) $8(2\sqrt{2} - 1)/9$. c) $(1 - \cos(\pi^2/4)) \cdot (2/\pi)$.
2. a) $E = \{(x, y) : 0 \le x \le 1, x \le y \le x^2 + 1\}$ and $\int_0^1 \int_x^{x^2+1} (x + y)\,dy\,dx = 71/60$. b) $E = \{(x, y, z) : 0 \le y \le 1, \sqrt{y} \le x \le 1, 0 \le z \le x^2 + y^2\}$ and $3\int_0^1 \int_{\sqrt{y}}^1 (x^2 + y^2)\,dx\,dy = 26/105$. c) $E = \{(x, y) : 0 \le x \le 1, 0 \le y \le x\}$ and $\int_0^1 \int_y^1 \sin(x^2)\,dx\,dy = (1 - \cos(1))/2$. d) $E = \{(x, y, z) : 0 \le x \le 1, 0 \le y \le x^2, x^3 \le z \le 1\}$ and $\int_0^1 \int_{\sqrt{y}}^1 \int_{x^3}^1 \sqrt{x^3 + z}\,dz\,dx\,dy = 4(2\sqrt{2} - 1)/45$.
3. a) $2/35$. b) 1. c) $492/5$. d) $1/8$.
4. a) 3π. b) $91/30$. c) $88/105$. d) $1/18$.
7. a) See Exercise 6.

7.4 Change of Variables.

2. a) $\pi(1 - \cos 4)/4$. b) $3/10$. c) $(\sqrt{2} + \log(1 + \sqrt{2}))(b^3 - a^3)/6$. (Recall that the indefinite integral of $\sec\theta$ is $\int \sec\theta = \log|\sec\theta + \tan\theta| + C$.)
3. a) $(\pi\sqrt{3}/3)\sin 3$. b) $16^2/(3 \cdot 5 \cdot 7)$.
4. a) $(6\sqrt{6} - 7)4\pi/5$. b) $\pi(4e^3 - 2(\sqrt{8} - 1)e^{\sqrt{8}})$. c) $16\sqrt{2}/15$.
5. b) $\pi r^2 d/a$.
6. a) $4/27$. b) $9/112$. c) $3(e - 1)/e$. d) 5. (Use the change of variables $x = u + v$, $y = u - v$.)
7. See Exercise 5 in Section 7.2.
9. See Exercise 7 in Section 5.1.
10. d) $\pi^{n/2}$.

7.5 Partitions of Unity.

3. See Theorem 4.46.

7.6 The Gamma Function and Volume.

5. Let ψ_n represent the spherical change of variables in $\mathbf{R}^n$ and observe that the cofactor $|A_1|$ of $-\rho \sin \varphi_1 \sin \varphi_2 \ldots \sin \varphi_{n-2} \sin \theta$ in the matrix $D\psi_n$ is identical to $\Delta_{\psi_{n-1}}$ if in $\Delta_{\psi_{n-1}}$, θ is replaced by φ_{n-2} and each entry in the last row of $D\psi_{n-1}$ is multiplied by $\sin \theta$.

8. $r^2 \text{Vol}(B_r(\mathbf{0}))/(n+2)$.

CHAPTER 8

8.1 Curves.

5. a) This curve spirals up the cone $x^2 + y^2 = z^2$ from $(0,1,1)$ to $(0, e^{2\pi}, e^{2\pi})$ and has arc length $\sqrt{3}(e^{2\pi} - 1)$. b) This curve coincides with the graph of $x = \pm y^{3/2}$, $0 \le y \le 1$ (looking like a stylized gull in flight), and has arc length $2(\sqrt{13^3}-1)/27$. c) This curve is a straight line segment from $(0,0,0)$ to $(4,4,4)$ and has arc length $4\sqrt{3}$. d) The arc length of the astroid is 6.

6. a) 27. b) $ab(a^2 + ab + b^2)/(3(a+b))$. c) 12π. d) $(5 + 3\sqrt{5})/2$.

7. b) Use Dini's Theorem.

9. Analyze what happens to (x,y) and $dy/dx := (dy/dt)/(dx/dt)$ as $t \to -\infty$, $t \to -1-$, $t \to -t+$, $t \to 0$, and $t \to \infty$. For example, prove that, as $t \to -1-$, the trace of $\phi(t)$ lies in the fourth quadrant and is asymptotic to the line $y = -x$.

11. b) Take the derivative of $\nu' \cdot \nu'$ using the Dot Product Rule. d) Observe that $\phi(t) = \nu(\ell(t))$ and use the Chain Rule to compute $\phi'(t)$ and $\phi''(t)$. Then calculate $\phi' \times \phi''$ directly.

8.2 Oriented Curves.

1. a) A spiral on the elliptic cylinder $y^2 + 9z^2 = 9$ oriented clockwise when viewed from far out the x axis. b) A cubical parabola (it looks like a stylized gull in flight) on the plane $z = x$ oriented from left to right when viewed from far out the plane $y = x$. c) A sine wave on the parabolic cylinder $y = x^2$ oriented from right to left when viewed from far out the y axis. d) An ellipse sliced by the plane $x = z$ out of the cylinder $y^2 + z^2 = 1$ oriented clockwise when viewed from far out the x axis. e) A sine wave traced vertically on the plane $y = x$ oriented from below to above when viewed from far out the x axis.

2. a) 128/3. b) $-\pi\sqrt{2}/2$. c) 0.

3. a) 5. b) $\pi(-1 + \sqrt{5})/2$. c) $|R|(2 - a - b)/2$. d) $-\sin(1) + 1/3$.

4. c) There exist functions ψ and τ on $[0,1]$ which are C^1 on $(0,1) \setminus \{j/N : j = 1, \ldots, N\}$ such that $\tau' > 0$ and $\psi = \phi_j \circ \tau$ on $((j-1)/N, j/N)$ for each $j = 1, \ldots, N$.

7. c) If F is conservative, consider the case when C is smooth first. If (*) holds, use parts a) and b) to prove that F is conservative.

8. Use Jensen's Inequality.

8.3 Surfaces.

1. a) $\sqrt{2}\pi(b^2 - a^2)$. b) $4\pi a^2$. c) $4\pi^2 ab$.

2. a) $\phi(u,v) = (u,v,u^2-v^2)$, $E = \{(u,v) : -1 \le u \le 1, -|u| \le v \le |u|\}$, $\psi_1(t) = (1,t,1-t^2)$, $\psi_2(t) = (-1,t,1-t^2)$, $\psi_3(t) = (t,t,0)$, $\psi_4(t) = (t,-t,0)$, $I_1 = I_2 = I_3 = I_4 = [-1,1]$, and $\iint_S g\,d\sigma = 22/3$. b) $\phi(u,v) = (u,u^3,v)$, $E = [0,2] \times [0,4]$, $\psi_1(t) = (t,t^3,4)$, $\psi_2(t) = (t,t^3,0)$, $\psi_3(t) = (0,0,t)$, $\psi_4(t) = (2,8,t)$, $I_1 = I_2 = [0,2]$, $I_3 = I_4 = [0,4]$, and $\iint_S g\,d\sigma = (4/27)(145^{3/2} - 1)$. c) $\phi(u,v) = (3\cos u \cos v, 3\sin u \cos v, 3\sin v)$, $E = [0,2\pi] \times [\pi/4, \pi/2]$, $\psi_1(t) = ((3/\sqrt{2})\cos t, (3/\sqrt{2})\sin t, 3/\sqrt{2})$, $\psi_2(t) = (3\cos t, 3\sin t, 0)$, $I = [0,2\pi]$, and $\iint_S g\,d\sigma = 27\pi/2$.
5. See Theorem 7.15.
6. If you got 52π, you gave up too much when you replaced $\|(x,y)\|$ by 3.

8.4 Oriented Surfaces.

1. a) Since the x axis lies to the left of the yz plane when viewed from far out the positive y axis, the boundary can be parametrized by $\phi(t) = (3\sin t, 0, 3\cos t)$, $I = [0,2\pi]$, and $\int_{\partial S} F \cdot T\,ds = -9\pi$. b) The boundary can be parametrized by $\phi_1(t) = (0,-t,1+2t)$, $I_1 = [-1/2,0]$; $\phi_2(t) = (t,0,1-t)$, $I_2 = [0,1]$; and $\phi_3(t) = (-t,(1+t)/2,0)$, $I_3 = [-1,0]$; and $\int_{\partial S} F \cdot T\,ds = -1/12$. c) The boundary can be parametrized by $\phi_1(t) = (2\sin t, 2\cos t, 4)$, $I_1 = [0,2\pi]$, and $\phi_2(t) = (\cos t, \sin t, 1)$, $I_2 = [0,2\pi]$, and $\int_{\partial S} F \cdot T\,ds = 3\pi$.
2. a) $\pi/2$. b) 16. c) $2\pi^2 ab^2$. d) $\pi/8$.
3. a) $-14/15$. b) $4\pi a^3/3$. c) $(3b^4 + 8a^3 - 8(a^2 - b^2)^{3/2})\pi/12$. d) $-2\pi/3$.
4. See Theorem 7.15.

8.5 Theorems of Green and Gauss.

1. a) $8/3$. b) $3\log 3 + 2(1-e^3)$. c) $-15\pi/4$.
2. a) $(b-a)(c-d)(c+d-2)/2$. b) $-1/6$. c) 0.
3. a) $2(5+e^3)$. b) π. c) 8. d) $\pi abc(a+b+c)/2$.
4. a) $224/3$. b) $2(8\sqrt{2}-7)/15$. c) 24π.
5. b) $3/2$. c) $\mathrm{Vol}(E) = (1/3)\int_{\partial E} x\,dy\,dz + y\,dz\,dx + z\,dx\,dy$. d) $2\pi^2 ab^2$.
9. c) Use Exercise 8 and Gauss's Theorem.
10. e) Use Green's Theorem and Exercise 5 in Section 7.2.

8.6 Stokes's Theorem.

1. a) $-\pi/4$. b) $27\pi/4$.
2. a) 0. b) -3π. c) -10π. d) $-1/12$.
3. a) $\pi^2/5$. b) $-\pi/(8\sqrt{2})$. c) 28π (not -28π because $\mathbf{i} \times \mathbf{k} = -\mathbf{j}$). d) 32π. e) $-\pi$.
4. a) 18π. b) 8π. c) $3(1-e) + 3\pi/2$. d) 0.
10. b) 2π.

CHAPTER 9

9.1 Introduction.

1. a) $a_0(x^2) = 2\pi^2/3$, $a_k(x^2) = 4(-1)^k/k^2$, and $b_k(x^2) = 0$ for $k = 1, 2, \ldots$ b) All Fourier coefficients of $\cos^2 x$ are zero except $a_0(\cos^2 x) = 1/2$ and $a_2(\cos^2 x) = 1/2$.

6. a) $a_k(f) = 0$ for $k = 0, 1, \ldots,$ $b_k(f) = 4/(k\pi)$ when k is odd and 0 when k is even.

   c) You may wish to use Theorem 5.32.

9.2 Summability of Fourier Series.

5. c) See Exercise 4b in Section 3.1.
7. See Theorem 5.33.

9.3 Growth of Fourier Coefficients.

4. See Exercise 4a in Section 9.2 and Theorem 4.25.

9.4 Convergence of Fourier Series.

1. Note: It is not assumed that f is periodic.
2. c) $\pi^2/8$.
4. a) Use Abel's Formula. For the first identity, you must show that $S_N \rho^N \to 0$ as $N \to \infty$ for all $\rho \in (0,1)$ if $\sum_{k=0}^{\infty} a_k r^k$ converges for all $r \in (0,1)$.
5. a) Prove that for each fixed h, $a_k(f(x+h)) = a_k(f)\cos kh + b_k(f)\sin kh$.

CHAPTER 10

10.1 Differential Forms on $\mathbf{R}^n$.

1. a) $dy\,dz - 3\,dz\,dx + 2\,dx\,dy$. b) $x^2\,dy\,dz + xy\,dz\,dx + yz\,dx\,dy$. c) $x^2\cos x\,dx\,dy\,dw - dy\,dz\,dw$.
2. a) $(2x+2y)\,dx\,dy$. b) $y\cos(xy)\,dx\,dz\,dw + x\cos(xy)\,dy\,dz\,dw - w\sin(zw)\,dx\,dy\,dz - z\sin(zw)\,dx\,dy\,dw$. c) $((x+y)/\sqrt{x^2+y^2})\,dx\,dy\,dz$.
   d) $(y\sin xe^{yz} - y\sin xe^{xy} + x\cos ye^{xy} - \cos xe^{xy})\,dx\,dy\,dz$.

10.2 Differentiable Manifolds.

5. a) Note: If $\mathcal{B}$ is another atlas of M, then $\mathcal{A}$ and $\mathcal{B}$ are compatible.

10.3 Stokes's Theorem on Manifolds.

1. $\pi^2 a^6/4$.
2. $\sum_{j=1}^{n}(-1)^{j-1}a_1 \ldots \widehat{a_j} \ldots a_n \cdot a_j^2$.

SUBJECT INDEX

NOTATION INDEX